* Applied Multivariate Statistic...
Richard Johnson and Dean ...
(0-13-092553-5)

* A Second Course in Statistics: Regression Analysis 6e
William Mendenhall and Terry Sincich
(0-13-022323-9)

* Biostatistics: How it Works
Steve Selvin
(0-13-046616-6)

* Statistics for the Life Sciences 3e
Jeff Witmer and Myra Samuels
(0-13-041316-X)

* Statistics with Applications to the Biological and Health Sciences 3e
Tony Schork and Richard Remington
(0-13-022327-1)

* Biostatistical Analysis 4e
Jerrold Zar
(0-13-081542-X)

* Fundamentals of Probability 3e
Saeed Ghahramani
(0-13-145340-8)

* A First Course in Probability 6e
Sheldon Ross
(0-13-033851-6)

* Introduction to Stochastic Processes with Biology Applications
Linda Allen
(0-13-035218-7)

* Mathematics of Medical Imaging
Charles Epstein
(0-13-067548-2)

Introduction
to
Mathematical Statistics

Sixth Edition

Robert V. Hogg
University of Iowa

Joseph W. McKean
Western Michigan University

Allen T. Craig
Late Professor of Statistics
University of Iowa

PEARSON
Prentice
Hall

Pearson Education International

1004595487.

Executive Acquisitions Editor: *George Lobell*
Executive Editor-in-Chief: *Sally Yagan*
Vice President/Director of Production and Manufacturing: *David W. Riccardi*
Production Editor: *Bayani Mendoza de Leon*
Senior Managing Editor: *Linda Mihatov Behrens*
Executive Managing Editor: *Kathleen Schiaparelli*
Assistant Manufacturing Manager/Buyer: *Michael Bell*
Manufacturing Manager: *Trudy Pisciotti*
Marketing Manager: *Halee Dinsey*
Marketing Assistant: *Rachael Beckman*
Art Director: *Jayne Conte*
Cover Designer: *Bruce Kenselaar*
Art Editor: *Thomas Benfatti*
Editorial Assistant: *Jennifer Brady*
Cover Image: *Tun shell (Tonna galea). David Roberts/Science Photo Library/Photo
Researchers, Inc.*

©2005, 1995, 1978, 1970, 1965, 1958 Pearson Education, Inc.
Pearson Prentice Hall
Pearson Education, Inc.
Upper Saddle River, NJ 07458

Printed in the United States of America
10 9 8 7 6 5 4 3

ISBN: 0-13-122605-3

Pearson Education, Ltd., *London*
Pearson Education Australia PTY. Limited, *Sydney*
Pearson Education Singapore, Pte., Ltd
Pearson Education North Asia Ltd, *Hong Kong*
Pearson Education Canada, Ltd., *Toronto*
Pearson Education de Mexico, S.A. de C.V.
Pearson Education - Japan, *Tokyo*
Pearson Education Malaysia, Pte. Ltd
Pearson Education, *Upper Saddle River, New Jersey*

To Ann and to Marge

Contents

     10.5.1 Efficacy 552
     10.5.2 Estimating Equations Based on General Scores 553
     10.5.3 Optimization: Best Estimates 554
  10.6 Adaptive Procedures 561
  10.7 Simple Linear Model 565
  10.8 Measures of Association 570
     10.8.1 Kendall's τ 571
     10.8.2 Spearman's Rho 574

11 Bayesian Statistics   **579**
  11.1 Subjective Probability 579
  11.2 Bayesian Procedures 582
     11.2.1 Prior and Posterior Distributions 583
     11.2.2 Bayesian Point Estimation 586
     11.2.3 Bayesian Interval Estimation 589
     11.2.4 Bayesian Testing Procedures 590
     11.2.5 Bayesian Sequential Procedures 592
  11.3 More Bayesian Terminology and Ideas 593
  11.4 Gibbs Sampler 600
  11.5 Modern Bayesian Methods 606
     11.5.1 Empirical Bayes 610

12 Linear Models   **615**
  12.1 Robust Concepts 615
     12.1.1 Norms and Estimating Equations 616
     12.1.2 Influence Functions 617
     12.1.3 Breakdown Point of an Estimator 621
  12.2 LS and Wilcoxon Estimators of Slope 624
     12.2.1 Norms and Estimating Equations 625
     12.2.2 Influence Functions 626
     12.2.3 Intercept 629
  12.3 LS Estimation for Linear Models 631
     12.3.1 Least Squares 633
     12.3.2 Basics of LS Inference under Normal Errors 635
  12.4 Wilcoxon Estimation for Linear Models 640
     12.4.1 Norms and Estimating Equations 641
     12.4.2 Influence Functions 641
     12.4.3 Asymptotic Distribution Theory 643
     12.4.4 Estimates of the Intercept Parameter 645
  12.5 Tests of General Linear Hypotheses 647
     12.5.1 Distribution Theory for the LS Test for Normal Errors 650
     12.5.2 Asymptotic Results 651
     12.5.3 Examples 652

Preface

Since Allen T. Craig's death in 1978, Bob Hogg has revised the later editions of this text. However, when Prentice Hall asked him to consider a sixth edition, he thought of his good friend, Joe McKean, and asked him to help. That was a great choice for Joe made many excellent suggestions on which we both could agree and these changes are outlined later in this preface.

In addition to Joe's ideas, our colleague Jon Cryer gave us his marked up copy of the fifth edition from which we changed a number of items. Moreover, George Woodworth and Kate Cowles made a number of suggestions concerning the new Bayesian chapter; in particular, Woodworth taught us about a "Dutch book" used in many Bayesian proofs. Of course, in addition to these three, we must thank others, both faculty and students, who have made worthwhile suggestions. However, our greatest debts of gratitude are for our special friend, Tom Hettmansperger of Penn State University, who used our revised notes in his mathematical statistics course during the 2002-2004 academic years and Suzanne Dubnicka of Kansas State University who used our notes in her mathematical statistics course during Fall of 2003. From these experiences, Tom and Suzanne and several of their students provided us with new ideas and corrections.

While in earlier editions, Hogg and Craig had resisted adding any "real" problems, Joe did insert a few among his more important changes. While the level of the book is aimed for beginning graduate students in Statistics, it is also suitable for senior undergraduate mathematics, statistics and actuarial science majors.

The major differences between this edition and the fifth edition are:

- It is easier to find various items because more definitions, equations, and theorems are given by chapter, section, and display numbers. Moreover, many theorems, definitions, and examples are given names in bold faced type for easier reference.

- Many of the distribution finding techniques, such as transformations and moment generating methods, are in the first three chapters. The concepts of expectation and conditional expectation are treated more thoroughly in the first two chapters.

- Chapter 3 on special distributions now includes contaminated normal distributions, the multivariate normal distribution, the t- and F-distributions, and a section on mixture distributions.

- Chapter 4 presents large sample theory on convergence in probability and distribution and ends with the Central Limit Theorem. In the first semester, if the instructor is pressed for time he or she can omit this chapter and proceed to Chapter 5.

- To enable the instructor to include some statistical inference in the first semester, Chapter 5 introduces sampling, confidence intervals and testing. These include many of the normal theory procedures for one and two sample location problems and the corresponding large sample procedures. The chapter concludes with an introduction to Monte Carlo techniques and bootstrap procedures for confidence intervals and testing. These procedures are used throughout the later chapters of the book.

- Maximum likelihood methods, Chapter 6, have been expanded. For illustration, the regularity conditions have been listed which allows us to provide better proofs of a number of associated theorems, such as the limiting distributions of the maximum likelihood procedures. This forms a more complete inference for these important methods. The EM algorithm is discussed and is applied to several maximum likelihood situations.

- Chapters 7-9 contain material on sufficient statistics, optimal tests of hypotheses, and inferences about normal models.

- Chapters 10-12 contain new material. Chapter 10 presents nonparametric procedures for the location models and simple linear regression. It presents estimation and confidence intervals as well as testing. Sections on optimal scores and adaptive methods are presented. Chapter 11 offers an introduction to Bayesian methods. This includes traditional Bayesian procedures as well as Markov Chain Monte Carlo procedures, including the Gibbs sampler, for hierarchical and empirical Bayes procedures. Chapter 12 offers a comparison of robust and traditional least squares methods for linear models. It introduces the concepts of influence functions and breakdown points for estimators.

  Not every instructor will include these new chapters in a two-semester course, but those interested in one of these areas will find their inclusion very worthwhile. These last three chapters are independent of one another.

- We have occasionally made use of the statistical softwares R, (Ihaka and Gentleman, 1996), and S-PLUS, (S-PLUS, 2000), in this edition; see Venables and Ripley (2002). Students do not need resource to these packages to use the text but the use of one (or that of another package) does add a computational flavor. The package R is freeware which can be downloaded for free at the site

  http://lib.stat.cmu.edu/R/CRAN/

There are versions of R for unix, pc and mac platforms. We have written some R functions for several procedures in the text. These we have listed in Appendix B but they can also be downloaded at the site

http://www.stat.wmich.edu/mckean/HMC/Rcode

These functions will run in S-PLUS also.

- The reference list has been expanded so that instructors and students can find the original sources better.

- The order of presentation has been greatly improved and more exercises have been added. As a matter of fact, there are now over one thousand exercises and, further, many new examples have been added.

Most instructors will find selections from the first nine chapters sufficient for a two-semester course. However, we hope that many will want to insert one of the three topic chapters into their course. As a matter of fact, there is really enough material for a three semester sequence, which at one time we taught at the University of Iowa. A few optional sections have been marked with an asterisk.

We would like to thank the following reviewers who read through earlier versions of the manuscript: Walter Freiberger, Brown University; John Leahy, University of Oregon; Bradford Crain, Portland State University; Joseph S. Verducci, Ohio State University. and Hosam M. Mahmoud, George Washington University. Their suggestions were helpful in editing the final version.

Finally, we would like to thank George Lobell and Prentice Hall who provided funds to have the fifth edition converted to LATEX 2_ε and Kimberly Crimin who carried out this work. It certainly helped us in writing the sixth edition in LATEX 2_ε. Also, a special thanks to Ash Abebe for technical assistance. Last, but not least, we must thank our wives, Ann and Marge, who provided great support for our efforts. Let's hope the readers approve of the results.

Bob Hogg
Joe McKean
joe@stat.wmich.edu

Chapter 1

Probability and Distributions

1.1 Introduction

Many kinds of investigations may be characterized in part by the fact that repeated experimentation, under essentially the same conditions, is more or less standard procedure. For instance, in medical research, interest may center on the effect of a drug that is to be administered; or an economist may be concerned with the prices of three specified commodities at various time intervals; or the agronomist may wish to study the effect that a chemical fertilizer has on the yield of a cereal grain. The only way in which an investigator can elicit information about any such phenomenon is to perform the experiment. Each experiment terminates with an *outcome*. But it is characteristic of these experiments that the outcome cannot be predicted with certainty prior to the performance of the experiment.

Suppose that we have such an experiment, the outcome of which cannot be predicted with certainty, but the experiment is of such a nature that a collection of every possible outcome can be described prior to its performance. If this kind of experiment can be repeated under the same conditions, it is called a *random experiment*, and the collection of every possible outcome is called the experimental space or the *sample space*.

Example 1.1.1. In the toss of a coin, let the outcome tails be denoted by T and let the outcome heads be denoted by H. If we assume that the coin may be repeatedly tossed under the same conditions, then the toss of this coin is an example of a random experiment in which the outcome is one of the two symbols T and H; that is, the sample space is the collection of these two symbols. ∎

Example 1.1.2. In the cast of one red die and one white die, let the outcome be the ordered pair (number of spots up on the red die, number of spots up on the white die). If we assume that these two dice may be repeatedly cast under the same conditions, then the cast of this pair of dice is a random experiment. The sample space consists of the 36 ordered pairs: $(1,1), \ldots, (1,6), (2,1), \ldots, (2,6), \ldots, (6,6)$. ∎

Let $\mathcal{C}$ denote a sample space, let c denote an element of $\mathcal{C}$, and let C represent a collection of elements of $\mathcal{C}$. If, upon the performance of the experiment, the outcome

1

is in C, we shall say that the *event* C has occurred. Now conceive of our having made N repeated performances of the random experiment. Then we can count the number f of times (the frequency) that the event C actually occurred throughout the N performances. The ratio f/N is called the *relative frequency* of the event C in these N experiments. A relative frequency is usually quite erratic for small values of N, as you can discover by tossing a coin. But as N increases, experience indicates that we associate with the event C a number, say p, that is equal or approximately equal to that number about which the relative frequency seems to stabilize. If we do this, then the number p can be interpreted as that number which, in future performances of the experiment, the relative frequency of the event C will either equal or approximate. Thus, although we *cannot* predict the outcome of a random experiment, we *can*, for a large value of N, predict approximately the relative frequency with which the outcome will be in C. The number p associated with the event C is given various names. Sometimes it is called the *probability* that the outcome of the random experiment is in C; sometimes it is called the *probability* of the event C; and sometimes it is called the *probability measure* of C. The context usually suggests an appropriate choice of terminology.

Example 1.1.3. Let $\mathcal{C}$ denote the sample space of Example 1.1.2 and let C be the collection of every ordered pair of $\mathcal{C}$ for which the sum of the pair is equal to seven. Thus C is the collection $(1,6), (2,5), (3,4), (4,3), (5,2)$, and $(6,1)$. Suppose that the dice are cast $N = 400$ times and let f, the frequency of a sum of seven, be $f = 60$. Then the relative frequency with which the outcome was in C is $f/N = \frac{60}{400} = 0.15$. Thus we might associate with C a number p that is close to 0.15, and p would be called the probability of the event C. ∎

Remark 1.1.1. The preceding interpretation of probability is sometimes referred to as the *relative frequency approach*, and it obviously depends upon the fact that an experiment can be repeated under essentially identical conditions. However, many persons extend probability to other situations by treating it as a rational measure of belief. For example, the statement $p = \frac{2}{5}$ would mean to them that their *personal* or *subjective* probability of the event C is equal to $\frac{2}{5}$. Hence, if they are not opposed to gambling, this could be interpreted as a willingness on their part to bet on the outcome of C so that the two possible payoffs are in the ratio $p/(1-p) = \frac{2}{5}/\frac{3}{5} = \frac{2}{3}$. Moreover, if they truly believe that $p = \frac{2}{5}$ is correct, they would be willing to accept either side of the bet: (a) win 3 units if C occurs and lose 2 if it does not occur, or (b) win 2 units if C does not occur and lose 3 if it does. However, since the mathematical properties of probability given in Section 1.3 are consistent with either of these interpretations, the subsequent mathematical development does not depend upon which approach is used. ∎

The primary purpose of having a mathematical theory of statistics is to provide mathematical models for random experiments. Once a model for such an experiment has been provided and the theory worked out in detail, the statistician may, within this framework, make inferences (that is, draw conclusions) about the random experiment. The construction of such a model requires a theory of probability. One of the more logically satisfying theories of probability is that based on the concepts of sets and functions of sets. These concepts are introduced in Section 1.2.

1.2   Set Theory

The concept of a *set* or a *collection* of objects is usually left undefined. However, a particular set can be described so that there is no misunderstanding as to what collection of objects is under consideration. For example, the set of the first 10 positive integers is sufficiently well described to make clear that the numbers $\frac{3}{4}$ and 14 are not in the set, while the number 3 is in the set. If an object belongs to a set, it is said to be an *element* of the set. For example, if C denotes the set of real numbers x for which $0 \leq x \leq 1$, then $\frac{3}{4}$ is an element of the set C. The fact that $\frac{3}{4}$ is an element of the set C is indicated by writing $\frac{3}{4} \in C$. More generally, $c \in C$ means that c is an element of the set C.

The sets that concern us will frequently be *sets of numbers*. However, the language of sets of *points* proves somewhat more convenient than that of sets of numbers. Accordingly, we briefly indicate how we use this terminology. In analytic geometry considerable emphasis is placed on the fact that to each point on a line (on which an origin and a unit point have been selected) there corresponds one and only one number, say x; and that to each number x there corresponds one and only one point on the line. This one-to-one correspondence between the numbers and points on a line enables us to speak, without misunderstanding, of the "point x" instead of the "number x." Furthermore, with a plane rectangular coordinate system and with x and y numbers, to each symbol (x, y) there corresponds one and only one point in the plane; and to each point in the plane there corresponds but one such symbol. Here again, we may speak of the "point (x, y)," meaning the "ordered number pair x and y." This convenient language can be used when we have a rectangular coordinate system in a space of three or more dimensions. Thus the "point $(x_1, x_2, \ldots, x_n)$" means the numbers $x_1, x_2, \ldots, x_n$ in the order stated. Accordingly, in describing our sets, we frequently speak of a set of points (a set whose elements are points), being careful, of course, to describe the set so as to avoid any ambiguity. The notation $C = \{x : 0 \leq x \leq 1\}$ is read "C is the one-dimensional set of points x for which $0 \leq x \leq 1$." Similarly, $C = \{(x, y) : 0 \leq x \leq 1, 0 \leq y \leq 1\}$ can be read "C is the two-dimensional set of points (x, y) that are interior to, or on the boundary of, a square with opposite vertices at $(0, 0)$ and $(1, 1)$." We now give some definitions (together with illustrative examples) that lead to an elementary algebra of sets adequate for our purposes.

Definition 1.2.1. *If each element of a set C_1 is also an element of set C_2, the set C_1 is called a* **subset** *of the set C_2. This is indicated by writing $C_1 \subset C_2$. If $C_1 \subset C_2$ and also $C_2 \subset C_1$, the two sets have the same elements, and this is indicated by writing $C_1 = C_2$.*

Example 1.2.1. Let $C_1 = \{x : 0 \leq x \leq 1\}$ and $C_2 = \{x : -1 \leq x \leq 2\}$. Here the one-dimensional set C_1 is seen to be a subset of the one-dimensional set C_2; that is, $C_1 \subset C_2$. Subsequently, when the dimensionality of the set is clear, we shall not make specific reference to it. ∎

Example 1.2.2. Define the two sets $C_1 = \{(x, y) : 0 \leq x = y \leq 1\}$ and $C_2 = \{(x, y) : 0 \leq x \leq 1, 0 \leq y \leq 1\}$. Because the elements of C_1 are the points on one diagonal of the square, then $C_1 \subset C_2$. ∎

Definition 1.2.2. *If a set C has no elements, C is called the **null set**. This is indicated by writing $C = \phi$.*

Definition 1.2.3. *The set of all elements that belong to at least one of the sets C_1 and C_2 is called the **union** of C_1 and C_2. The union of C_1 and C_2 is indicated by writing $C_1 \cup C_2$. The union of several sets $C_1, C_2, C_3, \ldots$ is the set of all elements that belong to at least one of the several sets, denoted by $C_1 \cup C_2 \cup C_3 \cup \cdots$ or by $C_1 \cup C_2 \cup \cdots \cup C_k$ if a finite number k of sets is involved.*

Example 1.2.3. Define the sets $C_1 = \{x : x = 8, 9, 10, 11, \text{ or } 11 < x \le 12\}$ and $C_2 = \{x : x = 0, 1, \ldots, 10\}$. Then

$$
\begin{aligned}
C_1 \cup C_2 &= \{x : x = 0, 1, \ldots, 8, 9, 10, 11, \text{ or } 11 < x \le 12\} \\
&= \{x : x = 0, 1, \ldots, 8, 9, 10 \text{ or } 11 \le x \le 12\}. \quad \blacksquare
\end{aligned}
$$

Example 1.2.4. Define C_1 and C_2 as in Example 1.2.1. Then $C_1 \cup C_2 = C_2$. $\blacksquare$

Example 1.2.5. Let $C_2 = \phi$. Then $C_1 \cup C_2 = C_1$, for every set C_1. $\blacksquare$

Example 1.2.6. For every set C, $C \cup C = C$. $\blacksquare$

Example 1.2.7. Let

$$
C_k = \left\{ x : \tfrac{1}{k+1} \le x \le 1 \right\}, \quad k = 1, 2, 3, \ldots
$$

Then $C_1 \cup C_2 \cup C_3 \cup \cdots = \{x : 0 < x \le 1\}$. Note that the number zero is not in this set, since it is not in one of the sets $C_1, C_2, C_3, \ldots$. $\blacksquare$

Definition 1.2.4. *The set of all elements that belong to each of the sets C_1 and C_2 is called the **intersection** of C_1 and C_2. The intersection of C_1 and C_2 is indicated by writing $C_1 \cap C_2$. The intersection of several sets $C_1, C_2, C_3, \ldots$ is the set of all elements that belong to each of the sets $C_1, C_2, C_3, \ldots$. This intersection is denoted by $C_1 \cap C_2 \cap C_3 \cap \cdots$ or by $C_1 \cap C_2 \cap \cdots \cap C_k$ if a finite number k of sets is involved.*

Example 1.2.8. Let $C_1 = \{(0,0), (0,1), (1,1)\}$ and $C_2 = \{(1,1), (1,2), (2,1)\}$. Then $C_1 \cap C_2 = \{(1,1)\}$. $\blacksquare$

Example 1.2.9. Let $C_1 = \{(x,y) : 0 \le x + y \le 1\}$ and $C_2 = \{(x,y) : 1 < x + y\}$. Then C_1 and C_2 have no points in common and $C_1 \cap C_2 = \phi$. $\blacksquare$

Example 1.2.10. For every set C, $C \cap C = C$ and $C \cap \phi = \phi$. $\blacksquare$

Example 1.2.11. Let

$$
C_k = \left\{ x : 0 < x < \tfrac{1}{k} \right\}, \quad k = 1, 2, 3, \ldots
$$

Then $C_1 \cap C_2 \cap C_3 \cap \cdots$ is the null set, since there is no point that belongs to each of the sets $C_1 \cap C_2 \cap C_3 \cap \cdots$. $\blacksquare$

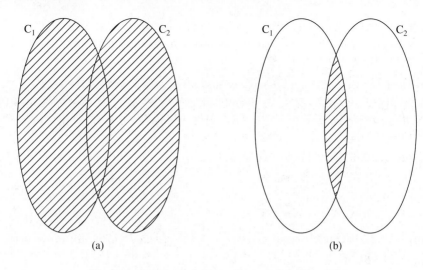

Figure 1.2.1: (a) $C_1 \cup C_2$ and (b) $C_1 \cap C_2$.

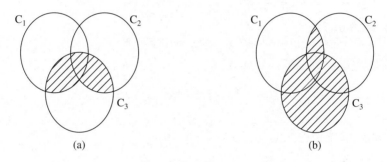

Figure 1.2.2: (a) $(C_1 \cup C_2) \cap C_3$ and (b) $(C_1 \cap C_2) \cup C_3$.

Example 1.2.12. Let C_1 and C_2 represent the sets of points enclosed, respectively, by two intersecting circles. Then the sets $C_1 \cup C_2$ and $C_1 \cap C_2$ are represented, respectively, by the shaded regions in the *Venn diagrams* in Figure 1.2.1. ∎

Example 1.2.13. Let C_1, C_2 and C_3 represent the sets of points enclosed, respectively, by three intersecting circles. Then the sets $(C_1 \cup C_2) \cap C_3$ and $(C_1 \cap C_2) \cup C_3$ are depicted in Figure 1.2.2. ∎

Definition 1.2.5. *In certain discussions or considerations, the totality of all elements that pertain to the discussion can be described. This set of all elements under consideration is given a special name. It is called the* **space.** *We shall often denote spaces by letters such as $\mathcal{C}$ and $\mathcal{D}$.*

Example 1.2.14. Let the number of heads, in tossing a coin four times, be denoted by x. Of necessity, the number of heads will be of the numbers $0, 1, 2, 3, 4$. Here, then, the space is the set $\mathcal{C} = \{0, 1, 2, 3, 4\}$. ∎

Example 1.2.15. Consider all nondegenerate rectangles of base x and height y. To be meaningful, both x and y must be positive. Then the space is given by the set $\mathcal{C} = \{(x, y) : x > 0, y > 0\}$. ∎

Definition 1.2.6. *Let $\mathcal{C}$ denote a space and let C be a subset of the set $\mathcal{C}$. The set that consists of all elements of $\mathcal{C}$ that are not elements of C is called the* **complement** *of C (actually, with respect to $\mathcal{C}$). The complement of C is denoted by C^c. In particular, $\mathcal{C}^c = \phi$.*

Example 1.2.16. Let $\mathcal{C}$ be defined as in Example 1.2.14, and let the set $C = \{0, 1\}$. The complement of C (with respect to $\mathcal{C}$) is $C^c = \{2, 3, 4\}$. ∎

Example 1.2.17. Given $C \subset \mathcal{C}$. Then $C \cup C^c = \mathcal{C}, C \cap C^c = \phi, C \cup \mathcal{C} = \mathcal{C}$, $C \cap \mathcal{C} = C$, and $(C^c)^c = C$. ∎

Example 1.2.18 (DeMorgan's Laws). A set of rules which will prove useful is known as DeMorgan's Laws. Let $\mathcal{C}$ denote a space and let $C_i \subset \mathcal{C}$, $i = 1, 2$. Then

$$(C_1 \cap C_2)^c = C_1^c \cup C_2^c \tag{1.2.1}$$
$$(C_1 \cup C_2)^c = C_1^c \cap C_2^c. \tag{1.2.2}$$

The reader is asked to prove these in Exercise 1.2.4. ∎

In the calculus, functions such as

$$f(x) = 2x, \quad -\infty < x < \infty$$

or

$$g(x, y) = \begin{cases} e^{-x-y} & 0 < x < \infty \quad 0 < y < \infty \\ 0 & \text{elsewhere,} \end{cases}$$

or possibly

$$h(x_1, x_2, \ldots, x_n) = \begin{cases} 3x_1 x_2 \cdots x_n & 0 \le x_i \le 1, \quad i = 1, 2, \ldots, n \\ 0 & \text{elsewhere,} \end{cases}$$

are of common occurrence. The value of $f(x)$ at the "point $x = 1$" is $f(1) = 2$; the value of $g(x, y)$ at the "point $(-1, 3)$" is $g(-1, 3) = 0$; the value of $h(x_1, x_2, \ldots, x_n)$ at the "point $(1, 1, \ldots, 1)$" is 3. Functions such as these are called functions of a point or, more simply, *point functions* because they are evaluated (if they have a value) at a point in a space of indicated dimension.

There is no reason why, if they prove useful, we should not have functions that can be evaluated, not necessarily at a point, but for an entire set of points. Such functions are naturally called functions of a set or, more simply, *set functions*. We shall give some examples of set functions and evaluate them for certain simple sets.

Example 1.2.19. Let C be a set in one-dimensional space and let $Q(C)$ be equal to the number of points in C which correspond to positive integers. Then $Q(C)$ is a function of the set C. Thus, if $C = \{x : 0 < x < 5\}$, then $Q(C) = 4$; if $C = \{-2, -1\}$, then $Q(C) = 0$; if $C = \{x : -\infty < x < 6\}$, then $Q(C) = 5$. ∎

Example 1.2.20. Let C be a set in two-dimensional space and let $Q(C)$ be the area of C, if C has a finite area; otherwise, let $Q(C)$ be undefined. Thus, if $C = \{(x, y) : x^2 + y^2 \leq 1\}$, then $Q(C) = \pi$; if $C = \{(0, 0), (1, 1), (0, 1)\}$, then $Q(C) = 0$; if $C = \{(x, y) : 0 \leq x, 0 \leq y, x + y \leq 1\}$, then $Q(C) = \frac{1}{2}$. ∎

Example 1.2.21. Let C be a set in three-dimensional space and let $Q(C)$ be the volume of C, if C has a finite volume; otherwise let $Q(C)$ be undefined. Thus, if $C = \{(x, y, z) : 0 \leq x \leq 2, 0 \leq y \leq 1, 0 \leq z \leq 3\}$, then $Q(C) = 6$; if $C = \{(x, y, z) : x^2 + y^2 + z^2 \geq 1\}$, then $Q(C)$ is undefined. ∎

At this point we introduce the following notations. The symbol

$$\int_C f(x)\, dx$$

will mean the ordinary (Riemann) integral of $f(x)$ over a prescribed one-dimensional set C; the symbol

$$\int\int_C g(x, y)\, dxdy$$

will mean the Riemann integral of $g(x, y)$ over a prescribed two-dimensional set C; and so on. To be sure, unless these sets C and these functions $f(x)$ and $g(x, y)$ are chosen with care, the integrals will frequently fail to exist. Similarly, the symbol

$$\sum_C f(x)$$

will mean the sum extended over all $x \in C$; the symbol

$$\sum\sum_C g(x, y)$$

will mean the sum extended over all $(x, y) \in C$; and so on.

Example 1.2.22. Let C be a set in one-dimensional space and let $Q(C) = \sum_C f(x)$, where

$$f(x) = \begin{cases} (\frac{1}{2})^x & x = 1, 2, 3, \ldots \\ 0 & \text{elsewhere.} \end{cases}$$

If $C = \{x : 0 \leq x \leq 3\}$, then

$$Q(C) = \frac{1}{2} + (\frac{1}{2})^2 + (\frac{1}{2})^3 = \frac{7}{8}. \quad ∎$$

Example 1.2.23. Let $Q(C) = \sum_C f(x)$, where

$$f(x) = \begin{cases} p^x (1 - p)^{1-x} & x = 0, 1 \\ 0 & \text{elsewhere.} \end{cases}$$

If $C = \{0\}$, then

$$Q(C) = \sum_{x=0}^{0} p^x (1-p)^{1-x} = 1 - p;$$

if $C = \{x : 1 \leq x \leq 2\}$, then $Q(C) = f(1) = p$. ∎

Example 1.2.24. Let C be a one-dimensional set and let

$$Q(C) = \int_C e^{-x} dx.$$

Thus, if $C = \{x : 0 \leq x < \infty\}$, then

$$Q(C) = \int_0^\infty e^{-x} dx = 1;$$

if $C = \{x : 1 \leq x \leq 2\}$, then

$$Q(C) = \int_1^2 e^{-x} dx = e^{-1} - e^{-2};$$

if $C_1 = \{x : 0 \leq x \leq 1\}$ and $C_2 = \{x : 1 < x \leq 3\}$, then

$$\begin{aligned}
Q(C_1 \cup C_2) &= \int_0^3 e^{-x} dx \\
&= \int_0^1 e^{-x} dx + \int_1^3 e^{-x} dx \\
&= Q(C_1) + Q(C_2);
\end{aligned}$$

if $C = C_1 \cup C_2$, where $C_1 = \{x : 0 \leq x \leq 2\}$ and $C_2 = \{x : 1 \leq x \leq 3\}$, then

$$\begin{aligned}
Q(C) = Q(C_1 \cup C_2) &= \int_0^3 e^{-x} dx \\
&= \int_0^2 e^{-x} dx + \int_1^3 e^{-x} dx - \int_1^2 e^{-x} dx \\
&= Q(C_1) + Q(C_2) - Q(C_1 \cap C_2). \quad ∎
\end{aligned}$$

Example 1.2.25. Let C be a set in n-dimensional space and let

$$Q(C) = \int \cdots \int_C dx_1 dx_2 \cdots dx_n.$$

If $C = \{(x_1, x_2, \ldots, x_n) : 0 \leq x_1 \leq x_2 \leq \cdots \leq x_n \leq 1\}$, then

$$\begin{aligned}
Q(C) &= \int_0^1 \int_0^{x_n} \cdots \int_0^{x_3} \int_0^{x_2} dx_1 dx_2 \cdots dx_{n-1} dx_n \\
&= \frac{1}{n!},
\end{aligned}$$

where $n! = n(n-1) \cdots 3 \cdot 2 \cdot 1$. ∎

EXERCISES

1.2.1. Find the union $C_1 \cup C_2$ and the intersection $C_1 \cap C_2$ of the two sets C_1 and C_2, where:

(a) $C_1 = \{0, 1, 2, \}$, $C_2 = \{2, 3, 4\}$.

(b) $C_1 = \{x : 0 < x < 2\}$, $C_2 = \{x : 1 \leq x < 3\}$.

(c) $C_1 = \{(x, y) : 0 < x < 2, 1 < y < 2\}$, $C_2 = \{(x, y) : 1 < x < 3, 1 < y < 3\}$.

1.2.2. Find the complement C^c of the set C with respect to the space $\mathcal{C}$ if:

(a) $\mathcal{C} = \{x : 0 < x < 1\}$, $C = \{x : \frac{5}{8} < x < 1\}$.

(b) $\mathcal{C} = \{(x, y, z) : x^2 + y^2 + z^2 \leq 1\}$, $C = \{(x, y, z) : x^2 + y^2 + z^2 = 1\}$.

(c) $\mathcal{C} = \{(x, y) : |x| + |y| \leq 2\}$, $C = \{(x, y) : x^2 + y^2 < 2\}$.

1.2.3. List all possible arrangements of the four letters m, a, r, and y. Let C_1 be the collection of the arrangements in which y is in the last position. Let C_2 be the collection of the arrangements in which m is in the first position. Find the union and the intersection of C_1 and C_2.

1.2.4. Referring to Example 1.2.18, verify DeMorgan's Laws (1.2.1) and (1.2.2) by using Venn diagrams and then prove that the laws are true. Generalize the laws to arbitrary unions and intersections.

1.2.5. By the use of Venn diagrams, in which the space $\mathcal{C}$ is the set of points enclosed by a rectangle containing the circles, compare the following sets. These laws are called the *distributive laws*.

(a) $C_1 \cap (C_2 \cup C_3)$ and $(C_1 \cap C_2) \cup (C_1 \cap C_3)$.

(b) $C_1 \cup (C_2 \cap C_3)$ and $(C_1 \cup C_2) \cap (C_1 \cup C_3)$.

1.2.6. If a sequence of sets $C_1, C_2, C_3, \ldots$ is such that $C_k \subset C_{k+1}$, $k = 1, 2, 3, \ldots$, the sequence is said to be a *nondecreasing sequence*. Give an example of this kind of sequence of sets.

1.2.7. If a sequence of sets $C_1, C_2, C_3, \ldots$ is such that $C_k \supset C_{k+1}$, $k = 1, 2, 3, \ldots$, the sequence is said to be a *nonincreasing sequence*. Give an example of this kind of sequence of sets.

1.2.8. If $C_1, C_2, C_3, \ldots$ are sets such that $C_k \subset C_{k+1}$, $k = 1, 2, 3, \ldots$, $\lim_{k \to \infty} C_k$ is defined as the union $C_1 \cup C_2 \cup C_3 \cup \cdots$. Find $\lim_{k \to \infty} C_k$ if:

(a) $C_k = \{x : 1/k \leq x \leq 3 - 1/k\}$, $k = 1, 2, 3, \ldots$.

(b) $C_k = \{(x, y) : 1/k \leq x^2 + y^2 \leq 4 - 1/k\}$, $k = 1, 2, 3, \ldots$.

1.2.9. If $C_1, C_2, C_3, \ldots$ are sets such that $C_k \supset C_{k+1}$, $k = 1, 2, 3, \ldots$, $\lim_{k \to \infty} C_k$ is defined as the intersection $C_1 \cap C_2 \cap C_3 \cap \cdots$. Find $\lim_{k \to \infty} C_k$ if:

(a) $C_k = \{x : 2 - 1/k < x \leq 2\}$, $k = 1, 2, 3, \ldots$.

(b) $C_k = \{x : 2 < x \leq 2 + 1/k\}$, $k = 1, 2, 3, \ldots$.

(c) $C_k = \{(x, y) : 0 \leq x^2 + y^2 \leq 1/k\}$, $k = 1, 2, 3, \ldots$.

1.2.10. For every one-dimensional set C, define the function $Q(C) = \sum_C f(x)$, where $f(x) = (\frac{2}{3})(\frac{1}{3})^x$, $x = 0, 1, 2, \ldots$, zero elsewhere. If $C_1 = \{x : x = 0, 1, 2, 3\}$ and $C_2 = \{x : x = 0, 1, 2, \ldots\}$, find $Q(C_1)$ and $Q(C_2)$.
Hint: Recall that $S_n = a + ar + \cdots + ar^{n-1} = a(1 - r^n)/(1 - r)$ and, hence, it follows that $\lim_{n \to \infty} S_n = a/(1 - r)$ provided that $|r| < 1$.

1.2.11. For every one-dimensional set C for which the integral exists, let $Q(C) = \int_C f(x)\,dx$, where $f(x) = 6x(1 - x)$, $0 < x < 1$, zero elsewhere; otherwise, let $Q(C)$ be undefined. If $C_1 = \{x : \frac{1}{4} < x < \frac{3}{4}\}$, $C_2 = \{\frac{1}{2}\}$, and $C_3 = \{x : 0 < x < 10\}$, find $Q(C_1), Q(C_2)$, and $Q(C_3)$.

1.2.12. For every two-dimensional set C contained in R^2 for which the integral exists, let $Q(C) = \int_C \int (x^2 + y^2)\,dxdy$. If $C_1 = \{(x, y) : -1 \leq x \leq 1, -1 \leq y \leq 1\}$, $C_2 = \{(x, y) : -1 \leq x = y \leq 1\}$, and $C_3 = \{(x, y) : x^2 + y^2 \leq 1\}$, find $Q(C_1), Q(C_2)$, and $Q(C_3)$.

1.2.13. Let $\mathcal{C}$ denote the set of points that are interior to, or on the boundary of, a square with opposite vertices at the points $(0, 0)$ and $(1, 1)$. Let $Q(C) = \int_C \int dy\,dx$.

(a) If $C \subset \mathcal{C}$ is the set $\{(x, y) : 0 < x < y < 1\}$, compute $Q(C)$.

(b) If $C \subset \mathcal{C}$ is the set $\{(x, y) : 0 < x = y < 1\}$, compute $Q(C)$.

(c) If $C \subset \mathcal{C}$ is the set $\{(x, y) : 0 < x/2 \leq y \leq 3x/2 < 1\}$, compute $Q(C)$.

1.2.14. Let $\mathcal{C}$ be the set of points interior to or on the boundary of a cube with edge of length 1. Moreover, say that the cube is in the first octant with one vertex at the point $(0, 0, 0)$ and an opposite vertex at the point $(1, 1, 1)$. Let $Q(C) = \int \int \int_C dx\,dydz$.

(a) If $C \subset \mathcal{C}$ is the set $\{(x, y, z) : 0 < x < y < z < 1\}$, compute $Q(C)$.

(b) If C is the subset $\{(x, y, z) : 0 < x = y = z < 1\}$, compute $Q(C)$.

1.2.15. Let C denote the set $\{(x, y, z) : x^2 + y^2 + z^2 \leq 1\}$. Evaluate $Q(C) = \int \int \int_C \sqrt{x^2 + y^2 + z^2}\,dxdydz$. *Hint:* Use spherical coordinates.

1.2.16. To join a certain club, a person must be either a statistician or a mathematician or both. Of the 25 members in this club, 19 are statisticians and 16 are mathematicians. How many persons in the club are both a statistician and a mathematician?

1.2.17. After a hard-fought football game, it was reported that, of the 11 starting players, 8 hurt a hip, 6 hurt an arm, 5 hurt a knee, 3 hurt both a hip and an arm, 2 hurt both a hip and a knee, 1 hurt both an arm and a knee, and no one hurt all three. Comment on the accuracy of the report.

1.3 The Probability Set Function

Let $\mathcal{C}$ denote the sample space. What should be our collection of events? As discussed in Section 2, we are interested in assigning probabilities to events, complements of events, and union and intersections of events (i.e., compound events). Hence, we want our collection of events to include these combinations of events. Such a collection of events is called a σ-field of subsets of $\mathcal{C}$, which is defined as follows.

Definition 1.3.1 (σ-Field). *Let $\mathcal{B}$ be a collection of subsets of $\mathcal{C}$. We say $\mathcal{B}$ is a σ-field if*

> *(1). $\phi \in \mathcal{B}$, ($\mathcal{B}$ is not empty).*
>
> *(2). If $C \in \mathcal{B}$ then $C^c \in \mathcal{B}$, ($\mathcal{B}$ is closed under complements).*
>
> *(3). If the sequence of sets $\{C_1, C_2, \ldots\}$ is in $\mathcal{B}$ then $\bigcup_{i=1}^{\infty} C_i \in \mathcal{B}$,*
>
> *($\mathcal{B}$ is closed under countable unions).*

Note by (1) and (2), a σ-field always contains ϕ and $\mathcal{C}$. By (2) and (3), it follows from DeMorgan's laws that a σ-field is closed under countable intersections, besides countable unions. This is what we need for our collection of events. To avoid confusion please note the equivalence: let $C \subset \mathcal{C}$. Then

> the statement C is an event is equivalent to the statement $C \in \mathcal{B}$.

We will use these expressions interchangeably in the text. Next, we present some examples of σ-fields.

1. Let $\mathcal{C}$ be any set and let $C \subset \mathcal{C}$. Then $\mathcal{B} = \{C, C^c, \phi, \mathcal{C}\}$ is a σ-field.

2. Let $\mathcal{C}$ be any set and let $\mathcal{B}$ be the power set of $\mathcal{C}$, (the collection of all subsets of $\mathcal{C}$). Then $\mathcal{B}$ is a σ-field.

3. Suppose $\mathcal{D}$ is a nonempty collection of subsets of $\mathcal{C}$. Consider the collection of events,
$$\mathcal{B} = \cap\{\mathcal{E} : \mathcal{D} \subset \mathcal{E} \text{ and } \mathcal{E} \text{ is a } \sigma\text{-field}\}. \tag{1.3.1}$$
As Exercise 1.3.20 shows, $\mathcal{B}$ is a σ-field. It is the smallest σ-field which contains $\mathcal{D}$; hence, it is sometimes referred to as the σ-field generated by $\mathcal{D}$.

4. Let $\mathcal{C} = R$, where R is the set of all real numbers. Let $\mathcal{I}$ be the set of all open intervals in R. Let
$$\mathcal{B}_0 = \cap\{\mathcal{E} : \mathcal{I} \subset \mathcal{E} \text{ and } \mathcal{E} \text{ is a } \sigma\text{-field}\}. \tag{1.3.2}$$

The σ-field, $\mathcal{B}_0$ is often referred to as the **Borel** σ-field on the real line. As Exercise 1.3.21 shows, it contains not only the open intervals, but the closed and half-open intervals of real numbers. This is an important σ-field.

Now that we have a sample space, $\mathcal{C}$, and our collection of events, $\mathcal{B}$, we can define the third component in our probability space, namely a probability set function. In order to motivate its definition, we consider the relative frequency approach to probability.

Remark 1.3.1. The definition of probability consists of three axioms which we will motivate by the following three intuitive properties of relative frequency. Let C be an event. Suppose we repeat the experiment N times. Then the relative frequency of C is $f_C = \#\{C\}/N$, where $\#\{C\}$ denotes the number of times C occurred in the N repetitions. Note that $f_C \geq 0$ and $f_C \leq 1$. These are the first two properties. For the third, suppose that C_1 and C_2 are disjoint events. Then $f_{C_1 \cup C_2} = f_{C_1} + f_{C_2}$. These three properties of relative frequencies form the axioms of a probability, except that the third axiom is in terms of countable unions. As with the axioms of probability, the readers should check that the theorems we prove below about probabilities agree with their intuition of relative frequency. ∎

Definition 1.3.2 (Probability). *Let $\mathcal{C}$ be a sample space and let $\mathcal{B}$ be a σ-field on $\mathcal{C}$. Let P be a real valued function defined on $\mathcal{B}$. Then P is a **probability set function** if P satisfies the following three conditions:*

1. *$P(C) \geq 0$, for all $C \in \mathcal{B}$.*

2. *$P(\mathcal{C}) = 1$.*

3. *If $\{C_n\}$ is a sequence of sets in $\mathcal{B}$ and $C_m \cap C_n = \phi$ for all $m \neq n$, then*

$$P\left(\bigcup_{n=1}^{\infty} C_n\right) = \sum_{n=1}^{\infty} P(C_n).$$

A probability set function tells us how the probability is distributed over the set of events, $\mathcal{B}$. In this sense we speak of a distribution of probability. We will often drop the word set and refer to P as a probability function.

The following theorems give us some other properties of a probability set function. In the statement of each of these theorems, $P(C)$ is taken, tacitly, to be a probability set function defined on a σ-field $\mathcal{B}$ of a sample space $\mathcal{C}$.

Theorem 1.3.1. *For each event $C \in \mathcal{B}$, $P(C) = 1 - P(C^c)$.*

Proof: We have $\mathcal{C} = C \cup C^c$ and $C \cap C^c = \phi$. Thus, from (2) and (3) of Definition 1.3.2, it follows that

$$1 = P(C) + P(C^c)$$

which is the desired result. ∎

Theorem 1.3.2. *The probability of the null set is zero; that is, $P(\phi) = 0$.*

Proof: In Theorem 1.3.1, take $C = \phi$ so that $C^c = \mathcal{C}$. Accordingly, we have

$$P(\phi) = 1 - P(\mathcal{C}) = 1 - 1 = 0$$

and the theorem is proved. ∎

Theorem 1.3.3. *If C_1 and C_2 are events such that $C_1 \subset C_2$, then $P(C_1) \le P(C_2)$.*

Proof: Now $C_2 = C_1 \cup (C_1^c \cap C_2)$ and $C_1 \cap (C_1^c \cap C_2) = \phi$. Hence, from (3) of Definition 1.3.2,
$$P(C_2) = P(C_1) + P(C_1^c \cap C_2).$$
From (1) of Definition 1.3.2, $P(C_1^c \cap C_2) \ge 0$. Hence, $P(C_2) \ge P(C_1)$. ∎

Theorem 1.3.4. *For each $C \in \mathcal{B}$, $0 \le P(C) \le 1$.*

Proof: Since $\phi \subset C \subset \mathcal{C}$, we have by Theorem 1.3.3 that

$$P(\phi) \le P(C) \le P(\mathcal{C}) \quad \text{or} \quad 0 \le P(C) \le 1$$

the desired result. ∎

Part (3) of the definition of probability says that $P(C_1 \cup C_2) = P(C_1) + P(C_2)$, if C_1 and C_2 are disjoint, i.e., $C_1 \cap C_2 = \phi$. The next theorem, gives the rule for any two events.

Theorem 1.3.5. *If C_1 and C_2 are events in $\mathcal{C}$, then*

$$P(C_1 \cup C_2) = P(C_1) + P(C_2) - P(C_1 \cap C_2).$$

Proof: Each of the sets $C_1 \cup C_2$ and C_2 can be represented, respectively, as a union of nonintersecting sets as follows:

$$C_1 \cup C_2 = C_1 \cup (C_1^c \cap C_2) \quad \text{and} \quad C_2 = (C_1 \cap C_2) \cup (C_1^c \cap C_2).$$

Thus, from (3) of Definition 1.3.2,

$$P(C_1 \cup C_2) = P(C_1) + P(C_1^c \cap C_2)$$

and

$$P(C_2) = P(C_1 \cap C_2) + P(C_1^c \cap C_2).$$

If the second of these equations is solved for $P(C_1^c \cap C_2)$ and this result substituted in the first equation, we obtain,

$$P(C_1 \cup C_2) = P(C_1) + P(C_2) - P(C_1 \cap C_2).$$

This completes the proof. ∎

Remark 1.3.2 (Inclusion-Exclusion Formula). It is easy to show (Exercise 1.3.9) that

$$P(C_1 \cup C_2 \cup C_3) = p_1 - p_2 + p_3,$$

where

$$
\begin{aligned}
p_1 &= P(C_1) + P(C_2) + P(C_3) \\
p_2 &= P(C_1 \cap C_2) + P(C_1 \cap C_3) + P(C_2 \cap C_3) \\
p_3 &= P(C_1 \cap C_2 \cap C_3).
\end{aligned}
\tag{1.3.3}
$$

This can be generalized to the **inclusion-exclusion formula**:

$$P(C_1 \cup C_2 \cup \cdots \cup C_k) = p_1 - p_2 + p_3 - \cdots + (-1)^{k+1} p_k, \tag{1.3.4}$$

where p_i equals the sum of the probabilities of all possible intersections involving i sets. It is clear in the case $k = 3$ that $p_1 \geq p_2 \geq p_3$, but more generally $p_1 \geq p_2 \geq \cdots \geq p_k$. As shown in Theorem 1.3.7,

$$p_1 = P(C_1) + P(C_2) + \cdots + P(C_k) \geq P(C_1 \cup C_2 \cup \cdots \cup C_k).$$

This is known as *Boole's inequality*. For $k = 2$, we have

$$1 \geq P(C_1 \cup C_2) = P(C_1) + P(C_2) - P(C_1 \cap C_2)$$

which gives **Bonferroni's Inequality**,

$$P(C_1 \cap C_2) \geq P(C_1) + P(C_2) - 1, \tag{1.3.5}$$

that is only useful when $P(C_1)$ and $P(C_2)$ are large. The inclusion-exclusion formula provides other inequalities that are useful; such as,

$$p_1 \geq P(C_1 \cup C_2 \cup \cdots \cup C_k) \geq p_1 - p_2$$

and

$$p_1 - p_2 + p_3 \geq P(C_1 \cup C_2 \cup \cdots \cup C_k) \geq p_1 - p_2 + p_3 - p_4.$$

Exercise 1.3.10 gives an interesting application of the inclusion-exclusion formula to the matching problem. ∎

Example 1.3.1. Let $\mathcal{C}$ denote the sample space of Example 1.1.2. Let the probability set function assign probability of $\frac{1}{36}$ to each of the 36 points in $\mathcal{C}$; that is the dice are fair. If $C_1 = \{(1,1), (2,1), (3,1), (4,1), (5,1)\}$ and $C_2 = \{(1,2), (2,2), (3,2)\}$, then $P(C_1) = \frac{5}{36}$, $P(C_2) = \frac{3}{36}$, $P(C_1 \cup C_2) = \frac{8}{36}$, and $P(C_1 \cap C_2) = 0$. ∎

Example 1.3.2. Two coins are to be tossed and the outcome is the ordered pair (face on the first coin, face on the second coin). Thus the sample space may be represented as $\mathcal{C} = \{(H,H), (H,T), (T,H), (T,T)\}$. Let the probability set function assign a probability of $\frac{1}{4}$ to each element of $\mathcal{C}$. Let $C_1 = \{(H,H), (H,T)\}$ and $C_2 = \{(H,H), (T,H)\}$. Then $P(C_1) = P(C_2) = \frac{1}{2}$, $P(C_1 \cap C_2) = \frac{1}{4}$, and, in accordance with Theorem 1.3.5, $P(C_1 \cup C_2) = \frac{1}{2} + \frac{1}{2} - \frac{1}{4} = \frac{3}{4}$. ∎

Let $\mathcal{C}$ denote a sample space and let $C_1, C_2, C_3, \ldots$ denote events of $\mathcal{C}$. If these events are such that no two have an element in common, they are called mutually disjoint sets and the corresponding events $C_1, C_2, C_3, \ldots$ are said to be *mutually exclusive events*. Then $P(C_1 \cup C_2 \cup C_3 \cup \cdots) = P(C_1) + P(C_2) + P(C_3) + \cdots$, in accordance with (3) of Definition 1.3.2. Moreover, if $\mathcal{C} = C_1 \cup C_2 \cup C_3 \cup \cdots$, the mutually exclusive events are further characterized as being *exhaustive* and the probability of their union is obviously equal to 1.

Example 1.3.3 (Equilikely Case). Let $\mathcal{C}$ be partitioned into k mutually disjoint subsets $C_1, C_2, \ldots, C_k$ in such a way that the union of these k mutually disjoint subsets is the sample space $\mathcal{C}$. Thus the events $C_1, C_2, \ldots, C_k$ are mutually exclusive and exhaustive. Suppose that the random experiment is of such a character that it is reasonable to *assume* that each of the mutually exclusive and exhaustive events $C_i, i = 1, 2, \ldots, k$, has the same probability. It is necessary, then, that $P(C_i) = 1/k$, $i = 1, 2, \ldots, k$; and we often say that the events $C_1, C_2, \ldots, C_k$ are *equally likely*. Let the event E be the union of r of these mutually exclusive events, say

$$E = C_1 \cup C_2 \cup \cdots \cup C_r, \quad r \leq k.$$

Then

$$P(E) = P(C_1) + P(C_2) + \cdots + P(C_r) = \frac{r}{k}.$$

Frequently, the integer k is called the total number of ways (for this particular partition of $\mathcal{C}$) in which the random experiment can terminate and the integer r is called the number of ways that are favorable to the event E. So, in this terminology, $P(E)$ is equal to the number of ways favorable to the event E divided by the total number of ways in which the experiment can terminate. It should be emphasized that in order to assign, *in this manner*, the probability r/k to the event E, we must assume that each of the mutually exclusive and exhaustive events $C_1, C_2, \ldots, C_k$ has the same probability $1/k$. This assumption of equally likely events then becomes a *part* of our probability model. Obviously, if this assumption is not realistic in an application, the probability of the event E cannot be computed in this way. ■

In order to illustrate the equilikely case, it is helpful to use some elementary counting rules. These are usually discussed in an elementary algebra course. In the next remark, we offer a brief review of these rules.

Remark 1.3.3 (Counting Rules). Suppose we have two experiments. The first experiment results in m outcomes while the second experiment results in n outcomes. The composite experiment, first experiment followed by second experiment, has mn outcomes which can be represented as mn ordered pairs. This is called the **multiplication rule** or the mn-**rule**. This is easily extended to more than two experiments.

Let A be a set with n elements. Suppose we are interested in k-tuples whose components are elements of A. Then by the extended multiplication rule, there are $n \cdot n \cdots n = n^k$ such k-tuples whose components are elements of A. Next, suppose $k \leq n$ and we are interested in k-tuples whose components are distinct (no repeats) elements of A. There are n elements from which to choose for the first component, $n-1$ for the second component, ..., $n-(k-1)$ for the kth. Hence, by the multiplication rule, there are $n(n-1)\cdots(n-(k-1))$ such k-tuples with distinct elements. We call each such k-tuple a **permutation** and use the symbol P_k^n to denote the number of k permutations taken from a set of n elements. Hence, we have the formula,

$$P_k^n = n(n-1)\cdots(n-(k-1)) = \frac{n!}{(n-k)!}. \tag{1.3.6}$$

Next suppose order is not important, so instead of counting the number of permutations we want to count the number of subsets of k elements taken from A. We will use the symbol $\binom{n}{k}$ to denote the total number of these subsets. Consider a subset of k elements from A. By the permutation rule it generates $P_k^k = k(k-1)\cdots 1$ permutations. Furthermore all these permutations are distinct from permutations generated by other subsets of k elements from A. Finally, each permutation of k distinct elements drawn form A, must be generated by one of these subsets. Hence, we have just shown that $P_k^n = \binom{n}{k}k!$; that is,

$$\binom{n}{k} = \frac{n!}{k!(n-k)!}. \tag{1.3.7}$$

We often use the terminology combinations instead of subsets. So we say that there are $\binom{n}{k}$ **combinations** of k things taken from a set of n things. Another common symbol for $\binom{n}{k}$ is C_k^n.

It is interesting to note that if we expand the binomial,

$$(a+b)^n = (a+b)(a+b)\cdots(a+b),$$

we get

$$(a+b)^n = \sum_{k=0}^{n} \binom{n}{k} a^k b^{n-k}; \tag{1.3.8}$$

because we can select the k factors from which to take a in $\binom{n}{k}$ ways. So $\binom{n}{k}$ is also referred to as a **binomial coefficient**. ∎

Example 1.3.4 (Poker Hands). Let a card be drawn at random from an ordinary deck of 52 playing cards which has been well shuffled. The sample space $\mathcal{C}$ is the union of $k = 52$ outcomes, and it is reasonable to assume that each of these outcomes has the same probability $\frac{1}{52}$. Accordingly, if E_1 is the set of outcomes that are spades, $P(E_1) = \frac{13}{52} = \frac{1}{4}$ because there are $r_1 = 13$ spades in the deck; that is, $\frac{1}{4}$ is the probability of drawing a card that is a spade. If E_2 is the set of outcomes that are kings, $P(E_2) = \frac{4}{52} = \frac{1}{13}$ because there are $r_2 = 4$ kings in the deck; that is, $\frac{1}{13}$ is the probability of drawing a card that is a king. These computations are very easy because there are no difficulties in the determination of the appropriate values of r and k.

However, instead of drawing only one card, suppose that five cards are taken, at random and without replacement, from this deck; i.e, a 5-card poker hand. In this instance, order is not important. So a hand is a subset of 5 elements drawn from a set of 52 elements. Hence, by (1.3.7) there are $\binom{52}{5}$ poker hands. If the deck is well shuffled, each hand should be equilikely; i.e., each hand has probability $1/\binom{52}{5}$. We can now compute the probabilities of some interesting poker hands. Let E_1 be the event of a flush, all 5 cards of the same suit. There are $\binom{4}{1} = 4$ suits to choose for the flush and in each suit there are $\binom{13}{5}$ possible hands; hence, using the multiplication rule, the probability of getting a flush is

$$P(E_1) = \frac{\binom{4}{1}\binom{13}{5}}{\binom{52}{5}} = \frac{4 \cdot 1287}{2598960} = 0.00198.$$

Real poker players note that this includes the probability of obtaining a straight flush.

Next, consider the probability of the event E_2 of getting exactly 3 of a kind, (the other two cards are distinct and are of different kinds). Choose the kind for the 3, in $\binom{13}{1}$ ways; choose the 3, in $\binom{4}{3}$ ways; choose the other 2 kinds, in $\binom{12}{2}$ ways; and choose 1 card from each of these last two kinds, in $\binom{4}{1}\binom{4}{1}$ ways. Hence the probability of exactly 3 of a kind is

$$P(E_2) = \frac{\binom{13}{1}\binom{4}{3}\binom{12}{2}\binom{4}{1}^2}{\binom{52}{5}} = 0.0211.$$

Now suppose that E_3 is the set of outcomes in which exactly three cards are kings and exactly two cards are queens. Select the kings, in $\binom{4}{3}$ ways and select the queens, in $\binom{4}{2}$ ways. Hence, the probability of E_3 is,

$$P(E_3) = \binom{4}{3}\binom{4}{2}\bigg/\binom{52}{5} = 0.0000093.$$

The event E_3 is an example of a full house: 3 of one kind and 2 of another kind. Exercise 1.3.19 asks for the determination of the probability of a full house. ∎

Example 1.3.4 and the previous discussion allow us to see one way in which we can define a probability set function, that is, a set function that satisfies the requirements of Definition 1.3.2. Suppose that our space $\mathcal{C}$ consists of k distinct points, which, for this discussion, we take to be in a one-dimensional space. If the random experiment that ends in one of those k points is such that it is reasonable to assume that these points are equally likely, we could assign $1/k$ to each point and let, for $C \subset \mathcal{C}$,

$$P(C) = \frac{\text{number of points in } C}{k}$$

$$= \sum_{x \in C} f(x), \quad \text{where} \quad f(x) = \frac{1}{k}, \quad x \in \mathcal{C}.$$

For illustration, in the cast of a die, we could take $\mathcal{C} = \{1, 2, 3, 4, 5, 6\}$ and $f(x) = \frac{1}{6}$, $x \in \mathcal{C}$, if we believe the die to be unbiased. Clearly, such a set function satisfies Definition 1.3.2.

The word *unbiased* in this illustration suggests the possibility that all six points might *not*, in all such cases, be equally likely. As a matter of fact, *loaded* dice do exist. In the case of a loaded die, some numbers occur more frequently than others in a sequence of casts of that die. For example, suppose that a die has been loaded so that the relative frequencies of the numbers in $\mathcal{C}$ *seem to stabilize* proportional to the number of spots that are on the *up* side. Thus we might assign $f(x) = x/21$, $x \in \mathcal{C}$, and the corresponding

$$P(C) = \sum_{x \in C} f(x)$$

would satisfy Definition 1.3.2. For illustration, this means that if $C = \{1, 2, 3\}$, then

$$P(C) = \sum_{x=1}^{3} f(x) = \frac{1}{21} + \frac{2}{21} + \frac{3}{21} = \frac{6}{21} = \frac{2}{7}.$$

Whether this probability set function is realistic can only be checked by performing the random experiment a large number of times.

We end this section with another property of probability which will be useful in the sequel. Recall in Exercise 1.2.8 we said that a sequence of events $\{C_n\}$ is an increasing sequence if $C_n \subset C_{n+1}$, for all n, in which case we wrote $\lim_{n \to \infty} C_n = \cup_{n=1}^{\infty} C_n$. Consider, $\lim_{n \to \infty} P(C_n)$. The question is: can we interchange the limit and P? As the following theorem shows the answer is yes. The result also holds for a decreasing sequence of events. Because of this interchange, this theorem is sometimes referred to as the continuity theorem of probability.

Theorem 1.3.6. *Let $\{C_n\}$ be an increasing sequence of events. Then*

$$\lim_{n \to \infty} P(C_n) = P(\lim_{n \to \infty} C_n) = P\left(\bigcup_{n=1}^{\infty} C_n\right). \tag{1.3.9}$$

Let $\{C_n\}$ be a decreasing sequence of events. Then

$$\lim_{n \to \infty} P(C_n) = P(\lim_{n \to \infty} C_n) = P\left(\bigcap_{n=1}^{\infty} C_n\right). \tag{1.3.10}$$

Proof. We prove the result (1.3.9) and leave the second result as Exercise 1.3.22. Define the sets, called rings as: $R_1 = C_1$ and for $n > 1$, $R_n = C_n \cap C_{n-1}^c$. It follows that $\cup_{n=1}^{\infty} C_n = \cup_{n=1}^{\infty} R_n$ and that $R_m \cap R_n = \phi$, for $m \neq n$. Also, $P(R_n) = P(C_n) - P(C_{n-1})$. Applying the third axiom of probability yields the following string of equalities:

$$P\left[\lim_{n \to \infty} C_n\right] = P\left(\bigcup_{n=1}^{\infty} C_n\right) = P\left(\bigcup_{n=1}^{\infty} R_n\right) = \sum_{n=1}^{\infty} P(R_n) = \lim_{n \to \infty} \sum_{j=1}^{n} P(R_j)$$

$$= \lim_{n \to \infty} \left\{P(C_1) + \sum_{j=2}^{n} [P(C_j) - P(C_{j-1})]\right\} = \lim_{n \to \infty} P(C_n). \tag{1.3.11}$$

This is the desired result. ∎

Another useful result for arbitrary unions is given by

Theorem 1.3.7 (Boole's Inequality). *Let $\{C_n\}$ be an arbitrary sequence of events. Then*

$$P\left(\bigcup_{n=1}^{\infty} C_n\right) \leq \sum_{n=1}^{\infty} P(C_n). \tag{1.3.12}$$

Proof: Let $D_n = \bigcup_{i=1}^{n} C_i$. Then $\{D_n\}$ is an increasing sequence of events which go up to $\bigcup_{n=1}^{\infty} C_n$. Also, for all j, $D_j = D_{j-1} \cup C_j$. Hence, by Theorem 1.3.5

$$P(D_j) \leq P(D_{j-1}) + P(C_j),$$

that is,

$$P(D_j) - P(D_{j-1}) \leq P(C_j).$$

In this case, the C_is are replaced by the D_is in expression (1.3.11). Hence, using the above inequality in this expression and the fact that $P(C_1) = P(D_1)$ we have

$$P\left(\bigcup_{n=1}^{\infty} C_n\right) = P\left(\bigcup_{n=1}^{\infty} D_n\right) = \lim_{n \to \infty} \left\{ P(D_1) + \sum_{j=2}^{n} [P(D_j) - P(D_{j-1})] \right\}$$

$$\leq \lim_{n \to \infty} \sum_{j=1}^{n} P(C_j) = \sum_{n=1}^{\infty} P(C_n),$$

which was to be proved. ∎

EXERCISES

1.3.1. A positive integer from one to six is to be chosen by casting a die. Thus the elements c of the sample space $\mathcal{C}$ are $1, 2, 3, 4, 5, 6$. Suppose $C_1 = \{1, 2, 3, 4\}$ and $C_2 = \{3, 4, 5, 6\}$. If the probability set function P assigns a probability of $\frac{1}{6}$ to each of the elements of $\mathcal{C}$, compute $P(C_1)$, $P(C_2)$, $P(C_1 \cap C_2)$, and $P(C_1 \cup C_2)$.

1.3.2. A random experiment consists of drawing a card from an ordinary deck of 52 playing cards. Let the probability set function P assign a probability of $\frac{1}{52}$ to each of the 52 possible outcomes. Let C_1 denote the collection of the 13 hearts and let C_2 denote the collection of the 4 kings. Compute $P(C_1)$, $P(C_2)$, $P(C_1 \cap C_2)$, and $P(C_1 \cup C_2)$.

1.3.3. A coin is to be tossed as many times as necessary to turn up one head. Thus the elements c of the sample space $\mathcal{C}$ are H, TH, TTH, $TTTH$, and so forth. Let the probability set function P assign to these elements the respective probabilities $\frac{1}{2}, \frac{1}{4}, \frac{1}{8}, \frac{1}{16}$, and so forth. Show that $P(\mathcal{C}) = 1$. Let $C_1 = \{c : c$ is $H, TH, TTH, TTTH,$ or $TTTTH\}$. Compute $P(C_1)$. Next, suppose that $C_2 = \{c : c$ is $TTTTH$ or $TTTTTH\}$. Compute $P(C_2)$, $P(C_1 \cap C_2)$, and $P(C_1 \cup C_2)$.

1.3.4. If the sample space is $\mathcal{C} = C_1 \cup C_2$ and if $P(C_1) = 0.8$ and $P(C_2) = 0.5$, find $P(C_1 \cap C_2)$.

1.3.5. Let the sample space be $\mathcal{C} = \{c : 0 < c < \infty\}$. Let $C \subset \mathcal{C}$ be defined by $C = \{c : 4 < c < \infty\}$ and take $P(C) = \int_C e^{-x} \, dx$. Evaluate $P(C)$, $P(C^c)$, and $P(C \cup C^c)$.

1.3.6. If the sample space is $\mathcal{C} = \{c : -\infty < c < \infty\}$ and if $C \subset \mathcal{C}$ is a set for which the integral $\int_C e^{-|x|} \, dx$ exists, show that this set function is not a probability set function. What constant do we multiply the integrand by to make it a probability set function?

1.3.7. If C_1 and C_2 are subsets of the sample space $\mathcal{C}$, show that

$$P(C_1 \cap C_2) \leq P(C_1) \leq P(C_1 \cup C_2) \leq P(C_1) + P(C_2).$$

1.3.8. Let C_1, C_2, and C_3 be three mutually disjoint subsets of the sample space $\mathcal{C}$. Find $P[(C_1 \cup C_2) \cap C_3]$ and $P(C_1^c \cup C_2^c)$.

1.3.9. Consider Remark 1.3.2.

(a) If C_1, C_2, and C_3 are subsets of $\mathcal{C}$, show that

$$\begin{aligned}
P(C_1 \cup C_2 \cup C_3) &= P(C_1) + P(C_2) + P(C_3) - P(C_1 \cap C_2) \\
&\quad - P(C_1 \cap C_3) - P(C_2 \cap C_3) + P(C_1 \cap C_2 \cap C_3),
\end{aligned}$$

(b) Now prove the general inclusion-exclusion formula given by the expression (1.3.4).

1.3.10. Suppose we turn over cards simultaneously from two well shuffled decks of ordinary playing cards. We say we obtain an exact match on a particular turn if the same card appears from each deck; for example, the queen of spades against the queen of spades. Let p_M equal the probability of at least one exact match.

(a) Show that

$$p_M = 1 - \frac{1}{2!} + \frac{1}{3!} - \frac{1}{4!} + \cdots - \frac{1}{52!}.$$

Hint: Let C_i denote the event of an exact match on the *ith* turn. Then $p_M = P(C_1 \cup C_2 \cup \cdots \cup C_{52})$. Now use the the general inclusion-exclusion formula given by (1.3.4). In this regard note that: $P(C_i) = 1/52$ and hence $p_1 = 52(1/52) = 1$. Also, $P(C_i \cap C_j) = 50!/52!$ and, hence, $p_2 = \binom{52}{2}/(52 \cdot 51)$.

(b) Show that p_m is approximately equal to $1 - e^{-1} = 0.632$.

Remark 1.3.4. In order to solve a number of exercises, like (1.3.11) - (1.3.19), certain reasonable assumptions must be made. ∎

1.3.11. A bowl contains 16 chips, of which 6 are red, 7 are white, and 3 are blue. If four chips are taken at random and without replacement, find the probability that: (a) each of the 4 chips is red; (b) none of the 4 chips is red; (c) there is at least 1 chip of each color.

1.3.12. A person has purchased 10 of 1000 tickets sold in a certain raffle. To determine the five prize winners, 5 tickets are to be drawn at random and without replacement. Compute the probability that this person will win at least one prize. *Hint:* First compute the probability that the person does not win a prize.

1.3.13. Compute the probability of being dealt at random and without replacement a 13-card bridge hand consisting of: (a) 6 spades, 4 hearts, 2 diamonds, and 1 club; (b) 13 cards of the same suit.

1.3.14. Three distinct integers are chosen at random from the first 20 positive integers. Compute the probability that: (a) their sum is even; (b) their product is even.

1.3.15. There are 5 red chips and 3 blue chips in a bowl. The red chips are numbered 1,2,3,4,5, respectively, and the blue chips are numbered 1,2,3, respectively. If 2 chips are to be drawn at random and without replacement, find the probability that these chips have either the same number or the same color.

1.3.16. In a lot of 50 light bulbs, there are 2 bad bulbs. An inspector examines 5 bulbs, which are selected at random and without replacement.

(a) Find the probability of at least 1 defective bulb among the 5.

(b) How many bulbs should be examined so that the probability of finding at least 1 bad bulb exceeds $\frac{1}{2}$?

1.3.17. If $C_1, \ldots, C_k$ are k events in the sample space $\mathcal{C}$, show that the probability that at least one of the events occurs is one minus the probability that none of them occur; i.e.,

$$P(C_1 \cup \cdots \cup C_k) = 1 - P(C_1^c \cap \cdots \cap C_k^c). \qquad (1.3.13)$$

1.3.18. A secretary types three letters and the three corresponding envelopes. In a hurry, he places at random one letter in each envelope. What is the probability that at least one letter is in the correct envelope? *Hint:* Let C_i be the event that the *ith* letter is in the correct envelope. Expand $P(C_1 \cup C_2 \cup C_3)$ to determine the probability.

1.3.19. Consider poker hands drawn form a well shuffled deck as described in Example 1.3.4. Determine the probability of a full house; i.e, three of one kind and two of another.

1.3.20. Suppose $\mathcal{D}$ is a nonempty collection of subsets of $\mathcal{C}$. Consider the collection of events,

$$\mathcal{B} = \cap\{\mathcal{E} \ : \ \mathcal{D} \subset \mathcal{E} \text{ and } \mathcal{E} \text{ is a } \sigma\text{-field}\}.$$

Note that $\phi \in \mathcal{B}$ because it is in each σ-field, and, hence, in particular, it is in each σ-field $\mathcal{E} \supset \mathcal{D}$. Continue in this way to show that $\mathcal{B}$ is a σ-field.

1.3.21. Let $\mathcal{C} = R$, where R is the set of all real numbers. Let $\mathcal{I}$ be the set of all open intervals in R. Recall from (1.3.2) the Borel σ-field on the real line; i.e, the σ-field $\mathcal{B}_0$ given by

$$\mathcal{B}_0 = \cap \{\mathcal{E} : \mathcal{I} \subset \mathcal{E} \text{ and } \mathcal{E} \text{ is a } \sigma\text{-field}\}.$$

By definition $\mathcal{B}_0$ contains the open intervals. Because $[a, \infty) = (-\infty, a)^c$ and $\mathcal{B}_0$ is closed under complements, it contains all intervals of the form $[a, \infty)$, for $a \in R$. Continue in this way and show that $\mathcal{B}_0$ contains all the closed and half-open intervals of real numbers.

1.3.22. Prove expression (1.3.10).

1.3.23. Suppose the experiment is to choose a real number at random in the interval $(0, 1)$. For any subinterval $(a, b) \subset (0, 1)$, it seems reasonable to assign the probability $P[(a, b)] = b - a$; i.e., the probability of selecting the point from a subinterval is directly proportional to the length of the subinterval. If this is the case, choose an appropriate sequence of subintervals and use expression (1.3.10) to show that $P[\{a\}] = 0$, for all $a \in (0, 1)$.

1.3.24. Consider the events C_1, C_2, C_3.

(a) Suppose C_1, C_2, C_3 are mutually exclusive events. If $P(C_i) = p_i$, $i = 1, 2, 3$, what is the restriction on the sum $p_1 + p_2 + p_3$?

(b) In the notation of Part (a), if $p_1 = 4/10$, $p_2 = 3/10$, and $p_3 = 5/10$ are C_1, C_2, C_3 mutually exclusive?

1.4 Conditional Probability and Independence

In some random experiments, we are interested only in those outcomes that are elements of a subset C_1 of the sample space $\mathcal{C}$. This means, for our purposes, that the sample space is effectively the subset C_1. We are now confronted with the problem of defining a probability set function with C_1 as the "new" sample space.

Let the probability set function $P(C)$ be defined on the sample space $\mathcal{C}$ and let C_1 be a subset of $\mathcal{C}$ such that $P(C_1) > 0$. We agree to consider only those outcomes of the random experiment that are elements of C_1; in essence, then, we take C_1 to be a sample space. Let C_2 be another subset of $\mathcal{C}$. How, relative to the new sample space C_1, do we want to define the probability of the event C_2? Once defined, this probability is called the *conditional probability* of the event C_2, relative to the hypothesis of the event C_1; or, more briefly, the conditional probability of C_2, given C_1. Such a conditional probability is denoted by the symbol $P(C_2|C_1)$. We now return to the question that was raised about the definition of this symbol. Since C_1 is now the sample space, the only elements of C_2 that concern us are those, if any, that are also elements of C_1, that is, the elements of $C_1 \cap C_2$. It seems desirable, then, to define the symbol $P(C_2|C_1)$ in such a way that

$$P(C_1|C_1) = 1 \quad \text{and} \quad P(C_2|C_1) = P(C_1 \cap C_2|C_1).$$

Moreover, from a relative frequency point of view, it would seem logically inconsistent if we did not require that the ratio of the probabilities of the events $C_1 \cap C_2$ and C_1, relative to the space C_1, be the same as the ratio of the probabilities of these events relative to the space C; that is, we should have

$$\frac{P(C_1 \cap C_2 | C_1)}{P(C_1 | C_1)} = \frac{P(C_1 \cap C_2)}{P(C_1)}.$$

These three desirable conditions imply that the relation

$$P(C_2 | C_1) = \frac{P(C_1 \cap C_2)}{P(C_1)}$$

is a suitable *definition* of the conditional probability of the event C_2, given the event C_1, provided that $P(C_1) > 0$. Moreover, we have

1. $P(C_2 | C_1) \geq 0$.

2. $P(C_2 \cup C_3 \cup \cdots | C_1) = P(C_2 | C_1) + P(C_3 | C_1) + \cdots$, provided that $C_2, C_3, \ldots$ are mutually disjoint sets.

3. $P(C_1 | C_1) = 1$.

Properties (1) and (3) are evident; proof of property (2) is left as Exercise (1.4.1). But these are precisely the conditions that a probability set function must satisfy. Accordingly, $P(C_2 | C_1)$ is a probability set function, defined for subsets of C_1. It may be called the conditional probability set function, relative to the hypothesis C_1; or the conditional probability set function, given C_1. It should be noted that this conditional probability set function, given C_1, is defined at this time only when $P(C_1) > 0$.

Example 1.4.1. A hand of 5 cards is to be dealt at random without replacement from an ordinary deck of 52 playing cards. The conditional probability of an all-spade hand (C_2), relative to the hypothesis that there are at least 4 spades in the hand (C_1), is, since $C_1 \cap C_2 = C_2$,

$$P(C_2 | C_1) = \frac{P(C_2)}{P(C_1)} = \frac{\binom{13}{5} / \binom{52}{5}}{\left[\binom{13}{4}\binom{39}{1} + \binom{13}{5}\right] / \binom{52}{5}}$$

$$= \frac{\binom{13}{5}}{\binom{13}{4}\binom{39}{1} + \binom{13}{5}} = 0.0441.$$

Note that this is not the same as drawing for a spade to complete a flush in draw poker; see Exercise 1.4.3. ∎

From the definition of the conditional probability set function, we observe that

$$P(C_1 \cap C_2) = P(C_1)P(C_2 | C_1).$$

This relation is frequently called the *multiplication rule* for probabilities. Sometimes, after considering the nature of the random experiment, it is possible to make reasonable assumptions so that both $P(C_1)$ and $P(C_2|C_1)$ can be assigned. Then $P(C_1 \cap C_2)$ can be computed under these assumptions. This will be illustrated in Examples 1.4.2 and 1.4.3.

Example 1.4.2. A bowl contains eight chips. Three of the chips are red and the remaining five are blue. Two chips are to be drawn successively, at random and without replacement. We want to compute the probability that the first draw results in a red chip (C_1) and that the second draw results in a blue chip (C_2). It is reasonable to assign the following probabilities:

$$P(C_1) = \tfrac{3}{8} \quad \text{and} \quad P(C_2|C_1) = \tfrac{5}{7} \ .$$

Thus, under these assignments, we have $P(C_1 \cap C_2) = (\tfrac{3}{8})(\tfrac{5}{7}) = \tfrac{15}{56} = 0.2679$. ∎

Example 1.4.3. From an ordinary deck of playing cards, cards are to be drawn successively, at random and without replacement. The probability that the third spade appears on the sixth draw is computed as follows. Let C_1 be the event of two spades in the first five draws and let C_2 be the event of a spade on the sixth draw. Thus the probability that we wish to compute is $P(C_1 \cap C_2)$. It is reasonable to take

$$P(C_1) = \frac{\binom{13}{2}\binom{39}{3}}{\binom{52}{5}} = 0.2743 \quad \text{and} \quad P(C_2|C_1) = \frac{11}{47} = 0.2340.$$

The desired probability $P(C_1 \cap C_2)$ is then the product of these two numbers, which to four places is 0.0642. ∎

The multiplication rule can be extended to three or more events. In the case of three events, we have, by using the multiplication rule for two events,

$$
\begin{aligned}
P(C_1 \cap C_2 \cap C_3) &= P[(C_1 \cap C_2) \cap C_3] \\
&= P(C_1 \cap C_2)P(C_3|C_1 \cap C_2).
\end{aligned}
$$

But $P(C_1 \cap C_2) = P(C_1)P(C_2|C_1)$. Hence, provided $P(C_1 \cap C_2) > 0$,

$$P(C_1 \cap C_2 \cap C_3) = P(C_1)P(C_2|C_1)P(C_3|C_1 \cap C_2).$$

This procedure can be used to extend the multiplication rule to four or more events. The general formula for k events can be proved by mathematical induction.

Example 1.4.4. Four cards are to be dealt successively, at random and without replacement, from an ordinary deck of playing cards. The probability of receiving a spade, a heart, a diamond, and a club, in that order, is $(\tfrac{13}{52})(\tfrac{13}{51})(\tfrac{13}{50})(\tfrac{13}{49}) = 0.0044$. This follows from the extension of the multiplication rule. ∎

Consider k mutually exclusive and exhaustive events $C_1, C_2, \ldots, C_k$ such that $P(C_i) > 0$, $i = 1, 2, \ldots, k$. Suppose these events form a partition of $\mathcal{C}$. Here the events $C_1, C_2, \ldots, C_k$ do *not* need to be equally likely. Let C be another event. Thus C occurs with one and only one of the events $C_1, C_2, \ldots, C_k$; that is,

$$
\begin{aligned}
C &= C \cap (C_1 \cup C_2 \cup \cdots C_k) \\
&= (C \cap C_1) \cup (C \cap C_2) \cup \cdots \cup (C \cap C_k).
\end{aligned}
$$

Since $C \cap C_i$, $i = 1, 2, \ldots, k$, are mutually exclusive, we have

$$P(C) = P(C \cap C_1) + P(C \cap C_2) + \cdots + P(C \cap C_k).$$

However, $P(C \cap C_i) = P(C_i)P(C|C_i)$, $i = 1, 2, \ldots, k$; so

$$
\begin{aligned}
P(C) &= P(C_1)P(C|C_1) + P(C_2)P(C|C_2) + \cdots + P(C_k)P(C|C_k) \\
&= \sum_{i=1}^{k} P(C_i)P(C|C_i).
\end{aligned}
$$

This result is sometimes called the *law of total probability*.

Suppose, also, that $P(C) > 0$. From the definition of conditional probability, we have, using the law of total probability, that

$$P(C_j|C) = \frac{P(C \cap C_j)}{P(C)} = \frac{P(C_j)P(C|C_j)}{\sum_{i=1}^{k} P(C_i)P(C|C_i)}, \tag{1.4.1}$$

which is the well-known *Bayes' theorem*. This permits us to calculate the conditional probability of C_j, given C, from the probabilities of $C_1, C_2, \ldots, C_k$ and the conditional probabilities of C, given C_i, $i = 1, 2, \ldots, k$.

Example 1.4.5. Say it is known that bowl C_1 contains 3 red and 7 blue chips and bowl C_2 contains 8 red and 2 blue chips. All chips are identical in size and shape. A die is cast and bowl C_1 is selected if five or six spots show on the side that is up; otherwise, bowl C_2 is selected. In a notation that is fairly obvious, it seems reasonable to assign $P(C_1) = \frac{2}{6}$ and $P(C_2) = \frac{4}{6}$. The selected bowl is handed to another person and one chip is taken at random. Say that this chip is red, an event which we denote by C. By considering the contents of the bowls, it is reasonable to assign the conditional probabilities $P(C|C_1) = \frac{3}{10}$ and $P(C|C_2) = \frac{8}{10}$. Thus the conditional probability of bowl C_1, given that a red chip is drawn, is

$$
\begin{aligned}
P(C_1|C) &= \frac{P(C_1)P(C|C_1)}{P(C_1)P(C|C_1) + P(C_2)P(C|C_2)} \\
&= \frac{\left(\frac{2}{6}\right)\left(\frac{3}{10}\right)}{\left(\frac{2}{6}\right)\left(\frac{3}{10}\right) + \left(\frac{4}{6}\right)\left(\frac{8}{10}\right)} = \frac{3}{19}.
\end{aligned}
$$

In a similar manner, we have $P(C_2|C) = \frac{16}{19}$. ∎

In Example 1.4.5, the probabilities $P(C_1) = \frac{2}{6}$ and $P(C_2) = \frac{4}{6}$ are called *prior probabilities* of C_1 and C_2, respectively, because they are known to be due to the random mechanism used to select the bowls. After the chip is taken and observed to be red, the conditional probabilities $P(C_1|C) = \frac{3}{19}$ and $P(C_2|C) = \frac{16}{19}$ are called *posterior probabilities*. Since C_2 has a larger proportion of red chips than does C_1, it appeals to one's intuition that $P(C_2|C)$ should be larger than $P(C_2)$ and, of course, $P(C_1|C)$ should be smaller than $P(C_1)$. That is, intuitively the chances of having bowl C_2 are better once that a red chip is observed than before a chip is taken. Bayes' theorem provides a method of determining exactly what those probabilities are.

Example 1.4.6. Three plants, C_1, C_2 and C_3, produce respectively, 10, 50, and 40 percent of a company's output. Although plant C_1 is a small plant, its manager believes in high quality and only 1 percent of its products are defective. The other two, C_2 and C_3, are worse and produce items that are 3 and 4 percent defective, respectively. All products are sent to a central warehouse. One item is selected at random and observed to be defective, say event C. The conditional probability that it comes from plant C_1 is found as follows. It is natural to assign the respective prior probabilities of getting an item from the plants as $P(C_1) = 0.1$, $P(C_2) = 0.5$ and $P(C_3) = 0.4$, while the conditional probabilities of defective items are $P(C|C_1) = 0.01$, $P(C|C_2) = 0.03$, and $P(C|C_3) = 0.04$. Thus the posterior probability of C_1, given a defective, is

$$P(C_1|C) = \frac{P(C_1 \cap C)}{P(C)} = \frac{(0.10)(0.01)}{(0.1)(0.01) + (0.5)(0.03) + (0.4)(0.04)},$$

which equals $\frac{1}{32}$; this is much smaller than the prior probability $P(C_1) = \frac{1}{10}$. This is as it should be because the fact that the item is defective decreases the chances that it comes from the high-quality plant C_1. ∎

Example 1.4.7. Suppose we want to investigate the percentage of abused children in a certain population. The events of interest are: a child is abused (A) and its complement a child is not abused ($N = A^c$). For the purposes of this example, we will assume that $P(A) = 0.01$ and, hence, $P(N) = 0.99$. The classification as to whether a child is abused or not is based upon a doctor's examination. Because doctors are not perfect, they sometimes classify an abused child (A) as one that is not abused (N_D, where N_D means classified as not abused by a doctor). On the other hand, doctors sometimes classify a nonabused child (N) as abused (A_D). Suppose these error rates of misclassification are $P(N_D | A) = 0.04$ and $P(A_D | N) = 0.05$; thus the probabilities of correct decisions are $P(A_D | A) = 0.96$ and $P(N_D | N) = 0.95$. Let us compute the probability that a child taken at random is classified as abused by a doctor. Because this can happen in two ways, $A \cap A_D$ or $N \cap A_D$, we have

$$P(A_D) = P(A_D | A)P(A) + P(A_D | N)P(N) = (0.96)(0.01) + (0.05)(0.99) = 0.0591,$$

which is quite high relative to the probability that of an abused child, 0.01. Further, the probability that a child is abused when the doctor classified the child as abused is

$$P(A | A_D) = \frac{P(A \cap A_D)}{P(A_D)} = \frac{(0.96)(0.01)}{0.0591} = 0.1624,$$

which is quite low. In the same way, the probability that a child is not abused when the doctor classified the child as abused is 0.8376, which is quite high. The reason that these probabilities are so poor at recording the true situation is that the doctors' error rates are so high relative to the fraction 0.01 of the population that is abused. An investigation such as this would, hopefully, lead to better training of doctors for classifying abused children. See, also, Exercise 1.4.17. ∎

Sometimes it happens that the occurrence of event C_1 does not change the probability of event C_2; that is, when $P(C_1) > 0$,

$$P(C_2|C_1) = P(C_2).$$

In this case, we say that the events C_1 and C_2 are *independent*. Moreover, the multiplication rule becomes

$$P(C_1 \cap C_2) = P(C_1)P(C_2|C_1) = P(C_1)P(C_2). \qquad (1.4.2)$$

This, in turn, implies, when $P(C_2) > 0$, that

$$P(C_1|C_2) = \frac{P(C_1 \cap C_2)}{P(C_2)} = \frac{P(C_1)P(C_2)}{P(C_2)} = P(C_1).$$

Note that if $P(C_1) > 0$ and $P(C_2) > 0$ then by the above discussion independence is equivalent to

$$P(C_1 \cap C_2) = P(C_1)P(C_2). \qquad (1.4.3)$$

What if either $P(C_1) = 0$ or $P(C_2) = 0$? In either case, the right side of (1.4.3) is 0. However, the left side is 0 also because $C_1 \cap C_2 \subset C_1$ and $C_1 \cap C_2 \subset C_2$. Hence, we will take equation (1.4.3) as our formal definition of independence; that is,

Definition 1.4.1. *Let C_1 and C_2 be two events. We say that C_1 and C_2 are independent if equation (1.4.3) holds.*

Suppose C_1 and C_2 are independent events. Then the following three pairs of events are independent: C_1 and C_2^c, C_1^c and C_2, and C_1^c and C_2^c, (see Exercise 1.4.11).

Remark 1.4.1. Events that are *independent* are sometimes called *statistically independent*, *stochastically independent*, or *independent in a probability sense*. In most instances, we use independent without a modifier if there is no possibility of misunderstanding. ∎

Example 1.4.8. A red die and a white die are cast in such a way that the number of spots on the two sides that are up are independent events. If C_1 represents a four on the red die and C_2 represents a three on the white die, with an equally likely assumption for each side, we assign $P(C_1) = \frac{1}{6}$ and $P(C_2) = \frac{1}{6}$. Thus, from independence, the probability of the ordered pair (red = 4, white = 3) is

$$P[(4,3)] = (\tfrac{1}{6})(\tfrac{1}{6}) = \tfrac{1}{36}.$$

The probability that the sum of the up spots of the two dice equals seven is

$$P[(1,6),(2,5),(3,4),(4,3),(5,2),(6,1)]$$
$$= \left(\tfrac{1}{6}\right)\left(\tfrac{1}{6}\right) + \left(\tfrac{1}{6}\right)\left(\tfrac{1}{6}\right) + \left(\tfrac{1}{6}\right)\left(\tfrac{1}{6}\right) + \left(\tfrac{1}{6}\right)\left(\tfrac{1}{6}\right) + \left(\tfrac{1}{6}\right)\left(\tfrac{1}{6}\right) + \left(\tfrac{1}{6}\right)\left(\tfrac{1}{6}\right) = \tfrac{6}{36}.$$

In a similar manner, it is easy to show that the probabilities of the sums of $2, 3, 4, 5, 6, 7, 8, 9, 10, 11, 12$ are, respectively,

$$\tfrac{1}{36}, \tfrac{2}{36}, \tfrac{3}{36}, \tfrac{4}{36}, \tfrac{5}{36}, \tfrac{6}{36}, \tfrac{5}{36}, \tfrac{4}{36}, \tfrac{3}{36}, \tfrac{2}{36}, \tfrac{1}{36}. \quad ∎$$

Suppose now that we have three events, C_1, C_2, and C_3. We say that they are *mutually independent* if and only if they are *pairwise independent*:

$$P(C_1 \cap C_3) = P(C_1)P(C_3), \quad P(C_1 \cap C_2) = P(C_1)P(C_2),$$
$$P(C_2 \cap C_3) = P(C_2)P(C_3),$$

and

$$P(C_1 \cap C_2 \cap C_3) = P(C_1)P(C_2)P(C_3).$$

More generally, the n events $C_1, C_2, \ldots, C_n$ are *mutually independent* if and only if for every collection of k of these events, $2 \leq k \leq n$, the following is true:

Say that $d_1, d_2, \ldots, d_k$ are k distinct integers from $1, 2, \ldots, n$; then

$$P(C_{d_1} \cap C_{d_2} \cap \cdots \cap C_{d_k}) = P(C_{d_1})P(C_{d_2}) \cdots P(C_{d_k}).$$

In particular, if $C_1, C_2, \ldots, C_n$ are mutually independent, then

$$P(C_1 \cap C_2 \cap \cdots \cap C_n) = P(C_1)P(C_2) \cdots P(C_n).$$

Also, as with two sets, many combinations of these events and their complements are independent, such as

1. The events C_1^c and $C_2 \cup C_3^c \cup C_4$ are independent;

2. The events $C_1 \cup C_2^c$, C_3^c and $C_4 \cap C_5^c$ are mutually independent.

If there is no possibility of misunderstanding, *independent* is often used without the modifier *mutually* when considering more than two events.

We often perform a sequence of random experiments in such a way that the events associated with one of them are independent of the events associated with the others. For convenience, we refer to these events as *independent experiments*, meaning that the respective events are independent. Thus we often refer to independent flips of a coin or independent casts of a die or, more generally, independent trials of some given random experiment.

Example 1.4.9. A coin is flipped independently several times. Let the event C_i represent a head (H) on the ith toss; thus C_i^c represents a tail (T). Assume that C_i and C_i^c are equally likely; that is, $P(C_i) = P(C_i^c) = \frac{1}{2}$. Thus the probability of an ordered sequence like HHTH is, from independence,

$$P(C_1 \cap C_2 \cap C_3^c \cap C_4) = P(C_1)P(C_2)P(C_3^c)P(C_4) = (\tfrac{1}{2})^4 = \tfrac{1}{16}.$$

Similarly, the probability of observing the first head on the third flip is

$$P(C_1^c \cap C_2^c \cap C_3) = P(C_1^c)P(C_2^c)P(C_3) = (\tfrac{1}{2})^3 = \tfrac{1}{8}.$$

Also, the probability of getting at least one head on four flips is

$$\begin{aligned}
P(C_1 \cup C_2 \cup C_3 \cup C_4) &= 1 - P[(C_1 \cup C_2 \cup C_3 \cup C_4)^c] \\
&= 1 - P(C_1^c \cap C_2^c \cap C_3^c \cap C_4^c) \\
&= 1 - (\tfrac{1}{2})^4 = \tfrac{15}{16}.
\end{aligned}$$

See Exercise 1.4.13 to justify this last probability. ∎

Example 1.4.10. A computer system is built so that if component K_1 fails, it is bypassed and K_2 is used. If K_2 fails then K_3 is used. Suppose that the probability that K_1 fails is 0.01, that K_2 fails is 0.03, and that K_3 fails is 0.08. Moreover, we can assume that the failures are mutually independent events. Then the probability of failure of the system is

$$(0.01)(0.03)(0.08) = 0.000024,$$

as all three components would have to fail. Hence, the probability that the system does not fail is $1 - 0.000024 = 0.999976$. ∎

EXERCISES

1.4.1. If $P(C_1) > 0$ and if $C_2, C_3, C_4, \ldots$ are mutually disjoint sets, show that $P(C_2 \cup C_3 \cup \cdots |C_1) = P(C_2|C_1) + P(C_3|C_1) + \cdots$.

1.4.2. Assume that $P(C_1 \cap C_2 \cap C_3) > 0$. Prove that

$$P(C_1 \cap C_2 \cap C_3 \cap C_4) = P(C_1)P(C_2|C_1)P(C_3|C_1 \cap C_2)P(C_4|C_1 \cap C_2 \cap C_3).$$

1.4.3. Suppose we are playing draw poker. We are dealt (from a well shuffled deck) 5 cards which contain 4 spades and another card of a different suit. We decide to discard the card of a different suit and draw one card from the remaining cards to complete a flush in spades (all 5 cards spades). Determine the probability of completing the flush.

1.4.4. From a well shuffled deck of ordinary playing cards, four cards are turned over one at a time without replacement. What is the probability that the spades and red cards alternate?

1.4.5. A hand of 13 cards is to be dealt at random and without replacement from an ordinary deck of playing cards. Find the conditional probability that there are at least three kings in the hand given that the hand contains at least two kings.

1.4.6. A drawer contains eight different pairs of socks. If six socks are taken at random and without replacement, compute the probability that there is at least one matching pair among these six socks. *Hint:* Compute the probability that there is not a matching pair.

1.4.7. A pair of dice is cast until either the sum of seven or eight appears.

(a) Show that the probability of a seven before an eight is 6/11.

(b) Next, this pair of dice is cast until a seven appears twice or until each of a six and eight have appeared at least once. Show that the probability of the six and eight occurring before two sevens is 0.546.

1.4.8. In a certain factory, machines I, II, and III are all producing springs of the same length. Machines I, II, and III produce 1%, 4% and 2% defective springs, respectively. Of the total production of springs in the factory, Machine I produces 30%, Machine II produces 25%, and Machine III produces 45%.

(a) If one spring is selected at random from the total springs produced in a given day, determine the probability that it is defective.

(b) Given that the selected spring is defective, find the conditional probability that it was produced by Machine II.

1.4.9. Bowl I contains 6 red chips and 4 blue chips. Five of these 10 chips are selected at random and without replacement and put in bowl II, which was originally empty. One chip is then drawn at random from bowl II. Given that this chip is blue, find the conditional probability that 2 red chips and 3 blue chips are transferred from bowl I to bowl II.

1.4.10. A professor of statistics has two boxes of computer disks: box C_1 contains seven Verbatim disks and three Control Data disks and box C_2 contains two Verbatim disks and eight Control Data disks. She selects a box at random with probabilities $P(C_1) = \frac{2}{3}$ and $P(C_2) = \frac{1}{3}$ because of their respective locations. A disk is then selected at random and the event C occurs if it is from Control Data. Using an equally likely assumption for each disk in the selected box, compute $P(C_1|C)$ and $P(C_2|C)$.

1.4.11. If C_1 and C_2 are independent events, show that the following pairs of events are also independent: (a) C_1 and C_2^c, (b) C_1^c and C_2, and (c) C_1^c and C_2^c. *Hint:* In (a), write $P(C_1 \cap C_2^c) = P(C_1)P(C_2^c|C_1) = P(C_1)[1 - P(C_2|C_1)]$. From independence of C_1 and C_2, $P(C_2|C_1) = P(C_2)$.

1.4.12. Let C_1 and C_2 be independent events with $P(C_1) = 0.6$ and $P(C_2) = 0.3$. Compute (a) $P(C_1 \cap C_2)$; $(b)P(C_1 \cup C_2)$; $(c)P(C_1 \cup C_2^c)$.

1.4.13. Generalize Exercise 1.2.5 to obtain

$$(C_1 \cup C_2 \cup \cdots \cup C_k)^c = C_1^c \cap C_2^c \cap \cdots \cap C_k^c.$$

Say that $C_1, C_2, \ldots, C_k$ are independent events that have respective probabilities $p_1, p_2, \ldots, p_k$. Argue that the probability of at least one of $C_1, C_2, \ldots, C_k$ is equal to

$$1 - (1 - p_1)(1 - p_2) \cdots (1 - p_k).$$

1.4.14. Each of four persons fires one shot at a target. Let C_k denote the event that the target is hit by person k, $k = 1, 2, 3, 4$. If C_1, C_2, C_3, C_4 are independent and if $P(C_1) = P(C_2) = 0.7$, $P(C_3) = 0.9$, and $P(C_4) = 0.4$, compute the probability that (a) all of them hit the target; (b) exactly one hits the target; (c) no one hits the target; (d) at least one hits the target.

1.4.15. A bowl contains three red (R) balls and seven white (W) balls of exactly the same size and shape. Select balls successively at random and with replacement so that the events of white on the first trial, white on the second, and so on, can be assumed to be independent. In four trials, make certain assumptions and compute the probabilities of the following ordered sequences: (a) WWRW; (b) RWWW; (c) WWWR; and (d) WRWW. Compute the probability of exactly one red ball in the four trials.

1.4.16. A coin is tossed two independent times, each resulting in a tail (T) or a head (H). The sample space consists of four ordered pairs: TT, TH, HT, HH. Making certain assumptions, compute the probability of each of these ordered pairs. What is the probability of at least one head?

1.4.17. For Example 1.4.7, obtain the following probabilities. Explain what they mean in terms of the problem.

(a) $P(N_D)$.

(b) $P(N \mid A_D)$.

(c) $P(A \mid N_D)$.

(d) $P(N \mid N_D)$.

1.4.18. A die is cast independently until the first 6 appears. If the casting stops on an odd number of times, Bob wins; otherwise, Joe wins.

(a) Assuming the die is fair, what is the probability that Bob wins?

(b) Let p denote the probability of a 6. Show that the game favors Bob, for all p, $0 < p < 1$.

1.4.19. Cards are drawn at random and with replacement from an ordinary deck of 52 cards until a spade appears.

(a) What is the probability that at least 4 draws are necessary?

(b) Same as part (a), except the cards are drawn without replacement.

1.4.20. A person answers each of two multiple choice questions at random. If there are four possible choices on each question, what is the conditional probability that both answers are correct given that at least one is correct?

1.4.21. Suppose a fair 6-sided die is rolled 6 independent times. A match occurs if side i is observed on the ith trial, $i = 1, \ldots, 6$.

(a) What is the probability of at least one match on the 6 rolls? *Hint:* Let C_i be the event of a match on the ith trial and use Exercise 1.4.13 to determine the desired probability.

(b) Extend Part (a) to a fair n-sided die with n independent rolls. Then determine the limit of the probability as $n \to \infty$.

1.4.22. Players A and B play a sequence of independent games. Player A throws a die first and wins on a "six." If he fails, B throws and wins on a "five" or "six ." If he fails, A throws and wins on a "four," "five," or "six." And so on. Find the probability of each player winning the sequence.

1.4.23. Let C_1, C_2, C_3 be independent events with probabilities $\frac{1}{2}$, $\frac{1}{3}$, $\frac{1}{4}$, respectively. Compute $P(C_1 \cup C_2 \cup C_3)$.

1.4.24. From a bowl containing 5 red, 3 white, and 7 blue chips, select 4 at random and without replacement. Compute the conditional probability of 1 red, 0 white, and 3 blue chips, given that there are at least 3 blue chips in this sample of 4 chips.

1.4.25. Let the three mutually independent events C_1, C_2, and C_3 be such that $P(C_1) = P(C_2) = P(C_3) = \frac{1}{4}$. Find $P[(C_1^c \cap C_2^c) \cup C_3]$.

1.4.26. Person A tosses a coin and then person B rolls a die. This is repeated independently until a head or one of the numbers $1, 2, 3, 4$ appears, at which time the game is stopped. Person A wins with the head and B wins with one of the numbers $1, 2, 3, 4$. Compute the probability that A wins the game.

1.4.27. Each bag in a large box contains 25 tulip bulbs. It is known that 60% of the bags contain bulbs for 5 red and 20 yellow tulips while the remaining 40% of the bags contain bulbs for 15 red and 10 yellow tulips. A bag is selected at random and a bulb taken at random from this bag is planted.

(a) What is the probability that it will be a yellow tulip?

(b) Given that it is yellow, what is the conditional probability it comes from a bag that contained 5 red and 20 yellow bulbs?

1.4.28. A bowl contains ten chips numbered $1, 2, \ldots, 10$, respectively. Five chips are drawn at random, one at a time, and without replacement. What is the probability that two even-numbered chips are drawn and they occur on even-numbered draws?

1.4.29. A person bets 1 dollar to b dollars that he can draw two cards from an ordinary deck of cards without replacement and that they will be of the same suit. Find b so that the bet will be fair.

1.4.30 (Monte Hall Problem). Suppose there are three curtains. Behind one curtain there is a nice prize while behind the other two there are worthless prizes. A contestant selects one curtain at random, and then Monte Hall opens one of the other two curtains to reveal a worthless prize. Hall then expresses the willingness to trade the curtain that the contestant has chosen for the other curtain that has not been opened. Should the contestant switch curtains or stick with the one that she has? If she sticks with the curtain she has then the probability of winning the prize is $1/3$. Hence, to answer the question determine the probability that she wins the prize if she switches.

1.4.31. A French nobleman, Chevalier de Méré, had asked a famous mathematician, Pascal, to explain why the following two probabilities were different (the difference had been noted from playing the game many times): (1) at least one six in 4 independent casts of a six-sided die; (2) at least a pair of sixes in 24 independent casts of a pair of dice. From proportions it seemed to de Méré that the probabilities should be the same. Compute the probabilities of (1) and (2).

1.4.32. Hunters A and B shoot at a target; the probabilities of hitting the target are p_1 and p_2, respectively. Assuming independence, can p_1 and p_2 be selected so that

$$P(\text{zero hits}) = P(\text{one hit}) = P(\text{two hits}) ?$$

1.5   Random Variables

The reader will perceive that a sample space $\mathcal{C}$ may be tedious to describe if the elements of $\mathcal{C}$ are not numbers. We shall now discuss how we may formulate a rule, or a set of rules, by which the elements c of $\mathcal{C}$ may be represented by numbers. We begin the discussion with a very simple example. Let the random experiment be the toss of a coin and let the sample space associated with the experiment be $\mathcal{C} = \{c : \text{where } c \text{ is T or } c \text{ is H}\}$ and T and H represent, respectively, tails and heads. Let X be a function such that $X(c) = 0$ if c is T and $X(c) = 1$ if c is H. Thus X is a real-valued function defined on the sample space $\mathcal{C}$ which takes us from the sample space $\mathcal{C}$ to a space of real numbers $\mathcal{D} = \{0, 1\}$. We now formulate the definition of a random variable and its space.

Definition 1.5.1. *Consider a random experiment with a sample space $\mathcal{C}$. A function X, which assigns to each element $c \in \mathcal{C}$ one and only one number $X(c) = x$, is called a* **random variable**. *The* **space** *or* **range** *of X is the set of real numbers $\mathcal{D} = \{x : x = X(c), c \in \mathcal{C}\}$.*

In this text, $\mathcal{D}$ will generally be a countable set or an interval of real numbers. We call random variables of the first type **discrete** random variables while we call those of the second type **continuous** random variables. In this section, we present examples of discrete and continuous random variables and then in the next two sections we discuss them separately.

A random variable X induces a new sample space $\mathcal{D}$ on the real number line, R. What are the analogues of the class of events $\mathcal{B}$ and the probability P?

Consider the case where X is a discrete random variable with a finite space $\mathcal{D} = \{d_1, \ldots, d_m\}$. There are m events of interest in this case which are given by:

$$\{c \in \mathcal{C} : X(c) = d_i\}, \quad \text{for } i = 1, \ldots, m.$$

Hence, for this random variable, the σ-field on $\mathcal{D}$ can be the one generated by the collection of *simple* events $\{\{d_1\}, \ldots, \{d_m\}\}$ which is the set of all subsets of $\mathcal{D}$. Let $\mathcal{F}$ denote this σ-field.

Thus we have a sample space and a collection of events. What about a probability set function? For any event B in $\mathcal{F}$ define

$$P_X(B) = P[\{c \in \mathcal{C} : X(c) \in B\}]. \tag{1.5.1}$$

We need to show that P_X satisfies the three axioms of probability given by Definition 1.3.2.

Note first that $P_X(B) \geq 0$. Second, because the domain of X is $\mathcal{C}$, we have $P_X(\mathcal{D}) = P(\mathcal{C}) = 1$. Thus P_X satisfies the first two axioms of a probability, see Definition 1.3.2. Exercise 1.5.10 shows that the third axiom is true also. Hence, P_X is a probability on $\mathcal{D}$. We say that P_X is the probability *induced* on $\mathcal{D}$ by the random variable X.

This discussion can be simplified by noting that, because any event B in $\mathcal{F}$ is a subset of $\mathcal{D} = \{d_1, \ldots, d_m\}$, P_X satisfies,

$$P_X(B) = \sum_{d_i \in B} P[\{c \in \mathcal{C} : X(c) = d_i\}].$$

Hence, P_X is completely determined by the function

$$p_X(d_i) = P_X[\{d_i\}] \quad \text{for } i = 1, \ldots, m. \tag{1.5.2}$$

The function $p_X(d_i)$ is called the **probability mass function** of X, which we abbreviate by **pmf**. After a brief remark, we will consider a specific example.

Remark 1.5.1. In equations (1.5.1) and (1.5.2), the subscript X on P_X and p_X identify the induced probability set function and the pmf with the random variable. We will often use this notation, especially when there are several random variables in the discussion. On the other hand, if the identity of the random variable is clear then we will often suppress the subscripts. ■

Example 1.5.1 (First Roll in Craps). Let X be the sum of the upfaces on a roll of a pair of fair 6-sided dice, each with the numbers 1 through 6 on it. The sample space is $\mathcal{C} = \{(i, j) : 1 \le i, j \le 6\}$. Because the dice are fair, $P[\{(i, j)\}] = 1/36$. The random variable X is $X(i, j) = i + j$. The space of X is $\mathcal{D} = \{2, \ldots, 12\}$. By enumeration, the pmf of X is given by

Range value x	2	3	4	5	6	7	8	9	10	11	12
Probability $p_X(x)$	$\frac{1}{36}$	$\frac{2}{36}$	$\frac{3}{36}$	$\frac{4}{36}$	$\frac{5}{36}$	$\frac{6}{36}$	$\frac{5}{36}$	$\frac{4}{36}$	$\frac{3}{36}$	$\frac{2}{36}$	$\frac{1}{36}$

The σ-field for the the probability space on $\mathcal{C}$ would consist of 2^{36} subsets, (the number of subsets of elements in $\mathcal{C}$). But our interest here is with the random variable X and for it there are only 11 simple events of interest; i.e, the events $\{X = k\}$, for $k = 2, \ldots, 12$. To illustrate the computation of probabilities concerning X, suppose $B_1 = \{x : x = 7, 11\}$ and $B_2 = \{x : x = 2, 3, 12\}$, then

$$P_X(B_1) = \sum_{x \in B_1} p_X(x) = \frac{6}{36} + \frac{2}{36} = \frac{8}{36}$$

$$P_X(B_2) = \sum_{x \in B_2} p_X(x) = \frac{1}{36} + \frac{2}{36} + \frac{1}{36} = \frac{4}{36},$$

where $p_X(x)$ is given in the display. ■

For an example of a continuous random variable, consider the following simple experiment: choose a real number at random from the interval $(0, 1)$. Let X be the number chosen. In this case the space of X is $\mathcal{D} = (0, 1)$. It is not obvious as it was in the last example what the induced probability P_X is. But there are some intuitive probabilities. For instance, because the number is chosen at random, it is reasonable to assign

$$P_X[(a, b)] = b - a, \text{ for } 0 < a < b < 1. \tag{1.5.3}$$

For continuous random variables X, we want the probability model of X to be determined by probabilities of intervals. Hence, we take as our class of events on R the Borel σ-field $\mathcal{B}_0$, (1.3.2), generated by intervals. Note that this includes discrete random variables, also. For example, the event of interest $\{d_i\}$ can be expressed as an intersection of intervals; e.g., $\{d_i\} = \cap_n (d_i - (1/n), d_i]$.

In a more advanced course, we would say that X is a random variable provided the set $\{c : X(c) \in B\}$ is in $\mathcal{B}$, for every Borel set B in the Borel σ-field $\mathcal{B}_0$, (1.3.2), on R. Continuing in this vein for a moment, we can define P_X in general. For any $B \in \mathcal{B}_0$, this probability is given by

$$P_X(B) = P(\{c : X(c) \in B\}). \tag{1.5.4}$$

As for the discrete example, above, Exercise 1.5.10 shows that P_X is a probability set function on R. Because the Borel σ-field $\mathcal{B}_0$ on R is generated by intervals, it can shown in a more advanced class that P_X can be completely determined once we know its values on intervals. In fact, its values on semi-closed intervals of the form $(-\infty, x]$ uniquely determine $P_X(B)$. This defines a very important function which is given by:

Definition 1.5.2 (Cumulative Distribution Function). *Let X be a random variable. Then its* **cumulative distribution function** *, (cdf), is defined by,*

$$F_X(x) = P_X((-\infty, x]) = P(X \leq x). \tag{1.5.5}$$

Remark 1.5.2. Recall that P is a probability on the sample space $\mathcal{C}$, so the term on the far right-side of Equation (1.5.5) needs to be defined. We shall define it as

$$P(X \leq x) = P(\{c \in \mathcal{C} : X(c) \leq x\}). \tag{1.5.6}$$

This is a convenient abbreviation, one which we shall often use.

Also, $F_X(x)$ is often called simply the distribution function (df). However, in this text, we use the modifier *cumulative* as $F_X(x)$ accumulates the probabilities less than or equal to x. ∎

The next example discusses a cdf for a discrete random variable.

Example 1.5.2 (First Roll in Craps, Continued). From Example 1.5.1, the space of X is $\mathcal{D} = \{2, \ldots, 12\}$. If $x < 2$ then $F_X(x) = 0$. If $2 \leq x < 3$ then $F_X(x) = 1/36$. Continuing this way, we see that the cdf of X is an increasing step function which steps up by $P(X = i)$ at each i in the space of X. The graph of F_X is similar to that of Figure 1.5.1. Given $F_X(x)$, we can determine the pmf of X. ∎

The following example discusses the cdf of a continuous random variable.

Example 1.5.3. Let X denote a real number chosen at random between 0 and 1. We now obtain the cdf of X. First, if $x < 0$, then $P(X \leq x) = 0$. Next, if $X > 1$, then $P(X \leq x) = 1$. Finally, if $0 < x < 1$, it follows from expression (1.5.3) that $P(X \leq x) = P(0 < X \leq x) = x - 0 = x$. Hence the cdf of X is

$$F_X(x) = \begin{cases} 0 & \text{if } x < 0 \\ x & \text{if } 0 \leq x < 1 \\ 1 & \text{if } x \geq 1. \end{cases} \tag{1.5.7}$$

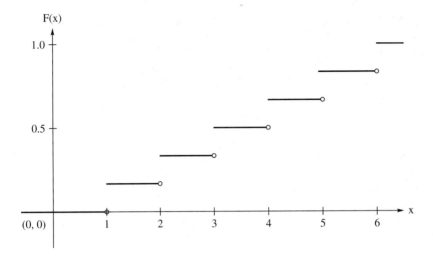

Figure 1.5.1: Distribution Function for the Upface of a Roll of a Fair Die.

A sketch of the cdf of X is given in Figure 1.5.2. Let $f_X(x)$ be given by,

$$f_X(x) = \begin{cases} 1 & 0 < x < 1 \\ 0 & \text{elsewhere.} \end{cases}$$

Then,

$$F_X(x) = \int_{-\infty}^{x} f_X(t)\,dt , \quad \text{for all } x \in R,$$

and $\frac{d}{dx} F_X(x) = f_X(x)$, for all $x \in R$, except for $x = 0$ and $x = 1$. The function $f_X(x)$ is defined as a **probability density function**, (pdf), of X in Section 1.7. To illustrate the computation of probabilities on X using the pdf, consider

$$P\left(\frac{1}{8} < X < \frac{3}{4}\right) = \int_{1/8}^{3/4} f_X(x)\,dx = \int_{1/8}^{3/4} 1\,dx = \frac{5}{8}. \quad \blacksquare$$

Let X and Y be two random variables. We say that X and Y are equal in distribution and write $X \overset{D}{=} Y$ if and only if $F_X(x) = F_Y(x)$, for all $x \in R$. It is important to note while X and Y may be equal in distribution they may be quite different. For instance, in the last example define the random variable Y as $Y = 1 - X$. Then $Y \neq X$. But the space of Y is the interval $(0, 1)$, the same as X. Further, the cdf of Y is 0 for $y < 0$; 1 for $y \geq 1$; and for $0 \leq y < 1$, it is

$$F_Y(y) = P(Y \leq y) = P(1 - X \leq y) = P(X \geq 1 - y) = 1 - (1 - y) = y.$$

Hence, Y has the same cdf as X, i.e., $Y \overset{D}{=} X$, but $Y \neq X$.

The cdfs displayed in Figures 1.5.1 and 1.5.2 show increasing functions with lower limits 0 and upper limits 1. In both figures, the cdfs are at least right continuous. As the next theorem proves, these properties are true in general for cdfs.

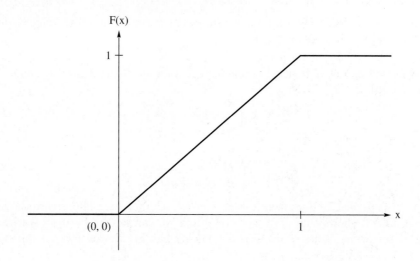

Figure 1.5.2: Distribution Function for Example 1.5.3.

Theorem 1.5.1. *Let X be a random variable with cumulative distribution function $F(x)$. Then*

(a). *For all a and b, if $a < b$ then $F(a) \leq F(b)$, (F is a nondecreasing function).*

(b). *$\lim_{x \to -\infty} F(x) = 0$, (the lower limit of F is 0).*

(c). *$\lim_{x \to \infty} F(x) = 1$, (the upper limit of F is 1).*

(d). *$\lim_{x \downarrow x_0} F(x) = F(x_0)$, (F is right continuous).*

Proof: We prove parts (a) and (d) and leave parts (b) and (c) for Exercise 1.5.11. Part (a): Because $a < b$, we have $\{X \leq a\} \subset \{X \leq b\}$. The result then follows from the monotonicity of P; see Theorem 1.3.3.

Part (d): Let $\{x_n\}$ be any sequence of real numbers such that $x_n \downarrow x_0$. Let $C_n = \{X \leq x_n\}$. Then the sequence of sets $\{C_n\}$ is decreasing and $\cap_{n=1}^{\infty} C_n = \{X \leq x_0\}$. Hence, by Theorem 1.3.6,

$$\lim_{n \to \infty} F(x_n) = P\left(\bigcap_{n=1}^{\infty} C_n\right) = F(x_0),$$

which is the desired result. ∎

The next theorem is helpful in evaluating probabilities using cdfs.

Theorem 1.5.2. *Let X be a random variable with cdf F_X. Then for $a < b$, $P[a < X \leq b] = F_X(b) - F_X(a)$.*

Proof: Note that,

$$\{-\infty < X \leq b\} = \{-\infty < X \leq a\} \cup \{a < X \leq b\}.$$

The proof of the result follows immediately because the union on the right side of this equation is a disjoint union. ∎

Example 1.5.4. Let X be the lifetime in years of a mechanical part. Assume that X has the cdf

$$F_X(x) = \begin{cases} 0 & x < 0 \\ 1 - e^{-x} & 0 \le x. \end{cases}$$

The pdf of X, $\frac{d}{dx}F_X(x)$, is

$$f_X(x) = \begin{cases} e^{-x} & 0 < x < \infty \\ 0 & \text{elsewhere.} \end{cases}$$

Actually the derivative does not exist at $x = 0$, but in the continuous case the next theorem (1.5.3) shows that $P(X = 0) = 0$ and we can assign $f_X(0) = 0$ without changing the probabilities concerning X. The probability that a part has a lifetime between 1 and 3 years is given by

$$P(1 < X \le 3) = F_X(3) - F_X(1) = \int_1^3 e^{-x}\, dx.$$

That is, the probability can be found by $F_X(3) - F_X(1)$ or evaluating the integral. In either case, it equals $e^{-1} - e^{-3} = 0.318$. ∎

Theorem 1.5.1 shows that cdfs are right continuous and monotone. Such functions can be shown to have only a countable number of discontinuities. As the next theorem shows, the discontinuities of a cdf have mass; that is, if x is a point of discontinuity of F_X then we have $P(X = x) > 0$.

Theorem 1.5.3. *For any random variable,*

$$P[X = x] = F_X(x) - F_X(x-), \tag{1.5.8}$$

for all $x \in R$, where $F_X(x-) = \lim_{z \uparrow x} F_X(z)$.

Proof: For any $x \in R$, we have

$$\{x\} = \bigcap_{n=1}^{\infty} \left(x - \frac{1}{n}, x\right],$$

that is, $\{x\}$ is the limit of a decreasing sequence of sets. Hence, by Theorem 1.3.6,

$$\begin{aligned}
P[X = x] &= P\left[\bigcap_{n=1}^{\infty}\left\{x - \frac{1}{n} < X \le x\right\}\right] \\
&= \lim_{n \to \infty} P\left[x - \frac{1}{n} < X \le x\right] \\
&= \lim_{n \to \infty}\left[F_X(x) - F_X(x - (1/n))\right] \\
&= F_X(x) - F_X(x-),
\end{aligned}$$

which is the desired result. ∎

Example 1.5.5. Let X have the discontinuous cdf

$$F_X(x) = \begin{cases} 0 & x < 0 \\ x/2 & 0 \le x < 1 \\ 1 & 1 \le x. \end{cases}$$

Then

$$P(-1 < X \le 1/2) = F_X(1/2) - F_X(-1) = \frac{1}{4} - 0 = \frac{1}{4},$$

and

$$P(X = 1) = F_X(1) - F_X(1-) = 1 - \frac{1}{2} = \frac{1}{2},$$

The value $1/2$ equals the value of the step of F_X at $x = 1$. ∎

Since the total probability associated with a random variable X of the discrete type with pmf $p_X(x)$ or of the continuous type with pdf $f_X(x)$ is 1, then it must be true that

$$\sum_{x \in D} p_X(x) = 1 \text{ and } \int_D f_X(x) \, dx = 1,$$

where D is the space of X. As the next two examples show, we can use this property to determine the pmf or pdf, if we know the pmf or pdf down to a constant of proportionality.

Example 1.5.6. Suppose X has the pmf

$$p_X(x) = \begin{cases} cx & x = 1, 2, \ldots, 10 \\ 0 & \text{elsewhere,} \end{cases}$$

then

$$1 = \sum_{x=1}^{10} p_X(x) = \sum_{x=1}^{10} cx = c(1 + 2 + \cdots + 10) = 55c,$$

and, hence, $c = 1/55$. ∎

Example 1.5.7. Suppose X has the pdf

$$f_X(x) = \begin{cases} cx^3 & 0 < x < 2 \\ 0 & \text{elsewhere,} \end{cases}$$

then

$$1 = \int_0^2 cx^3 \, dx = c\frac{x^4}{4} \Big|_0^2 = 4c,$$

and, hence, $c = 1/4$. For illustration of the computation of a probability involving X, we have

$$P\left(\frac{1}{4} < X < 1\right) = \int_{1/4}^1 \frac{x^3}{4} \, dx = \frac{255}{4096} = 0.06226. \quad ∎$$

EXERCISES

1.5.1. Let a card be selected from an ordinary deck of playing cards. The outcome c is one of these 52 cards. Let $X(c) = 4$ if c is an ace, let $X(c) = 3$ if c is a king, let $X(c) = 2$ if c is a queen, let $X(c) = 1$ if c is a jack, and let $X(c) = 0$ otherwise. Suppose that P assigns a probability of $\frac{1}{52}$ to each outcome c. Describe the induced probability $P_X(D)$ on the space $\mathcal{D} = \{0, 1, 2, 3, 4\}$ of the random variable X.

1.5.2. For each of the following, find the constant c so that $p(x)$ satisfies the condition of being a pmf of one random variable X.

(a) $p(x) = c(\frac{2}{3})^x$, $x = 1, 2, 3, \ldots$, zero elsewhere.

(b) $p(x) = cx$, $x = 1, 2, 3, 4, 5, 6$, zero elsewhere.

1.5.3. Let $p_X(x) = x/15$, $x = 1, 2, 3, 4, 5$, zero elsewhere, be the pmf of X. Find $P(X = 1 \text{ or } 2)$, $P(\frac{1}{2} < X < \frac{5}{2})$, and $P(1 \leq X \leq 2)$.

1.5.4. Let $p_X(x)$ be the pmf of a random variable X. Find the cdf $F(x)$ of X and sketch its graph along with that of $p_X(x)$ if:

(a) $p_X(x) = 1$, $x = 0$, zero elsewhere.

(b) $p_X(x) = \frac{1}{3}$, $x = -1, 0, 1$, zero elsewhere.

(c) $p_X(x) = x/15$, $x = 1, 2, 3, 4, 5$, zero elsewhere.

1.5.5. Let us select five cards at random and without replacement from an ordinary deck of playing cards.

(a) Find the pmf of X, the number of hearts in the five cards.

(b) Determine $P(X \leq 1)$.

1.5.6. Let the probability set function $P_X(D)$ of the random variable X be $P_X(D) = \int_D f(x)\, dx$, where $f(x) = 2x/9$, $x \in \mathcal{D} = \{x : 0 < x < 3\}$. Let $D_1 = \{x : 0 < x < 1\}$, $D_2 = \{x : 2 < x < 3\}$. Compute $P_X(D_1) = P(X \in D_1)$, $P_X(D_2) = P(X \in D_2)$, and $P_X(D_1 \cup D_2) = P(X \in D_1 \cup D_2)$.

1.5.7. Let the space of the random variable X be $\mathcal{D} = \{x : 0 < x < 1\}$. If $D_1 = \{x : 0 < x < \frac{1}{2}\}$ and $D_2 = \{x : \frac{1}{2} \leq x < 1\}$, find $P_X(D_2)$ if $P_X(D_1) = \frac{1}{4}$.

1.5.8. Given the cdf

$$F(x) = \begin{cases} 0 & x < -1 \\ \frac{x+2}{4} & -1 \leq x < 1 \\ 1 & 1 \leq x. \end{cases}$$

Sketch the graph of $F(x)$ and then compute: (a) $P(-\frac{1}{2} < X \leq \frac{1}{2})$; (b) $P(X = 0)$; (c) $P(X = 1)$; (d) $P(2 < X \leq 3)$.

1.5.9. Consider an urn which contains slips of paper each with one of the numbers $1, 2, \ldots, 100$ on it. Suppose there are i slips with the number i on it for $i = 1, 2, \ldots, 100$. For example, there are 25 slips of paper with the number 25. Assume that the slips are identical except for the numbers. Suppose one slip is drawn at random. Let X be the number on the slip.

(a) Show that X has the pmf $p(x) = x/5050$, $x = 1, 2, 3, \ldots, 100$, zero elsewhere.

(b) Compute $P(X \le 50)$.

(c) Show that the cdf of X is $F(x) = [x]([x] + 1)/10100$, for $1 \le x \le 100$, where $[x]$ is the greatest integer in x.

1.5.10. Let X be a random variable with space $\mathcal{D}$. For a sequence of sets $\{D_n\}$ in $\mathcal{D}$, show that

$$\{c : X(c) \in \cup_n D_n\} = \cup_n \{c : X(c) \in D_n\}.$$

Use this to show that the induced probability P_X, (1.5.1), satisfies the third axiom of probability.

1.5.11. Prove parts (b) and (c) of Theorem 1.5.1.

1.6 Discrete Random Variables

The first example of a random variable encountered in the last section was an example of a discrete random variable, which is defined next.

Definition 1.6.1 (Discrete Random Variable). *We say a random variable is a* **discrete random variable** *if its space is either finite or countable.*

A set $\mathcal{D}$ is said to be countable, if its elements can be listed; i.e., there is a one-to-one correspondence between $\mathcal{D}$ and the positive integers.

Example 1.6.1. Consider a sequence of independent flips of a coin, each resulting in a head (H) or a tail (T). Moreover, on each flip, we assume that H and T are equally likely, that is, $P(H) = P(T) = \frac{1}{2}$. The sample space $\mathcal{C}$ consists of sequences like TTHTHHT$\cdots$. Let the random variable X equal the number of flips needed to obtain the first head. For this given sequence, $X = 3$. Clearly, the space of X is $\mathcal{D} = \{1, 2, 3, 4, \ldots\}$. We see that $X = 1$ when the sequence begins with an H and thus $P(X = 1) = \frac{1}{2}$. Likewise, $X = 2$ when the sequence begins with TH, which has probability $P(X = 2) = (\frac{1}{2})(\frac{1}{2}) = \frac{1}{4}$ from the independence. More generally, if $X = x$, where $x = 1, 2, 3, 4, \ldots$, there must be a string of $x - 1$ tails followed by a head, that is, TT$\cdots$TH, where there are $x - 1$ tails in TT$\cdots$T. Thus, from independence, we have

$$P(X = x) = \left(\frac{1}{2}\right)^{x-1}\left(\frac{1}{2}\right) = \left(\frac{1}{2}\right)^x, \quad x = 1, 2, 3, \ldots, \tag{1.6.1}$$

the space of which is countable. An interesting event is that the first head appears on an odd number of flips; i.e., $X \in \{1, 3, 5, \ldots\}$. The probability of this event is

$$P[X \in \{1, 3, 5, \ldots\}] = \sum_{x=1}^{\infty} \left(\frac{1}{2}\right)^{2x-1} = \frac{1/2}{1 - (1/4)} = \frac{2}{3}. \quad \blacksquare$$

As the last example suggests, probabilities concerning a discrete random variable can be obtained in terms of the probabilities $P(X = x)$, for $x \in \mathcal{D}$. These probabilities determine an important function which we define as,

Definition 1.6.2 (Probability Mass Function (pmf)). *Let X be a discrete random variable with space $\mathcal{D}$. The* **probability mass function** *(pmf) of X is given by*

$$p_X(x) = P[X = x], \quad for \ x \in \mathcal{D}. \tag{1.6.2}$$

Note that pmfs satisfy the following two properties:

$$\text{(i). } 0 \leq p_X(x) \leq 1 \,, x \in \mathcal{D} \text{ and (ii). } \sum_{x \in \mathcal{D}} p_X(x) = 1. \tag{1.6.3}$$

In a more advanced class it can be shown that if a function satisfies properties (i) and (ii) for a discrete set $\mathcal{D}$ then this function uniquely determines the distribution of a random variable.

Let X be a discrete random variable with space $\mathcal{D}$. As Theorem 1.5.3 shows, discontinuities of $F_X(x)$ define a mass; that is, if x is a point of discontinuity of F_X then $P(X = x) > 0$. We now make a distinction between the space of a discrete random variable and these points of positive probability. We define the **support** of a discrete random variable X to be the points in the space of X which have positive probability. We will often use $\mathcal{S}$ to denote the support of X. Note that $\mathcal{S} \subset \mathcal{D}$, but it may be that $\mathcal{S} = \mathcal{D}$.

Also, we can use Theorem 1.5.3 to obtain a relationship between the pmf and cdf of a discrete random variable. If $x \in \mathcal{S}$ then $p_X(x)$ is equal to the size of the discontinuity of F_X at x. If $x \notin \mathcal{S}$ then $P[X = x] = 0$ and, hence, F_X is continuous at x.

Example 1.6.2. A lot, consisting of 100 fuses, is inspected by the following procedure. Five of these fuses are chosen at random and tested; if all 5 "blow" at the correct amperage, the lot is accepted. If, in fact, there are 20 defective fuses in the lot, the probability of accepting the lot is, under appropriate assumptions,

$$\frac{\binom{80}{5}}{\binom{100}{5}} = 0.32,$$

approximately. More generally, let the random variable X be the number of defective fuses among the 5 that are inspected. The pmf of X is given by

$$p_X(x) = \begin{cases} \dfrac{\binom{20}{x}\binom{80}{5-x}}{\binom{100}{5}} & \text{for } x = 0, 1, 2, 3, 4, 5 \\ 0 & \text{elsewhere.} \end{cases}$$

Clearly, the space of X is $\mathcal{D} = \{0, 1, 2, 3, 4, 5\}$. Thus this is an example of a random variable of the discrete type whose distribution is an illustration of a **hypergeometric distribution**. Based on the above discussion, it is easy to graph the cdf of X; see Exercise 1.6.5. ∎

1.6.1 Transformations

A problem often encountered in statistics is the following. We have a random variable X and we know its distribution. We are interested, though, in a random variable Y which is some **transformation** of X, say, $Y = g(X)$. In particular, we want to determine the distribution of Y. Assume X is discrete with space $\mathcal{D}_X$. Then the space of Y is $\mathcal{D}_Y = \{g(x) : x \in \mathcal{D}_X\}$. We will consider two cases.

In the first case, g is one-to-one. Then clearly the pmf of Y is obtained as,

$$p_Y(y) = P[Y = y] = P[g(X) = y] = P[X = g^{-1}(y)] = p_X(g^{-1}(y)). \quad (1.6.4)$$

Example 1.6.3 (Geometric Distribution). Consider the geometric random variable X of Example 1.6.1. Recall that X was the flip number on which the first head appeared. Let Y be the number of flips before the first head. Then $Y = X - 1$. In this case, the function g is $g(x) = x - 1$ whose inverse is given by $g^{-1}(y) = y + 1$. The space of Y is $\mathcal{D}_Y = \{0, 1, 2, \ldots\}$. The pmf of X is given by (1.6.1); hence, based on expression (1.6.4) the pmf of Y is

$$p_Y(y) = p_X(y + 1) = \left(\frac{1}{2}\right)^{y+1}, \quad \text{for } y = 0, 1, 2, \ldots. \quad ∎$$

Example 1.6.4. Let X have the pmf

$$p_X(x) = \begin{cases} \frac{3!}{x!(3-x)!} \left(\frac{2}{3}\right)^x \left(\frac{1}{3}\right)^{3-x} & x = 0, 1, 2, 3 \\ 0 & \text{elsewhere.} \end{cases}$$

We seek the pmf $p_Y(y)$ of the random variable $Y = X^2$. The transformation $y = g(x) = x^2$ maps $\mathcal{D}_X = \{x : x = 0, 1, 2, 3\}$ onto $\mathcal{D}_Y = \{y : y = 0, 1, 4, 9\}$. In general, $y = x^2$ does not define a one-to-one transformation; here, however, it does, for there are no negative value of x in $\mathcal{D}_X = \{x : x = 0, 1, 2, 3\}$. That is, we have the single-valued inverse function $x = g^{-1}(y) = \sqrt{y}$ (not $-\sqrt{y}$), and so

$$p_Y(y) = p_X(\sqrt{y}) = \frac{3!}{(\sqrt{y})!(3 - \sqrt{y})!} \left(\frac{2}{3}\right)^{\sqrt{y}} \left(\frac{1}{3}\right)^{3-\sqrt{y}}, \quad y = 0, 1, 4, 9. \quad ∎$$

The second case is where the transformation, $g(x)$, is not one-to-one. Instead of developing an overall rule, for most applications involving discrete random variables the pmf of Y can be obtained in a straightforward manner. We offer two examples as illustrations.

Consider the geometric random variable in Example 1.6.3. Suppose we are playing a game against the "house" (say, a gambling casino). If the first head appears on an odd number of flips we pay the house one dollar, while if it appears on an even number of flips we win one dollar from the house. Let Y denote our net gain. Then the space of Y is $\{-1, 1\}$. In Example 1.6.1, we showed that the probability that X is odd is $\frac{2}{3}$. Hence, the distribution of Y is given by $p_Y(-1) = 2/3$ and $p_Y(1) = 1/3$.

As a second illustration, let $Z = (X-2)^2$, where X is the geometric random variable of Example 1.6.1. Then the space of Z is $\mathcal{D}_Z = \{0, 1, 4, 9, 16, \ldots\}$. Note that $Z = 0$ if and only if $X = 2$; $Z = 1$ if and only if $X = 1$ or $X = 3$; while for the other values of the space there is a one-to-one correspondence given by $x = \sqrt{z}+2$, for $z \in \{4, 9, 16, \ldots\}$. Hence, the pmf of Z is:

$$p_Z(z) = \begin{cases} p_X(2) = \frac{1}{4} & \text{for } z = 0 \\ p_X(1) + p_X(3) = \frac{5}{8} & \text{for } z = 1 \\ p_X(\sqrt{z}+2) = \frac{1}{4}\left(\frac{1}{2}\right)^{\sqrt{z}} & \text{for } z = 4, 9, 16, \ldots. \end{cases} \tag{1.6.5}$$

For verification, the reader is asked to show in Exercise 1.6.9 that the pmf of Z sums to 1 over its space.

EXERCISES

1.6.1. Let X equal the number of heads in four independent flips of a coin. Using certain assumptions, determine the pmf of X and compute the probability that X is equal to an odd number.

1.6.2. Let a bowl contain 10 chips of the same size and shape. One and only one of these chips is red. Continue to draw chips from the bowl, one at a time and at random and without replacement, until the red chip is drawn.

   (a) Find the pmf of X, the number of trials needed to draw the red chip.

   (b) Compute $P(X \leq 4)$.

1.6.3. Cast a die a number of independent times until a six appears on the up side of the die.

   (a) Find the pmf $p(x)$ of X, the number of casts needed to obtain that first six.

   (b) Show that $\sum_{x=1}^{\infty} p(x) = 1$.

   (c) Determine $P(X = 1, 3, 5, 7, \ldots)$.

   (d) Find the cdf $F(x) = P(X \leq x)$.

1.6.4. Cast a die two independent times and let X equal the absolute value of the difference of the two resulting values (the numbers on the up sides). Find the pmf of X. *Hint:* It is not necessary to find a formula for the pmf.

1.6.5. For the random variable X defined in Example 1.6.2, graph the cdf of X.

1.6.6. For the random variable X defined in Example 1.6.1, graph the cdf of X.

1.6.7. Let X have a pmf $p(x) = \frac{1}{3}$, $x = 1, 2, 3$, zero elsewhere. Find the pmf of $Y = 2X + 1$.

1.6.8. Let X have the pmf $p(x) = (\frac{1}{2})^x$, $x = 1, 2, 3, \ldots$, zero elsewhere. Find the pmf of $Y = X^3$.

1.6.9. Show that the function given in expression (1.6.5) is a pmf.

1.7 Continuous Random Variables

In the last section, we discussed discrete random variables. Another class of random variables important in statistical applications is the class of continuous random variables which we define next.

Definition 1.7.1 (Continuous Random Variables). *We say a random variable is a **continuous random variable** if its cumulative distribution function $F_X(x)$ is a continuous function for all $x \in R$.*

Recall from Theorem 1.5.3 that $P(X = x) = F_X(x) - F_X(x-)$, for any random variable X. Hence, for a continuous random variable X there are no points of discrete mass; i.e., if X is continuous then $P(X = x) = 0$ for all $x \in R$. Most continuous random variables are absolutely continuous, that is,

$$F_X(x) = \int_{-\infty}^{x} f_X(t)\, dt, \tag{1.7.1}$$

for some function $f_X(t)$. The function $f_X(t)$ is called a **probability density function** (pdf) of X. If $f_X(x)$ is also continuous then the Fundamental Theorem of Calculus implies that,

$$\frac{d}{dx} F_X(x) = f_X(x). \tag{1.7.2}$$

The **support** of a continuous random variable X consists of all points x such that $f_X(x) > 0$. As in the discrete case, we will often denote the support of X by $\mathcal{S}$.

If X is a continuous random variable, then probabilities can be obtained by integration, i.e.,

$$P(a < X \leq b) = F_X(b) - F_X(a) = \int_{a}^{b} f_X(t)\, dt.$$

Also for continuous random variables, $P(a < X \leq b) = P(a \leq X \leq b) = P(a \leq X < b) = P(a < X < b)$. Because $f_X(x)$ is continuous over the support of X and $F_X(\infty) = 1$, pdfs satisfy the two properties,

$$\text{(i): } f_X(x) \geq 0 \text{ and (ii): } \int_{-\infty}^{\infty} f_X(t)\, dt = 1. \tag{1.7.3}$$

In an advanced course in probability, it is shown that if a function satisfies the above two properties, then it is a pdf for a continuous random variable; see, for example, Tucker (1967).

Recall in Example 1.5.3 the simple experiment where a number was chosen at random from the interval $(0,1)$. The number chosen, X, is an example of a continuous random variable. Recall that the cdf of X is $F_X(x) = x$, for $x \in (0,1)$. Hence, the pdf of X is given by

$$f_X(x) = \begin{cases} 1 & x \in (0,1) \\ 0 & \text{elsewhere.} \end{cases} \tag{1.7.4}$$

Any continuous or discrete random variable X whose pdf or pmf is constant on the support of X is said to have a **uniform** distribution.

Example 1.7.1 (Point Chosen at Random in the Unit Circle). Suppose we select a point at random in the interior of a circle of radius 1. Let X be the distance of the selected point from the origin. The sample space for the experiment is $\mathcal{C} = \{(w, y) : w^2 + y^2 < 1\}$. Because the point is chosen at random, it seems that subsets of $\mathcal{C}$ which have equal area are equilikely. Hence, the probability of the selected point lying in a set C interior to $\mathcal{C}$ is proportional to the area of C; i.e.,

$$P(C) = \frac{\text{area of } C}{\pi}.$$

For $0 < x < 1$, the event $\{X \leq x\}$ is equivalent to the point lying in a circle of radius x. By this probability rule $P(X \leq x) = \pi x^2/\pi = x^2$, hence, the cdf of X is

$$F_X(x) = \begin{cases} 0 & x < 0 \\ x^2 & 0 \leq x < 1 \\ 1 & 1 \leq x. \end{cases} \tag{1.7.5}$$

The pdf X is given by

$$f_X(x) = \begin{cases} 2x & 0 \leq x < 1 \\ 0 & \text{elsewhere.} \end{cases} \tag{1.7.6}$$

For illustration, the probability that the selected point falls in the ring with radii $1/4$ and $1/2$ is given by

$$P\left(\frac{1}{4} < X \leq \frac{1}{2}\right) = \int_{\frac{1}{4}}^{\frac{1}{2}} 2w \, dw = [w^2]_{\frac{1}{4}}^{\frac{1}{2}} = \frac{3}{16}. \quad \blacksquare$$

Example 1.7.2. Let the random variable be the time in seconds between incoming telephone calls at a busy switchboard. Suppose that a reasonable probability model for X is given by the pdf

$$f_X(x) = \begin{cases} \frac{1}{4}e^{-x/4} & 0 < x < \infty \\ 0 & \text{elsewhere.} \end{cases}$$

Note that f_X satisfies the two properties of a pdf, namely, (i) $f(x) \geq 0$ and (ii)

$$\int_0^\infty \frac{1}{4}e^{-x/4} \, dx = \left[-e^{-x/4}\right]_0^\infty = 1.$$

For illustration, the probability that the time between successive phone calls exceeds 4 seconds is given by

$$P(X > 4) = \int_4^\infty \frac{1}{4} e^{-x/4}\, dx = e^{-1} = .3679.$$

The pdf and the probability of interest are depicted in Figure 1.7.1. ∎

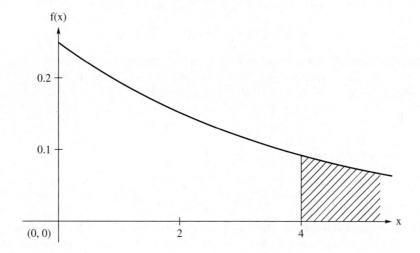

Figure 1.7.1: In Example 1.7.2, the area under the pdf to the right of 4 is $P(X > 4)$.

1.7.1 Transformations

Let X be a continuous random variable with a known pdf f_X. As in the discrete case, we are often interested in the distribution of a random variable Y which is some **transformation** of X, say, $Y = g(X)$. Often we can obtain the pdf of Y by first obtaining its cdf. We illustrate this with two examples.

Example 1.7.3. Let X be the random variable in Example 1.7.1. Recall that X was distance from the origin to the random point selected in the unit circle. Suppose instead, we are interested in the square of the distance; that is, let $Y = X^2$. The support of Y is the same as that of X, namely $\mathcal{S}_Y = (0, 1)$. What is the cdf of Y? By expression (1.7.5), the cdf of X is

$$F_X(x) = \begin{cases} 0 & x < 0 \\ x^2 & 0 \le x < 1 \\ 1 & 1 \le x. \end{cases} \tag{1.7.7}$$

Let y be in the support of Y; i.e., $0 < y < 1$. Then, using expression (1.7.7) and the fact that the support of X contains only positive numbers, the cdf of Y is

$$F_Y(y) = P(Y \le y) = P(X^2 \le y) = P(X \le \sqrt{y}) = F_X(\sqrt{y}) = \sqrt{y}^2 = y.$$

It follows that the pdf of Y is

$$f_Y(y) = \begin{cases} 1 & 0 < y < 1 \\ 0 & \text{elsewhere.} \end{cases} \quad \blacksquare$$

Example 1.7.4. Let $f_X(x) = \frac{1}{2}$, $-1 < x < 1$, zero elsewhere, be the pdf of a random variable X. Define the random variable Y by $Y = X^2$. We wish to find the pdf of Y. If $y \geq 0$, the probability $P(Y \leq y)$ is equivalent to

$$P(X^2 \leq y) = P(-\sqrt{y} \leq X \leq \sqrt{y}).$$

Accordingly, the cdf of Y, $F_Y(y) = P(Y \leq y)$, is given by

$$F_Y(y) = \begin{cases} 0 & y < 0 \\ \int_{-\sqrt{y}}^{\sqrt{y}} \frac{1}{2}\,dx = \sqrt{y} & 0 \leq y < 1 \\ 1 & 1 \leq y. \end{cases}$$

Hence, the pdf of Y is given by,

$$f_Y(y) = \begin{cases} \frac{1}{2\sqrt{y}} & 0 < y < 1 \\ 0 & \text{elsewhere.} \end{cases} \quad \blacksquare$$

These examples illustrate the *cumulative distribution function technique*. The transformation in the first example was one-to-one and in such cases we can obtain a simple formula for the pdf of Y in terms of the pdf of X, which we record in the next theorem.

Theorem 1.7.1. *Let X be a continuous random variable with pdf $f_X(x)$ and support $\mathcal{S}_X$. Let $Y = g(X)$, where $g(x)$ is a one-to-one differentiable function, on the support of X, $\mathcal{S}_X$. Denote the inverse of g by $x = g^{-1}(y)$ and let $dx/dy = d[g^{-1}(y)]/dy$. Then the pdf of Y is given by*

$$f_Y(y) = f_X(g^{-1}(y)) \left| \frac{dx}{dy} \right|, \quad \text{for } y \in \mathcal{S}_Y, \tag{1.7.8}$$

where the support of Y is the set $\mathcal{S}_Y = \{y = g(x) : x \in \mathcal{S}_X\}$.

Proof: Since $g(x)$ is one-to-one and continuous, it is either strictly monotonically increasing or decreasing. Assume that it is strictly monotonically increasing, for now. The cdf of Y is given by

$$F_Y(y) = P[Y \leq y] = P[g(X) \leq y] = P[X \leq g^{-1}(y)] = F_X(g^{-1}(y)). \tag{1.7.9}$$

Hence, the pdf of Y is

$$f_Y(y) = \frac{d}{dy} F_Y(y) = f_X(g^{-1}(y)) \frac{dx}{dy}, \tag{1.7.10}$$

where dx/dy is the derivative of the function $x = g^{-1}(y)$. In this case, because g is increasing, $dx/dy > 0$. Hence, we can write $dx/dy = |dx/dy|$.

Suppose $g(x)$ strictly monotonically decreasing. Then (1.7.9) becomes, $F_Y(y) = 1 - F_X(g^{-1}(y))$. Hence, the pdf of Y is $f_Y(y) = f_X(g^{-1}(y))(-dx/dy)$. But since g is decreasing $dx/dy < 0$ and, hence, $-dx/dy = |dx/dy|$. Thus equation (1.7.8) is true in both cases. ∎

Henceforth, we shall refer to $dx/dy = (d/dy)g^{-1}(y)$ as the **Jacobian** (denoted by J) of the transformation. In most mathematical areas, $J = dx/dy$ is referred to as the Jacobian of the inverse transformation $x = g^{-1}(y)$, but in this book it will be called the Jacobian of the transformation, simply for convenience.

Example 1.7.5. Let X have the pdf

$$f(x) = \begin{cases} 1 & 0 < x < 1 \\ 0 & \text{elsewhere.} \end{cases}$$

Consider the random variable $Y = -2\log X$. The support sets of X and Y are given by $(0, 1)$ and $(0, \infty)$, respectively. The transformation $g(x) = -2\log x$ is one-to-one between these sets. The inverse of the transformation is $x = g^{-1}(y) = e^{-y/2}$. The Jacobian of the transformation is

$$J = \frac{dx}{dy} = e^{-y/2} = -\frac{1}{2}e^{-y/2}.$$

Accordingly, the pdf of $Y = -2\log X$ is

$$f_Y(y) = \begin{cases} f_X(e^{-y/2})|J| = \frac{1}{2}e^{-y/2} & 0 < y < \infty \\ 0 & \text{elsewhere.} \end{cases} \quad ∎$$

We close this section by two examples of distributions that are neither of the discrete nor the continuous type.

Example 1.7.6. Let a distribution function be given by

$$F(x) = \begin{cases} 0 & x < 0 \\ \frac{x+1}{2} & 0 \le x < 1 \\ 1 & 1 \le x. \end{cases}$$

Then, for instance,

$$P\left(-3 < X \le \frac{1}{2}\right) = F\left(\frac{1}{2}\right) - F(-3) = \frac{3}{4} - 0 = \frac{3}{4}$$

and

$$P(X = 0) = F(0) - F(0-) = \frac{1}{2} - 0 = \frac{1}{2}.$$

The graph of $F(x)$ is shown in Figure 1.7.2. We see that $F(x)$ is not always continuous, nor is it a step function. Accordingly, the corresponding distribution is neither of the continuous type nor of the discrete type. It may be described as a mixture of those types. ∎

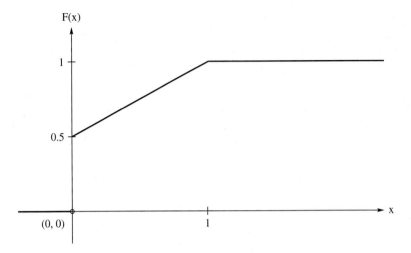

Figure 1.7.2: Graph of the cdf of Example 1.7.6.

Distributions that are mixtures of the continuous and discrete type do, in fact, occur frequently in practice. For illustration, in life testing, suppose we know that the length of life, say X, exceeds the number b, but the exact value of X is unknown. This is called *censoring*. For instance, this can happen when a subject in a cancer study simply disappears; the investigator knows that the subject has lived a certain number of months, but the exact length of life is unknown. Or it might happen when an investigator does not have enough time in an investigation to observe the moments of deaths of all the animals, say rats, in some study. Censoring can also occur in the insurance industry; in particular, consider a loss with a limited-pay policy in which the top amount is exceeded but is not known by how much.

Example 1.7.7. Reinsurance companies are concerned with large losses because they might agree, for illustration, to cover losses due to wind damages that are between \$2,000,000 and \$10,000,000. Say that X equals the size of a wind loss in millions of dollars, and suppose it has the cdf

$$F_X(x) = \begin{cases} 0 & -\infty < x < 0 \\ 1 - \left(\frac{10}{10+x}\right)^3 & 0 \le x < \infty. \end{cases}$$

If losses beyond \$10,000,000 are reported only as 10, then the cdf of this censored distribution is

$$F_Y(y) = \begin{cases} 0 & -\infty < y < 0 \\ 1 - \left(\frac{10}{10+y}\right)^3 & 0 \le y < 10, \\ 1 & 10 \le y < \infty, \end{cases}$$

which has a jump of $[10/(10+10)]^3 = \frac{1}{8}$ at $y = 10$. ∎

EXERCISES

1.7.1. Let a point be selected from the sample space $\mathcal{C} = \{c : 0 < c < 10\}$. Let $C \subset \mathcal{C}$ and let the probability set function be $P(C) = \int_C \frac{1}{10} \, dz$. Define the random variable X to be $X(c) = c^2$. Find the cdf and the pdf of X.

1.7.2. Let the space of the random variable X be $\mathcal{C} = \{x : 0 < x < 10\}$ and let $P_X(C_1) = \frac{3}{8}$, where $C_1 = \{x : 1 < x < 5\}$. Show that $P_X(C_2) \leq \frac{5}{8}$, where $C_2 = \{x : 5 \leq x < 10\}$.

1.7.3. Let the subsets $C_1 = \{\frac{1}{4} < x < \frac{1}{2}\}$ and $C_2 = \{\frac{1}{2} \leq x < 1\}$ of the space $\mathcal{C} = \{x : 0 < x < 1\}$ of the random variable X be such that $P_X(C_1) = \frac{1}{8}$ and $P_X(C_2) = \frac{1}{2}$. Find $P_X(C_1 \cup C_2)$, $P_X(C_1^c)$, and $P_X(C_1^c \cap C_2^c)$.

1.7.4. Given $\int_C [1/\pi(1 + x^2)] \, dx$, where $C \subset \mathcal{C} = \{x : -\infty < x < \infty\}$. Show that the integral could serve as a probability set function of a random variable X whose space is $\mathcal{C}$.

1.7.5. Let the probability set function of the random variable X be

$$P_X(C) = \int_C e^{-x} \, dx, \quad \text{where } \mathcal{C} = \{x : 0 < x < \infty\}.$$

Let $C_k = \{x : 2 - 1/k < x \leq 3\}$, $k = 1, 2, 3, \ldots$. Find $\lim_{k \to \infty} C_k$ and $P_X(\lim_{k \to \infty} C_k)$. Find $P_X(C_k)$ and $\lim_{k \to \infty} P_X(C_k) = P_X(\lim_{k \to \infty} C_k)$.

1.7.6. For each of the following pdfs of X, find $P(|X| < 1)$ and $P(X^2 < 9)$.

(a) $f(x) = x^2/18$, $-3 < x < 3$, zero elsewhere.

(b) $f(x) = (x + 2)/18$, $-2 < x < 4$, zero elsewhere.

1.7.7. Let $f(x) = 1/x^2$, $1 < x < \infty$, zero elsewhere, be the pdf of X. If $C_1 = \{x : 1 < x < 2\}$ and $C_2 = \{x : 4 < x < 5\}$, find $P_X(C_1 \cup C_2)$ and $P_X(C_1 \cap C_2)$.

1.7.8. A *mode* of a distribution of one random variable X is a value of x that maximizes the pdf or pmf. For X of the continuous type, $f(x)$ must be continuous. If there is only one such x, it is called the *mode of the distribution*. Find the mode of each of the following distributions:

(a) $p(x) = (\frac{1}{2})^x$, $x = 1, 2, 3, \ldots$, zero elsewhere.

(b) $f(x) = 12x^2(1 - x)$, $0 < x < 1$, zero elsewhere.

(c) $f(x) = (\frac{1}{2})x^2 e^{-x}$, $0 < x < \infty$, zero elsewhere.

1.7.9. A *median* of a distribution of one random variable X of the discrete or continuous type is a value of x such that $P(X < x) \leq \frac{1}{2}$ and $P(X \leq x) \geq \frac{1}{2}$. If there is only one such x, it is called the *median of the distribution*. Find the median of each of the following distributions:

(a) $p(x) = \frac{4!}{x!(4-x)!}(\frac{1}{4})^x(\frac{3}{4})^{4-x}$, $x = 0, 1, 2, 3, 4$, zero elsewhere.

(b) $f(x) = 3x^2$, $0 < x < 1$, zero elsewhere.

(c) $f(x) = \frac{1}{\pi(1+x^2)}$, $-\infty < x < \infty$.

Hint: In parts (b) and (c), $P(X < x) = P(X \le x)$ and thus that common value must equal $\frac{1}{2}$ if x is to be the median of the distribution.

1.7.10. Let $0 < p < 1$. A $(100p)th$ *percentile* (*quantile* of order p) of the distribution of a random variable X is a value ξ_p such that $P(X < \xi_p) \le p$ and $P(X \le \xi_p) \ge p$. Find the $20th$ percentile of the distribution that has pdf $f(x) = 4x^3$, $0 < x < 1$, zero elsewhere.

Hint: With a continuous-type random variable X, $P(X < \xi_p) = P(X \le \xi_p)$ and hence that common value must equal p.

1.7.11. Find the pdf $f(x)$, the $25th$ percentile, and the $60th$ percentile for each of the following cdfs: Sketch the graphs of $f(x)$ and $F(x)$.

(a) $F(x) = (1 + e^{-x})^{-1}$, $-\infty < x < \infty$.

(b) $F(x) = \exp\{-e^{-x}\}$, $-\infty < x < \infty$.

(c) $F(x) = \frac{1}{2} + \frac{1}{\pi}\tan^{-1}(x)$, $-\infty < x < \infty$.

1.7.12. Find the cdf $F(x)$ associated with each of the following probability density functions. Sketch the graphs of $f(x)$ and $F(x)$.

(a) $f(x) = 3(1 - x)^2$, $0 < x < 1$, zero elsewhere.

(b) $f(x) = 1/x^2$, $1 < x < \infty$, zero elsewhere.

(c) $f(x) = \frac{1}{3}$, $0 < x < 1$ or $2 < x < 4$, zero elsewhere.

Also find the median and the 25th percentile of each of these distributions.

1.7.13. Consider the cdf $F(x) = 1 - e^{-x} - xe^{-x}$, $0 \le x < \infty$, zero elsewhere. Find the pdf, the mode, and the median (by numerical methods) of this distribution.

1.7.14. Let X have the pdf $f(x) = 2x$, $0 < x < 1$, zero elsewhere. Compute the probability that X is at least $\frac{3}{4}$ given that X is at least $\frac{1}{2}$.

1.7.15. The random variable X is said to be **stochastically larger** than the random variable Y if

$$P(X > z) \ge P(Y > z), \qquad (1.7.11)$$

for all real z, with strict inequality holding for at least one z value. Show that this requires that the cdfs enjoy the following property

$$F_X(z) \le F_Y(z),$$

for all real z, with strict inequality holding for at least one z value.

1.7.16. Let X be a continuous random variable with support $(-\infty, \infty)$. If $Y = X + \Delta$ and $\Delta > 0$, using the definition in Exercise 1.7.15, that Y is stochastically larger than X.

1.7.17. Divide a line segment into two parts by selecting a point at random. Find the probability that the larger segment is at least three times the shorter. Assume a uniform distribution.

1.7.18. Let X be the number of gallons of ice cream that is requested at a certain store on a hot summer day. Assume that $f(x) = 12x(1000-x)^2/10^{12}$, $0 < x < 1000$, zero elsewhere, is the pdf of X. How many gallons of ice cream should the store have on hand each of these days, so that the probability of exhausting its supply on a particular day is 0.05?

1.7.19. Find the $25th$ percentile of the distribution having pdf $f(x) = |x|/4$, $-2 < x < 2$, zero elsewhere.

1.7.20. Let X have the pdf $f(x) = x^2/9$, $0 < x < 3$, zero elsewhere. Find the pdf of $Y = X^3$.

1.7.21. If the pdf of X is $f(x) = 2xe^{-x^2}$, $0 < x < \infty$, zero elsewhere, determine the pdf of $Y = X^2$.

1.7.22. Let X have the uniform pdf $f_X(x) = \frac{1}{\pi}$, for $-\frac{\pi}{2} < x < \frac{\pi}{2}$. Find the pdf of $Y = \tan X$. This is the pdf of a **Cauchy distribution**.

1.7.23. Let X have the pdf $f(x) = 4x^3$, $0 < x < 1$, zero elsewhere. Find the cdf and the pdf of $Y = -\ln X^4$.

1.7.24. Let $f(x) = \frac{1}{3}$, $-1 < x < 2$, zero elsewhere, be the pdf of X. Find the cdf and the pdf of $Y = X^2$.
Hint: Consider $P(X^2 \leq y)$ for two cases: $0 \leq y < 1$ and $1 \leq y < 4$.

1.8   Expectation of a Random Variable

In this section we introduce the expectation operator which we will use throughout the remainder of the text.

Definition 1.8.1 (Expectation). *Let X be a random variable. If X is a continuous random variable with pdf $f(x)$ and*

$$\int_{-\infty}^{\infty} |x| f(x)\, dx < \infty,$$

then the **expectation** *of X is*

$$E(X) = \int_{-\infty}^{\infty} x f(x)\, dx.$$

If X is a discrete random variable with pmf $p(x)$ and

$$\sum_x |x|\, p(x) < \infty,$$

then the **expectation** *of X is*

$$E(X) = \sum_x x\, p(x).$$

Sometimes the expectation $E(X)$ is called the **mathematical expectation** of X, the **expected value** of X, or the **mean** of X. When the mean designation is used, we often denote the $E(X)$ by μ; i.e, $\mu = E(X)$.

Example 1.8.1 (Expectation of a Constant). Consider a constant random variable, that is, a random variable with all its mass at a constant k. This is a discrete random variable with pmf $p(k) = 1$. Because $|k|$ is finite, we have by definition that

$$E(k) = kp(k) = k. \quad \blacksquare \qquad\qquad (1.8.1)$$

Remark 1.8.1. The terminology of expectation or expected value has its origin in games of chance. This can be illustrated as follows: Four small similar chips, numbered 1,1,1, and 2, respectively, are placed in a bowl and are mixed. A player is blindfolded and is to draw a chip from the bowl. If she draws one of the three chips numbered 1, she will receive one dollar. If she draws the chip numbered 2, she will receive two dollars. It seems reasonable to assume that the player has a "$\frac{3}{4}$ claim" on the \$1 and a "$\frac{1}{4}$ claim" on the \$2. Her "total claim" is $(1)(\frac{3}{4}) + 2(\frac{1}{4}) = \frac{5}{4}$, that is \$1.25. Thus the expectation of X is precisely the player's claim in this game. $\blacksquare$

Example 1.8.2. Let the random variable X of the discrete type have the pmf given by the table

x	1	2	3	4
$p(x)$	$\frac{4}{10}$	$\frac{1}{10}$	$\frac{3}{10}$	$\frac{2}{10}$

Here $p(x) = 0$ if x is not equal to one of the first four positive integers. This illustrates the fact that there is no need to have a formula to describe a pmf. We have

$$E(X) = (1)(\frac{4}{10}) + (2)(\frac{1}{10}) + (3)(\frac{3}{10}) + (4)(\frac{2}{10}) = \frac{23}{10} = 2.3. \quad \blacksquare$$

Example 1.8.3. Let X have the pdf

$$f(x) = \begin{cases} 4x^3 & 0 < x < 1 \\ 0 & \text{elsewhere.} \end{cases}$$

Then

$$E(X) = \int_0^1 x(4x^3)\, dx = \int_0^1 4x^4\, dx = \left[\frac{4x^5}{5}\right]_0^1 = \frac{4}{5}. \quad \blacksquare$$

Let us consider a function of a random variable X. Call this function $Y = g(X)$. Because Y is a random variable we could obtain its expectation by first finding the distribution of Y. However, as the following theorem states, we can use the distribution of X to determine the expectation of Y.

Theorem 1.8.1. *Let X be a random variable and let $Y = g(X)$ for some function g.*

(a). *Suppose X is continuous with pdf $f_X(x)$. If $\int_{-\infty}^{\infty} |g(x)| f_X(x)\, dx < \infty$, then the expectation of Y exists and it is given by*

$$E(Y) = \int_{-\infty}^{\infty} g(x) f_X(x)\, dx. \qquad (1.8.2)$$

(b). *Suppose X is discrete with pmf $p_X(x)$. Suppose the support of X is denoted by $\mathcal{S}_X$. If $\sum_{x \in \mathcal{S}_X} |g(x)| p_X(x) < \infty$, then the expectation of Y exists and it is given by*

$$E(Y) = \sum_{x \in \mathcal{S}_X} g(x) p_X(x). \qquad (1.8.3)$$

Proof: We give the proof in the discrete case. The proof for the continuous case requires some advanced results in analysis; see, also, Exercise 1.8.1. The assumption of absolute convergence,

$$\sum_{x \in \mathcal{S}_X} |g(x)| p_X(x) < \infty, \qquad (1.8.4)$$

implies that the following results are true:

(c). The series $\sum_{x \in \mathcal{S}_X} g(x) p_X(x)$ converges.

(d). Any rearrangement of either series (1.8.4) or (c) converges to the same value as the original series.

The rearrangement we need is through the support set $\mathcal{S}_Y$ of Y. Result (d) implies

$$
\begin{aligned}
\sum_{x \in \mathcal{S}_X} |g(x)| p_X(x) &= \sum_{y \in \mathcal{S}_Y} \sum_{\{x \in \mathcal{S}_X : g(x) = y\}} |g(x)| p_X(x) && (1.8.5) \\
&= \sum_{y \in \mathcal{S}_Y} |y| \sum_{\{x \in \mathcal{S}_X : g(x) = y\}} p_X(x) && (1.8.6) \\
&= \sum_{y \in \mathcal{S}_Y} |y| p_Y(y). && (1.8.7)
\end{aligned}
$$

By (1.8.4), the left side of (1.8.5) is finite; hence, the last term (1.8.7) is also finite. Thus $E(Y)$ exists. Using (d) we can then obtain another set of equations which are the same as (1.8.5) - (1.8.7) but without the absolute values. Hence,

$$\sum_{x \in \mathcal{S}_X} g(x) p_X(x) = \sum_{y \in \mathcal{S}_Y} y p_Y(y) = E(Y),$$

which is the desired result. ∎

Theorem 1.8.1 shows that the expectation operator E is a linear operator.

Theorem 1.8.2. *Let $g_1(X)$ and $g_2(X)$ be functions of a random variable X. Suppose the expectations of $g_1(X)$ and $g_2(X)$ exist. Then for any constants k_1 and k_2, the expectation of $k_1 g_1(X) + k_2 g_2(X)$ exists and it is given by,*

$$E[k_1 g_1(X) + k_2 g_2(X)] = k_1 E[g_1(X)] + k_2 E[g_2(X)]. \qquad (1.8.8)$$

Proof: For the continuous case existence follows from the hypothesis, the triangle inequality, and the linearity of the integral; i.e.,

$$\int_{-\infty}^{\infty} |k_1 g_1(x) + k_2 g_2(x)| f_X(x)\, dx \quad \le \quad |k_1| \int_{-\infty}^{\infty} |g_1(x)| f_X(x)\, dx$$

$$+ |k_2| \int_{-\infty}^{\infty} |g_2(x)| f_X(x)\, dx < \infty.$$

The result (1.8.8) follows similarly using the linearity of the integral. The proof for the discrete case follows likewise using the linearity of sums. ∎

The following examples illustrate these theorems.

Example 1.8.4. Let X have the pdf

$$f(x) = \begin{cases} 2(1-x) & 0 < x < 1 \\ 0 & \text{elsewhere.} \end{cases}$$

Then

$$E(X) = \int_{-\infty}^{\infty} x f(x)\, dx \quad = \quad \int_0^1 (x) 2(1-x)\, dx = \frac{1}{3},$$

$$E(X^2) = \int_{-\infty}^{\infty} x^2 f(x)\, dx \quad = \quad \int_0^1 (x^2) 2(1-x)\, dx = \frac{1}{6},$$

and, of course,

$$E(6X + 3X^2) = 6\left(\frac{1}{3}\right) + 3\left(\frac{1}{6}\right) = \frac{5}{2}. \quad ∎$$

Example 1.8.5. Let X have the pmf

$$p(x) = \begin{cases} \frac{x}{6} & x = 1, 2, 3 \\ 0 & \text{elsewhere.} \end{cases}$$

Then

$$E(X^3) \quad = \quad \sum_x x^3 p(x) = \sum_{x=1}^{3} x^3 \frac{x}{6}$$

$$= \quad \frac{1}{6} + \frac{16}{6} + \frac{81}{6} = \frac{98}{6}. \quad ∎$$

Example 1.8.6. Let us divide, at random, a horizontal line segment of length 5 into two parts. If X is the length of the left-hand part, it is reasonable to assume that X has the pdf

$$f(x) = \begin{cases} \frac{1}{5} & 0 < x < 5 \\ 0 & \text{elsewhere.} \end{cases}$$

The expected value of the length of X is $E(X) = \frac{5}{2}$ and the expected value of the length $5 - x$ is $E(5 - x) = \frac{5}{2}$. But the expected value of the product of the two lengths is equal to

$$E[X(5 - X)] = \int_0^5 x(5 - x)(\tfrac{1}{5})\,dx = \tfrac{25}{6} \neq (\tfrac{5}{2})^2.$$

That is, in general, the expected value of a product is not equal to the product of the expected values. ∎

Example 1.8.7. A bowl contains five chips, which cannot be distinguished by a sense of touch alone. Three of the chips are marked \$1 each and the remaining two are marked \$4 each. A player is blindfolded and draws, at random and without replacement, two chips from the bowl. The player is paid an amount equal to the sum of the values of the two chips that he draws and the game is over. If it costs \$4.75 to play the game, would we care to participate for any protracted period of time? Because we are unable to distinguish the chips by sense of touch, we assume that each of the 10 pairs that can be drawn has the same probability of being drawn. Let the random variable X be the number of chips, of the two to be chosen, that are marked \$1. Then, under our assumptions, X has the hypergeometric pmf

$$p(x) = \begin{cases} \dfrac{\binom{3}{x}\binom{2}{2-x}}{\binom{5}{2}} & x = 0, 1, 2 \\ 0 & \text{elsewhere.} \end{cases}$$

If $X = x$, the player receives $u(x) = x + 4(2 - x) = 8 - 3x$ dollars. Hence his mathematical expectation is equal to

$$E[8 - 3X] = \sum_{x=0}^{2} (8 - 3x)p(x) = \tfrac{44}{10},$$

or \$4.40. ∎

EXERCISES

1.8.1. Our proof of Theorem 1.8.1 was for the discrete case. The proof for the continuous case requires some advanced results in in analysis. If in addition, though, the function $g(x)$ is one-to-one, show that the result is true for the continuous case. *Hint:* First assume that $y = g(x)$ is strictly increasing. Then use the change of variable technique with Jacobian dx/dy on the integral $\int_{x \in S_X} g(x) f_X(x)\,dx$.

1.8.2. Let X be a random variable of either type. If $g(X) \equiv k$, where k is a constant, show that $E(g(X)) = k$.

1.8.3. Let X have the pdf $f(x) = (x + 2)/18$, $-2 < x < 4$, zero elsewhere. Find $E(X)$, $E[(X + 2)^3]$, and $E[6X - 2(X + 2)^3]$.

1.8.4. Suppose that $p(x) = \frac{1}{5}$, $x = 1, 2, 3, 4, 5$, zero elsewhere, is the pmf of the discrete type random variable X. Compute $E(X)$ and $E(X^2)$. Use these two results to find $E[(X + 2)^2]$ by writing $(X + 2)^2 = X^2 + 4X + 4$.

1.8.5. Let X be a number selected at random from a set of numbers $\{51, 52, \ldots, 100\}$. Approximate $E(1/X)$.

Hint: Find reasonable upper and lower bounds by finding integrals bounding $E(1/X)$.

1.8.6. Let the pmf $p(x)$ be positive at $x = -1, 0, 1$ and zero elsewhere.

(a) If $p(0) = \frac{1}{4}$, find $E(X^2)$.

(b) If $p(0) = \frac{1}{4}$ and if $E(X) = \frac{1}{4}$, determine $p(-1)$ and $p(1)$.

1.8.7. Let X have the pdf $f(x) = 3x^2$, $0 < x < 1$, zero elsewhere. Consider a random rectangle whose sides are X and $(1 - X)$. Determine the expected value of the area of the rectangle.

1.8.8. A bowl contains 10 chips, of which 8 are marked $2 each and 2 are marked $5 each. Let a person choose, at random and without replacement, 3 chips from this bowl. If the person is to receive the sum of the resulting amounts, find his expectation.

1.8.9. Let X be a random variable of the continuous type that has pdf $f(x)$. If m is the unique median of the distribution of X and b is a real constant, show that

$$E(|X - b|) = E(|X - m|) + 2 \int_m^b (b - x) f(x)\, dx,$$

provided that the expectations exist. For what value of b is $E(|X - b|)$ a minimum?

1.8.10. Let $f(x) = 2x$, $0 < x < 1$, zero elsewhere, be the pdf of X.

(a) Compute $E(1/X)$.

(b) Find the cdf and the pdf of $Y = 1/X$.

(c) Compute $E(Y)$ and compare this result with the answer obtained in Part (a).

1.8.11. Two distinct integers are chosen at random and without replacement from the first six positive integers. Compute the expected value of the absolute value of the difference of these two numbers.

1.8.12. Let X have the pdf $f(x) = \frac{1}{x^2}$, $1 < x < \infty$, zero elsewhere. Show that $E(X)$ does not exist.

1.8.13. Let X have a Cauchy distribution that is symmetric about zero. Why doesn't $E(X) = 0$?

1.8.14. Let X have the pdf $f(x) = 3x^2$, $0 < x < 1$, zero elsewhere.

(a) Compute $E(X^3)$.

(b) Show that $Y = X^3$ has a uniform$(0, 1)$ distribution.

(c) Compute $E(Y)$ and compare this result with the answer obtained in Part (a).

1.9 Some Special Expectations

Certain expectations, if they exist, have special names and symbols to represent them. First, let X be a random variable of the discrete type with pmf $p(x)$. Then

$$E(X) = \sum_x xp(x).$$

If the support of X is $\{a_1, a_2, a_3, \ldots\}$, it follows that

$$E(X) = a_1 p(a_1) + a_2 p(a_2) + a_3 p(a_3) + \cdots.$$

This sum of products is seen to be a "weighted average" of the values of $a_1, a_2, a_3, \ldots$, the "weight" associated with each a_i being $p(a_i)$. This suggests that we call $E(X)$ the arithmetic mean of the values of X, or, more simply, the *mean value* of X (or the mean value of the distribution).

Definition 1.9.1 (Mean). *Let X be a random variable whose expectation exists. The* **mean value** μ *of X is defined to be* $\mu = E(X)$.

The mean is the first moment (about 0) of a random variable. Another special expectation involves the second moment. Let X be a discrete random variable with support $\{a_1, a_2, \ldots\}$ and with pmf $p(x)$, then

$$
\begin{aligned}
E[(X - \mu)^2] &= \sum_x (x - \mu)^2 p(x) \\
&= (a_1 - \mu)^2 p(a_1) + (a_2 - \mu)^2 p(a_2) + \cdots.
\end{aligned}
$$

This sum of products may be interpreted as a "weighted average" of the squares of the deviations of the numbers $a_1, a_2, \ldots$ from the mean value μ of those numbers where the "weight" associated with each $(a_i - \mu)^2$ is $p(a_i)$. It can also be thought of as the second moment of X about μ. This will be an important expectation for all types of random variables and we shall usually refer to it as the variance.

Definition 1.9.2 (Variance). *Let X be a random variable with finite mean μ and such that $E[(X - \mu)^2]$ is finite. Then the* **variance** *of X is defined to be $E[(X - \mu)^2]$. It is usually denoted by σ^2 or by $Var(X)$.*

It is worthwhile to observe that the $Var(X)$ equals

$$\sigma^2 = E[(X - \mu)^2] = E(X^2 - 2\mu X + \mu^2);$$

and since E is a linear operator,

$$
\begin{aligned}
\sigma^2 &= E(X^2) - 2\mu E(X) + \mu^2 \\
&= E(X^2) - 2\mu^2 + \mu^2 \\
&= E(X^2) - \mu^2.
\end{aligned}
$$

This frequently affords an easier way of computing the variance of X.

It is customary to call σ (the positive square root of the variance) the **standard deviation** of X (or the standard deviation of the distribution). The number σ is sometimes interpreted as a measure of the dispersion of the points of the space relative to the mean value μ. If the space contains only one point k for which $p(k) > 0$, then $p(k) = 1$, $\mu = k$ and $\sigma = 0$.

Remark 1.9.1. Let the random variable X of the continuous type have the pdf $f_X(x) = 1/(2a)$, $-a < x < a$, zero elsewhere, so that $\sigma_X = a/\sqrt{3}$ is the standard deviation of the distribution of X. Next, let the random variable Y of the continuous type have the pdf $f_Y(y) = 1/4a$, $-2a < y < 2a$, zero elsewhere, so that $\sigma_Y = 2a/\sqrt{3}$ is the standard deviation of the distribution of Y. Here the standard deviation of Y is twice that of X; this reflects the fact that the probability for Y is spread out twice as much (relative to the mean zero) than is the probability for X. ∎

We next define a third special expectation.

Definition 1.9.3 (Moment Generating Function (mgf)). *Let X be a random variable such that for some $h > 0$, the expectation of e^{tX} exists for $-h < t < h$. The* **moment generating function** *of X is defined to be the function $M(t) = E(e^{tX})$, for $-h < t < h$. We will use the abbreviation* **mgf** *to denote moment generating function of a random variable.*

Actually all that is needed is that the mgf exists in an open neighborhood of 0. Such an interval, of course, will include an interval of the form $(-h, h)$ for some $h > 0$. Further, it is evident that if we set $t = 0$, we have $M(0) = 1$. But note that for a mgf to exist, it must exist in an open interval about 0. As will be seen by example, not every distribution has an mgf.

If we are discussing several random variables, it is often useful to subscript M as M_X to denote that this is the mgf of X.

Let X and Y be two random variables with mgfs. If X and Y have the same distribution, i.e, $F_X(z) = F_Y(z)$ for all z, then certainly $M_X(t) = M_Y(t)$ in a neighborhood of 0. But one of the most important properties of mgfs is that the converse of this statement is true too. That is, mgfs uniquely identify distributions. We state this as a theorem. The proof of this converse, though, is beyond the scope of this text; see Chung (1974). We will verify it for a discrete situation.

Theorem 1.9.1. *Let X and Y be random variables with moment generating functions M_X and M_Y, respectively, existing in open intervals about 0. Then $F_X(z) = F_Y(z)$ for all $z \in R$ if and only if $M_X(t) = M_Y(t)$ for all $t \in (-h, h)$ for some $h > 0$.*

Because of the importance of this theorem it does seem desirable to try to make the assertion plausible. This can be done if the random variable is of the discrete type. For example, let it be given that

$$M(t) = \tfrac{1}{10}e^t + \tfrac{2}{10}e^{2t} + \tfrac{3}{10}e^{3t} + \tfrac{4}{10}e^{4t}$$

is, for all real values of t, the mgf of a random variable X of the discrete type. If we let $p(x)$ be the pmf of X with support $\{a_1, a_2, a_3, \ldots\}$, then because

$$M(t) = \sum_x e^{tx} p(x),$$

we have

$$\tfrac{1}{10}e^t + \tfrac{2}{10}e^{2t} + \tfrac{3}{10}e^{3t} + \tfrac{4}{10}e^{4t} = p(a_1)e^{a_1 t} + p(a_2)e^{a_2 t} + \cdots.$$

Because this is an identity for all real values of t, it seems that the right-hand member should consist of but four terms and that each of the four should be equal, respectively, to one of those in the left-hand member; hence we may take $a_1 = 1$, $p(a_1) = \tfrac{1}{10}$; $a_2 = 2$, $p(a_2) = \tfrac{2}{10}$; $a_3 = 3$, $p(a_3) = \tfrac{3}{10}$; $a_4 = 4$, $p(a_4) = \tfrac{4}{10}$. Or, more simply, the pmf of X is

$$p(x) = \begin{cases} \frac{x}{10} & x = 1, 2, 3, 4 \\ 0 & \text{elsewhere.} \end{cases}$$

On the other hand, suppose X is a random variable of the continuous type. Let it be given that

$$M(t) = \frac{1}{1-t}, \quad t < 1,$$

is the mgf of X. That is, we are given

$$\frac{1}{1-t} = \int_{-\infty}^{\infty} e^{tx} f(x)\, dx, \quad t < 1.$$

It is not at all obvious how $f(x)$ is found. However, it is easy to see that a distribution with pdf

$$f(x) = \begin{cases} e^{-x} & 0 < x < \infty \\ 0 & \text{elsewhere,} \end{cases}$$

has the mgf $M(t) = (1-t)^{-1}$, $t < 1$. Thus the random variable X has a distribution with this pdf in accordance with the assertion of the uniqueness of the mgf.

Since a distribution that has an mgf $M(t)$ is completely determined by $M(t)$, it would not be surprising if we could obtain some properties of the distribution directly from $M(t)$. For example, the existence of $M(t)$ for $-h < t < h$ implies that derivatives of $M(t)$ of all orders exist at $t = 0$. Also, a theorem in analysis allows us to interchange the order of differentiation and integration (or summation in the discrete case). That is, if X is continuous,

$$M'(t) = \frac{dM(t)}{dt} = \frac{d}{dt}\int_{-\infty}^{\infty} e^{tx} f(x)\, dx = \int_{-\infty}^{\infty} \frac{d}{dt} e^{tx} f(x)\, dx, = \int_{-\infty}^{\infty} x e^{tx} f(x)\, dx.$$

Likewise, if X is a discrete random variable

$$M'(t) = \frac{dM(t)}{dt} = \sum_x x e^{tx} p(x).$$

Upon setting $t = 0$, we have in either case

$$M'(0) = E(X) = \mu.$$

The second derivative of $M(t)$ is

$$M''(t) = \int_{-\infty}^{\infty} x^2 e^{tx} f(x)\, dx \quad \text{or} \quad \sum_x x^2 e^{tx} p(x),$$

so that $M''(0) = E(X^2)$. Accordingly, the var(X) equals

$$\sigma^2 = E(X^2) - \mu^2 = M''(0) - [M'(0)]^2.$$

For example, if $M(t) = (1-t)^{-1}$, $t < 1$, as in the illustration above, then

$$M'(t) = (1-t)^{-2} \quad \text{and} \quad M''(t) = 2(1-t)^{-3}.$$

Hence

$$\mu = M'(0) = 1$$

and

$$\sigma^2 = M''(0) - \mu^2 = 2 - 1 = 1.$$

Of course, we could have computed μ and σ^2 from the pdf by

$$\mu = \int_{-\infty}^{\infty} x f(x)\, dx \quad \text{and} \quad \sigma^2 = \int_{-\infty}^{\infty} x^2 f(x)\, dx - \mu^2,$$

respectively. Sometimes one way is easier than the other.

In general, if m is a positive integer and if $M^{(m)}(t)$ means the mth derivative of $M(t)$, we have, by repeated differentiation with respect to t,

$$M^{(m)}(0) = E(X^m).$$

Now

$$E(X^m) = \int_{-\infty}^{\infty} x^m f(x)\, dx \quad \text{or} \quad \sum_x x^m f(x),$$

and the integrals (or sums) of this sort are, in mechanics, called *moments*. Since $M(t)$ generates the values of $E(X^m)$, $m = 1, 2, 3, \ldots$, it is called the moment-generating function (mgf). In fact, we shall sometimes call $E(X^m)$ the **mth moment** of the distribution, or the mth moment of X.

Example 1.9.1. Let X have the pdf

$$f(x) = \begin{cases} \frac{1}{2}(x+1) & -1 < x < 1 \\ 0 & \text{elsewhere.} \end{cases}$$

Then the mean value of X is

$$\mu = \int_{-\infty}^{\infty} x f(x)\, dx = \int_{-1}^{1} x \frac{x+1}{2}\, dx = \frac{1}{3}$$

while the variance of X is

$$\sigma^2 = \int_{-\infty}^{\infty} x^2 f(x)\, dx - \mu^2 = \int_{-1}^{1} x^2 \frac{x+1}{2}\, dx - \left(\frac{1}{3}\right)^2 = \frac{2}{9}. \quad \blacksquare$$

Example 1.9.2. If X has the pdf

$$f(x) = \begin{cases} \frac{1}{x^2} & 1 < x < \infty \\ 0 & \text{elsewhere,} \end{cases}$$

then the mean value of X does not exist, because

$$
\begin{aligned}
\int_1^\infty |x| \frac{1}{x^2}\, dx &= \lim_{b \to \infty} \int_1^b \frac{1}{x}\, dx \\
&= \lim_{b \to \infty} (\log b - \log 1)
\end{aligned}
$$

does not exist. ∎

Example 1.9.3. It is known that the series

$$\frac{1}{1^2} + \frac{1}{2^2} + \frac{1}{3^2} + \cdots$$

converges to $\pi^2/6$. Then

$$p(x) = \begin{cases} \frac{6}{\pi^2 x^2} & x = 1, 2, 3, \ldots \\ 0 & \text{elsewhere,} \end{cases}$$

is the pmf of a discrete type of random variable X. The mgf of this distribution, if it exists, is given by

$$
\begin{aligned}
M(t) &= E(e^{tX}) = \sum_x e^{tx} p(x) \\
&= \sum_{x=1}^\infty \frac{6 e^{tx}}{\pi^2 x^2}.
\end{aligned}
$$

The ratio test may be used to show that this series diverges if $t > 0$. Thus there does not exist a positive number h such that $M(t)$ exists for $-h < t < h$. Accordingly, the distribution has the pmf $p(x)$ of this example and does not have an mgf. ∎

Example 1.9.4. Let X have the mgf $M(t) = e^{t^2/2}$, $-\infty < t < \infty$. We can differentiate $M(t)$ any number of times to find the moments of X. However, it is instructive to consider this alternative method. The function $M(t)$ is represented by the following MacLaurin's series.

$$
\begin{aligned}
e^{t^2/2} &= 1 + \frac{1}{1!}\left(\frac{t^2}{2}\right) + \frac{1}{2!}\left(\frac{t^2}{2}\right)^2 + \cdots + \frac{1}{k!}\left(\frac{t^2}{2}\right)^k + \cdots \\
&= 1 + \frac{1}{2!}t^2 + \frac{(3)(1)}{4!}t^4 + \cdots + \frac{(2k-1)\cdots(3)(1)}{(2k)!}t^{2k} + \cdots.
\end{aligned}
$$

In general, the MacLaurin's series for $M(t)$ is

$$
\begin{aligned}
M(t) &= M(0) + \frac{M'(0)}{1!}t + \frac{M''(0)}{2!}t^2 + \cdots + \frac{M^{(m)}(0)}{m!}t^m + \cdots \\
&= 1 + \frac{E(X)}{1!}t + \frac{E(X^2)}{2!}t^2 + \cdots + \frac{E(X^m)}{m!}t^m + \cdots.
\end{aligned}
$$

Thus the coefficient of $(t^m/m!)$ in the MacLaurin's series representation of $M(t)$ is $E(X^m)$. So, for our particular $M(t)$, we have

$$E(X^{2k}) \quad = \quad (2k-1)(2k-3)\cdots(3)(1) = \frac{(2k)!}{2^k k!}, \quad k = 1, 2, 3, \ldots, \quad (1.9.1)$$

$$E(X^{2k-1}) \quad = \quad 0, \quad k = 1, 2, 3, \ldots \quad (1.9.2)$$

We will make use of this result in Section 3.4. ∎

Remark 1.9.2. In a more advanced course, we would not work with the mgf because so many distributions do not have moment-generating functions. Instead, we would let i denote the imaginary unit, t an arbitrary real, and we would define $\varphi(t) = E(e^{itX})$. This expectation exists for *every* distribution and it is called the *characteristic function* of the distribution. To see why $\varphi(t)$ exists for all real t, we note, in the continuous case, that its absolute value

$$|\varphi(t)| = \left| \int_{-\infty}^{\infty} e^{itx} f(x)\,dx \right| \le \int_{-\infty}^{\infty} |e^{itx} f(x)|\,dx.$$

However, $|f(x)| = f(x)$ since $f(x)$ is nonnegative and

$$|e^{itx}| = |\cos tx + i \sin tx| = \sqrt{\cos^2 tx + \sin^2 tx} = 1.$$

Thus

$$|\varphi(t)| \le \int_{-\infty}^{\infty} f(x)\,dx = 1.$$

Accordingly, the integral for $\varphi(t)$ exists for all real values of t. In the discrete case, a summation would replace the integral.

Every distribution has a unique characteristic function; and to each characteristic function there corresponds a unique distribution of probability. If X has a distribution with characteristic function $\varphi(t)$, then, for instance, if $E(X)$ and $E(X^2)$ exist, they are given, respectively, by $iE(X) = \varphi'(0)$ and $i^2 E(X^2) = \varphi''(0)$. Readers who are familiar with complex-valued functions may write $\varphi(t) = M(it)$ and, throughout this book, may prove certain theorems in complete generality.

Those who have studied Laplace and Fourier transforms will note a similarity between these transforms and $M(t)$ and $\varphi(t)$; it is the uniqueness of these transforms that allows us to assert the uniqueness of each of the moment-generating and characteristic functions. ∎

EXERCISES

1.9.1. Find the mean and variance, if they exist, of each of the following distributions.

(a) $p(x) = \frac{3!}{x!(3-x)!}(\frac{1}{2})^3$, $x = 0, 1, 2, 3$, zero elsewhere.

(b) $f(x) = 6x(1-x)$, $0 < x < 1$, zero elsewhere.

(c) $f(x) = 2/x^3$, $1 < x < \infty$, zero elsewhere.

1.9.2. Let $p(x) = (\frac{1}{2})^x$, $x = 1, 2, 3, \ldots$, zero elsewhere, be the pmf of the random variable X. Find the mgf, the mean, and the variance of X.

1.9.3. For each of the following distributions, compute $P(\mu - 2\sigma < X < \mu + 2\sigma)$.

(a) $f(x) = 6x(1 - x)$, $0 < x < 1$, zero elsewhere.

(b) $p(x) = (\frac{1}{2})^x$, $x = 1, 2, 3, \ldots$, zero elsewhere.

1.9.4. If the variance of the random variable X exists, show that

$$E(X^2) \geq [E(X)]^2.$$

1.9.5. Let a random variable X of the continuous type have a pdf $f(x)$ whose graph is symmetric with respect to $x = c$. If the mean value of X exists, show that $E(X) = c$.
Hint: Show that $E(X - c)$ equals zero by writing $E(X - c)$ as the sum of two integrals: one from $-\infty$ to c and the other from c to ∞. In the first, let $y = c - x$; and, in the second, $z = x - c$. Finally, use the symmetry condition $f(c-y) = f(c+y)$ in the first.

1.9.6. Let the random variable X have mean μ, standard deviation σ, and mgf $M(t)$, $-h < t < h$. Show that

$$E\left(\frac{X - \mu}{\sigma}\right) = 0, \quad E\left[\left(\frac{X - \mu}{\sigma}\right)^2\right] = 1,$$

and

$$E\left\{\exp\left[t\left(\frac{X - \mu}{\sigma}\right)\right]\right\} = e^{-\mu t/\sigma} M\left(\frac{t}{\sigma}\right), \quad -h\sigma < t < h\sigma.$$

1.9.7. Show that the moment generating function of the random variable X having the pdf $f(x) = \frac{1}{3}$, $-1 < x < 2$, zero elsewhere, is

$$M(t) = \begin{cases} \frac{e^{2t} - e^{-t}}{3t} & t \neq 0 \\ 1 & t = 0. \end{cases}$$

1.9.8. Let X be a random variable such that $E[(X - b)^2]$ exists for all real b. Show that $E[(X - b)^2]$ is a minimum when $b = E(X)$.

1.9.9. Let X denote a random variable for which $E[(X - a)^2]$ exists. Give an example of a distribution of a discrete type such that this expectation is zero. Such a distribution is called a *degenerate distribution*.

1.9.10. Let X denote a random variable such that $K(t) = E(t^X)$ exists for all real values of t in a certain open interval that includes the point $t = 1$. Show that $K^{(m)}(1)$ is equal to the mth *factorial moment* $E[X(X - 1) \cdots (X - m + 1)]$.

1.9.11. Let X be a random variable. If m is a positive integer, the expectation $E[(X - b)^m]$, if it exists, is called the mth moment of the distribution about the point b. Let the first, second, and third moments of the distribution about the point 7 be 3, 11, and 15, respectively. Determine the mean μ of X, and then find the first, second, and third moments of the distribution about the point μ.

1.9.12. Let X be a random variable such that $R(t) = E(e^{t(X-b)})$ exists for t such that $-h < t < h$. If m is a positive integer, show that $R^{(m)}(0)$ is equal to the mth moment of the distribution about the point b.

1.9.13. Let X be a random variable with mean μ and variance σ^2 such that the third moment $E[(X - \mu)^3]$ about the vertical line through μ exists. The value of the ratio $E[(X - \mu)^3]/\sigma^3$ is often used as a measure of *skewness*. Graph each of the following probability density functions and show that this measure is negative, zero, and positive for these respective distributions (which are said to be skewed to the left, not skewed, and skewed to the right, respectively).

(a) $f(x) = (x + 1)/2$, $-1 < x < 1$, zero elsewhere.

(b) $f(x) = \frac{1}{2}$, $-1 < x < 1$, zero elsewhere.

(c) $f(x) = (1 - x)/2$, $-1 < x < 1$, zero elsewhere.

1.9.14. Let X be a random variable with mean μ and variance σ^2 such that the fourth moment $E[(X - \mu)^4]$ exists. The value of the ratio $E[(X - \mu)^4]/\sigma^4$ is often used as a measure of *kurtosis*. Graph each of the following probability density functions and show that this measure is smaller for the first distribution.

(a) $f(x) = \frac{1}{2}$, $-1 < x < 1$, zero elsewhere.

(b) $f(x) = 3(1 - x^2)/4$, $-1 < x < 1$, zero elsewhere.

1.9.15. Let the random variable X have pmf

$$p(x) = \begin{cases} p & x = -1, 1 \\ 1 - 2p & x = 0 \\ 0 & \text{elsewhere,} \end{cases}$$

where $0 < p < \frac{1}{2}$. Find the measure of kurtosis as a function of p. Determine its value when $p = \frac{1}{3}$, $p = \frac{1}{5}$, $p = \frac{1}{10}$, and $p = \frac{1}{100}$. Note that the kurtosis increases as p decreases.

1.9.16. Let $\psi(t) = \log M(t)$, where $M(t)$ is the mgf of a distribution. Prove that $\psi'(0) = \mu$ and $\psi''(0) = \sigma^2$. The function $\psi(t)$ is called the **cumulant generating function**.

1.9.17. Find the mean and the variance of the distribution that has the cdf

$$F(x) = \begin{cases} 0 & x < 0 \\ \frac{x}{8} & 0 \le x < 2 \\ \frac{x^2}{16} & 2 \le x < 4 \\ 1 & 4 \le x. \end{cases}$$

1.9.18. Find the moments of the distribution that has mgf $M(t) = (1-t)^{-3}$, $t < 1$. *Hint:* Find the MacLaurin's series for $M(t)$.

1.9.19. Let X be a random variable of the continuous type with pdf $f(x)$, which is positive provided $0 < x < b < \infty$, and is equal to zero elsewhere. Show that

$$E(X) = \int_0^b [1 - F(x)]\, dx,$$

where $F(x)$ is the cdf of X.

1.9.20. Let X be a random variable of the discrete type with pmf $p(x)$ that is positive on the nonnegative integers and is equal to zero elsewhere. Show that

$$E(X) = \sum_{x=0}^{\infty} [1 - F(x)],$$

where $F(x)$ is the cdf of X.

1.9.21. Let X have the pmf $p(x) = 1/k$, $x = 1, 2, 3, \ldots, k$, zero elsewhere. Show that the mgf is

$$M(t) = \begin{cases} \frac{e^t(1-e^{kt})}{k(1-e^t)} & t \neq 0 \\ 1 & t = 0. \end{cases}$$

1.9.22. Let X have the cdf $F(x)$ that is a mixture of the continuous and discrete types, namely

$$F(x) = \begin{cases} 0 & x < 0 \\ \frac{x+1}{4} & 0 \leq x < 1 \\ 1 & 1 \leq x. \end{cases}$$

Determine reasonable definitions of $\mu = E(X)$ and $\sigma^2 = \text{var}(X)$ and compute each. *Hint:* Determine the parts of the pmf and the pdf associated with each of the discrete and continuous parts, and then sum for the discrete part and integrate for the continuous part.

1.9.23. Consider k continuous-type distributions with the following characteristics: pdf $f_i(x)$, mean μ_i, and variance σ_i^2, $i = 1, 2, \ldots, k$. If $c_i \geq 0$, $i = 1, 2, \ldots, k$, and $c_1 + c_2 + \cdots + c_k = 1$, show that the mean and the variance of the distribution having pdf $c_1 f_1(x) + \cdots + c_k f_k(x)$ are $\mu = \sum_{i=1}^k c_i \mu_i$ and $\sigma^2 = \sum_{i=1}^k c_i[\sigma_i^2 + (\mu_i - \mu)^2]$, respectively.

1.9.24. Let X be a random variable with a pdf $f(x)$ and mgf $M(t)$. Suppose f is symmetric about 0, $(f(-x) = f(x))$. Show that $M(-t) = M(t)$.

1.9.25. Let X have the exponential pdf, $f(x) = \beta^{-1} \exp\{-x/\beta\}$, $0 < x < \infty$, zero elsewhere. Find the mgf, the mean, and the variance of X.

1.10 Important Inequalities

In this section, we obtain the proofs of three famous inequalities involving expectations. We shall make use of these inequalities in the remainder of the text. We begin with a useful result.

Theorem 1.10.1. *Let X be a random variable and let m be a positive integer. Suppose $E[X^m]$ exists. If k is an integer and $k \leq m$, then $E[X^k]$ exists.*

Proof: We shall prove it for the continuous case; but the proof is similar for the discrete case if we replace integrals by sums. Let $f(x)$ be the pdf of X. Then

$$
\begin{aligned}
\int_{-\infty}^{\infty} |x|^k f(x)\, dx &= \int_{|x| \leq 1} |x|^k f(x)\, dx + \int_{|x| > 1} |x|^k f(x)\, dx \\
&\leq \int_{|x| \leq 1} f(x)\, dx + \int_{|x| > 1} |x|^m f(x)\, dx \\
&\leq \int_{-\infty}^{\infty} f(x)\, dx + \int_{-\infty}^{\infty} |x|^m f(x)\, dx \\
&\leq 1 + E[|X|^m] < \infty,
\end{aligned}
\tag{1.10.1}
$$

which is the the desired result. ∎

Theorem 1.10.2 (Markov's Inequality). *Let $u(X)$ be a nonnegative function of the random variable X. If $E[u(X)]$ exists, then for every positive constant c,*

$$
P[u(X) \geq c] \leq \frac{E[u(X)]}{c}.
$$

Proof. The proof is given when the random variable X is of the continuous type; but the proof can be adapted to the discrete case if we replace integrals by sums. Let $A = \{x : u(x) \geq c\}$ and let $f(x)$ denote the pdf of X. Then

$$
E[u(X)] = \int_{-\infty}^{\infty} u(x) f(x)\, dx = \int_A u(x) f(x)\, dx + \int_{A^c} u(x) f(x)\, dx.
$$

Since each of the integrals in the extreme right-hand member of the preceding equation is nonnegative, the left-hand member is greater than or equal to either of them. In particular,

$$
E[u(X)] \geq \int_A u(x) f(x)\, dx.
$$

However, if $x \in A$, then $u(x) \geq c$; accordingly, the right-hand member of the preceding inequality is not increased if we replace $u(x)$ by c. Thus

$$
E[u(X)] \geq c \int_A f(x)\, dx.
$$

Since

$$
\int_A f(x)\, dx = P(X \in A) = P[u(X) \geq c],
$$

it follows that

$$E[u(X)] \geq cP[u(X) \geq c],$$

which is the desired result. ∎

The preceding theorem is a generalization of an inequality that is often called *Chebyshev's inequality*. This inequality will now be established.

Theorem 1.10.3 (Chebyshev's Inequality). *Let the random variable X have a distribution of probability about which we assume only that there is a finite variance σ^2, (by Theorem 1.10.1 this implies the mean $\mu = E(X)$ exists). Then for every $k > 0$,*

$$P(|X - \mu| \geq k\sigma) \leq \frac{1}{k^2}, \tag{1.10.2}$$

or, equivalently,

$$P(|X - \mu| < k\sigma) \geq 1 - \frac{1}{k^2}.$$

Proof. In Theorem 1.10.2 take $u(X) = (X - \mu)^2$ and $c = k^2\sigma^2$. Then we have

$$P[(X - \mu)^2 \geq k^2\sigma^2] \leq \frac{E[(X - \mu)^2]}{k^2\sigma^2}.$$

Since the numerator of the right-hand member of the preceding inequality is σ^2, the inequality may be written

$$P(|X - \mu| \geq k\sigma) \leq \frac{1}{k^2},$$

which is the desired result. Naturally, we would take the positive number k to be greater than 1 to have an inequality of interest. ∎

A convenient form of Chebyshev's Inequality is found by taking $k\sigma = \epsilon$ for $\epsilon > 0$. Then equation (1.10.2) becomes

$$P(|X - \mu| \geq \epsilon) \leq \frac{\sigma^2}{\epsilon^2}, \quad \text{for all } \epsilon > 0. \tag{1.10.3}$$

Hence, the number $1/k^2$ is an upper bound for the probability $P(|X - \mu| \geq k\sigma)$. In the following example this upper bound and the exact value of the probability are compared in special instances.

Example 1.10.1. Let X have the pdf

$$f(x) = \begin{cases} \frac{1}{2\sqrt{3}} & -\sqrt{3} < x < \sqrt{3} \\ 0 & \text{elsewhere.} \end{cases}$$

Here $\mu = 0$ and $\sigma^2 = 1$. If $k = \frac{3}{2}$, we have the exact probability

$$P(|X - \mu| \geq k\sigma) = P\left(|X| \geq \frac{3}{2}\right) = 1 - \int_{-3/2}^{3/2} \frac{1}{2\sqrt{3}}\, dx = 1 - \frac{\sqrt{3}}{2}.$$

By Chebyshev's inequality, this probability has the upper bound $1/k^2 = \frac{4}{9}$. Since $1 - \sqrt{3}/2 = 0.134$, approximately, the exact probability in this case is considerably less than the upper bound $\frac{4}{9}$. If we take $k = 2$, we have the exact probability $P(|X - \mu| \geq 2\sigma) = P(|X| \geq 2) = 0$. This again is considerably less than the upper bound $1/k^2 = \frac{1}{4}$ provided by Chebyshev's inequality. ∎

In each of the instances in the preceding example, the probability $P(|X - \mu| \geq k\sigma)$ and its upper bound $1/k^2$ differ considerably. This suggests that this inequality might be made sharper. However, if we want an inequality that holds for every $k > 0$ and holds for all random variables having a finite variance, such an improvement is impossible, as is shown by the following example.

Example 1.10.2. Let the random variable X of the discrete type have probabilities $\frac{1}{8}, \frac{6}{8}, \frac{1}{8}$ at the points $x = -1, 0, 1$, respectively. Here $\mu = 0$ and $\sigma^2 = \frac{1}{4}$. If $k = 2$, then $1/k^2 = \frac{1}{4}$ and $P(|X - \mu| \geq k\sigma) = P(|X| \geq 1) = \frac{1}{4}$. That is, the probability $P(|X - \mu| \geq k\sigma)$ here attains the upper bound $1/k^2 = \frac{1}{4}$. Hence the inequality cannot be improved without further assumptions about the distribution of X. ∎

Definition 1.10.1.

A function ϕ defined on an interval (a, b), $-\infty \leq a < b \leq \infty$, is said to be a **convex** function if for all x, y in (a, b) and for all $0 < \gamma < 1$,

$$\phi[\gamma x + (1 - \gamma)y] \leq \gamma\phi(x) + (1 - \gamma)\phi(y). \tag{1.10.4}$$

We say ϕ is **strictly convex** if the above inequality is strict.

Depending on existence of first or second derivatives of ϕ, the following theorem can be proved.

Theorem 1.10.4. *If ϕ is differentiable on (a, b) then*

(a) ϕ is convex if and only if $\phi'(x) \leq \phi'(y)$, for all $a < x < y < b$,

(b) ϕ is strictly convex if and only if $\phi'(x) < \phi'(y)$, for all $a < x < y < b$.

If ϕ is twice differentiable on (a, b) then

(a) ϕ is convex if and only if $\phi''(x) \geq 0$, for all $a < x < b$,

(b) ϕ is strictly convex if $\phi''(x) > 0$, for all $a < x < b$.

Of course the second part of this theorem follows immediately from the first part. While the first part appeals to one's intuition, the proof of it can be found in most analysis books; see, for instance, Hewitt and Stromberg (1965). A very useful probability inequality follows from convexity.

Theorem 1.10.5 (Jensen's Inequality). *If ϕ is convex on an open interval I and X is a random variable whose support is contained in I and has finite expectation, then*

$$\phi[E(X)] \leq E[\phi(X)]. \tag{1.10.5}$$

If ϕ is strictly convex then the inequality is strict, unless X is a constant random variable.

Proof: For our proof we will assume that ϕ has a second derivative, but in general only convexity is required. Expand $\phi(x)$ into a Taylor series about $\mu = E[X]$ of order two:

$$\phi(x) = \phi(\mu) + \phi'(\mu)(x - \mu) + \frac{\phi''(\zeta)(x - \mu)^2}{2},$$

where ζ is between x and μ. Because the last term on the right side of the above equation is nonnegative, we have

$$\phi(x) \geq \phi(\mu) + \phi'(\mu)(x - \mu).$$

Taking expectations of both sides, leads to the result. The inequality will be strict if $\phi''(x) > 0$, for all $x \in (a, b)$, provided X is not a constant. ∎

Example 1.10.3. Let X be a nondegenerate random variable with mean μ and a finite second moment. Then $\mu^2 < E(X^2)$. This is obtained by Jensen's inequality using the strictly convex function $\phi(t) = t^2$. ∎

Example 1.10.4 (Harmonic and Geometric Means). Let $\{a_1, \ldots, a_n\}$ be a set of positive numbers. Create a distribution for a random variable X by placing weight $1/n$ on each of the numbers $a_1, \ldots, a_n$. Then the mean of X is the *arithmetic mean*, (AM), $E(X) = n^{-1} \sum_{i=1}^{n} a_i$. Then, since $-\log x$ is a convex function, we have by Jensen's inequality that

$$-\log\left(\frac{1}{n} \sum_{i=1}^{n} a_i\right) \leq E(-\log X) = -\frac{1}{n} \sum_{i=1}^{n} \log a_i = -\log(a_1 a_1 \cdots a_n)^{1/n}$$

or, equivalently,

$$\log\left(\frac{1}{n} \sum_{i=1}^{n} a_i\right) \geq \log(a_1 a_1 \cdots a_n)^{1/n},$$

and, hence,

$$(a_1 a_1 \cdots a_n)^{1/n} \leq \frac{1}{n} \sum_{i=1}^{n} a_i. \tag{1.10.6}$$

The quantity on the left side of this inequality is called the *geometric mean*, (GM). So (1.10.6) is equivalent to saying that GM $\leq$ AM for any finite set of positive numbers.

Now in (1.10.6) replace a_i by $1/a_i$, (which is positive, also). We then obtain,

$$\frac{1}{n} \sum_{i=1}^{n} \frac{1}{a_i} \geq \left(\frac{1}{a_1} \frac{1}{a_2} \cdots \frac{1}{a_n}\right)^{1/n}$$

or, equivalently,

$$\frac{1}{\frac{1}{n} \sum_{i=1}^{n} \frac{1}{a_i}} \leq (a_1 a_1 \cdots a_n)^{1/n}. \tag{1.10.7}$$

The left member of this inequality is called the *harmonic mean*, (HM). Putting (1.10.6) and (1.10.7) together we have shown the relationship

$$\text{HM} \le \text{GM} \le \text{AM}, \tag{1.10.8}$$

for any finite set of positive numbers. ∎

EXERCISES

1.10.1. Let X be a random variable with mean μ and let $E[(X - \mu)^{2k}]$ exist. Show, with $d > 0$, that $P(|X - \mu| \ge d) \le E[(X - \mu)^{2k}]/d^{2k}$. This is essentially Chebyshev's inequality when $k = 1$. The fact that this holds for all $k = 1, 2, 3, \ldots$, when those $(2k)$th moments exist, usually provides a much smaller upper bound for $P(|X - \mu| \ge d)$ than does Chebyshev's result.

1.10.2. Let X be a random variable such that $P(X \le 0) = 0$ and let $\mu = E(X)$ exist. Show that $P(X \ge 2\mu) \le \frac{1}{2}$.

1.10.3. If X is a random variable such that $E(X) = 3$ and $E(X^2) = 13$, use Chebyshev's inequality to determine a lower bound for the probability $P(-2 < X < 8)$.

1.10.4. Let X be a random variable with mgf $M(t)$, $-h < t < h$. Prove that

$$P(X \ge a) \le e^{-at} M(t), \quad 0 < t < h,$$

and that

$$P(X \le a) \le e^{-at} M(t), \quad -h < t < 0.$$

Hint: Let $u(x) = e^{tx}$ and $c = e^{ta}$ in Theorem 1.10.2. *Note.* These results imply that $P(X \ge a)$ and $P(X \le a)$ are less than the respective greatest lower bounds for $e^{-at} M(t)$ when $0 < t < h$ and when $-h < t < 0$.

1.10.5. The mgf of X exists for all real values of t and is given by

$$M(t) = \frac{e^t - e^{-t}}{2t}, \quad t \ne 0, \quad M(0) = 1.$$

Use the results of the preceding exercise to show that $P(X \ge 1) = 0$ and $P(X \le -1) = 0$. Note that here h is infinite.

1.10.6. Let X be a positive random variable; i.e., $P(X \le 0) = 0$. Argue that

(a) $E(1/X) \ge 1/E(X)$,

(b) $E[-\log X] \ge -\log[E(X)]$,

(c) $E[\log(1/X)] \ge \log[1/E(X)]$,

(d) $E[X^3] \ge [E(X)]^3$.

Chapter 2

Multivariate Distributions

2.1 Distributions of Two Random Variables

We begin the discussion of two random variables with the following example. A coin is to be tossed three times and our interest is in the ordered number pair (number of H's on first two tosses, number of H's on all three tosses), where H and T represent, respectively, heads and tails. Thus the sample space is $\mathcal{C} = \{c : c = c_i, \ i = 1, 2, \ldots, 8\}$, where c_1 is TTT, c_2 is TTH, c_3 is THT, c_4 is HTT, c_5 is THH, c_6 is HTH, c_7 is HHT, and c_8 is HHH. Let X_1 and X_2 be two functions such that $X_1(c_1) = X_1(c_2) = 0$, $X_1(c_3) = X_1(c_4) = X_1(c_5) = X_1(c_6) = 1$, $X_1(c_7) = X_1(c_8) = 2$; and $X_2(c_1) = 0$, $X_2(c_2) = X_2(c_3) = X_2(c_4) = 1$, $X_2(c_5) = X_2(c_6) = X_2(c_7) = 2$, and $X_2(c_8) = 3$. Thus X_1 and X_2 are real-valued functions defined on the sample space $\mathcal{C}$, which take us from the sample space to the space of ordered number pairs.

$$\mathcal{D} = \{(0, 0), (0, 1), (1, 1), (1, 2), (2, 2), (2, 3)\}.$$

Thus X_1 and X_2 are two random variables defined on the space $\mathcal{C}$, and, in this example, the space of these random variables is the two-dimensional set $\mathcal{D}$ which is a subset of two-dimensional Euclidean space R^2. Hence (X_1, X_2) is a vector function from $\mathcal{C}$ to $\mathcal{D}$. We now formulate the definition of a random vector.

Definition 2.1.1 (Random Vector). *Given a random experiment with a sample space $\mathcal{C}$. Consider two random variables X_1 and X_2, which assign to each element c of $\mathcal{C}$ one and only one ordered pair of numbers $X_1(c) = x_1$, $X_2(c) = x_2$. Then we say that (X_1, X_2) is a **random vector**. The space of (X_1, X_2) is the set of ordered pairs $\mathcal{D} = \{(x_1, x_2) : x_1 = X_1(c), x_2 = X_2(c), c \in \mathcal{C}\}$.*

We will often denote random vectors using vector notation $\mathbf{X} = (X_1, X_2)'$, where the $'$ denotes the transpose of the row vector (X_1, X_2).

Let $\mathcal{D}$ be the space associated with the random vector (X_1, X_2). Let A be a subset of $\mathcal{D}$. As in the case of one random variable, we shall speak of the event A. We wish to define the probability of the event A, which we denote by $P_{X_1, X_2}[A]$.

As with random variables in Section 1.5, we can uniquely define P_{X_1,X_2} in terms of the **cumulative distribution function**, (cdf), which is given by

$$F_{X_1,X_2}(x_1,x_2) = P[\{X_1 \le x_1\} \cap \{X_2 \le x_2\}], \tag{2.1.1}$$

for all $(x_1, x_2) \in R^2$. Because X_1 and X_2 are random variables, each of the events in the above intersection and the intersection of the events are events in the original sample space $\mathcal{C}$. Thus the expression is well defined. As with random variables, we will write $P[\{X_1 \le x_1\} \cap \{X_2 \le x_2\}]$ as $P[X_1 \le x_1, X_2 \le x_2]$. As Exercise 2.1.3 shows,

$$\begin{aligned} P[a_1 < X_1 \le b_1, a_2 < X_2 \le b_2] = \ & F_{X_1,X_2}(b_1, b_2) - F_{X_1,X_2}(a_1, b_2) \\ & -F_{X_1,X_2}(b_1, a_2) + F_{X_1,X_2}(a_1, a_2). \end{aligned} \tag{2.1.2}$$

Hence, all induced probabilities of sets of the form $(a_1, b_1] \times (a_2, b_2]$ can be formulated in terms of the cdf. Sets of this form in R^2 generate the Borel σ-field of subsets in R^2. This is the σ-field we will use in R^2. In a more advanced class it can be shown that the cdf uniquely determines a probability on R^2, (the induced probability distribution for the random vector (X_1, X_2)). We will often call this cdf the **joint cumulative distribution function** of (X_1, X_2).

As with random variables, we are mainly concerned with two types of random vectors, namely discrete and continuous. We will first discuss the discrete type.

A random vector (X_1, X_2) is a **discrete random vector** if its space $\mathcal{D}$ is finite or countable. Hence, X_1 and X_2 are both discrete, also. The **joint probability mass function** (pmf) of (X_1, X_2) is defined by,

$$p_{X_1,X_2}(x_1, x_2) = P[X_1 = x_1, X_2 = x_2], \tag{2.1.3}$$

for all $(x_1, x_2) \in \mathcal{D}$. As with random variables, the pmf uniquely defines the cdf. It also is characterized by the two properties:

$$\text{(i). } 0 \le p_{X_1,X_2}(x_1, x_2) \le 1 \text{ and (ii). } \sum_{\mathcal{D}}\sum p_{X_1,X_2}(x_1, x_2) = 1. \tag{2.1.4}$$

For an event $B \in \mathcal{D}$, we have

$$P[(X_1, X_2) \in B] = \sum_{B}\sum p_{X_1,X_2}(x_1, x_2).$$

Example 2.1.1. Consider the discrete random vector (X_1, X_2) defined in the example at the beginning of this section. We can conveniently table its pmf as:

		Support of X_2			
		0	1	2	3
	0	$\frac{1}{8}$	$\frac{1}{8}$	0	0
Support of X_1	1	0	$\frac{2}{8}$	$\frac{2}{8}$	0
	2	0	0	$\frac{1}{8}$	$\frac{1}{8}$

■

At times it will be convenient to speak of the **support** of a discrete random vector (X_1, X_2). These are all the points (x_1, x_2) in the space of (X_1, X_2) such that $p(x_1, x_2) > 0$. In the last example the support consists of the six points $\{(0,0), (0,1), (1,1), (1,2), (2,2), (2,3)\}$.

We say a random vector (X_1, X_2) with space $\mathcal{D}$ is of the **continuous** type if its cdf $F_{X_1, X_2}(x_1, x_2)$ is continuous. For the most part, the continuous random vectors in this book will have cdfs which can be represented as integrals of nonnegative functions. That is, $F_{X_1, X_2}(x_1, x_2)$ can be expressed as,

$$F_{X_1, X_2}(x_1, x_2) = \int_{-\infty}^{x_1} \int_{-\infty}^{x_2} f_{X_1, X_2}(w_1, w_2) \, dw_1 dw_2, \qquad (2.1.5)$$

for all $(x_1, x_2) \in R^2$. We call the integrand the **joint probability density function** (pdf) of (X_1, X_2). At points of continuity of $f_{X_1, X_2}(x_1, x_2)$, we have

$$\frac{\partial^2 F_{X_1, X_2}(x_1, x_2)}{\partial x_1 \, \partial x_2} = f_{X_1, X_2}(x_1, x_2).$$

A pdf is essentially characterized by the two properties:

$$\text{(i) } f_{X_1, X_2}(x_1, x_2) \geq 0 \text{ and (ii) } \int \int_{\mathcal{D}} f_{X_1, X_2}(x_1, x_2) \, dx_1 dx_2 = 1. \qquad (2.1.6)$$

For an event $A \in \mathcal{D}$, we have

$$P[(X_1, X_2) \in A] = \int \int_A f_{X_1, X_2}(x_1, x_2) \, dx_1 dx_2.$$

Note that the $P[(X_1, X_2) \in A]$ is just the volume under the surface $z = f_{X_1, X_2}(x_1, x_2)$ over the set A.

Remark 2.1.1. As with univariate random variables, we will often drop the subscript (X_1, X_2) from joint cdfs, pdfs, and pmfs, when it is clear from the context. We will also use notation such as f_{12} instead of f_{X_1, X_2}. Besides (X_1, X_2), we will often use (X, Y) to express random vectors. ∎

Example 2.1.2. Let

$$f(x_1, x_2) = \begin{cases} 6x_1^2 x_2 & 0 < x_1 < 1, \ 0 < x_2 < 1 \\ 0 & \text{elsewhere}, \end{cases}$$

be the pdf of two random variables X_1 and X_2 of the continuous type. We have, for instance,

$$\begin{aligned}
P(0 < X_1 < \tfrac{3}{4}, \tfrac{1}{3} < X_2 < 2) &= \int_{1/3}^{2} \int_{0}^{3/4} f(x_1, x_2) \, dx_1 dx_2 \\
&= \int_{1/3}^{1} \int_{0}^{3/4} 6x_1^2 x_2 \, dx_1 dx_2 + \int_{1}^{2} \int_{0}^{3/4} 0 \, dx_1 dx_2 \\
&= \tfrac{3}{8} + 0 = \tfrac{3}{8}.
\end{aligned}$$

Note that this probability is the volume under the surface $f(x_1, x_2) = 6x_1^2 x_2$ above the rectangular set $\{(x_1, x_2) : 0 < x_1 < \tfrac{3}{4}, \tfrac{1}{3} < x_2 < 1\} \in R^2$. ∎

For a continuous random vector (X_1, X_2), the **support** of (X_1, X_2) contains all points (x_1, x_2) for which $f(x_1, x_2) > 0$. We will denote the support of a random vector by $\mathcal{S}$. As in the univariate case $\mathcal{S} \subset \mathcal{D}$.

We may extend the definition of a pdf $f_{X_1, X_2}(x_1, x_2)$ over R^2 by using zero elsewhere. We shall do this consistently so that tedious, repetitious references to the space $\mathcal{D}$ can be avoided. Once this is done, we replace

$$\int\int_{\mathcal{D}} f_{X_1, X_2}(x_1, x_2)\, dx_1 dx_2 \quad \text{by} \quad \int_{-\infty}^{\infty}\int_{-\infty}^{\infty} f(x_1, x_2)\, dx_1\, dx_2.$$

Likewise we may extend the pmf $p_{X_1, X_2}(x_1, x_2)$ over a convenient set by using zero elsewhere. Hence, we replace

$$\sum\sum_{\mathcal{D}} p_{X_1, X_2}(x_1, x_2) \quad \text{by} \quad \sum_{x_2}\sum_{x_1} p(x_1, x_2).$$

Finally, if a pmf or a pdf in one or more variables is explicitly defined, we can see by inspection whether the random variables are of the continuous or discrete type. For example, it seems obvious that

$$p(x, y) = \begin{cases} \frac{9}{4^{x+y}} & x = 1, 2, 3, \ldots, \quad y = 1, 2, 3, \ldots \\ 0 & \text{elsewhere}, \end{cases}$$

is a pmf of two discrete-type random variables X and Y, whereas

$$f(x, y) = \begin{cases} 4xy e^{-x^2 - y^2} & 0 < x < \infty, \ 0 < y < \infty \\ 0 & \text{elsewhere}, \end{cases}$$

is clearly a pdf of two continuous-type random variables X and Y. In such cases it seems unnecessary to specify which of the two simpler types of random variables is under consideration.

Let (X_1, X_2) be a random vector. Each of X_1 and X_2 are then random variables. We can obtain their distributions in terms of the joint distribution of (X_1, X_2) as follows. Recall that the event which defined the cdf of X_1 at x_1 is $\{X_1 \le x_1\}$. However,

$$\{X_1 \le x_1\} = \{X_1 \le x_1\} \cap \{-\infty < X_2 < \infty\} = \{X_1 \le x_1, -\infty < X_2 < \infty\}.$$

Taking probabilities we have

$$F_{X_1}(x_1) = P[X_1 \le x_1, -\infty < X_2 < \infty], \tag{2.1.7}$$

for all $x_1 \in R$. By Theorem 1.3.6 we can write this equation as $F_{X_1}(x_1) = \lim_{x_2 \uparrow \infty} F(x_1, x_2)$. Thus we have a relationship between the cdfs, which we can extend to either the pmf or pdf depending on whether (X_1, X_2) is discrete or continuous.

First consider the discrete case. Let $\mathcal{D}_{X_1}$ be the support of X_1. For $x_1 \in \mathcal{D}_{X_1}$, equation (2.1.7) is equivalent to

$$F_{X_1}(x_1) = \sum_{w_1 \le x_1, -\infty < x_2 < \infty}\sum p_{X_1, X_2}(w_1, x_2) = \sum_{w_1 \le x_1}\left\{\sum_{x_2 < \infty} p_{X_1, X_2}(w_1, x_2)\right\}.$$

By the uniqueness of cdfs, the quantity in braces must be the pmf of X_1 evaluated at w_1; that is,

$$p_{X_1}(x_1) = \sum_{x_2 < \infty} p_{X_1, X_2}(x_1, x_2), \qquad (2.1.8)$$

for all $x_1 \in \mathcal{D}_{X_1}$.

Note what this says. To find the probability that X_1 is x_1, keep x_1 fixed and sum p_{X_1, X_2} over all of x_2. In terms of a tabled joint pmf with rows comprised of X_1 support values and columns comprised of X_2 support values, this says that the distribution of X_1 can be obtained by the marginal sums of the rows. Likewise, the pmf of X_2 can be obtained by marginal sums of the columns. For example, consider the joint distribution discussed in Example 2.1.1. We have added these marginal sums to the table:

		Support of X_2				
		0	1	2	3	$p_{X_1}(x_1)$
Support of X_1	0	$\frac{1}{8}$	$\frac{1}{8}$	0	0	$\frac{2}{8}$
	1	0	$\frac{2}{8}$	$\frac{2}{8}$	0	$\frac{4}{8}$
	2	0	0	$\frac{1}{8}$	$\frac{1}{8}$	$\frac{2}{8}$
	$p_{X_2}(x_2)$	$\frac{1}{8}$	$\frac{3}{8}$	$\frac{3}{8}$	$\frac{1}{8}$	

Hence, the final row of this table is the pmf of X_2 while the final column is the pmf of X_1. In general, because these distributions are recorded in the margins of the table, we often refer to them as **marginal** pmfs.

Example 2.1.3. Consider a random experiment that consists of drawing at random one chip from a bowl containing 10 chips of the same shape and size. Each chip has an ordered pair of numbers on it: one with $(1, 1)$, one with $(2, 1)$, two with $(3, 1)$, one with $(1, 2)$, two with $(2, 2)$, and three with $(3, 2)$. Let the random variables X_1 and X_2 be defined as the respective first and second values of the ordered pair. Thus the joint pmf $p(x_1, x_2)$ of X_1 and X_2 can be given by the following table, with $p(x_1, x_2)$ equal to zero elsewhere.

	x_1			
x_2	1	2	3	$p_2(x_2)$
1	$\frac{1}{10}$	$\frac{1}{10}$	$\frac{2}{10}$	$\frac{4}{10}$
2	$\frac{1}{10}$	$\frac{2}{10}$	$\frac{3}{10}$	$\frac{6}{10}$
$p_1(x_1)$	$\frac{2}{10}$	$\frac{3}{10}$	$\frac{5}{10}$	

The joint probabilities have been summed in each row and each column and these sums recorded in the margins to give the marginal probability density functions of X_1 and X_2, respectively. Note that it is not necessary to have a formula for $p(x_1, x_2)$ to do this. ∎

We next consider the continuous case. Let $\mathcal{D}_{X_1}$ be the support of X_1. For $x_1 \in \mathcal{D}_{X_1}$, equation (2.1.7) is equivalent to

$$F_{X_1}(x_1) = \int_{-\infty}^{x_1} \int_{-\infty}^{\infty} f_{X_1,X_2}(w_1, x_2)\, dx_2 dw_1 = \int_{-\infty}^{x_1} \left\{ \int_{-\infty}^{\infty} f_{X_1,X_2}(w_1, x_2)\, dx_2 \right\} dw_1.$$

By the uniqueness of cdfs, the quantity in braces must be the pdf of X_1, evaluated at w_1; that is,

$$f_{X_1}(x_1) = \int_{-\infty}^{\infty} f_{X_1,X_2}(x_1, x_2)\, dx_2 \qquad (2.1.9)$$

for all $x_1 \in \mathcal{D}_{X_1}$. Hence, in the continuous case the marginal pdf of X_1 is found by integrating out x_2. Similarly the marginal pdf of X_2 is found by integrating out x_1.

Example 2.1.4. Let X_1 and X_2 have the joint pdf

$$f(x_1, x_2) = \begin{cases} x_1 + x_2 & 0 < x_1 < 1,\ 0 < x_2 < 1 \\ 0 & \text{elsewhere.} \end{cases}$$

The marginal pdf of X_1 is

$$f_1(x_1) = \int_0^1 (x_1 + x_2)\, dx_2 = x_1 + \tfrac{1}{2}, \quad 0 < x_1 < 1,$$

zero elsewhere, and the marginal pdf of X_2 is

$$f_2(x_2) = \int_0^1 (x_1 + x_2)\, dx_1 = \tfrac{1}{2} + x_2, \quad 0 < x_2 < 1,$$

zero elsewhere. A probability like $P(X_1 \leq \tfrac{1}{2})$ can be computed from either $f_1(x_1)$ or $f(x_1, x_2)$ because

$$\int_0^{1/2} \int_0^1 f(x_1, x_2)\, dx_2 dx_1 = \int_0^{1/2} f_1(x_1)\, dx_1 = \tfrac{3}{8}.$$

However, to find a probability like $P(X_1 + X_2 \leq 1)$, we must use the joint pdf $f(x_1, x_2)$ as follows:

$$\int_0^1 \int_0^{1-x_1} (x_1 + x_2)\, dx_2 dx_1 = \int_0^1 \left[x_1(1 - x_1) + \frac{(1 - x_1)^2}{2} \right] dx_1$$

$$= \int_0^1 \left(\frac{1}{2} - \frac{1}{2}x_1^2 \right) dx_1 = \frac{1}{3}.$$

This latter probability is the volume under the surface $f(x_1, x_2) = x_1 + x_2$ above the set $\{(x_1, x_2) : 0 < x_1,\ x_1 + x_2 \leq 1\}$. ∎

2.1.1 Expectation

The concept of expectation extends in a straightforward manner. Let (X_1, X_2) be a random vector and let $Y = g(X_1, X_2)$ for some real valued function, i.e., $g : R^2 \to R$. Then Y is a random variable and we could determine its expectation by obtaining the distribution of Y. But Theorem 1.8.1 is true for random vectors, also. Note the proof we gave for this theorem involved the discrete case and Exercise 2.1.11 shows its extension to the random vector case.

Suppose (X_1, X_2) is of the continuous type. Then $E(Y)$ exists if

$$\int_{-\infty}^{\infty} \int_{-\infty}^{\infty} |g(x_1, x_2)| f_{X_1, X_2}(x_1, x_2) \, dx_1 dx_2 < \infty.$$

Then

$$E(Y) = \int_{-\infty}^{\infty} \int_{-\infty}^{\infty} g(x_1, x_2) f_{X_1, X_2}(x_1, x_2) \, dx_1 dx_2. \tag{2.1.10}$$

Likewise if (X_1, X_2) is discrete, then $E(Y)$ exists if

$$\sum_{x_1} \sum_{x_2} |g(x_1, x_2)| p_{X_1, X_2}(x_1, x_2) < \infty.$$

Then

$$E(Y) = \sum_{x_1} \sum_{x_2} g(x_1, x_2) p_{X_1, X_2}(x_1, x_2). \tag{2.1.11}$$

We can now show that E is a linear operator.

Theorem 2.1.1. *Let* (X_1, X_2) *be a random vector. Let* $Y_1 = g_1(X_1, X_2)$ *and* $Y_2 = g_2(X_1, X_2)$ *be random variables whose expectations exist. Then for any real numbers* k_1 *and* k_2,

$$E(k_1 Y_1 + k_2 Y_2) = k_1 E(Y_1) + k_2 E(Y_2). \tag{2.1.12}$$

Proof: We shall prove it for the continuous case. Existence of the expected value of $k_1 Y_1 + k_2 Y_2$ follows directly from the triangle inequality and linearity of integrals, i.e.,

$$\int_{-\infty}^{\infty} \int_{-\infty}^{\infty} |k_1 g_1(x_1, x_2) + k_2 g_1(x_1, x_2)| f_{X_1, X_2}(x_1, x_2) \, dx_1 dx_2 \leq$$

$$|k_1| \int_{-\infty}^{\infty} \int_{-\infty}^{\infty} |g_1(x_1, x_2)| f_{X_1, X_2}(x_1, x_2) \, dx_1 dx_2$$

$$+ |k_2| \int_{-\infty}^{\infty} \int_{-\infty}^{\infty} |g_2(x_1, x_2)| f_{X_1, X_2}(x_1, x_2) \, dx_1 dx_2 < \infty.$$

By once again using linearity of the integral we have,

$$
\begin{aligned}
E(k_1 Y_1 + k_2 Y_2) &= \int_{-\infty}^{\infty} \int_{-\infty}^{\infty} [k_1 g_1(x_1, x_2) + k_2 g_2(x_1, x_2)] f_{X_1, X_2}(x_1, x_2)\, dx_1 dx_2 \\
&= k_1 \int_{-\infty}^{\infty} \int_{-\infty}^{\infty} g_1(x_1, x_2) f_{X_1, X_2}(x_1, x_2)\, dx_1 dx_2 \\
&\quad + k_2 \int_{-\infty}^{\infty} \int_{-\infty}^{\infty} g_2(x_1, x_2) f_{X_1, X_2}(x_1, x_2)\, dx_1 dx_2 \\
&= k_1 E(Y_1) + k_2 E(Y_2),
\end{aligned}
$$

i.e., the desired result. ∎

We also note that the expected value of any function $g(X_2)$ of X_2 can be found in two ways:

$$
E(g(X_2)) = \int_{-\infty}^{\infty} \int_{-\infty}^{\infty} g(x_2) f(x_1, x_2)\, dx_1 dx_2 = \int_{-\infty}^{\infty} g(x_2) f_{X_2}(x_2)\, dx_2,
$$

the latter single integral being obtained from the double integral by integrating on x_1 first. The following example illustrates these ideas.

Example 2.1.5. Let X_1 and X_2 have the pdf

$$
f(x_1, x_2) = \begin{cases} 8x_1 x_2 & 0 < x_1 < x_2 < 1 \\ 0 & \text{elsewhere.} \end{cases}
$$

Then

$$
\begin{aligned}
E(X_1 X_2^2) &= \int_{-\infty}^{\infty} \int_{-\infty}^{\infty} x_1 x_2^2 f(x_1, x_2)\, dx_1 dx_2 \\
&= \int_0^1 \int_0^{x_2} 8 x_1^2 x_2^3\, dx_1 dx_2 \\
&= \int_0^1 \tfrac{8}{3} x_2^6\, dx_2 = \tfrac{8}{21}.
\end{aligned}
$$

In addition,

$$
E(X_2) = \int_0^1 \int_0^{x_2} x_2(8 x_1 x_2)\, dx_1 dx_2 = \tfrac{4}{5}.
$$

Since X_2 has the pdf $f_2(x_2) = 4x_2^3$, $0 < x_2 < 1$, zero elsewhere, the latter expectation can be found by

$$
E(X_2) = \int_0^1 x_2(4 x_2^3)\, dx_2 = \tfrac{4}{5}.
$$

Thus,

$$
\begin{aligned}
E(7 X_1 X_2^2 + 5 X_2) &= 7 E(X_1 X_2^2) + 5 E(X_2) \\
&= (7)(\tfrac{8}{21}) + (5)(\tfrac{4}{5}) = \tfrac{20}{3}. \quad \blacksquare
\end{aligned}
$$

Example 2.1.6. Continuing with Example 2.1.5, suppose the random variable Y is defined by $Y = X_1/X_2$. We determine $E(Y)$ in two ways. The first way is by definition, i.e., find the distribution of Y and then determine its expectation. The cdf of Y, for $0 < y \le 1$, is

$$
\begin{aligned}
F_Y(y) &= P(Y \le y) = P(X_1 \le yX_2) = \int_0^1 \int_0^{yx_2} 8x_1x_2 \, dx_1 dx_2 \\
&= \int_0^1 4y^2 x_2^3 \, dx_2 = y^2.
\end{aligned}
$$

Hence, the pdf of Y is

$$
f_Y(y) = F_Y'(y) = \begin{cases} 2y & 0 < y < 1 \\ 0 & \text{elsewhere,} \end{cases}
$$

which leads to

$$
E(Y) = \int_0^1 y(2y) \, dy = \frac{2}{3}.
$$

For the second way, we make use of expression (2.1.10) and find $E(Y)$ directly by

$$
\begin{aligned}
E(Y) &= E\left(\frac{X_1}{X_2}\right) = \int_0^1 \left\{ \int_0^{x_2} \left(\frac{x_1}{x_2}\right) 8x_1x_2 \, dx_1 \right\} dx_2 \\
&= \int_0^1 \frac{8}{3} x_2^3 \, dx_2 = \frac{2}{3}. \quad \blacksquare
\end{aligned}
$$

We next define the moment generating function of a random vector.

Definition 2.1.2 (Moment Generating Function of a Random Vector). *Let $\mathbf{X} = (X_1, X_2)'$ be a random vector. If $E(e^{t_1 X_1 + t_2 X_2})$ exists for $|t_1| < h_1$ and $|t_2| < h_2$, where h_1 and h_2 are positive, it is denoted by $M_{X_1,X_2}(t_1, t_2)$ and is called the* **moment-generating function** *(mgf) of $\mathbf{X}$.*

As with random variables, if it exists, the mgf of a random vector uniquely determines the distribution of the random vector.

Let $\mathbf{t} = (t_1, t_2)'$, Then we can write the mgf of $\mathbf{X}$ as,

$$
M_{X_1,X_2}(\mathbf{t}) = E\left[e^{\mathbf{t}'\mathbf{X}}\right], \tag{2.1.13}
$$

so it is quite similar to the mgf of a random variable. Also, the mgfs of X_1 and X_2 are immediately seen to be $M_{X_1,X_2}(t_1, 0)$ and $M_{X_1,X_2}(0, t_2)$, respectively. If there is no confusion, we often drop the subscripts on M.

Example 2.1.7. Let the continuous-type random variables X and Y have the joint pdf

$$
f(x, y) = \begin{cases} e^{-y} & 0 < x < y < \infty \\ 0 & \text{elsewhere.} \end{cases}
$$

The mgf of this joint distribution is

$$M(t_1, t_2) = \int_0^\infty \int_x^\infty \exp(t_1 x + t_2 y - y)\, dy\, dx$$

$$= \frac{1}{(1 - t_1 - t_2)(1 - t_2)},$$

provided that $t_1 + t_2 < 1$ and $t_2 < 1$. Furthermore, the moment-generating functions of the marginal distributions of X and Y are, respectively,

$$M(t_1, 0) = \frac{1}{1 - t_1}, \quad t_1 < 1,$$

$$M(0, t_2) = \frac{1}{(1 - t_2)^2}, \quad t_2 < 1.$$

These moment-generating functions are, of course, respectively, those of the marginal probability density functions,

$$f_1(x) = \int_x^\infty e^{-y}\, dy = e^{-x}, \quad 0 < x < \infty,$$

zero elsewhere, and

$$f_2(y) = e^{-y} \int_0^y dx = y e^{-y}, \quad 0 < y < \infty,$$

zero elsewhere. ∎

We will also need to define the expected value of the random vector itself, but this is not a new concept because it is defined in terms of componentwise expectation:

Definition 2.1.3 (Expected Value of a Random Vector). *Let* $\mathbf{X} = (X_1, X_2)'$ *be a random vector. Then the* **expected value** *of* $\mathbf{X}$ *exists if the expectations of* X_1 *and* X_2 *exist. If it exists, then the* **expected value** *is given by*

$$E[\mathbf{X}] = \left[\begin{array}{c} E(X_1) \\ E(X_2) \end{array} \right]. \tag{2.1.14}$$

EXERCISES

2.1.1. Let $f(x_1, x_2) = 4x_1 x_2$, $0 < x_1 < 1$, $0 < x_2 < 1$, zero elsewhere, be the pdf of X_1 and X_2. Find $P(0 < X_1 < \frac{1}{2}, \frac{1}{4} < X_2 < 1)$, $P(X_1 = X_2)$, $P(X_1 < X_2)$, and $P(X_1 \leq X_2)$.
Hint: Recall that $P(X_1 = X_2)$ would be the volume under the surface $f(x_1, x_2) = 4x_1 x_2$ and above the line segment $0 < x_1 = x_2 < 1$ in the $x_1 x_2$-plane.

2.1.2. Let $A_1 = \{(x, y) : x \leq 2, \ y \leq 4\}$, $A_2 = \{(x, y) : x \leq 2, \ y \leq 1\}$, $A_3 = \{(x, y) : x \leq 0, \ y \leq 4\}$, and $A_4 = \{(x, y) : x \leq 0 \ y \leq 1\}$ be subsets of the space $\mathcal{A}$ of two random variables X and Y, which is the entire two-dimensional plane. If $P(A_1) = \frac{7}{8}$, $P(A_2) = \frac{4}{8}$, $P(A_3) = \frac{3}{8}$, and $P(A_4) = \frac{2}{8}$, find $P(A_5)$, where $A_5 = \{(x, y) : 0 < x \leq 2, \ 1 < y \leq 4\}$.

2.1.3. Let $F(x, y)$ be the distribution function of X and Y. For all real constants $a < b$, $c < d$, show that $P(a < X \le b, \ c < Y \le d) = F(b, d) - F(b, c) - F(a, d) + F(a, c)$.

2.1.4. Show that the function $F(x, y)$ that is equal to 1 provided that $x + 2y \ge 1$, and that is equal to zero provided that $x + 2y < 1$, cannot be a distribution function of two random variables.

Hint: Find four numbers $a < b$, $c < d$, so that

$$F(b, d) - F(a, d) - F(b, c) + F(a, c)$$

is less than zero.

2.1.5. Given that the nonnegative function $g(x)$ has the property that

$$\int_0^\infty g(x) \, dx = 1.$$

Show that

$$f(x_1, x_2) = \frac{2g(\sqrt{x_1^2 + x_2^2})}{\pi \sqrt{x_1^2 + x_2^2}}, \ 0 < x_1 < \infty \ \ 0 < x_2 < \infty,$$

zero elsewhere, satisfies the conditions for a pdf of two continuous-type random variables X_1 and X_2.

Hint: Use polar coordinates.

2.1.6. Let $f(x, y) = e^{-x-y}$, $0 < x < \infty$, $0 < y < \infty$, zero elsewhere, be the pdf of X and Y. Then if $Z = X + Y$, compute $P(Z \le 0)$, $P(Z \le 6)$, and, more generally, $P(Z \le z)$, for $0 < z < \infty$. What is the pdf of Z?

2.1.7. Let X and Y have the pdf $f(x, y) = 1$, $0 < x < 1$, $0 < y < 1$, zero elsewhere. Find the cdf and pdf of the product $Z = XY$.

2.1.8. Let 13 cards be taken, at random and without replacement, from an ordinary deck of playing cards. If X is the number of spades in these 13 cards, find the pmf of X. If, in addition, Y is the number of hearts in these 13 cards, find the probability $P(X = 2, \ Y = 5)$. What is the joint pmf of X and Y?

2.1.9. Let the random variables X_1 and X_2 have the joint pmf described as follows:

(x_1, x_2)	$(0, 0)$	$(0, 1)$	$(0, 2)$	$(1, 0)$	$(1, 1)$	$(1, 2)$
$f(x_1, x_2)$	$\frac{2}{12}$	$\frac{3}{12}$	$\frac{2}{12}$	$\frac{2}{12}$	$\frac{2}{12}$	$\frac{1}{12}$

and $f(x_1, x_2)$ is equal to zero elsewhere.

(a) Write these probabilities in a rectangular array as in Example 2.1.3, recording each marginal pdf in the "margins".

(b) What is $P(X_1 + X_2 = 1)$?

2.1.10. Let X_1 and X_2 have the joint pdf $f(x_1, x_2) = 15x_1^2 x_2$, $0 < x_1 < x_2 < 1$, zero elsewhere. Find the marginal pdfs and compute $P(X_1 + X_2 \leq 1)$.

Hint: Graph the space X_1 and X_2 and carefully choose the limits of integration in determining each marginal pdf.

2.1.11. Let X_1, X_2 be two random variables with joint pmf $p(x_1, x_2)$, $(x_1, x_2) \in \mathcal{S}$, where $\mathcal{S}$ is the support of X_1, X_2. Let $Y = g(X_1, X_2)$ be a function such that

$$\sum\sum_{(x_1, x_2) \in \mathcal{S}} |g(x_1, x_2)| p(x_1, x_2) < \infty.$$

By following the proof of Theorem 1.8.1, show that

$$E(Y) = \sum\sum_{(x_1, x_2) \in \mathcal{S}} g(x_1, x_2) p(x_1, x_2) < \infty.$$

2.1.12. Let X_1, X_2 be two random variables with joint pmf $p(x_1, x_2) = (x_1 + x_2)/12$, for $x_1 = 1, 2$, $x_2 = 1, 2$, zero elsewhere. Compute $E(X_1)$, $E(X_1^2)$, $E(X_2)$, $E(X_2^2)$, and $E(X_1 X_2)$. Is $E(X_1 X_2) = E(X_1)E(X_2)$? Find $E(2X_1 - 6X_2^2 + 7X_1 X_2)$.

2.1.13. Let X_1, X_2 be two random variables with joint pdf $f(x_1, x_2) = 4x_1 x_2$, $0 < x_1 < 1$, $0 < x_2 < 1$, zero elsewhere. Compute $E(X_1)$, $E(X_1^2)$, $E(X_2)$, $E(X_2^2)$, and $E(X_1 X_2)$. Is $E(X_1 X_2) = E(X_1)E(X_2)$? Find $E(3X_2 - 2X_1^2 + 6X_1 X_2)$.

2.1.14. Let X_1, X_2 be two random variables with joint pmf $p(x_1, x_2) = (1/2)^{x_1 + x_2}$, for $1 \leq x_i < \infty, i = 1, 2$, where x_1 and x_2 are integers, zero elsewhere. Determine the joint mgf of X_1, X_2. Show that $M(t_1, t_2) = M(t_1, 0)M(0, t_2)$.

2.1.15. Let X_1, X_2 be two random variables with joint pdf $f(x_1, x_2) = x_1 \exp\{-x_2\}$, for $0 < x_1 < x_2 < \infty$, zero elsewhere. Determine the joint mgf of X_1, X_2. Does $M(t_1, t_2) = M(t_1, 0)M(0, t_2)$?

2.1.16. Let X and Y have the joint pdf $f(x, y) = 6(1 - x - y)$, $x + y < 1$, $0 < x$, $0 < y$, zero elsewhere. Compute $P(2X + 3Y < 1)$ and $E(XY + 2X^2)$.

2.2 Transformations: Bivariate Random Variables

Let (X_1, X_2) be a random vector. Suppose we know the joint distribution of (X_1, X_2) and we seek the distribution of a transformation of (X_1, X_2), say, $Y = g(X_1, X_2)$. We may be able to obtain the cdf of Y. Another way is to use a transformation. We considered transformation theory for random variables in Sections 1.6 and 1.7. In this section, we extend this theory to random vectors. It is best to discuss the discrete and continuous cases separately. We begin with the discrete case.

There are no essential difficulties involved in a problem like the following. Let $p_{X_1, X_2}(x_1, x_2)$ be the joint pmf of two discrete-type random variables X_1 and X_2 with $\mathcal{S}$ the (two-dimensional) set of points at which $p_{X_1, X_2}(x_1, x_2) > 0$, i.e., $\mathcal{S}$ is the support of (X_1, X_2). Let $y_1 = u_1(x_1, x_2)$ and $y_2 = u_2(x_1, x_2)$ define a one-to-one transformation that maps $\mathcal{S}$ onto $\mathcal{T}$. The joint pmf of the two new random variables $Y_1 = u_1(X_1, X_2)$ and $Y_2 = u_2(X_1, X_2)$ is given by

$$p_{Y_1, Y_2}(y_1, y_2) = \begin{cases} p_{X_1, X_2}[w_1(y_1, y_2), w_2(y_1, y_2)] & (y_1, y_2) \in \mathcal{T} \\ 0 & \text{elsewhere,} \end{cases}$$

where $x_1 = w_1(y_1, y_2)$, $x_2 = w_2(y_1, y_2)$ is the single-valued inverse of $y_1 = u_1(x_1, x_2)$, $y_2 = u_2(x_1, x_2)$. From this joint pmf $p_{Y_1, Y_2}(y_1, y_2)$ we may obtain the marginal pmf of Y_1 by summing on y_2 or the marginal pmf of Y_2 by summing on y_1.

In using this change of variable technique, it should be emphasized that we need two "new" variables to replace the two "old" variables. An example will help explain this technique.

Example 2.2.1. Let X_1 and X_2 have the joint pmf

$$p_{X_1, X_2}(x_1, x_2) = \frac{\mu_1^{x_1} \mu_2^{x_2} e^{-\mu_1} e^{-\mu_2}}{x_1! x_2!}, \quad x_1 = 0, 1, 2, 3, \ldots, \quad x_2 = 0, 1, 2, 3, \ldots,$$

and is zero elsewhere, where μ_1 and μ_2 are fixed positive real numbers. Thus the space $\mathcal{S}$ is the set of points (x_1, x_2), where each of x_1 and x_2 is a nonnegative integer. We wish to find the pmf of $Y_1 = X_1 + X_2$. If we use the change of variable technique, we need to define a second random variable Y_2. Because Y_2 is of no interest to us, let us choose it in such a way that we have a simple one-to-one transformation. For example, take $Y_2 = X_2$. Then $y_1 = x_1 + x_2$ and $y_2 = x_2$ represent a one-to-one transformation that maps $\mathcal{S}$ onto

$$\mathcal{T} = \{(y_1, y_2) : y_2 = 0, 1, \ldots, y_1 \quad \text{and} \quad y_1 = 0, 1, 2, \ldots\}.$$

Note that, if $(y_1, y_2) \in \mathcal{T}$, then $0 \le y_2 \le y_1$. The inverse functions are given by $x_1 = y_1 - y_2$ and $x_2 = y_2$. Thus the joint pmf of Y_1 and Y_2 is

$$p_{Y_1, Y_2}(y_1, y_2) = \frac{\mu_1^{y_1 - y_2} \mu_2^{y_2} e^{-\mu_1 - \mu_2}}{(y_1 - y_2)! y_2!}, \quad (y_1, y_2) \in \mathcal{T},$$

and is zero elsewhere. Consequently, the marginal pmf of Y_1 is given by

$$\begin{aligned} p_{Y_1}(y_1) &= \sum_{y_2=0}^{y_1} p_{Y_1, Y_2}(y_1, y_2) \\ &= \frac{e^{-\mu_1 - \mu_2}}{y_1!} \sum_{y_2=0}^{y_1} \frac{y_1!}{(y_1 - y_2)! y_2!} \mu_1^{y_1 - y_2} \mu_2^{y_2} \\ &= \frac{(\mu_1 + \mu_2)^{y_1} e^{-\mu_1 - \mu_2}}{y_1!}, \quad y_1 = 0, 1, 2, \ldots, \end{aligned}$$

and is zero elsewhere. ∎

For the continuous case we begin with an example which illustrates the cdf technique.

Example 2.2.2. Consider an experiment in which a person chooses at random a point (X, Y) from the unit square $\mathcal{S} = \{(x, y) : 0 < x < 1, \ 0 < y < 1\}$. Suppose that our interest is not in X or in Y but in $Z = X + Y$. Once a suitable probability model has been adopted, we shall see how to find the pdf of Z. To be specific, let the nature of the random experiment be such that it is reasonable to *assume* that the distribution of probability over the unit square is uniform. Then the pdf of X and Y may be written

$$f_{X,Y}(x, y) = \begin{cases} 1 & 0 < x < 1, \ 0 < y < 1 \\ 0 & \text{elsewhere,} \end{cases}$$

and this describes the probability model. Now let the cdf of Z be denoted by $F_Z(z) = P(X + Y \leq z)$. Then

$$F_Z(z) = \begin{cases} 0 & z < 0 \\ \int_0^z \int_0^{z-x} dy dx = \frac{z^2}{2} & 0 \leq z < 1 \\ 1 - \int_{z-1}^1 \int_{z-x}^1 dy dx = 1 - \frac{(2-z)^2}{2} & 1 \leq z < 2 \\ 1 & 2 \leq z. \end{cases}$$

Since $F_Z'(z)$ exists for all values of z, the pmf of Z may then be written

$$f_Z(z) = \begin{cases} z & 0 < z < 1 \\ 2 - z & 1 \leq z < 2 \\ 0 & \text{elsewhere.} \end{cases} \blacksquare$$

We now discuss in general the transformation technique for the continuous case. Let (X_1, X_2) have a jointly continuous distribution with pdf $f_{X_1, X_2}(x_1, x_2)$ and support set $\mathcal{S}$. Suppose the random variables Y_1 and Y_2 are given by $Y_1 = u_1(X_1, X_2)$ and $Y_2 = u_2(X_1, X_2)$, where the functions $y_1 = u_1(x_1, x_2)$ and $y_2 = u_2(x_1, x_2)$ define a one-to-one transformation that maps the set $\mathcal{S}$ in R^2 onto a (two-dimensional) set $\mathcal{T}$ in R^2 where $\mathcal{T}$ is the support of (Y_1, Y_2). If we express each of x_1 and x_2 in terms of y_1 and y_2, we can write $x_1 = w_1(y_1, y_2)$, $x_2 = w_2(y_1, y_2)$. The determinant of order 2,

$$J = \begin{vmatrix} \frac{\partial x_1}{\partial y_1} & \frac{\partial x_1}{\partial y_2} \\ \frac{\partial x_2}{\partial y_1} & \frac{\partial x_2}{\partial y_2} \end{vmatrix},$$

is called the **Jacobian** of the transformation and will be denoted by the symbol J. It will be assumed that these first-order partial derivatives are continuous and that the Jacobian J is not identically equal to zero in $\mathcal{T}$.

We can find, by use of a theorem in analysis, the joint pdf of (Y_1, Y_2). Let A be a subset of $\mathcal{S}$, and let B denote the mapping of A under the one-to-one transformation (see Figure 2.2.1).

Because the transformation is one-to-one, the events $\{(X_1, X_2) \in A\}$ and $\{(Y_1, Y_2) \in B\}$ are equivalent. Hence

$$\begin{aligned} P[(Y_1, Y_2) \in B] &= P[(X_1, X_2) \in A] \\ &= \int\int_A f_{X_1, X_2}(x_1, x_2) \, dx_1 dx_2. \end{aligned}$$

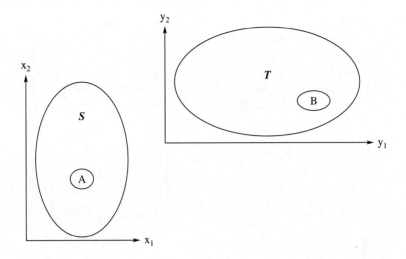

Figure 2.2.1: A general sketch of the supports of (X_1, X_2), $(\mathcal{S})$, and (Y_1, Y_2), $(\mathcal{T})$.

We wish now to change variables of integration by writing $y_1 = u_1(x_1, x_2)$, $y_2 = u_2(x_1, x_2)$, or $x_1 = w_1(y_1, y_2)$, $x_2 = w_2(y_1, y_2)$. It has been proven in analysis, (see, e.g., page 304 of Buck, 1965), that this change of variables requires

$$\int\int_A f_{X_1,X_2}(x_1, x_2)\, dx_1 dx_2 = \int\int_B f_{X_1,X_2}[w_1(y_1, y_2), w_2(y_1, y_2)]|J|\, dy_1 dy_2.$$

Thus, for every set B in $\mathcal{T}$,

$$P[(Y_1, Y_2) \in B] = \int\int_B f_{X_1,X_2}[w_1(y_1, y_2), w_2(y_1, y_2)]|J|\, dy_1 dy_2,$$

which implies that the joint pdf $f_{Y_1,Y_2}(y_1, y_2)$ of Y_1 and Y_2 is

$$f_{Y_1,Y_2}(y_1, y_2) = \begin{cases} f_{X_1,X_2}[w_1(y_1, y_2), w_2(y_1, y_2)]|J| & (y_1, y_2) \in \mathcal{T} \\ 0 & \text{elsewhere.} \end{cases}$$

Accordingly, the marginal pdf $f_{Y_1}(y_1)$ of Y_1 can be obtained from the joint pdf $f_{Y_1,Y_2}(y_1, y_2)$ in the usual manner by integrating on y_2. Several examples of this result will be given.

Example 2.2.3. Suppose (X_1, X_2) have the joint pdf,

$$f_{X_1,X_2}(x_1, x_2) = \begin{cases} 1 & 0 < x_1 < 1, 0 < x_2 < 1 \\ 0 & \text{elsewhere.} \end{cases}$$

The support of (X_1, X_2) is then the set $\mathcal{S} = \{(x_1, x_2) : 0 < x_1 < 1, 0 < x_2 < 1\}$ depicted in Figure 2.2.2.

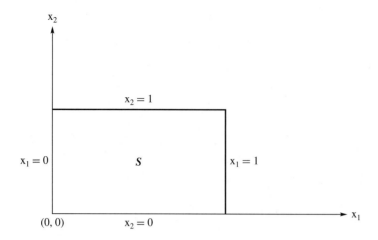

Figure 2.2.2: The support of (X_1, X_2) of Example 2.2.3.

Suppose $Y_1 = X_1 + X_2$ and $Y_2 = X_1 - X_2$. The transformation is given by

$$y_1 = u_1(x_1, x_2) = x_1 + x_2,$$
$$y_2 = u_2(x_1, x_2) = x_1 - x_2,$$

This transformation is one-to-one. We first determine the set $\mathcal{T}$ in the $y_1 y_2$-plane that is the mapping of $\mathcal{S}$ under this transformation. Now

$$x_1 = w_1(y_1, y_2) = \tfrac{1}{2}(y_1 + y_2),$$
$$x_2 = w_2(y_1, y_2) = \tfrac{1}{2}(y_1 - y_2).$$

To determine the set $\mathcal{S}$ in the $y_1 y_2$-plane onto which $\mathcal{T}$ is mapped under the transformation, note that the boundaries of $\mathcal{S}$ are transformed as follows into the boundaries of $\mathcal{T}$;

$$
\begin{array}{lll}
x_1 = 0 & \text{into} & 0 = \tfrac{1}{2}(y_1 + y_2), \\
x_1 = 1 & \text{into} & 1 = \tfrac{1}{2}(y_1 + y_2), \\
x_2 = 0 & \text{into} & 0 = \tfrac{1}{2}(y_1 - y_2), \\
x_2 = 1 & \text{into} & 1 = \tfrac{1}{2}(y_1 - y_2).
\end{array}
$$

Accordingly, $\mathcal{T}$ is shown in Figure 2.2.3. Next, the Jacobian is given by

$$
J = \begin{vmatrix} \dfrac{\partial x_1}{\partial y_1} & \dfrac{\partial x_1}{\partial y_2} \\ \dfrac{\partial x_2}{\partial y_1} & \dfrac{\partial x_2}{\partial y_2} \end{vmatrix} = \begin{vmatrix} \tfrac{1}{2} & \tfrac{1}{2} \\ \tfrac{1}{2} & -\tfrac{1}{2} \end{vmatrix} = -\dfrac{1}{2}.
$$

Although we suggest transforming the boundaries of $\mathcal{S}$, others might want to use the inequalities

$$0 < x_1 < 1 \quad \text{and} \quad 0 < x_2 < 1$$

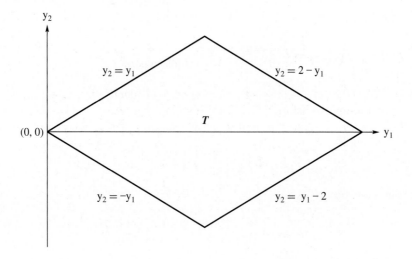

Figure 2.2.3: The support of (Y_1, Y_2) of Example 2.2.3.

directly. These four inequalities become

$$0 < \tfrac{1}{2}(y_1 + y_2) < 1 \quad \text{and} \quad 0 < \tfrac{1}{2}(y_1 - y_2) < 1.$$

It is easy to see that these are equivalent to

$$-y_1 < y_2, \quad y_2 < 2 - y_1, \quad y_2 < y_1 \quad y_1 - 2 < y_2;$$

and they define the set $\mathcal{T}$.

Hence, the joint pdf of (Y_1, Y_2) is given by,

$$f_{Y_1, Y_2}(y_1, y_2) = \begin{cases} f_{X_1, X_2}[\tfrac{1}{2}(y_1 + y_2), \tfrac{1}{2}(y_1 - y_2)]|J| = \tfrac{1}{2} & (y_1, y_2) \in \mathcal{T} \\ 0 & \text{elsewhere.} \end{cases}$$

The marginal pdf of Y_1 is given by

$$f_{Y_1}(y_1) = \int_{-\infty}^{\infty} f_{Y_1, Y_2}(y_1, y_2)\, dy_2.$$

If we refer to Figure 2.2.3, it is seen that

$$f_{Y_1}(y_1) = \begin{cases} \int_{-y_1}^{y_1} \tfrac{1}{2}\, dy_2 = y_1 & 0 < y_1 \leq 1 \\ \int_{y_1-2}^{2-y_1} \tfrac{1}{2}\, dy_2 = 2 - y_1 & 1 < y_1 < 2 \\ 0 & \text{elsewhere.} \end{cases}$$

In a similar manner, the marginal pdf $f_{Y_2}(y_2)$ is given by

$$f_{Y_2}(y_2) = \begin{cases} \int_{-y_2}^{y_2+2} \tfrac{1}{2}\, dy_1 = y_2 + 1 & -1 < y_2 \leq 0 \\ \int_{y_2}^{2-y_2} \tfrac{1}{2}\, dy_1 = 1 - y_2 & 0 < y_2 < 1 \\ 0 & \text{elsewhere.} \quad \blacksquare \end{cases}$$

Example 2.2.4. Let $Y_1 = \frac{1}{2}(X_1 - X_2)$, where X_1 and X_2 have the joint pdf,

$$f_{X_1,X_2}(x_1,x_2) = \begin{cases} \frac{1}{4}\exp\left(-\frac{x_1+x_2}{2}\right) & 0 < x_1 < \infty, \ 0 < x_2 < \infty \\ 0 & \text{elsewhere.} \end{cases}$$

Let $Y_2 = X_2$ so that $y_1 = \frac{1}{2}(x_1-x_2)$, $y_2 = x_2$ or, equivalently, $x_1 = 2y_1+y_2$, $x_2 = y_2$ define a one-to-one transformation from $\mathcal{S} = \{(x_1,x_2) : 0 < x_1 < \infty, 0 < x_2 < \infty\}$ onto $\mathcal{T} = \{(y_1,y_2) : -2y_1 < y_2 \text{ and } 0 < y_2, \ -\infty < y_1 < \infty\}$. The Jacobian of the transformation is

$$J = \begin{vmatrix} 2 & 1 \\ 0 & 1 \end{vmatrix} = 2;$$

hence the joint pdf of Y_1 and Y_2 is

$$f_{Y_1,Y_2}(y_1,y_2) = \begin{cases} \frac{|2|}{4} e^{-y_1-y_2} & (y_1,y_2) \in \mathcal{T} \\ 0 & \text{elsewhere.} \end{cases}$$

Thus the pdf of Y_1 is given by

$$f_{Y_1}(y_1) = \begin{cases} \int_{-2y_1}^{\infty} \frac{1}{2}e^{-y_1-y_2}\,dy_2 = \frac{1}{2} e^{y_1} & -\infty < y_1 < 0 \\ \int_0^{\infty} \frac{1}{2} e^{-y_1-y_2}\,dy_2 = \frac{1}{2} e^{-y_1} & 0 \le y_1 < \infty, \end{cases}$$

or

$$f_{Y_1}(y_1) = \frac{1}{2} e^{-|y_1|}, \quad -\infty < y_1 < \infty.$$

This pdf is frequently called the **double exponential** or **Laplace** pdf. ∎

Example 2.2.5. Let X_1 and X_2 have the joint pdf

$$f_{X_1,X_2}(x_1,x_2) = \begin{cases} 10x_1x_2^2 & 0 < x_1 < x_2 < 1 \\ 0 & \text{elsewhere.} \end{cases}$$

Suppose $Y_1 = X_1/X_2$ and $Y_2 = X_2$. Hence, the inverse transformation is $x_1 = y_1y_2$ and $x_2 = y_2$ which has the Jacobian

$$J = \begin{vmatrix} y_2 & y_1 \\ 0 & 1 \end{vmatrix} = y_2.$$

The inequalities defining the support $\mathcal{S}$ of (X_1, X_2) become

$$0 < y_1y_2, \ y_1y_2 < y_2, \text{ and } y_2 < 1.$$

These inequalities are equivalent to

$$0 < y_1 < 1 \text{ and } 0 < y_2 < 1,$$

which defines the support set $\mathcal{T}$ of (Y_1, Y_2). Hence, the joint pdf of (Y_1, Y_2) is

$$f_{Y_1,Y_2}(y_1,y_2) = 10y_1y_2y_2^2|y_2| = 10y_1y_2^4, \quad (y_1,y_2) \in \mathcal{T}.$$

The marginal pdfs are:

$$f_{Y_1}(y_1) = \int_0^1 10 y_1 y_2^4 \, dy_2 = 2y_1, \quad 0 < y_1 < 1$$

zero elsewhere, and

$$f_{Y_2}(y_2) = \int_0^1 10 y_1 y_2^4 \, dy_1 = 5 y_2^4, \quad 0 < y_1 < 1$$

zero elsewhere. ■

In addition to the change-of-variable and cdf techniques for finding distributions of functions of random variables, there is another method, called the moment generating function (mgf) technique, which works well for linear functions of random variables. In subsection 2.1.1, we pointed out that if $Y = g(X_1, X_2)$, then $E(Y)$, if it exists, could be found by

$$E(Y) = \int_{-\infty}^{\infty} \int_{-\infty}^{\infty} g(x_1, x_2) f_{X_1, X_2}(x_1, x_2) \, dx_1 dx_2$$

in the continuous case, with summations replacing integrals in the discrete case. Certainly that function $g(X_1, X_2)$ could be $\exp\{tu(X_1, X_2)\}$, so that in reality we would be finding the mgf of the function $Z = u(X_1, X_2)$. If we could then recognize this mgf as belonging to a certain distribution, then Z would have that distribution. We give two illustrations that demonstrate the power of this technique by reconsidering Examples 2.2.1 and 2.2.4.

Example 2.2.6 (Continuation of Example 2.2.1). Here X_1 and X_2 have the joint pmf

$$p_{X_1, X_2}(x_1, x_2) = \begin{cases} \frac{\mu_1^{x_1} \mu_2^{x_2} e^{-\mu_1} e^{-\mu_2}}{x_1! x_2!} & x_1 = 0, 1, 2, 3, \ldots, \quad x_2 = 0, 1, 2, 3, \ldots \\ 0 & \text{elsewhere,} \end{cases}$$

where μ_1 and μ_2 are fixed positive real numbers. Let $Y = X_1 + X_2$ and consider

$$\begin{aligned}
E(e^{tY}) &= \sum_{x_1=0}^{\infty} \sum_{x_2=0}^{\infty} e^{t(x_1+x_2)} p_{X_1, X_2}(x_1, x_2) \\
&= \sum_{x_1=0}^{\infty} e^{tx_1} \frac{\mu^{x_1} e^{-\mu_1}}{x_1!} \sum_{x_2=0}^{\infty} e^{tx_2} \frac{\mu^{x_2} e^{-\mu_2}}{x_2!} \\
&= \left[e^{-\mu_1} \sum_{x_1=0}^{\infty} \frac{(e^t \mu_1)^{x_1}}{x_1!} \right] \left[e^{-\mu_2} \sum_{x_2=0}^{\infty} \frac{(e^t \mu_2)^{x_2}}{x_2!} \right] \\
&= \left[e^{\mu_1(e^t - 1)} \right] \left[e^{\mu_2(e^t - 1)} \right] \\
&= e^{(\mu_1 + \mu_2)(e^t - 1)}.
\end{aligned}$$

Notice that the factors in the brackets in the next to last equality are the mgfs of X_1 and X_2, respectively. Hence, the mgf of Y is the same as that of X_1 except μ_1 has been replaced by $\mu_1 + \mu_2$. Therefore, by the uniqueness of mgfs the pmf of Y must be

$$p_Y(y) = e^{-(\mu_1 + \mu_2)} \frac{(\mu_1 + \mu_2)^y}{y!}, \quad y = 0, 1, 2, \ldots,$$

which is the same pmf that was obtained in Example 2.2.1. ∎

Example 2.2.7 (Continuation of Example 2.2.4). Here X_1 and X_2 have the joint pdf

$$f_{X_1, X_2}(x_1, x_2) = \begin{cases} \frac{1}{4} \exp\left(-\frac{x_1 + x_2}{2}\right) & 0 < x_1 < \infty, \ 0 < x_2 < \infty \\ 0 & \text{elsewhere.} \end{cases}$$

So the mgf of $Y = (1/2)(X_1 - X_2)$ is given by

$$
\begin{aligned}
E(e^{tY}) &= \int_0^\infty \int_0^\infty e^{t(x_1 - x_2)/2} \frac{1}{4} e^{-(x_1 + x_2)/2} \, dx_1 dx_2 \\
&= \left[\int_0^\infty \frac{1}{2} e^{-x_1(1-t)/2} \, dx_1\right]\left[\int_0^\infty \frac{1}{2} e^{-x_2(1+t)/2} \, dx_2\right] \\
&= \left[\frac{1}{1-t}\right]\left[\frac{1}{1+t}\right] = \frac{1}{1-t^2}
\end{aligned}
$$

provided that $1 - t > 0$ and $1 + t > 0$; i.e., $-1 < t < 1$. However, the mgf of a double exponential distribution is,

$$
\begin{aligned}
\int_{-\infty}^\infty e^{tx} \frac{e^{-|x|}}{2} \, dx &= \int_{-\infty}^0 \frac{e^{(1+t)x}}{2} \, dx + \int_0^\infty \frac{e^{(t-1)x}}{2} \, dx \\
&= \frac{1}{2(1+t)} + \frac{1}{2(1-t)} = \frac{1}{1-t^2},
\end{aligned}
$$

provided $-1 < t < 1$. Thus, by the uniqueness of mgfs, Y has the double exponential distribution. ∎

EXERCISES

2.2.1. If $p(x_1, x_2) = \left(\frac{2}{3}\right)^{x_1 + x_2} \left(\frac{1}{3}\right)^{2 - x_1 - x_2}$, $(x_1, x_2) = (0, 0), (0, 1), (1, 0), (1, 1)$, zero elsewhere, is the joint pmf of X_1 and X_2, find the joint pmf of $Y_1 = X_1 - X_2$ and $Y_2 = X_1 + X_2$.

2.2.2. Let X_1 and X_2 have the joint pmf $p(x_1, x_2) = x_1 x_2 / 36$, $x_1 = 1, 2, 3$ and $x_2 = 1, 2, 3$, zero elsewhere. Find first the joint pmf of $Y_1 = X_1 X_2$ and $Y_2 = X_2$, and then find the marginal pmf of Y_1.

2.2.3. Let X_1 and X_2 have the joint pdf $h(x_1, x_2) = 2e^{-x_1 - x_2}$, $0 < x_1 < x_2 < \infty$, zero elsewhere. Find the joint pdf of $Y_1 = 2X_1$ and $Y_2 = X_2 - X_1$.

2.2.4. Let X_1 and X_2 have the joint pdf $h(x_1, x_2) = 8x_1 x_2$, $0 < x_1 < x_2 < 1$, zero elsewhere. Find the joint pdf of $Y_1 = X_1/X_2$ and $Y_2 = X_2$.

Hint: Use the inequalities $0 < y_1 y_2 < y_2 < 1$ in considering the mapping from $\mathcal{S}$ onto $\mathcal{T}$.

2.2.5. Let X_1 and X_2 be continuous random variables with the joint probability density function, $f_{X_1, X_2}(x_1, x_2)$, $-\infty < x_i < \infty$, $i = 1, 2$. Let $Y_1 = X_1 + X_2$ and $Y_2 = X_2$.

(a) Find the joint pdf f_{Y_1, Y_2}.

(b) Show that

$$f_{Y_1}(y_1) = \int_{-\infty}^{\infty} f_{X_1, X_2}(y_1 - y_2, y_2)\, dy_2, \tag{2.2.1}$$

which is sometimes called the *convolution formula*.

2.2.6. Suppose X_1 and X_2 have the joint pdf $f_{X_1, X_2}(x_1, x_2) = e^{-(x_1 + x_2)}$, $0 < x_i < \infty$, $i = 1, 2$, zero elsewhere.

(a) Use formula (2.2.1) to find the pdf of $Y_1 = X_1 + X_2$.

(b) Find the mgf of Y_1.

2.2.7. Use the formula (2.2.1) to find the pdf of $Y_1 = X_1 + X_2$, where X_1 and X_2 have the joint pdf $f_{X_1, X_2}(x_1, x_2) = 2e^{-(x_1 + x_2)}$, $0 < x_1 < x_2 < \infty$, zero elsewhere.

2.3 Conditional Distributions and Expectations

In Section 2.1 we introduced the joint probability distribution of a pair of random variables. We also showed how to recover the individual (marginal) distributions for the random variables from the joint distribution. In this section, we discuss conditional distributions, i.e., the distribution of one of the random variables when the other has assumed a specific value. We discuss this first for the discrete case which follows easily from the concept of conditional probability presented in Section 1.4.

Let X_1 and X_2 denote random variables of the discrete type which have the joint pmf $p_{X_1, X_2}(x_1, x_2)$ which is positive on the support set $\mathcal{S}$ and is zero elsewhere. Let $p_{X_1}(x_1)$ and $p_{X_2}(x_2)$ denote, respectively, the marginal probability density functions of X_1 and X_2. Let x_1 be a point in the support of X_1; hence, $p_{X_1}(x_1) > 0$. Using the definition of conditional probability we have,

$$P(X_2 = x_2 | X_1 = x_1) = \frac{P(X_1 = x_1, X_2 = x_2)}{P(X_1 = x_1)} = \frac{p_{X_1, X_2}(x_1, x_2)}{p_{X_1}(x_1)}$$

for all x_2 in the support $\mathcal{S}_{X_2}$ of X_2. Define this function as,

$$p_{X_2 | X_1}(x_2 | x_1) = \frac{p_{X_1, X_2}(x_1, x_2)}{p_{X_1}(x_1)}, \quad x_2 \in \mathcal{S}_{X_2}. \tag{2.3.1}$$

For any fixed x_1 with $p_{X_1}(x_1) > 0$, this function $p_{X_2|X_1}(x_2|x_1)$ satisfies the conditions of being a pmf of the discrete type because $p_{X_2|X_1}(x_2|x_1)$ is nonnegative and

$$\sum_{x_2} p_{X_2|X_1}(x_2|x_1) = \sum_{x_2} \frac{p_{X_1,X_2}(x_1,x_2)}{p_{X_1}(x_1)} = \frac{1}{p_{X_1}(x_1)} \sum_{x_2} p_{X_1,X_2}(x_1,x_2) = \frac{p_{X_1}(x_1)}{p_{X_1}(x_1)} = 1.$$

We call $p_{X_2|X_1}(x_2|x_1)$ the **conditional pmf** of the discrete type of random variable X_2, given that the discrete type of random variable $X_1 = x_1$. In a similar manner, provided $x_2 \in \mathcal{S}_{X_2}$, we define the symbol $p_{X_1|X_2}(x_1|x_2)$ by the relation

$$p_{X_1|X_2}(x_1|x_2) = \frac{p_{X_1,X_2}(x_1,x_2)}{p_{X_2}(x_2)}, \quad x_1 \in \mathcal{S}_{X_1},$$

and we call $p_{X_1|X_2}(x_1|x_2)$ the conditional pmf of the discrete type of random variable X_1, given that the discrete type of random variable $X_2 = x_2$. We will often abbreviate $p_{X_1|X_2}(x_1|x_2)$ by $p_{1|2}(x_1|x_2)$ and $p_{X_2|X_1}(x_2|x_1)$ by $p_{2|1}(x_2|x_1)$. Similarly $p_1(x_1)$ and $p_2(x_2)$ will be used to denote the respective marginal pmfs.

Now let X_1 and X_2 denote random variables of the continuous type and have the joint pdf $f_{X_1,X_2}(x_1,x_2)$ and the marginal probability density functions $f_{X_1}(x_1)$ and $f_{X_2}(x_2)$, respectively. We shall use the results of the preceding paragraph to motivate a definition of a conditional pdf of a continuous type of random variable. When $f_{X_1}(x_1) > 0$, we define the symbol $f_{X_2|X_1}(x_2|x_1)$ by the relation

$$f_{X_2|X_1}(x_2|x_1) = \frac{f_{X_1,X_2}(x_1,x_2)}{f_{X_1}(x_1)}. \tag{2.3.2}$$

In this relation, x_1 is to be thought of as having a fixed (but any fixed) value for which $f_{X_1}(x_1) > 0$. It is evident that $f_{X_2|X_1}(x_2|x_1)$ is nonnegative and that

$$\begin{aligned}
\int_{-\infty}^{\infty} f_{X_2|X_1}(x_2|x_1)\, dx_2 &= \int_{-\infty}^{\infty} \frac{f_{X_1,X_2}(x_1,x_2)}{f_{X_1}(x_1)}\, dx_2 \\
&= \frac{1}{f_{X_1}(x_1)} \int_{-\infty}^{\infty} f_{X_1,X_2}(x_1,x_2)\, dx_2 \\
&= \frac{1}{f_{X_1}(x_1)} f_{X_1}(x_1) = 1.
\end{aligned}$$

That is, $f_{X_2|X_1}(x_1|x_1)$ has the properties of a pdf of one continuous type of random variable. It is called the **conditional pdf** of the continuous type of random variable X_2, given that the continuous type of random variable X_1 has the value x_1. When $f_{X_2}(x_2) > 0$, the conditional pdf of the continuous random variable X_1, given that the continuous type of random variable X_2 has the value x_2, is defined by

$$f_{X_1|X_2}(x_1|x_2) = \frac{f_{X_1,X_2}(x_1,x_2)}{f_{X_2}(x_2)}, \quad f_{X_2}(x_2) > 0.$$

We will often abbreviate these conditional pdfs by $f_{1|2}(x_1|x_2)$ and $f_{2|1}(x_2|x_1)$, respectively. Similarly $f_1(x_1)$ and $f_2(x_2)$ will be used to denote the respective marginal pdfs.

Since each of $f_{2|1}(x_2|x_1)$ and $f_{1|2}(x_1|x_2)$ is a pdf of one random variable, each has all the properties of such a pdf. Thus we can compute probabilities and mathematical expectations. If the random variables are of the continuous type, the probability

$$P(a < X_2 < b | X_1 = x_1) = \int_a^b f_{2|1}(x_2|x_1)\,dx_2$$

is called "the conditional probability that $a < X_2 < b$, given that $X_1 = x_1$." If there is no ambiguity, this may be written in the form $P(a < X_2 < b | x_1)$. Similarly, the conditional probability that $c < X_1 < d$, given $X_2 = x_2$, is

$$P(c < X_1 < d | X_2 = x_2) = \int_c^d f_{1|2}(x_1|x_2)\,dx_1.$$

If $u(X_2)$ is a function of X_2, the conditional expectation of $u(X_2)$, given that $X_1 = x_1$, if it exists, is given by

$$E[u(X_2)|x_1] = \int_{-\infty}^{\infty} u(x_2)f_{2|1}(x_2|x_1)\,dx_2.$$

In particular, if they do exist, then $E(X_2|x_1)$ is the mean and $E\{[X_2 - E(X_2|x_1)]^2 | x_1\}$ is the variance of the conditional distribution of X_2, given $X_1 = x_1$, which can be written more simply as $\mathrm{var}(X_2|x_1)$. It is convenient to refer to these as the "conditional mean" and the "conditional variance" of X_2, given $X_1 = x_1$. Of course, we have

$$\mathrm{var}(X_2|x_1) = E(X_2^2|x_1) - [E(X_2|x_1)]^2$$

from an earlier result. In like manner, the conditional expectation of $u(X_1)$, given $X_2 = x_2$, if it exists, is given by

$$E[u(X_1)|x_2] = \int_{-\infty}^{\infty} u(x_1)f_{1|2}(x_1|x_2)\,dx_1.$$

With random variables of the discrete type, these conditional probabilities and conditional expectations are computed by using summation instead of integration. An illustrative example follows.

Example 2.3.1. Let X_1 and X_2 have the joint pdf

$$f(x_1, x_2) = \begin{cases} 2 & 0 < x_1 < x_2 < 1 \\ 0 & \text{elsewhere.} \end{cases}$$

Then the marginal probability density functions are, respectively,

$$f_1(x_1) = \begin{cases} \int_{x_1}^1 2\,dx_2 = 2(1 - x_1) & 0 < x_1 < 1 \\ 0 & \text{elsewhere,} \end{cases}$$

and

$$f_2(x_2) = \begin{cases} \int_0^{x_2} 2\,dx_1 = 2x_2 & 0 < x_2 < 1 \\ 0 & \text{elsewhere.} \end{cases}$$

The conditional pdf of X_1, given $X_2 = x_2$, $0 < x_2 < 1$, is

$$f_{1|2}(x_1|x_2) = \begin{cases} \frac{2}{2x_2} = \frac{1}{x_2} & 0 < x_1 < x_2 \\ 0 & \text{elsewhere.} \end{cases}$$

Here the conditional mean and the conditional variance of X_1, given $X_2 = x_2$, are respectively,

$$\begin{aligned} E(X_1|x_2) &= \int_{-\infty}^{\infty} x_1 f_{1|2}(x_1|x_2)\, dx_1 \\ &= \int_0^{x_2} x_1 \left(\frac{1}{x_2}\right) dx_1 \\ &= \frac{x_2}{2}, \quad 0 < x_2 < 1, \end{aligned}$$

and

$$\begin{aligned} \text{var}(X_1|x_2) &= \int_0^{x_2} \left(x_1 - \frac{x_2}{2}\right)^2 \left(\frac{1}{x_2}\right) dx_1 \\ &= \frac{x_2^2}{12}, \quad 0 < x_2 < 1. \end{aligned}$$

Finally, we shall compare the values of

$$P(0 < X_1 < \tfrac{1}{2}|X_2 = \tfrac{3}{4}) \quad \text{and } P(0 < X_1 < \tfrac{1}{2}).$$

We have

$$P(0 < X_1 < \tfrac{1}{2}|X_2 = \tfrac{3}{4}) = \int_0^{1/2} f_{1|2}(x_1|\tfrac{3}{4})\, dx_1 = \int_0^{1/2} (\tfrac{4}{3})\, dx_1 = \tfrac{2}{3},$$

but

$$P(0 < X_1 < \tfrac{1}{2}) = \int_0^{1/2} f_1(x_1)\, dx_1 = \int_0^{1/2} 2(1 - x_1)\, dx_1 = \tfrac{3}{4}. \quad \blacksquare$$

Since $E(X_2|x_1)$ is a function of x_1, then $E(X_2|X_1)$ is a random variable with its own distribution, mean, and variance. Let us consider the following illustration of this.

Example 2.3.2. Let X_1 and X_2 have the joint pdf.

$$f(x_1, x_2) = \begin{cases} 6x_2 & 0 < x_2 < x_1 < 1 \\ 0 & \text{elsewhere.} \end{cases}$$

Then the marginal pdf of X_1 is

$$f_1(x_1) = \int_0^{x_1} 6x_2\, dx_2 = 3x_1^2, \quad 0 < x_1 < 1,$$

zero elsewhere. The conditional pdf of X_2, given $X_1 = x_1$, is

$$f_{2|1}(x_2|x_1) = \frac{6x_2}{3x_1^2} = \frac{2x_2}{x_1^2}, \quad 0 < x_2 < x_1,$$

zero elsewhere, where $0 < x_1 < 1$. The conditional mean of X_2, given $X_1 = x_1$, is

$$E(X_2|x_1) = \int_0^{x_1} x_2 \left(\frac{2x_2}{x_1^2}\right) dx_2 = \frac{2}{3}x_1, \quad 0 < x_1 < 1.$$

Now $E(X_2|X_1) = 2X_1/3$ is a random variable, say Y. The cdf of $Y = 2X_1/3$ is

$$G(y) = P(Y \le y) = P\left(X_1 \le \frac{3y}{2}\right), \quad 0 \le y < \frac{2}{3}.$$

From the pdf $f_1(x_1)$, we have

$$G(y) = \int_0^{3y/2} 3x_1^2 \, dx_1 = \frac{27y^3}{8}, \quad 0 \le y < \frac{2}{3}.$$

Of course, $G(y) = 0$, if $y < 0$, and $G(y) = 1$, if $\frac{2}{3} < y$. The pdf, mean, and variance of $Y = 2X_1/3$ are

$$g(y) = \frac{81y^2}{8}, \quad 0 \le y < \frac{2}{3},$$

zero elsewhere,

$$E(Y) = \int_0^{2/3} y\left(\frac{81y^2}{8}\right) dy = \frac{1}{2},$$

and

$$\text{var}(Y) = \int_0^{2/3} y^2\left(\frac{81y^2}{8}\right) dy - \frac{1}{4} = \frac{1}{60}.$$

Since the marginal pdf of X_2 is

$$f_2(x_2) = \int_{x_2}^1 6x_2 \, dx_1 = 6x_2(1 - x_2), \quad 0 < x_2 < 1,$$

zero elsewhere, it is easy to show that $E(X_2) = \frac{1}{2}$ and $\text{var}(X_2) = \frac{1}{20}$. That is, here

$$E(Y) = E[E(X_2|X_1)] = E(X_2)$$

and

$$\text{var}(Y) = \text{var}[E(X_2|X_1)] \le \text{var}(X_2). \quad \blacksquare$$

Example 2.3.2 is excellent, as it provides us with the opportunity to apply many of these new definitions as well as review the cdf technique for finding the distribution of a function of a random variable, name $Y = 2X_1/3$. Moreover, the two observations at the end of this example are no accident because they are true in general.

Theorem 2.3.1. *Let (X_1, X_2) be a random vector such that the variance of X_2 is finite. Then,*

(a). $E[E(X_2|X_1)] = E(X_2)$.

(b). $Var[E(X_2|X_1)] \leq Var(X_2)$.

Proof: The proof is for the continuous case. To obtain it for the discrete case, exchange summations for integrals. We first prove (a). Note that

$$
\begin{aligned}
E(X_2) &= \int_{-\infty}^{\infty} \int_{-\infty}^{\infty} x_2 f(x_1, x_2)\, dx_2 dx_1 \\
&= \int_{-\infty}^{\infty} \left[\int_{-\infty}^{\infty} x_2 \frac{f(x_1, x_2)}{f_1(x_1)}\, dx_2 \right] f_1(x_1)\, dx_1 \\
&= \int_{-\infty}^{\infty} E(X_2|x_1) f_1(x_1)\, dx_1 \\
&= E[E(X_2|X_1)],
\end{aligned}
$$

which is the first result.

Next we show (b). Consider with $\mu_2 = E(X_2)$,

$$
\begin{aligned}
\mathrm{var}(X_2) &= E[(X_2 - \mu_2)^2] \\
&= E\{[X_2 - E(X_2|X_1) + E(X_2|X_1) - \mu_2]^2\} \\
&= E\{[X_2 - E(X_2|X_1)]^2\} + E\{[E(X_2|X_1) - \mu_2]^2\} \\
&\quad + 2E\{[X_2 - E(X_2|X_1)][E(X_2|X_1) - \mu_2]\}.
\end{aligned}
$$

We shall show that the last term of the right-hand member of the immediately preceding equation is zero. It is equal to

$$
2 \int_{-\infty}^{\infty} \int_{-\infty}^{\infty} [x_2 - E(X_2|x_1)][E(X_2|x_1) - \mu_2] f(x_1, x_2)\, dx_2 dx_1
$$

$$
= 2 \int_{-\infty}^{\infty} [E(X_2|x_1) - \mu_2] \left\{ \int_{-\infty}^{\infty} [x_2 - E(X_2|x_1)] \frac{f(x_1, x_2)}{f_1(x_1)}\, dx_2 \right\} f_1(x_1)\, dx_1.
$$

But $E(X_2|x_1)$ is the conditional mean of X_2, given $X_1 = x_1$. Since the expression in the inner braces is equal to

$$
E(X_2|x_1) - E(X_2|x_1) = 0,
$$

the double integral is equal to zero. Accordingly, we have

$$
\mathrm{var}(X_2) = E\{[X_2 - E(X_2|X_1)]^2\} + E\{[E(X_2|X_1) - \mu_2]^2\}.
$$

The first term in the right-hand member of this equation is nonnegative because it is the expected value of a nonnegative function, namely $[X_2 - E(X_2|X_1)]^2$. Since $E[E(X_2|X_1)] = \mu_2$, the second term will be the $\mathrm{var}[E(X_2|X_1)]$. Hence we have

$$
\mathrm{var}(X_2) \geq \mathrm{var}[E(X_2|X_1)],
$$

which completes the proof. ∎

Intuitively, this result could have this useful interpretation. Both the random variables X_2 and $E(X_2|X_1)$ have the same mean μ_2. If we did not know μ_2, we could use either of the two random variables to guess at the unknown μ_2. Since, however, $\text{var}(X_2) \geq \text{var}[E(X_2|X_1)]$ we would put more reliance in $E(X_2|X_1)$ as a guess. That is, if we observe the pair (X_1, X_2) to be (x_1, x_2), we could prefer to use $E(X_2|x_1)$ to x_2 as a guess at the unknown μ_2. When studying the use of sufficient statistics in estimation in Chapter 6, we make use of this famous result, attributed to C. R. Rao and David Blackwell.

EXERCISES

2.3.1. Let X_1 and X_2 have the joint pdf $f(x_1, x_2) = x_1 + x_2$, $0 < x_1 < 1$, $0 < x_2 < 1$, zero elsewhere. Find the conditional mean and variance of X_2, given $X_1 = x_1$, $0 < x_1 < 1$.

2.3.2. Let $f_{1|2}(x_1|x_2) = c_1 x_1/x_2^2$, $0 < x_1 < x_2$, $0 < x_2 < 1$, zero elsewhere, and $f_2(x_2) = c_2 x_2^4$, $0 < x_2 < 1$, zero elsewhere, denote, respectively, the conditional pdf of X_1, given $X_2 = x_2$, and the marginal pdf of X_2. Determine:

(a) The constants c_1 and c_2.

(b) The joint pdf of X_1 and X_2.

(c) $P(\frac{1}{4} < X_1 < \frac{1}{2}|X_2 = \frac{5}{8})$.

(d) $P(\frac{1}{4} < X_1 < \frac{1}{2})$.

2.3.3. Let $f(x_1, x_2) = 21x_1^2 x_2^3$, $0 < x_1 < x_2 < 1$, zero elsewhere, be the joint pdf of X_1 and X_2.

(a) Find the conditional mean and variance of X_1, given $X_2 = x_2$, $0 < x_2 < 1$.

(b) Find the distribution of $Y = E(X_1|X_2)$.

(c) Determine $E(Y)$ and $\text{var}(Y)$ and compare these to $E(X_1)$ and $\text{var}(X_1)$, respectively.

2.3.4. Suppose X_1 and X_2 are random variables of the discrete type which have the joint pmf $p(x_1, x_2) = (x_1 + 2x_2)/18$, $(x_1, x_2) = (1,1), (1,2), (2,1), (2,2)$, zero elsewhere. Determine the conditional mean and variance of X_2, given $X_1 = x_1$, for $x_1 = 1$ or 2. Also compute $E(3X_1 - 2X_2)$.

2.3.5. Let X_1 and X_2 be two random variables such that the conditional distributions and means exist. Show that:

(a) $E(X_1 + X_2 \,|\, X_2) = E(X_1 \,|\, X_2) + X_2$

(b) $E(u(X_2) \,|\, X_2) = u(X_2)$.

2.3.6. Let the joint pdf of X and Y be given by

$$f(x,y) = \begin{cases} \frac{2}{(1+x+y)^3} & 0 < x < \infty, \ 0 < y < \infty \\ 0 & \text{elsewhere.} \end{cases}$$

(a) Compute the marginal pdf of X and the conditional pdf of Y, given $X = x$.

(b) For a fixed $X = x$, compute $E(1 + x + Y|x)$ and use the result to compute $E(Y|x)$.

2.3.7. Suppose X_1 and X_2 are discrete random variables which have the joint pmf $p(x_1, x_2) = (3x_1 + x_2)/24$, $(x_1, x_2) = (1, 1), (1, 2), (2, 1), (2, 2)$, zero elsewhere. Find the conditional mean $E(X_2|x_1)$, when $x_1 = 1$.

2.3.8. Let X and Y have the joint pdf $f(x, y) = 2\exp\{-(x + y)\}$, $0 < x < y < \infty$, zero elsewhere. Find the conditional mean $E(Y|x)$ of Y, given $X = x$.

2.3.9. Five cards are drawn at random and without replacement from an ordinary deck of cards. Let X_1 and X_2 denote, respectively, the number of spades and the number of hearts that appear in the five cards.

(a) Determine the joint pmf of X_1 and X_2.

(b) Find the two marginal pmfs.

(c) What is the conditional pmf of X_2, given $X_1 = x_1$?

2.3.10. Let X_1 and X_2 have the joint pmf $p(x_1, x_2)$ described as follows:

(x_1, x_2)	$(0, 0)$	$(0, 1)$	$(1, 0)$	$(1, 1)$	$(2, 0)$	$(2, 1)$
$p(x_1, x_2)$	$\frac{1}{18}$	$\frac{3}{18}$	$\frac{4}{18}$	$\frac{3}{18}$	$\frac{6}{18}$	$\frac{1}{18}$

and $p(x_1, x_2)$ is equal to zero elsewhere. Find the two marginal probability density functions and the two conditional means.

Hint: Write the probabilities in a rectangular array.

2.3.11. Let us choose at random a point from the interval $(0, 1)$ and let the random variable X_1 be equal to the number which corresponds to that point. Then choose a point at random from the interval $(0, x_1)$, where x_1 is the experimental value of X_1; and let the random variable X_2 be equal to the number which corresponds to this point.

(a) Make assumptions about the marginal pdf $f_1(x_1)$ and the conditional pdf $f_{2|1}(x_2|x_1)$.

(b) Compute $P(X_1 + X_2 \geq 1)$.

(c) Find the conditional mean $E(X_1|x_2)$.

2.3.12. Let $f(x)$ and $F(x)$ denote, respectively, the pdf and the cdf of the random variable X. The conditional pdf of X, given $X > x_0$, x_0 a fixed number, is defined by $f(x|X > x_0) = f(x)/[1 - F(x_0)]$, $x_0 < x$, zero elsewhere. This kind of conditional pdf finds application in a problem of time until death, given survival until time x_0.

(a) Show that $f(x|X > x_0)$ is a pdf.

(b) Let $f(x) = e^{-x}$, $0 < x < \infty$, and zero elsewhere. Compute $P(X > 2|X > 1)$.

2.4   The Correlation Coefficient

Because the result that we obtain in this section is more familiar in terms of X and Y, we use X and Y rather than X_1 and X_2 as symbols for our two random variables. Rather than discussing these concepts separately for continuous and discrete cases, we use continuous notation in our discussion. But the same properties hold for the discrete case also. Let X and Y have joint pdf $f(x, y)$. If $u(x, y)$ is a function of x and y, then $E[u(X, Y)]$ was defined, subject to its existence, in Section 2.1. The existence of all mathematical expectations will be assumed in this discussion. The means of X and Y, say μ_1 and μ_2, are obtained by taking $u(x, y)$ to be x and y, respectively; and the variances of X and Y, say σ_1^2 and σ_2^2, are obtained by setting the function $u(x, y)$ equal to $(x - \mu_1)^2$ and $(y - \mu_2)^2$, respectively. Consider the mathematical expectation

$$
\begin{aligned}
E[(X - \mu_1)(Y - \mu_2)] &= E(XY - \mu_2 X - \mu_1 Y + \mu_1 \mu_2) \\
&= E(XY) - \mu_2 E(X) - \mu_1 E(Y) + \mu_1 \mu_2 \\
&= E(XY) - \mu_1 \mu_2.
\end{aligned}
$$

This number is called the **covariance** of X and Y and is often denoted by $\operatorname{cov}(X, Y)$. If each of σ_1 and σ_2 is positive, the number

$$
\rho = \frac{E[(X - \mu_1)(Y - \mu_2)]}{\sigma_1 \sigma_2} = \frac{\operatorname{cov}(X, Y)}{\sigma_1 \sigma_2}
$$

is called the **correlation coefficient** of X and Y. It should be noted that the expected value of the product of two random variables is equal to the product of their expectations plus their covariance; that is $E(XY) = \mu_1 \mu_2 + \rho \sigma_1 \sigma_2 = \mu_1 \mu_2 + \operatorname{cov}(X, Y)$.

Example 2.4.1. Let the random variables X and Y have the joint pdf

$$
f(x, y) = \begin{cases} x + y & 0 < x < 1, \ 0 < y < 1 \\ 0 & \text{elsewhere.} \end{cases}
$$

We shall compute the correlation coefficient ρ of X and Y. Now

$$
\mu_1 = E(X) = \int_0^1 \int_0^1 x(x + y) \, dx \, dy = \frac{7}{12}
$$

and

$$
\sigma_1^2 = E(X^2) - \mu_1^2 = \int_0^1 \int_0^1 x^2(x + y) \, dx \, dy - \left(\frac{7}{12}\right)^2 = \frac{11}{144}.
$$

Similarly,

$$
\mu_2 = E(Y) = \frac{7}{12} \quad \text{and} \quad \sigma_2^2 = E(Y^2) - \mu_2^2 = \frac{11}{144}.
$$

The covariance of X and Y is

$$
E(XY) - \mu_1 \mu_2 = \int_0^1 \int_0^1 xy(x + y) \, dx \, dy - \left(\frac{7}{12}\right)^2 = -\frac{1}{144}.
$$

Accordingly, the correlation coefficient of X and Y is

$$\rho = \frac{-\frac{1}{144}}{\sqrt{\left(\frac{11}{144}\right)\left(\frac{11}{144}\right)}} = -\frac{1}{11}. \quad \blacksquare$$

Remark 2.4.1. For certain kinds of distributions of two random variables, say X and Y, the correlation coefficient ρ proves to be a very useful characteristic of the distribution. Unfortunately, the formal definition of ρ does not reveal this fact. At this time we make some observations about ρ, some of which will be explored more fully at a later stage. It will soon be seen that if a joint distribution of two variables has a correlation coefficient (that is, if both of the variances are positive), then ρ satisfies $-1 \leq \rho \leq 1$. If $\rho = 1$, there is a line with equation $y = a + bx$, $b > 0$, the graph of which contains all of the probability of the distribution of X and Y. In this extreme case, we have $P(Y = a + bX) = 1$. If $\rho = -1$, we have the same state of affairs except that $b < 0$. This suggests the following interesting question: When ρ does not have one of its extreme values, is there a line in the xy-plane such that the probability for X and Y tends to be concentrated in a band about this line? Under certain restrictive conditions this is in fact the case, and under those conditions we can look upon ρ as a measure of the intensity of the concentration of the probability for X and Y about that line. $\blacksquare$

Next, let $f(x, y)$ denote the joint pdf of two random variables X and Y and let $f_1(x)$ denote the marginal pdf of X. Recall from Section 2.3 that the conditional pdf of Y, given $X = x$, is

$$f_{2|1}(y|x) = \frac{f(x, y)}{f_1(x)}$$

at points where $f_1(x) > 0$, and the conditional mean of Y, given $X = x$, is given by

$$E(Y|x) = \int_{-\infty}^{\infty} y f_{2|1}(y|x) \, dy = \frac{\int_{-\infty}^{\infty} y f(x, y) \, dy}{f_1(x)},$$

when dealing with random variables of the continuous type. This conditional mean of Y, given $X = x$, is of course, a function of x, say $u(x)$. In like vein, the conditional mean of X, given $Y = y$, is a function of y, say $v(y)$.

In case $u(x)$ is a linear function of x, say $u(x) = a + bx$, we say the conditional mean of Y is linear in x; or that Y is a linear conditional mean. When $u(x) = a + bx$, the constants a and b have simple values which we will summarize in the following theorem.

Theorem 2.4.1. *Suppose (X, Y) have a joint distribution with the variances of X and Y finite and positive. Denote the means and variances of X and Y by μ_1, μ_2 and σ_1^2, σ_2^2, respectively, and let ρ be the correlation coefficient between X and Y. If $E(Y|X)$ is linear in X then*

$$E(Y|X) = \mu_2 + \rho \frac{\sigma_2}{\sigma_1}(X - \mu_1) \qquad (2.4.1)$$

and

$$E(Var(Y|X)) = \sigma_2^2(1 - \rho^2). \tag{2.4.2}$$

Proof: The proof will be given in the continuous case. The discrete case follows similarly by changing integrals to sums. Let $E(Y|x) = a + bx$. From

$$E(Y|x) = \frac{\int_{-\infty}^{\infty} yf(x,y)\,dy}{f_1(x)} = a + bx,$$

we have

$$\int_{-\infty}^{\infty} yf(x,y)\,dy = (a + bx)f_1(x). \tag{2.4.3}$$

If both members of Equation (2.4.3) are integrated on x, it is seen that

$$E(Y) = a + bE(X)$$

or

$$\mu_2 = a + b\mu_1, \tag{2.4.4}$$

where $\mu_1 = E(X)$ and $\mu_2 = E(Y)$. If both members of Equation 2.4.3 are first multiplied by x and then integrated on x, we have

$$E(XY) = aE(X) + bE(X^2),$$

or

$$\rho\sigma_1\sigma_2 + \mu_1\mu_2 = a\mu_1 + b(\sigma_1^2 + \mu_1^2), \tag{2.4.5}$$

where $\rho\sigma_1\sigma_2$ is the covariance of X and Y. The simultaneous solution of Equations 2.4.4 and 2.4.5 yields

$$a = \mu_2 - \rho\frac{\sigma_2}{\sigma_1}\mu_1 \quad \text{and} \quad b = \rho\frac{\sigma_2}{\sigma_1}.$$

These values give the first result (2.4.1).
The conditional variance of Y is given by

$$\begin{aligned}
\text{var}(Y|x) &= \int_{-\infty}^{\infty}\left[y - \mu_2 - \rho\frac{\sigma_2}{\sigma_1}(x - \mu_1)\right]^2 f_{2|1}(y|x)\,dy \\
&= \frac{\int_{-\infty}^{\infty}\left[(y - \mu_2) - \rho\frac{\sigma_2}{\sigma_1}(x - \mu_1)\right]^2 f(x,y)\,dy}{f_1(x)}.
\end{aligned} \tag{2.4.6}$$

This variance is nonnegative and is at most a function of x alone. If then, it is multiplied by $f_1(x)$ and integrated on x, the result obtained will be nonnegative.

This result is

$$\int_{-\infty}^{\infty} \int_{-\infty}^{\infty} \left[(y - \mu_2) - \rho\frac{\sigma_2}{\sigma_1}(x - \mu_1) \right]^2 f(x, y)\, dy\, dx$$

$$= \int_{-\infty}^{\infty} \int_{-\infty}^{\infty} \left[(y - \mu_2)^2 - 2\rho\frac{\sigma_2}{\sigma_1}(y - \mu_2)(x - \mu_1) + \rho^2\frac{\sigma_2^2}{\sigma_1^2}(x - \mu_1)^2 \right] f(x, y)\, dy\, dx$$

$$= E[(Y - \mu_2)^2] - 2\rho\frac{\sigma_2}{\sigma_1}E[(X - \mu_1)(Y - \mu_2)] + \rho^2\frac{\sigma_2^2}{\sigma_1^2}E[(X - \mu_1)^2]$$

$$= \sigma_2^2 - 2\rho\frac{\sigma_2}{\sigma_1}\rho\sigma_1\sigma_2 + \rho^2\frac{\sigma_2^2}{\sigma_1^2}\sigma_1^2$$

$$= \sigma_2^2 - 2\rho^2\sigma_2^2 + \rho^2\sigma_2^2 = \sigma_2^2(1 - \rho^2),$$

which is the desired result. ∎

Note that if the variance, Equation 2.4.6, is denoted by $k(x)$, then $E[k(X)] = \sigma_2^2(1 - \rho^2) \geq 0$. Accordingly, $\rho^2 \leq 1$, or $-1 \leq \rho \leq 1$. It is left as an exercise to prove that $-1 \leq \rho \leq 1$ whether the conditional mean is or is not linear; see Exercise 2.4.7.

Suppose that the variance, Equation 2.4.6, is positive but not a function of x; that is, the variance is a constant $k > 0$. Now if k is multiplied by $f_1(x)$ and integrated on x, the result is k, so that $k = \sigma_2^2(1 - \rho^2)$. Thus, in this case, the variance of each conditional distribution of Y, given $X = x$, is $\sigma_2^2(1 - \rho^2)$. If $\rho = 0$, the variance of each conditional distribution of Y, given $X = x$, is σ_2^2, the variance of the marginal distribution of Y. On the other hand, if ρ^2 is near one, the variance of each conditional distribution of Y, given $X = x$, is relatively small, and there is a high concentration of the probability for this conditional distribution near the mean $E(Y|x) = \mu_2 + \rho(\sigma_2/\sigma_1)(x - \mu_1)$. Similar comments can be made about $E(X|y)$ if it is linear. In particular, $E(X|y) = \mu_1 + \rho(\sigma_1/\sigma_2)(y - \mu_2)$ and $E[\mathrm{Var}(X|y)] = \sigma_1^2(1 - \rho^2)$.

Example 2.4.2. Let the random variables X and Y have the linear conditional means $E(Y|x) = 4x + 3$ and $E(X|y) = \frac{1}{16}y - 3$. In accordance with the general formulas for the linear conditional means, we see that $E(Y|x) = \mu_2$ if $x = \mu_1$ and $E(X|y) = \mu_1$ if $y = \mu_2$. Accordingly, in this special case, we have $\mu_2 = 4\mu_1 + 3$ and $\mu_1 = \frac{1}{16}\mu_2 - 3$ so that $\mu_1 = -\frac{15}{4}$ and $\mu_2 = -12$. The general formulas for the linear conditional means also show that the product of the coefficients of x and y, respectively, is equal to ρ^2 and that the quotient of these coefficients is equal to σ_2^2/σ_1^2. Here $\rho^2 = 4(\frac{1}{16}) = \frac{1}{4}$ with $\rho = \frac{1}{2}$ (not $-\frac{1}{2}$), and $\sigma_2^2/\sigma_1^2 = 64$. Thus, from the two linear conditional means, we are able to find the values of μ_1, μ_2, ρ, and σ_2/σ_1, but not the values of σ_1 and σ_2. ∎

Example 2.4.3. To illustrate how the correlation coefficient measures the intensity of the concentration of the probability for X and Y about a line, let these random variables have a distribution that is uniform over the area depicted in Figure 2.4.1. That is, the joint pdf of X and Y is

$$f(x, y) = \begin{cases} \frac{1}{4ah} & -a + bx < y < a + bx, \quad -h < x < h \\ 0 & \text{elsewhere.} \end{cases}$$

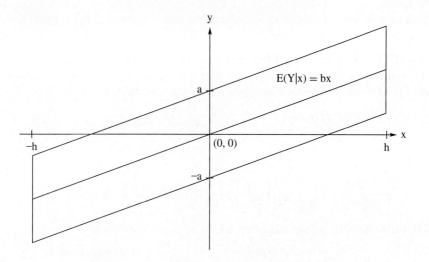

Figure 2.4.1: Illustration for Example 2.4.3.

We assume here that $b \geq 0$, but the argument can be modified for $b \leq 0$. It is easy to show that the pdf of X is uniform, namely

$$f_1(x) = \begin{cases} \int_{-a+bx}^{a+bx} \frac{1}{4ah} \, dy = \frac{1}{2h} & -h < x < h \\ 0 & \text{elsewhere.} \end{cases}$$

The conditional mean and variance are

$$E(Y|x) = bx \quad \text{and} \quad \text{var}(Y|x) = \frac{a^2}{3}.$$

From the general expressions for those characteristics we know that

$$b = \rho \frac{\sigma_2}{\sigma_1} \quad \text{and} \quad \frac{a^2}{3} = \sigma_2^2(1 - \rho^2).$$

Additionally, we know that $\sigma_1^2 = h^2/3$. If we solve these three equations, we obtain an expression for the correlation coefficient, namely

$$\rho = \frac{bh}{\sqrt{a^2 + b^2 h^2}}.$$

Referring to Figure 2.4.1, we note:

1. As a gets small (large), the straight line effect is more (less) intense and ρ is closer to one (zero).

2. As h gets large (small), the straight line effect is more (less) intense and ρ is closer to one (zero).

3. As b gets large (small), the straight line effect is more (less) intense and ρ is closer to one (zero). ∎

Recall that in Section 2.1 we introduced the mgf for the random vector (X, Y). As for random variables, the joint mgf also gives explicit formulas for certain moments. In the case of random variables of the continuous type,

$$\frac{\partial^{k+m} M(t_1, t_2)}{\partial t_1^k \partial t_2^m} = \int_{-\infty}^{\infty} \int_{-\infty}^{\infty} x^k y^m e^{t_1 x + t_2 y} f(x, y) \, dx dy,$$

so that

$$\frac{\partial^{k+m} M(t_1, t_2)}{\partial t_1^k \partial t_2^m} \bigg|_{t_1=t_2=0} = \int_{-\infty}^{\infty} \int_{-\infty}^{\infty} x^k y^m f(x, y) \, dx dy = E(X^k Y^m).$$

For instance, in a simplified notation which appears to be clear,

$$\mu_1 = E(X) = \frac{\partial M(0,0)}{\partial t_1}, \quad \mu_2 = E(Y) = \frac{\partial M(0,0)}{\partial t_2},$$

$$\sigma_1^2 = E(X^2) - \mu_1^2 = \frac{\partial^2 M(0,0)}{\partial t_1^2} - \mu_1^2,$$

$$\sigma_2^2 = E(Y^2) - \mu_2^2 = \frac{\partial^2 M(0,0)}{\partial t_2^2} - \mu_2^2, \tag{2.4.7}$$

$$E[(X - \mu_1)(Y - \mu_2)] = \frac{\partial^2 M(0,0)}{\partial t_1 \partial t_2} - \mu_1 \mu_2,$$

and from these we can compute the correlation coefficient ρ.

It is fairly obvious that the results of Equations 2.4.7 hold if X and Y are random variables of the discrete type. Thus the correlation coefficients may be computed by using the mgf of the joint distribution if that function is readily available. An illustrative example follows.

Example 2.4.4 (Example 2.1.7 Continued). In Example 2.1.7, we considered the joint density

$$f(x, y) = \begin{cases} e^{-y} & 0 < x < y < \infty \\ 0 & \text{elsewhere,} \end{cases}$$

and showed that the mgf was

$$M(t_1, t_2) = \frac{1}{(1 - t_1 - t_2)(1 - t_2)},$$

for $t_1 + t_2 < 1$ and $t_2 < 1$. For this distribution, Equations 2.4.7 become

$$\mu_1 = 1, \quad \mu_2 = 2,$$
$$\sigma_1^2 = 1, \quad \sigma_2^2 = 2, \tag{2.4.8}$$
$$E[(X - \mu_1)(Y - \mu_2)] = 1.$$

Verification of (2.4.8) is left as an exercise; see Exercise 2.4.5. If, momentarily, we accept these results, the correlation coefficient of X and Y is $\rho = 1/\sqrt{2}$. ∎

EXERCISES

2.4.1. Let the random variables X and Y have the joint pmf

(a) $p(x,y) = \frac{1}{3}$, $(x,y) = (0,0), (1,1), (2,2)$, zero elsewhere.

(b) $p(x,y) = \frac{1}{3}$, $(x,y) = (0,2), (1,1), (2,0)$, zero elsewhere.

(c) $p(x,y) = \frac{1}{3}$, $(x,y) = (0,0), (1,1), (2,0)$, zero elsewhere.

In each case compute the correlation coefficient of X and Y.

2.4.2. Let X and Y have the joint pmf described as follows:

(x,y)	$(1,1)$	$(1,2)$	$(1,3)$	$(2,1)$	$(2,2)$	$(2,3)$
$p(x,y)$	$\frac{2}{15}$	$\frac{4}{15}$	$\frac{3}{15}$	$\frac{1}{15}$	$\frac{1}{15}$	$\frac{4}{15}$

and $p(x,y)$ is equal to zero elsewhere.

(a) Find the means μ_1 and μ_2, the variances σ_1^2 and σ_2^2, and the correlation coefficient ρ.

(b) Compute $E(Y|X=1)$, $E(Y|X=2)$, and the line $\mu_2 + \rho(\sigma_2/\sigma_1)(x - \mu_1)$. Do the points $[k, E(Y|X=k)]$, $k = 1, 2$, lie on this line?

2.4.3. Let $f(x,y) = 2$, $0 < x < y$, $0 < y < 1$, zero elsewhere, be the joint pdf of X and Y. Show that the conditional means are, respectively, $(1+x)/2$, $0 < x < 1$, and $y/2$, $0 < y < 1$. Show that the correlation coefficient of X and Y is $\rho = \frac{1}{2}$.

2.4.4. Show that the variance of the conditional distribution of Y, given $X = x$, in Exercise 2.4.3, is $(1-x)^2/12$, $0 < x < 1$, and that the variance of the conditional distribution of X, given $Y = y$, is $y^2/12$, $0 < y < 1$.

2.4.5. Verify the results of Equations 2.4.8 of this section.

2.4.6. Let X and Y have the joint pdf $f(x,y) = 1$, $-x < y < x$, $0 < x < 1$, zero elsewhere. Show that, on the set of positive probability density, the graph of $E(Y|x)$ is a straight line, whereas that of $E(X|y)$ is not a straight line.

2.4.7. If the correlation coefficient ρ of X and Y exists, show that $-1 \leq \rho \leq 1$. *Hint:* Consider the discriminant of the nonnegative quadratic function

$$h(v) = E\{[(X - \mu_1) + v(Y - \mu_2)]^2\},$$

where v is real and is not a function of X nor of Y.

2.4.8. Let $\psi(t_1, t_2) = \log M(t_1, t_2)$, where $M(t_1, t_2)$ is the mgf of X and Y. Show that

$$\frac{\partial \psi(0,0)}{\partial t_i}, \quad \frac{\partial^2 \psi(0,0)}{\partial t_i^2}, \quad i = 1, 2,$$

and

$$\frac{\partial^2 \psi(0,0)}{\partial t_1 \partial t_2}$$

yield the means, the variances and the covariance of the two random variables. Use this result to find the means, the variances, and the covariance of X and Y of Example 2.4.4.

2.4.9. Let X and Y have the joint pmf $p(x,y) = \frac{1}{7}$, $(0,0), (1,0), (0,1), (1,1), (2,1)$, $(1,2), (2,2)$, zero elsewhere. Find the correlation coefficient ρ.

2.4.10. Let X_1 and X_2 have the joint pmf described by the following table:

(x_1, x_2)	$(0,0)$	$(0,1)$	$(0,2)$	$(1,1)$	$(1,2)$	$(2,2)$
$p(x_1, x_2)$	$\frac{1}{12}$	$\frac{2}{12}$	$\frac{1}{12}$	$\frac{3}{12}$	$\frac{4}{12}$	$\frac{1}{12}$

Find $p_1(x_1), p_2(x_2), \mu_1, \mu_2, \sigma_1^2, \sigma_2^2$, and ρ.

2.4.11. Let $\sigma_1^2 = \sigma_2^2 = \sigma^2$ be the common variance of X_1 and X_2 and let ρ be the correlation coefficient of X_1 and X_2. Show that

$$P[|(X_1 - \mu_1) + (X_2 - \mu_2)| \geq k\sigma] \leq \frac{2(1+\rho)}{k^2}.$$

2.5 Independent Random Variables

Let X_1 and X_2 denote the random variables of the continuous type which have the joint pdf $f(x_1, x_2)$ and marginal probability density functions $f_1(x_1)$ and $f_2(x_2)$, respectively. In accordance with the definition of the conditional pdf $f_{2|1}(x_2|x_1)$, we may write the joint pdf $f(x_1, x_2)$ as

$$f(x_1, x_2) = f_{2|1}(x_2|x_1)f_1(x_1).$$

Suppose that we have an instance where $f_{2|1}(x_2|x_1)$ does not depend upon x_1. Then the marginal pdf of X_2 is, for random variables of the continuous type,

$$\begin{aligned} f_2(x_2) &= \int_{-\infty}^{\infty} f_{2|1}(x_2|x_1)f_1(x_1)\, dx_1 \\ &= f_{2|1}(x_2|x_1) \int_{-\infty}^{\infty} f_1(x_1)\, dx_1 \\ &= f_{2|1}(x_2|x_1). \end{aligned}$$

Accordingly,

$$f_2(x_2) = f_{2|1}(x_2|x_1) \quad \text{and} \quad f(x_1, x_2) = f_1(x_1)f_2(x_2),$$

when $f_{2|1}(x_2|x_1)$ does not depend upon x_1. That is, if the conditional distribution of X_2, given $X_1 = x_1$, is independent of any assumption about x_1, then $f(x_1, x_2) = f_1(x_1)f_2(x_2)$.

The same discussion applies to the discrete case too, which we summarize in parentheses in the following definition.

Definition 2.5.1 (Independence). *Let the random variables X_1 and X_2 have the joint pdf $f(x_1, x_2)$ (joint pmf $p(x_1, x_2)$) and the marginal pdfs (pmfs) $f_1(x_1)$ ($p_1(x_1)$) and $f_2(x_2)$ ($p_2(x_2)$), respectively. The random variables X_1 and X_2 are said to be* **independent** *if, and only if, $f(x_1, x_2) \equiv f_1(x_1)f_2(x_2)$ ($p(x_1, x_2) \equiv p_1(x_1)p_2(x_2)$). Random variables that are not independent are said to be* **dependent***.*

Remark 2.5.1. Two comments should be made about the preceding definition. First, the product of two positive functions $f_1(x_1)f_2(x_2)$ means a function that is positive on the product space. That is, if $f_1(x_1)$ and $f_2(x_2)$ are positive on, and only on, the respective spaces $\mathcal{S}_1$ and $\mathcal{S}_2$, then the product of $f_1(x_1)$ and $f_2(x_2)$ is positive on, and only on, the product space $\mathcal{S} = \{(x_1, x_2) : x_1 \in \mathcal{S}_1, \ x_2 \in \mathcal{S}_2\}$. For instance, if $\mathcal{S}_1 = \{x_1 : 0 < x_1 < 1\}$ and $\mathcal{S}_2 = \{x_2 : 0 < x_2 < 3\}$, then $\mathcal{S} = \{(x_1, x_2) : 0 < x_1 < 1, \ 0 < x_2 < 3\}$. The second remark pertains to the identity. The identity in Definition 2.5.1 should be interpreted as follows. There may be certain points $(x_1, x_2) \in \mathcal{S}$ at which $f(x_1, x_2) \neq f_1(x_1)f_2(x_2)$. However, if A is the set of points (x_1, x_2) at which the equality does not hold, then $P(A) = 0$. In subsequent theorems and the subsequent generalizations, a product of nonnegative functions and an identity should be interpreted in an analogous manner. ∎

Example 2.5.1. Let the joint pdf of X_1 and X_2 be

$$f(x_1, x_2) = \begin{cases} x_1 + x_2 & 0 < x_1 < 1, \ 0 < x_2 < 1 \\ 0 & \text{elsewhere.} \end{cases}$$

It will be shown that X_1 and X_2 are dependent. Here the marginal probability density functions are

$$f_1(x_1) = \begin{cases} \int_{-\infty}^{\infty} f(x_1, x_2)\, dx_2 = \int_0^1 (x_1 + x_2)\, dx_2 = x_1 + \frac{1}{2} & 0 < x_1 < 1 \\ 0 & \text{elsewhere,} \end{cases}$$

and

$$f_2(x_2) = \begin{cases} \int_{-\infty}^{\infty} f(x_1, x_2)\, dx_1 = \int_0^1 (x_1 + x_2)\, dx_1 = \frac{1}{2} + x_2 & 0 < x_2 < 1 \\ 0 & \text{elsewhere.} \end{cases}$$

Since $f(x_1, x_2) \not\equiv f_1(x_1)f_2(x_2)$, the random variables X_1 and X_2 are dependent. ∎

The following theorem makes it possible to assert, without computing the marginal probability density functions, that the random variables X_1 and X_2 of Example 2.4.1 are dependent.

Theorem 2.5.1. *Let the random variables X_1 and X_2 have supports $\mathcal{S}_1$ and $\mathcal{S}_2$, respectively, and have the joint pdf $f(x_1, x_2)$. Then X_1 and X_2 are independent if and only if $f(x_1, x_2)$ can be written as a product of a nonnegative function of x_1 and a nonnegative function of x_2. That is,*

$$f(x_1, x_2) \equiv g(x_1)h(x_2),$$

where $g(x_1) > 0$, $x_1 \in \mathcal{S}_1$, zero elsewhere, and $h(x_2) > 0$, $x_2 \in \mathcal{S}_2$, zero elsewhere.

Proof. If X_1 and X_2 are independent, then $f(x_1, x_2) \equiv f_1(x_1)f_2(x_2)$, where $f_1(x_1)$ and $f_2(x_2)$ are the marginal probability density functions of X_1 and X_2, respectively. Thus the condition $f(x_1, x_2) \equiv g(x_1)h(x_2)$ is fulfilled.

Conversely, if $f(x_1, x_2) \equiv g(x_1)h(x_2)$, then, for random variables of the continuous type, we have

$$f_1(x_1) = \int_{-\infty}^{\infty} g(x_1)h(x_2)\, dx_2 = g(x_1) \int_{-\infty}^{\infty} h(x_2)\, dx_2 = c_1 g(x_1)$$

and

$$f_2(x_2) = \int_{-\infty}^{\infty} g(x_1)h(x_2)\, dx_1 = h(x_2) \int_{-\infty}^{\infty} g(x_1)\, dx_1 = c_2 h(x_2),$$

where c_1 and c_2 are constants, not functions of x_1 or x_2. Moreover, $c_1 c_2 = 1$ because

$$1 = \int_{-\infty}^{\infty}\int_{-\infty}^{\infty} g(x_1)h(x_2)\, dx_1 dx_2 = \left[\int_{-\infty}^{\infty} g(x_1)\, dx_1\right]\left[\int_{-\infty}^{\infty} h(x_2)\, dx_2\right] = c_2 c_1.$$

These results imply that

$$f(x_1, x_2) \equiv g(x_1)h(x_2) \equiv c_1 g(x_1) c_2 h(x_2) \equiv f_1(x_1)f_2(x_2).$$

Accordingly, X_1 and X_2 are independent. ∎

This theorem is true for the discrete case also. Simply replace the joint pdf by the joint pmf.

If we now refer to Example 2.5.1, we see that the joint pdf

$$f(x_1, x_2) = \begin{cases} x_1 + x_2 & 0 < x_1 < 1, \ 0 < x_2 < 1 \\ 0 & \text{elsewhere}, \end{cases}$$

cannot be written as the product of a nonnegative function of x_1 and a nonnegative function of x_2. Accordingly, X_1 and X_2 are dependent.

Example 2.5.2. Let the pdf of the random variable X_1 and X_2 be $f(x_1, x_2) = 8x_1 x_2$, $0 < x_1 < x_2 < 1$, zero elsewhere. The formula $8x_1 x_2$ might suggest to some that X_1 and X_2 are independent. However, if we consider the space $\mathcal{S} = \{(x_1, x_2) : 0 < x_1 < x_2 < 1\}$, we see that it is not a product space. This should make it clear that, in general, X_1 and X_2 must be dependent if the space of positive probability density of X_1 and X_2 is bounded by a curve that is neither a horizontal nor a vertical line. ∎

Instead of working with pdfs (or pmfs) we could have presented independence in terms of cumulative distribution functions. The following theorem shows the equivalence.

Theorem 2.5.2. *Let (X_1, X_2) have the joint cdf $F(x_1, x_2)$ and let X_1 and X_2 have the marginal cdfs $F_1(x_1)$ and $F_2(x_2)$, respectively. Then X_1 and X_2 are independent if and only if*

$$F(x_1, x_2) = F_1(x_1)F_2(x_2) \quad \text{for all } (x_1, x_2) \in R^2 . \tag{2.5.1}$$

Proof: We give the proof for the continuous case. Suppose expression (2.5.1) holds. Then the mixed second partial is

$$\frac{\partial^2}{\partial x_1 \partial x_2} F(x_1, x_2) = f_1(x_1) f_2(x_2).$$

Hence, X_1 and X_2 are independent. Conversely, suppose X_1 and X_2 are independent. Then by the definition of the joint cdf,

$$
\begin{aligned}
F(x_1, x_2) &= \int_{-\infty}^{x_1} \int_{-\infty}^{x_2} f_1(w_1) f_2(w_2) \, dw_2 dw_1 \\
&= \int_{-\infty}^{x_1} f_1(w_1) \, dw_1 \cdot \int_{-\infty}^{x_2} f_2(w_2) \, dw_2 = F_1(x_1) F_2(x_2).
\end{aligned}
$$

Hence, condition (2.5.1) is true. ∎

We now give a theorem that frequently simplifies the calculations of probabilities of events which involve independent variables.

Theorem 2.5.3. *The random variables X_1 and X_2 are independent random variables if and only if the following condition holds,*

$$P(a < X_1 \le b, c < X_2 \le d) = P(a < X_1 \le b)P(c < X_2 \le d) \qquad (2.5.2)$$

for every $a < b$ and $c < d$, where a, b, c, and d are constants.

Proof: If X_1 and X_2 are independent then an application of the last theorem and expression (2.1.2) shows that

$$
\begin{aligned}
P(a < X_1 \le b, c < X_2 \le d) &= F(b, d) - F(a, d) - F(b, c) + F(a, c) \\
&= F_1(b) F_2(d) - F_1(a) F_2(d) - F_1(b) F_2(c) \\
&\quad + F_1(a) F_2(c) \\
&= [F_1(b) - F_1(a)][F_2(d) - F_2(c)],
\end{aligned}
$$

which is the right side of expression (2.5.2). Conversely, condition (2.5.2) implies that the joint cdf of (X_1, X_2) factors into a product of the marginal cdfs, which in turn by Theorem 2.5.2 implies that X_1 and X_2 are independent. ∎

Example 2.5.3. (Example 2.5.1, continued) Independence is necessary for condition (2.5.2). For example consider the dependent variables X_1 and X_2 of Example 2.5.1. For these random variables, we have

$$P(0 < X_1 < \tfrac{1}{2}, 0 < X_2 < \tfrac{1}{2}) = \int_0^{1/2} \int_0^{1/2} (x_1 + x_2) \, dx_1 dx_2 = \tfrac{1}{8},$$

whereas

$$P(0 < X_1 < \tfrac{1}{2}) = \int_0^{1/2} (x_1 + \tfrac{1}{2}) \, dx_1 = \tfrac{3}{8}$$

and

$$P(0 < X_2 < \tfrac{1}{2}) = \int_0^{1/2} (\tfrac{1}{2} + x_1) \, dx_2 = \tfrac{3}{8}.$$

Hence, condition (2.5.2) does not hold. ∎

Not merely are calculations of some probabilities usually simpler when we have independent random variables, but many expectations, including certain moment-generating functions, have comparably simpler computations. The following result will prove so useful that we state it in the form of a theorem.

Theorem 2.5.4. *Suppose X_1 and X_2 are independent and that $E(u(X_1))$ and $E(v(X_2))$ exist. Then,*

$$E[u(X_1)v(X_2)] = E[u(X_1)]E[v(X_2)].$$

Proof. We give the proof in the continuous case. The independence of X_1 and X_2 implies that the joint pdf of X_1 and X_2 is $f_1(x_1)f_2(x_2)$. Thus we have, by definition of expectation,

$$
\begin{aligned}
E[u(X_1)v(X_2)] &= \int_{-\infty}^{\infty}\int_{-\infty}^{\infty} u(x_1)v(x_2)f_1(x_1)f_2(x_2)\,dx_1 dx_2 \\
&= \left[\int_{-\infty}^{\infty} u(x_1)f_1(x_1)\,dx_1\right]\left[\int_{-\infty}^{\infty} v(x_2)f_2(x_2)\,dx_2\right] \\
&= E[u(X_1)]E[v(X_2)].
\end{aligned}
$$

Hence, the result is true. ∎

Example 2.5.4. Let X and Y be two independent random variables with means μ_1 and μ_2 and positive variances σ_1^2 and σ_2^2, respectively. We shall show that the independence of X and Y implies that the correlation coefficient of X and Y is zero. This is true because the covariance of X and Y is equal to

$$E[(X-\mu_1)(Y-\mu_2)] = E(X-\mu_1)E(Y-\mu_2) = 0. \quad \blacksquare$$

We shall now prove a very useful theorem about independent random variables. The proof of the theorem relies heavily upon our assertion that an mgf, when it exists, is unique and that it uniquely determines the distribution of probability.

Theorem 2.5.5. *Suppose the joint mgf, $M(t_1, t_2)$, exists for the random variables X_1 and X_2. Then X_1 and X_2 are independent if and only if*

$$M(t_1, t_2) = M(t_1, 0)M(0, t_2),$$

that is, the joint mgf factors into the product of the marginal mgfs.

Proof. If X_1 and X_2 are independent, then

$$
\begin{aligned}
M(t_1, t_2) &= E(e^{t_1 X_1 + t_2 X_2}) \\
&= E(e^{t_1 X_1} e^{t_2 X_2}) \\
&= E(e^{t_1 X_1})E(e^{t_2 X_2}) \\
&= M(t_1, 0)M(0, t_2).
\end{aligned}
$$

Thus the independence of X_1 and X_2 implies that the mgf of the joint distribution factors into the product of the moment-generating functions of the two marginal distributions.

Suppose next that the mgf of the joint distribution of X_1 and X_2 is given by $M(t_1, t_2) = M(t_1, 0)M(0, t_2)$. Now X_1 has the unique mgf which, in the continuous case, is given by

$$M(t_1, 0) = \int_{-\infty}^{\infty} e^{t_1 x_1} f_1(x_1)\, dx_1.$$

Similarly, the unique mgf of X_2, in the continuous case, is given by

$$M(0, t_2) = \int_{-\infty}^{\infty} e^{t_2 x_2} f_2(x_2)\, dx_2.$$

Thus we have

$$
\begin{aligned}
M(t_1, 0)M(0, t_2) &= \left[\int_{-\infty}^{\infty} e^{t_1 x_1} f_1(x_1)\, dx_1\right]\left[\int_{-\infty}^{\infty} e^{t_2 x_2} f_2(x_2)\, dx_2\right] \\
&= \int_{-\infty}^{\infty}\int_{-\infty}^{\infty} e^{t_1 x_1 + t_2 x_2} f_1(x_1) f_2(x_2)\, dx_1 dx_2.
\end{aligned}
$$

We are given that $M(t_1, t_2) = M(t_1, 0)M(0, t_2)$; so

$$M(t_1, t_2) = \int_{-\infty}^{\infty}\int_{-\infty}^{\infty} e^{t_1 x_1 + t_2 x_2} f_1(x_1) f_2(x_2)\, dx_1 dx_2.$$

But $M(t_1, t_2)$ is the mgf of X_1 and X_2. Thus also

$$M(t_1, t_2) = \int_{-\infty}^{\infty}\int_{-\infty}^{\infty} e^{t_1 x_1 + t_2 x_2} f(x_1, x_2)\, dx_1 dx_2.$$

The uniqueness of the mgf implies that the two distributions of probability that are described by $f_1(x_1) f_2(x_2)$ and $f(x_1, x_2)$ are the same. Thus

$$f(x_1, x_2) \equiv f_1(x_1) f_2(x_2).$$

That is, if $M(t_1, t_2) = M(t_1, 0)M(0, t_2)$, then X_1 and X_2 are independent. This completes the proof when the random variables are of the continuous type. With random variables of the discrete type, the proof is made by using summation instead of integration. ∎

Example 2.5.5 (Example 2.1.7, Continued). Let (X, Y) be a pair of random variables with the joint pdf

$$f(x, y) = \begin{cases} e^{-y} & 0 < x < y < \infty \\ 0 & \text{elsewhere.} \end{cases}$$

In Example 2.1.7, we showed that the mgf of (X, Y) is

$$
\begin{aligned}
M(t_1, t_2) &= \int_0^{\infty}\int_x^{\infty} \exp(t_1 x + t_2 y - y)\, dy dx \\
&= \frac{1}{(1 - t_1 - t_2)(1 - t_2)},
\end{aligned}
$$

provided that $t_1 + t_2 < 1$ and $t_2 < 1$. Because $M(t_1, t_2) \neq M(t_1, 0)M(t_1, 0)$ the random variables are dependent. ∎

Example 2.5.6 (Exercise 2.1.14 continued). For the random variable X_1 and X_2 defined in Exercise 2.1.14, we showed that the joint mgf is

$$M(t_1, t_2) = \left[\frac{\exp\{t_1\}}{2 - \exp\{t_1\}}\right]\left[\frac{\exp\{t_2\}}{2 - \exp\{t_2\}}\right], \quad t_i < \log 2 \, , i = 1, 2.$$

We showed further that $M(t_1, t_2) = M(t_1, 0)M(0, t_2)$. Hence, X_1 and X_2 are independent random variables. ∎

EXERCISES

2.5.1. Show that the random variables X_1 and X_2 with joint pdf

$$f(x_1, x_2) = \begin{cases} 12x_1x_2(1 - x_2) & 0 < x_1 < 1, \ 0 < x_2 < 1 \\ 0 & \text{elsewhere} \end{cases}$$

are independent.

2.5.2. If the random variables X_1 and X_2 have the joint pdf $f(x_1, x_2) = 2e^{-x_1-x_2}$, $0 < x_1 < x_2$, $0 < x_2 < \infty$, zero elsewhere, show that X_1 and X_2 are dependent.

2.5.3. Let $p(x_1, x_2) = \frac{1}{16}$, $x_1 = 1, 2, 3, 4$, and $x_2 = 1, 2, 3, 4$, zero elsewhere, be the joint pmf of X_1 and X_2. Show that X_1 and X_2 are independent.

2.5.4. Find $P(0 < X_1 < \frac{1}{3}, 0 < X_2 < \frac{1}{3})$ if the random variables X_1 and X_2 have the joint pdf $f(x_1, x_2) = 4x_1(1 - x_2)$, $0 < x_1 < 1$, $0 < x_2 < 1$, zero elsewhere.

2.5.5. Find the probability of the union of the events $a < X_1 < b$, $-\infty < X_2 < \infty$, and $-\infty < X_1 < \infty$, $c < X_2 < d$ if X_1 and X_2 are two independent variables with $P(a < X_1 < b) = \frac{2}{3}$ and $P(c < X_2 < d) = \frac{5}{8}$.

2.5.6. If $f(x_1, x_2) = e^{-x_1-x_2}$, $0 < x_1 < \infty$, $0 < x_2 < \infty$, zero elsewhere, is the joint pdf of the random variables X_1 and X_2, show that X_1 and X_2 are independent and that $M(t_1, t_2) = (1 - t_1)^{-1}(1 - t_2)^{-1}$, $t_2 < 1$, $t_1 < 1$. Also show that

$$E(e^{t(X_1+X_2)}) = (1 - t)^{-2}, \quad t < 1.$$

Accordingly, find the mean and the variance of $Y = X_1 + X_2$.

2.5.7. Let the random variables X_1 and X_2 have the joint pdf $f(x_1, x_2) = 1/\pi$, for $(x_1 - 1)^2 + (x_2 + 2)^2 < 1$, zero elsewhere. Find $f_1(x_1)$ and $f_2(x_2)$. Are X_1 and X_2 independent?

2.5.8. Let X and Y have the joint pdf $f(x, y) = 3x$, $0 < y < x < 1$, zero elsewhere. Are X and Y independent? If not, find $E(X|y)$.

2.5.9. Suppose that a man leaves for work between 8:00 A.M.and 8:30 A.M. and takes between 40 and 50 minutes to get to the office. Let X denote the time of departure and let Y denote the time of travel. If we assume that these random variables are independent and uniformly distributed, find the probability that he arrives at the office before 9:00 A.M..

2.5.10. Let X and Y be random variables with the space consisting of the four points: $(0,0), (1,1), (1,0), (1,-1)$. Assign positive probabilities to these four points so that the correlation coefficient is equal to zero. Are X and Y independent?

2.5.11. Two line segments, each of length two units, are placed along the x-axis. The midpoint of the first is between $x = 0$ and $x = 14$ and that of the second is between $x = 6$ and $x = 20$. Assuming independence and uniform distributions for these midpoints, find the probability that the line segments overlap.

2.5.12. Cast a fair die and let $X = 0$ if $1, 2$, or 3 spots appear, let $X = 1$ if 4 or 5 spots appear, and let $X = 2$ if 6 spots appear. Do this two independent times, obtaining X_1 and X_2. Calculate $P(|X_1 - X_2| = 1)$.

2.5.13. For X_1 and X_2 in Example 2.5.6, show that the mgf of $Y = X_1 + X_2$ is $e^{2t}/(2 - e^t)^2$, $t < \log 2$, and then compute the mean and variance of Y.

2.6 Extension to Several Random Variables

The notions about two random variables can be extended immediately to n random variables. We make the following definition of the space of n random variables.

Definition 2.6.1. *Consider a random experiment with the sample space $\mathcal{C}$. Let the random variable X_i assign to each element $c \in \mathcal{C}$ one and only one real number $X_i(c) = x_i$, $i = 1, 2, \ldots, n$. We say that $(X_1, \ldots, X_n)$ is an n-dimensional* **random vector**. *The* **space** *of this random vector is the set of ordered n-tuples $\mathcal{D} = \{(x_1, x_2, \ldots, x_n) : x_1 = X_1(c), \ldots, x_n = X_n(c), c \in \mathcal{C}\}$. Furthermore, let A be a subset of the space $\mathcal{D}$. Then $P[(X_1, \ldots, X_n) \in A] = P(C)$, where $C = \{c : c \in \mathcal{C} \text{ and } (X_1(c), X_2(c), \ldots, X_n(c)) \in A\}$.*

In this section, we will often use vector notation. For example, we denote $(X_1, \ldots, X_n)'$ by the n dimensional column vector $\mathbf{X}$ and the observed values $(x_1, \ldots, x_n)'$ of the random variables by $\mathbf{x}$. The joint cdf is defined to be

$$F_{\mathbf{X}}(\mathbf{x}) = P[X_1 \leq x_1, \ldots, X_n \leq x_n]. \tag{2.6.1}$$

We say that the n random variables $X_1, X_2, \ldots, X_n$ are of the discrete type or of the continuous type and have a distribution of that type accordingly as the joint cdf can be expressed as

$$F_{\mathbf{X}}(\mathbf{x}) = \sum_{w_1 \leq x_1, \ldots, w_n \leq x_n} \cdots \sum p(w_1, \ldots, w_n),$$

or as

$$F_{\mathbf{X}}(\mathbf{x}) = \int_{w_1 \leq x_1, \ldots, w_n \leq x_n} \cdots \int f(w_1, \ldots, w_n) \, dw_1 \cdots dw_n.$$

For the continuous case,

$$\frac{\partial^n}{\partial x_1 \cdots \partial x_n} F_{\mathbf{X}}(\mathbf{x}) = f(\mathbf{x}). \tag{2.6.2}$$

In accordance with the convention of extending the definition of a joint pdf, it is seen that a point function f essentially satisfies the conditions of being a pdf if (a) f is defined and is nonnegative for all real values of its argument(s) and if (b) its integral over all real values of its argument(s) is 1. Likewise, a point function p essentially satisfies the conditions of being a joint pmf if (a) p is defined and is nonnegative for all real values of its argument(s) and if (b) its sum over all real values of its argument(s) is 1. As in previous sections, it is sometimes convenient to speak of the support set of a random vector. For the discrete case, this would be all points in $\mathcal{D}$ which have positive mass, while for the continuous case these would be all points in $\mathcal{D}$ which can be embedded in an open set of positive probability. We will use $\mathcal{S}$ to denote support sets.

Example 2.6.1. Let

$$f(x, y, z) = \begin{cases} e^{-(x+y+z)} & 0 < x, y, z < \infty \\ 0 & \text{elsewhere} \end{cases}$$

be the pdf of the random variables X, Y, and Z. Then the distribution function of X, Y, and Z is given by

$$\begin{aligned} F(x, y, z) &= P(X \le x, Y \le y, Z \le z) \\ &= \int_0^z \int_0^y \int_0^x e^{-u-v-w} \, du\, dv\, dw \\ &= (1 - e^{-x})(1 - e^{-y})(1 - e^{-z}), \quad 0 \le x, y, z < \infty, \end{aligned}$$

and is equal to zero elsewhere. The relationship (2.6.2) can easily be verified. ∎

Let $(X_1, X_2, \ldots, X_n)$ be a random vector and let $Y = u(X_1, X_2, \ldots, X_n)$ for some function u. As in the bivariate case, the expected value of the random variable exists if the n-fold integral

$$\int_{-\infty}^{\infty} \cdots \int_{-\infty}^{\infty} |u(x_1, x_2, \ldots, x_n)| f(x_1, x_2, \ldots, x_n) \, dx_1 dx_2 \cdots dx_n$$

exists when the random variables are of the continuous type, or if the n-fold sum

$$\sum_{x_n} \cdots \sum_{x_1} |u(x_1, x_2, \ldots, x_n)| f(x_1, x_2, \ldots, x_n)$$

exists when the random variables are of the discrete type. If the expected value of Y exists then its expectation is given by

$$E(Y) = \int_{-\infty}^{\infty} \cdots \int_{-\infty}^{\infty} u(x_1, x_2, \ldots, x_n) f(x_1, x_2, \ldots, x_n) \, dx_1 dx_2 \cdots dx_n \tag{2.6.3}$$

for the continuous case, and by

$$E(Y) = \sum_{x_n} \cdots \sum_{x_1} u(x_1, x_2, \ldots, x_n) f(x_1, x_2, \ldots, x_n) \tag{2.6.4}$$

for the discrete case. The properties of expectation discussed in Section 2.1 hold for the n-dimension case, also. In particular, E is a linear operator. That is, if $Y_j = u_j(X_1, \ldots, X_n)$ for $j = 1, \ldots, m$ and each $E(Y_i)$ exists then

$$E\left[\sum_{j=1}^{m} k_j Y_j\right] = \sum_{j=1}^{n} k_j E[Y_j], \tag{2.6.5}$$

where $k_1, \ldots, k_m$ are constants.

We shall now discuss the notions of marginal and conditional probability density functions from the point of view of n random variables. All of the preceding definitions can be directly generalized to the case of n variables in the following manner. Let the random variables $X_1, X_2, \ldots, X_n$ be of the continuous type with the joint pdf $f(x_1, x_2, \ldots, x_n)$. By an argument similar to the two-variable case, we have for every b,

$$F_{X_1}(b) = P(X_1 < b) = \int_{-\infty}^{b} f_1(x_1) \, dx_1,$$

where $f_1(x_1)$ is defined by the $(n-1)$-fold integral

$$f_1(x_1) = \int_{-\infty}^{\infty} \cdots \int_{-\infty}^{\infty} f(x_1, x_2, \ldots, x_n) \, dx_2 \cdots dx_n.$$

Therefore, $f_1(x_1)$ is the pdf of the random variable X_1 and $f_1(x_1)$ is called the marginal pdf of X_1. The marginal probability density functions $f_2(x_2), \ldots, f_n(x_n)$ of $X_2, \ldots, X_n$, respectively, are similar $(n-1)$-fold integrals.

Up to this point, each marginal pdf has been a pdf of one random variable. It is convenient to extend this terminology to joint probability density functions, which we shall do now. Let $f(x_1, x_2, \ldots, x_n)$ be the joint pdf of the n random variables $X_1, X_2, \ldots, X_n$, just as before. Now, however, let us take any group of $k < n$ of these random variables and let us find the joint pdf of them. This joint pdf is called the marginal pdf of this particular group of k variables. To fix the ideas, take $n = 6$, $k = 3$, and let us select the group X_2, X_4, X_5. Then the marginal pdf of X_2, X_4, X_5 is the joint pdf of this particular group of three variables, namely,

$$\int_{-\infty}^{\infty} \int_{-\infty}^{\infty} \int_{-\infty}^{\infty} f(x_1, x_2, x_3, x_4, x_5, x_6) \, dx_1 dx_3 dx_6,$$

if the random variables are of the continuous type.

Next we extend the definition of a conditional pdf. Suppose $f_1(x_1) > 0$. Then we define the symbol $f_{2,\ldots,n|1}(x_2, \ldots, x_n|x_1)$ by the relation

$$f_{2,\ldots,n|1}(x_2, \ldots, x_n|x_1) = \frac{f(x_1, x_2, \ldots, x_n)}{f_1(x_1)},$$

and $f_{2,\ldots,n|1}(x_2,\ldots,x_n|x_1)$ is called the **joint conditional pdf** of $X_2,\ldots,X_n$, given $X_1 = x_1$. The joint conditional pdf of any $n-1$ random variables, say $X_1,\ldots,X_{i-1},X_{i+1},\ldots,X_n$, given $X_i = x_i$, is defined as the joint pdf of $X_1,\ldots,X_n$ divided by the marginal pdf $f_i(x_i)$, provided that $f_i(x_i) > 0$. More generally, the joint conditional pdf of $n-k$ of the random variables, for given values of the remaining k variables, is defined as the joint pdf of the n variables divided by the marginal pdf of the particular group of k variables, provided that the latter pdf is positive. We remark that there are many other conditional probability density functions; for instance, see Exercise 2.3.12.

Because a conditional pdf is a pdf of a certain number of random variables, the expectation of a function of these random variables has been defined. To emphasize the fact that a conditional pdf is under consideration, such expectations are called conditional expectations. For instance, the conditional expectation of $u(X_2,\ldots,X_n)$ given $X_1 = x_1$, is, for random variables of the continuous type, given by

$$E[u(X_2,\ldots,X_n)|x_1] = \int_{-\infty}^{\infty} \cdots \int_{-\infty}^{\infty} u(x_2,\ldots,x_n) f_{2,\ldots,n|1}(x_2,\ldots,x_n|x_1)\, dx_2 \cdots dx_n$$

provided $f_1(x_1) > 0$ and the integral converges (absolutely). A useful random variable is given by $h(X_1) = E[u(X_2,\ldots,X_n)|X_1)]$.

The above discussion of marginal and conditional distributions generalizes to random variables of the discrete type by using pmfs and summations instead of integrals.

Let the random variables $X_1, X_2, \ldots, X_n$ have the joint pdf $f(x_1, x_2, \ldots, x_n)$ and the marginal probability density functions $f_1(x_1), f_2(x_2), \ldots, f_n(x_n)$, respectively. The definition of the independence of X_1 and X_2 is generalized to the mutual independence of $X_1, X_2, \ldots, X_n$ as follows: The random variables $X_1, X_2, \ldots, X_n$ are said to be **mutually independent** if and only if

$$f(x_1, x_2, \ldots, x_n) \equiv f_1(x_1) f_2(x_2) \cdots f_n(x_n),$$

for the continuous case. In the discrete case, $X_1, X_2, \ldots, X_n$ are said to be **mutually independent** if and only if

$$p(x_1, x_2, \ldots, x_n) \equiv p_1(x_1) p_2(x_2) \cdots p_n(x_n).$$

Suppose $X_1, X_2, \ldots, X_n$ are mutally independent. Then

$$P(a_1 < X_1 < b_1,\, a_2 < X_2 < b_2, \ldots, a_n < X_n < b_n)$$
$$= P(a_1 < X_1 < b_1) P(a_2 < X_2 < b_2) \cdots P(a_n < X_n < b_n)$$
$$= \prod_{i=1}^{n} P(a_i < X_i < b_i),$$

where the symbol $\prod_{i=1}^{n} \varphi(i)$ is defined to be

$$\prod_{i=1}^{n} \varphi(i) = \varphi(1)\varphi(2) \cdots \varphi(n).$$

The theorem that

$$E[u(X_1)v(X_2)] = E[u(X_1)]E[v(X_2)]$$

for independent random variables X_1 and X_2 becomes, for mutually independent random variables $X_1, X_2, \ldots, X_n$,

$$E[u_1(X_1)u_2(X_2) \cdots u_n(X_n)] = E[u_1(X_1)]E[u_2(X_2)] \cdots E[u_n(X_n)],$$

or

$$E\left[\prod_{i=1}^{n} u_i(X_i)\right] = \prod_{i=1}^{n} E[u_i(X_i)].$$

The moment-generating function (mgf) of the joint distribution of n random variables $X_1, X_2, \ldots, X_n$ is defined as follows. Let

$$E[\exp(t_1X_1 + t_2X_2 + \cdots + t_nX_n)]$$

exists for $-h_i < t_i < h_i$, $i = 1, 2, \ldots, n$, where each h_i is positive. This expectation is denoted by $M(t_1, t_2, \ldots, t_n)$ and it is called the mgf of the joint distribution of $X_1, \ldots, X_n$ (or simply the mgf of $X_1, \ldots, X_n$). As in the cases of one and two variables, this mgf is unique and uniquely determines the joint distribution of the n variables (and hence all marginal distributions). For example, the mgf of the marginal distributions of X_i is $M(0, \ldots, 0, t_i, 0, \ldots, 0)$, $i = 1, 2, \ldots, n$; that of the marginal distribution of X_i and X_j is $M(0, \ldots, 0, t_i, 0, \ldots, 0, t_j, 0, \ldots, 0)$; and so on. Theorem 2.5.5 of this chapter can be generalized, and the factorization

$$M(t_1, t_2, \ldots, t_n) = \prod_{i=1}^{n} M(0, \ldots, 0, t_i, 0, \ldots, 0) \tag{2.6.6}$$

is a necessary and sufficient condition for the mutual independence of $X_1, X_2, \ldots, X_n$. Note that we can write the joint mgf in vector notation as

$$M(\mathbf{t}) = E[\exp(\mathbf{t}'\mathbf{X})], \quad \text{for } \mathbf{t} \in B \subset R^n,$$

where $B = \{\mathbf{t} : -h_i < t_i < h_i, i = 1, \ldots, n\}$.

Example 2.6.2. Let X_1, X_2, and X_3 be three mutually independent random variables and let each have the pdf

$$f(x) = \begin{cases} 2x & 0 < x < 1 \\ 0 & \text{elsewhere.} \end{cases} \tag{2.6.7}$$

The joint pdf of X_1, X_2, X_3 is $f(x_1)f(x_2)f(x_3) = 8x_1x_2x_3$, $0 < x_i < 1$, $i = 1, 2, 3$, zero elsewhere. Then, for illustration, the expected value of $5X_1X_2^3 + 3X_2X_3^4$ is

$$\int_0^1 \int_0^1 \int_0^1 (5x_1x_2^3 + 3x_2x_3^4)8x_1x_2x_3 \, dx_1 dx_2 dx_3 = 2.$$

Let Y be the maximum of X_1, X_2, and X_3. Then, for instance, we have

$$
\begin{aligned}
P(Y \leq \tfrac{1}{2}) &= P(X_1 \leq \tfrac{1}{2}, X_2 \leq \tfrac{1}{2}, X_3 \leq \tfrac{1}{2}) \\
&= \int_0^{1/2} \int_0^{1/2} \int_0^{1/2} 8x_1 x_2 x_3 \, dx_1 dx_2 dx_3 \\
&= (\tfrac{1}{2})^6 = \tfrac{1}{64}.
\end{aligned}
$$

In a similar manner, we find that the cdf of Y is

$$
G(y) = P(Y \leq y) = \begin{cases} 0 & y < 0 \\ y^6 & 0 \leq y < 1 \\ 1 & 1 \leq y. \end{cases}
$$

Accordingly, the pdf of Y is

$$
g(y) = \begin{cases} 6y^5 & 0 < y < 1 \\ 0 & \text{elsewhere.} \end{cases} \quad \blacksquare
$$

Remark 2.6.1. If X_1, X_2, and X_3 are mutually independent, they are *pairwise independent* (that is, X_i and X_j, $i \neq j$, where $i, j = 1, 2, 3$, are independent). However, the following example, attributed to S. Bernstein, shows that pairwise independence does not necessarily imply mutual independence. Let X_1, X_2, and X_3 have the joint pmf

$$
f(x_1, x_2, x_3) = \begin{cases} \tfrac{1}{4} & (x_1, x_2, x_3) \in \{(1,0,0), (0,1,0), (0,0,1), (1,1,1)\} \\ 0 & \text{elsewhere.} \end{cases}
$$

The joint pmf of X_i and X_j, $i \neq j$, is

$$
f_{ij}(x_i, x_j) = \begin{cases} \tfrac{1}{4} & (x_i, x_j) \in \{(0,0), (1,0), (0,1), (1,1)\} \\ 0 & \text{elsewhere,} \end{cases}
$$

whereas the marginal pmf of X_i is

$$
f_i(x_i) = \begin{cases} \tfrac{1}{2} & x_i = 0, 1 \\ 0 & \text{elsewhere.} \end{cases}
$$

Obviously, if $i \neq j$, we have

$$
f_{ij}(x_i, x_j) \equiv f_i(x_i) f_j(x_j),
$$

and thus X_i and X_j are independent. However,

$$
f(x_1, x_2, x_3) \neq f_1(x_1) f_2(x_2) f_3(x_3).
$$

Thus X_1, X_2, and X_3 are not mutually independent.

Unless there is a possible misunderstanding between *mutual* and *pairwise* independence, we usually drop the modifier *mutual*. Accordingly, using this practice in Example 2.6.2, we say that X_1, X_2, X_3 are independent random variables, meaning that they are mutually independent. Occasionally, for emphasis, we use *mutually independent* so that the reader is reminded that this is different from *pairwise independence*.

In addition, if several random variables are mutually independent and have the same distribution, we say that they are **independent and identically distributed**, which we abbreviate as iid. So the random variables in Example 2.6.2 are iid with the common pdf given in expression (2.6.7). ∎

2.6.1 *Variance-Covariance

In Section 2.4 we discussed the covariance between two random variables. In this section we want to extend this discussion to the n-variate case. Let $\mathbf{X} = (X_1, \ldots, X_n)'$ be an n-dimensional random vector. Recall that we defined $E(\mathbf{X}) = (E(X_1), \ldots, E(X_n))'$, that is, the expectation of a random vector is just the vector of the expectations of its components. Now suppose $\mathbf{W}$ is an $m \times n$ matrix of random variables, say, $\mathbf{W} = [W_{ij}]$ for the random variables W_{ij}, $1 \le i \le m$ and $1 \le j \le n$. Note that we can always string out the matrix into an $mn \times 1$ random vector. Hence, we define the expectation of a random matrix

$$E[\mathbf{W}] = [E(W_{ij})]. \tag{2.6.8}$$

As the following theorem shows, linearity of the expectation operator easily follows from this definition:

Theorem 2.6.1. *Let $\mathbf{W}_1$ and $\mathbf{W}_2$ be $m \times n$ matrices of random variables, and let $\mathbf{A}_1$ and $\mathbf{A}_2$ be $k \times m$ matrices of constants, and let $\mathbf{B}$ be a $n \times l$ matrix of constants. Then*

$$E[\mathbf{A}_1\mathbf{W}_1 + \mathbf{A}_2\mathbf{W}_2] = \mathbf{A}_1 E[\mathbf{W}_1] + \mathbf{A}_2 E[\mathbf{W}_2] \tag{2.6.9}$$

$$E[\mathbf{A}_1\mathbf{W}_1\mathbf{B}] = \mathbf{A}_1 E[\mathbf{W}_1]\mathbf{B}. \tag{2.6.10}$$

Proof: Because of linearity of the operator E on random variables, we have for the $(i, j)th$ components of expression (2.6.9) that

$$E[\sum_{s=1}^{n} a_{1is}W_{1sj} + \sum_{s=1}^{n} a_{2is}W_{2sj}] = \sum_{s=1}^{n} a_{1is}E[W_{1sj}] + \sum_{s=1}^{n} a_{2is}E[W_{2sj}].$$

Hence by (2.6.8) expression (2.6.9) is true. The derivation of Expression (2.6.10) follows in the same manner. ∎

Let $\mathbf{X} = (X_1, \ldots, X_n)'$ be an n-dimensional random vector, such that $\sigma_i^2 = \mathrm{Var}(X_i) < \infty$. The **mean** of $\mathbf{X}$ is $\boldsymbol{\mu} = E[\mathbf{X}]$ and we define its **variance-covariance matrix** to be,

$$\mathrm{Cov}(\mathbf{X}) = E[(\mathbf{X} - \boldsymbol{\mu})(\mathbf{X} - \boldsymbol{\mu})'] = [\sigma_{ij}], \tag{2.6.11}$$

where σ_{ii} denotes σ_i^2. As Exercise 2.6.7 shows, the ith diagonal entry of $\mathrm{Cov}(\mathbf{X})$ is $\sigma_i^2 = \mathrm{Var}(X_i)$ and the $(i, j)th$ off diagonal entry is $\mathrm{cov}(X_i, X_j)$. So the name, variance-covariance matrix is appropriate.

Example 2.6.3 (Example 2.4.4, Continued). In Example 2.4.4, we considered the joint pdf

$$f(x, y) = \begin{cases} e^{-y} & 0 < x < y < \infty \\ 0 & \text{elsewhere,} \end{cases}$$

and showed that the first two moments are

$$\mu_1 = 1, \quad \mu_2 = 2,$$
$$\sigma_1^2 = 1, \quad \sigma_2^2 = 2, \tag{2.6.12}$$
$$E[(X - \mu_1)(Y - \mu_2)] = 1.$$

Let $\mathbf{Z} = (X, Y)'$. Then using the present notation, we have

$$E[\mathbf{Z}] = \begin{bmatrix} 1 \\ 2 \end{bmatrix} \text{ and } \text{cov}(\mathbf{Z}) = \begin{bmatrix} 1 & 1 \\ 1 & 2 \end{bmatrix}. \quad \blacksquare$$

Two properties of $\text{cov}(X_i, X_j)$ which we need later are summarized in the following theorem,

Theorem 2.6.2. *Let $\mathbf{X} = (X_1, \ldots, X_n)'$ be an n-dimensional random vector, such that $\sigma_i^2 = \sigma_{ii} = Var(X_i) < \infty$. Let $\mathbf{A}$ be an $m \times n$ matrix of constants. Then*

$$Cov(\mathbf{X}) = E[\mathbf{XX}'] - \boldsymbol{\mu\mu}' \tag{2.6.13}$$
$$Cov(\mathbf{AX}) = \mathbf{A}\, Cov(\mathbf{X})\mathbf{A}' \tag{2.6.14}$$

Proof: Use Theorem 2.6.1 to derive (2.6.13) ; i.e.,

$$\begin{aligned}
\text{Cov}(\mathbf{X}) &= E[(\mathbf{X} - \boldsymbol{\mu})(\mathbf{X} - \boldsymbol{\mu})'] \\
&= E[\mathbf{XX}' - \boldsymbol{\mu}\mathbf{X}' - \mathbf{X}\boldsymbol{\mu}' + \boldsymbol{\mu\mu}'] \\
&= E[\mathbf{XX}'] - \boldsymbol{\mu}E[\mathbf{X}'] - E[\mathbf{X}]\boldsymbol{\mu}' + \boldsymbol{\mu\mu}',
\end{aligned}$$

which is the desired result. The proof of (2.6.14) is left as an exercise. $\blacksquare$

All variance-covariance matrices are **positive semi-definite** (psd) matrices; that is, $\mathbf{a}'\text{Cov}(\mathbf{X})\mathbf{a} \geq 0$, for all vectors $\mathbf{a} \in R^n$. To see this let $\mathbf{X}$ be a random vector and let $\mathbf{a}$ be any $n \times 1$ vector of constants. Then $Y = \mathbf{a}'\mathbf{X}$ is a random variable and, hence, has nonnegative variance; i.e,

$$0 \leq \text{Var}(Y) = \text{Var}(\mathbf{a}'\mathbf{X}) = \mathbf{a}'\text{Cov}(\mathbf{X})\mathbf{a}; \tag{2.6.15}$$

hence, $\text{Cov}(\mathbf{X})$ is psd.

EXERCISES

2.6.1. Let X, Y, Z have joint pdf $f(x, y, z) = 2(x + y + z)/3$, $0 < x < 1$, $0 < y < 1$, $0 < z < 1$, zero elsewhere.

(a) Find the marginal probability density functions of $X, Y,$ and Z.

(b) Compute $P(0 < X < \frac{1}{2}, 0 < Y < \frac{1}{2}, 0 < Z < \frac{1}{2})$ and $P(0 < X < \frac{1}{2}) = P(0 < Y < \frac{1}{2}) = P(0 < Z < \frac{1}{2})$.

(c) Are X, Y, and Z independent?

(d) Calculate $E(X^2YZ + 3XY^4Z^2)$.

(e) Determine the cdf of X, Y and Z.

(f) Find the conditional distribution of X and Y, given $Z = z$, and evaluate $E(X + Y|z)$.

(g) Determine the conditional distribution of X, given $Y = y$ and $Z = z$, and compute $E(X|y, z)$.

2.6.2. Let $f(x_1, x_2, x_3) = \exp[-(x_1 + x_2 + x_3)]$, $0 < x_1 < \infty$, $0 < x_2 < \infty$, $0 < x_3 < \infty$, zero elsewhere, be the joint pdf of X_1, X_2, X_3.

(a) Compute $P(X_1 < X_2 < X_3)$ and $P(X_1 = X_2 < X_3)$.

(b) Determine the joint mgf of X_1, X_2, and X_3. Are these random variables independent?

2.6.3. Let X_1, X_2, X_3, and X_4 be four independent random variables, each with pdf $f(x) = 3(1 - x)^2$, $0 < x < 1$, zero elsewhere. If Y is the minimum of these four variables, find the cdf and the pdf of Y.
Hint: $P(Y > y) = P(X_i > y, i = 1, \ldots, 4)$.

2.6.4. A fair die is cast at random three independent times. Let the random variable X_i be equal to the number of spots that appear on the ith trial, $i = 1, 2, 3$. Let the random variable Y be equal to $\max(X_i)$. Find the cdf and the pmf of Y.
Hint: $P(Y \leq y) = P(X_i \leq y, i = 1, 2, 3)$.

2.6.5. Let $M(t_1, t_2, t_3)$ be the mgf of the random variables X_1, X_2, and X_3 of Bernstein's example, described in the remark following Example 2.6.2. Show that

$$M(t_1, t_2, 0) = M(t_1, 0, 0)M(0, t_2, 0), \quad M(t_1, 0, t_3) = M(t_1, 0, 0)M(0, 0, t_3)$$

and

$$M(0, t_2, t_3) = M(0, t_2, 0)M(0, 0, t_3)$$

are true, but that

$$M(t_1, t_2, t_3) \neq M(t_1, 0, 0)M(0, t_2, 0)M(0, 0, t_3).$$

Thus X_1, X_2, X_3 are pairwise independent but not mutually independent.

2.6.6. Let X_1, X_2, and X_3 be three random variables with means, variances, and correlation coefficients, denoted by $\mu_1, \mu_2, \mu_3; \sigma_1^2, \sigma_2^2, \sigma_3^2$; and $\rho_{12}, \rho_{13}, \rho_{23}$, respectively. For constants b_2 and b_3, suppose $E(X_1 - \mu_1 | x_2, x_3) = b_2(x_2 - \mu_2) + b_3(x_3 - \mu_3)$. Determine b_2 and b_3 in terms of the variances and the correlation coefficients.

2.6.7. Let $\mathbf{X} = (X_1, \ldots, X_n)'$ be an n-dimensional random vector, with variance-covariance matrix (2.6.11). Show that the *ith* diagonal entry of $\text{Cov}(\mathbf{X})$ is $\sigma_i^2 = \text{Var}(X_i)$ and that the $(i, j)th$ off diagonal entry is $\text{cov}(X_i, X_j)$.

2.6.8. Let X_1, X_2, X_3 be iid with common pdf $f(x) = \exp(-x)$, $0 < x < \infty$, zero elsewhere. Evaluate:

(a) $P(X_1 < X_2 | X_1 < 2X_2)$.

(b) $P(X_1 < X_2 < X_3 | X_3 < 1)$.

2.7 Transformations: Random Vectors

In Section 2.2 it was seen that the determination of the joint pdf of two functions of two random variables of the continuous type was essentially a corollary to a theorem in analysis having to do with the change of variables in a twofold integral. This theorem has a natural extension to n-fold integrals. This extension is as follows. Consider an integral of the form

$$\int \cdots \int_A h(x_1, x_2, \ldots, x_n) \, dx_1 \, dx_2 \cdots dx_n$$

taken over a subset A of an n-dimensional space $\mathcal{S}$. Let

$$y_1 = u_1(x_1, x_2, \ldots, x_n), \quad y_2 = u_2(x_1, x_2, \ldots, x_n), \ldots, y_n = u_n(x_1, x_2, \ldots, x_n),$$

together with the inverse functions

$$x_1 = w_1(y_1, y_2, \ldots, y_n), \quad x_2 = w_2(y_1, y_2, \ldots, y_n), \ldots, x_n = w_n(y_1, y_2, \ldots, y_n)$$

define a one-to-one transformation that maps $\mathcal{S}$ onto $\mathcal{T}$ in the $y_1, y_2, \ldots, y_n$ space and, hence, maps the subset A of $\mathcal{S}$ onto a subset B of $\mathcal{T}$. Let the first partial derivatives of the inverse functions be continuous and let the n by n determinant (called the Jacobian)

$$J = \begin{vmatrix} \frac{\partial x_1}{\partial y_1} & \frac{\partial x_1}{\partial y_2} & \cdots & \frac{\partial x_1}{\partial y_n} \\ \frac{\partial x_2}{\partial y_1} & \frac{\partial x_2}{\partial y_2} & \cdots & \frac{\partial x_2}{\partial y_n} \\ \vdots & \vdots & & \vdots \\ \frac{\partial x_n}{\partial y_1} & \frac{\partial x_n}{\partial y_2} & \cdots & \frac{\partial x_n}{\partial y_n} \end{vmatrix}$$

not be identically zero in $\mathcal{T}$. Then

$$\int \cdots \int_A h(x_1, x_2, \ldots, x_n) \, dx_1 dx_2 \cdots dx_n$$

$$= \int \cdots \int_B h[w_1(y_1, \ldots, y_n), w_2(y_1, \ldots, y_n), \ldots, w_n(y_1, \ldots, y_n)]|J| \, dy_1 dy_2 \cdots dy_n.$$

Whenever the conditions of this theorem are satisfied, we can determine the joint pdf of n functions of n random variables. Appropriate changes of notation in Section 2.2 (to indicate n-space as opposed to 2-space) are all that is needed to show that the joint pdf of the random variables $Y_1 = u_1(X_1, X_2, \ldots, X_n), \ldots, Y_n = u_n(X_1, X_2, \ldots, X_n)$, where the joint pdf of $X_1, \ldots, X_n$ is $h(x_1, \ldots, x_n)$ is given by

$$g(y_1, y_2, \ldots, y_n) = |J| h[w_1(y_1, \ldots, y_n), \ldots, w_n(y_1, \ldots, y_n)],$$

where $(y_1, y_2, \ldots, y_n) \in \mathcal{T}$, and is zero elsewhere.

Example 2.7.1. Let X_1, X_2, X_3 have the joint pdf

$$h(x_1, x_2, x_3) = \begin{cases} 48 x_1 x_2 x_3 & 0 < x_1 < x_2 < x_3 < 1 \\ 0 & \text{elsewhere.} \end{cases} \tag{2.7.1}$$

If $Y_1 = X_1/X_2$, $Y_2 = X_2/X_3$ and $Y_3 = X_3$, then the inverse transformation is given by

$$x_1 = y_1 y_2 y_3, \quad x_2 = y_2 y_3 \text{ and } x_3 = y_3 .$$

The Jacobian is given by

$$J = \begin{vmatrix} y_2 y_3 & y_1 y_3 & y_1 y_2 \\ 0 & y_3 & y_2 \\ 0 & 0 & 1 \end{vmatrix} = y_2 y_3^2.$$

Moreover inequalities defining the support are equivalent to

$$0 < y_1 y_2 y_3, \ y_1 y_2 y_3 < y_2 y_3, \ y_2 y_3 < y_3 \text{ and } y_3 < 1$$

which reduces to the support $\mathcal{T}$ of Y_1, Y_2, Y_3 of

$$\mathcal{T} = \{(y_1, y_2, y_3) : 0 < y_i < 1, i = 1, 2, 3\}.$$

Hence the joint pdf of Y_1, Y_2, Y_3 is

$$\begin{aligned} g(y_1, y_2, y_3) &= 48(y_1 y_2 y_3)(y_2 y_3) y_3 |y_2 y_3^2| \\ &= \begin{cases} 48 y_1 y_2^3 y_3^5 & 0 < y_i < 1, i = 1, 2, 3 \\ 0 & \text{elsewhere.} \end{cases} \end{aligned} \tag{2.7.2}$$

The marginal pdfs are

$$\begin{aligned} g_1(y_1) &= 2y_1, 0 < y_1 < 1, \text{zero elsewhere,} \\ g_2(y_2) &= 4y_2^3, 0 < y_2 < 1, \text{zero elsewhere,} \\ g_2(y_2) &= 6y_3^5, 0 < y_2 < 1, \text{zero elsewhere.} \end{aligned}$$

Because $g(y_1, y_2, y_3) = g_1(y_1) g_2(y_2) g_2(y_2)$, the random variables Y_1, Y_2, Y_3 are mutually independent. ■

Example 2.7.2. Let X_1, X_2, X_3 be iid with common pdf

$$f(x) = \begin{cases} e^{-x} & 0 < x < \infty \\ 0 & \text{elsewhere.} \end{cases}$$

Consequently the joint pdf of X_1, X_2, X_3 is

$$f_{X_1,X_2,X_3}(x_1, x_2, x_3) = \begin{cases} e^{-\sum_{i=1}^{3} x_i} & 0 < x_i < \infty, i = 1, 2, 3 \\ 0 & \text{elsewhere.} \end{cases}$$

Consider the random variables Y_1, Y_2, Y_3 defined by

$$Y_1 = \frac{X_1}{X_1+X_2+X_3}, \; Y_2 = \frac{X_2}{X_1+X_2+X_3} \text{ and } Y_3 = X_1 + X_2 + X_3.$$

Hence, the inverse transformation is given by,

$$x_1 = y_1 y_3, \; x_2 = y_2 y_3, \text{ and } x_3 = y_3 - y_1 y_3 - y_2 y_3,$$

with the Jacobian,

$$J = \begin{vmatrix} y_3 & 0 & y_1 \\ 0 & y_3 & y_2 \\ -y_3 & -y_3 & 1 - y_1 - y_2 \end{vmatrix} = y_3^2.$$

The support of X_1, X_2, X_3 maps onto

$$0 < y_1 y_3 < \infty, \, 0 < y_2 y_3 < \infty, \text{ and } 0 < y_3(1 - y_1 - y_2) < \infty,$$

which is equivalent to the support $\mathcal{T}$ given by

$$\mathcal{T} = \{(y_1, y_2, y_3) : \, 0 < y_1, 0 < y_2, 0 < 1 - y_1 - y_2, 0 < y_3 < \infty\}.$$

Hence the joint pdf of Y_1, Y_2, Y_3 is

$$g(y_1, y_2, y_3) = y_3^2 e^{-y_3}, \quad (y_1, y_2, y_3) \in \mathcal{T}.$$

The marginal pdf of Y_1 is

$$g_1(y_1) = \int_0^{1-y_1} \int_0^{\infty} y_3^2 e^{-y_3} \, dy_3 \, dy_2 = 2(1 - y_1), \quad 0 < y_1 < 1,$$

zero elsewhere. Likewise the marginal pdf of Y_2 is

$$g_2(y_2) = 2(1 - y_2), \quad 0 < y_2 < 1,$$

zero elsewhere, while the pdf of Y_3 is

$$g_3(y_3) = \int_0^1 \int_0^{1-y_1} y_3^2 e^{-y_3} \, dy_2 \, dy_1 = \frac{1}{2} y_3^2 e^{-y_3}, \quad 0 < y_3 < \infty,$$

zero elsewhere. Because $g(y_1, y_2, y_3) \neq g_1(y_1) g_2(y_2) g_3(y_3)$, Y_1, Y_2, Y_3 are dependent random variables.

Now however, that the joint pdf of Y_1 and Y_3 is

$$g_{13}(y_1, y_3) = \int_0^{1-y_1} y_1^2 e^{-y_3} \, dy_2 = (1 - y_1) y_3^2 e^{-y_3}, \quad 0 < y_1 < 1, 0 < y_3 < \infty,$$

zero elsewhere. Hence Y_1 and Y_3 are independent. In a similar manner, Y_2 and Y_3 are also independent. Because the joint pdf of Y_1 and Y_2 is

$$g_{12}(y_1, y_2) = \int_0^\infty y_3^2 e^{-y_3} \, dy_3 = 2, \quad 0 < y_1, 0 < y_2, y_1 + y_2 < 1,$$

zero elsewhere, Y_1 and Y_2 are seen to be dependent. ∎

We now consider some other problems that are encountered when transforming variables. Let X have the Cauchy pdf

$$f(x) = \frac{1}{\pi(1 + x^2)}, \quad -\infty < x < \infty,$$

and let $Y = X^2$. We seek the pdf $g(y)$ of Y. Consider the transformation $y = x^2$. This transformation maps the space of X, $\mathcal{S} = \{x : -\infty < x < \infty\}$, onto $\mathcal{T} = \{y : 0 \leq y < \infty\}$. However, the transformation is not one-to-one. To each $y \in \mathcal{T}$, with the exception of $y = 0$, there corresponds two points $x \in \mathcal{S}$. For example, if $y = 4$, we may have either $x = 2$ or $x = -2$. In such an instance, we represent $\mathcal{S}$ as the union of two disjoint sets A_1 and A_2 such that $y = x^2$ defines a one-to-one transformation that maps each of A_1 and A_2 onto $\mathcal{T}$. If we take A_1 to be $\{x : -\infty < x < 0\}$ and A_2 to be $\{x : 0 \leq x < \infty\}$, we see that A_1 is mapped onto $\{y : 0 < y < \infty\}$ whereas A_2 is mapped onto $\{y : 0 \leq y < \infty\}$, and these sets are not the same. Our difficulty is caused by the fact that $x = 0$ is an element of $\mathcal{S}$. Why, then, do we not return to the Cauchy pdf and take $f(0) = 0$? Then our new $\mathcal{S}$ is $\mathcal{S} = \{-\infty < x < \infty \text{ but } x \neq 0\}$. We then take $A_1 = \{x : -\infty < x < 0\}$ and $A_2 = \{x : 0 < x < \infty\}$. Thus $y = x^2$, with the inverse $x = -\sqrt{y}$, maps A_1 onto $\mathcal{T} = \{y : 0 < y < \infty\}$ and the transformation is one-to-one. Moreover, the transformation $y = x^2$, with inverse $x = \sqrt{y}$, maps A_2 onto $\mathcal{T} = \{y : 0 < y < \infty\}$ and the transformation is one-to-one. Consider the probability $P(Y \in B)$ where $B \subset \mathcal{T}$. Let $A_3 = \{x : x = -\sqrt{y}, y \in B\} \subset A_1$ and let $A_4 = \{x : x = \sqrt{y}, y \in B\} \subset A_2$. Then $Y \in B$ when and only when $X \in A_3$ or $X \in A_4$. Thus we have

$$
\begin{aligned}
P(Y \in B) &= P(X \in A_3) + P(X \in A_4) \\
&= \int_{A_3} f(x) \, dx + \int_{A_4} f(x) \, dx.
\end{aligned}
$$

In the first of these integrals, let $x = -\sqrt{y}$. Thus the Jacobian, say J_1, is $-1/2\sqrt{y}$; furthermore, the set A_3 is mapped onto B. In the second integral let $x = \sqrt{y}$. Thus the Jacobian, say J_2, is $1/2\sqrt{y}$; furthermore, the set A_4 is also mapped onto B.

Finally,

$$
\begin{aligned}
P(Y \in B) &= \int_B f(-\sqrt{y}) \left| -\frac{1}{2\sqrt{y}} \right| dy + \int_B f(\sqrt{y}) \frac{1}{2\sqrt{y}} \, dy \\
&= \int_B [f(-\sqrt{y}) + f(\sqrt{y})] \frac{1}{2\sqrt{y}} \, dy.
\end{aligned}
$$

Hence the pdf of Y is given by

$$
g(y) = \frac{1}{2\sqrt{y}} [f(-\sqrt{y}) + f(\sqrt{y})], \quad y \in \mathcal{T}.
$$

With $f(x)$ the Cauchy pdf we have

$$
g(y) = \begin{cases} \frac{1}{\pi(1+y)\sqrt{y}} & 0 < y < \infty \\ 0 & \text{elsewhere.} \end{cases}
$$

In the preceding discussion of a random variable of the continuous type, we had two inverse functions, $x = -\sqrt{y}$ and $x = \sqrt{y}$. That is why we sought to partition $\mathcal{S}$ (or a modification of $\mathcal{S}$) into two disjoint subsets such that the transformation $y = x^2$ maps each onto the same $\mathcal{T}$. Had there been three inverse functions, we would have sought to partition $\mathcal{S}$ (or a modified form of $\mathcal{S}$) into three disjoint subsets, and so on. It is hoped that this detailed discussion will make the following paragraph easier to read.

Let $h(x_1, x_2, \ldots, x_n)$ be the joint pdf of $X_1, X_2, \ldots, X_n$, which are random variables of the continuous type. Let $\mathcal{S}$ denote the n-dimensional space where this joint pdf $h(x_1, x_2, \ldots, x_n) > 0$, and consider the transformation $y_1 = u_1(x_1, x_2, \ldots, x_n)$, $\ldots, y_n = u_n(x_1, x_2, \ldots, x_n)$, which maps $\mathcal{S}$ onto $\mathcal{T}$ in the $y_1, y_2, \ldots, y_n$ space. To each point of $\mathcal{S}$ there will correspond, of course, only one point in $\mathcal{T}$; but to a point in $\mathcal{T}$ there may correspond more than one point in $\mathcal{S}$. That is, the transformation may not be one-to-one. Suppose, however, that we can represent $\mathcal{S}$ as the union of a finite number, say k, of mutually disjoint sets $A_1, A_2, \ldots, A_k$ so that

$$
y_1 = u_1(x_1, x_2, \ldots, x_n), \ldots, y_n = u_n(x_1, x_2, \ldots, x_n)
$$

define a one-to-one transformation of each A_i onto $\mathcal{T}$. Thus, to each point in $\mathcal{T}$ there will correspond exactly one point in each of $A_1, A_2, \ldots, A_k$. For $i = 1, \ldots, k$, let

$$
x_1 = w_{1i}(y_1, y_2, \ldots, y_n), x_2 = w_{2i}(y_1, y_2, \ldots, y_n), \ldots, x_n = w_{ni}(y_1, y_2, \ldots, y_n),
$$

denote the k groups of n inverse functions, one group for each of these k transformations. Let the first partial derivatives be continuous and let each

$$
J_i = \begin{vmatrix} \frac{\partial w_{1i}}{\partial y_1} & \frac{\partial w_{1i}}{\partial y_2} & \cdots & \frac{\partial w_{1i}}{\partial y_n} \\ \frac{\partial w_{2i}}{\partial y_1} & \frac{\partial w_{2i}}{\partial y_2} & \cdots & \frac{\partial w_{2i}}{\partial y_n} \\ \vdots & \vdots & & \vdots \\ \frac{\partial w_{ni}}{\partial y_1} & \frac{\partial w_{ni}}{\partial y_2} & \cdots & \frac{\partial w_{ni}}{\partial y_n} \end{vmatrix}, \quad i = 1, 2, \ldots, k,
$$

be not identically equal to zero in $\mathcal{T}$. Considering the probability of the union of k mutually exclusive events and by applying the change of variable technique to the probability of each of these events, it can be seen that the joint pdf of $Y_1 = u_1(X_1, X_2, \ldots, X_n)$, $Y_2 = u_2(X_1, X_2, \ldots, X_n), \ldots, Y_n = u_n(X_1, X_2, \ldots, X_n)$, is given by

$$g(y_1, y_2, \ldots, y_n) = \sum_{i=1}^{k} |J_i| h[w_{1i}(y_1, \ldots, y_n), \ldots, w_{ni}(y_1, \ldots, y_n)],$$

provided that $(y_1, y_2, \ldots, y_n) \in \mathcal{T}$, and equals zero elsewhere. The pdf of any Y_i, say Y_1 is then

$$g_1(y_1) = \int_{-\infty}^{\infty} \cdots \int_{-\infty}^{\infty} g(y_1, y_2, \ldots, y_n)\, dy_2 \cdots dy_n.$$

Example 2.7.3. Let X_1 and X_2 have the joint pdf defined over the unit circle given by

$$f(x_1, x_2) = \begin{cases} \frac{1}{\pi} & 0 < x_1^2 + x_2^2 < 1 \\ 0 & \text{elsewhere.} \end{cases}$$

Let $Y_1 = X_1^2 + X_2^2$ and $Y_2 = X_1^2/(X_1^2 + X_2^2)$. Thus, $y_1 y_2 = x_1^2$ and $x_2^2 = y_1(1 - y_2)$. The support S maps onto $\mathcal{T} = \{(y_1, y_2) : 0 < y_i < 1, i = 1, 2\}$. For each ordered pair $(y_1, y_2) \in \mathcal{T}$, there are four points in S given by,

$$(x_1, x_2) \quad \text{such that} \quad x_1 = \sqrt{y_1 y_2} \text{ and } x_2 = \sqrt{y_1(1 - y_2)};$$
$$(x_1, x_2) \quad \text{such that} \quad x_1 = \sqrt{y_1 y_2} \text{ and } x_2 = -\sqrt{y_1(1 - y_2)};$$
$$(x_1, x_2) \quad \text{such that} \quad x_1 = -\sqrt{y_1 y_2} \text{ and } x_2 = \sqrt{y_1(1 - y_2)};$$
$$\text{and } (x_1, x_2) \quad \text{such that} \quad x_1 = -\sqrt{y_1 y_2} \text{ and } x_2 = -\sqrt{y_1(1 - y_2)}.$$

The value of the first Jacobian is

$$J_1 = \begin{vmatrix} \frac{1}{2}\sqrt{y_2/y_1} & \frac{1}{2}\sqrt{y_1/y_2} \\ \frac{1}{2}\sqrt{(1 - y_2)/y_1} & -\frac{1}{2}\sqrt{y_1/(1 - y_2)} \end{vmatrix}$$
$$= \frac{1}{4}\left\{ -\sqrt{\frac{1 - y_2}{y_2}} - \sqrt{\frac{y_2}{1 - y_2}} \right\} = -\frac{1}{4}\frac{1}{\sqrt{y_2(1 - y_2)}}.$$

It is easy to see that the absolute value of each of the four Jacobians equals $1/4\sqrt{y_2(1 - y_2)}$. Hence, the joint pdf of Y_1 and Y_2 is the sum of four terms and can be written as

$$g(y_1, y_2) = 4\frac{1}{\pi}\frac{1}{4\sqrt{y_2(1 - y_2)}} = \frac{1}{\pi\sqrt{y_2(1 - y_2)}}, \quad (y_1, y_2) \in \mathcal{T}.$$

Thus Y_1 and Y_2 are independent random variables by Theorem 2.5.1. ∎

Of course, as in the bivariate case, we can use the mgf technique by noting that if $Y = g(X_1, X_2, \ldots, X_n)$ is a function of the random variables, then the mgf of Y is given by

$$E\left(e^{tY}\right) = \int_{-\infty}^{\infty} \int_{-\infty}^{\infty} \cdots \int_{-\infty}^{\infty} e^{tg(x_1, x_2, \ldots, x_n)} h(x_1, x_2, \ldots, x_n) \, dx_1 dx_2 \cdots dx_n,$$

in the continuous case, where $h(x_1, x_2, \ldots, x_n)$ is the joint pdf. In the discrete case, summations replace the integrals. This procedure is particularly useful in cases in which we are dealing with linear functions of independent random variables.

Example 2.7.4 (Extension of Example 2.2.6). Let X_1, X_2, X_3 be independent random variables with joint pmf

$$p(x_1, x_2, x_3) = \begin{cases} \frac{\mu_1^{x_1} \mu_2^{x_2} \mu_3^{x_3} e^{-\mu_1 - \mu_2 - \mu_3}}{x_1! x_2! x_3!} & x_i = 0, 1, 2, \ldots, i = 1, 2, 3 \\ 0 & \text{elsewhere.} \end{cases}$$

If $Y = X_1 + X_2 + X_3$, the mgf of Y is

$$\begin{aligned} E\left(e^{tY}\right) &= E\left(e^{t(X_1 + X_2 + X_3)}\right) \\ &= E\left(e^{tX_1} e^{tX_2} e^{tX_3}\right) \\ &= E\left(e^{tX_1}\right) E\left(e^{tX_2}\right) E\left(e^{tX_3}\right), \end{aligned}$$

because of the independence of X_1, X_2, X_3. In Example 2.2.6, we found that

$$E\left(e^{tX_i}\right) = \exp\{\mu_i(e^t - 1)\}, \quad i = 1, 2, 3.$$

Hence,

$$E\left(e^{tY}\right) = \exp\{(\mu_1 + \mu_2 + \mu_3)(e^t - 1)\}.$$

This, however, is the mgf of the pmf

$$p_Y(y) = \begin{cases} \frac{(\mu_1 + \mu_2 + \mu_3)^y e^{-(\mu_1 + \mu_2 + \mu_3)}}{y!} & y = 0, 1, 2 \ldots \\ 0 & \text{elsewhere,} \end{cases}$$

so $Y = X_1 + X_2 + X_3$ has this distribution. ∎

Example 2.7.5. Let X_1, X_2, X_3, X_4 be independent random variables with common pdf

$$f(x) = \begin{cases} e^{-x} & x > 0 \\ 0 & \text{elsewhere.} \end{cases}$$

If $Y = X_1 + X_2 + X_3 + X_4$ then, similar to the argument in the last example, the independence of X_1, X_2, X_3, X_4 implies that

$$E\left(e^{tY}\right) = E\left(e^{tX_1}\right) E\left(e^{tX_2}\right) E\left(e^{tX_3}\right) E\left(e^{tX_4}\right).$$

In Section 1.9, we saw that

$$E\left(e^{tX_i}\right) = (1 - t)^{-1}, \quad t < 1, \ i = 1, 2, 3, 4.$$

Hence,

$$E\left(e^{tY}\right) = (1-t)^{-4}.$$

In Section 3.3, we find that this is the mgf of a distribution with pdf

$$f_Y(y) = \begin{cases} \frac{1}{3!}y^3 e^{-y} & 0 < y < \infty \\ 0 & \text{elsewhere.} \end{cases}$$

Accordingly, Y has this distribution. ∎

EXERCISES

2.7.1. Let X_1, X_2, X_3 be iid, each with the distribution having pdf $f(x) = e^{-x}$, $0 < x < \infty$, zero elsewhere. Show that

$$Y_1 = \frac{X_1}{X_1 + X_2}, \quad Y_2 = \frac{X_1 + X_2}{X_1 + X_2 + X_3}, \quad Y_3 = X_1 + X_2 + X_3$$

are mutually independent.

2.7.2. If $f(x) = \frac{1}{2}$, $-1 < x < 1$, zero elsewhere, is the pdf of the random variable X, find the pdf of $Y = X^2$.

2.7.3. If X has the pdf of $f(x) = \frac{1}{4}$, $-1 < x < 3$, zero elsewhere, find the pdf of $Y = X^2$.
Hint: Here $\mathcal{T} = \{y : 0 \le y < 9\}$ and the event $Y \in B$ is the union of two mutually exclusive events if $B = \{y : 0 < y < 1\}$.

2.7.4. Let X_1, X_2, X_3 be iid with common pdf $f(x) = e^{-x}$, $x > 0$, 0 elsewhere. Find the joint pdf of $Y_1 = X_1$, $Y_2 = X_1 + X_2$, and $Y_3 = X_1 + X_2 + X_3$.

2.7.5. Let X_1, X_2, X_3 be iid with common pdf $f(x) = e^{-x}$, $x > 0$, 0 elsewhere. Find the joint pdf of $Y_1 = X_1/X_2$, $Y_2 = X_3/(X_1 + X_2)$, and $Y_3 = X_1 + X_2$. Are Y_1, Y_2, Y_3 mutually independent?

2.7.6. Let X_1, X_2 have the joint pdf $f(x_1, x_2) = 1/\pi$, $0 < x_1^2 + x_2^2 < 1$. Let $Y_1 = X_1^2 + X_2^2$ and $Y_2 = X_2$. Find the joint pdf of Y_1 and Y_2.

2.7.7. Let X_1, X_2, X_3, X_4 have the joint pdf $f(x_1, x_2, x_3, x_4) = 24$, $0 < x_1 < x_2 < x_3 < x_4 < 1$, 0 elsewhere. Find the joint pdf of $Y_1 = X_1/X_2$, $Y_2 = X_2/X_3, Y_3 = X_3/X_4, Y_4 = X_4$ and show that they are mutually independent.

2.7.8. Let X_1, X_2, X_3 be iid with common mgf $M(t) = ((3/4) + (1/4)e^t)^2$, for all $t \in R$.

(a) Determine the probabilities, $P(X_1 = k), k = 0, 1, 2$.

(b) Find the mgf of $Y = X_1 + X_2 + X_3$ and then determine the probabilities, $P(Y = k), k = 0, 1, 2, \ldots, 6$.

Chapter 3

Some Special Distributions

3.1 The Binomial and Related Distributions

In Chapter 1 we introduced the *uniform distribution* and the *hypergeometric distribution*. In this chapter we discuss some other important distributions of random variables frequently used in statistics. We begin with the binomial and related distributions.

A **Bernoulli experiment** is a random experiment, the outcome of which can be classified in but one of two mutually exclusive and exhaustive ways, for instance, success or failure (e.g., female or male, life or death, nondefective or defective). A sequence of **Bernoulli trials** occurs when a Bernoulli experiment is performed several independent times so that the probability of success, say p, remains the same from trial to trial. That is, in such a sequence, we let p denote the probability of success on each trial.

Let X be a random variable associated with a Bernoulli trial by defining it as follows:

$$X(\text{success}) = 1 \quad \text{and} \quad X(\text{failure}) = 0.$$

That is, the two outcomes, success and failure, are denoted by one and zero, respectively. The pmf of X can be written as

$$p(x) = p^x (1-p)^{1-x}, \quad x = 0, 1, \tag{3.1.1}$$

and we say that X has a *Bernoulli distribution*. The expected value of X is

$$\mu = E(X) = \sum_{x=0}^{1} x p^x (1-p)^{1-x} = (0)(1-p) + (1)(p) = p,$$

and the variance of X is

$$\sigma^2 = \text{var}(X) = \sum_{x=0}^{1} (x-p)^2 p^x (1-p)^{1-x}$$
$$= p^2(1-p) + (1-p)^2 p = p(1-p).$$

It follows that the standard deviation of X is $\sigma = \sqrt{p(1-p)}$.

In a sequence of n Bernoulli trials, we shall let X_i denote the Bernoulli random variable associated with the ith trial. An observed sequence of n Bernoulli trials will then be an n-tuple of zeros and ones. In such a sequence of Bernoulli trials, we are often interested in the total number of successes and not in the order of their occurrence. If we let the random variable X equal the number of observed successes in n Bernoulli trials, the possible values of X are $0, 1, 2, \ldots, n$. If x successes occur, where $x = 0, 1, 2, \ldots, n$, then $n - x$ failures occur. The number of ways of selecting the x positions for the x successes in the n trials is

$$\binom{n}{x} = \frac{n!}{x!(n-x)!}.$$

Since the trials are independent and the probabilities of success and failure on each trial are, respectively, p and $1 - p$, the probability of each of these ways is $p^x(1-p)^{n-x}$. Thus the pmf of X, say $p(x)$, is the sum of the probabilities of these $\binom{n}{x}$ mutually exclusive events; that is,

$$p(x) = \begin{cases} \binom{n}{x} p^x (1-p)^{n-x} & x = 0, 1, 2, \ldots, n \\ 0 & \text{elsewhere.} \end{cases}$$

Recall, if n is a positive integer, that

$$(a+b)^n = \sum_{x=0}^{n} \binom{n}{x} b^x a^{n-x}.$$

Thus it is clear that $p(x) \geq 0$ and that

$$\begin{aligned} \sum_x p(x) &= \sum_{x=0}^{n} \binom{n}{x} p^x (1-p)^{n-x} \\ &= [(1-p)+p]^n = 1. \end{aligned}$$

Therefore, $p(x)$ satisfies the conditions of being a pmf of a random variable X of the discrete type. A random variable X that has a pmf of the form of $p(x)$ is said to have a **binomial distribution**, and any such $p(x)$ is called a **binomial pmf**. A binomial distribution will be denoted by the symbol $b(n, p)$. The constants n and p are called the **parameters** of the binomial distribution. Thus, if we say that X is $b(5, \frac{1}{3})$, we mean that X has the binomial pmf

$$p(x) = \begin{cases} \binom{5}{x} \left(\frac{1}{3}\right)^x \left(\frac{2}{3}\right)^{5-x} & x = 0, 1, \ldots, 5 \\ 0 & \text{elsewhere.} \end{cases} \tag{3.1.2}$$

The mgf of a binomial distribution is easily obtained as follows,

$$\begin{aligned} M(t) &= \sum_x e^{tx} p(x) = \sum_{x=0}^{n} e^{tx} \binom{n}{x} p^x (1-p)^{n-x} \\ &= \sum_{x=0}^{n} \binom{n}{x} (pe^t)^x (1-p)^{n-x} \\ &= [(1-p) + pe^t]^n \end{aligned}$$

for all real values of t. The mean μ and the variance σ^2 of X may be computed from $M(t)$. Since

$$M'(t) = n[(1-p) + pe^t]^{n-1}(pe^t)$$

and

$$M''(t) = n[(1-p+pe^t]^{n-1}(pe^t) + n(n-1)[(1-p) + pe^t]^{n-2}(pe^t)^2,$$

if follows that

$$\mu = M'(0) = np$$

and

$$\sigma^2 = M''(0) - \mu^2 = np + n(n-1)p^2 - (np)^2 = np(1-p).$$

Example 3.1.1. Let X be the number of heads (successes) in $n = 7$ independent tosses of an unbiased coin. The pmf of X is

$$p(x) = \begin{cases} \binom{7}{x} \left(\frac{1}{2}\right)^x \left(1 - \frac{1}{2}\right)^{7-x} & x = 0, 1, 2, \ldots, 7 \\ 0 & \text{elsewhere.} \end{cases}$$

Then X has the mgf

$$M(t) = \left(\tfrac{1}{2} + \tfrac{1}{2}e^t\right)^7,$$

has mean $\mu = np = \frac{7}{2}$, and has variance $\sigma^2 = np(1-p) = \frac{7}{4}$. Furthermore, we have

$$P(0 \le X \le 1) = \sum_{x=0}^{1} p(x) = \frac{1}{128} + \frac{7}{128} = \frac{8}{128}$$

and

$$P(X = 5) = p(5) = \frac{7!}{5!2!} \left(\frac{1}{2}\right)^5 \left(\frac{1}{2}\right)^2 = \frac{21}{128}. \quad \blacksquare$$

Most computer packages have commands which obtain the binomial probabilities. To give the R (Ihaka and Gentleman, 1996) or S-PLUS (S-PLUS, 2000) commands, suppose X has a $b(n,p)$ distribution. Then the command `dbinom(k,n,p)` returns $P(X = k)$, while the command `pbinom(k,n,p)` returns the cumulative probability $P(X \le k)$.

Example 3.1.2. If the mgf of a random variable X is

$$M(t) = \left(\tfrac{2}{3} + \tfrac{1}{3}e^t\right)^5,$$

then X has a binomial distribution with $n = 5$ and $p = \frac{1}{3}$; that is, the pmf of X is

$$p(x) = \begin{cases} \binom{5}{x} \left(\frac{1}{3}\right)^x \left(\frac{2}{3}\right)^{5-x} & x = 0, 1, 2, \ldots, 5 \\ 0 & \text{elsewhere.} \end{cases}$$

Here $\mu = np = \frac{5}{3}$ and $\sigma^2 = np(1-p) = \frac{10}{9}$. $\blacksquare$

Example 3.1.3. If Y is $b(n, \frac{1}{3})$, then $P(Y \geq 1) = 1 - P(Y = 0) = 1 - (\frac{2}{3})^n$. Suppose that we wish to find the smallest value of n that yields $P(Y \geq 1) > 0.80$. We have $1 - (\frac{2}{3})^n > 0.80$ and $0.20 > (\frac{2}{3})^n$. Either by inspection or by use of logarithms, we see that $n = 4$ is the solution. That is, the probability of at least one success throughout $n = 4$ independent repetitions of a random experiment with probability of success $p = \frac{1}{3}$ is greater than 0.80. ∎

Example 3.1.4. Let the random variable Y be equal to the number of successes throughout n independent repetitions of a random experiment with probability p of success. That is, Y is $b(n, p)$. The ratio Y/n is called the relative frequency of success. Recall expression (1.10.3), the second version of Chebyshev's inequality (Theorem 1.10.3). Applying this result, we have for all $\epsilon > 0$ that

$$P\left(\left|\frac{Y}{n} - p\right| \geq \epsilon\right) \leq \frac{\text{Var}(Y/n)}{\epsilon^2} = \frac{p(1-p)}{n\epsilon^2}.$$

Now, for every fixed $\epsilon > 0$, the right-hand member of the preceding inequality is close to zero for sufficiently large n. That is,

$$\lim_{n \to \infty} P\left(\left|\frac{Y}{n} - p\right| \geq \epsilon\right) = 0$$

and

$$\lim_{n \to \infty} P\left(\left|\frac{Y}{n} - p\right| < \epsilon\right) = 1.$$

Since this is true for every fixed $\epsilon > 0$, we see, in a certain sense, that the relative frequency of success is for large values of n, close to the probability of p of success. This result is one form of the *Weak Law of Large Numbers*. It was alluded to in the initial discussion of probability in Chapter 1 and will be considered again, along with related concepts, in Chapter 4. ∎

Example 3.1.5. Let the independent random variables X_1, X_2, X_3 have the same cdf $F(x)$. Let Y be the middle value of X_1, X_2, X_3. To determine the cdf of Y, say $F_Y(y) = P(Y \leq y)$, we note that $Y \leq y$ if and only if at least two of the random variables X_1, X_2, X_3 are less than or equal to y. Let us say that the ith "trial" is a success if $X_i \leq y$, $i = 1, 2, 3$; here each "trial" has the probability of success $F(y)$. In this terminology, $F_Y(y) = P(Y \leq y)$ is then the probability of at least two successes in three independent trials. Thus

$$F_Y(y) = \binom{3}{2}[F(y)]^2[1 - F(y)] + [F(y)]^3.$$

If $F(x)$ is a continuous cdf so that the pdf of X is $F'(x) = f(x)$, then the pdf of Y is

$$f_Y(y) = F_Y'(y) = 6[F(y)][1 - F(y)]f(y). \quad ∎$$

Example 3.1.6. Consider a sequence of independent repetitions of a random experiment with constant probability p of success. Let the random variable Y denote the total number of failures in this sequence before the rth success, that is,

$Y + r$ is equal to the number of trials necessary to produce exactly r successes. Here r is a fixed positive integer. To determine the pmf of Y, let y be an element of $\{y : y = 0, 1, 2, \ldots\}$. Then, by the multiplication rule of probabilities, $P(Y = y) = g(y)$ is equal to the product of the probability

$$\binom{y + r - 1}{r - 1} p^{r-1}(1 - p)^y$$

of obtaining exactly $r - 1$ successes in the first $y + r - 1$ trials and the probability p of a success on the $(y + r)$th trial. Thus the pmf of Y is

$$p_Y(y) = \begin{cases} \binom{y+r-1}{r-1} p^r (1 - p)^y & y = 0, 1, 2, \ldots \\ 0 & \text{elsewhere.} \end{cases} \tag{3.1.3}$$

A distribution with a pmf of the form $p_Y(y)$ is called a **negative binomial distribution**; and any such $p_Y(y)$ is called a negative binomial pmf. The distribution derives its name from the fact that $p_Y(y)$ is a general term in the expansion of $p^r[1 - (1 - p)]^{-r}$. It is left as an exercise to show that the mgf of this distribution is $M(t) = p^r[1 - (1 - p)e^t]^{-r}$, for $t < -\ln(1 - p)$. If $r = 1$, then Y has the pmf

$$p_Y(y) = p(1 - p)^y, \quad y = 0, 1, 2, \ldots, \tag{3.1.4}$$

zero elsewhere, and the mgf $M(t) = p[1 - (1 - p)e^t]^{-1}$. In this special case, $r = 1$, we say that Y has a **geometric distribution** of the form. ∎

Suppose we have several independent binomial distributions with the same probability of success. Then it makes sense that the sum of these random variables is binomial, as shown in the following theorem. Note that the mgf technique gives a quick and easy proof.

Theorem 3.1.1. *Let $X_1, X_2, \ldots, X_m$ be independent random variables such that X_i has binomial $b(n_i, p)$ distribution, for $i = 1, 2, \ldots, m$. Let $Y = \sum_{i=1}^m X_i$. Then Y has a binomial $b(\sum_{i=1}^m n_i, p)$ distribution.*

Proof: Using independence of the X_is and the mgf of X_i, we obtain the mgf of Y as follows:

$$\begin{aligned} M_Y(t) &= E\left[\exp\{\sum_{i=1}^m t X_i\}\right] = E\left[\prod_{i=1}^m \exp\{t X_i\}\right] \\ &= \prod_{i=1}^m E\left[\exp\{t X_i\}\right] = \prod_{i=1}^m (1 - p + p e^t)^{n_i} = (1 - p + p e^t)^{\sum_{i=1}^m n_i}. \end{aligned}$$

Hence, Y has a binomial $b(\sum_{i=1}^m n_i, p)$ distribution. ∎

The binomial distribution is generalized to the multinomial distribution as follows. Let a random experiment be repeated n independent times. On each repetition, the experiment results in but one of k mutually exclusive and exhaustive ways, say $C_1, C_2, \ldots, C_k$. Let p_i be the probability that the outcome is an element of C_i

and let p_i remain constant throughout the n independent repetitions, $i = 1, 2, \ldots, k$. Define the random variable X_i to be equal to the number of outcomes that are elements of C_i, $i = 1, 2, \ldots, k-1$. Furthermore, let $x_1, x_2, \ldots, x_{k-1}$ be nonnegative integers so that $x_1 + x_2 + \cdots + x_{k-1} \le n$. Then the probability that exactly x_1 terminations of the experiment are in $C_1, \ldots$, exactly x_{k-1} terminations are in C_{k-1}, and hence exactly $n - (x_1 + \cdots + x_{k-1})$ terminations are in C_k is

$$\frac{n!}{x_1! \cdots x_{k-1}! x_k!} p_1^{x_1} \cdots p_{k-1}^{x_{k-1}} p_k^{x_k},$$

where x_k is merely an abbreviation for $n - (x_1 + \cdots + x_{k-1})$. This is the **multinomial pmf** of $k-1$ random variables $X_1, X_2, \ldots, X_{k-1}$ of the discrete type. To see that this is correct, note that the number of distinguishable arrangements of $x_1 \, C_1$s, $x_2 \, C_2$s, $\ldots, x_k \, C_k$s is

$$\binom{n}{x_1} \binom{n - x_1}{x_2} \cdots \binom{n - x_1 - \cdots - x_{k-2}}{x_{k-1}} = \frac{n!}{x_1! x_2! \cdots x_k!}$$

and the probability of each of these distinguishable arrangements is

$$p_1^{x_1} p_2^{x_2} \cdots p_k^{x_k}.$$

Hence the product of these two latter expressions gives the correct probability, which is an agreement with the formula for the multinomial pmf.

When $k = 3$, we often let $X = X_1$ and $Y = X_2$; then $n - X - Y = X_3$. We say that X and Y have a **trinomial distribution**. The joint pmf of X and Y is

$$p(x, y) = \frac{n!}{x! y! (n - x - y)!} p_1^x p_2^y p_3^{n-x-y},$$

where x and y are nonnegative integers with $x + y \le n$, and p_1, p_2, and p_3 are positive proper fractions with $p_1 + p_2 + p_3 = 1$; and let $p(x, y) = 0$ elsewhere. Accordingly, $p(x, y)$ satisfies the conditions of being a joint pmf of two random variables X and Y of the discrete type; that is, $p(x, y)$ is nonnegative and its sum over all points (x, y) at which $p(x, y)$ is positive is equal to $(p_1 + p_2 + p_3)^n = 1$.

If n is a positive integer and a_1, a_2, a_3 are fixed constants, we have

$$\sum_{x=0}^{n} \sum_{y=0}^{n-x} \frac{n!}{x! y! (n - x - y)!} a_1^x a_2^y a_3^{n-x-y}$$

$$= \sum_{x=0}^{n} \frac{n! a_1^x}{x! (n - x)!} \sum_{y=0}^{n-x} \frac{(n - x)!}{y! (n - x - y)!} a_2^y a_3^{n-x-y}$$

$$= \sum_{x=0}^{n} \frac{n!}{x! (n - x)!} a_1^x (a_2 + a_3)^{n-x}$$

$$= (a_1 + a_2 + a_3)^n. \tag{3.1.5}$$

Consequently, the mgf of a trinomial distribution, in accordance with Equation (3.1.5), is given by

$$
\begin{aligned}
M(t_1, t_2) &= \sum_{x=0}^{n} \sum_{y=0}^{n-x} \frac{n!}{x!y!(n-x-y)!} (p_1 e^{t_1})^x (p_2 e^{t_2})^y p_3^{n-x-y} \\
&= (p_1 e^{t_1} + p_2 e^{t_2} + p_3)^n,
\end{aligned}
$$

for all real values of t_1 and t_2. The moment-generating functions of the marginal distributions of X and Y are, respectively,

$$
M(t_1, 0) = (p_1 e^{t_1} + p_2 + p_3)^n = [(1 - p_1) + p_1 e^{t_1}]^n
$$

and

$$
M(0, t_2) = (p_1 + p_2 e^{t_2} + p_3)^n = [(1 - p_2) + p_2 e^{t_2}]^n.
$$

We see immediately, from Theorem 2.5.5 that X and Y are dependent random variables. In addition, X is $b(n, p_1)$ and Y is $b(n, p_2)$. Accordingly, the means and variances of X and Y are, respectively, $\mu_1 = np_1$, $\mu_2 = np_2$, $\sigma_1^2 = np_1(1 - p_1)$, and $\sigma_2^2 = np_2(1 - p_2)$.

Consider next the conditional pmf of Y, given $X = x$. We have

$$
p_{2|1}(y|x) = \begin{cases} \frac{(n-x)!}{y!(n-x-y)!} \left(\frac{p_2}{1-p_1}\right)^y \left(\frac{p_3}{1-p_1}\right)^{n-x-y} & y = 0, 1, \ldots, n-x \\ 0 & \text{elsewhere.} \end{cases}
$$

Thus the conditional distribution of Y, given $X = x$, is $b[n - x, p_2/(1 - p_1)]$. Hence the conditional mean of Y, given $X = x$, is the linear function

$$
E(Y|x) = (n - x) \left(\frac{p_2}{1 - p_1}\right).
$$

Also, the conditional distribution of X, given $Y = y$, is $b[n - y, p_1/(1 - p_2)]$ and thus

$$
E(X|y) = (n - y) \left(\frac{p_1}{1 - p_2}\right).
$$

Now recall from Example 2.4.2 that the square of the correlation coefficient ρ^2 is equal to the product of $-p_2/(1 - p_1)$ and $-p_1/(1 - p_2)$, the coefficients of x and y in the respective conditional means. Since both of these coefficients are negative (and thus ρ is negative), we have

$$
\rho = -\sqrt{\frac{p_1 p_2}{(1 - p_1)(1 - p_2)}}.
$$

In general, the mgf of a multinomial distribution is given by

$$
M(t_1, \ldots, t_{k-1}) = (p_1 e^{t_1} + \cdots + p_{k-1} e^{t_{k-1}} + p_k)^n
$$

for all real values of $t_1, t_2, \ldots, t_{k-1}$. Thus each one-variable marginal pmf is binomial, each two-variable marginal pmf is trinomial, and so on.

EXERCISES

3.1.1. If the mgf of a random variable X is $(\frac{1}{3} + \frac{2}{3}e^t)^5$, find $P(X = 2 \text{ or } 3)$.

3.1.2. The mgf of a random variable X is $(\frac{2}{3} + \frac{1}{3}e^t)^9$. Show that

$$P(\mu - 2\sigma < X < \mu + 2\sigma) = \sum_{x=1}^{5} \binom{9}{x} \left(\frac{1}{3}\right)^x \left(\frac{2}{3}\right)^{9-x}.$$

3.1.3. If X is $b(n, p)$, show that

$$E\left(\frac{X}{n}\right) = p \quad \text{and} \quad E\left[\left(\frac{X}{n} - p\right)^2\right] = \frac{p(1-p)}{n}.$$

3.1.4. Let the independent random variables X_1, X_2, X_3 have the same pdf $f(x) = 3x^2$, $0 < x < 1$, zero elsewhere. Find the probability that exactly two of these three variables exceed $\frac{1}{2}$.

3.1.5. Let Y be the number of successes in n independent repetitions of a random experiment having the probability of success $p = \frac{2}{3}$. If $n = 3$, compute $P(2 \leq Y)$; if $n = 5$, compute $P(3 \leq Y)$.

3.1.6. Let Y be the number of successes throughout n independent repetitions of a random experiment have probability of success $p = \frac{1}{4}$. Determine the smallest value of n so that $P(1 \leq Y) \geq 0.70$.

3.1.7. Let the independent random variables X_1 and X_2 have binomial distribution with parameters $n_1 = 3$, $p = \frac{2}{3}$ and $n_2 = 4$, $p = \frac{1}{2}$, respectively. Compute $P(X_1 = X_2)$.
Hint: List the four mutually exclusive ways that $X_1 = X_2$ and compute the probability of each.

3.1.8. For this exercise, the reader must have access to a statistical package that obtains the binomial distribution. Hints are given for R or S-PLUS code but other packages can be used too.

(a) Obtain the plot of the pmf for the $b(15, 0.2)$ distribution. Using either R or S-PLUS, the folllowing commands will return the plot:

```
x<-0:15
y<-dbinom(x,15,.2)
plot(x,y).
```

(b) Repeat Part (a) for the binomial distributions with $n = 15$ and with $p = 0.10, 0.20, \ldots, 0.90$. Comment on the plots.

(c) Obtain the plots of the pmfs with $p = .05$ and $n = 10, 20, 50, 200$. Comment on the plots, (What do the plots seem to be converging to?).

3.1.9. Toss two nickels and three dimes at random. Make appropriate assumptions and compute the probability that there are more heads showing on the nickels than on the dimes.

3.1.10. Let $X_1, X_2, \ldots, X_{k-1}$ have a multinomial distribution.

(a) Find the mgf of $X_2, X_3, \ldots, X_{k-1}$.

(b) What is the pmf of $X_2, X_3, \ldots, X_{k-1}$?

(c) Determine the conditional pmf of X_1 given that $X_2 = x_2, \ldots, X_{k-1} = x_{k-1}$.

(d) What is the conditional expectation $E(X_1 | x_2, \ldots, x_{k-1})$?

3.1.11. Let X be $b(2, p)$ and let Y be $b(4, p)$. If $P(X \geq 1) = \frac{5}{9}$, find $P(Y \geq 1)$.

3.1.12. If $x = r$ is the unique mode of a distribution that is $b(n, p)$, show that

$$(n+1)p - 1 < r < (n+1)p.$$

Hint: Determine the values of x for which the ratio $f(x+1)/f(x) > 1$.

3.1.13. Let X have a binomial distribution with parameters n and $p = \frac{1}{3}$. Determine the smallest integer n can be such that $P(X \geq 1) \geq 0.85$.

3.1.14. Let X have the pmf $p(x) = (\frac{1}{3})(\frac{2}{3})^x$, $x = 0, 1, 2, 3, \ldots$, zero elsewhere. Find the conditional pmf of X given that $X \geq 3$.

3.1.15. One of the numbers $1, 2, \ldots, 6$ is to be chosen by casting an unbiased die. Let this random experiment be repeated five independent times. Let the random variable X_1 be the number of terminations in the set $\{x : x = 1, 2, 3\}$ and let the random variable X_2 be the number of terminations in the set $\{x : x = 4, 5\}$. Compute $P(X_1 = 2, X_2 = 1)$.

3.1.16. Show that the moment generating function of the negative binomial distribution is $M(t) = p^r[1 - (1-p)e^t]^{-r}$. Find the mean and the variance of this distribution.
Hint: In the summation representing $M(t)$, make use of the MacLaurin's series for $(1 - w)^{-r}$.

3.1.17. Let X_1 and X_2 have a trinomial distribution. Differentiate the moment-generating function to show that their covariance is $-np_1 p_2$.

3.1.18. If a fair coin is tossed at random five independent times, find the conditional probability of five heads given that there are at least four heads.

3.1.19. Let an unbiased die be cast at random seven independent times. Compute the conditional probability that each side appears at least once given that side 1 appears exactly twice.

3.1.20. Compute the measures of skewness and kurtosis of the binomial distribution $b(n, p)$.

3.1.21. Let

$$p(x_1, x_2) = \binom{x_1}{x_2}\left(\frac{1}{2}\right)^{x_1}\left(\frac{x_1}{15}\right), \quad \begin{matrix} x_2 = 0, 1, \ldots, x_1, \\ x_1 = 1, 2, 3, 4, 5, \end{matrix}$$

zero elsewhere, be the joint pmf of X_1 and X_2. Determine:

(a) $E(X_2)$.

(b) $u(x_1) = E(X_2|x_1)$.

(c) $E[u(X_1)]$.

Compare the answers of Parts (a) and (c).
Hint: Note that $E(X_2) = \sum_{x_1=1}^{5}\sum_{x_2=0}^{x_1} x_2 p(x_1, x_2)$.

3.1.22. Three fair dice are cast. In 10 independent casts, let X be the number of times all three faces are alike and let Y be the number of times only two faces are alike. Find the joint pmf of X and Y and compute $E(6XY)$.

3.1.23. Let X have a geometric distribution. Show that

$$P(X \geq k + j \mid X \geq k) = P(X \geq j), \qquad (3.1.6)$$

where k and j are nonnegative integers. Note that we sometimes say in this situation that X is *memoryless*.

3.1.24. Let X equal the number of independent tosses of a fair coin that are required to observe heads on consecutive tosses. Let u_n equal the nth Fibonacci number, where $u_1 = u_2 = 1$ and $u_n = u_{n-1} + u_{n-2}$, $n = 3, 4, 5, \ldots$.

(a) Show that the pmf of X is

$$p(x) = \frac{u_{x-1}}{2^x}, \quad x = 2, 3, 4, \ldots.$$

(b) Use the fact that

$$u_n = \frac{1}{\sqrt{5}}\left[\left(\frac{1+\sqrt{5}}{2}\right)^n - \left(\frac{1-\sqrt{5}}{2}\right)^n\right]$$

to show that $\sum_{x=2}^{\infty} p(x) = 1$.

3.1.25. Let the independent random variables X_1 and X_2 have binomial distributions with parameters n_1, $p_1 = \frac{1}{2}$ and n_2, $p_2 = \frac{1}{2}$, respectively. Show that $Y = X_1 - X_2 + n_2$ has a binomial distribution with parameters $n = n_1 + n_2$, $p = \frac{1}{2}$.

3.2 The Poisson Distribution

Recall that the series

$$1 + m + \frac{m^2}{2!} + \frac{m^3}{3!} + \cdots = \sum_{x=0}^{\infty} \frac{m^x}{x!}$$

converges, for all values of m, to e^m. Consider the function $p(x)$ defined by

$$p(x) = \begin{cases} \frac{m^x e^{-m}}{x!} & x = 0, 1, 2, \ldots \\ 0 & \text{elsewhere,} \end{cases} \tag{3.2.1}$$

where $m > 0$. Since $m > 0$, then $p(x) \geq 0$ and

$$\sum_x p(x) = \sum_{x=0}^{\infty} \frac{m^x e^{-m}}{x!} = e^{-m} \sum_{x=0}^{\infty} \frac{m^x}{x!} = e^{-m} e^m = 1;$$

that is, $p(x)$ satisfies the conditions of being a pmf of a discrete type of random variable. A random variable that has a pmf of the form $p(x)$ is said to have a **Poisson distribution** with parameter m, and any such $p(x)$ is called a **Poisson pmf** with parameter m.

Remark 3.2.1. Experience indicates that the Poisson pmf may be used in a number of applications with quite satisfactory results. For example, let the random variable X denote the number of alpha particles emitted by a radioactive substance that enter a prescribed region during a prescribed interval of time. With a suitable value of m, it is found that X may be assumed to have a Poisson distribution. Again let the random variable X denote the number of defects on a manufactured article, such as a refrigerator door. Upon examining many of these doors, it is found, with an appropriate value of m, that X may be said to have a Poisson distribution. The number of automobile accidents in a unit of time (or the number of insurance claims in some unit of time) is often assumed to be a random variable which has a Poisson distribution. Each of these instances can be thought of as a process that generates a number of changes (accidents, claims, etc.) in a fixed interval (of time or space, etc.). If a process leads to a Poisson distribution, that process is called a *Poisson process*. Some assumptions that ensure a Poisson process will now be enumerated.

Let $g(x, w)$ denote the probability of x changes in each interval of length w. Furthermore, let the symbol $o(h)$ represent any function such that $\lim_{h \to 0} [o(h)/h] = 0$; for example, $h^2 = o(h)$ and $o(h) + o(h) = o(h)$. The Poisson postulates are the following:

1. $g(1, h) = \lambda h + o(h)$, where λ is a positive constant and $h > 0$.

2. $\sum_{x=2}^{\infty} g(x, h) = o(h)$.

3. The number of changes in nonoverlapping intervals are independent.

Postulates 1 and 3 state, in effect, that the probability of one change in a short interval h is independent of changes in other nonoverlapping intervals and is approximately proportional to the length of the interval. The substance of postulate 2 is that the probability of two or more changes in the same short interval h is essentially equal to zero. If $x = 0$, we take $g(0,0) = 1$. In accordance with postulates 1 and 2, the probability of at least one change in an interval h is $\lambda h + o(h) + o(h) = \lambda h + o(h)$. Hence the probability of zero changes in this interval of length h is $1 - \lambda h - o(h)$. Thus the probability $g(0, w + h)$ of zero changes in an interval of length $w + h$ is, in accordance with postulate 3, equal to the product of the probability $g(0, w)$ of zero changes in an interval of length w and the probability $[1 - \lambda h - o(h)]$ of zero changes in a nonoverlapping interval of length h. That is,

$$g(0, w + h) = g(0, w)[1 - \lambda h - o(h)].$$

Then

$$\frac{g(0, w + h) - g(0, w)}{h} = -\lambda g(0, w) - \frac{o(h)g(0, w)}{h}.$$

If we take the limit as $h \to 0$, we have

$$D_w[g(0, w)] = -\lambda g(0, w). \tag{3.2.2}$$

The solution of this differential equation is

$$g(0, w) = ce^{-\lambda w};$$

that is, the function $g(0, w) = ce^{-\lambda w}$ satisfies equation (3.2.2). The condition $g(0, 0) = 1$ implies that $c = 1$; thus

$$g(0, w) = e^{-\lambda w}.$$

If x is a positive integer, we take $g(x, 0) = 0$. The postulates imply that

$$g(x, w + h) = [g(x, w)][1 - \lambda h - o(h)] + [g(x - 1, w)][\lambda h + o(h)] + o(h).$$

Accordingly, we have

$$\frac{g(x, w + h) - g(x, w)}{h} = -\lambda g(x, w) + \lambda g(x - 1, w) + \frac{o(h)}{h}$$

and

$$D_w[g(x, w)] = -\lambda g(x, w) + \lambda g(x - 1, w),$$

for $x = 1, 2, 3, \ldots$. It can be shown, by mathematical induction, that the solutions to these differential equations, with boundary conditions $g(x, 0) = 0$ for $x = 1, 2, 3, \ldots$, are, respectively,

$$g(x, w) = \frac{(\lambda w)^x e^{-\lambda w}}{x!}, \quad x = 1, 2, 3, \ldots.$$

Hence the number of changes in X in an interval of length w has a Poisson distribution with parameter $m = \lambda w$. ∎

The mgf of a Poisson distribution is given by

$$M(t) = \sum_x e^{tx} p(x) = \sum_{x=0}^{\infty} e^{tx} \frac{m^x e^{-m}}{x!}$$

$$= e^{-m} \sum_{x=0}^{\infty} \frac{(me^t)^x}{x!}$$

$$= e^{-m} e^{me^t} = e^{m(e^t - 1)}$$

for all real values of t. Since

$$M'(t) = e^{m(e^t - 1)}(me^t)$$

and

$$M''(t) = e^{m(e^t - 1)}(me^t) + e^{m(e^t - 1)}(me^t)^2,$$

then

$$\mu = M'(0) = m$$

and

$$\sigma^2 = M''(0) - \mu^2 = m + m^2 - m^2 = m.$$

That is, a Poisson distribution has $\mu = \sigma^2 = m > 0$. On this account, a Poisson pmf is frequently written

$$p(x) = \begin{cases} \frac{\mu^x e^{-\mu}}{x!} & x = 0, 1, 2, \ldots \\ 0 & \text{elsewhere.} \end{cases}$$

Thus the parameter m in a Poisson pmf is the mean μ. Table I in Appendix C gives approximately the distribution for various values of the parameter $m = \mu$. On the other hand, if X has a Poisson distribution with parameter $m = \mu$ then the R or S-PLUS command dpois(k,m) returns the value that $P(X = k)$. The cumulative probability $P(X \leq k)$ is given by ppois(k,m).

Example 3.2.1. Suppose that X has a Poisson distribution with $\mu = 2$. Then the pmf of X is

$$p(x) = \begin{cases} \frac{2^x e^{-2}}{x!} & x = 0, 1, 2, \ldots \\ 0 & \text{elsewhere.} \end{cases}$$

The variance of this distribution is $\sigma^2 = \mu = 2$. If we wish to compute $P(1 \leq X)$, we have

$$P(1 \leq X) = 1 - P(X = 0)$$
$$= 1 - p(0) = 1 - e^{-2} = 0.865,$$

approximately, by Table I of Appendix C. ∎

Example 3.2.2. If the mgf of a random variable X is

$$M(t) = e^{4(e^t - 1)},$$

then X has a Poisson distribution with $\mu = 4$. Accordingly, by way of example,

$$P(X = 3) = \frac{4^3 e^{-4}}{3!} = \frac{32}{3} e^{-4},$$

or, by Table I,

$$P(X = 3) = P(X \le 3) - P(X \le 2) = 0.433 - 0.238 = 0.195. \quad \blacksquare$$

Example 3.2.3. Let the probability of exactly one blemish in 1 foot of wire be about $\frac{1}{1000}$ and let the probability of two or more blemishes in that length be, for all practical purposes, zero. Let the random variable X be the number of blemishes in 3000 feet of wire. If we assume the independence of the number of blemishes in nonoverlapping intervals, then the postulates of the Poisson process are approximated, with $\lambda = \frac{1}{1000}$ and $w = 3000$. Thus X has an approximate Poisson distribution with mean $3000(\frac{1}{1000}) = 3$. For example, the probability that there are five or more blemishes in 3000 feet of wire is

$$P(X \ge 5) = \sum_{k=5}^{\infty} \frac{3^k e^{-3}}{k!}$$

and by Table I,

$$P(X \ge 5) = 1 - P(X \le 4) = 1 - 0.815 = 0.185,$$

approximately. $\blacksquare$

The Poisson distribution satisfies the following important additive property.

Theorem 3.2.1. *Suppose $X_1, \ldots, X_n$ are independent random variables and suppose X_i has a Poisson distribution with parameter m_i. Then $Y = \sum_{i=1}^{n} X_i$ has a Poisson distribution with parameter $\sum_{i=1}^{n} m_i$.*

Proof: We shall obtain the result, by determining the mgf of Y. Using independence of the X_is and the mgf of each X_i, we have,

$$
\begin{aligned}
M_Y(t) &= E(e^{tY}) = E\left(e^{\sum_{i=1}^{n} tX_i}\right) \\
&= E\left(\prod_{i=1}^{n} e^{tX_i}\right) = \prod_{i=1}^{n} E\left(e^{tX_i}\right) \\
&= \prod_{i=1}^{n} e^{m_i(e^t - 1)} = e^{\sum_{i=1}^{n} m_i(e^t - 1)}.
\end{aligned}
$$

By the uniqueness of mgfs, we conclude that Y has a Poisson distribution with parameter $\sum_{i=1}^{n} m_i$. $\blacksquare$

Example 3.2.4 (Example 3.2.3, Continued). Suppose in Example 3.2.3 that a bail of wire consists of 3000 feet. Based on the information in the example, we expect 3 blemishes in a bail of wire and the probability of 5 or more blemishes is 0.185. Suppose in a sampling plan, three bails of wire are selected at random and we compute the mean number of blemishes in the wire. Now suppose we want to determine the probability that the mean of the three observations has 5 or more blemishes. Let X_i be the number of blemishes in the ith bail of wire for $i = 1, 2, 3$. Then X_i has a Poisson distribution with parameter 3. The mean of X_1, X_2, and X_3 is $\overline{X} = 3^{-1} \sum_{i=1}^{3} X_i$, which can also be expressed as $Y/3$ where $Y = \sum_{i=1}^{3} X_i$. By the last theorem, because the bails are independent of one another, Y has a Poisson distribution with parameter $\sum_{i=1}^{3} 3 = 9$. Hence, by Table 1 the desired probability is,

$$P(\overline{X} \geq 5) = P(Y \geq 15) = 1 - P(Y \leq 14) = 1 - 0.959 = 0.041.$$

Hence, while it is not too odd that a bail has 5 or more blemishes (probability is 0.185), it is unusual (probability is 0.041) that 3 independent bails of wire average 5 or more blemishes. ∎

EXERCISES

3.2.1. If the random variable X has a Poisson distribution such that $P(X = 1) = P(X = 2)$, find $P(X = 4)$.

3.2.2. The mgf of a random variable X is $e^{4(e^t-1)}$. Show that $P(\mu - 2\sigma < X < \mu + 2\sigma) = 0.931$.

3.2.3. In a lengthy manuscript, it is discovered that only 13.5 percent of the pages contain no typing errors. If we assume that the number of errors per page is a random variable with a Poisson distribution, find the percentage of pages that have exactly one error.

3.2.4. Let the pmf $p(x)$ be positive on and only on the nonnegative integers. Given that $p(x) = (4/x)p(x - 1)$, $x = 1, 2, 3, \ldots$. Find $p(x)$.
Hint: Note that $p(1) = 4p(0)$, $p(2) = (4^2/2!)p(0)$, and so on. That is, find each $p(x)$ in terms of $p(0)$ and then determine $p(0)$ from

$$1 = p(0) + p(1) + p(2) + \cdots.$$

3.2.5. Let X have a Poisson distribution with $\mu = 100$. Use Chebyshev's inequality to determine a lower bound for $P(75 < X < 125)$.

3.2.6. Suppose that $g(x, 0) = 0$ and that

$$D_w[g(x, w)] = -\lambda g(x, w) + \lambda g(x - 1, w)$$

for $x = 1, 2, 3, \ldots$. If $g(0, w) = e^{-\lambda w}$, show by mathematical induction that

$$g(x, w) = \frac{(\lambda w)^x e^{-\lambda w}}{x!}, \quad x = 1, 2, 3, \ldots.$$

3.2.7. Using the computer, obtain an overlay plot of the pmfs following two distributions:

(a) Poisson distribution with $\lambda = 2$.

(b) Binomial distribution with $n = 100$ and $p = 0.02$.

Why would these distributions be approximately the same? Discuss.

3.2.8. Let the number of chocolate drops in a certain type of cookie have a Poisson distribution. We want the probability that a cookie of this type contains at least two chocolate drops to be greater than 0.99. Find the smallest value of the mean that the distribution can take.

3.2.9. Compute the measures of skewness and kurtosis of the Poisson distribution with mean μ.

3.2.10. On the average a grocer sells 3 of a certain article per week. How many of these should he have in stock so that the chance of his running out within a week will be less than 0.01? Assume a Poisson distribution.

3.2.11. Let X have a Poisson distribution. If $P(X = 1) = P(X = 3)$, find the mode of the distribution.

3.2.12. Let X have a Poisson distribution with mean 1. Compute, if it exists, the expected value $E(X!)$.

3.2.13. Let X and Y have the joint pmf $p(x, y) = e^{-2}/[x!(y-x)!]$, $y = 0, 1, 2, \ldots$; $x = 0, 1, \ldots, y$, zero elsewhere.

(a) Find the mgf $M(t_1, t_2)$ of this joint distribution.

(b) Compute the means, the variances, and the correlation coefficient of X and Y.

(c) Determine the conditional mean $E(X|y)$.
   Hint: Note that

$$\sum_{x=0}^{y} [\exp(t_1 x)]y!/[x!(y-x)!] = [1 + \exp(t_1)]^y.$$

   Why?

3.2.14. Let X_1 and X_2 be two independent random variables. Suppose that X_1 and $Y = X_1 + X_2$ have Poisson distributions with means μ_1 and $\mu > \mu_1$, respectively. Find the distribution of X_2.

3.2.15. Let $X_1, X_2, \ldots, X_n$ denote n mutually independent random variables with the moment-generating functions $M_1(t)$, $M_2(t)$, $\ldots$, $M_n(t)$, respectively.

(a) Show that $Y = k_1 X_1 + k_2 X_2 + \cdots + k_n X_n$, where $k_1, k_2, \ldots, k_n$ are real constants, has the mgf $M(t) = \prod_{1}^{n} M_i(k_i t)$.

(b) If each $k_i = 1$ and if X_i is Poisson with mean μ_i, $i = 1, 2, \ldots, n$, using Part (a) prove that Y is Poisson with mean $\mu_1 + \cdots + \mu_n$. This is another proof of Theorem 3.2.1.

3.3 The Γ, χ^2, and β Distributions

In this section we introduce the gamma (Γ), chi-square (χ^2), and beta (β) distributions. It is proved in books on advanced calculus that the integral

$$\int_0^\infty y^{\alpha-1} e^{-y} \, dy$$

exists for $\alpha > 0$ and that the value of the integral is a positive number. The integral is called the gamma function of α, and we write

$$\Gamma(\alpha) = \int_0^\infty y^{\alpha-1} e^{-y} \, dy.$$

If $\alpha = 1$, clearly

$$\Gamma(1) = \int_0^\infty e^{-y} \, dy = 1.$$

If $\alpha > 1$, an integration by parts shows that

$$\Gamma(\alpha) = (\alpha - 1) \int_0^\infty y^{\alpha-2} e^{-y} \, dy = (\alpha - 1)\Gamma(\alpha - 1).$$

Accordingly, if α is a positive integer greater than 1,

$$\Gamma(\alpha) = (\alpha - 1)(\alpha - 2) \cdots (3)(2)(1)\Gamma(1) = (\alpha - 1)!.$$

Since $\Gamma(1) = 1$, this suggests we take $0! = 1$, as we have done.

In the integral that defines $\Gamma(\alpha)$, let us introduce a new variable by writing $y = x/\beta$, where $\beta > 0$. Then

$$\Gamma(\alpha) = \int_0^\infty \left(\frac{x}{\beta}\right)^{\alpha-1} e^{-x/\beta} \left(\frac{1}{\beta}\right) \, dx,$$

or, equivalently,

$$1 = \int_0^\infty \frac{1}{\Gamma(\alpha)\beta^\alpha} x^{\alpha-1} e^{-x/\beta} \, dx.$$

Since $\alpha > 0$, $\beta > 0$, and $\Gamma(\alpha) > 0$, we see that

$$f(x) = \begin{cases} \frac{1}{\Gamma(\alpha)\beta^\alpha} x^{\alpha-1} e^{-x/\beta} & 0 < x < \infty \\ 0 & \text{elsewhere,} \end{cases} \tag{3.3.1}$$

is a pdf of a random variable of the continuous type. A random variable X that has a pdf of this form is said to have a **gamma** distribution with parameters α and β. We will write that X has a $\Gamma(\alpha, \beta)$ distribution.

Remark 3.3.1. The gamma distribution is frequently a probability model for waiting times; for instance, in life testing, the waiting time until "death" is a random variable which is frequently modeled with a gamma distribution. To see this, let us assume the postulates of a Poisson process and let the interval of length w be a time interval. Specifically, let the random variable W be the time that is needed to obtain exactly k changes (possibly deaths), where k is a fixed positive integer. Then the cdf of W is

$$G(w) = P(W \leq w) = 1 - P(W > w).$$

However, the event $W > w$, for $w > 0$, is equivalent to the event in which there are less than k changes in a time interval of length w. That is, if the random variable X is the number of changes in an interval of length w, then

$$P(W > w) = \sum_{x=0}^{k-1} P(X = x) = \sum_{x=0}^{k-1} \frac{(\lambda w)^x e^{-\lambda w}}{x!}.$$

In Exercise 3.3.5, the reader is asked to prove that

$$\int_{\lambda w}^{\infty} \frac{z^{k-1} e^{-z}}{(k-1)!} \, dz = \sum_{x=0}^{k-1} \frac{(\lambda w)^x e^{-\lambda w}}{x!}.$$

If, momentarily, we accept this result, we have, for $w > 0$,

$$G(w) = 1 - \int_{\lambda w}^{\infty} \frac{z^{k-1} e^{-z}}{\Gamma(k)} \, dz = \int_{0}^{\lambda w} \frac{z^{k-1} e^{-z}}{\Gamma(k)} \, dz,$$

and for $w \leq 0$, $G(w) = 0$. If we change the variable of integration in the integral that defines $G(w)$ by writing $z = \lambda y$, then

$$G(w) = \int_{0}^{w} \frac{\lambda^k y^{k-1} e^{-\lambda y}}{\Gamma(k)} \, dy, \quad w > 0$$

and $G(w) = 0$ for $w \leq 0$. Accordingly, the pdf of W is

$$g(w) = G'(w) = \begin{cases} \frac{\lambda^k w^{k-1} e^{-\lambda w}}{\Gamma(k)} & 0 < w < \infty \\ 0 & \text{elsewhere.} \end{cases}$$

That is, W has a gamma distribution with $\alpha = k$ and $\beta = 1/\lambda$. If W is the waiting time until the first change, that is, if $k = 1$, the pdf of W is

$$g(w) = \begin{cases} \lambda e^{-\lambda w} & 0 < w < \infty \\ 0 & \text{elsewhere,} \end{cases} \tag{3.3.2}$$

and W is said to have an *exponential distribution* with parameter λ. ∎

We now find the mgf of a gamma distribution. Since

$$M(t) = \int_0^\infty e^{tx} \frac{1}{\Gamma(\alpha)\beta^\alpha} x^{\alpha-1} e^{-x/\beta}\, dx$$
$$= \int_0^\infty \frac{1}{\Gamma(\alpha)\beta^\alpha} x^{\alpha-1} e^{-x(1-\beta t)/\beta}\, dx,$$

we may set $y = x(1 - \beta t)/\beta$, $t < 1/\beta$, or $x = \beta y/(1 - \beta t)$, to obtain

$$M(t) = \int_0^\infty \frac{\beta/(1-\beta t)}{\Gamma(\alpha)\beta^\alpha} \left(\frac{\beta y}{1 - \beta t}\right)^{\alpha-1} e^{-y}\, dy.$$

That is,

$$M(t) = \left(\frac{1}{1-\beta t}\right)^\alpha \int_0^\infty \frac{1}{\Gamma(\alpha)} y^{\alpha-1} e^{-y}\, dy$$
$$= \frac{1}{(1-\beta t)^\alpha}, \quad t < \frac{1}{\beta}.$$

Now

$$M'(t) = (-\alpha)(1 - \beta t)^{-\alpha-1}(-\beta)$$

and

$$M''(t) = (-\alpha)(-\alpha - 1)(1 - \beta t)^{-\alpha-2}(-\beta)^2.$$

Hence, for a gamma distribution, we have

$$\mu = M'(0) = \alpha\beta$$

and

$$\sigma^2 = M''(0) - \mu^2 = \alpha(\alpha+1)\beta^2 - \alpha^2\beta^2 = \alpha\beta^2.$$

To calculate probabilities for gamma distributions with the program R or S-PLUS, suppose X has a gamma distribution with parameters $\alpha = a$ and $\beta = b$. Then the command `pgamma(x,shape=a,scale=b)` returns $P(X \leq x)$ while the value of the pdf of X at x is returned by the command `dgamma(x,shape=a,scale=b)`.

Example 3.3.1. Let the waiting time W have a gamma pdf with $\alpha = k$ and $\beta = 1/\lambda$. Accordingly, $E(W) = k/\lambda$. If $k = 1$, then $E(W) = 1/\lambda$; that is, the expected waiting time for $k = 1$ changes is equal to the reciprocal of λ. ∎

Example 3.3.2. Let X be a random variable such that

$$E(X^m) = \frac{(m+3)!}{3!} 3^m, \quad m = 1, 2, 3, \ldots.$$

Then the mgf of X is given by the series

$$M(t) = 1 + \frac{4!\, 3}{3!\, 1!} t + \frac{5!\, 3^2}{3!\, 2!} t^2 + \frac{6!\, 3^3}{3!\, 3!} t^3 + \cdots.$$

This, however, is the Maclaurin's series for $(1 - 3t)^{-4}$, provided that $-1 < 3t < 1$. Accordingly, X has a gamma distribution with $\alpha = 4$ and $\beta = 3$. ∎

Remark 3.3.2. The gamma distribution is not only a good model for waiting times, but one for many nonnegative random variables of the continuous type. For illustration, the distribution of certain incomes could be modeled satisfactorily by the gamma distribution, since the two parameters α and β provide a great deal of flexibility. Several gamma probability density functions are depicted in Figure 3.3.1. ∎

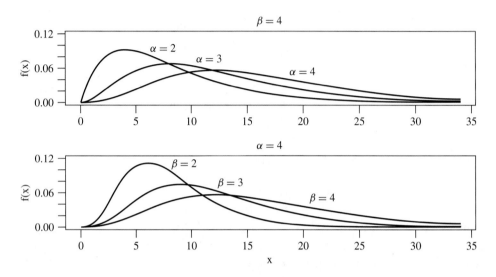

Figure 3.3.1: Several gamma densities

Let us now consider a special case of the gamma distribution in which $\alpha = r/2$, where r is a positive integer, and $\beta = 2$. A random variable X of the continuous type that has the pdf

$$f(x) = \begin{cases} \frac{1}{\Gamma(r/2)2^{r/2}}x^{r/2-1}e^{-x/2} & 0 < x < \infty \\ 0 & \text{elsewhere,} \end{cases} \tag{3.3.3}$$

and the mgf

$$M(t) = (1-2t)^{-r/2}, \quad t < \tfrac{1}{2},$$

is said to have a **chi-square distribution**, and any $f(x)$ of this form is called a **chi-square pdf**. The mean and the variance of a chi-square distribution are $\mu = \alpha\beta = (r/2)2 = r$ and $\sigma^2 = \alpha\beta^2 = (r/2)2^2 = 2r$, respectively. For no obvious reason, we call the parameter r the number of degrees of freedom of the chi-square distribution (or of the chi-square pdf). Because the chi-square distribution has an important role in statistics and occurs so frequently, we write, for brevity, that X is $\chi^2(r)$ to mean that the random variable X has a chi-square distribution with r degrees of freedom.

Example 3.3.3. If X has the pdf

$$f(x) = \begin{cases} \frac{1}{4}xe^{-x/2} & 0 < x < \infty \\ 0 & \text{elsewhere,} \end{cases}$$

then X is $\chi^2(4)$. Hence $\mu = 4$, $\sigma^2 = 8$, and $M(t) = (1 - 2t)^{-2}$, $t < \frac{1}{2}$. ■

Example 3.3.4. If X has the mgf $M(t) = (1 - 2t)^{-8}$, $t < \frac{1}{2}$, then X is $\chi^2(16)$. ■

If the random variable X is $\chi^2(r)$, then, with $c_1 < c_2$, we have

$$P(c_1 < X < c_2) = P(X \le c_2) - P(X \le c_1),$$

since $P(X = c_2) = 0$. To compute such a probability, we need the value of an integral like

$$P(X \le x) = \int_0^x \frac{1}{\Gamma(r/2)2^{r/2}} w^{r/2-1} e^{-w/2} \, dw.$$

Tables of this integral for selected values of r and x have been prepared and are partially reproduced in Table II in Appendix C. If, on the other hand, the package R or S-PLUS is available then the command `pchisq(x,r)` returns $P(X \le x)$ and the command `dchisq(x,r)` returns the value of the pdf of X at x when X has a chi-squared distribution with r degrees of freedom.

The following result will be used several times in the sequel; hence, we record it in a theorem.

Theorem 3.3.1. *Let X have a $\chi^2(r)$ distribution. If $k > -r/2$ then $E(X^k)$ exists and it is given by*

$$E(X^k) = \frac{2^k \Gamma\left(\frac{r}{2} + k\right)}{\Gamma\left(\frac{r}{2}\right)}, \quad \text{if } k > -r/2. \tag{3.3.4}$$

Proof. Note that

$$E(X^k) = \int_0^\infty \frac{1}{\Gamma\left(\frac{r}{2}\right) 2^{r/2}} x^{(r/2)+k-1} e^{-x/2} \, dx.$$

Make the change of variable $u = x/2$ in the above integral. This results in

$$E(X^k) = \int_0^\infty \frac{1}{\Gamma\left(\frac{r}{2}\right) 2^{(r/2)-1}} 2^{(r/2)+k-1} u^{(r/2)+k-1} e^{-u} \, du.$$

This yields the desired result provided that $k > -(r/2)$. ■

Notice that if k is a nonnegative integer then $k > -(r/2)$ is always true. Hence, all moments of a χ^2 distribution exist and the kth moment is given by (3.3.4).

Example 3.3.5. Let X be $\chi^2(10)$. Then, by Table II of Appendix C, with $r = 10$,

$$\begin{aligned}
P(3.25 \le X \le 20.5) &= P(X \le 20.5) - P(X \le 3.5) \\
&= 0.975 - 0.025 = 0.95.
\end{aligned}$$

Again, as an example, if $P(a < X) = 0.05$, then $P(X \le a) = 0.95$, and thus $a = 18.3$ from Table II with $r = 10$. ■

Example 3.3.6. Let X have a gamma distribution with $\alpha = r/2$, where r is a positive integer, and $\beta > 0$. Define the random variable $Y = 2X/\beta$. We seek the pdf of Y. Now the cdf of Y is

$$G(y) = P(Y \leq y) = P\left(X \leq \frac{\beta y}{2}\right).$$

If $y \leq 0$, then $G(y) = 0$; but if $y > 0$, then

$$G(y) = \int_0^{\beta y/2} \frac{1}{\Gamma(r/2)\beta^{r/2}} x^{r/2-1} e^{-x/\beta}\, dx.$$

Accordingly, the pdf of Y is

$$
\begin{aligned}
g(y) &= G'(y) = \frac{\beta/2}{\Gamma(r/2)\beta^{r/2}} (\beta y/2)^{r/2-1} e^{-y/2} \\
&= \frac{1}{\Gamma(r/2)2^{r/2}} y^{r/2-1} e^{-y/2}
\end{aligned}
$$

if $y > 0$. That is, Y is $\chi^2(r)$. ∎

One of the most important properties of the gamma distribution is its additive property.

Theorem 3.3.2. *Let $X_1, \ldots, X_n$ be independent random variables. Suppose, for $i = 1, \ldots, n$, that X_i has a $\Gamma(\alpha_i, \beta)$ distribution. Let $Y = \sum_{i=1}^n X_i$. Then Y has $\Gamma(\sum_{i=1}^n \alpha_i, \beta)$ distribution.*

Proof: Using the assumed independence and the mgf of a gamma distribution, we have for $t < 1/\beta$,

$$
\begin{aligned}
M_Y(t) &= E[\exp\{t\sum_{i=1}^n X_i\}] = \prod_{i=1}^n E[\exp\{tX_i\}] \\
&= \prod_{i=1}^n (1 - \beta t)^{-\alpha_i} = (1 - \beta t)^{-\sum_{i=1}^n \alpha_i},
\end{aligned}
$$

which is the mgf of a $\Gamma(\sum_{i=1}^n \alpha_i, \beta)$ distribution. ∎

In the sequel, we often will use this property for the χ^2 distribution. For convenience, we state the result as a corollary, since here $\beta = 2$ and $\sum \alpha_i = \sum r_i/2$.

Corollary 3.3.1. *Let $X_1, \ldots, X_n$ be independent random variables. Suppose, for $i = 1, \ldots, n$, that X_i has a $\chi^2(r_i)$ distribution. Let $Y = \sum_{i=1}^n X_i$. Then Y has $\chi^2(\sum_{i=1}^n r_i)$ distribution.*

We conclude this section with another important distribution called the **beta** distribution, which we will derive from a pair of independent Γ random variables.

Let X_1 and X_2 be two independent random variables that have Γ distributions and the joint pdf

$$h(x_1, x_2) = \frac{1}{\Gamma(\alpha)\Gamma(\beta)} x_1^{\alpha-1} x_2^{\beta-1} e^{-x_1-x_2}, \quad 0 < x_1 < \infty, \ 0 < x_2 < \infty,$$

zero elsewhere, where $\alpha > 0$, $\beta > 0$. Let $Y_1 = X_1 + X_2$ and $Y_2 = X_1/(X_1 + X_2)$. We shall show that Y_1 and Y_2 are independent.

The space $\mathcal{S}$ is, exclusive of the points on the coordinate axes, the first quadrant of the $x_1 x_2$-plane. Now

$$y_1 = u_1(x_1, x_2) = x_1 + x_2,$$
$$y_2 = u_2(x_1, x_2) = \frac{x_1}{x_1 + x_2}$$

may be written $x_1 = y_1 y_2$, $x_2 = y_1(1 - y_2)$, so

$$J = \begin{vmatrix} y_2 & y_1 \\ 1 - y_2 & -y_1 \end{vmatrix} = -y_1 \neq 0.$$

The transformation is one-to-one, and it maps $\mathcal{S}$ onto $\mathcal{T} = \{(y_1, y_2) : 0 < y_1 < \infty, \ 0 < y_2 < 1\}$ in the $y_1 y_2$-plane. The joint pdf of Y_1 and Y_2 is then

$$\begin{aligned} g(y_1, y_2) &= (y_1) \frac{1}{\Gamma(\alpha)\Gamma(\beta)} (y_1 y_2)^{\alpha-1} [y_1(1 - y_2)]^{\beta-1} e^{-y_1} \\ &= \begin{cases} \frac{y_2^{\alpha-1}(1-y_2)^{\beta-1}}{\Gamma(\alpha)\Gamma(\beta)} y_1^{\alpha+\beta-1} e^{-y_1} & 0 < y_1 < \infty, \ 0 < y_2 < 1 \\ 0 & \text{elsewhere.} \end{cases} \end{aligned}$$

In accordance with Theorem 2.5.1 the random variables are independent. The marginal pdf of Y_2 is

$$\begin{aligned} g_2(y_2) &= \frac{y_2^{\alpha-1}(1 - y_2)^{\beta-1}}{\Gamma(\alpha)\Gamma(\beta)} \int_0^\infty y_1^{\alpha+\beta-1} e^{-y_1} \, dy_1 \\ &= \begin{cases} \frac{\Gamma(\alpha+\beta)}{\Gamma(\alpha)\Gamma(\beta)} y_2^{\alpha-1}(1 - y_2)^{\beta-1} & 0 < y_2 < 1 \\ 0 & \text{elsewhere.} \end{cases} \end{aligned} \tag{3.3.5}$$

This pdf is that of the *beta distribution* with parameters α and β. Since $g(y_1, y_2) \equiv g_1(y_1) g_2(y_2)$, it must be that the pdf of Y_1 is

$$g_1(y_1) = \begin{cases} \frac{1}{\Gamma(\alpha+\beta)} y_1^{\alpha+\beta-1} e^{-y_1} & 0 < y_1 < \infty \\ 0 & \text{elsewhere,} \end{cases}$$

which is that of a gamma distribution with parameter values of $\alpha + \beta$ and 1.

It is an easy exercise to show that the mean and the variance of Y_2, which has a beta distribution with parameters α and β, are, respectively,

$$\mu = \frac{\alpha}{\alpha + \beta}, \quad \sigma^2 = \frac{\alpha\beta}{(\alpha + \beta + 1)(\alpha + \beta)^2}.$$

Either of the programs R or S-PLUS calculate probabilities for the beta distribution. If X has a beta distribution with parameters $\alpha = a$ and $\beta = b$ then the command `pbeta(x,a,b)` returns $P(X \leq x)$ and the command `dbeta(x,a,b)` returns the value of the pdf of X at x.

We close this section with another example of a random variable whose distribution is derived from a transfomation of gamma random variables.

Example 3.3.7 (Dirichlet Distribution). Let $X_1, X_2, \ldots, X_{k+1}$ be independent random variables, each having a gamma distribution with $\beta = 1$. The joint pdf of these variables may be written as

$$h(x_1, x_2, \ldots, x_{k+1}) = \begin{cases} \prod_{i=1}^{k+1} \frac{1}{\Gamma(\alpha_i)} x_i^{\alpha_i - 1} e^{-x_i} & 0 < x_i < \infty \\ 0 & \text{elsewhere.} \end{cases}$$

Let

$$Y_i = \frac{X_i}{X_1 + X_2 + \cdots + X_{k+1}}, \quad i = 1, 2, \ldots, k,$$

and $Y_{k+1} = X_1 + X_2 + \cdots + X_{k+1}$ denote $k+1$ new random variables. The associated transformation maps $\mathcal{A} = \{(x_1, \ldots, x_{k+1}) : 0 < x_i < \infty, \ i = 1, \ldots, k+1\}$ onto the space.

$$\mathcal{B} = \{(y_1, \ldots, y_k, y_{k+1}) : 0 < y_i, \ i = 1, \ldots, k, \ y_1 + \cdots + y_k < 1, \ 0 < y_{k+1} < \infty\}.$$

The single-valued inverse functions are $x_1 = y_1 y_{k+1}, \ldots, x_k = y_k y_{k+1}, x_{k+1} = y_{k+1}(1 - y_1 - \cdots - y_k)$, so that the Jacobian is

$$J = \begin{vmatrix} y_{k+1} & 0 & \cdots & 0 & y_1 \\ 0 & y_{k+1} & \cdots & 0 & y_2 \\ \vdots & \vdots & & \vdots & \vdots \\ 0 & 0 & \cdots & y_{k+1} & y_k \\ -y_{k+1} & -y_{k+1} & \cdots & -y_{k+1} & (1 - y_1 - \cdots - y_k) \end{vmatrix} = y_{k+1}^k.$$

Hence the joint pdf of $Y_1, \ldots, Y_k, Y_{k+1}$ is given by

$$\frac{y_{k+1}^{\alpha_1 + \cdots + \alpha_{k+1} - 1} y_1^{\alpha_1 - 1} \cdots y_k^{\alpha_k - 1} (1 - y_1 - \cdots - y_k)^{\alpha_{k+1} - 1} e^{-y_{k+1}}}{\Gamma(\alpha_1) \cdots \Gamma(\alpha_k) \Gamma(\alpha_{k+1})},$$

provided that $(y_1, \ldots, y_k, y_{k+1}) \in \mathcal{B}$ and is equal to zero elsewhere. The joint pdf of $Y_1, \ldots, Y_k$ is seen by inspection to be given by

$$g(y_1, \ldots, y_k) = \frac{\Gamma(\alpha_1 + \cdots + \alpha_{k+1})}{\Gamma(\alpha_1) \cdots \Gamma(\alpha_{k+1})} y_1^{\alpha_1 - 1} \cdots y_k^{\alpha_k - 1} (1 - y_1 - \cdots - y_k)^{\alpha_{k+1} - 1}, \quad (3.3.6)$$

when $0 < y_i, \ i = 1, \ldots, k, \ y_1 + \cdots + y_k < 1$, while the function g is equal to zero elsewhere. Random variables $Y_1, \ldots, Y_k$ that have a joint pdf of this form are said to have a *Dirichlet pdf*. It is seen , in the special case of $k = 1$, that the Dirichlet pdf becomes a beta pdf. Moreover, it is also clear from the joint pdf of $Y_1, \ldots, Y_k, Y_{k+1}$ that Y_{k+1} has a gamma distribution with parameters $\alpha_1 + \cdots + \alpha_k + \alpha_{k+1}$ and $\beta = 1$ and that Y_{k+1} is independent of $Y_1, Y_2, \ldots, Y_k$. ∎

EXERCISES

3.3.1. If $(1-2t)^{-6}$, $t < \frac{1}{2}$, is the mgf of the random variable X, find $P(X < 5.23)$.

3.3.2. If X is $\chi^2(5)$, determine the constants c and d so that $P(c < X < d) = 0.95$ and $P(X < c) = 0.025$.

3.3.3. Find $P(3.28 < X < 25.2)$, if X has a gamma distribution with $\alpha = 3$ and $\beta = 4$.
Hint: Consider the probability of the equivalent event $1.64 < Y < 12.6$, where $Y = 2X/4 = X/2$.

3.3.4. Let X be a random variable such that $E(X^m) = (m+1)!2^m$, $m = 1, 2, 3, \ldots$. Determine the mgf and the distribution of X.

3.3.5. Show that

$$\int_\mu^\infty \frac{1}{\Gamma(k)} z^{k-1} e^{-z}\, dz = \sum_{x=0}^{k-1} \frac{\mu^x e^{-\mu}}{x!}, \quad k = 1, 2, 3, \ldots.$$

This demonstrates the relationship between the cdfs of the gamma and Poisson distribution .
Hint: Either integrate by parts $k-1$ times or simply note that the "antiderivative" of $z^{k-1}e^{-z}$ is

$$-z^{k-1}e^{-z} - (k-1)z^{k-2}e^{-z} - \cdots - (k-1)!e^{-z}$$

by differentiating the latter expression.

3.3.6. Let X_1, X_2, and X_3 be iid random variables, each with pdf $f(x) = e^{-x}$, $0 < x < \infty$, zero elsewhere. Find the distribution of $Y = \text{minimum}(X_1, X_2, X_3)$.
Hint: $P(Y \le y) = 1 - P(Y > y) = 1 - P(X_i > y, i = 1, 2, 3)$.

3.3.7. Let X have a gamma distribution with pdf

$$f(x) = \frac{1}{\beta^2} x e^{-x/\beta}, \quad 0 < x < \infty,$$

zero elsewhere. If $x = 2$ is the unique mode of the distribution, find the parameter β and $P(X < 9.49)$.

3.3.8. Compute the measures of skewness and kurtosis of a gamma distribution which has parameters α and β.

3.3.9. Let X have a gamma distribution with parameters α and β. Show that $P(X \ge 2\alpha\beta) \le (2/e)^\alpha$.
Hint: Use the result of Exercise 1.10.4.

3.3.10. Give a reasonable definition of a chi-square distribution with zero degrees of freedom.
Hint: Work with the mgf of a distribution that is $\chi^2(r)$ and let $r = 0$.

3.3.11. Using the computer, obtain plots of the pdfs of chi-squared distributions with degrees of freedom $r = 1, 2, 5, 10, 20$. Comment on the plots.

3.3.12. Using the computer, plot the cdf of $\Gamma(5, 4)$ and use it to guess the median. Confirm it with a computer command which returns the median, (In R or S-PLUS, use the command `qgamma(.5,shape=5,scale=4)`).

3.3.13. Using the computer, obtain plots of beta pdfs for $\alpha = 5$ and $\beta = 1, 2, 5, 10, 20$.

3.3.14. In the Poisson postulates of Remark 3.2.1, let λ be a nonnegative function of w, say $\lambda(w)$, such that $D_w[g(0, w)] = -\lambda(w)g(0, w)$. Suppose that $\lambda(w) = krw^{r-1}$, $r \geq 1$.

(a) Find $g(0, w)$ noting that $g(0, 0) = 1$.

(b) Let W be the time that is needed to obtain exactly one change. Find the distribution function of W, i.e., $G(w) = P(W \leq w) = 1 - P(W > w) = 1 - g(0, w)$, $0 \leq w$, and then find the pdf of W. This pdf is that of the *Weibull distribution*, which is used in the study of breaking strengths of materials.

3.3.15. Let X have a Poisson distribution with parameter m. If m is an experimental value of a random variable having a gamma distribution with $\alpha = 2$ and $\beta = 1$, compute $P(X = 0, 1, 2)$.
Hint: Find an expression that represents the joint distribution of X and m. Then integrate out m to find the marginal distribution of X.

3.3.16. Let X have the uniform distribution with pdf $f(x) = 1$, $0 < x < 1$, zero elsewhere. Find the cdf of $Y = -\log X$. What is the pdf of Y?

3.3.17. Find the uniform distribution of the continuous type on the interval (b, c) that has the same mean and the same variance as those of a chi-square distribution with 8 degrees of freedom. That is, find b and c.

3.3.18. Find the mean and variance of the β distribution.
Hint: From the pdf, we know that

$$\int_0^1 y^{\alpha-1}(1 - y)^{\beta-1}\, dy = \frac{\Gamma(\alpha)\Gamma(\beta)}{\Gamma(\alpha + \beta)}$$

for all $\alpha > 0$, $\beta > 0$.

3.3.19. Determine the constant c in each of the following so that each $f(x)$ is a β pdf:

(a) $f(x) = cx(1 - x)^3$, $0 < x < 1$, zero elsewhere.

(b) $f(x) = cx^4(1 - x)^5$, $0 < x < 1$, zero elsewhere.

(c) $f(x) = cx^2(1 - x)^8$, $0 < x < 1$, zero elsewhere.

3.3.20. Determine the constant c so that $f(x) = cx(3 - x)^4$, $0 < x < 3$, zero elsewhere, is a pdf.

3.3.21. Show that the graph of the β pdf is symmetric about the vertical line through $x = \frac{1}{2}$ if $\alpha = \beta$.

3.3.22. Show, for $k = 1, 2, \ldots, n$, that

$$\int_p^1 \frac{n!}{(k-1)!(n-k)!} z^{k-1}(1-z)^{n-k}\,dz = \sum_{x=0}^{k-1} \binom{n}{x} p^x (1-p)^{n-x}.$$

This demonstrates the relationship between the cdfs of the β and binomial distributions.

3.3.23. Let X_1 and X_2 be independent random variables. Let X_1 and $Y = X_1 + X_2$ have chi-square distributions with r_1 and r degrees of freedom, respectively. Here $r_1 < r$. Show that X_2 has a chi-square distribution with $r - r_1$ degrees of freedom. *Hint:* Write $M(t) = E(e^{t(X_1+X_2)})$ and make use of the independence of X_1 and X_2.

3.3.24. Let X_1, X_2 be two independent random variables having gamma distributions with parameters $\alpha_1 = 3$, $\beta_1 = 3$ and $\alpha_2 = 5$, $\beta_2 = 1$, respectively.

(a) Find the mgf of $Y = 2X_1 + 6X_2$.

(b) What is the distribution of Y?

3.3.25. Let X have an exponential distribution.

(a) Show that

$$P(X > x + y \mid X > x) = P(X > y). \tag{3.3.7}$$

Hence, the exponential distribution has the *memoryless* property. Recall from (3.1.6) that the discrete geometric distribution had a similar property.

(b) Let $F(x)$ be the cdf of a continuous random variable Y. Assume that $F(0) = 0$ and $0 < F(y) < 1$ for $y > 0$. Suppose property (3.3.7) holds for Y. Show that $F_Y(y) = 1 - e^{-\lambda y}$ for $y > 0$.

Hint: Show that $g(y) = 1 - F_Y(y)$ satisfies the equation

$$g(y + z) = g(y)g(z),$$

3.3.26. Consider a random variable X of the continuous type with cdf $F(x)$ and pdf $f(x)$. The **hazard rate** (or **failure rate** or **force of mortality**) is defined by

$$r(x) = \lim_{\Delta \to 0} \frac{P(x \le X < x + \Delta \mid X \ge x)}{\Delta}. \tag{3.3.8}$$

In the case that X represents the failure time of an item, the above conditional probability represents the failure of an item in the interval $[x, x + \Delta]$ given that it has survived until time x. Viewed this way, $r(x)$ is the rate of instantaneous failure at time $x > 0$.

(a) Show that $r(x) = f(x)/(1 - F(x))$.

(b) If $r(x) = c$, where c is a positive constant, show that the underlying distribution is exponential. Hence, exponential distributions have constant failure rate over all time.

(c) If $r(x) = cx^b$, where c and b are positive constants, show that X has a **Weibull** distribution, i.e.,

$$f(x) = \begin{cases} cx^b \exp\left\{-\frac{cx^{b+1}}{b+1}\right\} & 0 < x < \infty \\ 0 & \text{elsewhere.} \end{cases} \tag{3.3.9}$$

(d) If $r(x) = ce^{bx}$, where c and b are positive constants, show that X has a **Gompertz** cdf given by

$$F(x) = \begin{cases} 1 - \exp\left\{\frac{c}{b}(1 - e^{bx})\right\} & 0 < x < \infty \\ 0 & \text{elsewhere.} \end{cases} \tag{3.3.10}$$

This is frequently used by actuaries as a distribution of "length of life."

3.3.27. Let $Y_1, \ldots, Y_k$ have a Dirichlet distribution with parameters $\alpha_1, \ldots, \alpha_k, \alpha_{k+1}$.

(a) Show that Y_1 has a beta distribution with parameters $\alpha = \alpha_1$ and $\beta = \alpha_2 + \cdots + \alpha_{k+1}$.

(b) Show that $Y_1 + \cdots + Y_r$, $r \le k$, has a beta distribution with parameters $\alpha = \alpha_1 + \cdots + \alpha_r$ and $\beta = \alpha_{r+1} + \cdots + \alpha_{k+1}$.

(c) Show that $Y_1 + Y_2$, $Y_3 + Y_4$, $Y_5, \ldots, Y_k$, $k \ge 5$, have a Dirichlet distribution with parameters $\alpha_1 + \alpha_2$, $\alpha_3 + \alpha_4$, $\alpha_5, \ldots, \alpha_k, \alpha_{k+1}$.
Hint: Recall the definition of Y_i in Example 3.3.7 and use the fact that the sum of several independent gamma variables with $\beta = 1$ is a gamma variable.

3.4 The Normal Distribution

Motivation for the normal distribution is found in the Central Limit Theorem which is presented in Section 4.4. This theorem shows that normal distributions provide an important family of distributions for applications and for statistical inference, in general. We will proceed by first introducing the standard normal distribution and through it the general normal distribution.

Consider the integral

$$I = \int_{-\infty}^{\infty} \frac{1}{\sqrt{2\pi}} \exp\left(\frac{-z^2}{2}\right) dz. \tag{3.4.1}$$

This integral exists because the integrand is a positive continuous function which is bounded by an integrable function; that is,

$$0 < \exp\left(\frac{-z^2}{2}\right) < \exp(-|z| + 1), \quad -\infty < z < \infty,$$

and

$$\int_{-\infty}^{\infty} \exp(-|z|+1)\,dz = 2e.$$

To evaluate the integral I, we note that $I > 0$ and that I^2 may be written

$$I^2 = \frac{1}{2\pi} \int_{-\infty}^{\infty} \int_{-\infty}^{\infty} \exp\left(-\frac{z^2+w^2}{2}\right)\,dz\,dw.$$

This iterated integral can be evaluated by changing to polar coordinates. If we set $z = r\cos\theta$ and $w = r\sin\theta$, we have

$$
\begin{aligned}
I^2 &= \frac{1}{2\pi} \int_0^{2\pi} \int_0^{\infty} e^{-r^2/2} r\,dr\,d\theta \\
&= \frac{1}{2\pi} \int_0^{2\pi} d\theta = 1.
\end{aligned}
$$

Because the integrand of display (3.4.1) is positive on R and integrates to 1 over R, it is a pdf of a continuous random variable with support R. We denote this random variable by Z. In summary, Z has the pdf,

$$f(z) = \frac{1}{\sqrt{2\pi}} \exp\left(\frac{-z^2}{2}\right), \quad -\infty < z < \infty. \qquad (3.4.2)$$

For $t \in R$, the mgf of Z can be derived by a completion of a square as follows:

$$
\begin{aligned}
E[\exp\{tZ\}] &= \int_{-\infty}^{\infty} \exp\{tz\} \frac{1}{\sqrt{2\pi}} \exp\left\{-\frac{1}{2}z^2\right\}\,dz \\
&= \exp\left\{\frac{1}{2}t^2\right\} \int_{-\infty}^{\infty} \frac{1}{\sqrt{2\pi}} \exp\left\{-\frac{1}{2}(z-t)^2\right\}\,dz \\
&= \exp\left\{\frac{1}{2}t^2\right\} \int_{-\infty}^{\infty} \frac{1}{\sqrt{2\pi}} \exp\left\{-\frac{1}{2}w^2\right\}\,dw, \qquad (3.4.3)
\end{aligned}
$$

where for the last integral we made the one-to-one change of variable $w = z - t$. By the identity (3.4.2), the integral in expression (3.4.3) has value 1. Thus the mgf of Z is:

$$M_Z(t) = \exp\left\{\frac{1}{2}t^2\right\}, \quad \text{for } -\infty < t < \infty. \qquad (3.4.4)$$

The first two derivatives of $M_Z(t)$ are easily shown to be:

$$
\begin{aligned}
M_Z'(t) &= t\exp\left\{\frac{1}{2}t^2\right\} \\
M_Z''(t) &= \exp\left\{\frac{1}{2}t^2\right\} + t^2\exp\left\{\frac{1}{2}t^2\right\}.
\end{aligned}
$$

Upon evaluation these derivatives at $t = 0$, the mean and variance of Z are:

$$E(Z) = 0 \quad \text{and} \quad \text{Var}(Z) = 1. \qquad (3.4.5)$$

Next, define the continuous random variable X by

$$X = bZ + a,$$

for $b > 0$. This is a one-to-one transformation. To derive the pdf of X, note that the inverse of the transformation and the Jacobian are: $z = b^{-1}(x-a)$ and $J = b^{-1}$. Because $b > 0$, it follows from (3.4.2) that the pdf of X is

$$f_X(x) = \frac{1}{\sqrt{2\pi}b} \exp\left\{ -\frac{1}{2}\left(\frac{x-a}{b}\right)^2 \right\}, \quad -\infty < x < \infty.$$

By (3.4.5) we immediately have, $E(X) = a$ and $\mathrm{Var}(X) = b^2$. Hence, in the expression for the pdf of X, we can replace a by $\mu = E(X)$ and b^2 by $\sigma^2 = \mathrm{Var}(X)$. We make this formal in the following definition,

Definition 3.4.1 (Normal Distribution). *We say a random variable X has a* **normal distribution** *if its pdf is*

$$f(x) = \frac{1}{\sqrt{2\pi}\sigma} \exp\left\{ -\frac{1}{2}\left(\frac{x-\mu}{\sigma}\right)^2 \right\}, \quad \textit{for } -\infty < x < \infty. \quad (3.4.6)$$

The parameters μ and σ^2 are the mean and variance of X, respectively. We will often write that X has a $N(\mu, \sigma^2)$ distribution.

In this notation, the random variable Z with pdf (3.4.2) has a $N(0,1)$ distribution. We call Z a **standard normal** random variable.

For the mgf of X use the relationship $X = \sigma Z + \mu$ and the mgf for Z, (3.4.4), to obtain:

$$
\begin{aligned}
E[\exp\{tX\}] &= E[\exp\{t(\sigma Z + \mu)\}] = \exp\{\mu t\} E[\exp\{t\sigma Z\}] \\
&= \exp\{\mu t\} \exp\left\{ \frac{1}{2}\sigma^2 t^2 \right\} = \exp\left\{ \mu t + \frac{1}{2}\sigma^2 t^2 \right\}, \quad (3.4.7)
\end{aligned}
$$

for $-\infty < t < \infty$.

We summarize the above discussion, by noting the relationship between Z and X:

X has a $N(\mu, \sigma^2)$ distribution if and only if $Z = \frac{X-\mu}{\sigma}$ has a $N(0,1)$ distribution.
$$(3.4.8)$$

Example 3.4.1. If X has the mgf

$$M(t) = e^{2t + 32t^2},$$

then X has a normal distribution with $\mu = 2$ and $\sigma^2 = 64$. Furthermore, the random variable $Z = \frac{X-2}{8}$ has a $N(0,1)$ distribution. ∎

Example 3.4.2. Recall Example 1.9.4. In that example we derived all the moments of a standard normal random variable by using its moment generating function. We can use this to obtain all the moments of X where X has a $N(\mu, \sigma^2)$ distribution. From above, we can write $X = \sigma Z + \mu$ where Z has a $N(0, 1)$ distribution. Hence, for all nonnegative integers k a simple application of the binomial theorem yields,

$$E(X^k) = E[(\sigma Z + \mu)^k] = \sum_{j=0}^{k} \binom{k}{j} \sigma^j E(Z^j) \mu^{k-j}. \qquad (3.4.9)$$

Recall from Example 1.9.4 that all the odd moments of Z are 0, while all the even moments are given by expression (1.9.1). These can be substituted into expression (3.4.9) to derive the moments of X. ∎

The graph of the normal pdf, (3.4.6), is seen in Figure 3.4.1 to have the characteristics: (1), symmetry about a vertical axis through $x = \mu$; (2), having its maximum of $1/(\sigma\sqrt{2\pi})$ at $x = \mu$; and (3), having the x-axis as a horizontal asymptote. It should also be verified that (4) there are points of inflection at $x = \mu \pm \sigma$; see Exercise 3.4.7.

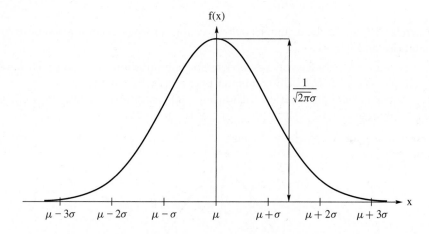

Figure 3.4.1: The Normal Density $f(x)$, (3.4.6).

As we discussed at the beginning of this section, many practical applications involve normal distributions. In particular, we need to be able to readily compute probabilities concerning them. Normal pdfs, however, contain some factor such as $\exp\{-s^2\}$. Hence, their antiderivatives cannot be obtained in closed form and numerical integration techniques must be used. Because of the relationship between normal and standard normal random variables, (3.4.8), we need only compute probabilities for standard normal random variables. To see this, denote the cdf of a standard normal random variable, Z, by

$$\Phi(z) = \int_{-\infty}^{z} \frac{1}{\sqrt{2\pi}} \exp\left\{\frac{-w^2}{2}\right\} dw. \qquad (3.4.10)$$

Let X have a $N(\mu, \sigma^2)$ distribution. Suppose we want to compute $F_X(x) = P(X \le x)$ for a specified x. For $Z = (X - \mu)/\sigma$, expression (3.4.8) implies

$$F_X(x) = P(X \le x) = P\left(Z \le \frac{x - \mu}{\sigma}\right) = \Phi\left(\frac{x - \mu}{\sigma}\right).$$

Thus we only need numerical integration computations for $\Phi(z)$. Normal quantiles can also be computed by using quantiles based on Z. For example, suppose we wanted the value x_p, such that $p = F_X(x_p)$, for a specified value of p. Take $z_p = \Phi^{-1}(p)$. Then by (3.4.8), $x_p = \sigma z_p + \mu$.

Figure 3.4.2 shows the standard normal density. The area under the density function to the left of z_p is p; that is, $\Phi(z_p) = p$. Table III in Appendix C offers an abbreviated table of probabilities for a standard normal distribution. Note that the table only gives probabilities for $z > 0$. Suppose we need to compute $\Phi(-z)$, where $z > 0$. Because the pdf of Z is symmetric about 0, we have

$$\Phi(-z) = 1 - \Phi(z), \tag{3.4.11}$$

see Exercise 3.4.24. In the examples below, we illustrate the computation of normal probabilities and quantiles.

Most computer packages offer functions for computation of these probabilities. For example, the R or S-PLUS command `pnorm(x,a,b)` calculates the $P(X \le x)$ when X has a normal distribution with mean a and standard deviation b, while the command `dnorm(x,a,b)` returns the value of the pdf of X at x.

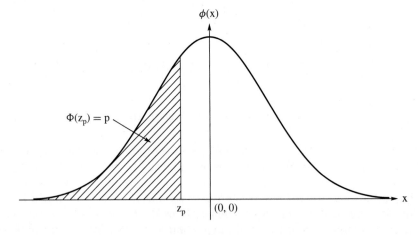

Figure 3.4.2: The Standard Normal Density: $p = \Phi(z_p)$ is the area under the curve to the left of z_p.

Example 3.4.3. Let X be $N(2, 25)$. Then, by Table III,

$$
\begin{aligned}
P(0 < X < 10) &= \Phi\left(\frac{10 - 2}{5}\right) - \Phi\left(\frac{0 - 2}{5}\right) \\
&= \Phi(1.6) - \Phi(-0.4) \\
&= 0.945 - (1 - 0.655) = 0.600
\end{aligned}
$$

and

$$
\begin{aligned}
P(-8 < X < 1) &= \Phi\left(\frac{1 - 2}{5}\right) - \Phi\left(\frac{-8 - 2}{5}\right) \\
&= \Phi(-0.2) - \Phi(-2) \\
&= (1 - 0.579) - (1 - 0.977) = 0.398. \quad \blacksquare
\end{aligned}
$$

Example 3.4.4. Let X be $N(\mu, \sigma^2)$. Then, by Table III,

$$
\begin{aligned}
P(\mu - 2\sigma < X < \mu + 2\sigma) &= \Phi\left(\frac{\mu + 2\sigma - \mu}{\sigma}\right) - \Phi\left(\frac{\mu - 2\sigma - \mu}{\sigma}\right) \\
&= \Phi(2) - \Phi(-2) \\
&= 0.977 - (1 - 0.977) = 0.954. \quad \blacksquare
\end{aligned}
$$

Example 3.4.5. Suppose that 10 percent of the probability for a certain distribution that is $N(\mu, \sigma^2)$ is below 60 and that 5 percent is above 90. What are the values of μ and σ? We are given that the random variable X is $N(\mu, \sigma^2)$ and that $P(X \leq 60) = 0.10$ and $P(X \leq 90) = 0.95$. Thus $\Phi[(60 - \mu)/\sigma] = 0.10$ and $\Phi[(90 - \mu)/\sigma] = 0.95$. From Table III we have

$$
\frac{60 - \mu}{\sigma} = -1.282, \quad \frac{90 - \mu}{\sigma} = 1.645.
$$

These conditions require that $\mu = 73.1$ and $\sigma = 10.2$ approximately. $\quad \blacksquare$

Remark 3.4.1. In this chapter we have illustrated three types of *parameters* associated with distributions. The mean μ of $N(\mu, \sigma^2)$ is called a *location parameter* because changing its value simply changes the location of the middle of the normal pdf; that is, the graph of the pdf looks exactly the same except for a shift in location. The *standard deviation* σ of $N(\mu, \sigma^2)$ is called a *scale parameter* because changing its value changes the spread of the distribution. That is, a small value of σ requires the graph of the normal pdf to be tall and narrow, while a large value of σ requires it to spread out and not be so tall. No matter what the values of μ and σ, however, the graph of the normal pdf will be that familiar "bell shape." Incidentally, the β of the gamma distribution is also a scale parameter. On the other hand, the α of the gamma distribution is called a *shape parameter*, as changing its value modifies the shape of the graph of the pdf as can be seen by referring to Figure 3.3.1. The parameters p and μ of the binomial and Poisson distributions, respectively, are also shape parameters. $\quad \blacksquare$

We close this part of the section with two important theorems.

Theorem 3.4.1. *If the random variable X is $N(\mu, \sigma^2)$, $\sigma^2 > 0$, then the random variable $V = (X - \mu)^2/\sigma^2$ is $\chi^2(1)$.*

Proof. Because $V = W^2$, where $W = (X - \mu)/\sigma$ is $N(0, 1)$, the cdf $G(v)$ for V is, for $v \geq 0$,

$$G(v) = P(W^2 \leq v) = P(-\sqrt{v} \leq W \leq \sqrt{v}).$$

That is,

$$G(v) = 2 \int_0^{\sqrt{v}} \frac{1}{\sqrt{2\pi}} e^{-w^2/2} \, dw, \quad 0 \leq v,$$

and

$$G(v) = 0, \quad v < 0.$$

If we change the variable of integration by writing $w = \sqrt{y}$, then

$$G(v) = \int_0^v \frac{1}{\sqrt{2\pi}\sqrt{y}} e^{-y/2} \, dy, \quad 0 \leq v.$$

Hence the pdf $g(v) = G'(v)$ of the continuous-type random variable V is

$$
\begin{aligned}
g(v) &= \frac{1}{\sqrt{\pi}\sqrt{2}} v^{1/2-1} e^{-v/2}, \quad 0 < v < \infty, \\
&= 0 \quad \text{elsewhere.}
\end{aligned}
$$

Since $g(v)$ is a pdf and hence

$$\int_0^\infty g(v) \, dv = 1,$$

it must be that $\Gamma(\frac{1}{2}) = \sqrt{\pi}$ and thus V is $\chi^2(1)$. ∎

One of the most important properties of the normal distribution is its additivity under independence.

Theorem 3.4.2. *Let $X_1, \ldots, X_n$ be independent random variables such that, for $i = 1, \ldots, n$, X_i has a $N(\mu_i, \sigma_i^2)$ distribution. Let $Y = \sum_{i=1}^n a_i X_i$, where $a_1, \ldots, a_n$ are constants. Then the distribution of Y is $N(\sum_{i=1}^n a_i \mu_i, \sum_{i=1}^n a_i^2 \sigma_i^2)$.*

Proof: Using independence and the mgf of normal distributions, for $t \in R$, the mgf of Y is,

$$
\begin{aligned}
M_Y(t) &= E[\exp\{tY\}] = E\left[\exp\left\{\sum_{i=1}^n t a_i X_i\right\}\right] \\
&= \prod_{i=1}^n E\left[\exp\{t a_i X_i\}\right] = \prod_{i=1}^n \exp\left\{t a_i \mu_i + (1/2)t^2 a_i^2 \sigma_i^2\right\} \\
&= \exp\left\{t \sum_{i=1}^n a_i \mu_i + (1/2)t^2 \sum_{i=1}^n a_i^2 \sigma_i^2\right\},
\end{aligned}
$$

which is the mgf of a $N(\sum_{i=1}^{n} a_i \mu_i, \sum_{i=1}^{n} a_i^2 \sigma_i^2)$ distribution. ∎

A simple corollary to this result gives the distribution of the mean $\overline{X} = n^{-1} \sum_{i=1}^{n} X_i$, when $X_1, X_2, \ldots X_n$ are iid normal random variables.

Corollary 3.4.1. *Let $X_1, \ldots, X_n$ be iid random variables with a common $N(\mu, \sigma^2)$ distribution. Let $\overline{X} = n^{-1} \sum_{i=1}^{n} X_i$. Then $\overline{X}$ has a $N(\mu, \sigma^2/n)$ distribution.*

To prove this corollary, simply take $a_i = (1/n)$, $\mu_i = \mu$, and $\sigma_i^2 = \sigma^2$, for $i = 1, 2, \ldots, n$, in Theorem 3.4.2.

3.4.1 Contaminated Normals

We next discuss a random variable whose distribution is a mixture of normals. As with the normal, we begin with a standardized random variable.

Suppose we are observing a random variable which most of the time follows a standard normal distribution but occasionally follows a normal distribution with a larger variance. In applications, we might say that most of the data are "good" but that there are occasional outliers. To make this precise let Z have a $N(0,1)$ distribution; let $I_{1-\epsilon}$ be a discrete random variable defined by,

$$I_{1-\epsilon} = \begin{cases} 1 & \text{with probability } 1 - \epsilon \\ 0 & \text{with probability } \epsilon, \end{cases}$$

and assume that Z and $I_{1-\epsilon}$ are independent. Let $W = ZI_{1-\epsilon} + \sigma_c Z(1 - I_{1-\epsilon})$. Then W is the random variable of interest.

The independence of Z and $I_{1-\epsilon}$ imply that the cdf of W is

$$
\begin{aligned}
F_W(w) = P[W \leq w] &= P[W \leq w, I_{1-\epsilon} = 1] + P[W \leq w, I_{1-\epsilon} = 0] \\
&= P[W \leq w | I_{1-\epsilon} = 1] P[I_{1-\epsilon} = 1] \\
&\quad + P[W \leq w | I_{1-\epsilon} = 0] P[I_{1-\epsilon} = 0] \\
&= P[Z \leq w](1 - \epsilon) + P[Z \leq w/\sigma_c]\epsilon \\
&= \Phi(w)(1 - \epsilon) + \Phi(w/\sigma_c)\epsilon \qquad (3.4.12)
\end{aligned}
$$

Therefore we have shown that the distribution of W is a mixture of normals. Further, because $W = ZI_{1-\epsilon} + \sigma_c Z(1 - I_{1-\epsilon})$, we have,

$$E(W) = 0 \text{ and } \text{Var}(W) = 1 + \epsilon(\sigma_c^2 - 1); \qquad (3.4.13)$$

see Exercise 3.4.25. Upon differentiating (3.4.12), the pdf of W is

$$f_W(w) = \phi(w)(1 - \epsilon) + \phi(w/\sigma_c)\frac{\epsilon}{\sigma_c}, \qquad (3.4.14)$$

where ϕ is the pdf of a standard normal.

Suppose, in general, that the random variable of interest is $X = a + bW$, where $b > 0$. Based on (3.4.13), the mean and variance of X are

$$E(X) = a \text{ and } \text{Var}(X) = b^2(1 + \epsilon(\sigma_c^2 - 1)). \qquad (3.4.15)$$

From expression (3.4.12), the cdf of X is

$$F_X(x) = \Phi\left(\frac{x-a}{b}\right)(1-\epsilon) + \Phi\left(\frac{x-a}{b\sigma_c}\right)\epsilon, \qquad (3.4.16)$$

which is a mixture of normal cdf's.

Based on expression (3.4.16) it is easy to obtain probabilities for contaminated normal distributions using R or S-PLUS. For example, suppose, as above, W has cdf (3.4.12). Then $P(W \leq w)$ is obtained by the R command `(1-eps)*pnorm(w) + eps*pnorm(w/sigc)`, where `eps` and `sigc` denote ϵ and σ_c, respectively. Similarly the pdf of W at w is returned by `(1-eps)*dnorm(w) + eps*dnorm(w/sigc)/sigc`. In Section 3.7, we explore mixture distributions in general.

EXERCISES

3.4.1. If

$$\Phi(x) = \int_{-\infty}^{z} \frac{1}{\sqrt{2\pi}} e^{-w^2/2}\, dw,$$

show that $\Phi(-z) = 1 - \Phi(z)$.

3.4.2. If X is $N(75, 100)$, find $P(X < 60)$ and $P(70 < X < 100)$ by using either Table III or if either R or S-PLUS is available the command `pnorm`.

3.4.3. If X is $N(\mu, \sigma^2)$, find b so that $P[-b < (X - \mu)/\sigma < b] = 0.90$, by using either Table III of Appendix C or if either R or S-PLUS is available the command `pnorm`.

3.4.4. Let X be $N(\mu, \sigma^2)$ so that $P(X < 89) = 0.90$ and $P(X < 94) = 0.95$. Find μ and σ^2.

3.4.5. Show that the constant c can be selected so that $f(x) = c2^{-x^2}$, $-\infty < x < \infty$, satisfies the conditions of a normal pdf.
Hint: Write $2 = e^{\log 2}$.

3.4.6. If X is $N(\mu, \sigma^2)$, show that $E(|X - \mu|) = \sigma\sqrt{2/\pi}$.

3.4.7. Show that the graph of a pdf $N(\mu, \sigma^2)$ has points of inflection at $x = \mu - \sigma$ and $x = \mu + \sigma$.

3.4.8. Evaluate $\int_2^3 \exp[-2(x-3)^2]\, dx$.

3.4.9. Determine the 90*th* percentile of the distribution, which is $N(65, 25)$.

3.4.10. If e^{3t+8t^2} is the mgf of the random variable X, find $P(-1 < X < 9)$.

3.4.11. Let the random variable X have the pdf

$$f(x) = \frac{2}{\sqrt{2\pi}} e^{-x^2/2}, \quad 0 < x < \infty, \quad \text{zero elsewhere.}$$

Find the mean and the variance of X.
Hint: Compute $E(X)$ directly and $E(X^2)$ by comparing the integral with the integral representing the variance of a random variable that is $N(0, 1)$.

3.4.12. Let X be $N(5, 10)$. Find $P[0.04 < (X - 5)^2 < 38.4]$.

3.4.13. If X is $N(1, 4)$, compute the probability $P(1 < X^2 < 9)$.

3.4.14. If X is $N(75, 25)$, find the conditional probability that X is greater than 80 given that X is greater than 77. See Exercise 2.3.12.

3.4.15. Let X be a random variable such that $E(X^{2m}) = (2m)!/(2^m m!)$, $m = 1, 2, 3, \ldots$ and $E(X^{2m-1}) = 0$, $m = 1, 2, 3, \ldots$. Find the mgf and the pdf of X.

3.4.16. Let the mutually independent random variables X_1, X_2, and X_3 be $N(0, 1)$, $N(2, 4)$, and $N(-1, 1)$, respectively. Compute the probability that exactly two of these three variables are less than zero.

3.4.17. Let X have a $N(\mu, \sigma^2)$ distribution. Use expression (3.4.9) to derive the third and fourth moments of X.

3.4.18. Compute the measures of skewness and kurtosis of a distribution which is $N(\mu, \sigma^2)$. See Exercises 1.9.13 and 1.9.14 for the definitions of skewness and kurtosis, respectively.

3.4.19. Let the random variable X have a distribution that is $N(\mu, \sigma^2)$.

(a) Does the random variable $Y = X^2$ also have a normal distribution?

(b) Would the random variable $Y = aX + b$, a and b nonzero constants have a normal distribution?
  Hint: In each case, first determine $P(Y \leq y)$.

3.4.20. Let the random variable X be $N(\mu, \sigma^2)$. What would this distribution be if $\sigma^2 = 0$?
Hint: Look at the mgf of X for $\sigma^2 > 0$ and investigate its limit as $\sigma^2 \to 0$.

3.4.21. Let Y have a *truncated* distribution with pdf $g(y) = \phi(y)/[\Phi(b) - \Phi(a)]$, for $a < y < b$, zero elsewhere, where $\phi(x)$ and $\Phi(x)$ are respectively the pdf and distribution function of a standard normal distribution. Show then that $E(Y)$ is equal to $[\phi(a) - \phi(b)]/[\Phi(b) - \Phi(a)]$.

3.4.22. Let $f(x)$ and $F(x)$ be the pdf and the cdf of a distribution of the continuous type such that $f'(x)$ exists for all x. Let the mean of the truncated distribution that has pdf $g(y) = f(y)/F(b)$, $-\infty < y < b$, zero elsewhere, be equal to $-f(b)/F(b)$ for all real b. Prove that $f(x)$ is a pdf of a standard normal distribution.

3.4.23. Let X and Y be independent random variables, each with a distribution that is $N(0, 1)$. Let $Z = X + Y$. Find the integral that represents the cdf $G(z) = P(X + Y \leq z)$ of Z. Determine the pdf of Z.
Hint: We have that $G(z) = \int_{-\infty}^{\infty} H(x, z)\, dx$, where

$$H(x, z) = \int_{-\infty}^{z-x} \frac{1}{2\pi} \exp[-(x^2 + y^2)/2]\, dy.$$

Find $G'(z)$ by evaluating $\int_{-\infty}^{\infty} [\partial H(x, z)/\partial z]\, dx$.

3.4.24. Suppose X is a random variable with the pdf $f(x)$ which is symmetric about 0, $(f(-x) = f(x))$. Show that $F(-x) = 1 - F(x)$, for all x in the support of X.

3.4.25. Derive the mean and variance of a contaminated normal random variable which is given in expression (3.4.13).

3.4.26. Assuming a computer is available, investigate the probabilities of an "outlier" for a contaminated normal random variable and a normal random variable. Specifically, determine the probability of observing the event $\{|X| \geq 2\}$ for the following random variables:

(a) X has a standard normal distribution.

(b) X has a contaminated normal distribution with cdf (3.4.12) where $\epsilon = 0.15$ and $\sigma_c = 10$.

(c) X has a contaminated normal distribution with cdf (3.4.12) where $\epsilon = 0.15$ and $\sigma_c = 20$.

(d) X has a contaminated normal distribution with cdf (3.4.12) where $\epsilon = 0.25$ and $\sigma_c = 20$.

3.4.27. Assuming a computer is available, plot the pdfs of the random variables defined in parts (a)-(d) of the last exercise. Obtain an overlay plot of all four pdfs, also. In eithet R or S-PLUS the domain values of the pdfs can easily be obtained by using the `seq` command. For instance, the command `x<-seq(-6,6,.1)` will return a vector of values between -6 and 6 in jumps of 0.1.

3.4.28. Let X_1 and X_2 be independent with normal distributions $N(6, 1)$ and $N(7, 1)$, respectively. Find $P(X_1 > X_2)$.
Hint: Write $P(X_1 > X_2) = P(X_1 - X_2 > 0)$ and determine the distribution of $X_1 - X_2$.

3.4.29. Compute $P(X_1 + 2X_2 - 2X_3 > 7)$, if X_1, X_2, X_3 are iid with common distribution $N(1, 4)$.

3.4.30. A certain job is completed in three steps in series. The means and standard deviations for the steps are (in minutes):

Step	Mean	Standard Deviation
1	17	2
2	13	1
3	13	2

Assuming independent steps and normal distributions, compute the probability that the job will take less than 40 minutes to complete.

3.4.31. Let X be $N(0,1)$. Use the moment-generating-function technique to show that $Y = X^2$ is $\chi^2(1)$.

Hint: Evaluate the integral that represents $E(e^{tX^2})$ by writing $w = x\sqrt{1-2t}$, $t < \frac{1}{2}$.

3.4.32. Suppose X_1, X_2 are iid with a common standard normal distribution. Find the joint pdf of $Y_1 = X_1^2 + X_2^2$ and $Y_2 = X_2$ and the marginal pdf of Y_1.

Hint: Note that the space of Y_1 and Y_2 is given by $-\sqrt{y_1} < y_2 < \sqrt{y_1}, 0 < y_1 < \infty$.

3.5 The Multivariate Normal Distribution

In this section we will present the multivariate normal distribution. We introduce it in general for an n-dimensional random vector, but we offer detailed examples for the bivariate case when $n = 2$. As with Section 3.4 on the normal distribution, the derivation of the distribution is simplified by first discussing the standard case and then proceeding to the general case. Also, vector and matrix notation will be used.

Consider the random vector $\mathbf{Z} = (Z_1, \ldots, Z_n)'$ where $Z_1, \ldots, Z_n$ are iid $N(0,1)$ random variables. Then the density of $\mathbf{Z}$ is

$$
\begin{aligned}
f_{\mathbf{Z}}(\mathbf{z}) &= \prod_{i=1}^{n} \frac{1}{\sqrt{2\pi}} \exp\left\{ -\frac{1}{2} z_i^2 \right\} = \left(\frac{1}{2\pi} \right)^{n/2} \exp\left\{ -\frac{1}{2} \sum_{i=1}^{n} z_i^2 \right\} \\
&= \left(\frac{1}{2\pi} \right)^{n/2} \exp\left\{ -\frac{1}{2} \mathbf{z}'\mathbf{z} \right\},
\end{aligned}
\tag{3.5.1}
$$

for $\mathbf{z} \in R^n$. Because the Z_i's have mean 0, variance 1, and are uncorrelated, the mean and covariance matrix of $\mathbf{Z}$ are

$$
E[\mathbf{Z}] = \mathbf{0} \text{ and } \text{Cov}[\mathbf{Z}] = \mathbf{I}_n,
\tag{3.5.2}
$$

where $\mathbf{I}_n$ denotes the identity matrix of order n. Recall that the mgf of Z_i is $\exp\{t_i^2/2\}$. Hence, because the Z_i's are independent, the mgf of $\mathbf{Z}$ is

$$
\begin{aligned}
M_{\mathbf{Z}}(\mathbf{t}) = E\left[\exp\{\mathbf{t}'\mathbf{Z}\}\right] &= E\left[\prod_{i=1}^{n} \exp\{t_i Z_i\} \right] = \prod_{i=1}^{n} E\left[\exp\{t_i Z_i\}\right] \\
&= \exp\left\{ \frac{1}{2} \sum_{i=1}^{n} t_i^2 \right\} = \exp\left\{ \frac{1}{2} \mathbf{t}'\mathbf{t} \right\},
\end{aligned}
\tag{3.5.3}
$$

for all $\mathbf{t} \in R^n$. We say that $\mathbf{Z}$ has a **multivariate normal distribution** with mean vector $\mathbf{0}$ and covariance matrix $\mathbf{I}_n$. We abbreviate this by saying that $\mathbf{Z}$ has an $N_n(\mathbf{0}, \mathbf{I}_n)$ distribution.

For the general case, suppose $\boldsymbol{\Sigma}$ is an $n \times n$, symmetric, and positive semi-definite matrix (psd). Then from linear algebra, we can always decompose $\boldsymbol{\Sigma}$ as

$$
\boldsymbol{\Sigma} = \boldsymbol{\Gamma}'\boldsymbol{\Lambda}\boldsymbol{\Gamma},
\tag{3.5.4}
$$

where $\mathbf{\Lambda}$ is the diagonal matrix $\mathbf{\Lambda} = \text{diag}(\lambda_1, \lambda_2, \ldots, \lambda_n)$, $\lambda_1 \geq \lambda_2 \geq \cdots \geq \lambda_n \geq 0$ are the eigenvalues of $\mathbf{\Sigma}$, and the columns of $\mathbf{\Gamma}'$, $\mathbf{v}_1, \mathbf{v}_2, \ldots, \mathbf{v}_n$, are the corresponding eigenvectors. This decomposition is called the **spectral decomposition** of $\mathbf{\Sigma}$. The matrix $\mathbf{\Gamma}$ is orthogonal, i.e., $\mathbf{\Gamma}^{-1} = \mathbf{\Gamma}'$, and, hence, $\mathbf{\Gamma}\mathbf{\Gamma}' = \mathbf{I}$. As Exercise 3.5.19 shows, we can write the spectral decomposition in another way, as

$$\mathbf{\Sigma} = \mathbf{\Gamma}'\mathbf{\Lambda}\mathbf{\Gamma} = \sum_{i=1}^{n} \lambda_i \mathbf{v}_i \mathbf{v}_i'. \tag{3.5.5}$$

Because the λ_i's are nonnegative, we can define the diagonal matrix $\mathbf{\Lambda}^{1/2} = (\sqrt{\lambda_1}, \ldots, \sqrt{\lambda_n})$. Then the orthogonality of $\mathbf{\Gamma}$ implies

$$\mathbf{\Sigma} = \mathbf{\Gamma}'\mathbf{\Lambda}^{1/2}\mathbf{\Gamma}\mathbf{\Gamma}'\mathbf{\Lambda}^{1/2}\mathbf{\Gamma}.$$

Define the **square root** of the psd matrix $\mathbf{\Sigma}$ as

$$\mathbf{\Sigma}^{1/2} = \mathbf{\Gamma}'\mathbf{\Lambda}^{1/2}\mathbf{\Gamma}, \tag{3.5.6}$$

where $\mathbf{\Lambda}^{1/2} = \text{diag}(\sqrt{\lambda_1}, \ldots, \sqrt{\lambda_n})$. Note that $\mathbf{\Sigma}^{1/2}$ is symmetric and psd. Suppose $\mathbf{\Sigma}$ is positive definite (pd); i.e., all of its eigenvalues are strictly positive. Then it is easy to show that

$$\left(\mathbf{\Sigma}^{1/2}\right)^{-1} = \mathbf{\Gamma}'\mathbf{\Lambda}^{-1/2}\mathbf{\Gamma}; \tag{3.5.7}$$

see Exercise 3.5.11. We write the left side of this equation as $\mathbf{\Sigma}^{-1/2}$. These matrices enjoy many additional properties of the law of exponents for numbers; see, for example, Arnold (1981). Here, though, all we need are the properties given above.

Let $\mathbf{Z}$ have a $N_n(\mathbf{0}, \mathbf{I}_n)$ distribution. Let $\mathbf{\Sigma}$ be a positive semi-definite, symmetric matrix and let $\boldsymbol{\mu}$ be an $n \times 1$ vector of constants. Define the random vector $\mathbf{X}$ by

$$\mathbf{X} = \mathbf{\Sigma}^{1/2}\mathbf{Z} + \boldsymbol{\mu}. \tag{3.5.8}$$

By (3.5.2) and Theorem 2.6.2, we immediately have

$$E[\mathbf{X}] = \boldsymbol{\mu} \text{ and } \text{Cov}[\mathbf{X}] = \mathbf{\Sigma}^{1/2}\mathbf{\Sigma}^{1/2} = \mathbf{\Sigma}. \tag{3.5.9}$$

Further the mgf of $\mathbf{X}$ is given by

$$
\begin{aligned}
M_{\mathbf{X}}(\mathbf{t}) = E\left[\exp\{\mathbf{t}'\mathbf{X}\}\right] &= E\left[\exp\{\mathbf{t}'\mathbf{\Sigma}^{1/2}\mathbf{Z} + \mathbf{t}'\boldsymbol{\mu}\}\right] \\
&= \exp\{\mathbf{t}'\boldsymbol{\mu}\}E\left[\exp\{\left(\mathbf{\Sigma}^{1/2}\mathbf{t}\right)'\mathbf{Z}\}\right] \\
&= \exp\{\mathbf{t}'\boldsymbol{\mu}\}\exp\{(1/2)\left(\mathbf{\Sigma}^{1/2}\mathbf{t}\right)'\mathbf{\Sigma}^{1/2}\mathbf{t}\} \\
&= \exp\{\mathbf{t}'\boldsymbol{\mu}\}\exp\{(1/2)\mathbf{t}'\mathbf{\Sigma}\mathbf{t}\}. \tag{3.5.10}
\end{aligned}
$$

This leads to the following definition:

Definition 3.5.1 (Multivariate Normal). *We say an n-dimensional random vector* **X** *has a* **multivariate normal distribution** *if its mgf is*

$$M_{\mathbf{X}}(\mathbf{t}) = \exp\left\{\mathbf{t}'\boldsymbol{\mu} + (1/2)\mathbf{t}'\boldsymbol{\Sigma}\mathbf{t}\right\}, \tag{3.5.11}$$

for all $\mathbf{t} \in R^n$ *and where* $\boldsymbol{\Sigma}$ *is a symmetric, positive semi-definite matrix and* $\boldsymbol{\mu} \in R^n$. *We abbreviate this by saying that* **X** *has a* $N_n(\boldsymbol{\mu}, \boldsymbol{\Sigma})$ *distribution.*

Note that our definition is for positive semi-definite matrices $\boldsymbol{\Sigma}$. Usually $\boldsymbol{\Sigma}$ is positive definite, in which case, we can further obtain the density of **X**. If $\boldsymbol{\Sigma}$ is positive definite then so is $\boldsymbol{\Sigma}^{1/2}$ and, as discussed above, its inverse is given by the expression (3.5.7). Thus the transformation between **X** and **Z**, (3.5.8), is one-to-one with the inverse transformation

$$\mathbf{Z} = \boldsymbol{\Sigma}^{-1/2}(\mathbf{X} - \boldsymbol{\mu}),$$

with Jacobian $|\boldsymbol{\Sigma}^{-1/2}| = |\boldsymbol{\Sigma}|^{-1/2}$. Hence, upon simplification, the pdf of **X** is given by

$$f_{\mathbf{X}}(\mathbf{x}) = \frac{1}{(2\pi)^{n/2}|\boldsymbol{\Sigma}|^{1/2}} \exp\left\{-\frac{1}{2}(\mathbf{x} - \boldsymbol{\mu})'\boldsymbol{\Sigma}^{-1}(\mathbf{x} - \boldsymbol{\mu})\right\}, \quad \text{for } \mathbf{x} \in R^n. \tag{3.5.12}$$

The following two theorems are very useful. The first says that a linear transformation of a multivariate normal random vector has a multivariate normal distribution.

Theorem 3.5.1. *Suppose* **X** *has a* $N_n(\boldsymbol{\mu}, \boldsymbol{\Sigma})$ *distribution. Let* $\mathbf{Y} = \mathbf{A}\mathbf{X} + \mathbf{b}$, *where* **A** *is an* $m \times n$ *matrix and* $\mathbf{b} \in R^m$. *Then* **Y** *has a* $N_m(\mathbf{A}\boldsymbol{\mu} + \mathbf{b}, \mathbf{A}\boldsymbol{\Sigma}\mathbf{A}')$ *distribution.*

Proof: From (3.5.11), for $\mathbf{t} \in R^m$, the mgf of **Y** is

$$
\begin{aligned}
M_{\mathbf{Y}}(\mathbf{t}) &= E\left[\exp\left\{\mathbf{t}'\mathbf{Y}\right\}\right] \\
&= E\left[\exp\left\{\mathbf{t}'(\mathbf{A}\mathbf{X} + \mathbf{b})\right\}\right] \\
&= \exp\left\{\mathbf{t}'\mathbf{b}\right\} E\left[\exp\left\{(\mathbf{A}'\mathbf{t})'\mathbf{X}\right\}\right] \\
&= \exp\left\{\mathbf{t}'\mathbf{b}\right\} \exp\left\{(\mathbf{A}'\mathbf{t})'\boldsymbol{\mu} + (1/2)(\mathbf{A}'\mathbf{t})'\boldsymbol{\Sigma}(\mathbf{A}'\mathbf{t})\right\} \\
&= \exp\left\{\mathbf{t}'(\mathbf{A}\boldsymbol{\mu} + \mathbf{b}) + (1/2)\mathbf{t}'\mathbf{A}\boldsymbol{\Sigma}\mathbf{A}'\mathbf{t}\right\},
\end{aligned}
$$

which is the mgf of an $N_m(\mathbf{A}\boldsymbol{\mu} + \mathbf{b}, \mathbf{A}\boldsymbol{\Sigma}\mathbf{A}')$ distribution. ∎

A simple corollary to this theorem gives marginal distributions of a multivariate normal random variable. Let $\mathbf{X}_1$ be any subvector of **X**, say of dimension $m < n$. Because we can always rearrange means and correlations, there is no loss in generality in writing **X** as

$$\mathbf{X} = \left[\begin{array}{c} \mathbf{X}_1 \\ \mathbf{X}_2 \end{array}\right], \tag{3.5.13}$$

where $\mathbf{X}_2$ is of dimension $p = n - m$. In the same way, partition the mean and covariance matrix of **X**; that is,

$$\boldsymbol{\mu} = \left[\begin{array}{c} \boldsymbol{\mu}_1 \\ \boldsymbol{\mu}_2 \end{array}\right] \text{ and } \boldsymbol{\Sigma} = \left[\begin{array}{cc} \boldsymbol{\Sigma}_{11} & \boldsymbol{\Sigma}_{12} \\ \boldsymbol{\Sigma}_{21} & \boldsymbol{\Sigma}_{22} \end{array}\right] \tag{3.5.14}$$

with the same dimensions as in expression (3.5.13). Note, for instance, that $\boldsymbol{\Sigma}_{11}$ is the covariance matrix of $\mathbf{X}_1$ and $\boldsymbol{\Sigma}_{12}$ contains all the covariances between the components of $\mathbf{X}_1$ and $\mathbf{X}_2$. Now define $\mathbf{A}$ to be the matrix,

$$\mathbf{A} = [\mathbf{I}_m \vdots \mathbf{O}_{mp}],$$

where $\mathbf{O}_{mp}$ is a $m \times p$ matrix of zeroes. Then $\mathbf{X}_1 = \mathbf{A}\mathbf{X}$. Hence, applying Theorem 3.5.1 to this transformation, along with some matrix algebra, we have the following corollary:

Corollary 3.5.1. *Suppose* $\mathbf{X}$ *has a* $N_n(\boldsymbol{\mu}, \boldsymbol{\Sigma})$ *distribution, partitioned as in expressions (3.5.13) and (3.5.14). Then* $\mathbf{X}_1$ *has a* $N_m(\boldsymbol{\mu}_1, \boldsymbol{\Sigma}_{11})$ *distribution.*

This is a useful result because it says that any marginal distribution of $\mathbf{X}$ is also normal and, further, its mean and covariance matrix are those associated with that partial vector.

Example 3.5.1. In this example, we explore the multivariate normal case when $n = 2$. The distribution in this case is called the bivariate normal. We will also use the customary notation of (X, Y) instead of (X_1, X_2). So, suppose (X, Y) has a $N_2(\boldsymbol{\mu}, \boldsymbol{\Sigma})$ distribution, where

$$\boldsymbol{\mu} = \begin{bmatrix} \mu_1 \\ \mu_2 \end{bmatrix} \text{ and } \boldsymbol{\Sigma} = \begin{bmatrix} \sigma_1^2 & \sigma_{12} \\ \sigma_{12} & \sigma_2^2 \end{bmatrix}. \tag{3.5.15}$$

Hence, μ_1 and σ_1^2 are the mean and variance, respectively, of X; μ_2 and σ_2^2 are the mean and variance, respectively, of Y; and σ_{12} is the covariance between X and Y. Recall that $\sigma_{12} = \rho\sigma_1\sigma_2$, where ρ is the correlation coefficient between X and Y. Substituting $\rho\sigma_1\sigma_2$ for σ_{12} in $\boldsymbol{\Sigma}$, it is easy to see that the determinant of $\boldsymbol{\Sigma}$ is $\sigma_1^2\sigma_2^2(1 - \rho^2)$. Recall that $\rho^2 \leq 1$. For the remainder of this example, assume that $\rho^2 < 1$. In this case, $\boldsymbol{\Sigma}$ is invertible (it is also positive-definite). Further, since $\boldsymbol{\Sigma}$ is a 2×2 matrix, its inverse can easily be determined to be

$$\boldsymbol{\Sigma}^{-1} = \frac{1}{\sigma_1^2\sigma_2^2(1 - \rho^2)} \begin{bmatrix} \sigma_2^2 & -\rho\sigma_1\sigma_2 \\ -\rho\sigma_1\sigma_2 & \sigma_1^2 \end{bmatrix}. \tag{3.5.16}$$

Using this expression, the pdf of (X, Y), expression (3.5.12), can be written as

$$f(x, y) = \frac{1}{2\pi\sigma_1\sigma_2\sqrt{1 - \rho^2}} e^{-q/2}, \quad -\infty < x < \infty, \quad -\infty < y < \infty, \tag{3.5.17}$$

where,

$$q = \frac{1}{1 - \rho^2} \left[\left(\frac{x - \mu_1}{\sigma_1} \right)^2 - 2\rho \left(\frac{x - \mu_1}{\sigma_1} \right) \left(\frac{y - \mu_2}{\sigma_2} \right) + \left(\frac{y - \mu_2}{\sigma_2} \right)^2 \right]; \tag{3.5.18}$$

see Exercise 3.5.12.

Recall in general, that if X and Y are independent random variable then their correlation coefficient is 0. If they are normal, by Corollary 3.5.1, X has a $N(\mu_1, \sigma_1^2)$

distribution and Y has a $N(\mu_2, \sigma_2^2)$ distribution. Further, based on the expression (3.5.17) for the joint pdf of (X, Y), we see that if the correlation coefficient is 0, then X and Y are independent. That is, for the bivariate normal case, independence is equivalent to $\rho = 0$. That this is true for the multivariate normal is shown by Theorem 3.5.2. ■

Recall in Section 2.5, Example 2.5.4, that if two random variables are independent then their covariance is 0. In general the converse is not true. However, as the following theorem shows, it is true for the multivariate normal distribution.

Theorem 3.5.2. *Suppose* $\mathbf{X}$ *has a* $N_n(\boldsymbol{\mu}, \boldsymbol{\Sigma})$ *distribution, partioned as in the expressions (3.5.13) and (3.5.14). Then* $\mathbf{X}_1$ *and* $\mathbf{X}_2$ *are independent if and only if* $\boldsymbol{\Sigma}_{12} = \mathbf{O}$.

Proof: First note that $\boldsymbol{\Sigma}_{21} = \boldsymbol{\Sigma}_{12}'$. The joint mgf of $\mathbf{X}_1$ and $\mathbf{X}_2$ is given by,

$$M_{\mathbf{X}_1, \mathbf{X}_2}(\mathbf{t}_1, \mathbf{t}_2) = \exp\left\{\mathbf{t}_1'\boldsymbol{\mu}_1 + \mathbf{t}_2'\boldsymbol{\mu}_2 + \frac{1}{2}(\mathbf{t}_1'\boldsymbol{\Sigma}_{11}\mathbf{t}_1 + \mathbf{t}_2'\boldsymbol{\Sigma}_{22}\mathbf{t}_2 + \mathbf{t}_2'\boldsymbol{\Sigma}_{21}\mathbf{t}_1 + \mathbf{t}_1'\boldsymbol{\Sigma}_{12}\mathbf{t}_2)\right\}$$
$$(3.5.19)$$

where $\mathbf{t}' = (\mathbf{t}_1', \mathbf{t}_2')$ is partitioned the same as $\boldsymbol{\mu}$. By Corollary 3.5.1, $\mathbf{X}_1$ has a $N_m(\boldsymbol{\mu}_1, \boldsymbol{\Sigma}_{11})$ distribution and $\mathbf{X}_2$ has a $N_p(\boldsymbol{\mu}_2, \boldsymbol{\Sigma}_{22})$ distribution. Hence, the product of their marginal mgfs is:

$$M_{\mathbf{X}_1}(\mathbf{t}_1) M_{\mathbf{X}_2}(\mathbf{t}_2) = \exp\left\{\mathbf{t}_1'\boldsymbol{\mu}_1 + \mathbf{t}_2'\boldsymbol{\mu}_2 + \frac{1}{2}(\mathbf{t}_1'\boldsymbol{\Sigma}_{11}\mathbf{t}_1 + \mathbf{t}_2'\boldsymbol{\Sigma}_{22}\mathbf{t}_2)\right\}. \qquad (3.5.20)$$

By (2.6.6) of Section 2.6, $\mathbf{X}_1$ and $\mathbf{X}_2$ are independent if and only if the expressions (3.5.19) and (3.5.20) are the same. If $\boldsymbol{\Sigma}_{12} = \mathbf{O}$ and, hence, $\boldsymbol{\Sigma}_{21} = \mathbf{O}$, then the expressions are the same and $\mathbf{X}_1$ and $\mathbf{X}_2$ are independent. If $\mathbf{X}_1$ and $\mathbf{X}_2$ are independent, then the covariances between their components are all 0; i.e., $\boldsymbol{\Sigma}_{12} = \mathbf{O}$ and $\boldsymbol{\Sigma}_{21} = \mathbf{O}$. ■

Corollary 3.5.1 showed that the marginal distributions of a multivariate normal are themselves normal. This is true for conditional distributions, too. As the following proof shows, we can combine the results of Theorems 3.5.1 and 3.5.2 to obtain the following theorem.

Theorem 3.5.3. *Suppose* $\mathbf{X}$ *has a* $N_n(\boldsymbol{\mu}, \boldsymbol{\Sigma})$ *distribution, which is partioned as in expressions (3.5.13) and (3.5.14). Assume that* $\boldsymbol{\Sigma}$ *is positive definite. Then the conditional distribution of* $\mathbf{X}_1 \,|\, \mathbf{X}_2$ *is*

$$N_m(\boldsymbol{\mu}_1 + \boldsymbol{\Sigma}_{12}\boldsymbol{\Sigma}_{22}^{-1}(\mathbf{X}_2 - \boldsymbol{\mu}_2), \boldsymbol{\Sigma}_{11} - \boldsymbol{\Sigma}_{12}\boldsymbol{\Sigma}_{22}^{-1}\boldsymbol{\Sigma}_{21}). \qquad (3.5.21)$$

Proof: Consider first the joint distribution of the random vector $\mathbf{W} = \mathbf{X}_1 - \boldsymbol{\Sigma}_{12}\boldsymbol{\Sigma}_{22}^{-1}\mathbf{X}_2$ and $\mathbf{X}_2$. This distribution is obtained from the transformation,

$$\begin{bmatrix} \mathbf{W} \\ \mathbf{X}_2 \end{bmatrix} = \begin{bmatrix} \mathbf{I}_m & -\boldsymbol{\Sigma}_{12}\boldsymbol{\Sigma}_{22}^{-1} \\ \mathbf{O} & \mathbf{I}_p \end{bmatrix} \begin{bmatrix} \mathbf{X}_1 \\ \mathbf{X}_2 \end{bmatrix}.$$

Because this is a linear transformation, it follows from Theorem 3.5.1 that joint distribution is multivariate normal, with $E[\mathbf{W}] = \boldsymbol{\mu}_1 - \boldsymbol{\Sigma}_{12}\boldsymbol{\Sigma}_{22}^{-1}\boldsymbol{\mu}_2$, $E[\mathbf{X}_2] = \boldsymbol{\mu}_2$, and covariance matrix

$$\begin{bmatrix} \mathbf{I}_m & -\boldsymbol{\Sigma}_{12}\boldsymbol{\Sigma}_{22}^{-1} \\ \mathbf{O} & \mathbf{I}_p \end{bmatrix} \begin{bmatrix} \boldsymbol{\Sigma}_{11} & \boldsymbol{\Sigma}_{12} \\ \boldsymbol{\Sigma}_{21} & \boldsymbol{\Sigma}_{22} \end{bmatrix} \begin{bmatrix} \mathbf{I}_m & \mathbf{O} \\ -\boldsymbol{\Sigma}_{22}^{-1}\boldsymbol{\Sigma}_{21} & \mathbf{I}_p \end{bmatrix} =$$
$$\begin{bmatrix} \boldsymbol{\Sigma}_{11} - \boldsymbol{\Sigma}_{12}\boldsymbol{\Sigma}_{22}^{-1}\boldsymbol{\Sigma}_{21} & \mathbf{O} \\ \mathbf{O} & \boldsymbol{\Sigma}_{22} \end{bmatrix}.$$

Hence, by Theorem 3.5.2 the random vectors $\mathbf{W}$ and $\mathbf{X}_2$ are independent. Thus, the conditional distribution of $\mathbf{W} \,|\, \mathbf{X}_2$ is the same as the marginal distribution of $\mathbf{W}$; that is

$$\mathbf{W} \,|\, \mathbf{X}_2 \text{ is } N_m(\boldsymbol{\mu}_1 - \boldsymbol{\Sigma}_{12}\boldsymbol{\Sigma}_{22}^{-1}\boldsymbol{\mu}_2, \boldsymbol{\Sigma}_{11} - \boldsymbol{\Sigma}_{12}\boldsymbol{\Sigma}_{22}^{-1}\boldsymbol{\Sigma}_{21}).$$

Further, because of this independence, $\mathbf{W} + \boldsymbol{\Sigma}_{12}\boldsymbol{\Sigma}_{22}^{-1}\mathbf{X}_2$ given $\mathbf{X}_2$ is distributed as

$$N_m(\boldsymbol{\mu}_1 - \boldsymbol{\Sigma}_{12}\boldsymbol{\Sigma}_{22}^{-1}\boldsymbol{\mu}_2 + \boldsymbol{\Sigma}_{12}\boldsymbol{\Sigma}_{22}^{-1}\mathbf{X}_2, \boldsymbol{\Sigma}_{11} - \boldsymbol{\Sigma}_{12}\boldsymbol{\Sigma}_{22}^{-1}\boldsymbol{\Sigma}_{21}), \qquad (3.5.22)$$

which is the desired result. ∎

Example 3.5.2 (Continuation of Example 3.5.1). Consider once more the bivariate normal distribution which was given in Example 3.5.1. For this case, reversing the roles so that $Y = X_1$ and $X = X_2$, expression (3.5.21) shows that the conditional distribution of Y given $X = x$ is

$$N\left[\mu_2 + \rho\frac{\sigma_2}{\sigma_1}(x - \mu_1), \sigma_2^2(1 - \rho^2)\right]. \qquad (3.5.23)$$

Thus, with a bivariate normal distribution, the conditional mean of Y, given that $X = x$, is linear in x and is given by

$$E(Y|x) = \mu_2 + \rho\frac{\sigma_2}{\sigma_1}(x - \mu_1).$$

Since the coefficient of x in this linear conditional mean $E(Y|x)$ is $\rho\sigma_2/\sigma_1$, and since σ_1 and σ_2 represent the respective standard deviations, ρ is the correlation coefficient of X and Y. This follows from the result, established in Section 2.4, that the coefficient of x in a general linear conditional mean $E(Y|x)$ is the product of the correlation coefficient and the ratio σ_2/σ_1.

  Although the mean of the conditional distribution of Y, given $X = x$, depends upon x (unless $\rho = 0$), the variance $\sigma_2^2(1 - \rho^2)$ is the same for all real values of x. Thus, by way of example, given that $X = x$, the conditional probability that Y is within $(2.576)\sigma_2\sqrt{1 - \rho^2}$ units of the conditional mean is 0.99, whatever the value of x may be. In this sense, most of the probability for the distribution of X and Y lies in the band

$$\mu_2 + \rho\frac{\sigma_2}{\sigma_1}(x - \mu_1) \pm (2.576)\sigma_2\sqrt{1 - \rho^2}$$

about the graph of the linear conditional mean. For every fixed positive σ_2, the width of this band depends upon ρ. Because the band is narrow when ρ^2 is nearly

1, we see that ρ does measure the intensity of the concentration of the probability for X and Y about the linear conditional mean. We alluded to this fact in the remark of Section 2.4.

In a similar manner we can show that the conditional distribution of X, given $Y = y$, is the normal distribution

$$N\left[\mu_1 + \rho\frac{\sigma_1}{\sigma_2}(y - \mu_2),\ \sigma_1^2(1 - \rho^2)\right].\quad\blacksquare$$

Example 3.5.3. Let us assume that in a certain population of married couples the height X_1 of the husband and the height X_2 of the wife have a bivariate normal distribution with parameters $\mu_1 = 5.8$ feet, $\mu_2 = 5.3$ feet, $\sigma_1 = \sigma_2 = 0.2$ foot, and $\rho = 0.6$. The conditional pdf of X_2 given $X_1 = 6.3$, is normal with mean $5.3 + (0.6)(6.3 - 5.8) = 5.6$ and standard deviation $(0.2)\sqrt{(1 - 0.36)} = 0.16$. Accordingly, given that the height of the husband is 6.3 feet, the probability that his wife has a height between 5.28 and 5.92 feet is

$$P(5.28 < X_2 < 5.92 | X_1 = 6.3) = \Phi(2) - \Phi(-2) = 0.954.$$

The interval $(5.28, 5.92)$ could be thought of as a 95.4 percent *prediction interval* for the wife's height, given $X_1 = 6.3$. $\blacksquare$

Recall that if the random variable X has a $N(\mu, \sigma^2)$ distribution then the random variable $[(X - \mu)/\sigma]^2$ has a $\chi^2(1)$ distribution. The multivariate analogue of this fact is given in the next theorem.

Theorem 3.5.4. *Suppose* $\mathbf{X}$ *has a* $N_n(\boldsymbol{\mu}, \boldsymbol{\Sigma})$ *distribution where* $\boldsymbol{\Sigma}$ *is positive definite. Then the random variable* $W = (\mathbf{X} - \boldsymbol{\mu})'\boldsymbol{\Sigma}^{-1}(\mathbf{X} - \boldsymbol{\mu})$ *has a* $\chi^2(n)$ *distribution.*

Proof: Write $\boldsymbol{\Sigma} = \boldsymbol{\Sigma}^{1/2}\boldsymbol{\Sigma}^{1/2}$ where $\boldsymbol{\Sigma}^{1/2}$ is defined as in (3.5.6). Then $\mathbf{Z} = \boldsymbol{\Sigma}^{-1/2}(\mathbf{X} - \boldsymbol{\mu})$ is $N_n(\mathbf{0}, \mathbf{I}_n)$. Let $W = \mathbf{Z}'\mathbf{Z} = \sum_{i=1}^{n} Z_i^2$. Because, for $i = 1, 2, \ldots, n$, Z_i has a $N(0, 1)$ distribution, it follows from Theorem 3.4.1 that Z_i^2 has a $\chi^2(1)$ distribution. Because $Z_1, \ldots, Z_n$ are independent standard normal random variables, by Corollary 3.3.1 $\sum_{i=1} Z_i^2 = W$ has a $\chi^2(n)$ distribution. $\blacksquare$

3.5.1 *Applications

In this section, we consider several applications of the multivariate normal distribution. These the reader may have already encountered in an applied course in statistics. The first is *principal components* which results in a linear function of a multivariate normal random vector which has independent components and preserves the "total" variation in the problem.

Let the random vector $\mathbf{X}$ have the multivariate normal distribution $N_n(\boldsymbol{\mu}, \boldsymbol{\Sigma})$ where $\boldsymbol{\Sigma}$ is positive definite. As in (3.5.4), write the spectral decomposition of $\boldsymbol{\Sigma}$ as $\boldsymbol{\Sigma} = \boldsymbol{\Gamma}'\boldsymbol{\Lambda}\boldsymbol{\Gamma}$. Recall that the columns, $\mathbf{v}_1, \mathbf{v}_2, \ldots, \mathbf{v}_n$, of $\boldsymbol{\Gamma}'$ are the eigenvectors corresponding to the eigenvalues $\lambda_1, \lambda_2, \ldots, \lambda_n$ which form the main diagonal of the matrix $\boldsymbol{\Lambda}$. Assume without loss of generality that the eigenvalues are decreasing, i.e., $\lambda_1 \geq \lambda_2 \geq \cdots \geq \lambda_n > 0$. Define the random vector $\mathbf{Y} = \boldsymbol{\Gamma}(\mathbf{X} - \boldsymbol{\mu})$. Since

$\mathbf{\Gamma\Sigma\Gamma'} = \mathbf{\Lambda}$, by Theorem 3.5.1 $\mathbf{Y}$ has a $N_n(\mathbf{0}, \mathbf{\Lambda})$ distribution. Hence the components $Y_1, Y_2, \ldots, Y_n$ are independent random variables and, for $i = 1, 2, \ldots, n$, Y_i has a $N(0, \lambda_i)$ distribution. The random vector $\mathbf{Y}$ is called the vector of **principal components**.

We say the **total variation**, (TV), of a random vector is the sum of the variances of its components. For the random vector $\mathbf{X}$, because $\mathbf{\Gamma}$ is an orthogonal matrix

$$TV(\mathbf{X}) = \sum_{i=1}^{n} \sigma_i^2 = \mathrm{tr}\mathbf{\Sigma} = \mathrm{tr}\mathbf{\Gamma'\Lambda\Gamma} = \mathrm{tr}\mathbf{\Lambda\Gamma\Gamma'} = \sum_{i=1}^{n} \lambda_i = TV(\mathbf{Y}).$$

Hence, $\mathbf{X}$ and $\mathbf{Y}$ have the same total variation.

Next, consider the first component of $\mathbf{Y}$ which is given by $Y_1 = \mathbf{v}_1'(\mathbf{X} - \boldsymbol{\mu})$. This is a linear combination of the components of $\mathbf{X} - \boldsymbol{\mu}$ with the property $\|\mathbf{v}_1\|^2 = \sum_{j=1}^{n} v_{1j}^2 = 1$, because $\mathbf{\Gamma'}$ is orthogonal. Consider any other linear combination of $(\mathbf{X} - \boldsymbol{\mu})$, say $\mathbf{a}'(\mathbf{X} - \boldsymbol{\mu})$ such that $\|\mathbf{a}\|^2 = 1$. Because $\mathbf{a} \in R^n$ and $\{\mathbf{v}_1, \ldots, \mathbf{v}_n\}$ forms a basis for R^n, we must have $\mathbf{a} = \sum_{j=1}^{n} a_j \mathbf{v}_j$ for some set of scalars $a_1, \ldots, a_n$. Furthermore, because the basis $\{\mathbf{v}_1, \ldots, \mathbf{v}_n\}$ is orthonormal

$$\mathbf{a}'\mathbf{v}_i = \left(\sum_{j=1}^{n} a_j \mathbf{v}_j\right)' \mathbf{v}_i = \sum_{j=1}^{n} a_j \mathbf{v}_j' \mathbf{v}_i = a_i.$$

Using (3.5.5) and the fact that $\lambda_i > 0$, we have the inequality

$$
\begin{aligned}
\mathrm{Var}(\mathbf{a}'\mathbf{X}) &= \mathbf{a}'\mathbf{\Sigma}\mathbf{a} \\
&= \sum_{i=1}^{n} \lambda_i (\mathbf{a}'\mathbf{v}_i)^2 \\
&= \sum_{i=1}^{n} \lambda_i a_i^2 \leq \lambda_1 \sum_{i=1}^{n} a_i^2 = \lambda_1 = \mathrm{Var}(Y_1). \quad (3.5.24)
\end{aligned}
$$

Hence, Y_1 has the maximum variance of any linear combination $\mathbf{a}'(\mathbf{X} - \boldsymbol{\mu})$, such that $\|\mathbf{a}\| = 1$. For this reason, Y_1 is called the **first principal component** of $\mathbf{X}$.

What about the other components, $Y_2, \ldots, Y_n$? As the following theorem shows, they share a similar property relative to the order of their associated eigenvalue. For this reason, they are called the **second, third**, through the **nth principal components**, respectively.

Theorem 3.5.5. *Consider the situation described above. For $j = 2, \ldots, n$ and $i = 1, 2, \ldots, j - 1$, $Var[\mathbf{a}'\mathbf{X}] \leq \lambda_j = Var(Y_j)$, for all vectors $\mathbf{a}$ such that $\mathbf{a} \perp \mathbf{v}_i$ and $\|\mathbf{a}\| = 1$.*

The proof of this theorem is similar to that for the first principal component and is left as Exercise 3.5.20. A second application concerning linear regression is offered in Exercise 3.5.22.

EXERCISES

3.5.1. Let X and Y have a bivariate normal distribution with respective parameters $\mu_x = 2.8$, $\mu_y = 110$, $\sigma_x^2 = 0.16$, $\sigma_y^2 = 100$, and $\rho = 0.6$. Compute:

(a) $P(106 < Y < 124)$.

(b) $P(106 < Y < 124|X = 3.2)$.

3.5.2. Let X and Y have a bivariate normal distribution with parameters $\mu_1 = 3$, $\mu_2 = 1$, $\sigma_1^2 = 16$, $\sigma_2^2 = 25$, and $\rho = \frac{3}{5}$. Determine the following probabilities:

(a) $P(3 < Y < 8)$.

(b) $P(3 < Y < 8|X = 7)$.

(c) $P(-3 < X < 3)$.

(d) $P(-3 < X < 3|Y = -4)$.

3.5.3. If $M(t_1, t_2)$ is the mgf of a bivariate normal distribution, compute the covariance by using the formula

$$\frac{\partial^2 M(0,0)}{\partial t_1 \partial t_2} - \frac{\partial M(0,0)}{\partial t_1} \frac{\partial M(0,0)}{\partial t_2}.$$

Now let $\psi(t_1, t_2) = \log M(t_1, t_2)$. Show that $\partial^2 \psi(0,0)/\partial t_1 \partial t_2$ gives this covariance directly.

3.5.4. Let U and V be independent random variables, each having a standard normal distribution. Show that the mgf $E(e^{t(UV)})$ of the random variable UV is $(1 - t^2)^{-1/2}$, $-1 < t < 1$.
Hint: Compare $E(e^{tUV})$ with the integral of a bivariate normal pdf that has means equal to zero.

3.5.5. Let X and Y have a bivariate normal distribution with parameters $\mu_1 = 5$, $\mu_2 = 10$, $\sigma_1^2 = 1$, $\sigma_2^2 = 25$, and $\rho > 0$. If $P(4 < Y < 16|X = 5) = 0.954$, determine ρ.

3.5.6. Let X and Y have a bivariate normal distribution with parameters $\mu_1 = 20$, $\mu_2 = 40$, $\sigma_1^2 = 9$, $\sigma_2^2 = 4$, and $\rho = 0.6$. Find the shortest interval for which 0.90 is the conditional probability that Y is in the interval, given that $X = 22$.

3.5.7. Say the correlation coefficient between the heights of husbands and wives is 0.70 and the mean male height is 5 feet 10 inches with standard deviation 2 inches, and the mean female height is 5 feet 4 inches with standard deviation $1\frac{1}{2}$ inches. Assuming a bivariate normal distribution, what is the best guess of the height of a woman whose husband's height is 6 feet? Find a 95 percent prediction interval for her height.

3.5.8. Let

$$f(x,y) = (1/2\pi) \exp\left[-\frac{1}{2}(x^2 + y^2)\right]\left\{1 + xy\exp\left[-\frac{1}{2}(x^2 + y^2 - 2)\right]\right\},$$

where $-\infty < x < \infty$, $-\infty < y < \infty$. If $f(x,y)$ is a joint pdf it is not a normal bivariate pdf. Show that $f(x,y)$ actually is a joint pdf and that each marginal pdf is normal. Thus the fact that each marginal pdf is normal does not imply that the joint pdf is bivariate normal.

3.5.9. Let X, Y, and Z have the joint pdf

$$\left(\frac{1}{2\pi}\right)^{3/2} \exp\left(-\frac{x^2 + y^2 + z^2}{2}\right)\left[1 + xyz\exp\left(-\frac{x^2 + y^2 + z^2}{2}\right)\right],$$

where $-\infty < x < \infty$, $-\infty < y < \infty$, and $-\infty < z < \infty$. While X, Y, and Z are obviously dependent, show that X, Y, and Z are pairwise independent and that each pair has a bivariate normal distribution.

3.5.10. Let X and Y have a bivariate normal distribution with parameters $\mu_1 = \mu_2 = 0$, $\sigma_1^2 = \sigma_2^2 = 1$, and correlation coefficient ρ. Find the distribution of the random variable $Z = aX + bY$ in which a and b are nonzero constants.

3.5.11. Establish formula (3.5.7) by a direct multiplication.

3.5.12. Show that the expression (3.5.12) becomes that of (3.5.17) in the bivariate case.

3.5.13. Show that expression (3.5.21) simplifies to expression (3.5.23) for the bivariate normal case.

3.5.14. Let $\mathbf{X} = (X_1, X_2, X_3)$ have a multivariate normal distribution with mean vector $\mathbf{0}$ and variance-covariance matrix

$$\boldsymbol{\Sigma} = \begin{bmatrix} 1 & 0 & 0 \\ 0 & 2 & 1 \\ 0 & 1 & 2 \end{bmatrix}.$$

Find $P(X_1 > X_2 + X_3 + 2)$.
Hint: Find the vector $\mathbf{a}$ so that $\mathbf{aX} = X_1 - X_2 - X_3$ and make use of Theorem 3.5.1.

3.5.15. Suppose X is distributed $N_n(\boldsymbol{\mu}, \boldsymbol{\Sigma})$. Let $\overline{X} = n^{-1}\sum_{i=1}^{n} X_i$.

(a) Write $\overline{X}$ as $\mathbf{aX}$ for an appropriate vector $\mathbf{a}$ and apply Theorem 3.5.1 to find the distribution of $\overline{X}$.

(b) Determine the distribution of $\overline{X}$, if all of its component random variables X_i have the same mean μ.

3.5.16. Suppose $\mathbf{X}$ is distributed $N_2(\boldsymbol{\mu}, \boldsymbol{\Sigma})$. Determine the distribution of the random vector $(X_1 + X_2, X_1 - X_2)$. Show that $X_1 + X_2$ and $X_1 - X_2$ are independent if $\mathrm{Var}(X_1) = \mathrm{Var}(X_2)$.

3.5.17. Suppose X is distributed $N_3(\mathbf{0}, \boldsymbol{\Sigma})$, where

$$\boldsymbol{\Sigma} = \begin{bmatrix} 3 & 2 & 1 \\ 2 & 2 & 1 \\ 1 & 1 & 3 \end{bmatrix}.$$

Find $P((X_1 - 2X_2 + X_3)^2 > 15.36)$.

3.5.18. Let X_1, X_2, X_3 be iid random variables each having a standard normal distribution. Let the random variables Y_1, Y_2, Y_3 be defined by

$$X_1 = Y_1 \cos Y_2 \sin Y_3, \quad X_2 = Y_1 \sin Y_2 \sin Y_3, \quad X_3 = Y_1 \cos Y_3,$$

where $0 \le Y_1 < \infty$, $0 \le Y_2 < 2\pi$, $0 \le Y_3 \le \pi$. Show that Y_1, Y_2, Y_3 are mutually independent.

3.5.19. Show that expression (3.5.5) is true.

3.5.20. Prove Theorem 3.5.5.

3.5.21. Suppose $\mathbf{X}$ has a multivariate normal distribution with mean $\mathbf{0}$ and covariance matrix

$$\boldsymbol{\Sigma} = \begin{bmatrix} 283 & 215 & 277 & 208 \\ 215 & 213 & 217 & 153 \\ 277 & 217 & 336 & 236 \\ 208 & 153 & 236 & 194 \end{bmatrix}$$

(a) Find the total variation of $\mathbf{X}$

(b) Find the principal component vector $\mathbf{Y}$.

(c) Show that the first principal component accounts for 90% of the total variation.

(d) Show that the first principal component Y_1 is essentially a rescaled $\overline{X}$. Determine the variance of $(1/2)\overline{X}$ and compare it to that of Y_1.

Note if either R or S-PLUS is available, the command `eigen(amat)` obtains the spectral decomposition of the matrix `amat`.

3.5.22. Readers may have encountered the multiple regression model in a previous course in statistics. We can briefly write it as follows. Suppose we have a vector of n observations $\mathbf{Y}$ which has the distribution $N_n(\mathbf{X}\boldsymbol{\beta}, \sigma^2\mathbf{I})$, where $\mathbf{X}$ is an $n \times p$ matrix of known values, which has full column rank p, and $\boldsymbol{\beta}$ is a $p \times 1$ vector of unknown parameters. The least squares estimator of $\boldsymbol{\beta}$ is

$$\widehat{\boldsymbol{\beta}} = (\mathbf{X}'\mathbf{X})^{-1}\mathbf{X}'\mathbf{Y}.$$

(a) Determine the distribution of $\widehat{\boldsymbol{\beta}}$.

(b) Let $\widehat{\mathbf{Y}} = \mathbf{X}\widehat{\boldsymbol{\beta}}$. Determine the distribution of $\widehat{\mathbf{Y}}$.

(c) Let $\widehat{\mathbf{e}} = \mathbf{Y} - \widehat{\mathbf{Y}}$. Determine the distribution of $\widehat{\mathbf{e}}$.

(d) By writing the random vector $(\widehat{\mathbf{Y}}', \widehat{\mathbf{e}}')'$ as a linear function of $\mathbf{Y}$, show that the random vectors $\widehat{\mathbf{Y}}$ and $\widehat{\mathbf{e}}$ are independent.

(e) Show that $\widehat{\boldsymbol{\beta}}$ solves the least squares problem, that is,

$$\|\mathbf{Y} - \mathbf{X}\widehat{\boldsymbol{\beta}}\|^2 = \min_{\mathbf{b} \in R^p} \|\mathbf{Y} - \mathbf{X}\mathbf{b}\|^2.$$

3.6 t and F-Distributions

It is the purpose of this section to define two additional distributions that are quite useful in certain problems of statistical inference. These are called, respectively, the (Student's) t-distribution and the F-distribution.

3.6.1   The t-distribution

Let W denote a random variable that is $N(0, 1)$; let V denote a random variable that is $\chi^2(r)$; and let W and V be independent. Then the joint pdf of W and V, say $h(w, v)$, is the product of the pdf of W and that of V or

$$h(w, v) = \begin{cases} \frac{1}{\sqrt{2\pi}} e^{-w^2/2} \frac{1}{\Gamma(r/2)2^{r/2}} v^{r/2-1} e^{-v/2} & -\infty < w < \infty, \quad 0 < v < \infty \\ 0 & \text{elsewhere.} \end{cases}$$

Define a new random variable T by writing

$$T = \frac{W}{\sqrt{V/r}}.$$

The change-of-variable technique will be used to obtain the pdf $g_1(t)$ of T. The equations

$$t = \frac{w}{\sqrt{v/r}} \quad \text{and} \quad u = v$$

define a transformation that maps $\mathcal{S} = \{(w, v) : -\infty < w < \infty, 0 < v < \infty\}$ one-to-one and onto $\mathcal{T} = \{(t, u) : -\infty < t < \infty, 0 < u < \infty\}$. Since $w = t\sqrt{u}/\sqrt{r}$, $v = u$, the absolute value of the Jacobian of the transformation is $|J| = \sqrt{u}/\sqrt{r}$. Accordingly, the joint pdf of T and $U = V$ is given by

$$\begin{aligned} g(t, u) &= h\left(\frac{t\sqrt{u}}{\sqrt{r}}, u\right)|J| \\ &= \begin{cases} \frac{1}{\sqrt{2\pi}\Gamma(r/2)2^{r/2}} u^{r/2-1} \exp\left[-\frac{u}{2}\left(1 + \frac{t^2}{r}\right)\right] \frac{\sqrt{u}}{\sqrt{r}} & |t| < \infty, \quad 0 < u < \infty \\ 0 & \text{elsewhere.} \end{cases} \end{aligned}$$

The marginal pdf of T is then

$$
\begin{aligned}
g_1(t) &= \int_{-\infty}^{\infty} g(t, u)\, du \\
&= \int_0^{\infty} \frac{1}{\sqrt{2\pi r}\,\Gamma(r/2)2^{r/2}}\, u^{(r+1)/2-1} \exp\left[-\frac{u}{2}\left(1 + \frac{t^2}{r}\right)\right] du.
\end{aligned}
$$

In this integral let $z = u[1 + (t^2/r)]/2$, and it is seen that

$$
\begin{aligned}
g_1(t) &= \int_0^{\infty} \frac{1}{\sqrt{2\pi r}\,\Gamma(r/2)2^{r/2}} \left(\frac{2z}{1 + t^2/r}\right)^{(r+1)/2-1} e^{-z} \left(\frac{2}{1 + t^2/r}\right) dz \\
&= \frac{\Gamma[(r+1)/2]}{\sqrt{\pi r}\,\Gamma(r/2)} \frac{1}{(1 + t^2/r)^{(r+1)/2}}, \qquad -\infty < t < \infty . \tag{3.6.1}
\end{aligned}
$$

Thus, if W is $N(0, 1)$, if V is $\chi^2(r)$, and if W and V are independent, then

$$
T = \frac{W}{\sqrt{V/r}} \tag{3.6.2}
$$

has the immediately preceding pdf $g_1(t)$. The distribution of the random variable T is usually called a *t-distribution*. It should be observed that a t-distribution is completely determined by the parameter r, the number of degrees of freedom of the random variable that has the chi-square distribution. Some approximate values of

$$
P(T \le t) = \int_{-\infty}^t g_1(w)\, dw
$$

for selected values of r and t can be found in Table IV in Appendix C.

The R or S-PLUS computer package can also be used to obtain critical values as well as probabilities concerning the t-distribution. For instance the command `qt(.975,15)` returns the 97.5th percentile of the t-distribution with 15 degrees of freedom, while the command `pt(2.0,15)` returns the probability that a t-distributed random variable with 15 degrees of freedom is less that 2.0 and the command `dt(2.0,15)` returns the value of the pdf of this distribution at 2.0.

Remark 3.6.1. This distribution was first discovered by W.S. Gosset when he was working for an Irish brewery. Gosset published under the pseudonym Student. Thus this distribution is often known as Student's t-distribution. ■

Example 3.6.1 (Mean and Variance of the t-distribution). Let T have a t-distribution with r degrees of freedom. Then, as in (3.6.2), we can write $T = W(V/r)^{-1/2}$, where W has a $N(0, 1)$ distribution, V has $\chi^2(r)$ distribution, and W and V are independent random variables. Independence of W and V and expression (3.3.4), provided $(r/2) - (k/2) > 0$ (i.e., $k < r$), implies the following,

$$
E(T^k) = E\left[W^k \left(\frac{V}{r}\right)^{-k/2}\right] = E(W^k) E\left[\left(\frac{V}{r}\right)^{-k/2}\right] \tag{3.6.3}
$$

$$
= E(W^k)\frac{2^{-k/2}\Gamma\left(\frac{r}{2} - \frac{k}{2}\right)}{\Gamma\left(\frac{r}{2}\right) r^{-k/2}}, \qquad \text{if } k < r. \tag{3.6.4}
$$

For the mean of T, use $k = 1$. Because $E(W) = 0$, as long as the degrees of freedom of T exceed 1, the mean of T is 0. For the variance, use $k = 2$. In this case the condition becomes $r > 2$. Since $E(W^2) = 1$ by expression (3.6.4) the variance of T is given by

$$\text{Var}(T) = E(T^2) = \frac{r}{r-2}. \tag{3.6.5}$$

Therefore, a t-distribution with $r > 2$ degrees of freedom has a mean of 0 and a variance of $r/(r-2)$. ∎

3.6.2   The F-distribution

Next consider two independent chi-square random variables U and V having r_1 and r_2 degrees of freedom, respectively. The joint pdf $h(u, v)$ of U and V is then

$$h(u, v) = \begin{cases} \frac{1}{\Gamma(r_1/2)\Gamma(r_2/2)2^{(r_1+r_2)/2}} u^{r_1/2-1} v^{r_2/2-1} e^{-(u+v)/2} & 0 < u, v < \infty \\ 0 & \text{elsewhere.} \end{cases}$$

We define the new random variable

$$W = \frac{U/r_1}{V/r_2}$$

and we propose finding the pdf $g_1(w)$ of W. The equations

$$w = \frac{u/r_1}{v/r_2}, \quad z = v,$$

define a one-to-one transformation that maps the set $\mathcal{S} = \{(u, v) : 0 < u < \infty,\ 0 < v < \infty\}$ onto the set $\mathcal{T} = \{(w, z) : 0 < w < \infty,\ 0 < z < \infty\}$. Since $u = (r_1/r_2)zw$, $v = z$, the absolute value of the Jacobian of the transformation is $|J| = (r_1/r_2)z$. The joint pdf $g(w, z)$ of the random variables W and $Z = V$ is then

$$g(w, z) = \frac{1}{\Gamma(r_1/2)\Gamma(r_2/2)2^{(r_1+r_2)/2}} \left(\frac{r_1 zw}{r_2}\right)^{\frac{r_1-2}{2}} z^{\frac{r_2-2}{2}} \exp\left[-\frac{z}{2}\left(\frac{r_1 w}{r_2} + 1\right)\right] \frac{r_1 z}{r_2},$$

provided that $(w, z) \in \mathcal{T}$, and zero elsewhere. The marginal pdf $g_1(w)$ of W is then

$$\begin{aligned} g_1(w) &= \int_{-\infty}^{\infty} g(w, z)\, dz \\ &= \int_0^{\infty} \frac{(r_1/r_2)^{r_1/2}(w)^{r_1/2-1}}{\Gamma(r_1/2)\Gamma(r_2/2)2^{(r_1+r_2)/2}} z^{(r_1+r_2)/2-1} \exp\left[-\frac{z}{2}\left(\frac{r_1 w}{r_2} + 1\right)\right] dz. \end{aligned}$$

If we change the variable of integration by writing

$$y = \frac{z}{2}\left(\frac{r_1 w}{r_2} + 1\right),$$

it can be seen that

$$g_1(w) = \int_0^\infty \frac{(r_1/r_2)^{r_1/2}(w)^{r_1/2-1}}{\Gamma(r_1/2)\Gamma(r_2/2)2^{(r_1+r_2)/2}} \left(\frac{2y}{r_1w/r_2+1}\right)^{(r_1+r_2)/2-1} e^{-y}$$

$$\times \left(\frac{2}{r_1w/r_2+1}\right) dy$$

$$= \begin{cases} \frac{\Gamma[(r_1+r_2)/2](r_1/r_2)^{r_1/2}}{\Gamma(r_1/2)\Gamma(r_2/2)} \frac{(w)^{r_1/2-1}}{(1+r_1w/r_2)^{(r_1+r_2)/2}} & 0 < w < \infty \\ 0 & \text{elsewhere.} \end{cases}$$

Accordingly, if U and V are independent chi-square variables with r_1 and r_2 degrees of freedom, respectively, then

$$W = \frac{U/r_1}{V/r_2}$$

has the immediately preceding pdf $g(w)$. The distribution of this random variable is usually called an F-*distribution*; and we often call the ratio, which we have denoted by W, F. That is,

$$F = \frac{U/r_1}{V/r_2}. \tag{3.6.6}$$

It should be observed that an F-distribution is completely determined by the two parameters r_1 and r_2. Table V in Appendix C gives some approximate values of

$$P(F \le b) = \int_0^b g_1(w)\,dw$$

for selected values of r_1, r_2, and b.

The R or S-PLUS program can also be used to find critical values and probabilities for F-distributed random variables. Suppose we want the 0.025 upper critical point for an F random variable with a and b degrees of freedom. This can be obtained by the command `qf(.975,a,b)`. Also, the probability that this F-distributed random variable is less than x is returned by the command `pf(x,a,b)` while the command `df(x,a,b)` returns the value of its pdf at x.

Example 3.6.2 (Moments of F-distributions). Let F have an F-distribution with r_1 and r_2 degrees of freedom. Then, as in expression (3.6.6), we can write $F = (r_2/r_1)(U/V)$ where U and V are independent χ^2 random variables with r_1 and r_2 degrees of freedom, respectively. Hence, for the kth moment of F, by independence we have

$$E\left(F^k\right) = \left(\frac{r_2}{r_1}\right)^k E\left(U^k\right) E\left(V^{-k}\right),$$

provided of course that both expectations on the right-side exist. By Theorem 3.3.1, because $k > -(r_1/2)$ is always true, the first expectation always exists. The second expectation, however, exists if $r_2 > 2k$, i.e., the denominator degrees of freedom must exceed twice k. Assuming this is true, it follows from (3.3.4) that the mean of F is given by

$$E(F) = \frac{r_2}{r_1} r_1 \frac{2^{-1} \Gamma \left(\frac{r_2}{2} - 1 \right)}{\Gamma \left(\frac{r_2}{2} \right)} = \frac{r_2}{r_2 - 2}. \tag{3.6.7}$$

If r_2 is large then $E(F)$ is about 1. In Exercise 3.6.6, a general expression for $E(F^k)$ is derived. ∎

3.6.3   Student's Theorem

Our final note in this section concerns an important result for the later chapters on inference for normal random variables. It is a corollary to the t-distribution derived above and is often referred to as Student's Theorem.

Theorem 3.6.1. *Let $X_1, \ldots, X_n$ be iid random variables each having a normal distribution with mean μ and variance σ^2. Define the random variables,*

$$\overline{X} = \frac{1}{n} \sum_{i=1}^{n} X_i \text{ and } S^2 = \frac{1}{n-1} \sum_{i=1}^{n} (X_i - \overline{X})^2.$$

Then,

(a). *$\overline{X}$ has a $N \left(\mu, \frac{\sigma^2}{n} \right)$ distribution.*

(b). *$\overline{X}$ and S^2 are independent.*

(c). *$(n-1)S^2/\sigma^2$ has a $\chi^2(n-1)$ distribution.*

(d). *The random variable*

$$T = \frac{\overline{X} - \mu}{S/\sqrt{n}}, \tag{3.6.8}$$

has a Student t-distribution with $n-1$ degrees of freedom.

Proof: Note that we have proved Part (a) in Corollary 3.4.1. Let $\mathbf{X} = (X_1, \ldots, X_n)'$. Because $X_1, \ldots, X_n$ are iid $N(\mu, \sigma^2)$ random variables, $\mathbf{X}$ has a multivariate normal distribution $N(\mu \mathbf{1}, \sigma^2 \mathbf{I})$, where $\mathbf{1}$ denotes a vector whose components are all one. Let $\mathbf{v}' = (1/n, \ldots, 1/n)' = (1/n)\mathbf{1}'$. Note that $\overline{X} = \mathbf{v}'\mathbf{X}$. Define the random vector $\mathbf{Y}$ by $\mathbf{Y} = (X_1 - \overline{X}, \ldots, X_n - \overline{X})'$. Consider the following transfomation:

$$\mathbf{W} = \begin{bmatrix} \overline{X} \\ \mathbf{Y} \end{bmatrix} = \begin{bmatrix} \mathbf{v}' \\ \mathbf{I} - \mathbf{1}\mathbf{v}' \end{bmatrix} \mathbf{X}. \tag{3.6.9}$$

Because $\mathbf{W}$ is a linear transformation of multivariate normal random vector, by Theorem 3.5.1 it has a multivariate normal distribution with mean

$$E\left[\mathbf{W}\right] = \begin{bmatrix} \mathbf{v}' \\ \mathbf{I} - \mathbf{1}\mathbf{v}' \end{bmatrix} \mu \mathbf{1} = \begin{bmatrix} \mu \\ \mathbf{0}_n \end{bmatrix}, \tag{3.6.10}$$

where $\mathbf{0}_n$ denotes a vector whose components are all 0, and covariance matrix

$$\Sigma = \begin{bmatrix} \mathbf{v}' \\ \mathbf{I} - \mathbf{1v}' \end{bmatrix} \sigma^2 \mathbf{I} \begin{bmatrix} \mathbf{v}' \\ \mathbf{I} - \mathbf{1v}' \end{bmatrix}'$$

$$= \sigma^2 \begin{bmatrix} \frac{1}{n} & \mathbf{0}'_n \\ \mathbf{0}_n & \mathbf{I} - \mathbf{1v}' \end{bmatrix}. \tag{3.6.11}$$

Because $\overline{X}$ is the first component of $\mathbf{W}$, we can also obtain Part (a) by Theorem 3.5.1. Next, because the covariances are 0, $\overline{X}$ is independent of $\mathbf{Y}$. But $\mathbf{S}^2 = (n-1)^{-1}\mathbf{Y}'\mathbf{Y}$. Hence, $\overline{X}$ is independent of $\mathbf{S}^2$, also. Thus Part (b) is true.

Consider the random variable,

$$V = \sum_{i=1}^{n} \left(\frac{X_i - \mu}{\sigma} \right)^2.$$

Each term in this sum is the square of a $N(0,1)$ random variable and, hence, has a $\chi^2(1)$ distribution, (Theorem 3.4.1). Because the summands are independent by Corollary 3.3.1, we have that V is a $\chi^2(n)$ random variable. Note the following identity,

$$\begin{aligned} V &= \sum_{i=1}^{n} \left(\frac{(X_i - \overline{X}) + (\overline{X} - \mu)}{\sigma} \right)^2 \\ &= \sum_{i=1}^{n} \left(\frac{X_i - \overline{X}}{\sigma} \right)^2 + \left(\frac{\overline{X} - \mu}{\sigma/\sqrt{n}} \right)^2 \\ &= \frac{(n-1)S^2}{\sigma^2} + \left(\frac{\overline{X} - \mu}{\sigma/\sqrt{n}} \right)^2. \end{aligned} \tag{3.6.12}$$

By Part (b), the two terms on the right side of the last equation are independent. Further, the second term is the square of a standard normal random variable and, hence, has a $\chi^2(1)$ distribution. Taking mgf of both sides, we have,

$$(1 - 2t)^{-n/2} = E\left[\exp\{t(n-1)S^2/\sigma^2\}\right](1 - 2t)^{-1/2}. \tag{3.6.13}$$

Solving for the mgf of $(n-1)S^2/\sigma^2$ on the right side we obtain Part (c). Finally, Part (d) follows immediately from Parts (a)-(c) upon writing T, (3.6.8), as

$$T = \frac{(\overline{X} - \mu)/(\sigma/\sqrt{n})}{\sqrt{(n-1)S^2/(\sigma^2(n-1))}}. \quad \blacksquare$$

EXERCISES

3.6.1. Let T have a t-distribution with 10 degrees of freedom. Find $P(|T| > 2.228)$ from either Table IV or, if available, R or S-PLUS.

3.6.2. Let T have a t-distribution with 14 degrees of freedom. Determine b so that $P(-b < T < b) = 0.90$. Use either Table IV or, if available, R or S-PLUS.

3.6.3. Let T have a t-distribution with $r > 4$ degrees of freedom. Use expression (3.6.4) to determine the kurtosis of T. See Exercise 1.9.14 for the definition of kurtosis.

3.6.4. Assuming a computer is available, plot the pdfs of the random variables defined in Parts (a)-(e) below. Obtain an overlay plot of all four pdfs, also. In R or S-PLUS the domain values of the pdfs can easily be obtained by using the seq command. For instance, the command x<-seq(-6,6,.1) will return a vector of values between -6 and 6 in jumps of 0.1.

(a) X has a standard normal distribution.

(b) X has a t-distribution with 1 degree of freedom.

(c) X has a t-distribution with 3 degrees of freedom.

(d) X has a t-distribution with 10 degrees of freedom.

(e) X has a t-distribution with 30 degrees of freedom.

3.6.5. Assuming a computer is available, investigate the probabilities of an "outlier" for a t-random variable and a normal random variable. Specifically, determine the probability of observing the event $\{|X| \geq 2\}$ for the following random variables:

(a) X has a standard normal distribution.

(b) X has a t-distribution with 1 degree of freedom.

(c) X has a t-distribution with 3 degrees of freedom.

(d) X has a t-distribution with 10 degrees of freedom.

(e) X has a t-distribution with 30 degrees of freedom.

3.6.6. Let F have an F-distribution with parameters r_1 and r_2. Assuming that $r_2 > 2k$, continue with Example 3.6.2 and derive the $E(F^k)$.

3.6.7. Let F have an F-distribution with parameters r_1 and r_2. Using the results of the last exercise, determine the kurtosis of F, assuming that $r_2 > 8$.

3.6.8. Let F have an F-distribution with parameters r_1 and r_2. Prove that $1/F$ has an F-distribution with parameters r_2 and r_1.

3.6.9. If F has an F-distribution with parameters $r_1 = 5$ and $r_2 = 10$, find a and b so that $P(F \leq a) = 0.05$ and $P(F \leq b) = 0.95$, and, accordingly, $P(a < F < b) = 0.90$.

Hint: Write $P(F \leq a) = P(1/F \geq 1/a) = 1 - P(1/F \leq 1/a)$, and use the result of Exercise 3.6.8 and Table V or, if available, use R or S-PLUS.

3.6.10. Let $T = W/\sqrt{V/r}$, where the independent variables W and V are, respectively, normal with mean zero and variance 1 and chi-square with r degrees of freedom. Show that T^2 has an F-distribution with parameters $r_1 = 1$ and $r_2 = r$. *Hint:* What is the distribution of the numerator of T^2?

3.6.11. Show that the t-distribution with $r = 1$ degree of freedom and the Cauchy distribution are the same.

3.6.12. Show that

$$Y = \frac{1}{1 + (r_1/r_2)W},$$

where W has an F-distribution with parameters r_1 and r_2, has a beta distribution.

3.6.13. Let X_1, X_2 be iid with common distribution having the pdf $f(x) = e^{-x}$, $0 < x < \infty$, zero elsewhere. Show that $Z = X_1/X_2$ has an F-distribution.

3.6.14. Let X_1, X_2, and X_3 be three independent chi-square variables with r_1, r_2, and r_3 degrees of freedom, respectively.

(a) Show that $Y_1 = X_1/X_2$ and $Y_2 = X_1 + X_2$ are independent and that Y_2 is $\chi^2(r_1 + r_2)$.

(b) Deduce that

$$\frac{X_1/r_1}{X_2/r_2} \quad \text{and} \quad \frac{X_3/r_3}{(X_1 + X_2)/(r_1 + r_2)}$$

are independent F-variables.

3.7 Mixture Distributions

Recall in Section 3.4.1, the discussion on the contaminated normal distribution. This was an example of a mixture of normal distributions. In this section, we extend this to mixtures of distributions in general. Generally, we will use continuous type notation for the discussion, but discrete pmfs can be handled the same way.

Suppose that we have k distributions with respective pdfs $f_1(x), f_2(x), \ldots, f_k(x)$, with supports $\mathcal{S}_1, \mathcal{S}_2, \ldots, \mathcal{S}_k$, means $\mu_1, \mu_2, \ldots, \mu_k$, and variances $\sigma_1^2, \sigma_2^2, \ldots, \sigma_k^2$, with positive mixing probabilities $p_1, p_2, \ldots, p_k$, where $p_1 + p_2 + \cdots + p_k = 1$. Let $\mathcal{S} = \cup_{i=1}^k \mathcal{S}_i$ and consider the function

$$f(x) = p_1 f_1(x) + p_2 f_2(x) + \cdots + p_k f_k(x) = \sum_{i=1}^k p_i f_i(x), \quad x \in \mathcal{S}. \tag{3.7.1}$$

Note that $f(x)$ is nonnegative and it is easy to see that it integrates to one over $(-\infty, \infty)$; hence, $f(x)$ is a pdf for some continuous type random variable X. The mean of X is given by

$$E(X) = \sum_{i=1}^k p_i \int_{-\infty}^{\infty} x f_i(x)\, dx = \sum_{i=1}^k p_i \mu_i = \bar{\mu}, \tag{3.7.2}$$

a weighted average of $\mu_1, \mu_2, \ldots, \mu_k$, and the variance equals

$$
\begin{aligned}
\text{var}(X) &= \sum_{i=1}^{k} p_i \int_{-\infty}^{\infty} (x - \overline{\mu})^2 f_i(x)\, dx \\
&= \sum_{i=1}^{k} p_i \int_{-\infty}^{\infty} [(x - \mu_i) + (\mu_i - \overline{\mu})]^2 f_i(x)\, dx \\
&= \sum_{i=1}^{k} p_i \int_{-\infty}^{\infty} (x - \mu_i)^2 f_i(x)\, dx + \sum_{i=1}^{k} p_i (\mu_i - \overline{\mu})^2 \int_{-\infty}^{\infty} f_i(x)\, dx,
\end{aligned}
$$

because the cross-product terms integrate to zero. That is,

$$
\text{var}(X) = \sum_{i=1}^{k} p_i \sigma_i^2 + \sum_{i=1}^{k} p_i (\mu_i - \overline{\mu})^2. \tag{3.7.3}
$$

Note that the variance is not simply the weighted average of the k variances but it also includes a positive term involving the weighted variance of the means.

Remark 3.7.1. It is extremely important to note these characteristics are associated with a mixture of k distributions and have nothing to do with a linear combination, say $\sum a_i X_i$, of k random variables. ∎

For the next example, we need the following distribution. We say that X has a *loggamma* pdf with parameters $\alpha > 0$ and $\beta > 0$ if it has pdf

$$
f_1(x) = \begin{cases} \frac{1}{\Gamma(\alpha)\beta^\alpha} x^{-(1+\beta)/\beta} (\log x)^{\alpha - 1} & x > 1 \\ 0 & \text{elsewhere.} \end{cases} \tag{3.7.4}
$$

The derivation of this pdf is given in Exercise 3.7.1, where its mean and variance are also derived. We will denote this distribution of X by $\log \Gamma(\alpha, \beta)$.

Example 3.7.1. Actuaries have found that a mixture of the loggamma and gamma distributions is an important model for claim distributions. Suppose, then, that X_1 is $\log \Gamma(\alpha_1, \beta_1)$, X_2 is $\Gamma(\alpha_2, \beta_2)$, and the mixing probabilities are p and $(1 - p)$. Then the pdf of the mixture distribution is

$$
f(x) = \begin{cases} \frac{1-p}{\beta_2^{\alpha_2}\Gamma(\alpha_2)} x^{\alpha_2 - 1} e^{-x/\beta_2} & 0 < x \le 1 \\ \frac{p}{\beta_1^{\alpha_1}\Gamma(\alpha_1)} (\log x)^{\alpha_1 - 1} x^{-(\beta_1 + 1)/\beta_1} + \frac{1-p}{\beta_2^{\alpha_2}\Gamma(\alpha_2)} x^{\alpha_2 - 1} e^{-x/\beta_2} & 1 < x \\ 0 & \text{elsewhere.} \end{cases} \tag{3.7.5}
$$

Provided $\beta_1 < 2^{-1}$, the mean and the variance of this mixture distribution are

$$
\begin{aligned}
\mu &= p(1 - \beta_1)^{-\alpha_1} + (1 - p)\alpha_2 \beta_2 \tag{3.7.6} \\
\sigma^2 &= p[(1 - 2\beta_1)^{-\alpha_1} - (1 - \beta_1)^{-2\alpha_1}] \\
&\quad + (1 - p)\alpha_2 \beta_2^2 + p(1 - p)[(1 - \beta_1)^{-\alpha_1} - \alpha_2 \beta_2]^2; \tag{3.7.7}
\end{aligned}
$$

see Exercise 3.7.2. ∎

The mixture of distributions is sometimes called *compounding*. Moreover, it does not need to be restricted to a finite number of distributions. As demonstrated in the following example, a continuous weighting function, which is of course a pdf, can replace $p_1, p_2, \ldots, p_k$; i.e., integration replaces summation.

Example 3.7.2. Let X_θ be a Poisson random variable with parameter θ. We want to mix an infinite number of Poisson distributions, each with a different value of θ. We let the weighting function be a pdf of θ, namely, a gamma with parameters α and β. For $x = 0, 1, 2, \ldots$, the pmf of the compound distribution is

$$
\begin{aligned}
p(x) &= \int_0^\infty \left[\frac{1}{\beta^\alpha \Gamma(\alpha)} \theta^{\alpha-1} e^{-\theta/\beta} \right] \left[\frac{\theta^x e^{-\theta}}{x!} \right] d\theta \\
&= \frac{1}{\Gamma(\alpha)\beta^\alpha x!} \int_0^\infty \theta^{\alpha+x-1} e^{-\theta(1+\beta)/\beta} \, d\theta \\
&= \frac{\Gamma(\alpha+x)\beta^x}{\Gamma(\alpha)x!(1+\beta)^{\alpha+x}},
\end{aligned}
$$

where the third line follows from the change of variable $t = \theta(1+\beta)/\beta$ to solve the integral of the second line.

An interesting case of this compound occurs when $\alpha = r$, a positive integer, and $\beta = (1-p)/p$, where $0 < p < 1$. In this case the pmf becomes

$$
p(x) = \frac{(r+x-1)!}{(r-1)!} \frac{p^r(1-p)^x}{x!}, \quad x = 0, 1, 2, \ldots.
$$

That is, this compound distribution is the same as that of the number of excess trials needed to obtain r successes in a sequence of independent trials, each with probability p of success; this is one form of the *negative binomial distribution*. The negative binomial distribution has been used successfully as a model for the number of accidents (see Weber, 1971). ∎

In compounding, we can think of the original distribution of X as being a conditional distribution given θ, whose pdf is denoted by $f(x|\theta)$. Then the weighting function is treated as a pdf for θ, say $g(\theta)$. Accordingly, the joint pdf is $f(x|\theta)g(\theta)$, and the compound pdf can be thought of as the marginal (unconditional) pdf of X,

$$
h(x) = \int_\theta g(\theta) f(x|\theta) \, d\theta,
$$

where a summation replaces integration in case θ has a discrete distribution. For illustration, suppose we know that the mean of the normal distribution is zero but the variance σ^2 equals $1/\theta > 0$, where θ has been selected from some random model. For convenience, say this latter is a gamma distribution with parameters α and β. Thus, given that θ, X is conditionally $N(0, 1/\theta)$ so that the joint distribution of X and θ is

$$
f(x|\theta)g(\theta) = \left[\frac{\sqrt{\theta}}{\sqrt{2\pi}} \exp\left(\frac{-\theta x^2}{2} \right) \right] \left[\frac{1}{\beta^\alpha \Gamma(\alpha)} \theta^{\alpha-1} \exp(-\theta/\beta) \right],
$$

for $-\infty < x < \infty$, $\ 0 < \theta < \infty$. Therefore, the marginal (unconditional) pdf $h(x)$ of X is found by integrating out θ; that is,

$$h(x) = \int_0^\infty \frac{\theta^{\alpha+1/2-1}}{\beta^\alpha \sqrt{2\pi}\Gamma(\alpha)} \exp\left[-\theta\left(\frac{x^2}{2} + \frac{1}{\beta}\right)\right] d\theta.$$

By comparing this integrand with a gamma pdf with parameters $\alpha + \frac{1}{2}$ and $[(1/\beta) + (x^2/2)]^{-1}$, we see that the integral equals

$$h(x) = \frac{\Gamma(\alpha + \frac{1}{2})}{\beta^\alpha \sqrt{2\pi}\Gamma(\alpha)} \left(\frac{2\beta}{2 + \beta x^2}\right)^{\alpha+1/2}, \quad -\infty < x < \infty.$$

It is interesting to note that if $\alpha = r/2$ and $\beta = 2/r$, where r is a positive integer, then X has an unconditional distribution which is Student's t with r degrees of freedom. That is, we have developed a generalization of Student's distribution through this type of mixing or compounding. We note that the resulting distribution (a generalization of Student's t) has much thicker tails than those of the conditional normal with which we started.

The next two examples offer two additional illustrations of this type of compounding.

Example 3.7.3. Suppose that we have a binomial distribution, but we are not certain about the probability p of success on a given trial. Suppose p has been selected first by some random process which has a beta pdf with parameters α and β. Thus X, the number of successes on n independent trials, has a conditional binomial distribution so that the joint pdf of X and p is

$$p(x|p)g(p) = \frac{n!}{x!(n-x)!}p^x(1-p)^{n-x}\frac{\Gamma(\alpha+\beta)}{\Gamma(\alpha)\Gamma(\beta)}p^{\alpha-1}(1-p)^{\beta-1},$$

for $x = 0, 1, \ldots, n$, $\ 0 < p < 1$. Therefore, the unconditional pdf of X is given by the integral

$$\begin{aligned}
h(x) &= \int_0^1 \frac{n!\Gamma(\alpha+\beta)}{x!(n-x)!\Gamma(\alpha)\Gamma(\beta)}p^{x+\alpha-1}(1-p)^{n-x+\beta-1}\,dp \\
&= \frac{n!\Gamma(\alpha+\beta)\Gamma(x+\alpha)\Gamma(n-x+\beta)}{x!(n-x)!\Gamma(\alpha)\Gamma(\beta)\Gamma(n+\alpha+\beta)}, \quad x = 0, 1, 2, \ldots, n.
\end{aligned}$$

Now suppose α and β are positive integers; since $\Gamma(k) = (k-1)!$, this unconditional (marginal or compound) pdf can be written

$$h(x) = \frac{n!(\alpha+\beta-1)!(x+\alpha-1)!(n-x+\beta-1)!}{x!(n-x)!(\alpha-1)!(\beta-1)!(n+\alpha+\beta-1)!}, \quad x = 0, 1, 2, \ldots, n.$$

Because the conditional mean $E(X|p) = np$, the unconditional mean is $n\alpha/(\alpha+\beta)$ since $E(p)$ equals the mean $\alpha/(\alpha+\beta)$ of the beta distribution. ∎

Example 3.7.4. In this example, we develop by compounding a heavy-tailed skewed distribution. Assume X has a conditional gamma pdf with parameters k and θ^{-1}. The weighting function for θ is a gamma pdf with parameters α and β. Thus the unconditional (marginal or compounded) pdf of X is

$$
\begin{aligned}
h(x) &= \int_0^\infty \left[\frac{\theta^{\alpha-1} e^{-\theta/\beta}}{\beta^\alpha \Gamma(\alpha)} \right] \left[\frac{\theta^k x^{k-1} e^{-\theta x}}{\Gamma(k)} \right] d\theta \\
&= \int_0^\infty \frac{x^{k-1} \theta^{\alpha+k-1}}{\beta^\alpha \Gamma(\alpha) \Gamma(k)} e^{-\theta(1+\beta x)/\beta} \, d\theta.
\end{aligned}
$$

Comparing this integrand to the gamma pdf with parameters $\alpha+k$ and $\beta/(1+\beta x)$, we see that

$$
h(x) = \frac{\Gamma(\alpha+k)\beta^k x^{k-1}}{\Gamma(\alpha)\Gamma(k)(1+\beta x)^{\alpha+k}}, \quad 0 < x < \infty,
$$

which is the pdf of the *generalized Pareto distribution* (and a generalization of the F distribution). Of course, when $k = 1$ (so that X has a conditional exponential distribution), the pdf is

$$
h(x) = \alpha\beta(1+\beta x)^{-(\alpha+1)}, \quad 0 < x < \infty
$$

which is the *Pareto pdf*. Both of these compound pdfs have thicker tails than the original (conditional) gamma distribution.

While the cdf of the generalized Pareto distribution cannot be expressed in a simple closed form, that of the Pareto distribution is

$$
H(x) = \int_0^x \alpha\beta(1+\beta t)^{-(\alpha+1)} \, dt = 1 - (1+\beta x)^{-\alpha}, \quad 0 \le x < \infty.
$$

From this, we can create another useful long-tailed distribution by letting $X = Y^\tau$, $0 < \tau$. Thus Y has the cdf

$$
G(y) = P(Y \le y) = P[X^{1/\tau} \le y] = P[X \le y^\tau].
$$

Hence, this probability is equal to

$$
G(y) = H(y^\tau) = 1 - (1+\beta y^\tau)^{-\alpha}, \quad 0 \le y < \infty,
$$

with corresponding pdf.

$$
G'(y) = g(y) = \frac{\alpha\beta\tau y^{\tau-1}}{(1+\beta y^\tau)^{\alpha+1}}, 0 < y < \infty.
$$

We call the associated distribution the *transformed Pareto distribution or Burr distribution* (Burr, 1942) and it has proved to be a useful one in modeling thicker tailed distributions. ∎

EXERCISES

3.7.1. Suppose Y has a $\Gamma(\alpha, \beta)$ distribution. Let $X = e^Y$. Show that the pdf of X is given by expression (3.7.4). Derive the mean and variance of X.

3.7.2. In Example 3.7.1, derive the pdf of the mixture distribution given in expression (3.7.5), then obtain its mean and variance as given in expressions (3.7.6) and (3.7.7).

3.7.3. Consider the mixture distribution, $(9/10)N(0,1)+(1/10)N(0,9)$. Show that its kurtosis is 8.34.

3.7.4. Let X have the conditional geometric pmf $\theta(1-\theta)^{x-1}$, $x = 1, 2, \ldots$, where θ is a value of a random variable having a beta pdf with parameters α and β. Show that the marginal (unconditional) pmf of X is

$$\frac{\Gamma(\alpha+\beta)\Gamma(\alpha+1)\Gamma(\beta+x-1)}{\Gamma(\alpha)\Gamma(\beta)\Gamma(\alpha+\beta+x)}, \quad x = 1, 2, \ldots.$$

If $\alpha = 1$, we obtain

$$\frac{\beta}{(\beta+x)(\beta+x-1)}, \quad x = 1, 2, \ldots,$$

which is one form of *Zipf's law*.

3.7.5. Repeat Exercise 3.7.4, letting X have a conditional negative binomial distribution instead of the geometric one.

3.7.6. Let X have a generalized Pareto distribution with parameters k, α, and β. Show, by change of variables, that $Y = \beta X/(1 + \beta X)$ has a beta distribution.

3.7.7. Show that the failure rate (hazard function) of the Pareto distribution is

$$\frac{h(x)}{1 - H(x)} = \frac{\alpha}{\beta^{-1} + x}.$$

Find the failure rate (hazard function) of the Burr distribution with cdf

$$G(y) = 1 - \left(\frac{1}{1 + \beta y^\tau}\right)^\alpha, \quad 0 \le y < \infty.$$

In each of these two failure rates, note what happens as the value of the variable increases.

3.7.8. For the Burr distribution, show that

$$E(X^k) = \frac{1}{\beta^{k/\tau}} \Gamma\left(\alpha - \frac{k}{\tau}\right) \Gamma\left(\frac{k}{\tau} + 1\right) \Big/ \Gamma(\alpha),$$

provided $k < \alpha\tau$.

3.7.9. Let the number X of accidents have a Poisson distribution with mean $\lambda\theta$. Suppose λ, the liability to have an accident, has, given θ, a gamma pdf with parameters $\alpha = h$ and $\beta = h^{-1}$; and θ, an accident proneness factor, has a generalized Pareto pdf with parameters α, $\lambda = h$, and k. Show that the unconditional pdf of X is

$$\frac{\Gamma(\alpha+k)\Gamma(\alpha+h)\Gamma(\alpha+h+k)\Gamma(h+k)\Gamma(k+x)}{\Gamma(\alpha)\Gamma(\alpha+k+h)\Gamma(h)\Gamma(k)\Gamma(\alpha+h+k+x)x!}, \quad x = 0, 1, 2, \ldots,$$

sometimes called the *generalized Waring* pdf.

3.7.10. Let X have a conditional Burr distribution with fixed parameters β and τ, given parameter α.

(a) If α has the geometric pdf $p(1-p)^\alpha$, $\alpha = 0, 1, 2, \ldots$, show that the unconditional distribution of X is a Burr distribution.

(b) If α has the exponential pdf $\beta^{-1}e^{-\alpha/\beta}$, $\alpha > 0$, find the unconditional pdf of X.

3.7.11. Let X have the conditional Weibull pdf.

$$f(x|\theta) = \theta\tau x^{\tau-1}e^{-\theta x^\tau}, \quad 0 < x < \infty,$$

and let the pdf (weighting function) $g(\theta)$ be gamma with parameters α and β. Show that the compound (marginal) pdf of X is that of Burr.

3.7.12. If X has a Pareto distribution with parameters α and β and if c is a positive constant, show that $Y = cX$ has a Pareto distribution with parameters α and $c\lambda$.

Chapter 4

Unbiasedness, Consistency, and Limiting Distributions

In the previous chapters, we were concerned with probability models and distributions. In the next chapter, we begin discussing statistical inference which will remain our focus for the remainder of this book. In this chapter, we present some tools drawn from asymptotic theory. These are useful in statistics as well as in probability theory.

In our discussion, we use some examples from statistics so the concept of a random sample will prove helpful for this chapter. More details on sampling are given in Chapter 5. Suppose we have a random variable X which has pdf, (or pmf), $f(x; \theta)$, $(p(x; \theta))$, where θ is either a real number or a vector of real numbers. Assume that $\theta \in \Omega$ which is a subset of R^p, for $p \geq 1$. For example, θ could be the vector (μ, σ^2) when X has a $N(\mu, \sigma^2)$ distribution or θ could be the probability of success p when X has a binomial distribution. In the previous chapters, say to work a probability problem, we would know θ. In statistics, though, θ is unknown. Our information about θ comes from a sample $X_1, X_2, \ldots, X_n$. We often assume that this is a **random sample** which means that the random variables $X_1, X_2, \ldots, X_n$ are independent and have the same distribution as X; that is, $X_1, X_2, \ldots, X_n$ are iid. A **statistic** T is a function of the sample; i.e, $T = T(X_1, X_2, \ldots, X_n)$. We may use T to estimate θ. In which case, we would say that T is a **point estimator** of θ. For example, suppose $X_1, X_2, \ldots, X_n$ is a random sample from a distribution with mean μ and variance σ^2. Then the statistics $\overline{X}$ and S^2 are referred to as the **sample mean** and the **sample variance** of this random sample. They are point estimators of μ and σ^2, respectively.

As another illustration, consider the case where X is Bernoulli with probability of success p; that is, X assumes the values 1 or 0 with probabilities p or $1 - p$, respectively. Suppose we perform n Bernoulli trials. Recall that Bernoulli trials are performed independently of one another and under identical conditions. Let X_i be the outcome on the *ith* trial, $i = 1, 2, \ldots, n$. Then $X_1, X_2, \ldots, X_n$ form a random sample from the distribution X. A statistic of interest here is $\overline{X}$, which is the proportion of successes in the sample. It is a point estimator of p.

4.1   Expectations of Functions

Let $\mathbf{X} = (X_1, \ldots, X_n)'$ denote a random vector from some experiment. Often we are interested in a function of $\mathbf{X}$, say, $T = T(\mathbf{X})$. For example, if $\mathbf{X}$ is a sample, T may be a statistic of interest. We begin by considering linear functions of $\mathbf{X}$; i.e., functions of the form

$$T = \mathbf{a}'\mathbf{X} = \sum_{i=1}^{n} a_i X_i,$$

for a specified vector $\mathbf{a} = (a_1, \ldots, a_n)'$. We will obtain the mean and variance of such random variables.

The mean of T follows immediately from the linearity of the expectation operator, E, but for easy reference we state this as a theorem:

Theorem 4.1.1. *Let $T = \sum_{i=1}^{n} a_i X_i$. Provided $E[|X_i|] < \infty$, for $i = 1, \ldots, n$,*

$$E(T) = \sum_{i=1}^{n} a_i E(X_i).$$

For the variance of T, we first state a very general result involving covariances. Let $\mathbf{Y} = (Y_1, \ldots, Y_m)'$ denote another random vector and let $W = \mathbf{b}'\mathbf{Y}$ for a specified vector $\mathbf{b} = (b_1, \ldots, b_m)'$.

Theorem 4.1.2. *Let $T = \sum_{i=1}^{n} a_i X_i$ and let $W = \sum_{i=1}^{m} b_i Y_i$. If $E[X_i^2] < \infty$, and $E[Y_j^2] < \infty$ for $i = 1, \ldots, n$ and $j = 1, \ldots, m$, then*

$$cov(T, W) = \sum_{i=1}^{n} \sum_{j=1}^{m} a_i b_j \, cov(X_i, Y_j).$$

Proof: Using the definition of the covariance and Theorem 4.1.1 we have the first equality below, while the second equality follows from the linearity of E

$$
\begin{aligned}
cov(T, W) &= E[\sum_{i=1}^{n} \sum_{j=1}^{m} (a_i X_i - a_i E(X_i))(b_j Y_j - b_j E(Y_j))] \\
&= \sum_{i=1}^{n} \sum_{j=1}^{m} a_i b_j E[(X_i - E(X_i))(Y_j - E(Y_j))],
\end{aligned}
$$

which is the desired result. ∎

To obtain the variance of T, simply replace W by T in Theorem 4.1.2. We state the result as a theorem:

Corollary 4.1.1. *Let $T = \sum_{i=1}^{n} a_i X_i$. Provided $E[X_i^2] < \infty$, for $i = 1, \ldots, n$,*

$$Var(T) = cov(T, T) = \sum_{i=1}^{n} a_i^2 \, Var(X_i) + 2 \sum_{i<j} a_i a_j \, cov(X_i, X_j). \qquad (4.1.1)$$

Note that if $X_1, \ldots, X_n$ are independent random variables then the covariance $\text{cov}(X_i, X_j) = 0$; see Example 2.5.4. This leads to a simplification of (4.1.1) which we record in the following corollary.

Corollary 4.1.2. *If $X_1, \ldots, X_n$ are independent random variables with finite variances, then*

$$Var(T) = \sum_{i=1}^{n} a_i^2 \, Var(X_i). \tag{4.1.2}$$

Note that we need only X_i and X_j to be uncorrelated for all $i \neq j$ to obtain this result; for example, $\text{Cov}(X_i, X_j) = 0$, $i \neq j$, which is true when $X_1, \ldots, X_n$ are independent.

Let us now return to the discussion of sampling and statistics found at the beginning of this chapter. Consider the situation where we have a random variable X of interest whose density is given by $f(x; \theta)$, for $\theta \in \Omega$. The parameter θ is unknown and we seek a statistic based on a sample to estimate it. Our first property of an estimator concerns its expectation.

Definition 4.1.1. *Let X be a random variable with pdf $f(x; \theta)$ or pmf $p(x; \theta)$, $\theta \in \Omega$. Let $X_1, \ldots, X_n$ be a random sample from the distribution of X and let T denote a statistic. We say T is an* **unbiased** *estimator of θ if*

$$E(T) = \theta, \quad \text{for all } \theta \in \Omega. \tag{4.1.3}$$

If T is not unbiased (that is, $E(T) \neq \theta$), we say that T is a **biased** *estimator of θ.*

Example 4.1.1 (Sample Mean). Let $X_1, \ldots, X_n$ be a random sample from the distribution of a random variable X which has mean μ and variance σ^2. Recall that the **sample mean** is given by $\overline{X} = n^{-1} \sum_{i=1}^{n} X_i$. This is a linear combination of the sample observations with $a_i \equiv n^{-1}$; hence, by Theorem 4.1.1 and Corollary 4.1.2 we have,

$$E(\overline{X}) = \mu \text{ and } Var(\overline{X}) = \frac{\sigma^2}{n}. \tag{4.1.4}$$

Hence, $\overline{X}$ is an unbiased estimator of μ. Furthermore, the variance of $\overline{X}$ becomes small as n gets large. That is, it seems in the limit that the mass of the distribution of the sample mean $\overline{X}$ is converging to μ as n gets large. This is presented in the next section. ∎

Example 4.1.2 (Sample Variance). As in the last example, let $X_1, \ldots, X_n$ be a random sample from the distribution of a random variable X which has mean μ and variance σ^2. Define the **sample variance** by

$$S^2 = (n-1)^{-1} \sum_{i=1}^{n} (X_i - \overline{X})^2 = (n-1)^{-1} \left(\sum_{i=1}^{n} X_i^2 - n\overline{X}^2 \right), \tag{4.1.5}$$

where the second equality follows after some algebra; see Exercise 4.1.3. Using the above theorems, the results of the last example, and the fact that $E(X^2) = \sigma^2 + \mu^2$, we have the following

$$E(S^2) = (n-1)^{-1}\left(\sum_{i=1}^{n} E(X_i^2) - nE(\overline{X}^2)\right)$$

$$= (n-1)^{-1}\left\{n\sigma^2 + n\mu^2 - n[(\sigma^2/n) + \mu^2]\right\}$$

$$= \sigma^2. \tag{4.1.6}$$

Hence, the sample variance is an unbiased estimate of σ^2. If $V = n^{-1}\sum_{i=1}^{n}(X_i - \overline{X})^2$ then $E(V) = ((n-1)/n)\sigma^2$. That is V is a biased estimator of σ^2. This is one reason for dividing by $n-1$ instead of n in the definition of the sample variance. ∎

Example 4.1.3 (Maximum of a Sample from a Uniform Distribution). Let $X_1, \ldots, X_n$ be a random sample from a uniform$(0, \theta)$ distribution. Suppose θ is unknown. An intuitive estimate of θ is the maximum of the sample. Let $Y_n = \max\{X_1, \ldots, X_n\}$. Exercise 4.1.2 shows that the cdf of Y_n is

$$F_{Y_n}(t) = \begin{cases} 1 & t > \theta \\ \left(\frac{t}{\theta}\right)^n & 0 < t \le \theta \\ 0 & t \le 0. \end{cases} \tag{4.1.7}$$

Hence, the pdf of Y_n is

$$f_{Y_n}(t) = \begin{cases} \frac{n}{\theta^n} t^{n-1} & 0 < t \le \theta \\ 0 & \text{elsewhere.} \end{cases} \tag{4.1.8}$$

Based on its pdf, it is easy to show that $E(Y_n) = (n/(n+1))\theta$. Thus, Y_n is a biased estimator of θ. Note, however, that $((n+1)/n)Y_n$ is an unbiased estimator of θ. ∎

Example 4.1.4 (Sample Median). Let $X_1, X_2, \ldots, X_n$ be a random sample from the distribution of X, which has pdf $f(x)$. Suppose $\mu = E(X)$ exists and, further, that the pdf $f(x)$ is symmetric about μ. In Example 4.1.1, we showed that the sample mean was an unbiased estimator of μ. What about the sample median, $T = T(X_1, X_2, \ldots, X_n) = \text{med}\{X_1, X_2, \ldots, X_n\}$? The sample median satisfies two properties: (1), if we increase (or decrease) the sample items by b then the sample median increases (or decreases) by b, and (2), if we multiply each sample item by -1, then the median gets multiplied by -1. We can abbreviate these properties as:

$$T(X_1 + b, X_2 + b, \ldots, X_n + b) = T(X_1, X_2, \ldots, X_n) + b \tag{4.1.9}$$

$$T(-X_1, -X_2, \ldots, -X_n) = -T(X_1, X_2, \ldots, X_n). \tag{4.1.10}$$

As Exercise 4.1.1 shows, if X_i is symmetrically distributed about μ, the distribution of the random vector $(X_1 - \mu, \ldots, X_n - \mu)$ is the same as the distribution of the random vector $(-(X_1 - \mu), \ldots, -(X_n - \mu))$. In particular, expectations taken under these random vectors are the same. By this fact and (4.1.9) and (4.1.10), we have the following:

$$E[T] - \mu = E[T(X_1, \ldots, X_n)] - \mu = E[T(X_1 - \mu, \ldots, X_n - \mu)]$$

$$= E[T(-(X_1 - \mu), \ldots, -(X_n - \mu))]$$

$$= -E[T(X_1 - \mu, \ldots, X_n - \mu)]$$

$$= -E[T(X_1, \ldots, X_n)] + \mu = -E[T] + \mu. \tag{4.1.11}$$

That is, $2E(T) = 2\mu$, so we have $E[T] = \mu$. However, the sample median satisfies (4.1.9) and (4.1.10); thus, under these conditions the sample median is an unbiased estimator of θ. Which estimator, the sample mean or the sample median, is better? We will consider this question later. ∎

Note that the median is transparent to the argument in the last example. That is, if T is an estimator of μ which satisfies the conditions (4.1.9) and (4.1.10) and the pdf of X is symmetric about μ, then T is an unbiased estimator of μ.

EXERCISES

4.1.1. Suppose X has a pdf which is symmetric about b; i.e., $f(b + x) = f(b - x)$, for all $-\infty < x < \infty$. We say that X is symmetrically distributed about b.

(a) Show that $Y = X - b$ is symmetrically distributed about 0.

(b) Show that $Z = -(X - b)$ has the same distribution as Y in Part (a).

(c) Show that $(X_1 - \mu, \ldots, X_n - \mu)$ and $(-(X_1 - \mu), \ldots, -(X_n - \mu))$ as defined in Example 4.1.4 have the same distribution.

4.1.2. Derive the cdf given in expression (4.1.7).

4.1.3. Derive the second equality in expression (4.1.5).

4.1.4. Let X_1, X_2, X_3, X_4 be four iid random variables having the same pdf $f(x) = 2x$, $0 < x < 1$, zero elsewhere. Find the mean and variance of the sum Y of these four random variables.

4.1.5. Let X_1 and X_2 be two independent random variables so that the variances of X_1 and X_2 are $\sigma_1^2 = k$ and $\sigma_2^2 = 2$, respectively. Given that the variance of $Y = 3X_2 - X_1$ is 25, find k.

4.1.6. If the independent variables X_1 and X_2 have means μ_1, μ_2 and variances σ_1^2, σ_2^2, respectively, show that the mean and variance of the product $Y = X_1 X_2$ are $\mu_1 \mu_2$ and $\sigma_1^2 \sigma_2^2 + \mu_1^2 \sigma_2^2 + \mu_2^2 \sigma_1^2$, respectively.

4.1.7. Find the mean and variance of the sum Y of the observations of a random sample of size 5 from the distribution having pdf $f(x) = 6x(1 - x)$, $0 < x < 1$, zero elsewhere.

4.1.8. Determine the mean and variance of the mean $\overline{X}$ of a random sample of size 9 from a distribution having pdf $f(x) = 4x^3$, $0 < x < 1$, zero elsewhere.

4.1.9. Let X and Y be random variables with $\mu_1 = 1$, $\mu_2 = 4$, $\sigma_1^2 = 4$, $\sigma_2^2 = 6$, $\rho = \frac{1}{2}$. Find the mean and variance of $Z = 3X - 2Y$.

4.1.10. Let X and Y be independent random variables with means μ_1, μ_2 and variances σ_1^2, σ_2^2. Determine the correlation coefficient of X and $Z = X - Y$ in terms of μ_1, $\mu_2, \sigma_1^2, \sigma_2^2$.

4.1.11. Let μ and σ^2 denote the mean and variance of the random variable X. Let $Y = c + bX$, where b and c are real constants. Show that the mean and the variance of Y are, respectively, $c + b\mu$ and $b^2\sigma^2$.

4.1.12. Find the mean and the variance of $Y = X_1 - 2X_2 + 3X_3$, where X_1, X_2, X_3 are observations of a random sample from a chi-square distribution with 6 degrees of freedom.

4.1.13. Determine the correlation coefficient of the random variables X and Y if $\mathrm{var}(X) = 4$, $\mathrm{var}(Y) = 2$, and $\mathrm{var}(X + 2Y) = 15$.

4.1.14. Let X and Y be random variables with means μ_1, μ_2; variances σ_1^2, σ_2^2; and correlation coefficient ρ. Show that the correlation coefficient of $W = aX + b$, $a > 0$, and $Z = cY + d$, $c > 0$, is ρ.

4.1.15. A person rolls a die, tosses a coin, and draws a card from an ordinary deck. He receives \$3 for each point up on the die, \$10 for a head and \$0 for a tail, and \$1 for each spot on the card (jack $= 11$, queen $= 12$, king $= 13$). If we assume that the three random variables involved are independent and uniformly distributed, compute the mean and variance of the amount to be received.

4.1.16. Let X_1 and X_2 be independent random variables with nonzero variances. Find the correlation coefficient of $Y = X_1 X_2$ and X_1 in terms of the means and variances of X_1 and X_2.

4.1.17. Let X_1 and X_2 have a joint distribution with parameters μ_1, μ_2, σ_1^2, σ_2^2, and ρ. Find the correlation coefficient of the linear functions of $Y = a_1 X_1 + a_2 X_2$ and $Z = b_1 X_1 + b_2 X_2$ in terms of the real constants a_1, a_2, b_1, b_2, and the parameters of the distribution.

4.1.18. Let X_1, X_2, and X_3 be random variables with equal variances but with correlation coefficients $\rho_{12} = 0.3$, $\rho_{13} = 0.5$, and $\rho_{23} = 0.2$. Find the correlation coefficient of the linear functions $Y = X_1 + X_2$ and $Z = X_2 + X_3$.

4.1.19. Find the variance of the sum of 10 random variables if each has variance 5 and if each pair has correlation coefficient 0.5.

4.1.20. Let X and Y have the parameters μ_1, μ_2, σ_1^2, σ_2^2, and ρ. Show that the correlation coefficient of X and $[Y - \rho(\sigma_2/\sigma_1)X]$ is zero.

4.1.21. Let X_1 and X_2 have a bivariate normal distribution with parameters μ_1, μ_2, σ_1^2, σ_2^2, and ρ. Compute the means, the variances, and the correlation coefficient of $Y_1 = \exp(X_1)$ and $Y_2 = \exp(X_2)$.
Hint: Various moments of Y_1 and Y_2 can be found by assigning appropriate values to t_1 and t_2 in $E[\exp(t_1 X_1 + t_2 X_2)]$.

4.1.22. Let X be $N(\mu, \sigma^2)$ and consider the transformation $X = \log(Y)$ or, equivalently, $Y = e^X$.

 (a) Find the mean and the variance of Y by first determining $E(e^X)$ and $E[(e^X)^2]$, by using the mgf of X.

(b) Find the pdf of Y. This is the pdf of the *lognormal distribution*.

4.1.23. Let X_1 and X_2 have a trinomial distribution with parameters n, p_1, p_2.

(a) What is the distribution of $Y = X_1 + X_2$?

(b) From the equality $\sigma_Y^2 = \sigma_1^2 + \sigma_2^2 + 2\rho\sigma_1\sigma_2$, once again determine the correlation coefficient ρ of X_1 and X_2.

4.1.24. Let $Y_1 = X_1 + X_2$ and $Y_2 = X_2 + X_3$, where X_1, X_2, and X_3 are three independent random variables. Find the joint mgf and the correlation coefficient of Y_1 and Y_2 provided that:

(a) X_i has a Poisson distribution with mean μ_i, $i = 1, 2, 3$.

(b) X_i is $N(\mu_i, \sigma_i^2)$, $i = 1, 2, 3$.

4.1.25. Let S^2 be the sample variance of a random sample from a distribution with variance $\sigma^2 > 0$. Since $E(S^2) = \sigma^2$, why isn't $E(S) = \sigma$? *Hint:* Use Jensen's inequality to show that $E(S) < \sigma$.

4.1.26. For the last exercise, suppose that the sample is drawn from a $N(\mu, \sigma^2)$ distribution. Recall that $(n-1)S^2/\sigma^2$ has a $\chi^2(n-1)$ distribution. Use Theorem 3.3.1 to determine an unbiased estimator of σ.

4.1.27. Let S^2 be the sample variance of a random sample drawn from a $N(\mu, \sigma^2)$ distribution. Show that the constant $c = (n-1)/(n+1)$ minimizes $E[(cS^2 - \sigma^2)^2]$. Hence, the estimator $(n+1)^{-1}\sum_{i=1}^{n}(X_i - \overline{X})^2$ of σ^2 minimizes the mean square error among estimators of the form cS^2.

4.2 Convergence in Probability

In this section, we formalize a way of saying that a sequence of random variables is getting "close" to another random variable. We will use this concept throughout the book.

Definition 4.2.1. *Let $\{X_n\}$ be a sequence of random variables and let X be a random variable defined on a sample space. We say that X_n **converges in probability** to X if for all $\epsilon > 0$*

$$\lim_{n\to\infty} P[|X_n - X| \geq \epsilon] = 0,$$

or equivalently,

$$\lim_{n\to\infty} P[|X_n - X| < \epsilon] = 1.$$

If so, we write

$$X_n \xrightarrow{P} X.$$

If $X_n \xrightarrow{P} X$, we often say that the mass of the difference $X_n - X$ is converging to 0. In statistics, often the limiting random variable X is a constant; i.e., X is a degenerate random variable with all its mass at some constant a. In this case, we write $X_n \xrightarrow{P} a$. Also, as Exercise 4.2.1 shows, convergence of real sequence $a_n \to a$ is equivalent to $a_n \xrightarrow{P} a$.

One way of showing convergence in probability is to use Chebyshev's Theorem (1.10.3). An illustration of this is given in the following proof. To emphasize the fact that we are working with sequences of random variables, we may place a subscript n on random variables, like $\overline{X}$ to read $\overline{X}_n$.

Theorem 4.2.1 (Weak Law of Large Numbers). *Let $\{X_n\}$ be a sequence of iid random variables having common mean μ and variance $\sigma^2 < \infty$. Let $\overline{X}_n = n^{-1} \sum_{i=1}^{n} X_i$. Then*

$$\overline{X}_n \xrightarrow{P} \mu.$$

Proof: Recall from Example 4.1.1 that mean and variance of $\overline{X}_n$ is μ and σ^2/n, respectively. Hence, by Chebyshev's Theorem we have for any $\epsilon > 0$,

$$P[|\overline{X}_n - \mu| \geq \epsilon] = P[|\overline{X}_n - \mu| \geq (\epsilon\sqrt{n}/\sigma)(\sigma/\sqrt{n})] \leq \frac{\sigma^2}{n\epsilon^2} \to 0. \quad \blacksquare$$

This theorem says that all the mass of the distribution of $\overline{X}_n$ is converging to μ, as n converges to ∞. In a sense, for n large, $\overline{X}_n$ is close to μ. But how close? For instance, if we were to estimate μ by $\overline{X}_n$, what can we say about the error of estimation? We will answer this in Section 4.3.

Actually in a more advanced course a Strong Law of Large Numbers is proven; see page 124 of Chung (1974). One result of this theorem is that we can weaken the hypothesis of Theorem 4.2.1 to the assumption that the random variables X_i are independent and each has finite mean μ. Thus the Strong Law of Large Numbers is a first moment theorem, while the Weak Law requires the existence of the second moment.

There are several theorems concerning convergence in probability which will be useful in the sequel. Together the next two theorems say that convergence in probability is closed under linearity.

Theorem 4.2.2. *Suppose $X_n \xrightarrow{P} X$ and $Y_n \xrightarrow{P} Y$. Then $X_n + Y_n \xrightarrow{P} X + Y$.*

Proof: Let $\epsilon > 0$ be given. Using the triangle inequality we can write

$$|X_n - X| + |Y_n - Y| \geq |(X_n + Y_n) - (X + Y)| \geq \epsilon.$$

Since P is monotone relative to set containment, we have

$$
\begin{aligned}
P[|(X_n + Y_n) - (X + Y)| \geq \epsilon] &\leq P[|X_n - X| + |Y_n - Y| \geq \epsilon] \\
&\leq P[|X_n - X| \geq \epsilon/2] + P[|Y_n - Y| \geq \epsilon/2].
\end{aligned}
$$

By the hypothesis of the theorem, the last two terms converge to 0 which gives us the desired result. $\blacksquare$

Theorem 4.2.3. *Suppose* $X_n \xrightarrow{P} X$ *and* a *is a constant. Then* $aX_n \xrightarrow{P} aX$.

Proof: If $a = 0$, the result is immediate. Suppose $a \neq 0$. Let $\epsilon > 0$. The result follows from these equalities:

$$P[|aX_n - aX| \geq \epsilon] = P[|a||X_n - X| \geq \epsilon] = P[|X_n - X| \geq \epsilon/|a|],$$

and by hypotheses the last term goes to 0. ∎

Theorem 4.2.4. *Suppose* $X_n \xrightarrow{P} a$ *and the real function* g *is continuous at* a. *Then* $g(X_n) \xrightarrow{P} g(a)$.

Proof: Let $\epsilon > 0$. Then since g is continuous at a, there exists a $\delta > 0$ such that if $|x - a| < \delta$, then $|g(x) - g(a)| < \epsilon$. Thus

$$|g(x) - g(a)| \geq \epsilon \Rightarrow |x - a| \geq \delta.$$

Substituting X_n for x in the above implication, we obtain

$$P[|g(X_n) - g(a)| \geq \epsilon] \leq P[|X_n - a| \geq \delta].$$

By the hypothesis, the last term goes to 0 as $n \to \infty$, which gives us the result. ∎

This theorem gives us many useful results. For instance, if $X_n \xrightarrow{P} a$, then

$$X_n^2 \xrightarrow{P} a^2,$$
$$1/X_n \xrightarrow{P} 1/a, \quad \text{provided } a \neq 0,$$
$$\sqrt{X_n} \xrightarrow{P} \sqrt{a}, \quad \text{provided } a \geq 0.$$

Actually, in a more advanced class, it is shown that if $X_n \xrightarrow{P} X$ and g is a continuous function then $g(X_n) \xrightarrow{P} g(X)$; see page 104 of Tucker (1967). We make use of this in the next theorem.

Theorem 4.2.5. *Suppose* $X_n \xrightarrow{P} X$ *and* $Y_n \xrightarrow{P} Y$. *Then* $X_n Y_n \xrightarrow{P} XY$.

Proof: Using the above results, we have

$$
\begin{aligned}
X_n Y_n &= \frac{1}{2} X_n^2 + \frac{1}{2} Y_n^2 - \frac{1}{2}(X_n - Y_n)^2 \\
&\xrightarrow{P} \frac{1}{2} X^2 + \frac{1}{2} Y^2 - \frac{1}{2}(X - Y)^2 = XY. \quad \blacksquare
\end{aligned}
$$

Let us return to our discussion of sampling and statistics. Consider the situation where we have a random variable X whose distribution has an unknown parameter $\theta \in \Omega$. We seek a statistic based on a sample to estimate θ. In the last section, we introduced the property of unbiasedness for an estimator. We now introduce consistency:

Definition 4.2.2. *Let X be a random variable with cdf $F(x, \theta)$, $\theta \in \Omega$. Let $X_1, \ldots, X_n$ be a sample from the distribution of X and let T_n denote a statistic. We say T_n is a* **consistent** *estimator of θ if*

$$T_n \xrightarrow{P} \theta.$$

If $X_1, \ldots, X_n$ is a random sample from a distribution with finite mean μ and variance σ^2, then by the Weak Law of Large Numbers, the sample mean, $\overline{X}$, is a consistent estimator of μ.

Example 4.2.1 (Sample Variance). Let $X_1, \ldots, X_n$ denote a random sample from a distribution with mean μ and variance σ^2. Theorem 4.2.1 showed that $\overline{X}_n \xrightarrow{P} \mu$. To show that the sample variance converges in probability to σ^2, assume further that $E[X_1^4] < \infty$, so that $\mathrm{Var}(S^2) < \infty$. Using the preceding results, we can show the following:

$$S_n^2 = \frac{1}{n-1} \sum_{i=1}^{n} (X_i - \overline{X}_n)^2 \quad = \quad \frac{n}{n-1} \left(\frac{1}{n} \sum_{i=1}^{n} X_i^2 - \overline{X}_n^2 \right)$$

$$\xrightarrow{P} \quad 1 \cdot [E(X_1^2) - \mu^2] = \sigma^2.$$

Hence, the sample variance is a consistent estimator of σ^2. ∎

Unlike the last example, sometimes we can obtain the convergence by using the distribution function. We illustrate this with the following example:

Example 4.2.2 (Maximum of a Sample from a Uniform Distribution). Reconsider Example 4.1.3, where $X_1, \ldots, X_n$ is a random sample from a uniform$(0, \theta)$ distribution. Let $Y_n = \max\{X_1, \ldots, X_n\}$. The cdf of Y_n is given by expression (4.1.7), from which it is easily seen that $Y_n \xrightarrow{P} \theta$ and the sample maximum is a consistent estimate of θ. Note that the unbiased estimator, $((n+1)/n)Y_n$, is also consistent. ∎

To expand on Example 4.2.2, by the Weak Law of Large Numbers, Theorem 4.2.1, it follows that $\overline{X}_n$ is a consistent estimator of $\theta/2$ so $2\overline{X}_n$ is a consistent estimator of θ. Note the difference in how we showed that Y_n and $2\overline{X}_n$ converge to θ in probability. For Y_n we used the cdf of Y_n but for $2\overline{X}_n$ we appealed to the Weak Law of Large Numbers. In fact, the cdf of $2\overline{X}_n$ is quite complicated for the uniform model. In many situations, the cdf of the statistic cannot be obtained but we can appeal to asymptotic theory to establish the result. There are other estimators of θ. Which is the "best" estimator? In future chapters we will be concerned with such questions.

Consistency is a very important property for an estimator to have. It is a poor estimator that does not approach its target as the sample size gets large. Note that the same cannot be said for the property of unbiasedness. For example, instead of using the sample variance to estimate σ^2, suppose we use $V = n^{-1} \sum_{i=1}^{n} (X_i - \overline{X})^2$. Then V is consistent for σ^2, but it is biased, because $E(V) = (n-1)\sigma^2/n$. Thus the bias of V is σ^2/n, which vanishes as $n \to \infty$.

EXERCISES

4.2.1. Let $\{a_n\}$ be a sequence of real numbers. Hence, we can also say that $\{a_n\}$ is a sequence of constant (degenerate) random variables. Let a be a real number. Show that $a_n \to a$ is equivalent to $a_n \overset{P}{\to} a$.

4.2.2. Let the random variable Y_n have a distribution that is $b(n, p)$.

(a) Prove that Y_n/n converges in probability p. This result is one form of the weak law of large numbers.

(b) Prove that $1 - Y_n/n$ converges in probability to $1 - p$.

(c) Prove that $(Y_n/n)(1 - Y_n/n)$ converges in probability to $p(1 - p)$.

4.2.3. Let W_n denote a random variable with mean μ and variance b/n^p, where $p > 0$, μ, and b are constants (not functions of n). Prove that W_n converges in probability to μ.
Hint: Use Chebyshev's inequality.

4.2.4. Let $X_1, \ldots, X_n$ be iid random variables with common pdf

$$f(x) = \begin{cases} e^{-(x-\theta)} & x > \theta \ -\infty < \theta < \infty \\ 0 & \text{elsewhere.} \end{cases} \tag{4.2.1}$$

This pdf is called the **shifted exponential**. Let $Y_n = \min\{X_1, \ldots, X_n\}$. Prove that $Y_n \to \theta$ in probability, by obtaining the cdf and the pdf of Y_n.

4.2.5. For Exercise 4.2.4, obtain the mean of Y_n. Is Y_n an unbiased estimator of θ? Obtain an unbiased estimator of θ based on Y_n.

4.3 Convergence in Distribution

In the last section, we introduced the concept of convergence in probability. With this concept, we can formally say, for instance, that a statistic converges to a parameter and, furthermore, in many situations we can show this without having to obtain the distribution function of the statistic. But how close is the statistic to the estimator? For instance, can we obtain the error of estimation with some credence? The method of convergence discussed in this section, in conjunction with earlier results, gives us affirmative answers to these questions.

Definition 4.3.1 (Convergence in Distribution). *Let $\{X_n\}$ be a sequence of random variables and let X be a random variable. Let F_{X_n} and F_X be, respectively, the cdfs of X_n and X. Let $C(F_X)$ denote the set of all points where F_X is continuous. We say that X_n **converges in distribution** to X if*

$$\lim_{n \to \infty} F_{X_n}(x) = F_X(x), \quad \text{for all } x \in C(F_X).$$

We denote this convergence by

$$X_n \overset{D}{\to} X.$$

Remark 4.3.1. This material on convergence in probability and in distribution comes under what statisticians and probabilists refer to as *asymptotic theory.* Often, we say that the ditsribution of X is the **asymptotic distribution** or the **limiting distribution** of the sequence $\{X_n\}$. We might even refer informally to the asymptotics of certain situations. Moreover, for illustration, instead of saying $X_n \overset{D}{\to} X$, where X has a standard normal random, we may write

$$X_n \overset{D}{\to} N(0,1),$$

as an abbreviated way of saying the same thing. Clearly the right-hand member of this last expression is a distribution and not a random variable as it should be, but we will make use of this convention. In addition, we may say that X_n has a *limiting* standard normal distribution to mean that $X_n \overset{D}{\to} X$, where X has a standard normal random, or equivalently $X_n \overset{D}{\to} N(0,1)$. ∎

Motivation for only considering points of continuity of F_X is given by the following simple example. Let X_n be a random variable with all its mass at $\frac{1}{n}$ and let X be a random variable with all its mass at 0. Then as Figure 4.3.1 shows all the mass of X_n is converging to 0, i.e., the distribution of X. At the point of discontinuity of F_X, $\lim F_{X_n}(0) = 0 \neq 1 = F_X(0)$; while at continuity points x of F_X, (i.e., , $x \neq 0$), $\lim F_{X_n}(x) = F_X(x)$. Hence, according to the definition, $X_n \overset{D}{\to} X$.

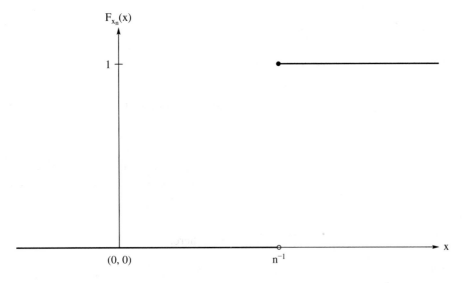

Figure 4.3.1: Cdf of X_n which has all its mass at n^{-1}

Convergence in probability is a way of saying that a sequence of random variables X_n is getting close to another random variable X. On the other hand, convergence in distribution is only concerned with the cdfs F_{X_n} and F_X. A simple example illustrates this. Let X be a continuous random variable with a pdf $f_X(x)$ which is symmetric about 0; i.e., $f_X(-x) = f_X(x)$. Then is easy to show that the density of the random variable $-X$ is also $f_X(x)$. Thus X and $-X$ have the same distributions. Define the sequence of random variables X_n as

$$X_n = \begin{cases} X & \text{if } n \text{ is odd} \\ -X & \text{if } n \text{ is even.} \end{cases} \tag{4.3.1}$$

Clearly $F_{X_n}(x) = F_X(x)$ for all x in the support of X, so that $X_n \xrightarrow{D} X$. On the other hand, the sequence X_n does not get close to X. In particular, $X_n \nrightarrow X$ in probability.

Example 4.3.1. Let $\overline{X}_n$ have the cdf

$$F_n(\overline{x}) = \int_{-\infty}^{\overline{x}} \frac{1}{\sqrt{1/n}\sqrt{2\pi}} e^{-nw^2/2} \, dw.$$

If the change of variable $v = \sqrt{n}w$ is made, we have

$$F_n(\overline{x}) = \int_{-\infty}^{\sqrt{n}\overline{x}} \frac{1}{\sqrt{2\pi}} e^{-v^2/2} \, dv.$$

It is clear that

$$\lim_{n\to\infty} F_n(\overline{x}) = \begin{cases} 0 & \overline{x} < 0 \\ \frac{1}{2} & \overline{x} = 0 \\ 1 & \overline{x} > 0. \end{cases}$$

Now the function

$$F(\overline{x}) = \begin{cases} 0 & \overline{x} < 0 \\ 1 & \overline{x} \geq 0, \end{cases}$$

is a cdf and $\lim_{n\to\infty} F_n(\overline{x}) = F(\overline{x})$ at every point of continuity of $F(\overline{x})$. To be sure, $\lim_{n\to\infty} F_n(0) \neq F(0)$, but $F(\overline{x})$ is not continuous at $\overline{x} = 0$. Accordingly, the sequence $\overline{X}_1, \overline{X}_2, \overline{X}_3, \ldots$ converges in distribution to a random variable that has a degenerate distribution at $\overline{x} = 0$. ∎

Example 4.3.2. Even if a sequence $X_1, X_2, X_3, \ldots$ converges in distribution to a random variable X, we cannot in general determine the distribution of X by taking the limit of the pmf of X_n. This is illustrated by letting X_n have the pmf

$$p_n(x) = \begin{cases} 1 & x = 2 + n^{-1} \\ 0 & \text{elsewhere.} \end{cases}$$

Clearly, $\lim_{n\to\infty} p_n(x) = 0$ for all values of x. This may suggest that X_n, for $n = 1, 2, 3, \ldots$, does not converge in distribution. However, the cdf of X_n is

$$F_n(x) = \begin{cases} 0 & x < 2 + n^{-1} \\ 1 & x \geq 2 + n^{-1}, \end{cases}$$

and

$$\lim_{n\to\infty} F_n(x) = \begin{cases} 0 & x \leq 2 \\ 1 & x > 2. \end{cases}$$

Since

$$F(x) = \begin{cases} 0 & x < 2 \\ 1 & x \geq 2, \end{cases}$$

is a cdf, and since $\lim_{n\to\infty} F_n(x) = F(x)$ at all points of continuity of $F(x)$, the sequence $X_1, X_2, X_3, \ldots$ converges in distribution to a random variable with cdf $F(x)$. ∎

The last example showed in general that we cannot determine limiting distributions by considering pmfs or pdfs. But under certain conditions we can determine convergence in distribution by considering the sequence of pdfs as the following example shows.

Example 4.3.3. Let T_n have a t-distribution with n degrees of freedom, $n = 1, 2, 3, \ldots$. Thus its cdf is

$$F_n(t) = \int_{-\infty}^{t} \frac{\Gamma[(n+1))/2]}{\sqrt{\pi n}\, \Gamma(n/2)} \frac{1}{(1 + y^2/n)^{(n+1)/2}}\, dy,$$

where the integrand is the pdf $f_n(y)$ of T_n. Accordingly,

$$\lim_{n\to\infty} F_n(t) = \lim_{n\to\infty} \int_{-\infty}^{t} f_n(y)\, dy = \int_{-\infty}^{t} \lim_{n\to\infty} f_n(y)\, dy.$$

By a result in analysis, (Lebesgue Dominated Convergence Theorem), that allows us to interchange the order of the limit and integration provided $|f_n(y)|$ is dominated by a function which is integrable. This is true because

$$|f_n(y)| \leq 10 f_1(y)$$

and

$$\int_{-\infty}^{t} 10 f_1(y)\, dy = \frac{10}{\pi} \arctan t < \infty,$$

for all real t. Hence, we can find the limiting distribution by finding the limit of the pdf of T_n. It is

$$\lim_{n\to\infty} f_n(y) = \lim_{n\to\infty} \left\{ \frac{\Gamma[(n+1)/2]}{\sqrt{n/2}\, \Gamma(n/2)} \right\} \lim_{n\to\infty} \left\{ \frac{1}{(1 + y^2/n)^{1/2}} \right\}$$

$$\times \lim_{n\to\infty} \left\{ \frac{1}{\sqrt{2\pi}} \left[\left(1 + \frac{y^2}{n} \right) \right]^{-n/2} \right\}.$$

Using the fact from elementary calculus that

$$\lim_{n\to\infty} \left(1 + \frac{y^2}{n} \right)^{n} = e^{y^2},$$

the limit associated with the third factor is clearly the pdf of the standard normal distribution. The second limit obviously equals 1. By Remark 4.3.2, the first limit also equals 1. Thus we have

$$\lim_{n \to \infty} F_n(t) = \int_{-\infty}^{t} \frac{1}{\sqrt{2\pi}} e^{-y^2/2} \, dy,$$

and hence T_n has a limiting standard normal distribution. ∎

Remark 4.3.2 (Stirling's Formula). In advanced calculus the following approximation is derived,

$$\Gamma(k+1) \doteq \sqrt{2\pi} k^{k+1/2} e^{-k}. \tag{4.3.2}$$

This is known as *Stirling's formula* and it is an excellent approximation when k is large. Because $\Gamma(k+1) = k!$, for k an integer, this formula gives one an idea of how fast $k!$ grows. As Exercise 4.3.20 shows, this approximation can be used to show that the first limit in Example 4.3.3 is 1. ∎

Example 4.3.4 (Maximum of a Sample from a Uniform Distribution, Continued). Recall Example 4.1.3, where $X_1, \ldots, X_n$ is a random sample from a uniform$(0, \theta)$ distribution. Again let $Y_n = \max\{X_1, \ldots, X_n\}$, but now consider the random variable $Z_n = n(\theta - Y_n)$. Let $t \in (0, n\theta)$. Then, using the cdf of Y_n, (4.1.7), the cdf of Z_n is

$$
\begin{aligned}
P[Z_n \leq t] &= P[Y_n \geq \theta - (t/n)] \\
&= 1 - \left(\frac{\theta - (t/n)}{\theta} \right)^n \\
&= 1 - \left(1 - \frac{t/\theta}{n} \right)^n \\
&\to 1 - e^{-t/\theta}.
\end{aligned}
$$

Note that the last quantity is the cdf of an exponential random variable with mean θ, (3.3.2). So we would say that $Z_n \xrightarrow{D} Z$, where Z is distributed exp(θ). ∎

Remark 4.3.3. To simplify several of the proofs of this section, we make use of the $\underline{\lim}$ and $\overline{\lim}$ of a sequence. For readers who are unfamilar with these concepts, we discuss them in Appendix A. In this brief remark, we highlight the properties needed for understanding the proofs. Let $\{a_n\}$ be a sequence of real numbers and define the two subsequences,

$$
\begin{aligned}
b_n &= \sup\{a_n, a_{n+1}, \ldots\}, \quad n = 1, 2, 3 \ldots, \tag{4.3.3} \\
c_n &= \inf\{a_n, a_{n+1}, \ldots\}, \quad n = 1, 2, 3 \ldots. \tag{4.3.4}
\end{aligned}
$$

While $\{c_n\}$ is a nondecreasing sequence, $\{b_n\}$ is a nonincreasing sequence. Hence, their limits always exist, (may be $\pm\infty$). Denote them respectively by $\underline{\lim}_{n\to\infty} a_n$ and $\overline{\lim}_{n\to\infty} a_n$. Further, $c_n \leq a_n \leq b_n$, for all n. Hence, by the Sandwich Theorem (see Theorem A.2.1 of Appendix A), if $\underline{\lim}_{n\to\infty} a_n = \overline{\lim}_{n\to\infty} a_n$, then $\lim_{n\to\infty} a_n$ exists and is given by $\lim_{n\to\infty} a_n = \overline{\lim}_{n\to\infty} a_n$.

As discussed in the appendix, several other properties of these concepts are useful. For example, suppose $\{p_n\}$ is a sequence of probabilities and $\overline{\lim}_{n\to\infty} p_n = 0$. Then by the Sandwich Theorem, since $0 \leq p_n \leq \sup\{p_n, p_{n+1}, \ldots\}$, for all n, we have $\lim_{n\to\infty} p_n = 0$. Also, for any two sequences $\{a_n\}$ and $\{b_n\}$, it easily follows that $\overline{\lim}_{n\to\infty}(a_n + b_n) \leq \overline{\lim}_{n\to\infty} a_n + \overline{\lim}_{n\to\infty} b_n$. ∎

As the following theorem shows, convergence in distribution is weaker than convergence in probability. Thus convergence in distribution is often called weak convergence.

Theorem 4.3.1. *If X_n converges to X in probability, then X_n converges to X in distribution.*

Proof: Let x be a point of continuity of $F_X(x)$. Let $\epsilon > 0$. We have,

$$
\begin{aligned}
F_{X_n}(x) &= P[X_n \leq x] \\
&= P[\{X_n \leq x\} \cap \{|X_n - X| < \epsilon\}] + P[\{X_n \leq x\} \cap \{|X_n - X| \geq \epsilon\}] \\
&\leq P[X \leq x + \epsilon] + P[|X_n - X| \geq \epsilon].
\end{aligned}
$$

Based on this inequality and the fact that $X_n \xrightarrow{P} X$, we see that

$$
\overline{\lim_{n\to\infty}} F_{X_n}(x) \leq F_X(x + \epsilon). \tag{4.3.5}
$$

To get a lower bound, we proceed similarly with the complement to show that

$$
P[X_n > x] \leq P[X \geq x - \epsilon] + P[|X_n - X| \geq \epsilon].
$$

Hence,

$$
\underline{\lim_{n\to\infty}} F_{X_n}(x) \geq F_X(x - \epsilon). \tag{4.3.6}
$$

Using a relationship between $\overline{\lim}$ and $\underline{\lim}$, it follows from (4.3.5) and (4.3.6) that

$$
F_X(x - \epsilon) \leq \underline{\lim_{n\to\infty}} F_{X_n}(x) \leq \overline{\lim_{n\to\infty}} F_{X_n}(x) \leq F_X(x + \epsilon).
$$

Letting $\epsilon \downarrow 0$ gives us the desired result. ∎

Reconsider the sequence of random variables $\{X_n\}$ defined by expression (4.3.1). Here, $X_n \xrightarrow{D} X$ but $X_n \xrightarrow{P}\!\!\!\!\!/ \ X$. So in general the converse of the above theorem is not true. However, it is true if X is degenerate as shown by the following theorem.

Theorem 4.3.2. *If X_n converges to the constant b in distribution, then X_n converges to b in probability.*

Proof: Let $\epsilon > 0$ be given. Then,

$$
\lim_{n\to\infty} P[|X_n - b| \leq \epsilon] = \lim_{n\to\infty} F_{X_n}(b + \epsilon) - \lim_{n\to\infty} F_{X_n}(b - \epsilon - 0) = 1 - 0 = 1,
$$

which is the desired result. ∎

A result that will prove quite useful is the following:

Theorem 4.3.3. *Suppose X_n converges to X in distribution and Y_n converges in probability to 0. Then $X_n + Y_n$ converges to X in distribution.*

The proof is similar to that of Theorem 4.3.2 and is left to Exercise 4.3.12. We often use this last result as follows. Suppose it is difficult to show that X_n converges to X in distribution; but it is easy to show that Y_n converges in distribution to X and that $X_n - Y_n$ converges to 0 in probability. Hence, by this last theorem, $X_n = Y_n + (X_n - Y_n) \xrightarrow{D} X$, as desired.

The next two theorems state general results. A proof of the first result can be found in a more advanced text while the second, Slutsky's Theorem, follows similar to that of Theorem 4.3.1.

Theorem 4.3.4. *Suppose X_n converges to X in distribution and g is a continuous function on the support of X. Then $g(X_n)$ converges to $g(X)$ in distribution.*

Theorem 4.3.5 (Slutsky's Theorem). *Let X_n, X, A_n, B_n be random variables and let a and b be constants. If $X_n \xrightarrow{D} X$, $A_n \xrightarrow{P} a$, and $B_n \xrightarrow{P} b$, then*

$$A_n + B_n X_n \xrightarrow{D} a + bX.$$

4.3.1 Bounded in Probability

Another useful concept, related to convergence in distribution, is boundedness in probability of a sequence of random variables.

First consider any random variable X with cdf $F_X(x)$. Then given $\epsilon > 0$, we can bound X in the following way. Because the lower limit of F_X is 0 and its upper limit is 1 we can find η_1 and η_2 such that

$$F_X(x) < \epsilon/2 \text{ for } x \leq \eta_1 \text{ and } F_X(x) > 1 - (\epsilon/2) \text{ for } x \geq \eta_2.$$

Let $\eta = \max\{|\eta_1|, |\eta_2|\}$ then

$$P[|X| \leq \eta] = F_X(\eta) - F_X(-\eta - 0) \geq 1 - (\epsilon/2) - (\epsilon/2) = 1 - \epsilon. \tag{4.3.7}$$

Thus random variables which are not bounded (e.g., X is $N(0,1)$) are still bounded in the above way. This is a useful concept for sequences of random variables which we define next.

Definition 4.3.2 (Bounded in Probability). *We say that the sequence of random variables $\{X_n\}$ is bounded in probability, if for all $\epsilon > 0$ there exists a constant $B_\epsilon > 0$ and an integer N_ϵ such that*

$$n \geq N_\epsilon \Rightarrow P[|X_n| \leq B_\epsilon] \geq 1 - \epsilon.$$

Next, consider a sequence of random variables $\{X_n\}$ which converge in distribution to a random variable X which has cdf F. Let $\epsilon > 0$ be given and choose η so that (4.3.7) holds for X. We can always choose η so that η and $-\eta$ are continuity points of F. We then have,

$$\lim_{n \to \infty} P[|X_n| \leq \eta] \geq \lim_{n \to \infty} F_{X_n}(\eta) - \lim_{n \to \infty} F_{X_n}(-\eta - 0) = F_X(\eta) - F_X(-\eta) \geq 1 - \epsilon.$$

To be precise, we can then choose N so large that $P[|X_n| \leq \eta] \geq 1 - \epsilon$, for $n \geq N$. We have thus proved the following theorem

Theorem 4.3.6. *Let $\{X_n\}$ be a sequence of random variables and let X be a random variable. If $X_n \to X$ in distribution, then $\{X_n\}$ is bounded in probability.*

As the following example shows, the converse of this theorem is not true.

Example 4.3.5. Take $\{X_n\}$ to be the following sequence of degenerate random variables. For $n = 2m$ even, $X_{2m} = 2 + (1/(2m))$ with probability one. For $n = 2m - 1$ odd, $X_{2m-1} = 1 + (1/(2m))$ with probability one. Then the sequence $\{X_2, X_4, X_6, \ldots\}$ converges in distribution to the degenerate random variable $Y = 2$, while the sequence $\{X_1, X_3, X_5, \ldots\}$ converges in distribution to the degenerate random variable $W = 1$. Since the distributions of Y and W are not the same, the sequence $\{X_n\}$ does not converge in distribution. Because all of the mass of the sequence $\{X_n\}$ is in the interval $[1, 5/2]$, however, the sequence $\{X_n\}$ is bounded in probability. ∎

One way of thinking of a sequence which is bounded in probability (or one which is converging to a random variable in distribution) is that the probability mass of $|X_n|$ is not escaping to ∞. At times we can use boundedness in probability instead of convergence in distribution. A property we will need later is given in the following theorem

Theorem 4.3.7. *Let $\{X_n\}$ be a sequence of random variables bounded in probability and let $\{Y_n\}$ be a sequence of random variables which converge to 0 in probability. Then*

$$X_n Y_n \xrightarrow{P} 0.$$

Proof: Let $\epsilon > 0$ be given. Choose $B_\epsilon > 0$ and an integer N_ϵ such that

$$n \geq N_\epsilon \Rightarrow P[|X_n| \leq B_\epsilon] \geq 1 - \epsilon.$$

Then,

$$
\begin{aligned}
\varlimsup_{n \to \infty} P[|X_n Y_n| \geq \epsilon] \quad &\leq \quad \varlimsup_{n \to \infty} P[|X_n Y_n| \geq \epsilon, |X_n| \leq B_\epsilon] \\
&\quad + \varlimsup_{n \to \infty} P[|X_n Y_n| \geq \epsilon, |X_n| > B_\epsilon] \\
&\leq \quad \varlimsup_{n \to \infty} P[|Y_n| \geq \epsilon/B_\epsilon] + \epsilon = \epsilon. \quad (4.3.8)
\end{aligned}
$$

From which the desired result follows. ∎

4.3.2   Δ-Method

Recall a common problem discussed in the last three chapters is the situation where we know the distribution of a random variable, but we want to determine the distribution of a function of it. This is also true in asymptotic theory and Theorems 4.3.4 and 4.3.5 are illustrations of this. Another such result is called the Δ-method. To establish this result we need a convenient form of the mean value theorem with remainder, sometimes called Young's Theorem; see Hardy (1992) or Lehmann (1999). Suppose $g(x)$ is differentiable at x. Then we can write,

$$g(y) = g(x) + g'(x)(y - x) + o(|y - x|), \tag{4.3.9}$$

where the notation o means

$$a = o(b) \text{ if and only if } \tfrac{a}{b} \to 0, \text{ as } b \to 0.$$

The *little o* notation is used in terms of convergence in probability, also. We often write $o_p(X_n)$, which means

$$Y_n = o_p(X_n) \text{ if and only if } \tfrac{Y_n}{X_n} \xrightarrow{P} 0, \text{ as } n \to \infty. \tag{4.3.10}$$

There is a corresponding *big O* notation which is given by

$$Y_n = O_p(X_n) \text{ if and only if } \tfrac{Y_n}{X_n} \text{ is bounded in probability as } n \to \infty. \tag{4.3.11}$$

The following theorem illustrates the little-o notation, but it is also serves as a lemma for Theorem 4.3.9.

Theorem 4.3.8. *Suppose $\{Y_n\}$ is a sequence of random variables which is bounded in probability. Suppose $X_n = o_p(Y_n)$. Then $X_n \xrightarrow{P} 0$, as $n \to \infty$.*

Proof: Let $\epsilon > 0$ be given. Because the sequence $\{Y_n\}$ is bounded in probability, there exists positive constants N_ϵ and B_ϵ such that

$$n \geq N_\epsilon \implies P[|Y_n| \leq B_\epsilon] \geq 1 - \epsilon. \tag{4.3.12}$$

Also, because $X_n = o_p(Y_n)$, we have

$$\frac{X_n}{Y_n} \xrightarrow{P} 0, \tag{4.3.13}$$

as $n \to \infty$. We then have,

$$\begin{aligned} P[|X_n| \geq \epsilon] &= P[|X_n| \geq \epsilon, |Y_n| \leq B_\epsilon] + P[|X_n| \geq \epsilon, |Y_n| > B_\epsilon] \\ &\leq P\left[\frac{X_n}{|Y_n|} \geq \frac{\epsilon}{B_\epsilon}\right] + P[|Y_n| > B_\epsilon]. \end{aligned}$$

By (4.3.13) and (4.3.12), respectively, the first and second terms on the right-side can be made arbitrarily small by choosing n sufficiently large. Hence, the result is true. ■

Theorem 4.3.9. *Let $\{X_n\}$ be a sequence of random variables such that*

$$\sqrt{n}(X_n - \theta) \xrightarrow{D} N(0, \sigma^2). \tag{4.3.14}$$

Suppose the function $g(x)$ is differentiable at θ and $g'(\theta) \neq 0$. Then

$$\sqrt{n}(g(X_n) - g(\theta)) \xrightarrow{D} N(0, \sigma^2(g'(\theta))^2). \tag{4.3.15}$$

Proof: Using expression (4.3.9), we have,

$$g(X_n) = g(\theta) + g'(\theta)(X_n - \theta) + o_p(|X_n - \theta|),$$

where o_p is interpreted as in (4.3.10). Rearranging, we have

$$\sqrt{n}(g(X_n) - g(\theta)) = g'(\theta)\sqrt{n}(X_n - \theta) + o_p(\sqrt{n}|X_n - \theta|).$$

Because (4.3.14) holds, Theorem 4.3.6 implies that $\sqrt{n}|X_n - \theta|$ is bounded in probability. Therefore by Theorem 4.3.8 $o_p(\sqrt{n}|X_n - \theta|) \to 0$, in probability. Hence by (4.3.14) and Theorem 4.3.1 the result follows. ∎

Illustrations of the Δ- method can be found in Example 4.3.8 and the exercises.

4.3.3 Moment Generating Function Technique

To find the limiting distribution function of a random variable X_n by using the definition obviously requires that we know $F_{X_n}(x)$ for each positive integer n. But it is often difficult to obtain $F_{X_n}(x)$ in closed form. Fortunately, if it exists, the mgf that corresponds to the cdf $F_{X_n}(x)$ often provides a convenient method of determining the limiting cdf.

The following theorem, which is essentially Curtiss' modification of a theorem of Lévy and Cramér, explains how the mgf may be used in problems of limiting distributions. A proof of the theorem is beyond of the scope of this book. It can readily be found in more advanced books; see, for instance, page 171 of Breiman (1968).

Theorem 4.3.10. *Let $\{X_n\}$ be a sequence of random variables with mgf $M_{X_n}(t)$ that exists for $-h < t < h$ for all n. Let X be a random variable with mgf $M(t)$, which exists for $|t| \leq h_1 \leq h$. If $\lim_{n \to \infty} M_{X_n}(t) = M(t)$ for $|t| \leq h_1$, then $X_n \xrightarrow{D} X$.*

In this and the subsequent sections are several illustrations of the use of Theorem 4.3.10. In some of these examples it is convenient to use a certain limit that is established in some courses in advanced calculus. We refer to a limit of the form

$$\lim_{n \to \infty} \left[1 + \frac{b}{n} + \frac{\psi(n)}{n} \right]^{cn},$$

where b and c do not depend upon n and where $\lim_{n \to \infty} \psi(n) = 0$. Then

$$\lim_{n \to \infty} \left[1 + \frac{b}{n} + \frac{\psi(n)}{n} \right]^{cn} = \lim_{n \to \infty} \left(1 + \frac{b}{n} \right)^{cn} = e^{bc}. \qquad (4.3.16)$$

For example,

$$\lim_{n \to \infty} \left(1 - \frac{t^2}{n} + \frac{t^3}{n^{3/2}} \right)^{-n/2} = \lim_{n \to \infty} \left(1 - \frac{t^2}{n} + \frac{t^2/\sqrt{n}}{n} \right)^{-n/2}.$$

Here $b = -t^2, c = -\frac{1}{2}$, and $\psi(n) = t^2/\sqrt{n}$. Accordingly, for every fixed value of t, the limit is $e^{t^2/2}$.

Example 4.3.6. Let Y_n have a distribution that is $b(n, p)$. Suppose that the mean $\mu = np$ is the same for every n; that is, $p = \mu/n$, where μ is a constant. We shall find the limiting distribution of the binomial distribution, when $p = \mu/n$, by finding the limit of $M(t; n)$. Now

$$M(t; n) = E(e^{tY_n}) = [(1 - p) + pe^t]^n = \left[1 + \frac{\mu(e^t - 1)}{n}\right]^n$$

for all real values of t. Hence we have

$$\lim_{n \to \infty} M(t; n) = e^{\mu(e^t - 1)}$$

for all real values of t. Since there exists a distribution, namely the Poisson distribution with mean μ, that has mgf $e^{\mu(e^t - 1)}$, then, in accordance with the theorem and under the conditions stated, it is seen that Y_n has a limiting Poisson distribution with mean μ.

Whenever a random variable has a limiting distribution, we may, if we wish, use the limiting distribution as an approximation to the exact distribution function. The result of this example enables us to use the Poisson distribution as an approximation to the binomial distribution when n is large and p is small. To illustrate the use of the approximation, let Y have a binomial distribution with $n = 50$ and $p = \frac{1}{25}$. Then

$$Pr(Y \leq 1) = (\tfrac{24}{25})^{50} + 50(\tfrac{1}{25})(\tfrac{24}{25})^{49} = 0.400,$$

approximately. Since $\mu = np = 2$, the Poisson approximation to this probability is

$$e^{-2} + 2e^{-2} = 0.406. \quad \blacksquare$$

Example 4.3.7. Let Z_n be $\chi^2(n)$. Then the mgf of Z_n is $(1 - 2t)^{-n/2}$, $t < \frac{1}{2}$. The mean and the variance of Z_n are, respectively, n and $2n$. The limiting distribution of the random variable $Y_n = (Z_n - n)/\sqrt{2n}$ will be investigated. Now the mgf of Y_n is

$$
\begin{aligned}
M(t; n) &= E\left\{\exp\left[t\left(\frac{Z_n - n}{\sqrt{2n}}\right)\right]\right\} \\
&= e^{-tn/\sqrt{2n}} E(e^{tZ_n/\sqrt{2n}}) \\
&= \exp\left[-\left(t\sqrt{\frac{2}{n}}\right)\left(\frac{n}{2}\right)\right]\left(1 - 2\frac{t}{\sqrt{2n}}\right)^{-n/2}, \quad t < \frac{\sqrt{2n}}{2}.
\end{aligned}
$$

This may be written in the form

$$M(t; n) = \left(e^{t\sqrt{2/n}} - t\sqrt{\frac{2}{n}}e^{t\sqrt{2/n}}\right)^{-n/2}, \quad t < \sqrt{\frac{n}{2}}.$$

In accordance with Taylor's formula, there exists a number $\xi(n)$, between 0 and $t\sqrt{2/n}$, such that

$$e^{t\sqrt{2/n}} = 1 + t\sqrt{\frac{2}{n}} + \frac{1}{2}\left(t\sqrt{\frac{2}{n}}\right)^2 + \frac{e^{\xi(n)}}{6}\left(t\sqrt{\frac{2}{n}}\right)^3.$$

If this sum is substituted for $e^{t\sqrt{2/n}}$ in the last expression for $M(t; n)$, it is seen that

$$M(t; n) = \left(1 - \frac{t^2}{n} + \frac{\psi(n)}{n}\right)^{-n/2},$$

where

$$\psi(n) = \frac{\sqrt{2}t^3 e^{\xi(n)}}{3\sqrt{n}} - \frac{\sqrt{2}t^3}{\sqrt{n}} - \frac{2t^4 e^{\xi(n)}}{3n}.$$

Since $\xi(n) \to 0$ as $n \to \infty$, then $\lim \psi(n) = 0$ for every fixed value of t. In accordance with the limit proposition cited earlier in this section, we have

$$\lim_{n \to \infty} M(t; n) = e^{t^2/2}$$

for all real values of t. That is, the random variable $Y_n = (Z_n - n)/\sqrt{2n}$ has a limiting standard normal distribution. ∎

Example 4.3.8 (Example 4.3.7 Continued). In the notation of the last example, we showed that

$$\sqrt{n}\left[\frac{1}{\sqrt{2n}}Z_n - \frac{1}{\sqrt{2}}\right] \xrightarrow{D} N(0, 1). \qquad (4.3.17)$$

For this situation, though, there are times when we are interested in the square-root of Z_n. Let $g(t) = \sqrt{t}$ and let $W_n = g(Z_n/(\sqrt{2}n)) = (Z_n/(\sqrt{2}n))^{1/2}$. Note that $g(1/\sqrt{2}) = 1/2^{1/4}$ and $g'(1/\sqrt{2}) = 2^{-3/4}$. Therefore, by the Δ-method, Theorem 4.3.9, and (4.3.17), we have

$$\sqrt{n}\left[W_n - 1/2^{1/4}\right] \xrightarrow{D} N(0, 2^{-3/2}). \quad ∎ \qquad (4.3.18)$$

EXERCISES

4.3.1. Let $\overline{X}_n$ denote the mean of a random sample of size n from a distribution that is $N(\mu, \sigma^2)$. Find the limiting distribution of $\overline{X}_n$.

4.3.2. Let Y_1 denote the minimum of a random sample of size n from a distribution that has pdf $f(x) = e^{-(x-\theta)}$, $\theta < x < \infty$, zero elsewhere. Let $Z_n = n(Y_1 - \theta)$. Investigate the limiting distribution of Z_n.

4.3.3. Let Y_n denote the maximum of a random sample from a distribution of the continuous type that has cdf $F(x)$ and pdf $f(x) = F'(x)$. Find the limiting distribution of $Z_n = n[1 - F(Y_n)]$.

4.3.4. Let Y_2 denote the second smallest item of a random sample of size n from a distribution of the continuous type that has cdf $F(x)$ and pdf $f(x) = F'(x)$. Find the limiting distribution of $W_n = nF(Y_2)$.

4.3.5. Let the pmf of Y_n be $p_n(y) = 1$, $y = n$, zero elsewhere. Show that Y_n does not have a limiting distribution. (In this case, the probability has "escaped" to infinity.)

4.3.6. Let $X_1, X_2, \ldots, X_n$ be a random sample of size n from a distribution that is $N(\mu, \sigma^2)$, where $\sigma^2 > 0$. Show that the sum $Z_n = \sum_1^n X_i$ does not have a limiting distribution.

4.3.7. Let X_n have a gamma distribution with parameter $\alpha = n$ and β, where β is not a function of n. Let $Y_n = X_n/n$. Find the limiting distribution of Y_n.

4.3.8. Let Z_n be $\chi^2(n)$ and let $W_n = Z_n/n^2$. Find the limiting distribution of W_n.

4.3.9. Let X be $\chi^2(50)$. Approximate $P(40 < X < 60)$.

4.3.10. Let $p = 0.95$ be the probability that a man, in a certain age group, lives at least 5 years.

(a) If we are to observe 60 such men and if we assume independence, find the probability that at least 56 of them live 5 or more years.

(b) Find an approximation to the result of part (a) by using the Poisson distribution.
Hint: Redefine p to be 0.05 and $1 - p = 0.95$.

4.3.11. Let the random variable Z_n have a Poisson distribution with parameter $\mu = n$. Show that the limiting distribution of the random variable $Y_n = (Z_n - n)/\sqrt{n}$ is normal with mean zero and variance 1.

4.3.12. Prove Theorem 4.3.3

4.3.13. Let X_n and Y_n have a bivariate normal distribution with parameters $\mu_1, \mu_2, \sigma_1^2, \sigma_2^2$ (free of n) but $\rho = 1 - 1/n$. Consider the conditional distribution of Y_n, given $X_n = x$. Investigate the limit of this conditional distribution as $n \to \infty$. What is the limiting distribution if $\rho = -1 + 1/n$? Reference to these facts is made in the Remark in Section 2.4.

4.3.14. Let $\overline{X}_n$ denote the mean of a random sample of size n from a Poisson distribution with parameter $\mu = 1$.

(a) Show that the mgf of $Y_n = \sqrt{n}(\overline{X}_n - \mu)/\sigma = \sqrt{n}(\overline{X}_n - 1)$ is given by $\exp[-t\sqrt{n} + n(e^{t/\sqrt{n}} - 1)]$.

(b) Investigate the limiting distribution of Y_n as $n \to \infty$.
Hint: Replace, by its MacLaurin's series, the expression $e^{t/\sqrt{n}}$, which is in the exponent of the mgf of Y_n.

4.3.15. Using Exercise 4.3.14, find the limiting distribution of $\sqrt{n}(\sqrt{\overline{X}_n} - 1)$.

4.3.16. Let $\overline{X}_n$ denote the mean of a random sample of size n from a distribution that has pdf $f(x) = e^{-x}$, $0 < x < \infty$, zero elsewhere.

(a) Show that the mgf $M(t; n)$ of $Y_n = \sqrt{n}(\overline{X}_n - 1)$ is

$$M(t; n) = [e^{t/\sqrt{n}} - (t/\sqrt{n})e^{t/\sqrt{n}}]^{-n}, \quad t < \sqrt{n}.$$

(b) Find the limiting distribution of Y_n as $n \to \infty$.

Exercises 4.3.14 and 4.3.16 are special instances of an important theorem that will be proved in the next section.

4.3.17. Using Exercise 4.3.16, find the limiting distribution of $\sqrt{n}(\sqrt{\overline{X}_n} - 1)$.

4.3.18. Let $Y_1 < Y_2 < \cdots < Y_n$ be the order statistics of a random sample from a distribution with pdf $f(x) = e^{-x}, 0 < x < \infty$, zero elsewhere. Determine the limiting distribution of $Z_n = (Y_n - \log n)$.

4.3.19. Let $Y_1 < Y_2 < \cdots < Y_n$ be the order statistics of a random sample from a distribution with pdf $f(x) = 5x^4, 0 < x < 1$, zero elsewhere. Find p so that $Z_n = n^p Y_1$ converges in distribution.

4.3.20. Use Stirling's formula, (4.3.2), to show that the first limit in Example 4.3.3 is 1.

4.4 Central Limit Theorem

It was seen (Section 3.4) that, if $X_1, X_2, \ldots, X_n$ is a random sample from a normal distribution with mean μ and variance σ^2, the random variable

$$\frac{\sum_1^n X_i - n\mu}{\sigma\sqrt{n}} = \frac{\sqrt{n}(\overline{X}_n - \mu)}{\sigma}$$

is, for every positive integer n, normally distributed with zero mean and unit variance. In probability theory there is a very elegant theorem called the *central limit theorem*. A special case of this theorem asserts the remarkable and important fact that if $X_1, X_2, \ldots, X_n$ denote the observations of a random sample of size n from any distribution having finite variance $\sigma^2 > 0$ (and hence finite mean μ), then the random variable $\sqrt{n}(\overline{X}_n - \mu)/\sigma$ converges in distribution to a random variable having a standard normal distribution. Thus, whenever the conditions of the theorem are satisfied, for large n the random variable $\sqrt{n}(\overline{X}_n - \mu)/\sigma$ has an approximate normal distribution with mean zero and variance 1. It will then be possible to use this approximate normal distribution to compute approximate probabilities concerning $\overline{X}$. In the statistical problem where μ is unknown, we will use this approximate distribution of $\overline{X}_n$ to establish approximate confidence intervals for μ; see Section 5.4.

We will often use the notation Y_n has a limiting standard normal distribution to mean that Y_n converges in distribution to a standard normal random variable; see Remark 4.3.1.

The more general form of the theorem is stated, but it is proved only in the modified case. However, this is exactly the proof of the theorem that would be given if we could use the characteristic function in place of the mgf.

Theorem 4.4.1. *Let $X_1, X_2, \ldots, X_n$ denote the observations of a random sample from a distribution that has mean μ and positive variance σ^2. Then the random variable $Y_n = (\sum_1^n X_i - n\mu)/\sqrt{n}\sigma = \sqrt{n}(\overline{X}_n - \mu)/\sigma$ converges in distribution to a random variable which has a normal distribution with mean zero and variance 1.*

Proof: For this proof, additionaly assume that the mgf $M(t) = E(e^{tX})$ exists for $-h < t < h$. If one replaces the mgf by the characteristic function $\varphi(t) = E(e^{itX})$, which always exists, then our proof is essentially the same as the proof in a more advanced course which uses characteristic functions.

The function

$$m(t) = E[e^{t(X-\mu)}] = e^{-\mu t} M(t)$$

also exists for $-h < t < h$. Since $m(t)$ is the mgf for $X - \mu$, it must follow that $m(0) = 1$, $m'(0) = E(X - \mu) = 0$, and $m''(0) = E[(X - \mu)^2] = \sigma^2$. By Taylor's formula there exists a number ξ between 0 and t such that

$$\begin{aligned} m(t) &= m(0) + m'(0)t + \frac{m''(\xi)t^2}{2} \\ &= 1 + \frac{m''(\xi)t^2}{2}. \end{aligned}$$

If $\sigma^2 t^2/2$ is added and subtracted, then

$$m(t) = 1 + \frac{\sigma^2 t^2}{2} + \frac{[m''(\xi) - \sigma^2]t^2}{2} \tag{4.4.1}$$

Next consider $M(t; n)$, where

$$\begin{aligned} M(t; n) &= E\left[\exp\left(t\frac{\sum X_i - n\mu}{\sigma\sqrt{n}}\right)\right] \\ &= E\left[\exp\left(t\frac{X_1 - \mu}{\sigma\sqrt{n}}\right)\exp\left(t\frac{X_2 - \mu}{\sigma\sqrt{n}}\right)\cdots\exp\left(t\frac{X_n - \mu}{\sigma\sqrt{n}}\right)\right] \\ &= E\left[\exp\left(t\frac{X_1 - \mu}{\sigma\sqrt{n}}\right)\right]\cdots E\left[\exp\left(t\frac{X_n - \mu}{\sigma\sqrt{n}}\right)\right] \\ &= \left\{E\left[\exp\left(t\frac{X - \mu}{\sigma\sqrt{n}}\right)\right]\right\}^n \\ &= \left[m\left(\frac{t}{\sigma\sqrt{n}}\right)\right]^n, \quad -h < \frac{t}{\sigma\sqrt{n}} < h. \end{aligned}$$

In Equation 4.4.1 replace t by $t/\sigma\sqrt{n}$ to obtain

$$m\left(\frac{t}{\sigma\sqrt{n}}\right) = 1 + \frac{t^2}{2n} + \frac{[m''(\xi) - \sigma^2]t^2}{2n\sigma^2},$$

where now ξ is between 0 and $t/\sigma\sqrt{n}$ with $-h\sigma\sqrt{n} < t < h\sigma\sqrt{n}$. Accordingly,

$$M(t; n) = \left\{1 + \frac{t^2}{2n} + \frac{[m''(\xi) - \sigma^2]t^2}{2n\sigma^2}\right\}^n.$$

Since $m''(t)$ is continuous at $t = 0$ and since $\xi \to 0$ as $n \to \infty$, we have

$$\lim_{n\to\infty}[m''(\xi) - \sigma^2] = 0.$$

The limit proposition (4.3.16) cited in Section 4.3 shows that

$$\lim_{n \to \infty} M(t; n) = e^{t^2/2},$$

for all real values of t. This proves that the random variable $Y_n = \sqrt{n}(\overline{X}_n - \mu)/\sigma$ has a limiting standard normal distribution. ∎

As cited in Remark 4.3.1, we say that Y_n has a limiting standard normal distribution. We interpret this theorem as saying that, when n is a large, fixed positive integer, the random variable $\overline{X}$ has an approximate normal distribution with mean μ and variance σ^2/n; and in applications we use the approximate normal pdf as though it were the exact pdf of $\overline{X}$.

Some illustrative examples, here and below, will help show the importance of this version of the central limit theorem.

Example 4.4.1. Let $\overline{X}$ denote the mean of a random sample of size 75 from the distribution that has the pdf

$$f(x) = \begin{cases} 1 & 0 < x < 1 \\ 0 & \text{elsewhere.} \end{cases}$$

For this situation, it can be shown that $g(\overline{x})$ has a graph when $0 < \overline{x} < 1$ that is composed of arcs of 75 different polynomials of degree 74. The computation of such a probability as $P(0.45 < \overline{X} < 0.55)$ would be extremely laborious. The conditions of the theorem are satisfied, since $M(t)$ exists for all real values of t. Moreover, $\mu = \frac{1}{2}$ and $\sigma^2 = \frac{1}{12}$, so that we have approximately

$$
\begin{aligned}
P(0.45 < \overline{X} < 0.55) &= P\left[\frac{\sqrt{n}(0.45 - \mu)}{\sigma} < \frac{\sqrt{n}(\overline{X} - \mu)}{\sigma} < \frac{\sqrt{n}(0.55 - \mu)}{\sigma} \right] \\
&= P[-1.5 < 30(\overline{X} - 0.5) < 1.5] \\
&= 0.866,
\end{aligned}
$$

from Table III in Appendix B. ∎

Example 4.4.2. Let $X_1, X_2, \ldots, X_n$ denote a random sample from a distribution that is $b(1, p)$. Here $\mu = p$, $\sigma^2 = p(1 - p)$, and $M(t)$ exists for all real values of t. If $Y_n = X_1 + \cdots + X_n$, it is known that Y_n is $b(n, p)$. Calculation of probabilities for Y_n, when we do not use the Poisson approximation, are simplified by making use of the fact that $(Y_n - np)/\sqrt{np(1 - p)} = \sqrt{n}(\overline{X}_n - p)/\sqrt{p(1 - p)} = \sqrt{n}(\overline{X}_n - \mu)/\sigma$ has a limiting distribution that is normal with mean zero and variance 1. Frequently, statisticians say that Y_n, or more simply Y, has an approximate normal distribution with mean np and variance $np(1 - p)$. Even with n as small as 10, with $p = \frac{1}{2}$ so that the binomial distribution is symmetric about $np = 5$, we note in Figure 4.4.1 how well the normal distribution, $N(5, \frac{5}{2})$, fits the binomial distribution, $b(10, \frac{1}{2})$, where the heights of the rectangles represent the probabilities of the respective integers $0, 1, 2, \ldots, 10$. Note that the area of the rectangle whose base is $(k - 0.5, k + 0.5)$ and the area under the normal pdf between $k - 0.5$ and $k + 0.5$ are *approximately* equal for each $k = 0, 1, 2, \ldots, 10$, even with $n = 10$. This example should help the reader understand Example 4.4.3. ∎

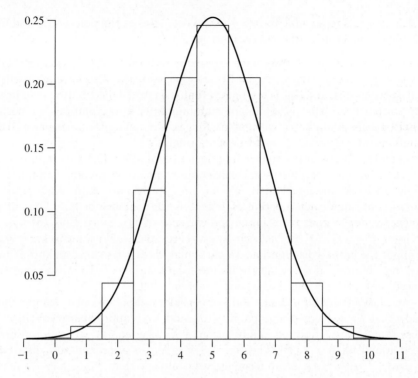

Figure 4.4.1: The $b\left(10, \frac{1}{2}\right)$ pmf overlaid by the $N\left(5, \frac{5}{2}\right)$ pdf

Example 4.4.3. With the background of Example 4.4.2, let $n = 100$ and $p = \frac{1}{2}$, and suppose that we wish to compute $P(Y = 48, 49, 50, 51, 52)$. Since Y is a random variable of the discrete type, $\{Y = 48, 49, 50, 51, 52\}$ and $\{47.5 < Y < 52.5\}$ are equivalent events. That is, $P(Y = 48, 49, 50, 51, 52) = P(47.5 < Y < 52.5)$. Since $np = 50$ and $np(1 - p) = 25$, the latter probability may be written

$$P(47.5 < Y < 52.5) \;=\; P\left(\frac{47.5 - 50}{5} < \frac{Y - 50}{5} < \frac{52.5 - 50}{5}\right)$$

$$=\; P\left(-0.5 < \frac{Y - 50}{5} < 0.5\right).$$

Since $(Y - 50)/5$ has an approximate normal distribution with mean zero and variance 1, Table III show this probability to be approximately 0.382.

The convention of selecting the event $47.5 < Y < 52.5$, instead of another event, say, $47.8 < Y < 52.3$, as the event equivalent to the event $Y = 48, 49, 50, 51, 52$ seems to have originated as: The probability, $P(Y = 48, 49, 50, 51, 52)$, can be interpreted as the sum of five rectangular areas where the rectangles have widths one but the heights are, respectively, $P(Y = 48), \ldots, P(Y = 52)$. If these rectangles are so located that the midpoints of their bases are, respectively, at the points $48, 49, \ldots, 52$ on a horizontal axis, then in approximating the sum of these areas by an area bounded by the horizontal axis, the graph of a normal pdf, and two ordinates, it seems reasonable to take the two ordinates at the points 47.5 and 52.5. This is called the **continuity correction**. ∎

We know that $\overline{X}$ and $\sum_1^n X_i$ have approximate normal distributions, provided that n is large enough. Later, we find that other statistics also have approximate normal distributions, and this is the reason that the normal distribution is so important to statisticians. That is, while not many underlying distributions are normal, the distributions of statistics calculated from random samples arising from these distributions are often very close to being normal.

Frequently, we are interested in functions of statistics that have approximate normal distributions. To illustrate, consider the sequence of random variable Y_n of Example 4.4.2. As discussed there, Y_n has an approximate $N[np, np(1-p)]$. So $np(1-p)$ is an important function of p as it is the variance of Y_n. Thus, if p is unknown, we might want to estimate the variance of Y_n. Since $E(Y_n/n) = p$, we might use $n(Y_n/n)(1 - Y_n/n)$ as such an estimator and would want to know something about the latter's distribution. In particular, does it also have an approximate normal distribution? If so, what are its mean and variance? To answer questions like these, we can apply the Δ-method, Theorem 4.3.9.

As an illustration of the Δ-method, we consider a function of the sample mean. We know that $\overline{X}_n$ converges in probability to μ and $\overline{X}_n$ is approximately $N(\mu, \sigma^2/n)$. Suppose that we are interested in a function of $\overline{X}_n$, say $u(\overline{X}_n)$, where u is differentiable at μ and $u'(\mu) \neq 0$. By Theorem 4.3.9, $u(\overline{X}_n)$ is approximately distributed as $N\{u(\mu), [u'(\mu)]^2\sigma^2/n\}$. More formally, we could say that

$$\frac{u(\overline{X}_n) - u(\mu)}{\sqrt{[u'(\mu)]^2\sigma^2/n}}$$

has a limiting standard normal distribution.

Example 4.4.4. Let Y_n (or Y for simplicity) be $b(n, p)$. Thus Y/n is approximately $N[p, p(1-p)/n]$. Statisticians often look for functions of statistics whose variances do not depend upon the parameter. Here the variance of Y/n depends upon p. Can we find a function, say $u(Y/n)$, whose variance is essentially free of p? Since Y/n converges in probability to p, we can approximate $u(Y/n)$ by the first two terms of its Taylor's expansion about p, namely by

$$u\left(\frac{Y}{n}\right) \doteq v\left(\frac{Y}{n}\right) = u(p) + \left(\frac{Y}{n} - p\right)u'(p).$$

Of course, $v(Y/n)$ is a linear function of Y/n and thus also has an approximate normal distribution; clearly, it has mean $u(p)$ and variance

$$[u'(p)]^2\frac{p(1-p)}{n}.$$

But it is the latter that we want to be essentially free of p; thus we set it equal to a constant, obtaining the differential equation

$$u'(p) = \frac{c}{\sqrt{p(1-p)}}.$$

A solution of this is

$$u(p) = (2c)\arcsin\sqrt{p}.$$

If we take $c = \frac{1}{2}$, we have, since $u(Y/n)$ is approximately equal to $v(Y/n)$, that

$$u\left(\frac{Y}{n}\right) = \arcsin\sqrt{\frac{Y}{n}}$$

has an approximate normal distribution with mean $\arcsin\sqrt{p}$ and variance $1/4n$, which is free of p. ∎

EXERCISES

4.4.1. Let $\overline{X}$ denote the mean of a random sample of size 100 from a distribution that is $\chi^2(50)$. Compute an approximate value of $P(49 < \overline{X} < 51)$.

4.4.2. Let $\overline{X}$ denote the mean of a random sample of size 128 from a gamma distribution with $\alpha = 2$ and $\beta = 4$. Approximate $P(7 < \overline{X} < 9)$.

4.4.3. Let Y be $b(72, \frac{1}{3})$. Approximate $P(22 \leq Y \leq 28)$.

4.4.4. Compute an approximate probability that the mean of a random sample of size 15 from a distribution having pdf $f(x) = 3x^2$, $0 < x < 1$, zero elsewhere, is between $\frac{3}{5}$ and $\frac{4}{5}$.

4.4.5. Let Y denote the sum of the observations of a random sample of size 12 from a distribution having pmf $p(x) = \frac{1}{6}$, $x = 1, 2, 3, 4, 5, 6$, zero elsewhere. Compute an approximate value of $P(36 \leq Y \leq 48)$.
Hint: Since the event of interest is $Y = 36, 37, \ldots, 48$, rewrite the probability as $P(35.5 < Y < 48.5)$.

4.4.6. Let Y be $b(400, \frac{1}{5})$. Compute an approximate value of $P(0.25 < Y/n)$.

4.4.7. If Y is $b(100, \frac{1}{2})$, approximate the value of $P(Y = 50)$.

4.4.8. Let Y be $b(n, 0.55)$. Find the smallest value of n which is such that (approximately) $P(Y/n > \frac{1}{2}) \geq 0.95$.

4.4.9. Let $f(x) = 1/x^2$, $1 < x < \infty$, zero elsewhere, be the pdf of a random variable X. Consider a random sample of size 72 from the distribution having this pdf. Compute approximately the probability that more than 50 of the observations of the random sample are less than 3.

4.4.10. Forty-eight measurements are recorded to several decimal places. Each of these 48 numbers is rounded off to the nearest integer. The sum of the original 48 numbers is approximated by the sum of these integers. If we assume that the errors made by rounding off are iid and have a uniform distribution over the interval $(-\frac{1}{2}, \frac{1}{2})$, compute approximately the probability that the sum of the integers is within two units of the true sum.

4.4.11. We know that $\overline{X}$ is approximately $N(\mu, \sigma^2/n)$ for large n. Find the approximate distribution of $u(\overline{X}) = \overline{X}^3$.

4.4.12. Let $X_1, X_2, \ldots, X_n$ be a random sample from a Poisson distribution with mean μ. Thus $Y = \sum_{i=1}^{n} X_i$ has a Poisson distribution with mean $n\mu$. Moreover, $\overline{X} = Y/n$ is approximately $N(\mu, \mu/n)$ for large n. Show that $u(Y/n) = \sqrt{Y/n}$ is a function of Y/n whose variance is essentially free of μ.

4.5   *Asymptotics for Multivariate Distributions

In this section, we briefly discuss asymptotic concepts for sequences of random vectors. The concepts introduced for univariate random variables generalize in a straightforward manner to the multivariate case. Our development is brief and the interested reader can consult more advanced texts for more depth; see Serfling (1980).

We need some notation. For a vector $\mathbf{v} \in R^p$, recall that Euclidean norm of $\mathbf{v}$ is defined to be

$$\|\mathbf{v}\| = \sqrt{\sum_{i=1}^{p} v_i^2}. \tag{4.5.1}$$

This norm satisfies the usual three properties given by

(a). For all $\mathbf{v} \in R^p$, $\|\mathbf{v}\| \geq 0$, and $\|\mathbf{v}\| = 0$ if and only if $\mathbf{v} = \mathbf{0}$.
(b). For all $\mathbf{v} \in R^p$ and $a \in R$, $\|a\mathbf{v}\| = |a|\|\mathbf{v}\|$. $\qquad$ (4.5.2)
(c). For all $\mathbf{v}, \mathbf{u} \in R^p$, $\|\mathbf{u} + \mathbf{v}\| \leq \|\mathbf{u}\| + \|\mathbf{v}\|$.

Denote the standard basis of R^p by the vectors $\mathbf{e}_1, \ldots, \mathbf{e}_p$, where all the components of $\mathbf{e}_i$ are 0 except for the *ith* component which is 1. Then we can always write any vector $\mathbf{v}' = (v_1, \ldots, v_p)$ as

$$\mathbf{v} = \sum_{i=1}^{p} v_i \mathbf{e}_i.$$

The following lemma will be useful:

Lemma 4.5.1. *Let* $\mathbf{v}' = (v_1, \ldots, v_p)$ *be any vector in* R^p. *Then,*

$$|v_j| \leq \|\mathbf{v}\| \leq \sum_{i=1}^{n} |v_i|, \quad \text{for all } j = 1, \ldots, p. \tag{4.5.3}$$

Proof: Note that for all j,

$$v_j^2 \leq \sum_{i=1}^{p} v_i^2 = \|\mathbf{v}\|^2;$$

hence, taking the square root of this equality leads to the first part of the desired inequality. The second part is

$$\|\mathbf{v}\| = \|\sum_{i=1}^{p} v_i \mathbf{e}_i\| \leq \sum_{i=1}^{p} |v_i| \|\mathbf{e}_i\| = \sum_{i=1}^{p} |v_i|. \quad \blacksquare$$

Let $\{\mathbf{X}_n\}$ denote a sequence of p dimensional vectors. Because the absolute value is the Euclidean norm in R^1, the definition of convergence in probability for random vectors is an immediate generalization:

Definition 4.5.1. *Let $\{\mathbf{X}_n\}$ be a sequence of p dimensional vectors and let $\mathbf{X}$ be a random vector, all defined on the same sample space. We say that $\{\mathbf{X}_n\}$ converges in probability to $\mathbf{X}$ if*

$$\lim_{n\to\infty} P[\|\mathbf{X}_n - \mathbf{X}\| \geq \epsilon] = 0, \qquad (4.5.4)$$

for all $\epsilon > 0$. As in the univariate case, we write $\mathbf{X}_n \overset{P}{\to} \mathbf{X}$.

As the next theorem shows, convergence in probability of vectors is equivalent to componentwise convergence in probability.

Theorem 4.5.1. *Let $\{\mathbf{X}_n\}$ be a sequence of p dimensional vectors and let $\mathbf{X}$ be a random vector, all defined on the same sample space. Then*

$$\mathbf{X}_n \overset{P}{\to} \mathbf{X} \text{ if and only if } X_{nj} \overset{P}{\to} X_j \text{ for all } j = 1, \ldots, p.$$

Proof: This follows immediately from Lemma 4.5.1. Suppose $\mathbf{X}_n \overset{P}{\to} \mathbf{X}$. For any j, from the first part of the inequality (4.5.3), we have, for $\epsilon > 0$,

$$\epsilon \leq |X_{nj} - X_j| \leq \|\mathbf{X}_n - \mathbf{X}\|.$$

Hence,

$$\overline{\lim}_{n\to\infty} P[|X_{nj} - X_j| \geq \epsilon] \leq \overline{\lim}_{n\to\infty} P[\|\mathbf{X}_n - \mathbf{X}\| \geq \epsilon] = 0,$$

which is the desired result.

Conversely, if $X_{nj} \overset{P}{\to} X_j$ for all $j = 1, \ldots, p$, then by the second part of the inequality (4.5.3),

$$\epsilon \leq \|\mathbf{X}_n - \mathbf{X}\| \leq \sum_{i=1}^{p} |X_{nj} - X_j|,$$

for any $\epsilon > 0$. Hence,

$$\overline{\lim}_{n\to\infty} P[\|\mathbf{X}_n - \mathbf{X}\| \geq \epsilon] \leq \overline{\lim}_{n\to\infty} P[\sum_{i=1}^{p} |X_{nj} - X_j| \geq \epsilon]$$

$$\leq \sum_{i=1}^{p} \overline{\lim}_{n\to\infty} P[|X_{nj} - X_j| \geq \epsilon/p] = 0. \quad \blacksquare$$

Based on this result many of the theorems involving convergence in probability can easily be extended to the multivariate setting. Some of these results are given in the exercises. This is true of statistical results, too. For example, in Section 4.2, we showed that if $X_1, \ldots, X_n$ is a random sample from the distribution of a random variable X with mean, μ, and variance, σ^2, then $\overline{X}_n$ and S_n^2 are consistent estimates of μ and σ^2. By the last theorem, we have that $(\overline{X}_n, S_n^2)$ is a consistent estimate of (μ, σ^2).

As another simple application, consider the multivariate analog of the sample mean and sample variance. Let $\{\mathbf{X}_n\}$ be a sequence of iid random vectors with common mean vector $\boldsymbol{\mu}$ and variance-covariance matrix $\boldsymbol{\Sigma}$. Denote the vector of means by

$$\overline{\mathbf{X}}_n = \frac{1}{n}\sum_{i=1}^{n}\mathbf{X}_i. \tag{4.5.5}$$

Of course, $\overline{\mathbf{X}}_n$ is just the vector of sample means, $(\overline{X}_1,\ldots,\overline{X}_p)'$. By the Weak Law of Large Numbers, Theorem 4.2.1, $\overline{X}_j \to \mu_j$, in probability, for each j. Hence, by Theorem 4.5.1, $\overline{\mathbf{X}}_n \to \boldsymbol{\mu}$, in probability.

How about the analog of the sample variances? Let $\mathbf{X}_i = (X_{i1},\ldots,X_{ip})'$. Define the sample variances and covariances by,

$$S_{n,j}^2 = \frac{1}{n-1}\sum_{i=1}^{n}(X_{ij}-\overline{X}_j)^2 \tag{4.5.6}$$

$$S_{n,jk} = \frac{1}{n-1}\sum_{i=1}^{n}(X_{ij}-\overline{X}_j)(X_{ik}-\overline{X}_k), \tag{4.5.7}$$

for $j,k=1,\ldots,p$. Assuming finite fourth moments, the Weak Law of Large Numbers shows that all these componentwise sample variances and sample covariances converge in probability to distribution variances and covariances, respectively. If we define the $p \times p$ matrix $\mathbf{S}$ to be the matrix with the jth diagonal entry $S_{n,j}^2$ and (j,k)th entry $S_{n,jk}$ then $\mathbf{S} \to \boldsymbol{\Sigma}$, in probability.

The definition of convergence in distribution remains the same. We state it here in terms of vector notation.

Definition 4.5.2. *Let $\{\mathbf{X}_n\}$ be a sequence of random vectors with $\mathbf{X}_n$ having distribution function $F_n(\mathbf{x})$ and $\mathbf{X}$ be a random vector with distribution function $F(\mathbf{x})$. Then $\{\mathbf{X}_n\}$ **converges in distribution** to $\mathbf{X}$ if*

$$\lim_{n\to\infty} F_n(\mathbf{x}) = F(\mathbf{x}), \tag{4.5.8}$$

for all points $\mathbf{x}$ at which $F(\mathbf{x})$ is continuous. We write $\mathbf{X}_n \overset{D}{\to} \mathbf{X}$.

In the multivariate case, there are analogs to many of the theorems in Section 4.3. We state two important theorems without proof.

Theorem 4.5.2. *Let $\{\mathbf{X}_n\}$ be a sequence of random vectors which converge in distribution to a random vector $\mathbf{X}$ and let $g(\mathbf{x})$ be a function which is continuous on the support of $\mathbf{X}$. Then $g(\mathbf{X}_n)$ converges in distribution to $g(\mathbf{X})$.*

We can apply this theorem to show that convergence in distribution implies marginal convergence. Simply take $g(\mathbf{x}) = x_j$ where $\mathbf{X} = (x_1,\ldots,x_p)'$. Since g is continuous, the desired result follows.

It is often difficult to determine convergence in distribution by using the definition. As in the univariate case, convergence in distribution is equivalent to convergence of moment generating functions, which we state in the following theorem.

Theorem 4.5.3. *Let $\{\mathbf{X}_n\}$ be a sequence of random vectors with $\mathbf{X}_n$ having distribution function $F_n(\mathbf{x})$ and moment generating function $M_n(\mathbf{t})$. Let $\mathbf{X}$ be a random vector with distribution function $F(\mathbf{x})$ and moment generating function $M(\mathbf{t})$. Then $\{\mathbf{X}_n\}$ converges in distribution to $\mathbf{X}$ if and only if for some $h > 0$,*

$$\lim_{n \to \infty} M_n(\mathbf{t}) = M(\mathbf{t}), \qquad (4.5.9)$$

for all $\mathbf{t}$ such that $\|\mathbf{t}\| < h$.

The proof of this theorem can be found in more advanced books; see, for instance, Tucker (1967). Also, the usual proof is for characteristic functions instead of moment generating functions. As we mentioned previously, characteristic functions always exist, so convergence in distribution is completely characterized by convergence of corresponding characteristic functions.

The moment generating function of $\mathbf{X}_n$ is $E[\exp\{\mathbf{t}'\mathbf{X}_n\}]$. Note that $\mathbf{t}'\mathbf{X}_n$ is a random variable. We can frequently use this and univariate theory to derive results in the multivariate case. A perfect example of this is the multivariate central limit theorem.

Theorem 4.5.4 (Multivariate Central Limit Theorem). *Let $\{\mathbf{X}_n\}$ be a sequence of iid random vectors with common mean vector $\boldsymbol{\mu}$ and variance-covariance matrix $\boldsymbol{\Sigma}$ which is positive definite. Assume the common moment generating function $M(\mathbf{t})$ exists in an open neighborhood of $\mathbf{0}$. Let*

$$\mathbf{Y}_n = \frac{1}{\sqrt{n}} \sum_{i=1}^{n} (\mathbf{X}_i - \boldsymbol{\mu}).$$

Then $\mathbf{Y}_n$ converges in distribution to a $N_p(\mathbf{0}, \boldsymbol{\Sigma})$ distribution.

Proof: Let $\mathbf{t} \in R^p$ be a vector in the stipulated neighbohood of $\mathbf{0}$. The moment generating function of $\mathbf{Y}_n$ is,

$$
\begin{aligned}
M_n(\mathbf{t}) &= E\left[\exp\left\{\mathbf{t}'\frac{1}{\sqrt{n}}\sum_{i=1}^{n}(\mathbf{X}_i - \boldsymbol{\mu})\right\}\right] \\
&= E\left[\exp\left\{\frac{1}{\sqrt{n}}\sum_{i=1}^{n}\mathbf{t}'(\mathbf{X}_i - \boldsymbol{\mu})\right\}\right] \\
&= E\left[\exp\left\{\frac{1}{\sqrt{n}}\sum_{i=1}^{n}W_i\right\}\right], \qquad (4.5.10)
\end{aligned}
$$

where $W_i = \mathbf{t}'(\mathbf{X}_i - \boldsymbol{\mu})$. Note that $W_1, \ldots, W_n$ are iid with mean 0 and variance $\mathrm{Var}(W_i) = \mathbf{t}'\boldsymbol{\Sigma}\mathbf{t}$. Hence, by the simple Central Limit Theorem

$$\frac{1}{\sqrt{n}}\sum_{i=1}^{n}W_i \xrightarrow{D} N(0, \mathbf{t}'\boldsymbol{\Sigma}\mathbf{t}). \qquad (4.5.11)$$

Expression (4.5.10), though, is the mgf of $(1/\sqrt{n})\sum_{i=1}^{n}W_i$ evaluated at 1. Therefore by (4.5.11), we must have

$$M_n(\mathbf{t}) = E\left[\exp\left\{1\frac{1}{\sqrt{n}}\sum_{i=1}^{n}W_i\right\}\right] \to e^{1^2\mathbf{t}'\boldsymbol{\Sigma}\mathbf{t}/2} = e^{\mathbf{t}'\boldsymbol{\Sigma}\mathbf{t}/2}.$$

Because the last quantity is the moment generating function of a $N_p(\mathbf{0}, \boldsymbol{\Sigma})$ distribution, we have the desired result. ∎

Suppose $\mathbf{X}_1, \mathbf{X}_2, \ldots, \mathbf{X}_n$ is a random sample from a distribution with mean vector $\boldsymbol{\mu}$ and variance-covariance matrix $\boldsymbol{\Sigma}$. Let $\overline{\mathbf{X}}_n$ be the vector of sample means. Then, from the Central Limit Theorem, we say that

$$\overline{\mathbf{X}}_n \text{ has an approximate } N_p\left(\boldsymbol{\mu}, \tfrac{1}{n}\boldsymbol{\Sigma}\right) \text{ distribution.} \tag{4.5.12}$$

A result that we use frequently concerns linear transformations. Its proof is obtained by using moment generating functions and is left as an exercise.

Theorem 4.5.5. *Let $\{\mathbf{X}_n\}$ be a sequence of p-dimensional random vectors. Suppose $\mathbf{X}_n \overset{D}{\to} N(\boldsymbol{\mu}, \boldsymbol{\Sigma})$. Let $\mathbf{A}$ be an $m \times p$ matrix of constants and let $\mathbf{b}$ be an m-dimensional vector of constants. Then $\mathbf{A}\mathbf{X}_n + \mathbf{b} \overset{D}{\to} N(\mathbf{A}\boldsymbol{\mu} + \mathbf{b}, \mathbf{A}\boldsymbol{\Sigma}\mathbf{A}')$.*

A result that will prove to be quite useful is the extension of the Δ-method; see Theorem 4.3.9. A proof can be found in Chapter 3 of Serfling (1980).

Theorem 4.5.6. *Let $\{\mathbf{X}_n\}$ be a sequence of p-dimensional random vectors. Suppose*

$$\sqrt{n}(\mathbf{X}_n - \boldsymbol{\mu}_0) \overset{D}{\to} N_p(\mathbf{0}, \boldsymbol{\Sigma}).$$

Let $\mathbf{g}$ be a transformation $\mathbf{g}(\mathbf{x}) = (g_1(\mathbf{x}), \ldots, g_k(\mathbf{x}))'$ such that $1 \leq k \leq p$ and the $k \times p$ matrix of partial derivatives,

$$\mathbf{B} = \left[\frac{\partial g_i}{\partial \mu_j}\right], \quad i = 1, \ldots k; \; j = 1, \ldots, p,$$

are continuous and do not vanish in a neighborhood of $\boldsymbol{\mu}_0$. Let $\mathbf{B}_0 = \mathbf{B}$ at $\boldsymbol{\mu}_0$. Then

$$\sqrt{n}(\mathbf{g}(\mathbf{X}_n) - \mathbf{g}(\boldsymbol{\mu}_0)) \overset{D}{\to} N_k(\mathbf{0}, \mathbf{B}_0\boldsymbol{\Sigma}\mathbf{B}_0'). \tag{4.5.13}$$

EXERCISES

4.5.1. Let $\{\mathbf{X}_n\}$ be a sequence of p dimensional random vectors. Show that

$$\mathbf{X}_n \overset{D}{\to} N_p(\boldsymbol{\mu}, \boldsymbol{\Sigma}) \text{ if and only if } \mathbf{a}'\mathbf{X}_n \overset{D}{\to} N_1(\mathbf{a}'\boldsymbol{\mu}, \mathbf{a}'\boldsymbol{\Sigma}\mathbf{a}),$$

for all vectors $\mathbf{a} \in R^p$.

4.5.2. Let $X_1, \ldots, X_n$ be a random sample from a uniform(a, b) distribution. Let $Y_1 = \min X_i$ and let $Y_2 = \max X_i$. Show that $(Y_1, Y_2)'$ converges in probability to the vector $(a, b)'$.

4.5.3. Let $\mathbf{X}_n$ and $\mathbf{Y}_n$ be p dimensional random vectors. Show that if

$$\mathbf{X}_n - \mathbf{Y}_n \overset{P}{\to} \mathbf{0} \text{ and } \mathbf{X}_n \overset{D}{\to} \mathbf{X},$$

where $\mathbf{X}$ is a p dimensional random vector, then $\mathbf{Y}_n \overset{D}{\to} \mathbf{X}$.

4.5.4. Let $\mathbf{X}_n$ and $\mathbf{Y}_n$ be p dimensional random vectors such that $\mathbf{X}_n$ and $\mathbf{Y}_n$ are independent for each n and their mgfs exist. Show that if

$$\mathbf{X}_n \overset{D}{\to} \mathbf{X} \text{ and } \mathbf{Y}_n \overset{D}{\to} \mathbf{Y},$$

where $\mathbf{X}$ and $\mathbf{Y}$ are p dimensional random vectors, then $(\mathbf{X}_n, \mathbf{Y}_n) \overset{D}{\to} (\mathbf{X}, \mathbf{Y})$.

4.5.5. Suppose $\mathbf{X}_n$ has a $N_p(\boldsymbol{\mu}_n, \boldsymbol{\Sigma}_n)$ distribution. Show that

$$\mathbf{X}_n \overset{D}{\to} N_p(\boldsymbol{\mu}, \boldsymbol{\Sigma}) \text{ iff } \boldsymbol{\mu}_n \to \boldsymbol{\mu} \text{ and } \boldsymbol{\Sigma}_n \to \boldsymbol{\Sigma}.$$

Chapter 5

Some Elementary Statistical Inferences

5.1 Sampling and Statistics

In last chapter, we introduced the concepts of samples and statistics. We continue with this development in this chapter while introducing the main tools of inference: confidence intervals and tests of hypotheses.

In a typical statistical problem, we have a random variable X of interest but its pdf $f(x)$ or pmf $p(x)$ is not known. Our ignorance about $f(x)$ or $p(x)$ can roughly be classified in one of two ways:

(1) $f(x)$ or $p(x)$ is completely unknown.

(2) The form of $f(x)$ or $p(x)$ is known down to a parameter θ, where θ may be a vector.

For now, we will consider the second classification. Some examples are:

(a) X has an exponential distribution, $\text{Exp}(\theta)$, (3.3.2), where θ is unknown.

(b) X has a binomial distribution $b(n, p)$, (3.1.2), where n is known but p is unknown.

(c) X has a gamma distribution $\Gamma(\alpha, \beta)$, (3.3.1), where both α and β are unknown.

We will often denote this problem, by saying that the random variable X has a density or mass function of the form $f(x; \theta)$ or $p(x; \theta)$, where $\theta \in \Omega$ for a specified set Ω. For example in (a) above, $\Omega = \{\theta \,|\, \theta > 0\}$. We call θ a parameter of the distribution. Because θ is unknown, we want to estimate it. In this chapter, we discuss certain desirable properties of estimators, while in later chapters, we will present general methodologies for estimation. Estimation is based on a sample and, in this section, we formalize this sampling process.

To illustrate these ideas, think of an urn which contains m balls, labeled from $1, \ldots, m$ and which are identical except for the number. The experiment is to choose a ball at random and record the number. Let X denote the number. Then the distribution of X is given by

$$P(X = x) = \frac{1}{m}, \quad \text{for } x = 1, \ldots, m. \tag{5.1.1}$$

Consider the situation where there are many balls in the urn and we do not know how many, i.e., m is unknown. So in this case $\theta = m$ and Ω is the set of positive integers. To obtain information on m we take a sample of n balls, which we will denote as $\mathbf{X} = (X_1, \ldots, X_n)'$, where X_i is the number on the ith ball.

Now the sample can be drawn in several ways. Two ways of interest are:

1. **Sampling with Replacement:** Here a ball is selected at random, its number is recorded, the ball is replaced in the urn, the balls are then remixed, and the next ball is chosen. In this case it is easy to see that the $X_1, \ldots, X_n$ are mutually independent random variables and each has the same distribution as X. This we will define, below, as a random sample.

2. **Sampling without Replacement:** Here n balls are selected at random. If the balls are selected one-at-a-time, they are not replaced after each draw. As Exercise 5.1.1 shows, the $X_1, \ldots, X_n$ are not independent, but each X_i has the same distribution. This type of sampling is often called simple random sampling.

If m is much greater than n, the sampling schemes are practically the same.

For a second illustration, suppose we are interested in the lifetime X of some electronic part. Suppose we are willing to assume that X has an exponential distribution with parameter θ. Hopefully, a sample of n trials of the experiment will provide useful information concerning θ. The ith trial consists of putting a part on test and recording its lifetime, say, X_i. The outcomes of the n trials , $X_1, \ldots, X_n$, constitute our sample. In this case, a sample is considered random if the parts were made independent of one another under identical conditions and the trials are run independently of one another under identical conditions. We next formally define a random sample.

Definition 5.1.1 (Random Sample). *The random variables $X_1, \ldots, X_n$ constitute a* **random sample** *on a random variable X if they are independent and each has the same distribution as X. We will abbreviate this by saying that $X_1, \ldots, X_n$ are* **iid***; i.e., independent and identically distributed.*

Let $F(x)$ and $f(x)$ be, respectively, the cdf and pdf of X. It follows from Section 2.6 that the joint cdf of a random sample $X_1, \ldots, X_n$ is given by

$$F_{X_1, \ldots, X_n}(x_1, \ldots, x_n) = \prod_{i=1}^{n} F(x_i),$$

while the joint pdf is given by

$$f_{X_1,\ldots,X_n}(x_1,\ldots,x_n) = \prod_{i=1}^{n} f(x_i).$$

A similar statement can be made about a discrete random variable X with pmf $p(x)$. We often use vector notation to denote a sample, $\mathbf{X} = (X_1,\ldots,X_n)'$. We next define a statistic.

Definition 5.1.2 (Statistic). *Suppose the n random variables $X_1, X_2, X_3, \ldots, X_n$ constitute a sample from the distribution of a random variable X. Then any function $T = T(X_1,\ldots,X_n)$ of the sample is called a* **statistic**.

In a more advanced class, we would require the function to be Borel measurable.

Since a statistic is a function of the sample, it is also a random variable. Statistics are frequently summaries of the data. A statistic, $T = T(X_1,\ldots,X_n)$, may convey information about the unknown parameter θ. In such cases, we may call the statistic a **point estimator** of θ. Recall from Chapter 4, that T is an *unbiased estimator* of θ if $E(T) = \theta$, and that T is a *consistent* estimator of θ if $T \to \theta$ in probability. Once the random sample is drawn, $X_1,\ldots,X_n$ is observed to be $x_1,\ldots,x_n$, the computed value $T(x_1,\ldots,x_n)$ is called a *point estimate* of θ. What characterizes "good" point estimators? In the sequel, we will discuss properties of estimators. The following simple example illustrates some of the problems.

Example 5.1.1 (Simple Urn Problem). Consider the urn problem discussed above. Recall that the urn contains m balls labeled from $1,\ldots,m$ and that the balls are identical except for the number. Suppose m is unknown. To estimate m we draw a random sample of the balls, $X_1,\ldots,X_n$ with replacement. The distribution of each X_i is $P(X = x) = 1/m$, for $x = 1,\ldots,m$. An intuitive point estimator of m is the statistic $T = \max\{X_1,\ldots,X_n\}$. This would appear to be a "good" estimator of m. But how far is T from m? One way of answering this is to consider the distribution of T. The support of T is $\{1,\ldots,m\}$. To determine the cdf of T, note that because T is the maximum of the X observations, the event $T \leq t$, for $1 \leq t \leq m$, is characterized as,

$$\{T \leq t\} = \{X_1 \leq t, \ldots X_n \leq t\} = \cap_{i=1}^{n}\{X_i \leq t\}.$$

Hence, using the fact that $X_1,\ldots,X_n$ are iid, the cdf of T is,

$$P[T \leq t] = \prod_{i=1}^{n} P[X_i \leq t] = [P(X_1 \leq t)]^n = \left(\frac{[t]}{m}\right)^n, \qquad (5.1.2)$$

where $[t]$ denotes the greatest integer less than or equal to t. Hence for $0 \leq t \leq m$,

$$P[T_n \leq t] = \left(\frac{[t]}{m}\right)^n \to \begin{cases} 0 & \text{if } t < m \\ 1 & \text{if } t = m. \end{cases}$$

Thus, $T_n \xrightarrow{D} m$ and so by the Theorem 4.3.2, $T_n \xrightarrow{P} m$. Thus T_n is a consistent estimate of m.

Note that in this problem, $E(X) = (m+1)/2$. Hence $E(2\overline{X} - 1) = m$, where $\overline{X} = n^{-1} \sum_{i=1}^{n} X_i$ denotes the **sample mean**. Perhaps, $2\overline{X} - 1$, is also a good estimator of m as it is unbiased. While it is satisfactory, we will show later that T is a better estimator in this case. ∎

In the sequel, we will be studying properties of estimators. Based on these properties, we will attempt to classify statistics into their usefulness for estimating certain parameters. This will enable us to give precise answers to questions such as those raised at the end of Example 5.1.1.

Example 5.1.2. Suppose X is a random variable with unknown mean θ. Let $X_1, \ldots, X_n$ be a random sample from the distribution of X and let $\overline{X} = n^{-1} \sum_{i=1}^{n} X_i$ be the sample mean. Then because $E(\overline{X}) = \theta$, the statistic $\overline{X}$ is an unbiased point estimator of θ. But how close is $\overline{X}$ to θ? We will answer this for general situations later in this chapter, but for now we look at a specific case. Suppose that X has a normal $N(\theta, \sigma^2)$ distribution and that σ^2 is known. By Section 3.4, the distribution of $\overline{X}$ is $N(\theta, \sigma^2/n)$. We can then use the knowledge of the distribution of $\overline{X}$ to answer the above question. Because $(\overline{X} - \theta)/(\sigma/\sqrt{n})$ has a standard normal, $N(0, 1)$, distribution, we have by simple algebra

$$
\begin{aligned}
0.954 &= P\left(-2 < \frac{\overline{X} - \theta}{\sigma/\sqrt{n}} < 2\right) \\
&= P\left(\overline{X} - 2\frac{\sigma}{\sqrt{n}} < \theta < \overline{X} + 2\frac{\sigma}{\sqrt{n}}\right). \tag{5.1.3}
\end{aligned}
$$

Expression (5.1.3) says that before the sample is drawn the probability that the random interval $\left(\overline{X} - 2\frac{\sigma}{\sqrt{n}}, \overline{X} + 2\frac{\sigma}{\sqrt{n}}\right)$ traps θ is 0.954. After the sample is drawn the realized interval

$$
\left(\overline{x} - 2\frac{\sigma}{\sqrt{n}}, \overline{x} + 2\frac{\sigma}{\sqrt{n}}\right) \tag{5.1.4}
$$

has either trapped θ or it has not. But because of the high probability of success, namely 0.954, before the sample is drawn, we call the interval (5.1.4) a 95.4% **confidence interval** for θ. We can say, with some confidence, that $\overline{x}$ is within $2\frac{\sigma}{\sqrt{n}}$ from θ. The number $0.954 = 95.4\%$ is called a **confidence coefficient**. Instead of using 2, we could use, say, 1.282, 1.645, or 2.576 to obtain 80%, 90% or 99% confidence intervals for θ. Note that the lengths of these confidence intervals increase as the confidence increases; i.e., the increase in confidence implies a loss in precision. On the other hand, for any confidence coefficient, an increase in sample size leads to shorter confidence intervals. We will return to confidence intervals for the mean in Section 5.4. ∎

EXERCISES

5.1.1. In the second sampling procedure (sampling without replacement) for the urn problem at the beginning of this section, show that

(a) The random variables X_i and X_j, $i \neq j$, are dependent. *Hint*: Find their joint pmf.

(b) The distribution of X_i is given by (5.1.1).

5.1.2. Let $X_1, \ldots, X_n$ be a random sample from the Bernoulli distribution, $b(1, p)$, where p is unknown. Let $Y = \sum_{i=1}^{n} X_i$.

(a) Find the distribution of Y.

(b) Show that Y/n is an unbiased estimator of p.

(c) What is the variance of Y/n?

5.1.3. Let $\overline{X}$ be the mean of a random sample from the exponential distribution, $\text{Exp}(\theta)$.

(a) Show that $\overline{X}$ is an unbiased point estimator of θ.

(b) Using the mgf technique determine the distribution of $\overline{X}$.

(c) Use (b) to show that $Y = 2n\overline{X}/\theta$ has a χ^2 distribution with $2n$ degrees of freedom.

(d) Based on Part (c), find a 95% confidence interval for θ if $n = 10$. *Hint:* Find c and d such that $P\left(c < \frac{2n\overline{X}}{\theta} < d\right) = 0.95$ and solve the inequalities for θ.

5.1.4. Let $X_1, \ldots, X_n$ be a random sample from the $\Gamma(2, \theta)$ distribution, where θ is unknown. Let $Y = \sum_{i=1}^{n} X_i$.

(a) Find the distribution of Y and determine c so that cY is an unbiased estimator of θ.

(b) If $n = 5$, show that

$$P\left(9.59 < \frac{2Y}{\theta} < 34.2\right) = 0.95.$$

(c) Using Part (b), show that if y is the value of Y once the sample is drawn, then the interval

$$\left(\frac{2y}{34.2}, \frac{2y}{9.59}\right)$$

is a 95% confidence interval for θ.

(d) Suppose the sample results in the values,

$$44.8079 \quad 1.5215 \quad 12.1929 \quad 12.5734 \quad 43.2305$$

Based on these data, obtain the point estimate of θ as determined in Part (a) and the computed 95% confidence interval in Part (c). What does the confidence interval mean?

5.1.5. Suppose the number of customers X that enter a store between the hours 9:00 AM and 10:00 AM follows a Poisson distribution with parameter θ. Suppose a random sample of the number of customers for 10 days results in the values,

$$9 \quad 7 \quad 9 \quad 15 \quad 10 \quad 13 \quad 11 \quad 7 \quad 2 \quad 12$$

Based on these data, obtain an unbiased point estimate of θ. Explain the meaning of this estimate in terms of the number of customers.

5.1.6. Let $X_1, X_2, \ldots, X_n$ be a random sample from a continuous type distribution.

(a) Find $P(X_1 \leq X_2), P(X_1 \leq X_2, X_1 \leq X_3), \ldots, P(X_1 \leq X_i, i = 2, 3, \ldots, n)$.

(b) Suppose the sampling continues until X_1 is no longer the smallest observation, (i.e., $X_j < X_1 \leq X_i, i = 2, 3, \ldots, j - 1$). Let Y equal the number of trials until X_1 is no longer the smallest observation, (i.e., $Y = j - 1$). Show that the distribution of Y is

$$P(Y = y) = \frac{1}{y(y+1)}, \quad y = 1, 2, 3, \ldots.$$

(c) Compute the mean and variance of Y if they exist.

5.2 Order Statistics

In this section the notion of an order statistic will be defined and we shall investigate some of the simple properties of such a statistic. These statistics have in recent times come to play an important role in statistical inference partly because some of their properties do not depend upon the distribution from which the random sample is obtained.

Let $X_1, X_2, \ldots, X_n$ denote a random sample from a distribution of the *continuous type* having a pdf $f(x)$ that has support $\mathcal{S} = (a, b)$, where $-\infty \leq a < b \leq \infty$. Let Y_1 be the smallest of these X_i, Y_2 the next X_i in order of magnitude, ..., and Y_n the largest of X_i. That is, $Y_1 < Y_2 < \cdots < Y_n$ represent $X_1, X_2, \ldots, X_n$ when the latter are arranged in ascending order of magnitude. We call Y_i, $i = 1, 2, \ldots, n$, the ith order statistic of the random sample $X_1, X_2, \ldots, X_n$. Then the joint pdf of $Y_1, Y_2, \ldots, Y_n$ is given in the following theorem.

Theorem 5.2.1. *Using the above notation, let $Y_1 < Y_2 < \cdots < Y_n$ denote the n order statistics based on the random sample $X_1, X_2, \ldots, X_n$ from a continuous distribution with pdf $f(x)$ and support (a, b). Then the joint pdf of $Y_1, Y_2, \ldots, Y_n$ is given by,*

$$g(y_1, y_2, \ldots, y_n) = \begin{cases} n! f(y_1) f(y_2) \cdots f(y_n) & a < y_1 < y_2 < \cdots < y_n < b \\ 0 & elsewhere. \end{cases} \quad (5.2.1)$$

Proof. Note that the support of $X_1, X_2, \ldots, X_n$ can be partitioned into $n!$ mutually disjoint sets which map onto the support of $Y_1, Y_2, \ldots, Y_n$, namely $\{(y_1, y_2, \ldots, y_n) :$

$a < y_1 < y_2 < \cdots < y_n < b\}$. One of these $n!$ sets is $a < x_1 < x_2 < \cdots < x_n < b$, and the others can be found by permuting the n x values in all possible ways. The transformation associated with the one listed is $x_1 = y_1, x_2 = y_2, \ldots, x_n = y_n$ which has a Jacobian equal to one. However, the Jacobian of each of the other transformations is either ± 1. Thus,

$$
\begin{aligned}
g(y_1, y_2, \ldots, y_n) &= \sum_{i=1}^{n!} |J_i| f(y_1) f(y_2) \cdots f(y_n) \\
&= \begin{cases} n! f(y_1) f(y_2) \cdots f(y_n) & a < y_1 < y_2 < \cdots < y_n < b \\ 0 & \text{elsewhere.} \end{cases}
\end{aligned}
$$

as was to be proved. ∎

Example 5.2.1. Let X denote a random variable of the continuous type with a pdf $f(x)$ that is positive and continuous, with support $\mathcal{S} = (a, b)$, $-\infty \le a < b \le \infty$. The distribution function $F(x)$ of X may be written

$$
F(x) = \int_a^x f(w)\,dw, \quad a < x < b.
$$

If $x \le a$, $F(x) = 0$; and if $b \le x$, $F(x) = 1$. Thus there is a unique median m of the distribution with $F(m) = \frac{1}{2}$. Let X_1, X_2, X_3 denote a random sample from this distribution and let $Y_1 < Y_2 < Y_3$ denote the order statistics of the sample. We shall compute the probability that $Y_2 \le m$. The joint pdf of the three order statistics is

$$
g(y_1, y_2, y_3) = \begin{cases} 6 f(y_1) f(y_2) f(y_3) & a < y_1 < y_2 < y_3 < b \\ 0 & \text{elsewhere.} \end{cases}
$$

The pdf of Y_2 is then

$$
\begin{aligned}
h(y_2) &= 6 f(y_2) \int_{y_2}^b \int_a^{y_2} f(y_1) f(y_3)\,dy_1 dy_3 \\
&= \begin{cases} 6 f(y_2) F(y_2)[1 - F(y_2)] & a < y_2 < b \\ 0 & \text{elsewhere.} \end{cases}
\end{aligned}
$$

Accordingly,

$$
\begin{aligned}
P(Y_2 \le m) &= 6 \int_a^m \{F(y_2) f(y_2) - [F(y_2)]^2 f(y_2)\}\,dy_2 \\
&= 6 \left\{ \frac{[F(y_2)]^2}{2} - \frac{[F(y_2)]^3}{3} \right\}_a^m = \frac{1}{2}. \ \blacksquare
\end{aligned}
$$

Once it is observed that

$$
\int_a^x [F(w)]^{\alpha-1} f(w)\,dw = \frac{[F(x)]^\alpha}{\alpha}, \quad \alpha > 0
$$

and that

$$\int_y^b [1 - F(w)]^{\beta-1} f(w)\,dw = \frac{[1 - F(y)]^\beta}{\beta}, \quad \beta > 0,$$

it is easy to express the marginal pdf of any order statistic, say Y_k, in terms of $F(x)$ and $f(x)$. This is done by evaluating the integral

$$g_k(y_k) = \int_a^{y_k} \cdots \int_a^{y_2} \int_{y_k}^b \cdots \int_{y_{n-1}}^b n! f(y_1) f(y_2) \cdots f(y_n)\,dy_n \cdots dy_{k+1}\,dy_1 \cdots dy_{k-1}.$$

The result is

$$g_k(y_k) = \begin{cases} \frac{n!}{(k-1)!(n-k)!} [F(y_k)]^{k-1} [1 - F(y_k)]^{n-k} f(y_k) & a < y_k < b \\ 0 & \text{elsewhere.} \end{cases} \quad (5.2.2)$$

Example 5.2.2. Let $Y_1 < Y_2 < Y_3 < Y_4$ denote the order statistics of a random sample of size 4 from a distribution having pdf

$$f(x) = \begin{cases} 2x & 0 < x < 1 \\ 0 & \text{elsewhere.} \end{cases}$$

We express the pdf of Y_3 in terms of $f(x)$ and $F(x)$ and then compute $P(\frac{1}{2} < Y_3)$. Here $F(x) = x^2$, provided that $0 < x < 1$, so that

$$g_3(y_3) = \begin{cases} \frac{4!}{2!\,1!} (y_3^2)^2 (1 - y_3^2)(2y_3) & 0 < y_3 < 1 \\ 0 & \text{elsewhere.} \end{cases}$$

Thus

$$
\begin{aligned}
P(\tfrac{1}{2} < Y_3) &= \int_{1/2}^\infty g_3(y_3)\,dy_3 \\
&= \int_{1/2}^1 24(y_3^5 - y_3^7)\,dy_3 = \frac{243}{256}.
\end{aligned}
$$

Finally, the joint pdf of any two order statistics, say $Y_i < Y_j$, is easily expressed in terms of $F(x)$ and $f(x)$. We have

$$g_{ij}(y_i, y_j) = \int_a^{y_i} \cdots \int_a^{y_2} \int_{y_i}^{y_j} \cdots \int_{y_{j-2}}^{y_j} \int_{y_j}^b \cdots \int_{y_{n-1}}^b n! f(y_1) \cdots$$

$$f(y_n)\,dy_n \cdots dy_{j+1} dy_{j-1} \cdots dy_{i+1} dy_1 \cdots dy_{i-1}.$$

Since, for $\gamma > 0$,

$$
\begin{aligned}
\int_x^y [F(y) - F(w)]^{\gamma-1} f(w)\,dw &= \left. -\frac{[F(y) - F(w)]^\gamma}{\gamma} \right|_x^y \\
&= \frac{[F(y) - F(x)]^\gamma}{\gamma},
\end{aligned}
$$

it is found that

$$
g_{ij}(y_i, y_j) = \begin{cases} \frac{n!}{(i-1)!(j-i-1)!(n-j)!}[F(y_i)]^{i-1}[F(y_j) - F(y_i)]^{j-i-1} \\ \quad \times [1 - F(y_j)]^{n-j} f(y_i) f(y_j) & a < y_i < y_j < b \\ 0 & \text{elsewhere.} \quad \blacksquare \end{cases}
$$
$$(5.2.3)$$

Remark 5.2.1 (Heuristic Derivation). There is an easy method of remembering the pdf of a vector of order statistics such as the one given in Formula (5.2.3). The probability $P(y_i < Y_i < y_i + \Delta_i, \; y_j < Y_j < y_j + \Delta_j)$, where Δ_i and Δ_j are small, can be approximated by the following multinomial probability. In n independent trials, $i - 1$ outcomes must be less than y_i [an event that has probability $p_1 = F(y_i)$ on each trial]; $j - i - 1$ outcomes must be between $y_i + \Delta_i$ and y_j [an event with approximate probability $p_2 = F(y_j) - F(y_i)$ on each trial]; $n - j$ outcomes must be greater than $y_j + \Delta_j$ [an event with approximate probability $p_3 = 1 - F(y_j)$ on each trial]; one outcome must be between y_i and $y_i + \Delta_i$ [an event with approximate probability $p_4 = f(y_i)\Delta_i$ on each trial]; and finally one outcome must be between y_j and $y_j + \Delta_j$ [an event with approximate probability $p_5 = f(y_j)\Delta_j$ on each trial]. This multinomial probability is

$$
\frac{n!}{(i-1)!(j-i-1)!(n-j)! \; 1! \; 1!} p_1^{i-1} p_2^{j-i-1} p_3^{n-j} p_4 p_5,
$$

which is $g_{i,j}(y_i, y_j)\Delta_i\Delta_j$. $\blacksquare$

Certain functions of the order statistics $Y_1, Y_2, \ldots, Y_n$ are important statistics themselves. A few of these are: (a) $Y_n - Y_1$, which is called the **range** of the random sample; (b) $(Y_1 + Y_n)/2$, which is called the **midrange** of the random sample; and (c) if n is odd, $Y_{(n+1)/2}$, which is called the **median** of the random sample.

Example 5.2.3. Let Y_1, Y_2, Y_3 be the order statistics of a random sample of size 3 from a distribution having pdf

$$
f(x) = \begin{cases} 1 & 0 < x < 1 \\ 0 & \text{elsewhere.} \end{cases}
$$

We seek the pdf of the sample range $Z_1 = Y_3 - Y_1$. Since $F(x) = x$, $0 < x < 1$, the joint pdf of Y_1 and Y_3 is

$$
g_{13}(y_1, y_3) = \begin{cases} 6(y_3 - y_1) & 0 < y_1 < y_3 < 1 \\ 0 & \text{elsewhere.} \end{cases}
$$

In addition to $Z_1 = Y_3 - Y_1$, let $Z_2 = Y_3$. The functions $z_1 = y_3 - y_1$, $z_2 = y_3$, have respective inverses $y_1 = z_2 - z_1$, $y_3 = z_2$, so that the corresponding Jacobian of the one-to-one transformation is

$$
J = \begin{vmatrix} \frac{\partial y_1}{\partial z_1} & \frac{\partial y_1}{\partial z_2} \\ \frac{\partial y_3}{\partial z_1} & \frac{\partial y_3}{\partial z_2} \end{vmatrix} = \begin{vmatrix} -1 & 1 \\ 0 & 1 \end{vmatrix} = -1.
$$

Thus the joint pdf of Z_1 and Z_2 is

$$h(z_1, z_2) = \begin{cases} |-1|6z_1 = 6z_1 & 0 < z_1 < z_2 < 1 \\ 0 & \text{elsewhere.} \end{cases}$$

Accordingly, the pdf of the range $Z_1 = Y_3 - Y_1$ of the random sample of size 3 is

$$h_1(z_1) = \begin{cases} \int_{z_1}^1 6z_1 \, dz_2 = 6z_1(1 - z_1) & 0 < z_1 < 1 \\ 0 & \text{elsewhere.} \end{cases} \quad \blacksquare$$

5.2.1   Quantiles

Let X be a random variable with a continuous cdf $F(x)$. For $0 < p < 1$, define the **pth quantile** of X to be $\xi_p = F^{-1}(p)$. For example, $\xi_{0.5}$, the median of X, is the 0.5 quantile. Let $X_1, X_2, \ldots, X_n$ be a random sample from the distribution of X and let $Y_1 < Y_2 < \cdots < Y_n$ be the corresponding order statistics. Let $k = [p(n+1)]$. We next define an estimator of ξ_p after making the following observation. The area under the pdf $f(x)$ to the left of Y_k is $F(Y_k)$. The expected value of this area is

$$E(F(Y_k)) = \int_a^b F(y_k)g_k(y_k) \, dy_k,$$

where $g_k(y_k)$ is the pdf of Y_k given in expression (5.2.2). If, in this integral, we make a change of variables through the transformation $z = F(y_k)$, we have

$$E(F(Y_k)) = \int_0^1 \frac{n!}{(k-1)!(n-k)!} z^k (1 - z)^{n-k} \, dz.$$

Comparing this to the integral of a beta pdf, we see that it is equal to

$$E(F(Y_k)) = \frac{n!k!(n-k)!}{(k-1)!(n-k)!(n+1)!} = \frac{k}{n+1}.$$

On the average, there is $k/(n + 1)$ of the total area to the left of Y_k. Because $p \doteq k/(n + 1)$, it seems reasonable to take Y_k as an estimator of the quantile ξ_p. Hence, we call Y_k the *pth* **sample quantile** It is also called the $100p$th **percentile of the sample**.

Remark 5.2.2. Some statisticians define sample quantiles slightly differently from what we have. For one modification with $1/(n + 1) < p < n/(n + 1)$, if $(n + 1)/p$ is not equal to an integer then the *pth* quantile of the sample may be defined as follows. Write $(n + 1)p = k + r$, where $k = [(n + 1)p]$ and r is a proper fraction and use the weighted average, then the *pth* quantile of the sample is the weighted average

$$(1 - r)Y_k + rY_{k+1},$$

as an estimator of the *pth* quantile. As n becomes large, however, all these modified definitions are essentially the same. $\blacksquare$

Sample quantiles are useful descriptive statistics. For instance, if Y_k is the *pth* quantile of the sample, then we know that approximately $p100\%$ of the data are less than or equal to Y_k and approximately $(1-p)100\%$ of the data are greater than or equal to Y_k. Next we discuss two statistical applications of quantiles.

A **five number** summary of the data consist of the following five sample quantiles: the minimum (Y_1), the first quartile $(Y_{[.25(n+1)]})$, the median $(Y_{[.50(n+1)]})$, the third quartile $(Y_{[.75(n+1)]})$, and the maximum (Y_n). Note that the median given is for odd sample sizes. In the case that n is even, we use the traditional $(Y_{(n/2)} + (Y_{(n/2+1)})/2$ as our estimator of the median $\xi_{.5}$. For this section, we will use the notation Q_1, Q_2, and Q_3 to denote, respectively, the first quartile, median, and third quartile of the sample.

The five number summary divides the data into their quartiles, offering a simple and easily interpretable description of the data. Five number summaries were made popular by the work of the late Professor John Tukey (see Tukey , 1977 and Mosteller and Tukey, 1977). He used slightly different quantities in place of the first and third quartiles which he called *hinges*. We prefer to use the sample quartiles.

Example 5.2.4. The following data are the ordered realizations of a random sample of size 15 on a random variable X.

$$\begin{array}{cccccccc} 56 & 70 & 89 & 94 & 96 & 101 & 102 & 102 \\ 102 & 105 & 106 & 108 & 110 & 113 & 116 \end{array}$$

For these data since $n + 1 = 16$, the realizations of the five number summary are $y_1 = 56$, $Q_1 = y_4 = 94$, $Q_2 = y_8 = 102$, $Q_3 = y_{12} = 108$, and $y_{15} = 116$. Hence, based on the five number summary, the data range from 56 to 118; the middle 50% of the data range from 94 to 108; and the middle of the data occurred at 102. ∎

The five number summary is the basis for a useful and quick plot of the data. This is called a **boxplot** of the data. The box encloses the middle 50% of the data and a line segment is usually used to indicate the median. The extreme order statistics, however, are very sensitive to outlying points. So care must be used in placing these on the plot. We will make use of the *box and whisker* plots defined by John Tukey. In order to define this plot, we need to define a potential outlier. Let $h = 1.5(Q_3 - Q_1)$ and define the *lower fence, (LF)*, and the *upper fence, (UF)*, by

$$LF = Q_1 - h \text{ and } UF = Q_3 + h. \tag{5.2.4}$$

Points that lie outside the fences; i.e., outside of the interval (LF, UF) are called *potential outliers* and they are denoted by the symbol "0" on the boxplot. The whiskers then protrude from the sides of the box to what are called the *adjacent points*, which are the points within the fences but closest to the fences. Exercise 5.2.2 shows that the probability of an observation from a normal distribution being a potential outlier is 0.006977.

Example 5.2.5 (Example 5.2.4 Continued). Consider the data given in Example 5.2.4. For these data, $h = 1.5(108 - 94) = 21$, $LF = 73$, and $UF = 129$. Hence, the observations 56 and 70 are potential outliers. There are no outliers on the high side of the data. The lower adjacent point is 89. Hence, the boxplot of data is given by Panel A of Figure 5.2.1,

Note that the point 56 is over $2h$ from Q_1. Some statisticians call such a point an "outlier" and label it with a symbol other than "O" but we will not make this distinction. ∎

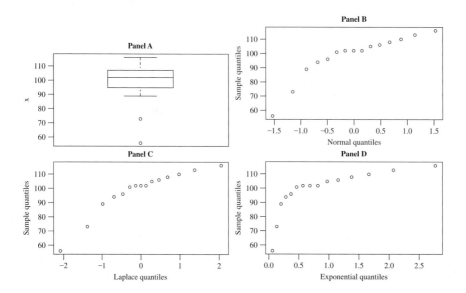

Figure 5.2.1: Boxplot and Quantile Plots for the data of Example 5.2.4

In practice, we often assume that the data follow a certain distribution. For example, we may assume that $X_1, \ldots, X_n$ are a random sample from a normal distribution with unknown mean and variance. Thus the form of the distribution of X is known but the specific parameters are not. Such an assumption needs to be checked and there are many statistical tests which do so; see D'Agostino and Stephens (1986) for a thorough discussion of such tests. As our second statistical application of quantiles, we discuss one such diagnostic plot in this regard.

We will consider the location and scale family. Suppose X is a random variable with cdf $F((x - a)/b)$ where $F(x)$ is known but a and $b > 0$ may not be. Let $Z = (X - a)/b$, then Z has cdf $F(z)$. Let $0 < p < 1$ and let $\xi_{X,p}$ be the pth quantile of X. Let $\xi_{Z,p}$ be the pth quantile of $Z = (X - a)/b$. Because $F(z)$ is known, $\xi_{Z,p}$ is known. But

$$p = P[X \leq \xi_{X,p}] = P\left[Z \leq \frac{\xi_{X,p} - a}{b}\right];$$

from which we have the linear relationship,

$$\xi_{X,p} = b\xi_{Z,p} + a. \tag{5.2.5}$$

Thus if X has a cdf of the form of $F((x - a)/b)$, then the quantiles of X are linearly related to the quantiles of Z. Of course, in practice, we do not know the quantiles of X, but we can estimate them. Let $X_1, \ldots, X_n$ be a random sample from the distribution of X and let $Y_1 < \cdots < Y_n$ be the order statistics. For $k = 1, \ldots, n$, let $p_k = k/(n+1)$. Then Y_k is an estimator of ξ_{X,p_k}. Denote the corresponding quantiles of the cdf $F(z)$ by $\xi_{Z,p_k} = F^{-1}(p_k)$. The plot of Y_k versus ξ_{Z,p_k} is called a **q−q plot**, as it plots one set of quantiles from the sample against another set from the theoretical cdf $F(z)$. Based on the above discussion, linearity of such a plot indicates that the cdf of X is of the form $F((x-a)/b)$.

Example 5.2.6 (Example 5.2.5 Continued). Panels B, C and D of Figure 5.2.1 contain $q-q$ plots of the data of Example 5.2.4 for three different distributions. The quantiles of a standard normal random variable are used for the plot in Panel B. Hence as described above, this is the plot of Y_k versus $\Phi^{-1}(k/(n+1))$, for $k = 1, 2, \ldots, n$. For Panel C, the population quantiles of the standard **Laplace** distribution are used; that is, the density of Z is $f(z) = (1/2)e^{-|z|}$, $-\infty < z < \infty$. For Panel D, the quantiles were generated from an exponential distribution with density $f(z) = e^{-z}$, $0 < z < \infty$, zero elsewhere. The generation of these quantiles are discussed in Exercise 5.2.1.

The plot farthest from linearity is that of Panel D. Note that this plot gives an indication of a more correct distribution. For the points to lie on a line, the lower quantiles of Z must be spread out as are the higher quantiles; i.e, symmetric distributions may be more appropriate. The plots in Panels B and C are more linear than that of Panel D, but they still contain some curvature. Of the two, Panel C appears to be more linear. Actually the data were generated from a Laplace distribution. ■

The $q-q$ plot using normal quantiles is often called a *normal q−q plot*.

5.2.2 Confidence Intervals of Quantiles

The following data are the realizations of a random sample of size 15 on a random variable X. Let X be a continuous random variable with cdf $F(x)$. For $0 < p < 1$, define the $100pth$ distribution percentile to be ξ_p where $F(\xi_p) = p$. For a sample of size n on X, let $Y_1 < Y_2 < \cdots < Y_n$ be the order statistics. Let $k = [(n+1)p]$. Then the $100pth$ sample percentile Y_k is a point estimate of ξ_p.

We now derive a **distribution free** confidence interval for ξ_p, meaning it is a confidence interval for ξ_p which is free of any assumptions about $F(x)$ other than it is of the continuous type. Let $i < [(n+1)p] < j$, and consider the order statistics $Y_i < Y_j$ and the event $Y_i < \xi_p < Y_j$. For the ith order statistic Y_i to be less than ξ_p it must be true that at least i of the X values are less than ξ_p. Moreover, for the jth order statistic to be greater than ξ_p, fewer than j of the X values are less than ξ_p. To put this in the context of a binomial distribution, the probability of success is $P(X < \xi_p) = F(\xi_p) = p$. Further the event $Y_i < \xi_p < Y_j$ is equivalent to obtaining between i (inclusive) and j (exclusive) successes in n independent trials. Therefore,

$$P(Y_i < \xi_p < Y_j) = \sum_{w=i}^{j-1} \binom{n}{w} p^w (1-p)^{n-w}, \qquad (5.2.6)$$

the probability of having at least i, but less than j, successes. When particular values of n, i, and j are specified, this probability can be computed. By this procedure, suppose that it has been found that $\gamma = P(Y_i < \xi_p < Y_j)$. Then the probability is γ that the random interval (Y_i, Y_j) includes the quantile of order p. If the experimental values of Y_i and Y_j are, respectively, y_i and y_j, the interval (y_i, y_j) serves as a 100γ percent confidence interval for ξ_p, the quantile of order p. We use this in the next example to find a confidence interval for the median.

Example 5.2.7 (Confidence Interval for the Median). Let X be a continuous random variable with cdf $F(x)$. Let $\xi_{1/2}$ denote the median of $F(x)$; i.e., $\xi_{1/2}$ solves $F(\xi_{1/2}) = 1/2$. Suppose $X_1, X_2, \ldots, X_n$ is a random sample from the distribution of X with corresponding order statistics $Y_1 < Y_2 < \cdots < Y_n$. As before, let Q_2 denote the sample median which is a point estimator of $\xi_{1/2}$. Select α, so that $0 < \alpha < 1$. Take $c_{\alpha/2}$ to be the $\alpha/2th$ quantile of a binomial $b(n, 1/2)$ distribution; that is, $P[S \leq c_{\alpha/2}] = \alpha/2$, where S is distributed $b(n, 1/2)$. Then note also that $P[S \geq n - c_{\alpha/2}] = \alpha/2$. Thus, it follows from expression(5.2.6) that

$$P[Y_{c_{\alpha/2}+1} < \xi_{1/2} < Y_{n-c_{\alpha/2}}] = 1 - \alpha. \qquad (5.2.7)$$

Hence when ths sample is drawn, if $y_{c_{\alpha/2}+1}$ and $y_{n-c_{\alpha/2}}$ are the realized values of the order statistics $Y_{c_{\alpha/2}+1}$ and $Y_{n-c_{\alpha/2}}$ then the interval $(y_{c_{\alpha/2}+1}, y_{n-c_{\alpha/2}})$ is a $(1-\alpha)100\%$ confidence interval for $\xi_{1/2}$.

To illustrate this confidence interval, consider the data of Example 5.2.4. Suppose we want a 88% confidence interval for ξ. Then $\alpha/2 = 0.060$. Then $c_{\alpha/2} = 4$ because $P[S \leq 4] = 0.059$, for S distributed as a $b(15, .5)$. Therefore, an 88% confidence interval for ξ is $(y_5, y_{11}) = (96, 106)$. ∎

Note that because of the discreteness of the binomial distribution only certain confidence levels are possible for this confidence interval for the median. If we further assume that $f(x)$ is symmetric about ξ, Chapter 10 presents other distribution free confidence intervals where this discreteness is much less of a problem.

EXERCISES

5.2.1. Obtain closed form expressions for the distribution quantiles based on the exponential and Laplace distributions as discussed in Example 5.2.6.

5.2.2. Obtain the probability that an observation is a potential outlier for the following distributions.

(a) The underlying distribution is normal.

(b) The underlying distribution is the *logistic*; that is, the pdf is given by

$$f(x) = \frac{e^{-x}}{(1 + e^{-x})^2}, \quad -\infty < x < \infty. \qquad (5.2.8)$$

(c) The underlying distribution is Laplace. the pdf is given by

$$f(x) = \frac{1}{2}e^{-|x|}, \quad -\infty < x < \infty. \qquad (5.2.9)$$

5.2.3. Consider the sample of data:

$$
\begin{array}{cccccccccc}
13 & 5 & 202 & 15 & 99 & 4 & 67 & 83 & 36 & 11 & 301 \\
23 & 213 & 40 & 66 & 106 & 78 & 69 & 166 & 84 & 64 &
\end{array}
$$

(a) Obtain the five number summary of these data.

(b) Determine if there are any outliers.

(c) Boxplot the data. Comment on the plot.

(d) Obtain a 92% confidence interval for the median $\xi_{1/2}$.

5.2.4. Consider the data in Exercise 5.2.3. Obtain the normal $q-q$ plot for these data. Does the plot suggest that the underlying distribution is normal? Use the plot to determine, if any, what quantiles associated with a different theoretical distribution would lead to a more linear plot. Then obtain the plot.

5.2.5. Let $Y_1 < Y_2 < Y_3 < Y_4$ be the order statistics of a random sample of size 4 from the distribution having pdf $f(x) = e^{-x}$, $0 < x < \infty$, zero elsewhere. Find $P(3 \leq Y_4)$.

5.2.6. Let X_1, X_2, X_3 be a random sample from a distribution of the continuous type having pdf $f(x) = 2x$, $0 < x < 1$, zero elsewhere.

(a) Compute the probability that the smallest of X_1, X_2, X_3 exceeds the median of the distribution.

(b) If $Y_1 < Y_2 < Y_3$ are the order statistics, find the correlation between Y_2 and Y_3.

5.2.7. Let $f(x) = \frac{1}{6}$, $x = 1, 2, 3, 4, 5, 6$, zero elsewhere, be the pmf of a distribution of the discrete type. Show that the pmf of the smallest observation of a random sample of size 5 from this distribution is

$$
g_1(y_1) = \left(\frac{7 - y_1}{6}\right)^5 - \left(\frac{6 - y_1}{6}\right)^5, \quad y_1 = 1, 2, \ldots, 6,
$$

zero elsewhere. Note that in this exercise the random sample is from a distribution of the discrete type. All formulas in the text were derived under the assumption that the random sample is from a distribution of the continuous type and are not applicable. Why?

5.2.8. Let $Y_1 < Y_2 < Y_3 < Y_4 < Y_5$ denote the order statistics of a random sample of size 5 from a distribution having pdf $f(x) = e^{-x}$, $0 < x < \infty$, zero elsewhere. Show that $Z_1 = Y_2$ and $Z_2 = Y_4 - Y_2$ are independent.
Hint: First find the joint pdf of Y_2 and Y_4.

5.2.9. Let $Y_1 < Y_2 < \cdots < Y_n$ be the order statistics of a random sample of size n from a distribution with pdf $f(x) = 1$, $0 < x < 1$, zero elsewhere. Show that the kth order statistic Y_k has a beta pdf with parameters $\alpha = k$ and $\beta = n - k + 1$.

5.2.10. Let $Y_1 < Y_2 < \cdots < Y_n$ be the order statistics from a Weibull distribution, Exercise 3.3.26. Find the distribution function and pdf of Y_1.

5.2.11. Find the probability that the range of a random sample of size 4 from the uniform distribution having the pdf $f(x) = 1$, $0 < x < 1$, zero elsewhere, is less than $\frac{1}{2}$.

5.2.12. Let $Y_1 < Y_2 < Y_3$ be the order statistics of a random sample of size 3 from a distribution having the pdf $f(x) = 2x$, $0 < x < 1$, zero elsewhere. Show that $Z_1 = Y_1/Y_2$, $Z_2 = Y_2/Y_3$, and $Z_3 = Y_3$ are mutually independent.

5.2.13. Suppose a random sample of size 2 is obtained from a distribution which has pdf $f(x) = 2(1 - x)$, $0 < x < 1$, zero elsewhere. Compute the probability that one sample observation is at least twice as large as the other.

5.2.14. Let $Y_1 < Y_2 < Y_3$ denote the order statistics of a random sample of size 3 from a distribution with pdf $f(x) = 1$, $0 < x < 1$, zero elsewhere. Let $Z = (Y_1 + Y_3)/2$ be the midrange of the sample. Find the pdf of Z.

5.2.15. Let $Y_1 < Y_2$ denote the order statistics of a random sample of size 2 from $N(0, \sigma^2)$.

   (a) Show that $E(Y_1) = -\sigma/\sqrt{\pi}$.
      Hint: Evaluate $E(Y_1)$ by using the joint pdf of Y_1 and Y_2 and first integrating on y_1.

   (b) Find the covariance of Y_1 and Y_2.

5.2.16. Let $Y_1 < Y_2$ be the order statistics of a random sample of size 2 from a distribution of the continuous type which has pdf $f(x)$ such that $f(x) > 0$, provided that $x \geq 0$, and $f(x) = 0$ elsewhere. Show that the independence of $Z_1 = Y_1$ and $Z_2 = Y_2 - Y_1$ characterizes the gamma pdf $f(x)$, which has parameters $\alpha = 1$ and $\beta > 0$.
Hint: Use the change-of-variable technique to find the joint pdf of Z_1 and Z_2 from that of Y_1 and Y_2. Accept the fact that the functional equation $h(0)h(x + y) \equiv h(x)h(y)$ has the solution $h(x) = c_1 e^{c_2 x}$, where c_1 and c_2 are constants.

5.2.17. Let $Y_1 < Y_2 < Y_3 < Y_4$ be the order statistics of a random sample of size $n = 4$ from a distribution with pdf $f(x) = 2x$, $0 < x < 1$, zero elsewhere.

   (a) Find the joint pdf of Y_3 and Y_4.

   (b) Find the conditional pdf of Y_3, given $Y_4 = y_4$.

   (c) Evaluate $E(Y_3|y_4)$.

5.2.18. Two numbers are selected at random from the interval $(0, 1)$. If these values are uniformly and independently distributed, by cutting the interval at these numbers compute the probability that the three resulting line segments form a triangle.

5.2.19. Let X and Y denote independent random variables with respective probability density functions $f(x) = 2x$, $0 < x < 1$, zero elsewhere, and $g(y) = 3y^2$, $0 < y < 1$, zero elsewhere. Let $U = \min(X, Y)$ and $V = \max(X, Y)$. Find the joint pdf of U and V.

Hint: Here the two inverse transformations are given by $x = u$, $y = v$ and $x = v$, $y = u$.

5.2.20. Let the joint pdf of X and Y be $f(x, y) = \frac{12}{7}x(x+y)$, $0 < x < 1$, $0 < y < 1$, zero elsewhere. Let $U = \min(X, Y)$ and $V = \max(X, Y)$. Find the joint pdf of U and V.

5.2.21. Let $X_1, X_2, \ldots, X_n$ be random sample from a distribution of either type. A measure of spread is *Gini's mean difference*

$$G = \sum_{j=2}^{n} \sum_{i=1}^{j-1} |X_i - X_j| / \binom{n}{2}. \qquad (5.2.10)$$

(a) If $n = 10$, find $a_1, a_2, \ldots, a_{10}$ so that $G = \sum_{i=1}^{10} a_i Y_i$, where $Y_1, Y_2, \ldots, Y_{10}$ are the order statistics of the sample.

(b) Show that $E(G) = 2\sigma/\sqrt{\pi}$ if the sample arises from the normal distribution $N(\mu, \sigma^2)$.

5.2.22. Let $Y_1 < Y_2 < \cdots < Y_n$ be the order statistics of a random sample of size n from the exponential distribution with pdf $f(x) = e^{-x}$, $0 < x < \infty$, zero elsewhere.

(a) Show that $Z_1 = nY_1$, $Z_2 = (n-1)(Y_2 - Y_1)$, $Z_3 = (n-2)(Y_3 - Y_2), \ldots, Z_n = Y_n - Y_{n-1}$ are independent and that each Z_i has the exponential distribution.

(b) Demonstrate that all linear functions of $Y_1, Y_2, \ldots, Y_n$, such as $\sum_{1}^{n} a_i Y_i$, can be expressed as linear functions of independent random variables.

5.2.23. In the Program Evaluation and Review Technique (PERT) we are interested in the total time to complete a project that is comprised of a large number of subprojects. For illustration, let X_1, X_2, X_3 be three independent random times for three subprojects. If these subprojects are in series (the first one must be completed before the second starts, etc.), then we are interested in the sum $Y = X_1 + X_2 + X_3$. If these are in parallel (can be worked on simultaneously), then we are interested in $Z = \max(X_1, X_2, X_3)$. In the case each of these random variables has the uniform distribution with pdf $f(x) = 1$, $0 < x < 1$, zero elsewhere, find (a) the pdf of Y and (b) the pdf of Z.

5.2.24. Let Y_n denote the nth order statistic of a random sample of size n from a distribution of the continuous type. Find the smallest value of n for which the inequality $P(\xi_{0.9} < Y_n) \geq 0.75$ is true.

5.2.25. Let $Y_1 < Y_2 < Y_3 < Y_4 < Y_5$ denote the order statistics of a random sample of size 5 from a distribution of the continuous type. Compute:

(a) $P(Y_1 < \xi_{0.5} < Y_5)$.

(b) $P(Y_1 < \xi_{0.25} < Y_3)$.

(c) $P(Y_4 < \xi_{0.80} < Y_5)$.

5.2.26. Compute $P(Y_3 < \xi_{0.5} < Y_7)$ if $Y_1 < \cdots < Y_9$ are the order statistics of a random sample of size 9 from a distribution of the continuous type.

5.2.27. Find the smallest value of n for which $P(Y_1 < \xi_{0.5} < Y_n) \geq 0.99$, where $Y_1 < \cdots < Y_n$ are the order statistics of a random sample of size n from a distribution of the continuous type.

5.2.28. Let $Y_1 < Y_2$ denote the order statistics of a random sample of size 2 from a distribution which is $N(\mu, \sigma^2)$, where σ^2 is known.

(a) Show that $P(Y_1 < \mu < Y_2) = \frac{1}{2}$ and compute the expected value of the random length $Y_2 - Y_1$.

(b) If $\overline{X}$ is the mean of this sample, find the constant c which solves the equation $P(\overline{X} - c\sigma < \mu < \overline{X} + c\sigma) = \frac{1}{2}$, and compare the length of this random interval with the expected value of that of part (a).

5.2.29. Let $y_1 < y_2 < y_3$ be the observed values of the order statistics of a random sample of size $n = 3$ from a continuous type distribution. Without knowing these values, a statistician is given these values in a random order, and she wants to select the largest; but once she refuses an observation, she cannot go back. Clearly if she selects the first one, her probability of getting the largest is $1/3$. Instead, she decides to use the following algorithm: She looks at the first but refuses it and then takes the second if it is larger than the first, or else she takes the third. Show that this algorithm has probability of $1/2$ in selecting the largest.

5.3   *Tolerance Limits for Distributions

We propose now to investigate a problem that has something of the same flavor as that treated in Section 5.2. Specifically, can we compute the probability that a certain random interval includes (or *covers*) a preassigned percentage of the probability of the distribution under consideration? And, by appropriate selection of the random interval, can we be led to an additional distribution-free method of statistical inference?

Let X be a random variable with distribution function $F(x)$ of the continuous type. Let $Z = F(X)$. Then, as shown in Exercise 5.3.1 Z has a uniform$(0, 1)$ distribution. That is, $Z = F(X)$ has the pdf

$$h(z) = \begin{cases} 1 & 0 < z < 1 \\ 0 & \text{elsewhere,} \end{cases}$$

then, if $0 < p < 1$, we have

$$P[F(X) \leq p] = \int_0^p dz = p.$$

Now $F(x) = P(X \leq x)$. Since $P(X = x) = 0$, then $F(x)$ is the fractional part of the probability for the distribution of X that is between $-\infty$ and x. If $F(x) \leq p$, then no more than $100p$ percent of the probability for the distribution of X is between $-\infty$ and x. But recall $P[F(X) \leq p] = p$. That is, the probability that the random variable $Z = F(X)$ is less than or equal to p is precisely the probability that the random interval $(-\infty, X)$ contains no more than $100p$ percent of the probability for the distribution. For example, if $p = 0.70$, the probability that the random interval $(-\infty, X)$ contains no more than 70 percent of the probability for the distribution is 0.70; and the probability that the random interval $(-\infty, X)$ contains more than 70 percent of the probability for the distribution is $1 - 0.70 = 0.30$.

We now consider certain functions of the order statistics. Let $X_1, X_2, \ldots, X_n$ denote a random sample of size n from a distribution that has a positive and continuous pdf $f(x)$ if and only if $a < x < b$ and let $F(x)$ denote the associated distribution function. Consider the random variables $F(X_1), F(X_2), \ldots, F(X_n)$. These random variables are independent and each, in accordance with Exercise 5.3.1, has a uniform distribution on the interval $(0, 1)$. Thus $F(X_1), F(X_2), \ldots, F(X_n)$ is a random sample of size n from a uniform distribution on the interval $(0, 1)$. Consider the order statistics of this random sample $F(X_1), F(X_2), \ldots, F(X_n)$. Let Z_1 be the smallest of these $F(X_i)$, Z_2 the next $F(X_i)$ in order of magnitude, $\ldots$, and Z_n the largest of $F(X_i)$. If $Y_1, Y_2, \ldots, Y_n$ are the order statistics of the initial random sample $X_1, X_2, \ldots, X_n$, the fact that $F(x)$ is a nondecreasing (here, strictly increasing) function of x implies that $Z_1 = F(Y_1), Z_2 = F(Y_2), \ldots, Z_n = F(Y_n)$. Hence, it follows from (5.2.1) that the joint pdf of $Z_1, Z_2, \ldots, Z_n$ is given by

$$h(z_1, z_2, \ldots, z_n) = \begin{cases} n! & 0 < z_1 < z_2 < \cdots < z_n < 1 \\ 0 & \text{elsewhere.} \end{cases} \tag{5.3.1}$$

This proves a special case of the following theorem.

Theorem 5.3.1. *Let $Y_1, Y_2, \ldots, Y_n$ denote the order statistics of a random sample of size n from a distribution of the continuous type that has pdf $f(x)$ and cdf $F(x)$. The joint pdf of the random variables $Z_i = F(Y_i)$, $i = 1, 2, \ldots, n$, is given by expression (5.3.1).*

Because the distribution function of $Z = F(X)$ is given by z, $0 < z < 1$, it follows from 5.2.2 that the marginal pdf of $Z_k = F(Y_k)$ is the following beta pdf:

$$h_k(z_k) = \begin{cases} \frac{n!}{(k-1)!(n-k)!} z_k^{k-1} (1 - z_k)^{n-k} & 0 < z_k < 1 \\ 0 & \text{elsewhere.} \end{cases} \tag{5.3.2}$$

Moreover, from (5.2.3), the joint pdf of $Z_i = F(Y_i)$ and $Z_j = F(Y_j)$ is, with $i < j$, given by

$$h(z_i, z_j) = \begin{cases} \frac{n! z_i^{i-1}(z_j - z_i)^{j-i-1}(1 - z_j)^{n-j}}{(i-1)!(j-i-1)!(n-j)!} & 0 < z_i < z_j < 1 \\ 0 & \text{elsewhere.} \end{cases} \tag{5.3.3}$$

Consider the difference $Z_j - Z_i = F(Y_j) - F(Y_i)$, $i < j$. Now $F(y_j) = P(X \le y_j)$ and $F(y_i) = P(X \le y_i)$. Since $P(X = y_i) = P(X = y_j) = 0$, then the difference $F(y_j) - F(y_i)$ is that fractional part of the probability for the distribution of X that is between y_i and y_j. Let p denote a positive proper fraction. If $F(y_j) - F(y_i) \ge p$, then at least $100p$ percent of the probability for the distribution of X is between y_i and y_j. Let it be given that $\gamma = P[F(Y_j) - F(Y_i) \ge p]$. Then the random interval (Y_i, Y_j) has probability γ of containing at least $100p$ percent of the probability for the distribution of X. Now if y_i and y_j denote, respectively, observational values of Y_i and Y_j, the interval (y_i, y_j) either does or does not contain at least $100p$ percent of the probability for the distribution of X. However, we refer to the interval (y_i, y_j) as a 100γ percent *tolerance interval* for $100p$ percent of the probability for the distribution of X. In like vein, y_i and y_j are called the 100γ percent *tolerance limits* for $100p$ percent of the probability for the distribution of X.

One way to compute the probability $\gamma = P[F(Y_j) - F(Y_i) \ge p]$ is to use Equation (5.3.3), which gives the joint pdf of $Z_i = F(Y_i)$ and $Z_j = F(Y_j)$. The required probability is then given by

$$\gamma = P(Z_j - Z_i \ge p) = \int_0^{1-p} \left[\int_{p+z_i}^1 h_{ij}(z_i, z_j)\, dz_j \right] dz_i.$$

Sometimes, this is a rather tedious computation. For this reason and also for the reason that *converages* are important in distribution-free statistical inference, we choose to introduce at this time the concepts of a coverage.

Consider the random variables $W_1 = F(Y_1) = Z_1$, $W_2 = F(Y_2) - F(Y_1) = Z_2 - Z_1$, $W_3 = F(Y_3) - F(Y_2) = Z_3 - Z_2, \ldots, W_n = F(Y_n) - F(Y_{n-1}) = Z_n - Z_{n-1}$. The random variable W_1 is called a *coverage* of the random interval $\{x : -\infty < x < Y_1\}$ and the random variable W_i, $i = 2, 3, \ldots, n$, is called a *coverage* of the random interval $\{x : Y_{i-1} < x < Y_i\}$. We shall find the joint pdf of the n coverages $W_1, W_2, \ldots, W_n$. First we note that the inverse functions of the associated transformation are given by

$$z_i = \textstyle\sum_{j=1}^i w_j, \text{ for } i = 1, 2, \ldots, n.$$

We also note that the Jacobian is equal to one and that the space of positive probability density is

$$\{(w_1, w_2, \ldots, w_n) : 0 < w_i,\ i = 1, 2, \ldots, n,\ w_1 + \cdots + w_n < 1\}.$$

Since the joint pdf of $Z_1, Z_2, \ldots, Z_n$ is $n!$, $0 < z_1 < z_2 < \cdots < z_n < 1$, zero elsewhere, the joint pdf of the n coverages is

$$k(w_1, \ldots, w_n) = \begin{cases} n! & 0 < w_i,\ i = 1, \ldots, n,\ w_1 + \cdots w_n < 1 \\ 0 & \text{elsewhere.} \end{cases}$$

Because the pdf $k(w_1, \ldots, w_n)$ is symmetric in $w_1, w_2, \ldots, w_n$, it is evident that the distribution of every sum of r, $r < n$, of these coverages $W_1, \ldots, W_n$ is exactly the same for each fixed value of r. For instance, if $i < j$ and $r = j - i$, the distribution of $Z_j - Z_i = F(Y_j) - F(Y_i) = W_{i+1} + W_{i+2} + \cdots + W_j$ is exactly the same as that of $Z_{j-i} = F(Y_{j-i}) = W_1 + W_2 + \cdots + W_{j-i}$. But we know that the pdf of Z_{j-i} is the beta pdf of the form

$$h_{j-i}(v) = \begin{cases} \frac{\Gamma(n+1)}{\Gamma(j-i)\Gamma(n-j+i+1)} v^{j-i-1}(1-v)^{n-j+i} & 0 < v < 1 \\ 0 & \text{elsewhere.} \end{cases}$$

Consequently, $F(Y_j) - F(Y_i)$ has this pdf and

$$P[F(Y_j) - F(Y_i) \geq p] = \int_p^1 h_{j-i}(v)\, dv.$$

Example 5.3.1. Let $Y_1 < Y_2 < \cdots < Y_6$ be the order statistics of a random sample of size 6 from a distribution of the continuous type. We want to use the observed interval (y_1, y_6) as a tolerance interval for 80 percent of the distribution. Then

$$\gamma = P[F(Y_6) - F(Y_1) \geq 0.8]$$
$$= 1 - \int_0^{0.8} 30v^4(1-v)\, dv,$$

because the integrand is the pdf of $F(Y_6) - F(Y_1)$. Accordingly,

$$\gamma = 1 - 6(0.8)^5 + 5(0.8)^6 = 0.34,$$

approximately. That is, the observed values of Y_1 and Y_6 will define a 34 percent tolerance interval for 80 percent of the probability for the distribution. ∎

Remark 5.3.1. Tolerance intervals are extremely important and often they are more desirable than confidence intervals. For illustration, consider a "fill" problem in which a manufacturer says that each container has at least 12 ounces of the product. Let X be the amount in a container. The company would be pleased to note that 12.1 to 12.3, for instance, is a 95% tolerance interval for 99% of the distribution of X. This would be true in this case, because the federal agency, FDA, allows a very small fraction of the containers to be less than 12 ounces. ∎

EXERCISES

5.3.1. Let X be a random variable with a continuous cdf $F(x)$. Assume that $F(x)$ is strictly increasing on the space of X. Consider the random variable $Z = F(X)$. Show that Z has a uniform distribution on the interval $(0, 1)$.

5.3.2. Let Y_1 and Y_n be, respectively, the first and the nth order statistic of a random sample of size n from a distribution of the continuous type having cdf $F(x)$. Find the smallest value of n such that $P[F(Y_n) - F(Y_1) \geq 0.5]$ is at least 0.95.

5.3.3. Let Y_2 and Y_{n-1} denote the second and the $(n-1)$st order statistics of a random sample of size n from a distribution of the continuous type having a distribution function $F(x)$. Compute $P[F(Y_{n-1}) - F(Y_2) \geq p]$, where $0 < p < 1$.

5.3.4. Let $Y_1 < Y_2 < \cdots < Y_{48}$ be the order statistics of a random sample of size 48 from a distribution of the continuous type. We want to use the observed interval (y_4, y_{45}) as a 100γ percent tolerance interval for 75 percent of the distribution.

(a) What is the value of γ?

(b) Approximate the integral in Part (a) by noting that it can be written as a Partial sum of a binomial pdf, which in turn can be approximated by probabilities associated with a normal distribution, (see Section 4.4).

5.3.5. Let $Y_1 < Y_2 < \cdots < Y_n$ be the order statistics of a random sample of size n from a distribution of the continuous type having distribution function $F(x)$.

(a) What is the distribution of $U = 1 - F(Y_j)$?

(b) Determine the distribution of $V = F(Y_n) - F(Y_j) + F(Y_i) - F(Y_1)$, where $i < j$.

5.3.6. Let $Y_1 < Y_2 < \cdots < Y_{10}$ be the order statistics of a random sample from a continuous-type distribution with distribution function $F(x)$. What is the joint distribution of $V_1 = F(Y_4) - F(Y_2)$ and $V_2 = F(Y_{10}) - F(Y_6)$?

5.4   More on Confidence Intervals

Let us return to the statistical problem that we were discussing in Section 5.1. Recall that the random variable of interest X has density $f(x; \theta), \theta \in \Omega$, where θ is unknown. In that section, we discussed estimating θ by a statistic $T = T(X_1, \ldots, X_n)$, where $X_1, \ldots, X_n$ is a sample from the distribution of X. When the sample is drawn, it is unlikely that the value of T is the true value of the parameter. In fact, if T has a continuous distribution then $P_\theta(T = \theta) = 0$. What is needed is an estimate of the error of the estimation; i.e., by how much did T miss θ. We addressed this briefly in Examples 5.1.2 and 5.2.7 when we introduced confidence intervals for μ and $\xi_{1/2}$, repectively. In this section, based on the Central Limit Theorem (Theorem 4.4.1), we discuss these issues further. At the end of this section, we will consider the case when the sample is drawn from a normal distribution.

To avoid confusion, let θ_0 denote the true, unknown value of the parameter θ. Suppose T is an estimator of θ_0 such that

$$\sqrt{n}(T - \theta_0) \xrightarrow{D} N(0, \sigma_T^2). \tag{5.4.1}$$

The parameter σ_T^2 is the asymptotic variance of $\sqrt{n}T$ and, in practice, it is usually unknown. For the present, though, we will think of σ_T^2 as known.

Let $Z = \sqrt{n}(T - \theta_0)/\sigma_T$ be the standardized random variable. Then Z is asymptotically $N(0, 1)$. Hence, $P(-1.96 < Z < 1.96) \doteq 0.95$. This leads to the following algebraic derivation:

$$
\begin{aligned}
0.95 &\doteq P(-1.96 < Z < 1.96) \\
&= P\left(-1.96 < \frac{\sqrt{n}(T - \theta_0)}{\sigma_T} < 1.96\right) \\
&= P\left(T - 1.96\frac{\sigma_T}{\sqrt{n}} < \theta_0 < T + 1.96\frac{\sigma_T}{\sqrt{n}}\right).
\end{aligned}
\tag{5.4.2}
$$

Because the interval $(T - 1.96\sigma_T/\sqrt{n}, T + 1.96\sigma_T/\sqrt{n})$ is a function of the random variable T, we will call it a **random interval**. By this derivation, the probability that the random interval contains θ is approximately 0.95. Let t be the value of the statistic T when the sample is drawn. Then the value of the random interval is given by

$$
(t - 1.96\sigma_T/\sqrt{n}, t + 1.96\sigma_T/\sqrt{n}).
\tag{5.4.3}
$$

This interval will either contain θ_0 or it will not. It is the outcome of a Bernoulli trial where success occurs when the interval traps θ_0. Based on the random interval (5.4.2), the probability of success of this Bernoulli trial is approximately 0.95. So even though the interval (5.4.3) either traps θ_0 or it does not, because of the high probability of success, we are fairly confident of a successful interval. Thus, using the terminology given in Section 5.1, we call the interval (5.4.3) a **95% confidence interval** for θ_0 and refer to the initial probability $0.95 = 95\%$ as a **confidence coefficient**.

Of course in practice, we often do not know σ_T. Suppose that the statistic S_T is a consistent estimator of σ_T. It then follows from Theorem 4.3.5 (Slutsky's Theorem) that

$$
\frac{\sqrt{n}(T - \theta_0)}{S_T} \xrightarrow{D} N(0, 1).
$$

Hence by the same chain of reasoning, the interval $(T - 1.96S_T/\sqrt{n}, T + 1.96S_T/\sqrt{n})$ would be a random interval with approximate probability 0.95 of covering θ_0. Once the sample is drawn, denote the observed values of T and S_T by t and s_t, respectively. Then an approximate 95% confidence interval for θ_0 is

$$
(t - 1.96s_t/\sqrt{n}, t + 1.96s_t/\sqrt{n}).
\tag{5.4.4}
$$

This interval would be used in practice and often $s_t/\sqrt{n}$ is called the **standard error** of T.

Example 5.4.1 (Confidence Interval for the Mean μ). Let $X_1, \ldots, X_n$ be a random sample from the distribution of a random variable X which has unknown mean μ and unknown variance σ^2. Let $\overline{X}$ and S^2 denote sample mean and variance, respectively. By the Central Limit Theorem and Theorem 4.3.5, then $\sqrt{n}(\overline{X} - \mu)/S$ has an approximate $N(0, 1)$ distribution. Hence, an approximate 95% confidence interval for μ is

$$
(\overline{x} - 1.96s/\sqrt{n}, \overline{x} + 1.96s/\sqrt{n}). \quad \blacksquare
\tag{5.4.5}
$$

There is nothing special about the confidence coefficient 0.95. Consider a general confidence coefficient $(1 - \alpha)$, where $0 < \alpha < 1$. Define $z_{\alpha/2}$ as the upper $\alpha/2$ quantile of a standard normal random variable; i.e., $1 - \Phi(z_{\alpha/2}) = \alpha/2$, where $\Phi(z)$ is the standard normal distribution function given by (3.4.10). The same chain of reasoning works for this general $1 - \alpha$ confidence coefficient as it did for the confidence coefficient 95%. That is, as above, let S_T be a consistent estimate of σ_T. A similar argument, leads to the following interval being an approximate $(1 - \alpha)100\%$ confidence interval for θ_0:

$$(t - z_{\alpha/2}s_T/\sqrt{n}, t + z_{\alpha/2}s_T/\sqrt{n}). \qquad (5.4.6)$$

In the case of Example 5.4.1,

$$(\overline{x} - z_{\alpha/2}s/\sqrt{n}, \overline{x} + z_{\alpha/2}s/\sqrt{n}) \qquad (5.4.7)$$

is an approximate $(1 - \alpha)100\%$ confidence interval for μ.

We often write these intervals as $t \pm z_{\alpha/2}s_T/\sqrt{n}$ and $\overline{x} \pm z_{\alpha/2}s/\sqrt{n}$ and think of the terms, $z_{\alpha/2}s_T/\sqrt{n}$ and $z_{\alpha/2}s/\sqrt{n}$, as the error parts of the confidence intervals. These errors are the maxima of by how much the estimates can miss the unknown parameters with confidence $(1 - \alpha)100\%$.

Note that the parts of this error term make sense intuitively:

1. Because $\alpha < \alpha^*$ implies that $z_{\alpha/2} > z_{\alpha^*/2}$, selection of higher values for confidence coefficients leads to larger error terms and hence, longer confidence intervals, assuming all else remains the same.

2. Choosing a larger sample size decreases the error part and hence, leads to shorter confidence intervals, assuming all else stays the same.

3. Usually the parameter σ_T is some type of scale parameter of the underlying distribution. In these situations, assuming all else remains the same, an increase in scale (noise level), generally results in larger error terms and, hence, longer confidence intervals.

Example 5.4.2 (Confidence Interval for p). Let X be a Bernoulli random variable with probability of success p. Suppose $X_1, \ldots, X_n$ is a random sample from the distribution of X. Let $\widehat{p} = \overline{X}$ be the sample proportion of successes. By the Central Limit Theorem, $\widehat{p}$ has an approximate $N(p, p(1 - p)/n)$ distribution. By the Weak Law of Large Numbers $\widehat{p}$ is a consistent estimate of p and, further, by Theorem 4.2.4, $\widehat{p}(1 - \widehat{p})$ is a consistent estimate of $p(1 - p)$. Hence, an approximate $(1 - \alpha)100\%$ confidence interval for p is given by

$$(\widehat{p} - z_{\alpha/2}\sqrt{\widehat{p}(1 - \widehat{p})/n}, \widehat{p} + z_{\alpha/2}\sqrt{\widehat{p}(1 - \widehat{p})/n}), \qquad (5.4.8)$$

where $\sqrt{\widehat{p}(1 - \widehat{p})/n}$ is called the standard error of $\widehat{p}$. ∎

In general, the confidence intervals developed so far in this section are approximate. They are based on the Central Limit Theorem and also, often require a consistent estimate of σ_T. In our next example, we develop an exact confidence interval for the mean when sampling from a normal distribution.

Example 5.4.3 (Confidence Interval for μ under Normality). Suppose the random variables $X_1, \ldots, X_n$ are a random sample from a $N(\mu, \sigma^2)$ distribution. Let $\overline{X}$ and S^2 denote sample mean and sample variance, respectively. By Part (d) of Theorem 3.6.1, the random variable $T = (\overline{X} - \mu)/(S/\sqrt{n})$ has a t distribution with $n-1$ degrees of freedom. For $0 < \alpha < 1$, define $t_{\alpha/2, n-1}$ to be the upper $\alpha/2$ critical point of a t distribution with $n-1$ degrees of freedom; i.e., $\alpha/2 = P(T > t_{\alpha/2, n-1})$. Using a similar algebraic derivation as in (5.4.3), we obtain,

$$
\begin{aligned}
1 - \alpha &= P(-t_{\alpha/2, n-1} < T < t_{\alpha/2, n-1}) = P\left(-t_{\alpha/2, n-1} < \frac{\overline{X} - \mu}{S/\sqrt{n}} < t_{\alpha/2, n-1}\right) \\
&= P\left(\overline{X} - t_{\alpha/2, n-1}\frac{S}{\sqrt{n}} < \mu < \overline{X} + t_{\alpha/2, n-1}\frac{S}{\sqrt{n}}\right). \qquad (5.4.9)
\end{aligned}
$$

Letting $\overline{x}$ and s denote the realized values of the statistics $\overline{x}$ and s, respectively, after the sample is drawn, a $(1 - \alpha)100\%$ confidence interval for μ is given by,

$$
(\overline{x} - t_{\alpha/2, n-1}s/\sqrt{n}, \overline{x} + t_{\alpha/2, n-1}s/\sqrt{n}). \qquad (5.4.10)
$$

For $\alpha = 0.05$, note that the only difference between this confidence interval and the large sample 95% confidence interval (5.4.3), is that $t_{.025, n-1}$ replaces 1.96. This one is exact while (5.4.3) is approximate. Of course, we have to assume we are sampling a normal population to get the exactness. Also this one uses $t_{\alpha/2, n-1}$ while the former uses $z_{\alpha/2}$.

In practice, we often do not know if the population is normal. Which confidence interval should we use? Generally, for the same α, the intervals based on $t_{\alpha/2, n-1}$ are larger than those based on $z_{\alpha/2}$. Hence, the interval (5.4.10) is generally more conservative than the interval (5.4.7). So in practice, statisticians generally prefer the interval (5.4.10). ∎

In later chapters of this book, we will develop other exact confidence intervals for certain specific situations. We will also consider bootstrap confidence intervals. These generally avoid the estimation of σ_T.

5.4.1   Confidence Intervals for Differences in Means

A practical problem of interest is the comparison of two distributions; that is, comparing the distributions of two random variables, say X and Y. In this section, we will compare the means of X and Y. Denote the means of X and Y by μ_1 and μ_2, respectively. In particular, we shall obtain confidence intervals for the difference $\Delta = \mu_1 - \mu_2$. Assume that the variances of X and Y are finite and denote them as $\sigma_1^2 = \text{Var}(X)$ and let $\sigma_2^2 = \text{Var}(Y)$. Let $X_1, \ldots, X_{n_1}$ be a random sample from the distribution of X and let $Y_1, \ldots, Y_{n_2}$ be a random sample from the distribution of Y. Assume that the samples were gathered independently of one another. Let $\overline{X} = n_1^{-1}\sum_{i=1}^{n_1} X_i$ and $\overline{Y} = n_2^{-1}\sum_{i=1}^{n_2} Y_i$ be the sample means. Let $\widehat{\Delta} = \overline{X} - \overline{Y}$. The statistic $\widehat{\Delta}$ is an unbiased estimator of Δ.

Next we obtain a large sample confidence interval for Δ based on the asymptotic distribution of $\widehat{\Delta}$. Denote the total sample size by $n = n_1 + n_2$. We need one further assumption concerning the sample sizes. Assume that

$$\frac{n_1}{n} \to \lambda_1 \text{ and } \frac{n_2}{n} \to \lambda_2 \text{ where } \lambda_1 + \lambda_2 = 1. \tag{5.4.11}$$

This says that the sample sizes are of the same relative order. Recall from the Central Limit Theorem (4.4.1) that

$$\sqrt{n_1}(\overline{X} - \mu_1) \xrightarrow{D} N(0, \sigma_1^2).$$

Because (5.4.11) holds, we then have

$$\sqrt{n}(\overline{X} - \mu_1) = \sqrt{\frac{n}{n_1}} \sqrt{n_1}(\overline{X} - \mu_1) \xrightarrow{D} N\left(0, \frac{1}{\lambda_1}\sigma_1^2\right). \tag{5.4.12}$$

Likewise,

$$\sqrt{n}(\overline{Y} - \mu_2) \xrightarrow{D} N\left(0, \frac{1}{\lambda_2}\sigma_2^2\right). \tag{5.4.13}$$

Since the samples are independent of one another, we can combine these results, (5.4.12) and (5.4.13), to obtain

$$\sqrt{n}[(\overline{X} - \overline{Y}) - (\mu_1 - \mu_2)] \xrightarrow{D} N\left(0, \frac{1}{\lambda_1}\sigma_1^2 + \frac{1}{\lambda_2}\sigma_2^2\right). \tag{5.4.14}$$

This last result is the basis for the confidence interval for Δ. In terms of the actual sample sizes, note that we can rewrite this result as

$$\frac{(\overline{X} - \overline{Y}) - (\mu_1 - \mu_2)}{\sqrt{\frac{\sigma_1^2}{n_1} + \frac{\sigma_2^2}{n_2}}} \text{ has a limiting } N(0, 1) \text{ distribution.} \tag{5.4.15}$$

From this last result, it is easy to see that

$$\left((\overline{x} - \overline{y}) - z_{\alpha/2}\sqrt{\frac{\sigma_1^2}{n_1} + \frac{\sigma_2^2}{n_2}}, (\overline{x} - \overline{y}) + z_{\alpha/2}\sqrt{\frac{\sigma_1^2}{n_1} + \frac{\sigma_2^2}{n_2}}\right), \tag{5.4.16}$$

is an approximate $(1 - \alpha)100\%$ confidence interval for $\Delta = \mu_1 - \mu_2$. In practice, though, the variances σ_1^2 and σ_2^2 would not be known. We can estimate these variances by the respective sample variances, $S_1^2 = (n_1 - 1)^{-1}\sum_{i=1}^{n_1}(X_i - \overline{X})^2$ and $S_2^2 = (n_2 - 1)^{-1}\sum_{i=1}^{n_2}(Y_i - \overline{Y})^2$. Because S_1^2 and S_2^2 are consistent estimators of σ_1^2 and σ_2^2, the result in expression (5.4.14) is still true if σ_1^2 and σ_2^2 are replaced by S_1^2 and S_2^2, respectively; see Exercise 5.4.18. This leads to the approximate $(1-\alpha)100\%$ confidence interval for $\Delta = \mu_1 - \mu_2$ given by

$$\left((\overline{x} - \overline{y}) - z_{\alpha/2}\sqrt{\frac{s_1^2}{n_1} + \frac{s_2^2}{n_2}}, (\overline{x} - \overline{y}) + z_{\alpha/2}\sqrt{\frac{s_1^2}{n_1} + \frac{s_2^2}{n_2}}\right), \tag{5.4.17}$$

where $\sqrt{(s_1^2/n_1) + (s_2^2/n_2)}$ is the standard error of $\overline{X} - \overline{Y}$.

The above confidence interval is approximate. In this situation we can obtain exact confidence intervals if we assume that the distributions of X and Y are the same except for a difference in means; i.e., a *location* model. In particular, the variances of X and Y are assumed to be the same. Assume further that X is distributed $N(\mu_1, \sigma^2)$ and Y is distributed $N(\mu_2, \sigma^2)$, where σ^2 is the common variance of X and Y. As above, let $X_1, \ldots, X_{n_1}$ be a random sample from the distribution of X and let $Y_1, \ldots, Y_{n_2}$ be a random sample from the distribution of Y. Assume that the samples are independent of one another and let $n = n_1 + n_2$ be the total sample size. Our estimator of Δ remains $\overline{X} - \overline{Y}$. Our goal is to show that a random variable, defined below, has a t-distribution, which is defined in Section 3.6.

Because $\overline{X}$ is distributed $N(\mu_1, \sigma^2/n_1)$, $\overline{Y}$ is distributed $N(\mu_2, \sigma^2/n_2)$, and $\overline{X}$ and $\overline{Y}$ are independent we have the result

$$\frac{(\overline{X}-\overline{Y})-(\mu_1-\mu_2)}{\sigma\sqrt{\frac{1}{n_1}+\frac{1}{n_2}}} \text{ has a } N(0,1) \text{ distribution.} \tag{5.4.18}$$

This will serve as the numerator of our T-statistic.

Let

$$S_p^2 = \frac{(n_1-1)S_1^2 + (n_2-1)S_2^2}{n_1+n_2-2}. \tag{5.4.19}$$

Note that S_p^2 is a weighted average of S_1^2 and S_2^2. It is easy to see that S_p^2 is an unbiased estimator of σ^2. It is called the *pooled estimator* of σ^2. Also because $(n_1-1)S_1^2/\sigma^2$ has a $\chi^2(n_1-1)$ distribution, $(n_2-1)S_2^2/\sigma^2$ has a $\chi^2(n_2-1)$ distribution, and S_1^2 and S_2^2 are independent, we have that $(n-2)S_p^2/\sigma^2$ has a $\chi^2(n-2)$ distribution; see Corollary 3.3.1. Finally, because S_1^2 is independent of $\overline{X}$ and S_2^2 is independent of $\overline{Y}$, and the random samples are independent of each other, it follows that S_p^2 is independent of expression (5.4.18). Therefore, by Student's t, we have

$$\begin{aligned} T &= \frac{[(\overline{X}-\overline{Y})-(\mu_1-\mu_2)]/\sigma\sqrt{n_1^{-1}+n_2^{-1}}}{\sqrt{(n-2)S_p^2/(n-2)\sigma^2}} \\ &= \frac{(\overline{X}-\overline{Y})-(\mu_1-\mu_2)}{S_p\sqrt{\frac{1}{n_1}+\frac{1}{n_2}}} \end{aligned} \tag{5.4.20}$$

has a t-distribution with $n-2$ degrees of freedom. From this last result, it is easy to see that the following interval is an exact $(1-\alpha)100\%$ confidence interval for $\Delta = \mu_1 - \mu_2$:

$$\left((\overline{x}-\overline{y}) - t_{\alpha/2,n-2}s_p\sqrt{\frac{1}{n_1}+\frac{1}{n_2}}, (\overline{x}-\overline{y}) + t_{\alpha/2,n-2}s_p\sqrt{\frac{1}{n_1}+\frac{1}{n_2}} \right). \tag{5.4.21}$$

A consideration of the difficulty encountered when the unknown variances of the two normal distributions are not equal is assigned to one of the exercises.

Example 5.4.4. Suppose $n_1 = 10$, $n_2 = 7$, $\overline{x} = 4.2$, $\overline{y} = 3.4$, $s_1^2 = 49$, $s_2^2 = 32$. Then using (5.4.21) a 90% confidence interval for $\mu_1 - \mu_2$ is $(-5.16, 6.76)$. ■

Remark 5.4.1. Suppose X and Y are not normally distributed but still their distributions only differ in location. It is easy to see that S_p^2 will be a consistent, unbiased estimator of the common variance σ^2. The above interval, (5.4.21), is then approximate and not exact. ∎

5.4.2 Confidence Interval for Difference in Proportions

Let X and Y be two independent random variables with Bernoulli distributions $b(1, p_1)$ and $b(1, p_2)$, respectively. Let us now turn to the problem of finding a confidence interval for the difference $p_1 - p_2$. Let $X_1, \ldots, X_{n_1}$ be a random sample from the distribution of X and let $Y_1, \ldots, Y_{n_2}$ be a random sample from the distribution of Y. As above, assume that the samples are independent of one another and let $n = n_1 + n_2$ be the total sample size. Our estimator of $p_1 - p_2$ is the difference in sample proportions which, of course, is given by $\overline{X} - \overline{Y}$. We will use the traditional notation and write $\hat{p}_1$ and $\hat{p}_2$ instead of $\overline{X}$ and $\overline{Y}$, respectively. Hence, from the above discussion the interval like (5.4.16) serves as an approximate confidence interval for $p_1 - p_2$. Here, $\sigma_1^2 = p_1(1 - p_1)$ and $\sigma_2^2 = p_2(1 - p_2)$. These variances are unknown, but it is easily shown that $\hat{p}_1(1 - \hat{p}_1)$ and $\hat{p}_2(1 - \hat{p}_2)$ are consistent estimators of $p_1(1 - p_1)$ and $p_2(1 - p_2)$, respectively. Similar to expression (5.4.17), we can substitute the estimators for the parameters. Thus, our approximate $(1 - \alpha)100\%$ confidence interval for $p_1 - p_2$ is

$$\hat{p}_1 - \hat{p}_2 \pm z_{\alpha/2} \sqrt{\frac{\hat{p}_1(1 - \hat{p}_1)}{n_1} + \frac{\hat{p}_2(1 - \hat{p}_2)}{n_2}} \qquad (5.4.22)$$

Example 5.4.5. If, in the preceding discussion, we take $n_1 = 100$, $n_2 = 400$, $y_1 = 30$, $y_2 = 80$, then the observed values of $Y_1/n_1 - Y_2/n_2$ and its standard error are 0.1 and $\sqrt{(0.3)(0.7)/100 + (0.2)(0.8)/400} = 0.05$, respectively. Thus the interval $(0, 0.2)$ is an approximate 95.4 percent confidence interval for $p_1 - p_2$. ∎

EXERCISES

5.4.1. Let the observed value of the mean $\overline{X}$ of a random sample of size 20 from a distribution that is $N(\mu, 80)$ be 81.2. Find a 95 percent confidence interval for μ.

5.4.2. Let $\overline{X}$ be the mean of a random sample of size n from a distribution that is $N(\mu, 9)$. Find n such that $P(\overline{X} - 1 < \mu < \overline{X} + 1) = 0.90$, approximately.

5.4.3. Let a random sample of size 17 from the normal distribution $N(\mu, \sigma^2)$ yield $\overline{x} = 4.7$ and $s^2 = 5.76$. Determine a 90 percent confidence interval for μ.

5.4.4. Let $\overline{X}$ denote the mean of a random sample of size n from a distribution that has mean μ and variance $\sigma^2 = 10$. Find n so that the probability is approximately 0.954 that the random interval $(\overline{X} - \frac{1}{2}, \overline{X} + \frac{1}{2})$ includes μ.

5.4.5. Let $X_1, X_2, \ldots, X_9$ be a random sample of size 9 from a distribution that is $N(\mu, \sigma^2)$.

(a) If σ is known, find the length of a 95 percent confidence interval for μ if this interval is based on the random variable $\sqrt{9}(\overline{X} - \mu)/\sigma$.

(b) If σ is unknown, find the expected value of the length of a 95 percent confidence interval for μ if this interval is based on the random variable $\sqrt{9}(\overline{X} - \mu)/S$. *Hint:* Write $E(S) = (\sigma/\sqrt{n-1})E[((n-1)S^2/\sigma^2)^{1/2}]$.

(c) Compare these two answers.

5.4.6. Let $X_1, X_2, \ldots, X_n, X_{n+1}$ be a random sample of size $n+1$, $n > 1$, from a distribution that is $N(\mu, \sigma^2)$. Let $\overline{X} = \sum_1^n X_i/n$ and $S^2 = \sum_1^n (X_i - \overline{X})^2/(n-1)$. Find the constant c so that the statistic $c(\overline{X} - X_{n+1})/S$ has a t-distribution. If $n = 8$, determine k such that $P(\overline{X} - kS < X_9 < \overline{X} + kS) = 0.80$. The observed interval $(\overline{x} - ks, \overline{x} + ks)$ is often called an 80 percent *prediction interval* for X_9.

5.4.7. Let Y be $b(300, p)$. If the observed value of Y is $y = 75$, find an approximate 90 percent confidence interval for p.

5.4.8. Let $\overline{X}$ be the mean of a random sample of size n from a distribution that is $N(\mu, \sigma^2)$, where the positive variance σ^2 is known. Because $\Phi(2) - \Phi(-2) = 0.954$, find, for each μ, $c_1(\mu)$ and $c_2(\mu)$ such that $P[c_1(\mu) < \overline{X} < c_2(\mu)] = 0.954$. Note that $c_1(\mu)$ and $c_2(\mu)$ are increasing functions of μ. Solve for the respective functions $d_1(\overline{x})$ and $d_2(\overline{x})$; thus we also have that $P[d_2(\overline{X}) < \mu < d_1(\overline{X})] = 0.954$. Compare this with the answer obtained previously in the text.

5.4.9. Let $\overline{X}$ denote the mean of a random sample of size 25 from a gamma-type distribution with $\alpha = 4$ and $\beta > 0$. Use the Central Limit Theorem to find an approximate 0.954 confidence interval for μ, the mean of the gamma distribution. *Hint:* Use the random variable $(\overline{X} - 4\beta)/(4\beta^2/25)^{1/2} = 5\overline{X}/2\beta - 10$.

5.4.10. Let $\overline{x}$ be the observed mean of a random sample of size n from a distribution having mean μ and known variance σ^2. Find n so that $\overline{x} - \sigma/4$ to $\overline{x} + \sigma/4$ is an approximate 95 percent confidence interval for μ.

5.4.11. Assume a binomial model for a certain random variable. If we desire a 90 percent confidence interval for p that is at most 0.02 in length, find n. *Hint:* Note that $\sqrt{(y/n)(1 - y/n)} \leq \sqrt{(\frac{1}{2})(1 - \frac{1}{2})}$.

5.4.12. It is known that a random variable X has a Poisson distribution with parameter μ. A sample of 200 observations from this distribution has a mean equal to 3.4. Construct an approximate 90 percent confidence interval for μ.

5.4.13. Let $Y_1 < Y_2 < \cdots < Y_n$ denote the order statistics of a random sample of size n from a distribution that has pdf $f(x) = 3x^2/\theta^3$, $0 < x < \theta$, zero elsewhere.

(a) Show that $P(c < Y_n/\theta < 1) = 1 - c^{3n}$, where $0 < c < 1$.

(b) If n is 4 and if the observed value of Y_4 is 2.3, what is a 95 percent confidence interval for θ?

5.4.14. Let $X_1, X_2, \ldots, X_n$ be a random sample from $N(\mu, \sigma^2)$, where both parameters μ and σ^2 are unknown. A *confidence interval* for σ^2 can be found as follows. We know that $(n-1)S^2/\sigma^2$ is a random variable with a $\chi^2(n-1)$ distribution. Thus we can find constants a and b so that $P((n-1)S^2/\sigma^2 < b) = 0.975$ and $P(a < (n-1)S^2/\sigma^2 < b) = 0.95$.

(a) Show that this second probability statement can be written as

$$P((n-1)S^2/b < \sigma^2 < (n-1)S^2/a) = 0.95.$$

(b) If $n = 9$ and $s^2 = 7.93$, find a 95 percent confidence interval for σ^2.

(c) If μ is known, how would you modify the preceding procedure for finding a confidence interval for σ^2?

5.4.15. Let $X_1, X_2, \ldots, X_n$ be a random sample from a gamma distributions with known parameter $\alpha = 3$ and unknown $\beta > 0$. Discuss the construction of a confidence interval for β.

Hint: What is the distribution of $2\sum_1^n X_i/\beta$? Follow the procedure outlined in Exercise 5.4.14.

5.4.16. When 100 tacks were thrown on a table, 60 of them landed point up. Obtain a 95 percent confidence interval for the probability that a tack of this type will land point up. Assume independence.

5.4.17. Complete the proof of expression (5.4.14) in the text.

5.4.18. Using the assumptions behind the confidence interval given in expression (5.4.17), show that

$$\sqrt{\frac{S_1^2}{n_1} + \frac{S_2^2}{n_2}} \bigg/ \sqrt{\frac{\sigma_1^2}{n_1} + \frac{\sigma_2^2}{n_2}} \xrightarrow{P} 1.$$

5.4.19. Let two independent random samples, each of size 10, from two normal distributions $N(\mu_1, \sigma^2)$ and $N(\mu_2, \sigma^2)$ yield $\bar{x} = 4.8$, $s_1^2 = 8.64$, $\bar{y} = 5.6$, $s_2^2 = 7.88$. Find a 95 percent confidence interval for $\mu_1 - \mu_2$.

5.4.20. Let two independent random variables, Y_1 and Y_2, with binomial distributions that have parameters $n_1 = n_2 = 100$, p_1, and p_2, respectively, be observed to be equal to $y_1 = 50$ and $y_2 = 40$. Determine an approximate 90 percent confidence interval for $p_1 - p_2$.

5.4.21. Discuss the problem of finding a confidence interval for the difference $\mu_1 - \mu_2$ between the two means of two normal distributions if the variances σ_1^2 and σ_2^2 are known but not necessarily equal.

5.4.22. Discuss Exercise 5.4.21 when it is assumed that the variances are unknown and unequal. This is a very difficult problem, and the discussion should point out exactly where the difficulty lies. If, however, the variances are unknown but their ratio σ_1^2/σ_2^2 is a known constant k, then a statistic that is a T random variable can again be used. Why?

5.4.23. To illustrate Exercise 5.4.22, let $X_1, X_2, \ldots, X_9$ and $Y_1, Y_2, \ldots, Y_{12}$ represent two independent random samples from the respective normal distributions $N(\mu_1, \sigma_1^2)$ and $N(\mu_2, \sigma_2^2)$. It is given that $\sigma_1^2 = 3\sigma_2^2$, but σ_2^2 is unknown. Define a random variable which has a t-distribution that can be used to find a 95 percent confidence interval for $\mu_1 - \mu_2$.

5.4.24. Let $\overline{X}$ and $\overline{Y}$ be the means of two independent random samples, each of size n, from the respective distributions $N(\mu_1, \sigma^2)$ and $N(\mu_2, \sigma_2)$, where the common variance is known. Find n such that

$$P(\overline{X} - \overline{Y} - \sigma/5 < \mu_1 - \mu_2 < \overline{X} - \overline{Y} + \sigma/5) = 0.90.$$

5.4.25. Let $X_1, X_2, \ldots, X_n$ and $Y_1, Y_2, \ldots, Y_m$ be two independent random samples from the respective normal distributions $N(\mu_1, \sigma_1^2)$ and $N(\mu_2, \sigma_2^2)$, where the four parameters are unknown. To construct a *confidence interval for the ratio, σ_1^2/σ_2^2*, of the variances, form the quotient of the two independent chi-square variables, each divided by its degrees of freedom, namely

$$F = \frac{\frac{(m-1)S_2^2}{\sigma_2^2}/(m-1)}{\frac{(n-1)S_1^2}{\sigma_1^2}/(n-1)} = \frac{S_2^2/\sigma_2^2}{S_1^2/\sigma_1^2},$$

where S_1^2 and S_2^2 are the respective sample variances.

(a) What kind of distribution does F have?

(b) From the appropriate table, a and b can be found so that $P(F < b) = 0.975$ and $P(a < F < b) = 0.95$.

(c) Rewrite the second probability statement as

$$P\left[a\frac{S_1^2}{S_2^2} < \frac{\sigma_1^2}{\sigma_2^2} < b\frac{S_1^2}{S_2^2}\right] = 0.95.$$

The observed values, s_1^2 and s_2^2, can be inserted in these inequalities to provide a 95 percent confidence interval for σ_1^2/σ_2^2.

5.5 Introduction to Hypothesis Testing

Point estimation and confidence intervals are useful statistical inference procedures. Another type of inference that is frequently used concerns tests of hypotheses. As in the last section, suppose our interest centers on a random variable X which has density function $f(x; \theta)$ where $\theta \in \Omega$. Suppose we think, due to theory or a preliminary experiment, that $\theta \in \omega_0$ or $\theta \in \omega_1$ where ω_0 and ω_1 are subsets of Ω and $\omega_0 \cup \omega_1 = \Omega$. We label these hypotheses as

$$H_0 : \theta \in \omega_0 \text{ versus } H_1 : \theta \in \omega_1. \tag{5.5.1}$$

The hypothesis H_0 is referred to as the **null hypothesis** while H_1 is referred to as the **alternative hypothesis**. Often the null hypothesis represents no change or no difference from the past while the alternative represents change or difference. The alternative is often referred to as the research worker's hypothesis. The decision rule to take H_0 or H_1 is based on a sample $X_1, \ldots, X_n$ from the distribution of X and hence, the decision could be wrong. For instance, we could decide that $\theta \in \omega_0$ when really $\theta \in \omega_1$. Table 5.5.1 displays the various situations. As we show in Chapter 8 a careful analysis of these errors can lead in certain situations to optimal decision rules. In this section, though, we simply want to introduce the elements of hypothesis testing. To set ideas, consider the following example.

Example 5.5.1 (Zea Mays Data). In 1878 Charles Darwin recorded some data on the heights of *zea mays* plants to determine what effect cross-fertilized or self-fertilized had on the height of *zea mays*. The experiment was to select one cross-fertilized plant and one self-fertiled plant, grow them in the same pot, and then later to measure their heights. An interesting hypotheses for this example would be that the cross-fertilized plants are generally taller than the self-fertilized plants. This will be the alternative hypothesis; i.e, the research worker's hypothesis. The null hypothesis is that the plants generally grow to the same height regardless of whether they were self or cross-fertilized. Data for 15 pots were recorded.

We will represent the data as $(X_1, Y_1), \ldots, (X_{15}, Y_{15})$ where X_i and Y_i are the heights of the cross-fertilized and self-fertilized plants, respectively, in the ith pot. Let $W_i = X_i - Y_i$. Due to growing in the same pot, X_i and Y_i may be dependent random variables but it seems appropriate to assume independence between pots, i.e., independence between the paired random vectors. So we will assume that $W_1, \ldots, W_{15}$ form a random sample. As a tentative model, consider

$$W_i = \mu + e_i, \quad i = 1, \ldots, 15,$$

where the random variables e_i are iid with continuous density $f(x)$. For this model, there is no loss in generality in assuming that the mean of e_i is 0 for, otherwise, we can simply redefine μ. Hence, $E(W_i) = \mu$. Further, the density of W_i is $f_W(x; \mu) = f(x - \mu)$. In practice the goodness of the model is always a concern and diagnostics based on the data would be run to confirm the quality of the model.

If $\mu = E(W_i) = 0$ then $E(X_i) = E(Y_i)$; i.e., on the average the cross-fertilized plants grow to the same height as the self-fertilized plants. While, if $\mu > 0$ then $E(X_i) > E(Y_i)$; i.e., on the average the cross-fertilized plants are taller than the self-fertilized plants. Under this model, our hypotheses are:

$$H_0: \ \mu = 0 \text{ versus } H_1: \ \mu > 0. \tag{5.5.2}$$

Hence, $\omega_0 = \{0\}$ represents no difference in the treatments and $\omega_1 = (0, \infty)$ represents a difference in the treatments. ∎

To complete the testing structure for the general problem described at the beginning of this section, we need to discuss decision rules. Recall that $X_1, \ldots, X_n$ is a random sample from the distribution of a random variable X which has density $f(x; \theta)$ where $\theta \in \Omega$. Consider testing the hypotheses $H_0: \ \theta \in \omega_0$ versus

Table 5.5.1: 2×2 Decision Table for a Test of Hypothesis

Decision	True State of Nature	
	H_0 is True	H_1 is True
Reject H_0	Type I Error	Correct Decision
Accept H_0	Correct Decision	Type II Error

$H_1 :$ $\theta \in \omega_1$, where $\omega_0 \cup \omega_1 = \Omega$. Denote the space of the sample by $\mathcal{D}$; that is, $\mathcal{D} = \text{space} \{(X_1, \ldots, X_n)\}$. A **test** of H_0 versus H_1 is based on a subset C of $\mathcal{D}$. This set C is called the **critical region** and its corresponding decision rule (test) is:

$$\text{Reject } H_0, (\text{Accept } H_1), \quad \text{if } (X_1, \ldots, X_n) \in C \quad (5.5.3)$$
$$\text{Retain } H_0, (\text{Reject } H_1), \quad \text{if } (X_1, \ldots, X_n) \in C^c.$$

For a given critical region, the 2×2 Decision Table 5.5.1, summarizes the results of the hypothesis test in terms of the true state of nature. Besides the correct decisions, two errors can occur. A **Type I** error occurs if H_0 is rejected when it is true while a **Type II** error occurs if H_0 is accepted when H_1 is true.

The goal, of course, is to select a critical region from all possible critical regions which minimizes the probabilities of these errors. In general, this is not possible. The probabilities of these errors often have a see-saw effect. This can be seen immediately in an extreme case. Simply let $C = \phi$. With this critical region, we would never reject H_0, so the probability of Type I error would be 0, but the probability of Type II error is 1. Often we consider Type I error to be the worse of the two errors. We then proceed by selecting critical regions which bound the probability of Type I error and then among these critical regions we try to select one which minimizes the probability of Type II error.

Definition 5.5.1. *We say a critical region C is of* **size** α *if*

$$\alpha = \max_{\theta \in \omega_0} P_\theta[(X_1, \ldots, X_n) \in C]. \quad (5.5.4)$$

Over all critical regions of size α, we want to consider critical regions which have lower probabilities of Type II error. We also can look at the complement of a Type II error, namely rejecting H_0 when H_1 is true which is a correct decision, as marked in Table 5.5.1. Since we desire to maximize the probability of this latter decision, we want the probability of it to be as large as possible. That is, for $\theta \in \omega_1$, we want to maximize

$$1 - P_\theta[\text{Type II Error}] = P_\theta[(X_1, \ldots, X_n) \in C].$$

The probability on the right side of this equation is called the **power** of the test at θ. It is the probability that the test detects the alternative θ when $\theta \in \omega_1$ is the true parameter. So minimizing the probability of Type II error is equivalent to maximizing power.

We define the **power function** of a critical region to be

$$\gamma_C(\theta) = P_\theta[(X_1, \ldots, X_n) \in C]; \quad \theta \in \omega_1. \qquad (5.5.5)$$

Hence, given two critical regions C_1 and C_2 which are both of size α, C_1 is better than C_2 if $\gamma_{C_1}(\theta) \geq \gamma_{C_2}(\theta)$ for all $\theta \in \omega_1$. In Chapter 8, we will obtain optimal critical regions for specific situations. In this section, we want to illustrate these concepts of hypotheses testing with several examples.

Example 5.5.2 (Test for a Binomial Proportion of Success). Let X be a Bernoulli random variable with probability of success p. Suppose we want to test at size α,

$$H_0 : p = p_0 \text{ versus } H_1 : p < p_0, \qquad (5.5.6)$$

where p_0 is specified. As an illustration, suppose "success" is dying from a certain disease and p_0 is the probability of dying with some standard treatment. A new treatment is used on several (randomly chosen) patients, and it is hoped that the probability of dying under this new treatment is less than p_0. Let $X_1, \ldots, X_n$ be a random sample from the distribution of X and let $S = \sum_{i=1}^n X_i$ be the total number of successes in the sample. An intuitive decision rule (critical region) is:

$$\text{Reject } H_0 \text{ in favor of } H_1 \text{ if } S \leq k, \qquad (5.5.7)$$

where k is such that $\alpha = P_{H_0}[S \leq k]$. Since S has a $b(n, p_0)$ distribution under H_0, k is determined by $\alpha = P_{p_0}[S \leq k]$. Because the binomial distribution is discrete, however, it is likely that there is no integer k which solves this equation. For example, suppose $n = 20$, $p_0 = 0.7$, and $\alpha = 0.15$. Then under H_0, S has a binomial $b(20, 0.7)$ distribution. Hence, computationally, $P_{H_0}[S \leq 11] = 0.1133$ and $P_{H_0}[S \leq 12] = 0.2277$. Hence, erring on the conservative side, we would probably choose k to be 11 and $\alpha = 0.1133$. As n increases this is less of a problem; see, also, the later discussion on p-values. In general, the power of the test for the hypotheses (5.5.6) is

$$\gamma(p) = P_p[S \leq k], \quad p < p_0. \qquad (5.5.8)$$

The curve labeled Test 1 in Figure 5.5.1 is the power function for the case $n = 20$, $p = 0.7$ and $\alpha = 0.1133$, (the R function which produced this plot is given in the appendix). Notice that the function is decreasing. The power is higher to detect the alternative $p = 0.2$ than $p = 0.6$. We will prove in general the monotonicity of the power function for binomial tests of these hypotheses in Section 8.2. It allows us to extend our test to the more general null hypothesis $H_0 : p \geq p_0$ rather than simply $H_0 : p = p_0$. Using the same decision rule as we used for the hypotheses (5.5.6), the definition of the size of a test (5.5.4) and the monotonicity of the power curve, we have

$$\max_{p \geq p_0} P_p[S \leq k] = P_{p_0}[S \leq k] = \alpha,$$

i.e., the same size as for the original null hypothesis.

Denote by Test 1 the test for the situation with $n = 20$, $p_0 = 0.70$, and size $\alpha = 0.1133$. Suppose we have a second test (Test 2) with an increased size. How

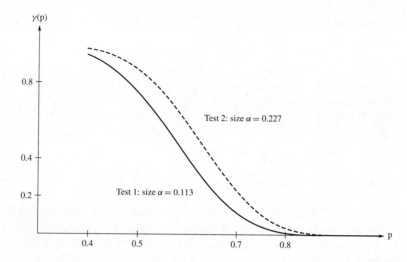

Figure 5.5.1: Power Curves for Tests 1 and 2; see Example 5.5.2

does the power function of Test 2 compare to Test 1? As an example, suppose for Test 2, we select $\alpha = 0.2277$. Hence, for Test 2, we reject H_0 if $S \leq 12$. Figure 5.5.1 displays the resulting power function. Note that while Test 2 has a higher probability of committing a Type I error it also has a higher power at each alternative. Exercise 5.5.7 shows this is true for these binomial tests. It is true in general; that is, if size of the test increases, power does too. ∎

Remark 5.5.1 (Nomenclature). Since in Example 5.5.2, the first null hypothesis $H_0 : p = p_0$ completely specifies the underlying distribution, it is called a **simple** hypothesis. Most hypotheses, such as $H_1 : p < p_0$, are **composite** hypotheses, because they are composed of many simple hypotheses and hence do not completely specify the distribution.

As we study more and more statistics, we find out that often other names are used for the size, α, of the critical region. Frequently, α is also called the **significance level** of the test associated with that critical region. Moreover, sometimes α is called the "maximum of probabilities of committing an error of Type I" and the "maximum of the power of the test when H_0 is true." It is disconcerting to the student to discover that there are so many names for the same thing. However, all of them are used in the statistical literature, and we feel obligated to point out this fact. ∎

The test in the last example is based on the exact distribution of its test statistic, i.e., the binomial distribution. Often we cannot obtain the distribution of the test statistic in closed form. But we can frequently appeal to the Central Limit Theorem to obtain an approximate test. Such is the case for the next example.

Example 5.5.3 (Large Sample Test for the Mean). Let X be a random variable with mean μ and finite variance σ^2. We want to test the hypotheses

$$H_0 : \mu = \mu_0 \text{ versus } H_1 : \mu > \mu_0, \tag{5.5.9}$$

where μ_0 is specified. To illustrate, suppose μ_0 is the mean level on a standardized test of students who have been taught a course by a standard method of teaching. Suppose it is hoped that a new method which incorporates computers will have a mean level $\mu > \mu_0$, where $\mu = E(X)$ and X is the score of a student taught by the new method. This conjecture will be tested by having n students (randomly selected) to be taught under this new method.

Let $X_1, \ldots, X_n$ be a random sample from the distribution of X and denote the sample mean and variance by $\overline{X}$ and S^2, respectively. Because $\overline{X} \to \mu$, in probability, an intuitive decision rule is given by

$$\text{Reject } H_0 \text{ in favor of } H_1 \text{ if } \overline{X} \text{ is much larger than } \mu_0. \tag{5.5.10}$$

In general, the distribution of the sample mean cannot be obtained in closed form. At the end of this section, under the strong assumption of normality for the distribution of X, we will obtain an exact test. For now, we will appeal to the Central Limit Theorem to find the critical region. We know from Section 4.4 that

$$\frac{\overline{X} - \mu}{S/\sqrt{n}} \xrightarrow{D} Z,$$

where Z has a standard normal distribution. Using this, we obtain a test with an approximate size α, if

$$\text{Reject } H_0 \text{ in favor of } H_1 \text{ if } \frac{\overline{X} - \mu_0}{S/\sqrt{n}} \geq z_\alpha. \tag{5.5.11}$$

The test is intuitive. To reject H_0, $\overline{X}$ must exceed μ_0 by at least $z_\alpha S/\sqrt{n}$. To approximate the power function of the test, we use the Central Limit Theorem. Upon substituting σ for S, it readily follows that the approximate power function is

$$
\begin{aligned}
\gamma(\mu) &= P_\mu(\overline{X} \geq \mu_0 + z_\alpha \sigma/\sqrt{n}) \\
&= P_\mu\left(\frac{\overline{X} - \mu}{\sigma/\sqrt{n}} \geq \frac{\mu_0 - \mu}{\sigma/\sqrt{n}} + z_\alpha\right) \\
&\doteq 1 - \Phi\left(z_\alpha + \frac{\sqrt{n}(\mu_0 - \mu)}{\sigma}\right) \\
&= \Phi\left(-z_\alpha - \frac{\sqrt{n}(\mu_0 - \mu)}{\sigma}\right).
\end{aligned}
\tag{5.5.12}
$$

So if we have some reasonable idea of what σ equals, we can compute the approximate power function. As Exercise 5.5.1 shows, this approximate power function is strictly increasing in μ, so as in the last example, we can change the null hypotheses to

$$H_0 : \mu \leq \mu_0 \text{ versus } H_1 : \mu > \mu_0. \tag{5.5.13}$$

Our asymptotic test will have approximate size α for these hypotheses. ∎

Example 5.5.4 (Test for μ under Normality). Let X have a $N(\mu, \sigma^2)$ distribution. As in Example 5.5.3, consider the hypotheses

$$H_0 : \mu = \mu_0 \text{ versus } H_1 : \mu > \mu_0, \qquad (5.5.14)$$

where μ_0 is specified. Assume that the desired size of the test is α, for $0 < \alpha < 1$, Suppose $X_1, \ldots, X_n$ is a random sample from a $N(\mu, \sigma^2)$ distribution. Let $\overline{X}$ and S^2 denote sample mean and variance, respectively. Our intuitive rejection rule is to reject H_0 in favor of H_A, if $\overline{X}$ is much larger than μ_0. Unlike Example 5.5.3, we now know the distribution of the statistic $\overline{X}$. In particular, by Part (d) of Theorem 3.6.1, under H_0 the statistic $T = (\overline{X} - \mu_0)/(S/\sqrt{n})$ has a t distribution with $n - 1$ degrees of freedom. Using the distribution of T, it is easy to show that the following rejection rule has exact level α:

$$\text{Reject } H_0 \text{ in favor of } H_1 \text{ if } T = \frac{\overline{X} - \mu_0}{S/\sqrt{n}} \geq t_{\alpha, n-1}, \qquad (5.5.15)$$

where $t_{\alpha, n-1}$ is the upper α critical point of a t distribution with $n - 1$ degrees of freedom; i.e., $\alpha = P(T > t_{\alpha, n-1})$. This is often called the t test of H_0 $\mu = \mu_0$.

Note the differences between this rejection rule and the large sample rule, (5.5.11). The large sample rule has approximate level α, while this has exact level α. Of course, we now have to assume that X has a normal distribution. In practice, we may not be willing to assume that the population is normal. In general, t critical values are larger than z critical values. Hence, the t test is conservative relative to the large sample test. So in practice, many statisticians often use the t test. ∎

Example 5.5.5 (Example 5.5.1 Continued). The data for Darwin's experiment on *zea mays* are recorded in Table 5.5.2. A boxplot and a normal $q-q$ plot of the 15 differences, $w_i = x_i - y_i$ are found in Figure 5.5.2. Based on these plots, we can see that there seem to be two outliers, Pots 2 and 15. In these two pots, the self-fertilized *zea mays* are much taller than the their cross-fertilized pairs. Except for these two outliers, the differences, $x_i - y_i$, are positive, indicating that the cross-fertilization leads to taller plants. We proceed to conduct a test of hypotheses (5.5.2), as discussed in Example 5.5.1. We will use the decision rule given by (5.5.15) with $\alpha = 0.05$. As Exercise 5.5.2 shows, the values of the sample mean and standard deviation for the differences, w_i, are: $\overline{w} = 2.62$ and $s_w = 4.72$. Hence, the t-test statistic is 2.15 which exceeds the t-critical value, $t_{.05,14} = 1.76$. Thus, we reject H_0 and conclude that on the average cross-fertilization of *zea mays* are on the average taller than self-fertilized *zea mays*. Because of the outliers, normality of the error distribution is somewhat dubious, and we use the test in a conservative manner as discussed at the end of Example 5.5.4. ∎

EXERCISES

5.5.1. Show that the approximate power function given in expression (5.5.12) of Example 5.5.3 is a strictly increasing function of μ. Show then that the test discussed in this example has approximate size α for testing

$$H_0 : \mu \leq \mu_0 \text{ versus } H_1 : \mu > \mu_0.$$

Table 5.5.2: Plant Growth

Pot	1	2	3	4	5	6	7	8
Cross	23.500	12.000	21.000	22.000	19.125	21.500	22.125	20.375
Self	17.375	20.375	20.000	20.000	18.375	18.625	18.625	15.250
Pot	9	10	11	12	13	14	15	
Cross	18.250	21.625	23.250	21.000	22.125	23.000	12.000	
Self	16.500	18.000	16.250	18.000	12.750	15.500	18.000	

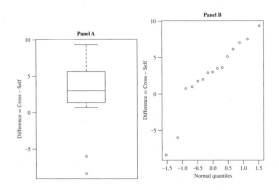

Figure 5.5.2: Boxplot and Normal $q-q$ Plot for the data of Example 5.2.4

5.5.2. For the Darwin data tabled in Example 5.5.5, verify that the Student t-test statistic is 2.15.

5.5.3. Let X have a pdf of the form $f(x; \theta) = \theta x^{\theta-1}$, $0 < x < 1$, zero elsewhere, where $\theta \in \{\theta : \theta = 1, 2\}$. To test the simple hypothesis $H_0 : \theta = 1$ against the alternative simple hypothesis $H_1 : \theta = 2$, use a random sample X_1, X_2 of size $n = 2$ and define the critical region to be $C = \{(x_1, x_2) : \frac{3}{4} \leq x_1 x_2\}$. Find the power function of the test.

5.5.4. Let X have a binomial distribution with the number of trials $n = 10$ and with p either $1/4$ or $1/2$. The simple hypothesis $H_0 : p = \frac{1}{2}$ is rejected, and the alternative simple hypothesis $H_1 : p = \frac{1}{4}$ is accepted, if the observed value of X_1, a random sample of size 1, is less than or equal to 3. Find the significance level and the power of the test.

5.5.5. Let X_1, X_2 be a random sample of size $n = 2$ from the distribution having pdf $f(x; \theta) = (1/\theta)e^{-x/\theta}$, $0 < x < \infty$, zero elsewhere. We reject $H_0 : \theta = 2$ and accept $H_1 : \theta = 1$ if the observed values of X_1, X_2, say x_1, x_2, are such that

$$\frac{f(x_1; 2)f(x_2; 2)}{f(x_1; 1)f(x_2; 1)} \leq \frac{1}{2}.$$

Here $\Omega = \{\theta : \theta = 1, 2\}$. Find the significance level of the test and the power of the test when H_0 is false.

5.5.6. Consider the tests Test 1 and Test 2 for the situation discussed in Example 5.5.2. Consider the test which rejects H_0 if $S \leq 10$. Find the level of significance for this test and sketch its power curve as in Figure 5.5.1.

5.5.7. Consider the situation described in Example 5.5.2. Suppose we have two tests A and B defined as follows. For Test A, H_0 is rejected if $S \leq k_A$ while for Test B, H_0 is rejected if $S \leq k_B$. If Test A has a higher level of significance than Test B, show that Test A has higher power than Test B at each alternative.

5.5.8. Let us say the life of a tire in miles, say X, is normally distributed with mean θ and standard deviation 5000. Past experience indicates that $\theta = 30,000$. The manufacturer claims that the tires made by a new process have mean $\theta > 30,000$. It is possible that $\theta = 35,000$. Check his claim by testing $H_0 : \theta = 30,000$ against $H_1 : \theta > 30,000$. We shall observe n independent values of X, say $x_1, \ldots, x_n$, and we shall reject H_0 (thus accept H_1) if and only if $\bar{x} \geq c$. Determine n and c so that the power function $\gamma(\theta)$ of the test has the values $\gamma(30,000) = 0.01$ and $\gamma(35,000) = 0.98$.

5.5.9. Let X have a Poisson distribution with mean θ. Consider the simple hypothesis $H_0 : \theta = \frac{1}{2}$ and the alternative composite hypothesis $H_1 : \theta < \frac{1}{2}$. Thus $\Omega = \{\theta : 0 < \theta \leq \frac{1}{2}\}$. Let $X_1, \ldots, X_{12}$ denote a random sample of size 12 from this distribution. We reject H_0 if and only if the observed value of $Y = X_1 + \cdots + X_{12} \leq 2$. If $\gamma(\theta)$ is the power function of the test, find the powers $\gamma(\frac{1}{2})$, $\gamma(\frac{1}{3})$, $\gamma(\frac{1}{4})$, $\gamma(\frac{1}{6})$, and $\gamma(\frac{1}{12})$. Sketch the graph of $\gamma(\theta)$. What is the significance level of the test?

5.5.10. Let Y have a binomial distribution with parameters n and p. We reject $H_0 : p = \frac{1}{2}$ and accept $H_1 : p > \frac{1}{2}$ if $Y \geq c$. Find n and c to give a power function $\gamma(p)$ which is such that $\gamma(\frac{1}{2}) = 0.10$ and $\gamma(\frac{2}{3}) = 0.95$, approximately.

5.5.11. Let $Y_1 < Y_2 < Y_3 < Y_4$ be the order statistics of a random sample of size $n = 4$ from a distribution with pdf $f(x; \theta) = 1/\theta$, $0 < x < \theta$, zero elsewhere, where $0 < \theta$. The hypothesis $H_0 : \theta = 1$ is rejected and $H_1 : \theta > 1$ is accepted if the observed $Y_4 \geq c$.

(a) Find the constant c so that the significance level is $\alpha = 0.05$.

(b) Determine the power function of the test.

5.5.12. Let $X_1, X_2, \ldots, X_8$ be a random sample of size $n = 8$ from a Poisson distribution with mean μ. Reject the simple null hypothesis $H_0 : \mu = 0.5$ and accept $H_1 : \mu > 0.5$ if the observed sum $\sum_{i=1}^{8} x_i \geq 8$.

(a) Compute the significance level α of the test.

(b) Find the power function $\gamma(\mu)$ of the test as a sum of Poisson probabilities.

(c) Using the Appendix, determine $\gamma(0.75)$, $\gamma(1)$, and $\gamma(1.25)$.

5.5.13. Let p denote the probability that, for a particular tennis player, the first serve is good. Since $p = 0.40$, this player decided to take lessons in order to increase p. When the lessons are completed, the hypothesis $H_0 : p = 0.40$ will be tested against $H_1 : p > 0.40$ based on $n = 25$ trials. Let y equal the number of first serves that are good, and let the critical region be defined by $C = \{y : y \geq 13\}$.

(a) Determine $\alpha = P(Y \geq 13; ; p = 0.40)$.

(b) Find $\beta = P(Y < 13)$ when $p = 0.60$; that is, $\beta = P(Y \leq 12; \ p = 0.60)$ so that $1 - \beta$ is the power at $p = 0.60$.

5.6   Additional Comments About Statistical Tests

All of the alternative hypotheses considered in Section 5.5 were *one-sided hypotheses*. For illustration, in Exercise 5.5.8 we tested $H_0 : \mu = 30,000$ against the one-sided alternative $H_1 : \mu > 30,000$, where μ is the mean of a normal distribution having standard deviation $\sigma = 5000$. Perhaps in this situation, though, we think the manufacturer's process has changed but are unsure of the direction. That is, we are interested in the alternative $H_1 : \ \mu \neq 30,000$. In this section, we further explore hypotheses testing and we begin with the construction of a test for a two sided alternative involving the mean of a random variable.

Example 5.6.1 (Large Sample Two-Sided Test for the Mean). In order to see how to construct a test for a two sided alternative, reconsider Example 5.5.3, where we constructed a large sample one-sided test for the mean of a random variable. As in Example 5.5.3, let X be a random variable with mean μ and finite variance σ^2. Here, though, we want to test

$$H_0 : \ \mu = \mu_0 \text{ versus } H_1 : \ \mu \neq \mu_0, \tag{5.6.1}$$

where μ_0 is specified. Let $X_1, \ldots, X_n$ be a random sample from the distribution of X and denote the sample mean and variance by $\overline{X}$ and S^2, respectively. For the one-sided test, we rejected H_0 if $\overline{X}$ was too large; hence, for the hypotheses (5.6.1), we use the decision rule

$$\text{Reject } H_0 \text{ in favor of } H_1 \text{ if } \overline{X} \leq h \text{ or } \overline{X} \geq k, \tag{5.6.2}$$

where h and k are such that $\alpha = P_{H_0}[\overline{X} \leq h \text{ or } \overline{X} \geq k]$. Clearly $h < k$, hence, we have

$$\alpha = P_{H_0}[\overline{X} \leq h \text{ or } \overline{X} \geq k] = P_{H_0}[\overline{X} \leq h] + P_{H_0}[\overline{X} \geq k].$$

Because, at least asymptotically, the distribution of $\overline{X}$ is symmetrically distributed about μ_0, under H_0, an intuitive rule is to divide α equally between the two terms on the right-side of the above expression; that is, h and k are chosen by

$$P_{H_0}[\overline{X} \leq h] = \alpha/2 \text{ and } P_{H_0}[\overline{X} \geq k] = \alpha/2. \tag{5.6.3}$$

By the Central Limit Theorem (Theorem 4.4.1) and the consistency of S^2 to σ^2, we have under H_0 that $(\overline{X} - \mu_0)/(S/\sqrt{n}) \xrightarrow{D} N(0,1)$. This and (5.6.3) leads to the approximate decision rule:

$$\text{Reject } H_0 \text{ in favor of } H_1 \text{ if } \left| \frac{\overline{X} - \mu_0}{S/\sqrt{n}} \right| \geq z_{\alpha/2}. \qquad (5.6.4)$$

To approximate the power function of the test, we use the Central Limit Theorem. Upon substituting σ for S, it readily follows that the approximate power function is

$$\begin{aligned}
\gamma(\mu) &= P_\mu(\overline{X} \leq \mu_0 - z_{\alpha/2}\sigma/\sqrt{n}) + P_\mu(\overline{X} \geq \mu_0 + z_{\alpha/2}\sigma/\sqrt{n}) \\
&= \Phi\left(\frac{\sqrt{n}(\mu_0 - \mu)}{\sigma} - z_{\alpha/2} \right) + 1 - \Phi\left(\frac{\sqrt{n}(\mu_0 - \mu)}{\sigma} + z_{\alpha/2} \right), \quad (5.6.5)
\end{aligned}$$

where $\Phi(z)$ is the cdf of a standard normal random variable; see (3.4.10). So if we have some reasonable idea of what σ equals, we can compute the approximate power function. Note that the derivative of the power function is

$$\gamma'(\mu) = \frac{\sqrt{n}}{\sigma}\left[\phi\left(\frac{\sqrt{n}(\mu_0 - \mu)}{\sigma} + z_{\alpha/2} \right) - \phi\left(\frac{\sqrt{n}(\mu_0 - \mu)}{\sigma} - z_{\alpha/2} \right) \right], \qquad (5.6.6)$$

where $\phi(z)$ is the pdf of a standard normal random variable. Note that $\gamma(\mu)$ has a critical value at μ_0. As Exercise 5.6.2, this gives the minimum of $\gamma(\mu)$. Further, $\gamma(\mu)$ is strictly decreasing for $\mu < \mu_0$ and strictly increasing for $\mu > \mu_0$. ∎

Consider again the situation at the beginning of this section. Suppose we want to test

$$H_0 : \mu = 30,000 \text{ versus } H_1 : \mu \neq 30,000. \qquad (5.6.7)$$

Suppose $n = 20$ and $\alpha = 0.01$. Then the rejection rule (5.6.4) becomes

$$\text{Reject } H_0 \text{ in favor of } H_1 \text{ if } \left| \frac{\overline{X} - 30,000}{S/\sqrt{20}} \right| \geq 2.575. \qquad (5.6.8)$$

Figure 5.6.1 shows the power curve for this test when $\sigma = 5,000$, as in Exercise 5.5.8, is substituted in for S. For comparison, the power curve for for the test with level $\alpha = .05$ is also shown; see Exercise 5.6.1.

The two sided test for the mean is approximate. If we assume that X has a normal distribution, then as Exercise 5.6.3 shows the following test has exact size α for testing $H_0 : \mu = \mu_0$ versus $H_A : \mu \neq \mu_0$:

$$\text{Reject } H_0 \text{ in favor of } H_1 \text{ if } \left| \frac{\overline{X} - \mu_0}{S/\sqrt{n}} \right| \geq t_{\alpha/2, n-1}. \qquad (5.6.9)$$

It too has a bowl shaped power curve similar to Figure 5.6.1; although, it is not as easy to show, see Lehmann (1986).

There exists a relationship between two sided tests and confidence intervals. Consider the two sided t-test (5.6.9). Here, we use the rejection rule absolutely (if and only if instead of if). Hence, in terms of acceptance we have

$$\text{Accept } H_0 \text{ if and only if } \mu_0 - t_{\alpha/2, n-1}S/\sqrt{n} < \overline{X} < \mu_0 + t_{\alpha/2, n-1}S/\sqrt{n}.$$

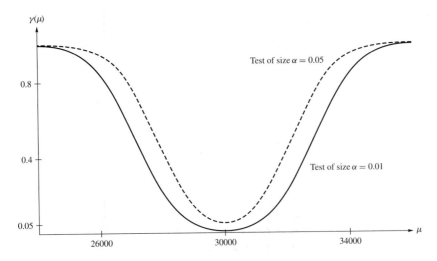

Figure 5.6.1: Power Curves for the Tests of Hypotheses 5.6.7

But this is easily shown to be

Accept H_0 if and only if $\mu_0 \in (\overline{X} - t_{\alpha/2,n-1}S/\sqrt{n}, \overline{X} + t_{\alpha/2,n-1}S/\sqrt{n})$,   (5.6.10)

that is, we accept H_0 at significance level α if and only if μ_0 is in the $(1-\alpha)100\%$ confidence interval for μ. Equivalently, we reject H_0 at significance level α if and only if μ_0 is not in the $(1-\alpha)100\%$ confidence interval for μ. This will be true for all the two sided tests and hypotheses discussed in this text. There is also a similar relationship between one sided tests and one sided confidence intervals.

Once we recognize this relationship between confidence intervals and tests of hypothesis, we can use all those statistics that we used to construct confidence intervals to test hypotheses, not only against two-sided alternatives but one-sided ones as well. Without listing all of these in a table, we present enough of them so that the principle can be understood.

Example 5.6.2. Let independent random samples be taken from $N(\mu_1, \sigma^2)$ and $N(\mu_2, \sigma^2)$, respectively. Say these have the respective sample characteristics n_1, $\overline{X}$, S_1^2 and n_2, $\overline{Y}$, S_2^2. Let $n = n_1 + n_2$ denote the combined sample size and let $S_p^2 = [(n_1-1)S_1^2 + (n_2-1)S_2^2]/(n-2)$, (5.4.19), be the pooled estimator of the common variance. At $\alpha = 0.05$, reject $H_0 : \mu_1 = \mu_2$ and accept the one-sided alternative $H_1 : \mu_1 > \mu_2$ if

$$T = \frac{\overline{X} - \overline{Y} - 0}{S_p\sqrt{\frac{1}{n_1} + \frac{1}{n_2}}} \geq t_{.05,n-2}.$$

■ A rigorous development of this test is given in Example 8.3.1. ■

Example 5.6.3. Say X is $b(1, p)$. Consider testing $H_0 : p = p_0$ against $H_1 : p < p_0$. Let $X_1 \ldots, X_n$ be a random sample from the distribution of X and let $\widehat{p} = \overline{X}$. To test H_0 versus H_A, we use either

$$Z_1 = \frac{\widehat{p} - p_0}{\sqrt{p_0(1 - p_0)/n}} \le c \quad \text{or} \quad Z_2 = \frac{\widehat{p} - p_0}{\sqrt{\widehat{p}(1 - \widehat{p})/n}} \le c.$$

If n is large, both Z_1 and Z_2 have approximate standard normal distributions provided that $H_0 : p = p_0$ is true. Hence, if c is set at -1.645 then the approximate significance level is $\alpha = 0.05$. Some statisticians use Z_1 and others Z_2. We do not have strong preferences one way or the other because the two methods provide about the same numerical results. As one might suspect, using Z_1 provides better probabilities for power calculations if the true p is close to p_0 while Z_2 is better if H_0 is clearly false. However, with a two-sided alternative hypothesis, Z_2 does provide a better relationship with the confidence interval for p. That is, $|Z_2| < z_{\alpha/2}$ is equivalent to p_0 being in the interval from

$$\widehat{p} - z_{\alpha/2}\sqrt{\frac{\widehat{p}(1 - \widehat{p})}{n}} \quad \text{to} \quad \widehat{p} + z_{\alpha/2}\sqrt{\frac{\widehat{p}(1 - \widehat{p})}{n}},$$

which is the interval that provides a $(1 - \alpha)100\%$ approximate confidence interval for p as considered in Section 5.5. ∎

In closing this section, we introduce the concepts of *randomized tests* and *p-values* through an example and remarks that follow the example.

Example 5.6.4. Let $X_1, X_2, \ldots, X_{10}$ be a random sample of size $n = 10$ from a Poisson distribution with mean θ. A critical region for testing $H_0 : \theta = 0.1$ against $H_1 : \theta > 0.1$ is given by $Y = \sum_1^{10} X_i \ge 3$. The statistic Y has a Poisson distribution with mean 10θ. Thus, with $\theta = 0.1$ so that the mean of Y is 1, the significance level of the test is

$$P(Y \ge 3) = 1 - P(Y \le 2) = 1 - 0.920 = 0.080.$$

If the critical region defined by $\sum_1^{10} x_i \ge 4$ is used, the significance level is

$$\alpha = P(Y \ge 4) = 1 - P(Y \le 3) = 1 - 0.981 = 0.019.$$

For instance, if a significance level of about $\alpha = 0.05$, say, is desired, most statisticians would use one of these tests; that is, they would adjust the significance level to that of one of these convenient tests. However, a significance level of $\alpha = 0.05$ can be achieved in the following way. Let W have a Bernoulli distribution with probability of success equal to

$$P(W = 1) = \frac{0.050 - 0.019}{0.080 - 0.019} = \frac{31}{61}.$$

Assume that W is selected independently of the sample. Consider the rejection rule

Reject H_0 if $\sum_1^{10} x_i \ge 4$ of if $\sum_1^{10} x_i = 3$ and $W = 1$.

The significance level of this rule is

$$P_{H_0}(Y \geq 4) + P_{H_0}(\{Y = 3\} \cap \{W = 1\}) = P_{H_0}(Y \geq 4)$$
$$+ P_{H_0}(Y = 3)P(W = 1)$$
$$= 0.019 + 0.061\frac{31}{61} = 0.05;$$

hence, the decision rule has exactly level 0.05. The process of performing the auxiliary experiment to decide whether to reject or not when $Y = 3$ is sometimes referred to as a **randomized test**. ∎

Remark 5.6.1 (Observed Significance Level). Not many statisticians like randomized tests in practice, because the use of them means that two statisticians could make the same assumptions, observe the same data, apply the same test, and yet make different decisions. Hence they usually adjust their significance level so as not to randomize. As a matter of fact, many statisticians report what are commonly called **observed significance level** or p-**values** (for *probability values*). For illustration, if in Example 5.6.4 the observed Y is $y = 4$, the p-value is 0.019; and if it is $y = 3$, the p-value is 0.080. That is, the p-value is the observed "tail" probability of a statistic being at least as extreme as the particular observed value when H_0 is true. Hence, more generally, if $Y = u(X_1, X_2, \ldots, X_n)$ is the statistic to be used in a test of H_0 and if the critical region is of the form

$$u(x_1, x_2, \ldots, x_n) \leq c,$$

an observed value $u(x_1, x_2, \ldots, x_n) = d$ would mean that the

$$p\text{-value} = P(Y \leq d;\ H_0).$$

That is, if $G(y)$ is the distribution function of $Y = u(X_1, X_2, \ldots, X_n)$, provided that H_0 is true, the p-value is equal to $G(d)$ in this case. However, $G(Y)$, in the continuous case, is uniformly distributed on the unit interval, so an observed value $G(d) \leq 0.05$ would be equivalent to selecting c, so that

$$P[u(X_1, X_2, \ldots, X_n) \leq c;\ H_0] = 0.05$$

and observing that $d \leq c$. Most computer programs automatically print out the p-value of a test. ∎

Example 5.6.5. Let $X_1, X_2, \ldots, X_{25}$ be a random sample from $N(\mu, \sigma^2 = 4)$. To test $H_0 : \mu = 77$ against the one-sided alternative hypothesis $H_1 : \mu < 77$, say we observe the 25 values and determine that $\bar{x} = 76.1$. The variance of $\overline{X}$ is $\sigma^2/n = 4/25 = 0.16$; so we know that $Z = (\overline{X} - 77)/0.4$ is $N(0, 1)$ provided that $\mu = 77$. Since the observed value of this test statistic is $z = (76.1 - 77)/0.4 = -2.25$, the p-value of the test is $\Phi(-2.25) = 1 - 0.988 = 0.012$. Accordingly, if we were using a significance level of $\alpha = 0.05$, we would reject H_0 and accept $H_1 : \mu < 77$ because $0.012 < 0.05$. ∎

EXERCISES

5.6.1. For the test at level 0.05 of the hypotheses given by (5.6.1) with $\mu_0 = 30,000$ and $n = 20$. Obtain the power function, (use $\sigma = 5,000$). Evaluate the power function for the following values: $\mu = 25,000; 27,500; 30,000; 32,500;$ and $35,000$. Then sketch this power function and see if it agrees with Figure 5.6.1.

5.6.2. Consider the power function $\gamma(\mu)$ and its derivative $\gamma'(\mu)$ given by (5.6.5) and (5.6.6). Show that $\gamma'(\mu)$ is strictly negative for $\mu < \mu_0$ and strictly positive for $\mu > \mu_0$.

5.6.3. Show that the test defined by 5.6.9 has exact size α for testing $H_0 : \mu = \mu_0$ versus $H_A : \mu \neq \mu_0$:

5.6.4. Consider the one-sided t-test for $H_0 : \mu = \mu_0$ versus $H_{A1} : \mu > \mu_0$ constructed in Example 5.5.4 and the two-sided t-test for t-test for $H_0 : \mu = \mu_0$ versus $H_{A1} : \mu \neq \mu_0$ given in (5.6.9). Assume that both tests are of size α. Show that for $\mu > \mu_0$, the power function of the one sided test is larger than the power function of the two sided test.

5.6.5. Assume that the weight of cereal in a "10-ounce box" is $N(\mu, \sigma^2)$. To test $H_0 : \mu = 10.1$ against $H_1 : \mu > 10.1$, we take a random sample of size $n = 16$ and observe that $\bar{x} = 10.4$ and $s = 0.4$.

(a) Do we accept or reject H_0 at the 5 percent significance level?

(b) What is the approximate p-value of this test?

5.6.6. Each of 51 golfers hit three golf balls of brand X and three golf balls of brand Y in a random order. Let X_i and Y_i equal the averages of the distances traveled by the brand X and brand Y golf balls hit by the ith golfer, $i = 1, 2, \ldots, 51$. Let $W_i = X_i - Y_i$, $i = 1, 2, \ldots, 51$. To test $H_0 : \mu_W = 0$ against $H_1 : \mu_W > 0$, where μ_W is the mean of the differences. If $\bar{w} = 2.07$ and $s_W^2 = 84.63$, would H_0 be accepted or rejected at an $\alpha = 0.05$ significance level? What is the p-value of this test?

5.6.7. Among the data collected for the World Health Organization air quality monitoring project is a measure of suspended particles in $\mu g/m^3$. Let X and Y equal the concentration of suspended particles in $\mu g/m^3$ in the city center (commercial district) for Melbourne and Houston, respectively. Using $n = 13$ observations of X and $m = 16$ observations of Y, we shall test $H_0 : \mu_X = \mu_Y$ against $H_1 : \mu_X < \mu_Y$.

(a) Define the test statistic and critical region, assuming that the unknown variances are equal. Let $\alpha = 0.05$.

(b) If $\bar{x} = 72.9$, $s_x = 25.6$, $\bar{y} = 81.7$, and $s_y = 28.3$, calculate the value of the test statistic and state your conclusion.

5.6.8. Let p equal the proportion of drivers who use a seat belt in a state that does not have a mandatory seat belt law. It was claimed that $p = 0.14$. An advertising campaign was conducted to increase this proportion. Two months after the campaign, $y = 104$ out of a random sample of $n = 590$ drivers were wearing their seat belts. Was the campaign successful?

(a) Define the null and alternative hypotheses.

(b) Define a critical region with an $\alpha = 0.01$ significance level.

(c) Determine the approximate p-value and state your conclusion.

5.6.9. In Exercise 5.4.14 we found a confidence interval for the variance σ^2 using the variance S^2 of a random sample of size n arising from $N(\mu, \sigma^2)$, where the mean μ is unknown. In testing $H_0 : \sigma^2 = \sigma_0^2$ against $H_1 : \sigma^2 > \sigma_0^2$, use the critical region defined by $(n-1)S^2/\sigma_0^2 \geq c$. That is, reject H_0 and accept H_1 if $S^2 \geq c\sigma_0^2/(n-1)$. If $n = 13$ and the significance level $\alpha = 0.025$, determine c.

5.6.10. In Exercise 5.4.25, in finding a confidence interval for the ratio of the variances of two normal distributions, we used a statistic S_1^2/S_2^2, which has an F-distribution when those two variances are equal. If we denote that statistic by F, we can test $H_0 : \sigma_1^2 = \sigma_2^2$ against $H_1 : \sigma_1^2 > \sigma_2^2$ using the critical region $F \geq c$. If $n = 13$, $m = 11$, and $\alpha = 0.05$, find c.

5.7 Chi-Square Tests

In this section we introduce tests of statistical hypotheses called *chi-square tests*. A test of this sort was originally proposed by Karl Pearson in 1900, and it provided one of the earlier methods of statistical inference.

Let the random variable X_i be $N(\mu_i, \sigma_i^2)$, $i = 1, 2, \ldots, n$, and let $X_1, X_2, \ldots, X_n$ be mutually independent. Thus the joint pdf of these variables is

$$\frac{1}{\sigma_1 \sigma_2 \cdots \sigma_n (2\pi)^{n/2}} \exp\left[-\frac{1}{2} \sum_1^n \left(\frac{x_i - \mu_i}{\sigma_i} \right)^2 \right], \quad -\infty < x_i < \infty.$$

The random variable that is defined by the exponent (apart from the coefficient $-\frac{1}{2}$) is $\sum_1^n (X_i - \mu_i)^2/\sigma_i^2$, and this random variable has a $\chi^2(n)$ distribution. In Section 3.5 we generalized this joint normal distribution of probability to n random variables that are *dependent* and we call the distribution a *multivariate normal distribution*. In Section 9.8, it will be shown that a certain exponent in the joint pdf (apart from a coefficient of $-\frac{1}{2}$) defines a random variable that is $\chi^2(n)$. This fact is the mathematical basis of the chi-square tests.

Let us now discuss some random variables that have approximate chi-square distributions. Let X_1 be $b(n, p_1)$. Consider the random variable

$$Y = \frac{X_1 - np_1}{\sqrt{np_1(1 - p_1)}}$$

which has, as $n \to \infty$, a limiting distribution that is $N(0, 1)$, we would strongly suspect that the limiting distribution of $Z = Y^2$ is $\chi^2(1)$. This is guaranteed by Theorem 4.3.4, but, as we show, it readily follows. If $G_n(y)$ represents the distribution function of Y, we know that

$$\lim_{n \to \infty} G_n(y) = \Phi(y), \quad -\infty < y < \infty,$$

where $\Phi(y)$ is the distribution function of a distribution that is $N(0, 1)$. For each positive integer n, let $H_n(z)$ represent the distribution function of $Z = Y^2$. Thus, if $z \geq 0$,

$$H_n(z) = P(Z \leq z) = P(-\sqrt{z} \leq Y \leq \sqrt{z}) = G_n(\sqrt{z}) - G_n[-\sqrt{z}].$$

Accordingly, since $\Phi(y)$ is continuous everywhere,

$$\lim_{n \to \infty} H_n(z) = \Phi(\sqrt{z}) - \Phi(-\sqrt{z}) = 2 \int_0^{\sqrt{z}} \frac{1}{\sqrt{2\pi}} e^{-w^2/2} \, dw.$$

If we change the variable of integration in this last integral by writing $w^2 = v$, then

$$\lim_{n \to \infty} H_n(z) = \int_0^z \frac{1}{\Gamma(\frac{1}{2})2^{1/2}} v^{1/2-1} e^{-v/2} \, dv,$$

provided that $z \geq 0$. If $z < 0$, then $\lim_{n \to \infty} H_n(z) = 0$. Thus $\lim_{n \to \infty} H_n(z)$ is equal to the distribution function of a random variable that is $\chi^2(1)$. This is the desired result.

Let us now return to the random variable X_1 which is $b(n, p_1)$. Let $X_2 = n - X_1$ and let $p_2 = 1 - p_1$. If we denote Y^2 by Q_1 instead of Z, we see that Q_1 may be written as

$$
\begin{aligned}
Q_1 = \frac{(X_1 - np_1)^2}{np_1(1 - p_1)} &= \frac{(X_1 - np_1)^2}{np_1} + \frac{(X_1 - np_1)^2}{n(1 - p_1)} \\
&= \frac{(X_1 - np_1)^2}{np_1} + \frac{(X_2 - np_2)^2}{np_2}
\end{aligned}
$$

because $(X_1 - np_1)^2 = (n - X_2 - n + np_2)^2 = (X_2 - np_2)^2$. Since Q_1 has a limiting chi-square distribution with 1 degree of freedom, we say, when n is a positive integer, that Q_1 has an approximate chi-square distribution with 1 degree of freedom. This result can be generalized as follows.

Let $X_1, X_2, \ldots, X_{k-1}$ have a multinomial distribution with the parameters n and $p_1, \ldots, p_{k-1}$, as in Section 3.1. Let $X_k = n - (X_1 + \cdots + X_{k-1})$ and let $p_k = 1 - (p_1 + \cdots + p_{k-1})$. Define Q_{k-1} by

$$Q_{k-1} = \sum_{i=1}^k \frac{(X_i - np_i)^2}{np_i}.$$

It is proved in a more advanced course that, as $n \to \infty$, Q_{k-1} has a limiting distribution that is $\chi^2(k - 1)$. If we accept this fact, we can say that Q_{k-1} has an approximate chi-square distribution with $k - 1$ degrees of freedom when n is a positive integer. Some writers caution the user of this approximation to be certain that n is large enough so that each np_i, $i = 1, 2, \ldots, k$, is at least equal to 5. In any case it is important to realize that Q_{k-1} does not have a chi-square distribution, only an approximate chi-square distribution.

The random variable Q_{k-1} may serve as the basis of the tests of certain statistical hypotheses which we now discuss. Let the sample space $\mathcal{A}$ of a random experiment be the union of a finite number k of mutually disjoint sets $A_1, A_2, \ldots, A_k$. Furthermore, let $P(A_i) = p_i$, $i = 1, 2, \ldots, k$, where $p_k = 1 - p_1 - \cdots - p_{k-1}$, so that p_i is the probability that the outcome of the random experiment is an element of the set A_i. The random experiment is to be repeated n independent times and X_i will represent the number of times the outcome is an element of set A_i. That is, $X_1, X_2, \ldots, X_k = n - X_1 - \cdots - X_{k-1}$ are the frequencies with which the outcome is, respectively, an element of $A_1, A_2, \ldots, A_k$. Then the joint pmf of $X_1, X_2, \ldots, X_{k-1}$ is the multinomial pmf with the parameters n, $p_1, \ldots, p_{k-1}$. Consider the simple hypothesis (concerning this multinomial pmf) $H_0 : p_1 = p_{10}$, $p_2 = p_{20}, \ldots, p_{k-1} = p_{k-1,0}$ $(p_k = p_{k0} = 1 - p_{10} - \cdots - p_{k-1,0})$, where $p_{10}, \ldots, p_{k-1,0}$ are specified numbers. It is desired to test H_0 against all alternatives.

If the hypothesis H_0 is true, the random variable

$$Q_{k-1} = \sum_{1}^{k} \frac{(X_i - np_{i0})^2}{np_{i0}}$$

has an approximate chi-square distribution with $k - 1$ degrees of freedom. Since, when H_0 is true, np_{i0} is the expected value of X_i, one would feel intuitively that observed values of Q_{k-1} should not be too large if H_0 is true. With this in mind, we may use Table II of Appendix B, with $k - 1$ degrees of freedom, and find c so that $P(Q_{k-1} \geq c) = \alpha$, where α is the desired significance level of the test. If, then, the hypothesis H_0 is rejected when the observed value of Q_{k-1} is at least as great as c, the test of H_0 will have a significance level that is approximately equal to α. This is frequently called a *goodness of fit test*.

Some illustrative examples follow.

Example 5.7.1. One of the first six positive integers is to be chosen by a random experiment (perhaps by the cast of a die). Let $A_i = \{x : x = i\}$, $i = 1, 2, \ldots, 6$. The hypothesis $H_0 : P(A_i) = p_{i0} = \frac{1}{6}$, $i = 1, 2, \ldots, 6$, will be tested, at the approximate 5 percent significance level, against all alternatives. To make the test, the random experiment will be repeated under the same conditions, 60 independent times. In this example $k = 6$ and $np_{i0} = 60(\frac{1}{6}) = 10$, $i = 1, 2, \ldots, 6$. Let X_i denote the frequency with which the random experiment terminates with the outcome in A_i, $i = 1, 2, \ldots, 6$, and let $Q_5 = \sum_{1}^{6} (X_i - 10)^2/10$. If H_0 is true, Table II, with $k - 1 = 6 - 1 = 5$ degrees of freedom, shows that we have $P(Q_5 \geq 11.1) = 0.05$. Now suppose that the experiment frequencies of $A_1, A_2, \ldots, A_6$ are, respectively, 13, 19, 11, 8, 5, and 4. The observed value of Q_5 is

$$\frac{(13 - 10)^2}{10} + \frac{(19 - 10)^2}{10} + \frac{(11 - 10)^2}{10} + \frac{(8 - 10)^2}{10} + \frac{(5 - 10)^2}{10} + \frac{(4 - 10)^2}{10} = 15.6.$$

Since $15.6 > 11.1$, the hypothesis $P(A_i) = \frac{1}{6}$, $i = 1, 2, \ldots, 6$, is rejected at the (approximate) 5 percent significance level. ∎

Example 5.7.2. A point is to be selected from the unit interval $\{x : 0 < x < 1\}$ by a random process. Let $A_1 = \{x : 0 < x \leq \frac{1}{4}\}$, $A_2 = \{x : \frac{1}{4} < x \leq \frac{1}{2}\}$, $A_3 = \{x : \frac{1}{2} < x \leq \frac{3}{4}\}$, and $A_4 = \{x : \frac{3}{4} < x < 1\}$. Let the probabilities p_i, $i = 1, 2, 3, 4$, assigned to these sets under the hypothesis be determined by the pdf $2x$, $0 < x < 1$, zero elsewhere. Then these probabilities are, respectively,

$$p_{10} = \int_0^{1/4} 2x\,dx = \frac{1}{16}, \quad p_{20} = \frac{3}{16}, \quad p_{30} = \frac{5}{16}, \quad p_{40} = \frac{7}{16}.$$

Thus the hypothesis to be tested is that p_1, p_2, p_3, and $p_4 = 1 - p_1 - p_2 - p_3$ have the preceding values in a multinomial distribution with $k = 4$. This hypothesis is to be tested at an approximate 0.025 significance level by repeating the random experiment $n = 80$ independent times under the same conditions. Here the np_{i0} for $i = 1, 2, 3, 4$, are, respectively, 5, 15, 25, and 35. Suppose the observed frequencies of A_1, A_2, A_3, and A_4 are 6, 18, 20, and 36, respectively. Then the observed value of $Q_3 = \sum_1^4 (X_i - np_{i0})^2/(np_{i0})$ is

$$\frac{(6-5)^2}{5} + \frac{(18-15)^2}{15} + \frac{(20-25)^2}{25} + \frac{(36-35)^2}{35} = \frac{64}{35} = 1.83,$$

approximately. From Table II, with $4 - 1 = 3$ degrees of freedom, the value corresponding to a 0.025 significance level is $c = 9.35$. Since the observed value of Q_3 is less than 9.35, the hypothesis is accepted at the (approximate) 0.025 level of significance. ∎

Thus far we have used the chi-square test when the hypothesis H_0 is a simple hypothesis. More often we encounter hypotheses H_0 in which the multinomial probabilities $p_1, p_2, \ldots, p_k$ are not completely specified by the hypothesis H_0. That is, under H_0, these probabilities are functions of unknown parameters. For an illustration, suppose that a certain random variable Y can take on any real value. Let us partition the space $\{y : -\infty < y < \infty\}$ into k mutually disjoint sets $A_1, A_2, \ldots, A_k$ so that the events $A_1, A_2, \ldots, A_k$ are mutually exclusive and exhaustive. Let H_0 be the hypothesis that Y is $N(\mu, \sigma^2)$ with μ and σ^2 unspecified. Then each

$$p_i = \int_{A_i} \frac{1}{\sqrt{2\pi}\sigma} \exp[-(y - \mu)^2/2\sigma^2]\,dy, \quad i = 1, 2, \ldots, k,$$

is a function of the unknown parameters μ and σ^2. Suppose that we take a random sample $Y_1, \ldots, Y_n$ of size n from this distribution. If we let X_i denote the frequency of A_i, $i = 1, 2, \ldots, k$, so that $X_1 + X_2 + \cdots + X_k = n$, the random variable

$$Q_{k-1} = \sum_{i=1}^k \frac{(X_i - np_i)^2}{np_i}$$

cannot be computed once $X_1, \ldots, X_k$ have been observed, since each p_i, and hence Q_{k-1}, is a function of μ and σ^2. Accordingly, choose the values of μ and σ^2 that minimize Q_{k-1}. These values depend upon the observed $X_1 = x_1, \ldots, X_k = x_k$ and are called *minimum chi-square estimates* of μ and σ^2. These point estimates of μ and σ^2 enable us to compute numerically the estimates of each p_i. Accordingly, if these values are used, Q_{k-1} can be computed once $Y_1, Y_2, \ldots, Y_n$, and hence $X_1, X_2, \ldots, X_k$, are observed. However, a very important aspect of the fact, which we accept without proof, is that now Q_{k-1} is approximately $\chi^2(k-3)$. That is, the number of degrees of freedom of the limiting chi-square distribution of Q_{k-1} is reduced by one for each parameter estimated by the observed data. This statement applies not only to the problem at hand but also to more general situations. Two examples will now be given. The first of these examples will deal with the test of the hypothesis that two multinomial distributions are the same.

Remark 5.7.1. In many instances, such as that involving the mean μ and the variance σ^2 of a normal distribution, minimum chi-square estimates are difficult to compute. Hence other estimates, such as the maximum likelihood estimates $\hat{\mu} = \overline{Y}$ and $\widehat{\sigma^2} = V = (n-1)S^2/n$, are used to evaluate p_i and Q_{k-1}. In general, Q_{k-1} is not minimized by maximum likelihood estimates, and thus its computed value is somewhat greater than it would be if minimum chi-square estimates are used. Hence, when comparing it to a critical value listed in the chi-square table with $k-3$ degrees of freedom, there is a greater chance of rejection than there would be if the actual minimum of Q_{k-1} is used. Accordingly, the approximate significance level of such a test will be somewhat higher than that value found in the table. This modification should be kept in mind and, if at all possible, each p_i should be estimated using the frequencies $X_1, \ldots, X_k$ rather than directly using the observations $Y_1, Y_2, \ldots, Y_n$ of the random sample. ∎

Example 5.7.3. In this example, we consider two multinomial distributions with parameters $n_j, p_{1j}, p_{2j}, \ldots, p_{kj}$ and $j = 1, 2$, respectively. Let X_{ij}, $i = 1, 2, \ldots, k$, $j = 1, 2$, represent the corresponding frequencies. If n_1 and n_2 are large and the observations from one distribution are independent of those from the other, the random variable

$$\sum_{j=1}^{2} \sum_{i=1}^{k} \frac{(X_{ij} - n_j p_{ij})^2}{n_j p_{ij}}$$

is the sum of two independent random variables each of which we treat as though it were $\chi^2(k-1)$; that is, the random variable is approximately $\chi^2(2k-2)$. Consider the hypothesis

$$H_0 : p_{11} = p_{12}, p_{21} = p_{22}, \ldots, p_{k1} = p_{k2},$$

where each $p_{i1} = p_{i2}$, $i = 1, 2, \ldots, k$, is unspecified. Thus we need point estimates of these parameters. The maximum likelihood estimator of $p_{i1} = p_{i2}$, based upon the frequencies X_{ij}, is $(X_{i1} + X_{i2})/(n_1 + n_2)$, $i = 1, 2, \ldots, k$. Note that we need only $k-1$ point estimates, because we have a point estimate of $p_{k1} = p_{k2}$ once we have point estimates of the first $k-1$ probabilities. In accordance with the fact that has been stated, the random variable

$$\sum_{j=1}^{2}\sum_{i=1}^{k} \frac{\{X_{ij} - n_j[(X_{i1} + X_{i2})/(n_1 + n_2)]\}^2}{n_j[(X_{i1} + X_{i2})/(n_1 + n_2)]}$$

has an approximate χ^2 distribution with $2k - 2 - (k-1) = k-1$ degrees of freedom. Thus we are able to test the hypothesis that two multinomial distributions are the same; this hypothesis is rejected when the computed value of this random variable is at least as great as an appropriate number from Table II, with $k - 1$ degrees of freedom. This test is often called the chi-square test for *homogeneity*, (the null is equivalent to homogeneous distributions). ∎

The second example deals with the subject of *contingency tables*.

Example 5.7.4. Let the result of a random experiment be classified by two attributes (such as the color of the hair and the color of the eyes). That is, one attribute of the outcome is one and only one of certain mutually exclusive and exhaustive events, say $A_1, A_2, \ldots, A_a$; and the attribute of the outcome is also one and only one of certain mutually exclusive and exhaustive events, say $B_1, B_2, \ldots, B_b$. The $p_{ij} = P(A_i \cap B_j)$, $i = 1, 2, \ldots, a$; $j = 1, 2, \ldots, b$. The random experiment is to be repeated n independent times and X_{ij} will denote the frequencies of the event $A_i \cap B_j$. Since there are $k = ab$ such events as $A_i \cap B_j$, the random variable

$$Q_{ab-1} = \sum_{j=1}^{b}\sum_{i=1}^{a} \frac{(X_{ij} - np_{ij})^2}{np_{ij}}$$

has an approximate chi-square distribution with $ab - 1$ degrees of freedom, provided that n is large. Suppose that we wish to test the independence of the A and the B attributes, i.e., the hypothesis $H_0 : P(A_i \cap B_j) = P(A_i)P(B_j)$, $i = 1, 2, \ldots, a$; $j = 1, 2, \ldots, b$. Let us denote $P(A_i)$ by $p_{i.}$ and $P(B_j)$ by $p_{.j}$. It follows that,

$$p_{i.} = \sum_{j=1}^{b} p_{ij}, \quad p_{.j} = \sum_{i=1}^{a} p_{ij}, \text{ and } 1 = \sum_{j=1}^{b}\sum_{i=1}^{a} p_{ij} = \sum_{j=1}^{b} p_{.j} = \sum_{i=1}^{a} p_{i.}.$$

Then the hypothesis can be formulated as $H_0 : p_{ij} = p_{i.}p_{.j}$, $i = 1, 2, \ldots, a$; $j = 1, 2, \ldots, b$. To test H_0, we can use Q_{ab-1} with p_{ij} replaced by $p_{i.}p_{.j}$. But if $p_{i.}$, $i = 1, 2, \ldots, a$, and $p_{.j}$, $j = 1, 2, \ldots, b$, are unknown, as they frequently are in applications, we cannot compute Q_{ab-1} once the frequencies are observed. In such a case we estimate these unknown parameters by

$$\hat{p}_{i.} = \frac{X_{i.}}{n}, \text{ where } X_{i.} = \sum_{j=1}^{b} X_{ij}, \text{ for } i = 1, 2, \ldots, a,$$

and

$$\hat{p}_{.j} = \frac{X_{.j}}{n}, \text{ where } X_{.j} = \sum_{i=1}^{a} X_{ij}, \text{ for } j = 1, 2, \ldots, b.$$

Since $\sum_{i} p_{i.} = \sum_{j} p_{.j} = 1$, we have estimated only $a - 1 + b - 1 = a + b - 2$ parameters.

So if these estimates are used in Q_{ab-1}, with $p_{ij} = p_{i.}p_{.j}$, then, according to the rule that has been stated in this section, the random variable

$$\sum_{j=1}^{b}\sum_{i=1}^{a} \frac{[X_{ij} - n(X_{i.}/n)(X_{.j}/n)]^2}{n(X_{i.}/n)(X_{.j}/n)}$$

has an approximate chi-square distribution with $ab - 1 - (a+b-2) = (a-1)(b-1)$ degrees of freedom provided that H_0 is true. The hypothesis H_0 is then rejected if the computed value of this statistic exceeds the constant c, where c is selected from Table II so that the test has the desired significance level α. This is the *chi-square test* for independence. ■

In each of the four examples of this section, we have indicated that the statistic used to test the hypothesis H_0 has an approximate chi-square distribution, provided that n is sufficiently large and H_0 is true. To compute the power of any of these tests for values of the parameters not described by H_0, we need the distribution of the statistic when H_0 is not true. In each of these cases, the statistic has an approximate distribution called a *noncentral chi-square distribution*. The noncentral chi-square distribution will be discussed in Section 9.3.

EXERCISES

5.7.1. A number is to be selected from the interval $\{x : 0 < x < 2\}$ by a random process. Let $A_i = \{x : (i-1)/2 < x \le i/2\}$, $i = 1, 2, 3$, and let $A_4 = \{x : \frac{3}{2} < x < 2\}$. For $i = 1, 2, 3, 4$, suppose a certain hypothesis assigns probabilities p_{i0} to these sets in accordance with $p_{i0} = \int_{A_i} (\frac{1}{2})(2 - x)\, dx$, $i = 1, 2, 3, 4$. This hypothesis (concerning the multinomial pdf with $k = 4$) is to be tested at the 5 percent level of significance by a chi-square test. If the observed frequencies of the sets A_i, $i = 1, 2, 3, 4$, are respectively, 30, 30, 10, 10, would H_0 be accepted at the (approximate) 5 percent level of significance?

5.7.2. Define the sets $A_1 = \{x : -\infty < x \le 0\}$, $A_i = \{x : i - 2 < x \le i - 1\}$, $i = 2, \ldots, 7$, and $A_8 = \{x : 6 < x < \infty\}$. A certain hypothesis assigns probabilities p_{i0} to these sets A_i in accordance with

$$p_{i0} = \int_{A_i} \frac{1}{2\sqrt{2\pi}} \exp\left[-\frac{(x-3)^2}{2(4)}\right] dx, \quad i = 1, 2, \ldots, 7, 8.$$

This hypothesis (concerning the multinomial pdf with $k = 8$) is to be tested, at the 5 percent level of significance, by a chi-square test. If the observed frequencies of the sets A_i, $i = 1, 2, \ldots, 8$, are, respectively, 60, 96, 140, 210, 172, 160, 88, and 74, would H_0 be accepted at the (approximate) 5 percent level of significance?

5.7.3. A die was cast $n = 120$ independent times and the following data resulted:

Spots Up	1	2	3	4	5	6
Frequency	b	20	20	20	20	40-b

If we use a chi-square test, for what values of b would the hypothesis that the die is unbiased be rejected at the 0.025 significance level?

5.7.4. Consider the problem from genetics of crossing two types of peas. The Mendelian theory states that the probabilities of the classifications (a) round and yellow, (b) wrinkled and yellow, (c) round and green, and (d) wrinkled and green are $\frac{9}{16}$, $\frac{3}{16}$, $\frac{3}{16}$, and $\frac{1}{16}$, respectively. If from 160 independent observations the observed frequencies of these respective classifications are 86, 35, 26, and 13, are these data consistent with the Mendelian theory? That is, test, with $\alpha = 0.01$, the hypothesis that the respective probabilities are $\frac{9}{16}$, $\frac{3}{16}$, $\frac{3}{16}$, and $\frac{1}{16}$.

5.7.5. Two different teaching procedures were used on two different groups of students. Each group contained 100 students of about the same ability. At the end of the term, an evaluating team assigned a letter grade to each student. The results were tabulated as follows.

Group	Grade					Total
	A	B	C	D	F	
I	15	25	32	17	11	100
II	9	18	29	28	16	100

If we consider these data to be independent observations from two respective multinomial distributions with $k = 5$, test at the 5 percent significance level the hypothesis that the two distributions are the same (and hence the two teaching procedures are equally effective).

5.7.6. Let the result of a random experiment be classified as one of the mutually exclusive and exhaustive ways A_1, A_2, A_3 and also as one of the mutually exclusive and exhaustive ways B_1, B_2, B_3, B_4. Two hundred independent trials of the experiment result in the following data:

	B_1	B_2	B_3	B_4
A_1	10	21	15	6
A_2	11	27	21	13
A_3	6	19	27	24

Test, at the 0.05 significance level, the hypothesis of independence of the A attribute and the B attribute, namely $H_0 : P(A_i \cap B_j) = P(A_i)P(B_j)$, $i = 1, 2, 3$ and $j = 1, 2, 3, 4$, against the alternative of dependence.

5.7.7. A certain genetic model suggests that the probabilities of a particular trinomial distribution are, respectively, $p_1 = p^2$, $p_2 = 2p(1-p)$, and $p_3 = (1-p)^2$, where $0 < p < 1$. If X_1, X_2, X_3 represent the respective frequencies in n independent trials, explain how we could check on the adequacy of the genetic model.

5.7.8. Let the result of a random experiment be classified as one of the mutually exclusive and exhaustive ways A_1, A_2, A_3 and also as one of the mutually exhaustive ways B_1, B_2, B_3, B_4. Say that 180 independent trials of the experiment result in the following frequencies:

	B_1	B_2	B_3	B_4
A_1	$15 - 3k$	$15 - k$	$15 + k$	$15 + 3k$
A_2	15	15	15	15
A_3	$15 + 3k$	$15 + k$	$15 - k$	$15 - 3k$

where k is one of the integers $0, 1, 2, 3, 4, 5$. What is the smallest value of k that will lead to the rejection of the independence of the A attribute and the B attribute at the $\alpha = 0.05$ significance level?

5.7.9. It is proposed to fit the Poisson distribution to the following data

x	0	1	2	3	$3 < x$
Frequency	20	40	16	18	6

(a) Compute the corresponding chi-square goodness-of-fit statistic.
   Hint: In computing the mean, treat $3 < x$ as $x = 4$.

(b) How many degrees of freedom are associated with this chi-square?

(c) Do these data result in the rejection of the Poisson model at the $\alpha = 0.05$ significance level?

5.8   The Method of Monte Carlo

In this section we introduce the concept of generating observations from a specified distribution or sample. This is often called **Monte Carlo** generation. This technique has been used for simulating complicated processes and investigating finite sample properties of statistical methodology for some time now. In the last 20 years, however, this has become a very important concept in modern statistics in the realm of inference based on the bootstrap (resampling) and modern Bayesian methods. We will repeatedly make use of this concept throughout the book.

As we will see, a generator of random uniform observations is, for the most part, all that is needed. It is not easy to construct a device which generates random uniform observations. However there has been considerable work done in this area, not only in the construction of such generators, but in the testing of their accuracy, as well.

Most statistical software packages have reliable uniform generators. We will make use of this fact in the text. For the examples and exercises we will often use *simple* algorithms written in R code, (Ihaka and Gentleman, 1996), which are almost *self-explanatory*. All code contained in this text is freely available at the web site www.stat.wmich.edu/mckean/HMC/. There are, however, other excellent statistical computing packages such as S-PLUS and Maple which can be easily used.

Suppose then we have a device capable of generating a stream of independent and identically distributed observations from a uniform $(0, 1)$ distribution. For example, the following command will generate 10 such observations in the languages

R: `runif(10)`. In this command the `r` stands for random, the `unif` stands for uniform, the 10 stands for the number of observations requested, and the lack of additional arguments means that the standard uniform $(0, 1)$ generator will be used.

For observations from a discrete distribution, often a uniform generator will suffice. For a simple example, consider an experiment where a fair six-sided die is rolled and the random variable X is 1 if the upface is a "low number," namely $\{1, 2\}$, otherwise $X = 0$. Note that the mean of X is $\mu = 1/3$. If U has a uniform $(0, 1)$ distribution, then X can be realized as

$$X = \begin{cases} 1 & \text{if } 0 < U \le 1/3 \\ 0 & \text{if } 1/3 < U < 1. \end{cases}$$

Using the command above, we generated 10 observations from this experiment. The following table displays the results.

u_i	0.4743	0.7891	0.5550	0.9693	0.0299
x_i	0	0	0	0	1
u_i	0.8425	0.6012	0.1009	0.0545	0.4677
x_i	0	0	1	1	0

Note that observations form a realization of a random sample $X_1, \ldots, X_{10}$ drawn from the distribution of X. From Section 4.2, $\overline{X}$ is a consistent estimate of μ, and for these 10 observations it is equal to $\overline{x} = 0.3$.

Example 5.8.1 (Estimation of π). Consider the experiment where a pair of numbers (U_1, U_2) are chosen at random in the unit square, as shown in Figure 5.8.1; that is, U_1 and U_2 are iid uniform $(0, 1)$ random variables. Since the point is chosen at random, the probability of (U_1, U_2) lying within the unit circle is $\pi/4$. Let X be the random variable,

$$X = \begin{cases} 1 & \text{if } U_1^2 + U_2^2 < 1 \\ 0 & \text{otherwise.} \end{cases}$$

Hence the mean of X is $\mu = \pi/4$. Now suppose π is unknown. One way of estimating π is to repeat the experiment n independent times; hence, obtaining a random sample $X_1, \ldots, X_n$ on X. The statistic $4\overline{X}$ is a consistent estimator of π. In Appendix C, a simple R routine, `piest`, is found which repeats the experiment n times and returns the estimate of π. Figure 5.8.1 shows 20 realizations of this experiment. Note that of the 20 points, 15 fall within the unit circle. Hence, our estimate of π is $4(15/20) = 3.00$. We ran this code for various values of n with the following results:

n	100	500	1000	10,000	100,000
$4\overline{x}$	3.24	3.072	3.132	3.138	3.13828
$1.96 \cdot 4\sqrt{\overline{x}(1 - \overline{x})/n}$	0.308	0.148	0.102	0.032	0.010

We can use the large sample confidence interval derived in Section 5.4 to estimate the error of estimation. The corresponding 95% confidence interval for π is

$$\left(4\overline{X} - 1.96 \cdot 4\sqrt{\overline{X}(1 - \overline{X})/n}, \; 4\overline{X} + 1.96 \cdot 4\sqrt{\overline{X}(1 - \overline{X})/n} \right). \tag{5.8.1}$$

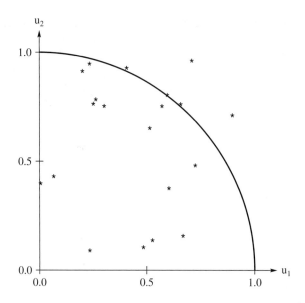

Figure 5.8.1: Unit Square with the First Quadrant of the Unit Circle, Example 5.8.1

The last row of the above table contains the error part of the confidence intervals. Notice that all five confidence intervals trapped the true value of π. ∎

What about continuous random variables? For these we have the following theorem:

Theorem 5.8.1. *Suppose the random variable U has a uniform $(0,1)$ distribution. Let F be a continuous distribution function. Then the random variable $X = F^{-1}(U)$ has distribution function F.*

Proof: Recall from the definition of a uniform distribution that U has the distribution function $F_U(u) = u$ for $u \in (0,1)$. Using this and the distribution-function technique and assuming that $F(x)$ is strictly monotone, the distribution function of X is,

$$
\begin{aligned}
P[X \leq x] &= P[F^{-1}(U) \leq x] \\
&= P[U \leq F(x)] \\
&= F(x),
\end{aligned}
$$

which proves the theorem. ∎

In the proof, we assumed that $F(x)$ was strictly monotone. As Exercise 5.8.12 shows, we can weaken this.

We can use this theorem to generate realizations (observations) of many different random variables. Consider the random variable X with an exponential distribution with $\theta = 1$, (3.3.2). Suppose we have a uniform generator and we want to generate a realization of X. The distribution function of X is

$$F(x) = 1 - e^{-x}, \quad x > 0.$$

Hence, the inverse of the distribution function is given by,

$$F^{-1}(u) = -\log(1 - u), \quad 0 < u < 1. \tag{5.8.2}$$

So if U has uniform $(0, 1)$ distribution, then $X = -\log(1 - U)$ has an exponential distribution. For instance, suppose our uniform generator generated the following stream of uniform observations,

$$0.473, \ 0.858, \ 0.501, \ 0.676, \ 0.240,$$

then the corresponding stream of exponential observations is

$$0.641, \ 1.95, \ 0.696, \ 1.13, \ 0.274.$$

In the exercises, the reader is asked to determine how to generate random observations for other distributions.

Example 5.8.2 (Monte Carlo Integration). Suppose we want to obtain the integral $\int_a^b g(x)\,dx$ for a continuous function g over the closed and bounded interval $[a, b]$. If the anti-derivative of g does not exist, then numerical integration is in order. A simple numerical technique is the method of Monte Carlo. We can write the integral as

$$\int_a^b g(x)\,dx = (b - a) \int_a^b g(x) \frac{1}{b - a}\,dx = (b - a)E[g(X)],$$

where X has the uniform (a, b) distribution. The Monte Carlo technique is then to generate a random sample $X_1, \ldots, X_n$ of size n from the uniform (a, b) distribution and compute $Y_i = (b - a)g(X_i)$. Then $\overline{Y}$ is a consistent estimate of $\int_a^b g(x)\,dx$. ∎

Example 5.8.3 (Estimation of π by Monte Carlo Integration). For a numerical example, reconsider the estimation of π. Instead of the experiment described in Example 5.8.1, we shall use the method of Monte Carlo integration. Let $g(x) = 4\sqrt{1 - x^2}$ for $0 < x < 1$. Then

$$\pi = \int_0^1 g(x)\,dx = E[g(X)],$$

where X has the uniform $(0, 1)$ distribution. Hence, we need to generate a random sample $X_1, \ldots, X_n$ from the uniform $(0, 1)$ distribution and form $Y_i = 4\sqrt{1 - X_i^2}$. Then $\overline{Y}$ is a consistent estimate of π. Note that $\overline{Y}$ is estimating a mean, so the large sample confidence interval (5.8.3) derived in Example 5.4.1 for means can be used to estimate the error of estimation. Recall that this 95% confidence interval is given by,

$$(\overline{y} - 1.96s/\sqrt{n}, \overline{y} + 1.96s/\sqrt{n}),$$

where s is the value of the sample standard deviation. The table below gives the results for estimates of π for various runs of different sample sizes along with the confidence intervals.

n	100	1000	10,000	100,000
$\bar{y}$	3.217849	3.103322	3.135465	3.142066
$\bar{y} - 1.96(s/\sqrt{n})$	3.054664	3.046330	3.118080	3.136535
$\bar{y} + 1.96(s/\sqrt{n})$	3.381034	3.160314	3.152850	3.147597

Note that for each experiment the confidence interval trapped π. See Appendix C, `piest2`, for the actual code used for the evaluation. ∎

Numerical integration techniques have made great strides over the last 20 years. But the simplicity of integration by Monte Carlo still makes it a powerful technique.

As Theorem 5.8.1 shows, if we can obtain $F_X^{-1}(u)$ in closed form then we can easily generate observations with cdf F_X. In many cases where this is not possible, techniques have been developed to generate observations. Note that the normal distribution serves as an example of such a case and, in the next example, we show how to generate normal observations. In Section 5.8.1, we discuss an algorithm which can be adapted for many of these cases.

Example 5.8.4 (Generating Normal Observations). To simulate normal variables, Box and Muller (1958) suggested the following procedure. Let Y_1, Y_2 be a random sample from the uniform distribution over $0 < y < 1$. Define X_1 and X_2 by

$$\begin{aligned} X_1 &= (-2 \log Y_1)^{1/2} \cos(2\pi Y_2), \\ X_2 &= (-2 \log Y_1)^{1/2} \sin(2\pi Y_2). \end{aligned}$$

This transformation is one-to-one and maps $\{(y_1, y_2) : 0 < y_1 < 1,\ 0 < y_2 < 1\}$ onto $\{(x_1, x_2) : -\infty < x_1 < \infty,\ -\infty < x_2 < \infty\}$ except for sets involving $x_1 = 0$ and $x_2 = 0$, which have probability zero. The inverse transformation is given by

$$y_1 = \exp\left(-\frac{x_1^2 + x_2^2}{2}\right),$$

$$y_2 = \frac{1}{2\pi} \arctan \frac{x_2}{x_1}.$$

This has the Jacobian

$$J = \begin{vmatrix} (-x_1)\exp\left(-\dfrac{x_1^2 + x_2^2}{2}\right) & (-x_2)\exp\left(-\dfrac{x_1^2 + x_2^2}{2}\right) \\[2ex] \dfrac{-x_2/x_1^2}{(2\pi)(1 + x_2^2/x_1^2)} & \dfrac{1/x_1}{(2\pi)(1 + x_2^2/x_1^2)} \end{vmatrix}$$

$$= \frac{-(1 + x_2^2/x_1^2)\exp\left(-\dfrac{x_1^2 + x_2^2}{2}\right)}{(2\pi)(1 + x_2^2/x_1^2)} = \frac{-\exp\left(-\dfrac{x_1^2 + x_2^2}{2}\right)}{2\pi}.$$

Since the joint pdf of Y_1 and Y_2 is 1 on $0 < y_1 < 1, 0 < y_2 < 1$, and zero elsewhere, the joint pdf of X_1 and X_2 is

$$\frac{\exp\left(-\dfrac{x_1^2 + x_2^2}{2}\right)}{2\pi}, \quad -\infty < x_1 < \infty, \quad -\infty < x_2 < \infty.$$

That is, X_1 and X_2 are independent, standard normal random variables. One of the most commonly used normal generators is a variant of the above procedure called the Marsaglia and Bray (1964) algorithm; see Exercise 5.8.20. ∎

Observations from a contaminated normal distribution, discussed in Section 3.4.1, can easily be generated using a normal generator and a uniform generator. We close this section by estimating via Monte Carlo the significance level of a t-test when the underlying distribution is a contaminated normal.

Example 5.8.5. Let X be a random variable with mean μ and consider the hypotheses:

$$H_0 : \ \mu = 0 \text{ versus } H_A : \ \mu > 0. \tag{5.8.3}$$

Suppose we decide to base this test on a sample of size $n = 20$ from the distribution of X, using the t-test with rejection rule,

$$\text{Reject } H_0 : \ \mu = 0 \text{ in favor of } H_A : \ \mu > 0 \text{ if } t > t_{.05,19} = 1.729, \tag{5.8.4}$$

where $t = \overline{x}/(s/\sqrt{20})$ and $\overline{x}$ and s are the sample mean and standard deviation, respectively. If X has a normal distribution, then this test will have level 0.05. But what if X does not have a normal distribution? In particular, for this example, suppose X has the contaminated normal distribution given by (3.4.14) with $\epsilon = 0.25$ and $\sigma_c = 25$; that is, 75% of the time an observation is generated by a standard normal distribution while 25% of the time it is generated by a normal distribution with mean 0 and standard deviation 25. Hence, the mean of X is 0, so H_0 is true. To obtain the exact significance level of the test would be quite complicated. We would have to obtain the distribution of t when X has this contaminated normal distribution. As an alternative, we will estimate the level (and the error of estimation) by simulation. Let N be the number of simulations. The following algorithm gives the steps of our simulation:

1. Set $k = 1$, $I = 0$.

2. Simulate a random sample of size 20 from the distribution of X.

3. Based on this sample, compute the test statistic t.

4. If $t > 1.729$, increase I by 1.

5. If $k = N$ go to Step 6, else increase k by 1 and go to Step 2.

6. Compute $\widehat{\alpha} = I/N$ and the approximate error $= 1.96\sqrt{\widehat{\alpha}(1 - \widehat{\alpha})/N}$.

Then $\widehat{\alpha}$ is our simulated estimate of α and the half width of a confidence interval for α serves as our estimate of error of estimation.

The routine empalphacn, found in Appendix C, provides R or S-PLUS code for this algorithm. When we ran it for $N = 10,000$ we obtained the results:

Numb. Simulat.	Empirical $\widehat{\alpha}$	Error	95% CI for α
10,000	0.0412	0.0039	$(0.0373, 0.0451)$

Based on these results the t test appears to be slightly conservative when the sample is drawn from this contaminated normal distribution. ∎

5.8.1   Accept-Reject Generation Algorithm

In this section, we develop the **accept-reject** procedure that can often be used to simulate random variables whose inverse cdf cannot be obtained in closed form. Let X be a continuous random variable with pdf $f(x)$. For this discussion, we will call this pdf the *target* pdf. Suppose it is relatively easy to generate an observation of the random variable Y which has pdf $g(x)$ and that for some constant M we have

$$f(x) \leq Mg(x), \quad -\infty < x < \infty. \tag{5.8.5}$$

We will call $g(x)$ the *instrumental* pdf. For clarity, we will write the accept-reject as an algorithm:

Algorithm 5.8.1. *(Accept-Reject Algorithm) Let $f(x)$ be a pdf. Suppose that Y is a random variable with pdf $g(y)$, U is a random variable with a uniform$(0,1)$ distribution, Y and U are independent and (5.8.5) holds. The following algorithm generates a random variable X with pdf $f(x)$.*

1. *Generate Y and U.*

2. *If $U \leq \frac{f(Y)}{Mg(Y)}$ then take $X = Y$. Else return to Step (1).*

3. *X has pdf $f(x)$.*

Proof of the validity of the algorithm: Let $-\infty < x < \infty$. Then

$$
\begin{aligned}
P[X \leq x] &= P\left[Y \leq x | U \leq \frac{f(Y)}{Mg(Y)}\right] \\
&= \frac{P\left[Y \leq x, U \leq \frac{f(Y)}{Mg(Y)}\right]}{P\left[U \leq \frac{f(Y)}{Mg(Y)}\right]} \\
&= \frac{\int_{-\infty}^{x}\left[\int_{0}^{f(y)/Mg(y)} du\right] g(y)dy}{\int_{-\infty}^{\infty}\left[\int_{0}^{f(y)/Mg(y)} du\right] g(y)dy} \\
&= \frac{\int_{-\infty}^{x} \frac{f(y)}{Mg(y)} g(y)dy}{\int_{-\infty}^{\infty} \frac{f(y)}{Mg(y)} g(y)dy} \tag{5.8.6} \\
&= \int_{-\infty}^{x} f(y)\, dy. \tag{5.8.7}
\end{aligned}
$$

Hence, by differentiating both sides, we find that the pdf of X is $f(x)$. ∎

As Exercise 5.8.13 shows, from step (5.8.6) of the proof, we can ignore normalizing constants of the two pdfs $f(x)$ and $g(x)$. For example, if $f(x) = kh(x)$ and $g(x) = ct(x)$ for constants c and k then we can use the rule

$$h(x) \leq M_2 t(x) , \quad -\infty < x < \infty, \tag{5.8.8}$$

and change the ratio in Step (2) of the algorithm to $U \leq \frac{h(Y)}{M_2 t(Y)}$. This often simplifies the use of the Accept-Reject Algorithm.

As an example of the Accept-Reject Algorithm, consider simulating a $\Gamma(\alpha, \beta)$ distribution. There are several approaches to generating gamma observations; see, for instance, Kennedy and Gentle (1980). We present the approach discussed in Robert and Casella (1999). Recall if X has a $\Gamma(\alpha, 1)$ distribution, then the random variable βX has a $\Gamma(\alpha, \beta)$ distribution. So without loss of generality, we can assume that $\beta = 1$. If α is an integer, then by Theorem 3.3.2, $X = \sum_{i=1}^{\alpha} Y_i$, where Y_i are iid $\Gamma(1, 1)$. In this case, by expression (5.8.2), we see that the inverse cdf of Y_i is easily written in closed form, and, hence X is easy to generate. Thus, the only remaining case is when α is not an integer.

Assume then that X has a $\Gamma(\alpha, 1)$ distribution, where α is not an integer. Let Y have a $\Gamma([\alpha], 1/b)$ distribution, where $b < 1$ will be chosen later and, as usual, $[\alpha]$ means the greatest integer less than or equal to α. To establish rule (5.8.8), consider the ratio, with $h(x)$ and $t(x)$ proportional to the pdfs of x and y respectively, given by

$$\frac{h(x)}{t(x)} = b^{-[\alpha]} x^{\alpha - [\alpha]} e^{-(1-b)x}, \tag{5.8.9}$$

where we have ignored some of the normalizing constants. We next determine the constant b.

As Exercise 5.8.14 shows the derivative of expression (5.8.9) is

$$\frac{d}{dx} b^{-[\alpha]} x^{\alpha - [\alpha]} e^{-(1-b)x} = b^{-[\alpha]} e^{-(1-b)x} [(\alpha - [\alpha]) - x(1 - b)] x^{\alpha - [\alpha] - 1}; , \tag{5.8.10}$$

which has a maximum critical value at $x = (\alpha - [\alpha])/(1 - b)$. Hence, using the maximum of $h(x)/t(x)$,

$$\frac{h(x)}{t(x)} \leq b^{-[\alpha]} \left[\frac{\alpha - [\alpha]}{(1 - b)e} \right]^{\alpha - [\alpha]} . \tag{5.8.11}$$

Now, we need to find our choice of b. As Exercise 5.8.15 shows,

$$\frac{d}{db} b^{-[\alpha]} (1 - b)^{[\alpha] - \alpha} = -b^{-[\alpha]} (1 - b)^{[\alpha] - \alpha} \left[\frac{[\alpha] - \alpha b}{b(1 - b)} \right], \tag{5.8.12}$$

which has a critical value at $b = [\alpha]/\alpha < 1$. As Exercise 5.8.15 shows this value of b provides a minimum of the right side of expression (5.8.11). Thus if we take $b = [\alpha]/\alpha < 1$, then equality (5.8.11) holds and it will be the tightest inequality possible. The final value of M is the right side of expression (5.8.11) evaluated at $b = [\alpha]/\alpha < 1$.

The following example offers a simpler derivation for a normal generator where the instrumental pdf is the pdf of a Cauchy random variable.

Example 5.8.6. Suppose that X is a normally distributed random variable with pdf $\phi(x) = (2\pi)^{-1/2} \exp\{-x^2/2\}$ and Y has a Cauchy distribution with pdf $g(x) = \pi^{-1}(1+x^2)^{-1}$. As Exercise 5.8.8 shows the Cauchy distribution is easy to simulate because its inverse cdf is a known function. Ignoring normalizing constants, the ratio to bound is

$$\frac{f(x)}{g(x)} \propto (1 + x^2) \exp\{-x^2/2\}, \quad -\infty < x < \infty. \tag{5.8.13}$$

As Exercise 5.8.16 shows, the derivative of this ratio is $-x \exp\{-x^2/2\}(x^2-1)$ which has critical values at ± 1. These values provide maximums to (5.8.13). Hence,

$$(1 + x^2) \exp\{-x^2/2\} \le 2 \exp\{-1/2\} = 1.213,$$

so $M = 1.213$. ∎

One result of the proof of Algorithm 5.8.1 is that the probability of acceptance in the algorithm is M^{-1}. This follows immediately from the denominator factor in Step (5.8.6) of the proof. Note, however, that this holds only for properly normed pdfs. For instance in the last example, the maximum value of the ratio of properly normed pdfs is

$$\frac{\pi}{\sqrt{2\pi}} 2 \exp\{-1/2\} = 1.52.$$

Hence, $1/M = 1.52^{-1} = 0.66$. Therefore the probability that the algorithm accepts is 0.66.

EXERCISES

5.8.1. Approximate log 2 by using a uniform$(0, 1)$ generator. Obtain an error of estimation in terms of a large sample 95% confidence interval. If you have access to the statistical package R, write an R function for the estimate and the error of estimation. Obtain your estimate for 10,000 simulations and compare it to the true value.

Hint: Recall that log $2 = \int_0^1 \frac{1}{x+1} \, dx$.

5.8.2. Similar to Exercise 5.8.1 but now approximate $\int_0^{1.96} \frac{1}{\sqrt{2\pi}} \exp\left\{-\frac{1}{2}t^2\right\} \, dt$.

5.8.3. Suppose X is a random variable with the pdf $f_X(x) = b^{-1}f((x - a)/b)$, where $b > 0$. Suppose we can generate observations from $f(z)$. Explain how we can generate observations from $f_X(x)$.

5.8.4. Determine a method to generate random observations for the logistic pdf, (5.2.8). If access is available, write an R function which returns a random sample of observations from a logistic distribution.

5.8.5. Determine a method to generate random observations for the following pdf,

$$f(x) = \begin{cases} 4x^3 & 0 < x < 1 \\ 0 & \text{elsewhere.} \end{cases}$$

If access is available, write an R function which returns a random sample of observations from this pdf.

5.8.6. Determine a method to generate random observations for the Laplace pdf, (5.2.9). If access is available, write an R function which returns a random sample of observations from a Laplace distribution.

5.8.7. Determine a method to generate random observations for the extreme valued pdf which is given by

$$f(x) = \exp\left\{x - e^x\right\}, \quad -\infty < x < \infty. \tag{5.8.14}$$

If access is available, write an R function which returns a random sample of observation from an extreme valued distribution.

5.8.8. Determine a method to generate random observations for the Cauchy distribution with pdf

$$f(x) = \frac{1}{\pi(1 + x^2)}, \quad -\infty < x < \infty. \tag{5.8.15}$$

If access is available, write an R function which returns a random sample of observation from a Cauchy distribution.

5.8.9. A Weibull distribution with pdf

$$f(x) = \begin{cases} \frac{1}{\theta^3} 3x^2 e^{-x^3/\theta^3} & 0 < x < \infty \\ 0 & \text{elsewhere.} \end{cases}$$

is often used to model the length of life times of certain products. Determine a method to generate random observations from this Weibull distribution. If access is available, write an R function which returns a random sample of observation from a Weibull distribution.

5.8.10. Consider the situation in Example 5.8.5 with the hypotheses (5.8.3). Write an algorithm which will simulate the power of the test (5.8.4) to detect the alternative $\mu = 0.5$ under the same contaminated normal distribution as in the example. If access is available, modify the R function `empalphacn(N)` to simulate this power and to obtain an estimate of the error of estimation.

5.8.11. For the last exercise, write an algorithm to simulate the significance level and power to detect the alternative $\mu = 0.5$ for the test (5.8.4) when the underling distribution is the logistic distribution (5.2.8).

5.8.12. For the proof of Theorem 5.8.1, we assumed that the cdf was strictly increasing over its support. Consider a random variable X with cdf $F(x)$ which is not strictly increasing. Define as the inverse of $F(x)$ the function

$$F^{-1}(u) = \inf\{x : F(x) \geq u\}, \quad 0 < u < 1.$$

Let U have a uniform $(0, 1)$ distribution. Prove that the random variable $F^{-1}(U)$ has cdf $F(x)$.

5.8.13. Show the discussion at the end of the proof of Algorithm 5.8.13 of not needing to use normalizing constants in the Accept-Reject Algorithm is true.

5.8.14. Verify the derivative in expression (5.8.10) and show that the function (5.8.9) attains a maximum at the critical value $x = (\alpha - [\alpha])/(1 - b)$.

5.8.15. Derive expression (5.8.12) and show that the resulting critical value $b = [\alpha]/\alpha < 1$ gives a minimum of the function which is the right side of expression (5.8.11).

5.8.16. Show that the derivative of the ratio in expression (5.8.13) is given by the function $-x \exp\{-x^2/2\}(x^2 - 1)$ with critical values ± 1. Show that the critical values provide maximums for expression (5.8.13).

5.8.17. Consider the pdf

$$f(x) = \begin{cases} \beta x^{\beta-1} & 0 < x < 1 \\ 0 & \text{elsewhere,} \end{cases}$$

for $\beta > 1$.

(a) Use Theorem 5.8.1 to generate an observation from this pdf.

(b) Use the Accept-Reject Algorithm to generate an observation from this pdf.

5.8.18. Proceeding similar to Example 5.8.6, use the Accept-Reject Algorithm to generate an observation from a t distribution with $r > 1$ degrees of freedom using the Cauchy distribution.

5.8.19. For $\alpha > 0$ and $\beta > 0$, consider the following accept/reject algorithm:

(1) Generate U_1 and U_2 iid uniform$(0, 1)$ random variables. Set $V_1 = U_1^{1/\alpha}$ and $V_2 = U_2^{1/\beta}$.

(2) Set $W = V_1 + V_2$. If $W \leq 1$, set $X = V_1/W$, else go to Step (1).

(3) Deliver X.

Show that X has a beta distribution with parameters α and β, (3.3.5). See Kennedy and Gentle (1980).

5.8.20. Consider the following algorithm:

(1) Generate U and V independent uniform $(-1, 1)$ random variables.

(2) Set $W = U^2 + V^2$.

(3) If $W > 1$ goto Step (1).

(4) Set $Z = \sqrt{(-2 \log W)/W}$ and let $X_1 = UZ$ and $X_2 = VZ$.

Show that the random variables X_1 and X_2 are iid with a common $N(0, 1)$ distribution. This algorithm was proposed by Marsaglia and Bray (1964).

5.9   Bootstrap Procedures

In the last section, we introduced the method of Monte Carlo and discussed several of its applications. In the last few years, however, Monte Carlo procedures have become increasingly used in statistical inference. In this section, we present the *bootstrap*, one of these procedures. We concentrate on confidence intervals and tests for one and two sample problems in this section.

5.9.1   Percentile Bootstrap Confidence Intervals

Let X be a random variable of the continuous type with pdf $f(x; \theta)$, for $\theta \in \Omega$. Suppose $\mathbf{X} = (X_1, X_2, \ldots, X_n)$ is a random sample on X and $\widehat{\theta} = \widehat{\theta}(\mathbf{X})$ is a point estimator of θ. The vector notation, $\mathbf{X}$, will prove useful in this section. In Sections 5.4, we discussed the problem of obtaining confidence intervals for θ in certain situations. In this section, we discuss a general method called the *percentile-bootstrap* procedure, which is a *resampling* procedure. Good discussions of such procedures can be found in the books by Efron and Tibshirani (1993) and Davison and Hinkley (1997).

To motivate the procedure, suppose for the moment that $\widehat{\theta}$ has a $N(\theta, \sigma_{\widehat{\theta}}^2)$ distribution. Then as in Section 5.4, a $(1 - \alpha)100\%$ confidence interval for θ is $(\widehat{\theta}_L, \widehat{\theta}_U)$ where

$$\widehat{\theta}_L = \widehat{\theta} - z^{(1-\alpha/2)} \sigma_{\widehat{\theta}} \quad \text{and} \quad \widehat{\theta}_U = \widehat{\theta} - z^{(\alpha/2)} \sigma_{\widehat{\theta}}, \tag{5.9.1}$$

and $z^{(\gamma)}$ denotes the $\gamma 100th$ percentile of a standard normal random variable; i.e., $z^{(\gamma)} = \Phi^{-1}(\gamma)$ where Φ is the cdf of a $N(0, 1)$ random variable, (see, also, Exercise 5.9.4). We have gone to a superscript notation here to avoid confusion with the usual subscript notation on critical values.

Now suppose $\widehat{\theta}$ and $\sigma_{\widehat{\theta}}$ are realizations from the sample and $\widehat{\theta}_L$ and $\widehat{\theta}_U$ are calculated. Say $\widehat{\theta}^*$ is a random variable with a $N(\widehat{\theta}, \sigma_{\widehat{\theta}}^2)$ distribution. Then by (5.9.1)

$$P(\widehat{\theta}^* \le \widehat{\theta}_L) = P\left(\frac{\widehat{\theta}^* - \widehat{\theta}}{\sigma_{\widehat{\theta}}} \le -z^{(1-\alpha/2)} \right) = \alpha/2. \tag{5.9.2}$$

Likewise, $P(\widehat{\theta}^* \le \widehat{\theta}_U) = 1 - (\alpha/2)$. Therefore, $\widehat{\theta}_L$ and $\widehat{\theta}_U$ are the $\frac{\alpha}{2}100th$ and $(1 - \frac{\alpha}{2})100th$ percentiles of the distribution of $\widehat{\theta}^*$. That is, the percentiles of the $N(\widehat{\theta}, \sigma_{\widehat{\theta}}^2)$ distribution form the $(1 - \alpha)100\%$ confidence interval for θ.

Before continuing, we show that the normal assumption is transparent to the argument. Suppose H is the cdf of $\widehat{\theta}$ and that H depends on θ. Then using Theorem 5.8.1 we can find an increasing transformation $\phi = m(\theta)$ such that the distribution of $\widehat{\phi} = m(\widehat{\theta})$ is $N(\phi, \sigma_c^2)$, where $\phi = m(\theta)$ and σ_c^2 is some variance. For example, take the transformation to be $m(\theta) = F_c^{-1}(H(\theta))$, where $F_c(x)$ is the cdf of a $N(\phi, \sigma_c^2)$ distribution. Then as above, $(\widehat{\phi} - z^{(1-\alpha/2)} \sigma_c, \widehat{\phi} - z^{(\alpha/2)} \sigma_c)$ is a

$(1 - \alpha)100\%$ confidence interval for ϕ. But note that

$$
\begin{aligned}
1 - \alpha &= P\left[\widehat{\phi} - z^{(1-\alpha/2)}\sigma_c < \phi < \widehat{\phi} - z^{(\alpha/2)}\sigma_c)\right] \\
&= P\left[m^{-1}(\widehat{\phi} - z^{(1-\alpha/2)}\sigma_c) < \theta < m^{-1}(\widehat{\phi} - z^{(\alpha/2)}\sigma_c)\right]. \quad (5.9.3)
\end{aligned}
$$

Hence, $(m^{-1}(\widehat{\phi} - z^{(1-\alpha/2)}\sigma_c), m^{-1}(\widehat{\phi} - z^{(\alpha/2)}\sigma_c))$ is a $(1-\alpha)100\%$ confidence interval for θ. Now suppose $\widehat{H}$ is the cdf H with a realization $\widehat{\theta}$ substituted in for θ; i.e, analogous to the $N(\widehat{\theta}, \sigma_{\widehat{\theta}}^2)$ distribution above. Suppose θ^* is a random variable with cdf $\widehat{H}$. Let $\widehat{\phi} = m(\widehat{\theta})$ and $\widehat{\phi}^* = m(\widehat{\theta}^*)$. We have

$$
\begin{aligned}
P\left[\widehat{\theta}^* \leq m^{-1}(\widehat{\phi} - z^{(1-\alpha/2)}\sigma_c)\right] &= P\left[\widehat{\phi}^* \leq \widehat{\phi} - z^{(1-\alpha/2)}\sigma_c\right] \\
&= P\left[\frac{\widehat{\phi}^* - \widehat{\phi}}{\sigma_c} \leq -z^{(1-\alpha/2)}\right] = \alpha/2,
\end{aligned}
$$

similar to (5.9.2). Therefore, $m^{-1}(\widehat{\phi} - z^{(1-\alpha/2)}\sigma_c)$ is the $\frac{\alpha}{2}100th$ percentile of the cdf $\widehat{H}$. Likewise, $m^{-1}(\widehat{\phi} - z^{(\alpha/2)}\sigma_c)$ is the $(1 - \frac{\alpha}{2})100th$ percentile of the cdf $\widehat{H}$. Therefore, in the general case too, the percentiles of the distribution of $\widehat{H}$ form the confidence interval for θ.

In practice, of course, we do not know the cdf $\widehat{H}$; hence, the above confidence interval defined by statement (5.9.3) cannot be obtained. But suppose we could take an infinite number of samples $\mathbf{X}_1, \mathbf{X}_2, \ldots$; obtain $\widehat{\theta}^* = \widehat{\theta}(\mathbf{X}^*)$ for each sample $\mathbf{X}^*$; and then form the histogram of these estimates $\widehat{\theta}^*$. The percentiles of this histogram would be the confidence interval (5.9.3). Since we only have one sample, this is impossible. It is, however, the idea behind bootstrap procedures.

Bootstrap procedures simply resample from the empirical distribution defined by the one sample. The sampling is done at random and with replacement and the resamples are all of size n, the size of the original sample. That is, suppose $\mathbf{x}' = (x_1, x_2, \ldots, x_n)$ denotes the realization of the sample. Let $\widehat{F}_n$ denote the empirical distribution function of the sample. Recall that $\widehat{F}_n$ is a discrete cdf which puts mass n^{-1} at each point x_i and that $\widehat{F}_n(x)$ is an estimator of $F(x)$. Then a bootstrap sample is a random sample, say $\mathbf{x}^{*\prime} = (x_1^*, x_2^*, \ldots, x_n^*)$, drawn from $\widehat{F}_n$. As Exercise 5.9.1 shows, $E(x_i^*) = \overline{x}$ and $V(x_i^*) = n^{-1}\sum_{i=1}^n (x_i - \overline{x})^2$. At first glance this resampling the sample seems like it would not work. But our only information on sampling variability is within the sample itself and by resampling the sample we are simulating this variability.

We now give an algorithm which obtains a bootstrap confidence interval. For clarity, we present a formal algorithm, which can be readily coded into languages such as R. Let $\mathbf{x}' = (x_1, x_2, \ldots, x_n)$ be the realization of a random sample drawn from a cdf $F(x; \theta)$, $\theta \in \Omega$. Let $\widehat{\theta}$ be a point estimator of θ. Let B, an integer, denote the number of bootstrap replications, i.e., the number of resamples. In practice B is often 3000 or more.

1. Set $j = 1$.

2. While $j \leq B$, do steps (2)-(5).

3. Let $\mathbf{x}_j^*$ be a random sample of size n drawn from the sample $\mathbf{x}$. That is, the observations $\mathbf{x}_j^*$ are drawn at random from $x_1, x_2, \ldots, x_n$, with replacement.

4. Let $\widehat{\theta}_j^* = \widehat{\theta}(\mathbf{x}_j^*)$.

5. Replace j by $j + 1$.

6. Let $\widehat{\theta}_{(1)}^* \leq \widehat{\theta}_{(2)}^* \leq \cdots \leq \widehat{\theta}_{(B)}^*$ denote the ordered values of $\widehat{\theta}_1^*, \widehat{\theta}_2^*, \ldots, \widehat{\theta}_B^*$. Let $m = [(\alpha/2)B]$, where $[\cdot]$ denotes the greatest integer function. Form the interval

$$(\widehat{\theta}_{(m)}^*, \widehat{\theta}_{(B+1-m)}^*); \tag{5.9.4}$$

that is, obtain the $\frac{\alpha}{2}100\%$ and $(1 - \frac{\alpha}{2})100\%$ percentiles of the sampling distribution of $\widehat{\theta}_1^*, \widehat{\theta}_2^*, \ldots, \widehat{\theta}_B^*$.

The interval in (5.9.4) is called the **percentile bootstrap** confidence interval for θ.

In Step 6, the subscripted parenthetical notation is a common notation for order statistics which will be handy in this section.

Example 5.9.1. In this example, we will sample form a known distribution, but, in practice, the distribution is usually unknown. Let $X_1, X_2, \ldots, X_n$ be a random sample from a $\Gamma(1, \beta)$ distribution. Since the mean of this distribution is β, the sample average $\overline{X}$ is an unbiased estimator of β. In this example, $\overline{X}$ will serve as our point estimator of β. The following 20 data points are the realizations (rounded) of a random sample of size $n = 20$ from a $\Gamma(1, 100)$ distribution:

131.7	182.7	73.3	10.7	150.4	42.3	22.2	17.9	264.0	154.4
4.3	265.6	61.9	10.8	48.8	22.5	8.8	150.6	103.0	85.9

The value of $\overline{X}$ for this sample is $\overline{x} = 90.59$, which is our point estimate of β. For illustration, we generated one bootstrap sample of these data. This ordered bootstrap sample is:

4.3	4.3	4.3	10.8	10.8	10.8	10.8	17.9	22.5	42.3
48.8	48.8	85.9	131.7	131.7	150.4	154.4	154.4	264.0	265.6

As Exercise 5.9.1 shows, in general, the sample mean of a bootstrap sample is an unbiased estimator of original sample mean $\overline{x}$. The sample mean of this particular bootstrap sample is $\overline{x}^* = 78.725$. We wrote an R-function to generate bootstrap samples and the percentile confidence interval above; see the program `percentciboot.s` of Appendix C. Figure 5.9.1 displays a histogram of 3000 $\overline{x}^*$s for the above sample. The sample mean of these 3000 values is 90.13, close to $\overline{x} = 90.59$. Our program also obtained a 90% (bootstrap percentile) confidence interval given by $(61.655, 120.48)$ which the reader can locate on the figure. It did trap $\mu = 100$.

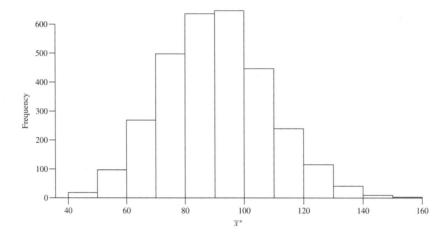

Figure 5.9.1: Histogram of the 3000 bootstrap $\overline{x}^*$s. The bootstrap confidence interval is $(61.655, 120.48)$.

Exercise 5.9.2 shows if we are sampling from a $\Gamma(1, \beta)$ distribution then the interval $(2n\overline{X}/[\chi_{2n}^2]^{(1-(\alpha/2))}, 2n\overline{X}/[\chi_{2n}^2]^{(\alpha/2)})$ is an exact $(1 - \alpha)100\%$ confidence interval for β. Note that, in keeping with our superscript notation for critical values, $[\chi_{2n}^2]^{(\gamma)}$ denotes the $\gamma100\%$ percentile of a χ^2 distribution with $2n$ degrees of freedom. The 90% confidence interval for our sample is $(64.99, 136.69)$. ∎

What about the validity of a bootstrap confidence interval? Davison and Hinkley (1997) discuss the theory behind the bootstrap in Chapter 2 of their book. We shall briefly outline it, using much of their notation. Consider a general situation, in which we are interested in a random variable of the form $Q_n(\mathbf{X}; F)$, where $\mathbf{X} = (X_1, X_2, \ldots, X_n)$ is a random sample from a distribution with cdf $F(x)$. For instance, $Q_n(\mathbf{X}; F)$ could be a point estimator $\widehat{\theta} = \widehat{\theta}(\mathbf{X})$, as above, or it could be a test statistic of the form $\sqrt{n}(\overline{X} - \theta_0)$. Our goal is to estimate the cdf of Q_n denoted by

$$G_{F,n}(x) = P_F[Q_n(\mathbf{X}; F) \leq x], \quad -\infty < x < \infty; \tag{5.9.5}$$

where the subscript F means that the probability is taken under $F(x)$, the true cdf of X_i. The bootstrap estimate of (5.9.5) is given by

$$G_{\hat{F}_n,n}(x) = P_{\hat{F}_n}[Q_n(\mathbf{X}^*; \hat{F}_n) \leq x], \quad -\infty < x < \infty. \tag{5.9.6}$$

The consistency of the bootstrap procedures follows, if we can establish, for all $\epsilon > 0$ and for $-\infty < x < \infty$,

$$\lim_{n \to \infty} P[|G_{\hat{F}_n,n}(x) - G_{F,\infty}(x)| > \epsilon] = 0, \tag{5.9.7}$$

where $G_{F,\infty}(x)$ denotes the limiting distribution of $G_{F,n}(x)$. Davison and Hinkley provide three regularity conditions to ensure this convergence, which we have not bothered to list. Example 5.9.1 shows the plausibility of this theory. The $\overline{x}^*$s have a bell-shaped histogram and the limiting distribution of $\sqrt{n}(\overline{X} - \theta)$ is normal by the Central Limit Theorem.

One way of improving the bootstrap is to use a pivot random variable, a variable whose distribution is free of other parameters. For instance, in the last example, instead of using $\overline{X}$ use $\overline{X}/\hat{\sigma}_{\overline{X}}$, where $\hat{\sigma}_{\overline{X}} = S/\sqrt{n}$ and $S = [\sum(X_i - \overline{X})^2/(n-1)]^{1/2}$; that is, adjust $\overline{X}$ by its standard error. This is discussed in Exercise 5.9.5. Other improvements are discussed in the two books cited earlier.

5.9.2 Bootstrap Testing Procedures

Bootstrap procedures can also be used effectively in testing hypotheses. We begin by discussing these procedures for two sample problems, which cover many of the nuances of the use of the bootstrap in testing.

Consider a two sample location problem; that is, $\mathbf{X}' = (X_1, X_2, \ldots, X_{n_1})$ is a random sample from a distribution with cdf $F(x)$ and $\mathbf{Y}' = (Y_1, Y_2, \ldots, Y_{n_2})$ is a random sample from a distribution with the cdf $F(x - \Delta)$, where $\Delta \in R$. The parameter Δ is the shift in locations between the two samples. Hence, Δ can be written as the difference in location parameters. In particular, assuming that the means μ_Y and μ_X exist, we have $\Delta = \mu_Y - \mu_X$. We consider the one sided hypotheses given by

$$H_0 : \Delta = 0 \text{ versus } H_A : \Delta > 0 . \tag{5.9.8}$$

As our test statistic, we take the difference in sample means, i.e.,

$$V = \overline{Y} - \overline{X}. \tag{5.9.9}$$

Our decision rule is to reject H_0 if $V \geq c$. As is often done in practice, we will base our decision on the p-value of the test. Recall if the samples result in the values $x_1, x_2, \ldots, x_{n_1}$ and $y_1, y_2, \ldots, y_{n_2}$ with realized sample means $\overline{x}$ and $\overline{y}$ respectively, then the p-value of the test is

$$\hat{p} = P_{H_0}[V \geq \overline{y} - \overline{x}]. \tag{5.9.10}$$

Our goal is a bootstrap estimate of the p-value. But, unlike the last section, the bootstraps here have to be performed when H_0 is true. An easy way to do this is to combine the samples into one large sample and then to resample at random and with replacement the combined sample into two samples, one of size n_1 (new xs) and one of size n_2 (new ys). Hence, the resampling is performed under one distribution; i.e, H_0 is true. Let B be an positive integer. Our bootstrap algorithm is

1. Combine the samples into one sample: $\mathbf{z}' = (\mathbf{x}', \mathbf{y}')$.

2. Set $j = 1$.

3. While $j \leq B$ do steps (3)-(6).

4. Obtain a random sample with replacement of size n_1 from $\mathbf{Z}$. Call the sample $\mathbf{x}^{*\prime} = (x_1^*, x_2^*, \ldots, x_{n_1}^*)$. Compute $\overline{x}_j^*$.

5. Obtain a random sample with replacement of size n_2 from **Z**. Call the sample $\mathbf{y}^{*\prime} = (y_1^*, y_2^*, \ldots, y_{n_2}^*)$. Compute $\bar{y}_j^*$.

6. Compute $v_j^* = \bar{y}_j^* - \bar{x}_j^*$.

7. The bootstrap estimated p-value is given by

$$\hat{p}^* = \frac{\#_{j=1}^B \{v_j^* \geq v\}}{B}. \tag{5.9.11}$$

Note that the theory discussed for the bootstrap confidence intervals covers this testing situation also. In particular, the situation we are considering (two sample location model with finite means) satisfies the regularity conditions. Hence, this bootstrap p-value is valid, (consistent).

Example 5.9.2. For illustration, we generated data sets from a contaminated normal distribution. Let W be a random variable with the contaminated normal distribution (3.4.14) with proportion of contamination $\epsilon = 0.20$ and $\sigma_c = 4$. Thirty independent observations $W_1, W_2, \ldots, W_{30}$ were generated from this distribution. Then we let $X_i = 10W_i + 100$ for $1 \leq i \leq 15$ and $Y_i = 10W_{i+15} + 120$ for $1 \leq i \leq 15$. Hence, the true shift parameter is $\Delta = 20$. The actual (rounded) data are:

			X variates				
94.2	111.3	90.0	99.7	116.8	92.2	166.0	95.7
109.3	106.0	111.7	111.9	111.6	146.4	103.9	
			Y variates				
125.5	107.1	67.9	98.2	128.6	123.5	116.5	143.2
120.3	118.6	105.0	111.8	129.3	130.8	139.8	

Based on the comparison boxplots below, the scales of the two data sets appear to be the same while the y-variates (Sample 2) appear to be shift to the right of x-variates (Sample 1).

```
                                    --------
Sample 1                    ----I       +I--              *           0
                                    --------


                                  ----------
Sample 2        *            ------I      +      I--------
                                  ----------
             +---------+---------+---------+---------+---------+------C3
             60        80        100       120       140       160
```

There are three outliers in the data sets.

Our test statistic for these data is $v = \bar{y} - \bar{x} = 117.74 - 111.11 = 6.63$. Computing with the R-program `boottesttwo.s` found in Appendix C, we performed the bootstrap algorithm given above for $B = 3000$ bootstrap replications. The bootstrap p-value was $\widehat{p}^* = 0.169$. This means that $(0.169)(3000) = 507$ of the bootstrap test statistics exceeded the value of the test statistic. Furthermore these bootstrap values were generated under H_0. In practice, H_0 would generally not be rejected for a p-value this high. In Figure 5.9.2, we display a histogram of the 3000 values of the bootstrap test statistic that were obtained. The relative area to the right of the value of the test statistic, 6.63, is approximately equal to $\widehat{p}^*$.

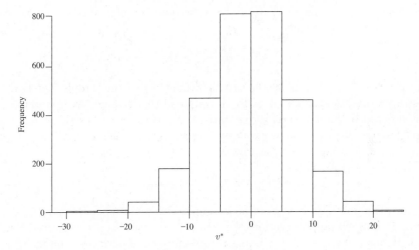

Figure 5.9.2: Histogram of the 3000 bootstrap v^*s. Locate the value of the test statistic $v = \overline{y} - \overline{x} = 6.63$ on the horizontal axis. The area (proportional to overall area) to the right is the p-value of the bootstrap test.

For comparison purposes, we used the two sample "pooled" t-test discussed in Example 5.6.2 to test these hypotheses. As the reader can obtain in Exercise 5.9.7, for these data, $t = 0.93$ with a p-value of 0.18, which is quite close to the bootstrap p-value. ∎

The above test uses the difference in sample means as the test statistic. Certainly other test statistics could be used. Exercise 5.9.6 asks the reader to obtain the bootstrap test based on the difference in sample medians. Often, as with confidence intervals, standardizing the test statistic by a scale estimator improves the bootstrap test.

The bootstrap test described above for the two sample problem is analogous to permutation tests. In the permutation test, the test statistic is calculated for all possible samples of xs and ys drawn without replacement from the combined data. Often, it is approximated by Monte Carlo methods in which case it is quite similar to the bootstrap test except, in the case of the bootstrap, the sampling is done with replacement; see Exercise 5.9.9. Usually the permutation tests and the bootstrap tests give very similar solutions; see Efron and Tibshirani (1993) for discussion.

As our second testing situation, consider a one sample location problem. Suppose $X_1, X_2, \ldots, X_n$ is a random sample from a continuous cdf $F(x)$ with finite mean μ. Suppose we want to test the hypotheses

$$H_0 : \mu = \mu_0 \text{ versus } H_A : \mu > \mu_0,$$

where μ_0 is specified. As a test statistic we use $\overline{X}$ with the decision rule:

Reject H_0 in favor of H_A if $\overline{X}$ is too large.

Let $x_1, x_2, \ldots, x_n$ be the realization of the random sample. We base our decision on p-value of the test, namely

$$\widehat{p} = P_{H_0}[\overline{X} \geq \overline{x}],$$

where $\overline{x}$ is the realized value of the sample average when the sample is drawn. Our bootstrap test is to obtain a bootstrap estimate of this p-value. At first glance, one might proceed by bootstrapping the statistic $\overline{X}$. But, note that the p-value must be estimated under H_0. One way of assuring H_0 is true is instead of bootstrapping $x_1, x_2, \ldots, x_n$ is to bootstrap the values:

$$z_i = x_i - \overline{x} + \mu_0, \quad i = 1, 2, \ldots, n. \tag{5.9.12}$$

Our bootstrap procedure is to randomly sample with replacement from $z_1, z_2, \ldots, z_n$. Letting z^* be such an observation, it is easy to see that $E(z^*) = \mu_0$; see Exercise 5.9.10. Hence using the z_is, the bootstrap resampling is performed under H_0.

To be precise, here is our algorithm to compute this bootstrap test. Let B be an positive integer.

1. Form the vector of shifted observations: $\mathbf{z}' = (z_1, z_2, \ldots, z_n)$, where $z_i = x_i - \overline{x} + \mu_0$.

2. Set $j = 1$.

3. While $j \leq B$ do steps (3)-(5).

4. Obtain a random sample with replacement of size n from $\mathbf{z}$. Call the sample $\mathbf{z}_j^*$. Compute its sample mean $\overline{z}_j^*$.

5. j is replaced by $j + 1$.

6. The bootstrap estimated p-value is given by

$$\widehat{p}^* = \frac{\#_{j=1}^B \{\overline{z}_j^* \geq \overline{x}\}}{B}. \tag{5.9.13}$$

The theory discussed for the bootstrap confidence intervals remains valid for this testing situation also.

Example 5.9.3. To illustrate the bootstrap test described in the last paragraph consider the following data set. We generated $n = 20$ observations $X_i = 10W_i + 100$

where W_i has a contaminated normal distribution with proportion of contamination 20% and $\sigma_c = 4$. Suppose we are interested in testing

$$H_0 : \mu = 90 \text{ versus } H_A : \mu > 90.$$

Because the true mean of X_i is 100, the null hypothesis is false. The data generated are:

119.7	104.1	92.8	85.4	108.6	93.4	67.1	88.4	101.0	97.2
95.4	77.2	100.0	114.2	150.3	102.3	105.8	107.5	0.9	94.1

The sample mean of these values is $\overline{x} = 95.27$ which exceeds 90, but is it significantly over 90? We wrote a R-function to perform the algorithm described above, bootstrapping the values $z_i = x_i - 95.27 + 90$; see program *boottestonemean* of Appendix C. We obtained 3000 values $\overline{z}_j^*$ which are displayed in the histogram in Figure 5.9.3. The mean of these 3000 values is 89.96, which is quite close to 90. Of these 3000 values, 563 exceeded $\overline{x} = 95.27$; hence, the p-value of the bootstrap test is 0.188. The fraction of the total area which is to the right of 95.27 in Figure 5.9.3 is approximately equal to 0.188. Such a high p-value is usually deemed nonsignificant; hence, the null hypothesis would not be rejected.

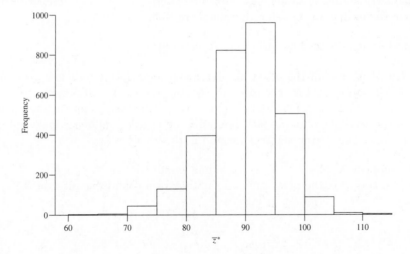

Figure 5.9.3: Histogram of the 3000 bootstrap $\overline{z}^*$s discussed Example 5.9.3. The bootstrap p-value is the area (relative to the total area) under the histogram to right of the 95.27.

For comparison, the reader is asked to show in Exercise 5.9.11 that the value of the one sample t-test is $t = 0.84$ which has a p-value of 0.20. A test based on the median is discussed in Exercise 5.9.12. ∎

EXERCISES

5.9.1. Let $x_1, x_2, \ldots, x_n$ be the values of a random sample. A bootstrap sample, $\mathbf{x}^{*\prime} = (x_1^*, x_2^*, \ldots, x_n^*)$, is a random sample of $x_1, x_2, \ldots, x_n$ drawn with replacement.

(a) Show that $x_1^*, x_2^*, \ldots, x_n^*$ are iid with common cdf $\widehat{F}_n$, the empirical cdf of $x_1, x_2, \ldots, x_n$.

(b) Show that $E(x_i^*) = \overline{x}$.

(c) If n is odd, show that median $\{x_i^*\} = x_{((n+1)/2)}$.

(d) Show that $V(x_i^*) = n^{-1} \sum_{i=1}^{n} (x_i - \overline{x})^2$.

5.9.2. Let $X_1, X_2, \ldots, X_n$ be a random sample from a $\Gamma(1, \beta)$ distribution.

(a) Show that the confidence interval $(2n\overline{X}/(\chi_{2n}^2)^{(1-(\alpha/2))}, 2n\overline{X}/(\chi_{2n}^2)^{(\alpha/2)})$ is an exact $(1 - \alpha)100\%$ confidence interval for β.

(b) Show that value of a 90% confidence interval for the data of the example is $(64.99, 136.69)$.

5.9.3. Consider the situation discussed in Example 5.9.1. Suppose we want to estimate the median of X_i using the sample median.

(a) Determine the median for a $\Gamma(1, \beta)$ distribution.

(b) The algorithm for the bootstrap percentile confidence intervals is general and hence can be used for the median. Rewrite the R-code in program *percentciboot.s* of Appendix C so the median is the estimator. Using the sample given in the example, obtain a 90% bootstrap percentile confidence interval for the median. Did it trap the true median in this case?

5.9.4. Suppose $X_1, X_2, \ldots, X_n$ is a random sample drawn from a $N(\mu, \sigma^2)$ distribution. In this case, the pivot random variable for a confidence interval is

$$t = \frac{\overline{X} - \mu}{S/\sqrt{n}}, \tag{5.9.14}$$

where $\overline{X}$ and S are the sample mean and standard deviation, respectfully. Recall by Theorem 3.6.1 that t has a Student t-distribution with $n - 1$ degrees of freedom; hence, its distribution is free of all parameters for this normal situation. In the notation of this section $t_{n-1}^{(\gamma)}$ denotes the $\gamma 100\%$ percentile of a t-distribution with $n - 1$ degrees of freedom. Using this notation show that a $(1 - \alpha)100\%$ confidence interval for μ is

$$\left(\overline{x} - t^{(1-\alpha/2)} \frac{s}{\sqrt{n}}, \overline{x} - t^{(\alpha/2)} \frac{s}{\sqrt{n}} \right). \tag{5.9.15}$$

5.9.5. Frequently the bootstrap percentile confidence interval can be improved if the estimator $\hat{\theta}$ is standardized by an estimate of scale. To illustrate this consider a bootstrap for a confidence interval for the mean. Let $x_1^*, x_2^*, \ldots, x_n^*$ be a bootstrap sample drawn from the sample $x_1, x_2, \ldots, x_n$. Consider the bootstrap pivot (analogue of (5.9.14):

$$t^* = \frac{\bar{x}^* - \bar{x}}{s^*/\sqrt{n}}, \tag{5.9.16}$$

where $\bar{x}^* = n^{-1} \sum_{i=1}^n x_i^*$ and

$$s^{*2} = (n-1)^{-1} \sum_{i=1}^n (x_i^* - \bar{x}^*)^2.$$

(a) Rewrite the percentile bootstrap confidence interval algorithm using the mean and collecting t_j^*, for $j = 1, 2, \ldots B$. Form the interval

$$\left(\bar{x} - t^{*(1-\alpha/2)} \frac{s}{\sqrt{n}}, \bar{x} - t^{*(\alpha/2)} \frac{s}{\sqrt{n}} \right), \tag{5.9.17}$$

where $t^{*(\gamma)} = t_{([\gamma*B])}^*$; that is, order the t_j^*s and pick off the quantiles.

(b) Rewrite the R-program `percentciboot.s` of Appendix C and use it to find a 90% confidence interval for μ for the data in Example 5.9.3. Use 3000 bootstraps.

(c) Compare your confidence interval in the last part with the nonstandardized bootstrap confidence interval based on program `percentciboot.s` of Appendix C.

5.9.6. Consider the algorithm for a two sample bootstrap test given in Section 5.9.2.

(a) Rewrite the algorithm for the bootstrap test based on the difference in medians.

(b) Consider the data in Example 5.9.2. By substituting the difference in medians for the difference in means in the R-program `boottesttwo.s` of Appendix C, obtain the bootstrap test for the algorithm of Part (a).

(c) Obtain the estimated p-value of your test for $B = 3000$ and compare it to the estimated p-value of 0.063 which the authors obtained.

5.9.7. Consider the data of Example 5.9.2. The two sample t-test of Example 5.6.2 can be used to test these hypotheses. The test is not exact here (Why?), but it is an approximate test. Show that the value of the test statistic is $t = 0.93$ with an approximate p-value of 0.18.

5.9.8. In Example 5.9.3, suppose we are testing the two sided hypotheses,

$$H_0 : \mu = 90 \text{ versus } H_A : \mu \neq 90.$$

(a) Determine the bootstrap p-value for this situation.

(b) Rewrite the R-program `boottestonemean` of Appendix C to obtain this p-value.

(c) Compute the p-value based on 3000 bootstraps.

5.9.9. Consider the following permutation test for the two sample problem with hypotheses (5.9.8). Let $\mathbf{x'} = (x_1, x_2, \ldots, x_{n_1})$ and $\mathbf{y'} = (y_1, y_2, \ldots, y_{n_2})$ be the realizations of the two random samples. The test statistic is the difference in sample means $\overline{y} - \overline{x}$. The estimated p-value of the test is calculated as follows:

1. Combine the data into one sample $\mathbf{z'} = (\mathbf{x'}, \mathbf{y'})$.

2. Obtain all possible samples of size n_1 drawn without replacement from $\mathbf{z}$. Each such sample automatically gives another sample of size n_2; i.e, all elements of $\mathbf{z}$ not in the sample of size n_1. There are $M = \binom{n_1 + n_2}{n_1}$ such samples.

3. For each such sample j:

    (a) Label the sample of size n_1 by $\mathbf{x^*}$ and label the sample of size n_2 by $\mathbf{y^*}$.

    (b) Calculate $v_j^* = \overline{y}^* - \overline{x}^*$.

4. The estimated p-value is $\widehat{p}^* = \#\{v_j^* \geq \overline{y} - \overline{x}\}/M$.

(a) Suppose we have two samples each of size 3 which result in the realizations: $\mathbf{x'} = (10, 15, 21)$ and $\mathbf{y'} = (20, 25, 30)$. Determine the test statistic and the permutation test described above along with the p-value.

(b) If we ignore distinct samples, then we can approximate the permutation test by using the bootstrap algorithm with resampling performed at random and without replacement. Modify the bootstrap program `boottesttwo.s` of Appendix C to do this and obtain this approximate permutation test based on 3000 resamples for the data of Example 5.9.2.

(c) In general what is the probability of having distinct samples in the approximate permutation test described in the last part? Assume that the original data are distinct values.

5.9.10. Let z^* be drawn at random from the discrete distribution which has mass n^{-1} at each point $z_i = x_i - \overline{x} + \mu_0$, where $(x_1, x_2, \ldots, x_n)$ is the realization of a random sample. Determine $E(z^*)$ and $V(z^*)$.

5.9.11. For the situation described in Example 5.9.3, show that the value of the one sample t-test is $t = 0.84$ and its associated p-value is 0.20.

5.9.12. For the situation described in Example 5.9.3, obtain the bootstrap test based on medians. Use the same hypotheses; i.e,

$$H_0 : \mu = 90 \text{ versus } H_A : \mu > 90.$$

5.9.13. Consider the Darwin's experiment on *Zea Mays* discussed in Examples 5.5.1 and 5.5.5.

(a) Obtain a bootstrap test for this experimental data. Keep in mind that the data are recorded in pairs. Hence, your resampling must keep this dependence in tact and still be under H_0.

(b) Provided computational facilities exist, write an R program that executes your bootstrap test and compare its p-value with that found in Example 5.5.5.

Chapter 6

Maximum Likelihood Methods

6.1 Maximum Likelihood Estimation

In this chapter we develop statistical inference (estimation and testing) based on likelihood methods. We show that these procedures are asymptotically optimal under certain conditions (regularity conditions). Suppose that $X_1, \ldots, X_n$ are iid random variables with common pdf $f(x; \theta), \theta \in \Omega$. In general, we will use pdf rather than pmf, $p(x; \theta)$, but the results extend to the discrete case, also. For now, we assume that θ is a scalar but we will extend the results to vectors in Sections 6.4 and 6.5. The parameter θ is unknown. The basis of our inferential procedures is the likelihood function given by,

$$L(\theta; \mathbf{x}) = \prod_{i=1}^{n} f(x_i; \theta), \quad \theta \in \Omega, \tag{6.1.1}$$

where $\mathbf{x} = (x_1, \ldots, x_n)'$. Because we will treat L as a function of θ in this chapter, we have transposed the x_i and θ in the argument of the likelihood function. In fact we will often write it as $L(\theta)$. Actually, the log of this function is usually more convenient to work with mathematically. Denote the $\log L(\theta)$ by

$$l(\theta) = \log L(\theta) = \sum_{i=1}^{n} \log f(x_i; \theta), \quad \theta \in \Omega. \tag{6.1.2}$$

Note that there is no loss of information in using $l(\theta)$ because the log is a one-to-one function. Most of our discussion in this chapter remains the same if X is a random vector. Although, we will generally consider X as a random variable, for several of our examples it will be a random vector.

To motivate the use of the likelihood function, we begin with a simple example and then provide a theoretical justification.

311

Example 6.1.1. Let $X_1, X_2, \ldots, X_n$ denote a random sample from the distribution with pmf

$$p(x) = \begin{cases} \theta^x (1-\theta)^{1-x} & x = 0, 1 \\ 0 & \text{elsewhere,} \end{cases}$$

where $0 \leq \theta \leq 1$. The probability that $X_1 = x_1$, $X_2 = x_2, \ldots, X_n = x_n$ is the joint pmf

$$\theta^{x_1}(1-\theta)^{1-x_1}\theta^{x_2}(1-\theta)^{1-x_2} \cdots \theta^{x_n}(1-\theta)^{1-x_n} = \theta^{\sum x_i}(1-\theta)^{n-\sum x_i},$$

where x_i equals zero or 1, $i = 1, 2, \ldots, n$. This probability, which is the joint pmf of $X_1, X_2, \ldots, X_n$, as a function of θ is the likelihood function $L(\theta)$ defined above. That is,

$$L(\theta) = \theta^{\sum x_i}(1-\theta)^{n-\sum x_i}, \quad 0 \leq \theta \leq 1.$$

We might ask what value of θ would maximize the probability $L(\theta)$ of obtaining this particular observed sample $x_1, x_2, \ldots, x_n$. Certainly, this maximizing value of θ would seemingly be a good estimate of θ because it would provide the largest probability of this particular sample. Since the likelihood function $L(\theta)$ and its logarithm, $l(\theta) = \log L(\theta)$, are maximized for the same value of θ, either $L(\theta)$ or $l(\theta)$ can be used. Here

$$l(\theta) = \log L(\theta) = \left(\sum_1^n x_i\right) \log \theta + \left(n - \sum_1^n x_i\right) \log(1-\theta);$$

so we have

$$\frac{dl(\theta)}{d\theta} = \frac{\sum x_i}{\theta} - \frac{n - \sum x_i}{1-\theta} = 0,$$

provided that θ is not equal to zero or 1. This is equivalent to the equation

$$(1-\theta)\sum_1^n x_i = \theta\left(n - \sum_1^n x_i\right),$$

whose solution for θ is $\sum_1^n x_i/n$. That $\sum_1^n x_i/n$ actually maximizes $L(\theta)$ and $\log L(\theta)$ can be easily checked, even in the cases in which all of $x_1, x_2, \ldots, x_n$ equal zero together or one together. That is, $\sum_1^n x_i/n$ is the value of θ that maximizes $L(\theta)$. The corresponding statistic

$$\hat{\theta} = \frac{1}{n}\sum_{i=1}^n X_i = \overline{X},$$

is called the *maximum likelihood estimator* of θ. As formally defined below, we will call $\sum_1^n x_i/n$ the *maximum likelihood estimate* of θ. For a simple example, suppose that $n = 3$, and $x_1 = 1$, $x_2 = 0$, $x_3 = 1$, then $L(\theta) = \theta^2(1-\theta)$ and the observed $\hat{\theta} = \frac{2}{3}$ is the maximum likelihood estimate of θ. ∎

Let θ_0 denote the *true value* of θ. Theorem 6.1.1 gives a theoretical reason for maximizing the likelihood function. It says that the maximum of $L(\theta)$ asymptotically separates the true model at θ_0 from models at $\theta \neq \theta_0$. To prove this theorem, we will assume certain assumptions, usually called *regularity conditions*.

Assumptions 6.1.1. (Regularity Conditions).

(R0): *The pdfs are distinct; i.e.,* $\theta \neq \theta' \Rightarrow f(x_i; \theta) \neq f(x_i; \theta')$.

(R1): *The pdfs have common support for all θ.*

(R2): *The point θ_0 is an interior point in Ω.*

The first assumption states that the parameters identify the pdfs. The second assumption implies that the support of X_i does not depend on θ. This is restrictive and some examples and exercises will cover models where (R1) is not true.

Theorem 6.1.1. *Let θ_0 be the true parameter. Under assumptions (R0) and (R1),*

$$\lim_{n \to \infty} P_{\theta_0}[L(\theta_0, \mathbf{X}) > L(\theta, \mathbf{X})] = 1, \quad \text{for all } \theta \neq \theta_0. \tag{6.1.3}$$

Proof: By taking logs, the inequality $L(\theta_0, \mathbf{X}) > L(\theta, \mathbf{X})$ is equivalent to

$$\frac{1}{n} \sum_{i=1}^{n} \log \left[\frac{f(X_i; \theta)}{f(X_i; \theta_0)} \right] < 0.$$

Because the summands are iid with finite expectation and the function $\phi(x) = -\log(x)$ is strictly convex, it follows from the Law of Large Numbers (Theorem 4.2.1) and Jensen's inequality, (Theorem 1.10.5), that, when θ_0 is the true parameter,

$$\frac{1}{n} \sum_{i=1}^{n} \log \left[\frac{f(X_i; \theta)}{f(X_i; \theta_0)} \right] \xrightarrow{P} E_{\theta_0} \left[\log \frac{f(X_1; \theta)}{f(X_1; \theta_0)} \right] < \log E_{\theta_0} \left[\frac{f(X_1; \theta)}{f(X_1; \theta_0)} \right].$$

But

$$E_{\theta_0} \left[\frac{f(X_1; \theta)}{f(X_1; \theta_0)} \right] = \int \frac{f(x; \theta)}{f(x; \theta_0)} f(x; \theta_0) \, dx = 1.$$

Because $\log 1 = 0$, the theorem follows. Note that common support is needed to obtain the last equalities. ∎

Theorem 6.1.1 says that asymptotically the likelihood function is maximized at the true value θ_0. So in considering estimates of θ_0, it seems natural to consider the value of θ which maximizes the likelihood.

Definition 6.1.1 (Maximum Likelihood Estimator). *We say that $\hat{\theta} = \hat{\theta}(\mathbf{X})$ is a* **maximum likelihood estimator** *(mle) of θ if*

$$\hat{\theta} = \operatorname{Argmax} L(\theta; \mathbf{X}); \tag{6.1.4}$$

The notation Argmax means that $L(\theta; \mathbf{X})$ achieves its maximum value at $\hat{\theta}$.

As in the example above, to determine the mle, we often take the log of the likelihood and determine its critical value; that is, letting $l(\theta) = \log L(\theta)$ the mle solves the equation

$$\frac{\partial l(\theta)}{\partial \theta} = 0. \tag{6.1.5}$$

This is an example of an **estimating equation** which we will often label as an EE. This is the first of several EEs in the text.

There is no guarantee that the mle exists or if it does whether it is unique. This is often clear from the application as in the next three examples. Other examples are given in the exercises.

Example 6.1.2 (Exponential Distribution). Suppose the common pdf is the exponential(θ) density given by (3.3.2). The log of the likelihood function is given by,

$$l(\theta) = -n \log \theta - \theta^{-1} \sum_{i=1}^{n} x_i.$$

For this example, differentiable calculus leads directly to the mle. The first partial of the log-likelihood with respect to θ is:

$$\frac{\partial l}{\partial \theta} = -n\theta^{-1} + \theta^{-2} \sum_{i=1}^{n} x_i.$$

Setting this partial to 0 and solving for θ we obtain the solution $\bar{x}$. There is only one critical value and, furthermore, the second partial of the log likelihood evaluated at $\bar{x}$ is strictly negative, verifying that it is indeed a maximum. Hence, for this example the statistic $\widehat{\theta} = \overline{X}$ is the mle of θ. ∎

Example 6.1.3 (Laplace Distribution). Let $X_1, \ldots, X_n$ be iid with density

$$f(x; \theta) = \frac{1}{2} e^{-|x-\theta|}, \quad -\infty < x < \infty, -\infty < \theta < \infty. \tag{6.1.6}$$

This pdf is referred to as either the *Laplace* or the *double exponential distribution*. The log of the likelihood simplifies to

$$l(\theta) = -n \log 2 - \sum_{i=1}^{n} |x_i - \theta|.$$

The first partial derivative is

$$l'(\theta) = \sum_{i=1}^{n} \text{sgn}(x_i - \theta), \tag{6.1.7}$$

where $\text{sgn}(t) = 1, 0,$ or -1 depending on whether $t > 0, t = 0,$ or $t < 0$. Note that we have used $\frac{d}{dt}|t| = \text{sgn}(t)$ which is true unless $t = 0$. Setting the equation (6.1.7) to 0, the solution for θ is $\text{med}\{x_1, x_2, \ldots, x_n\}$, because the median will make half the terms of the sum in expression (6.1.7) nonpositive and half nonnegative. Recall that we denote the median of a sample by Q_2 (the second quartile of the sample). Hence $\widehat{\theta} = Q_2$ is the mle of θ for the Laplace pdf (6.1.6). ∎

Example 6.1.4 (Logistic Distribution). Let $X_1, \ldots, X_n$ be iid with density

$$f(x; \theta) = \frac{\exp\{-(x - \theta)\}}{(1 + \exp\{-(x - \theta)\})^2}, \quad -\infty < x < \infty, \ -\infty < \theta < \infty. \quad (6.1.8)$$

The log of the likelihood simplifies to

$$l(\theta) = \sum_{i=1}^{n} \log f(x_i; \theta) = n\theta - n\bar{x} - 2 \sum_{i=1}^{n} \log(1 + \exp\{-(x_i - \theta)\}).$$

Using this, the first partial derivative is

$$l'(\theta) = n - 2 \sum_{i=1}^{n} \frac{\exp\{-(x_i - \theta)\}}{1 + \exp\{-(x_i - \theta)\}}. \quad (6.1.9)$$

Setting this equation to 0 and rearranging terms results in the equation,

$$\sum_{i=1}^{n} \frac{\exp\{-(x_i - \theta)\}}{1 + \exp\{-(x_i - \theta)\}} = \frac{n}{2}. \quad (6.1.10)$$

Although this does not simplify, we can show that equation (6.1.10) has a unique solution. The derivative of the left side of equation (6.1.10) simplifies to,

$$(\partial/\partial\theta) \sum_{i=1}^{n} \frac{\exp\{-(x_i - \theta)\}}{1 + \exp\{-(x_i - \theta)\}} = \sum_{i=1}^{n} \frac{\exp\{-(x_i - \theta)\}}{(1 + \exp\{-(x_i - \theta)\})^2} > 0.$$

Thus the left side of equation (6.1.10) is a strictly increasing function of θ. Finally, the left side of (6.1.10) approachs 0 as $\theta \to -\infty$ and approachs n as $\theta \to \infty$. Thus, the equation (6.1.10) has a unique solution. Also the second derivative of $l(\theta)$ is strictly negative for all θ; so the solution is a maximum.

Having shown that the mle exists and is unique, we can use a numerical method to obtain the solution. In this case, Newton's procedure is useful. We discuss this in general in the next section at which time we will reconsider this example. ∎

Even though we did not get the mle in closed form in the last example, in all three of these examples standard differential calculus methods led us to the solution. For the next example, the support of the random variable involves θ and, hence, does not satisfy the regularity conditions. For such cases, differential calculus may not be useful.

Example 6.1.5 (Uniform Distribution). Let $X_1, \ldots, X_n$ be iid with the uniform $(0, \theta)$ density, i.e., $f(x) = 1/\theta$ for $0 < x \leq \theta$, 0 elsewhere. Because θ is in the support, differentiation is not helpful here. The likelihood function can be written as

$$L(\theta) = \theta^{-n} I(\max\{x_i\}, \theta); \quad \text{for all } \theta > 0,$$

where $I(a, b)$ is 1 or 0 if $a \leq b$ or $a > b$, respectively. This function is a decreasing function of θ for all $\theta \geq \max\{x_i\}$ and is 0, otherwise, (please sketch it). Hence, the maximum occurs at the smallest value of θ; i.e., the mle is $\hat{\theta} = \max\{X_i\}$. ∎

Example 6.1.6. In Example 6.1.1, we discussed the mle of the probability of success θ for a random sample $X_1, X_2, \ldots, X_n$ from the Bernoulli distribution with pmf

$$p(x) = \begin{cases} \theta^x(1-\theta)^{1-x} & x = 0, 1 \\ 0 & \text{elsewhere,} \end{cases}$$

where $0 \le \theta \le 1$. Recall that the mle is $\overline{X}$, the proportion of sample successes. Now suppose that we know in advance that, instead of $0 \le \theta \le 1$, θ is restricted by the inequalities $0 \le \theta \le 1/3$. If the observations were such that $\overline{x} > 1/3$, then $\overline{x}$ would not be a satisfactory estimate. Since $\frac{\partial l(\theta)}{\partial \theta} > 0$, provided $\theta < \overline{x}$, under the restriction $0 \le \theta \le 1/3$, we can maximize $l(\theta)$ by taking $\widehat{\theta} = \min\left\{\overline{x}, \frac{1}{3}\right\}$. ∎

The following is an appealing property of maximum likelihood estimates.

Theorem 6.1.2. *Let $X_1, \ldots, X_n$ be iid with the pdf $f(x; \theta), \theta \in \Omega$. For a specified function g, let $\eta = g(\theta)$ be a parameter of interest. Suppose $\widehat{\theta}$ is the mle of θ. Then $g(\widehat{\theta})$ is the mle of $\eta = g(\theta)$.*

Proof: First suppose g is a one-to-one function. The likelihood of interest is $L(g(\theta))$, but because g is one-to-one,

$$\max L(g(\theta)) = \max_{\eta = g(\theta)} L(\eta) = \max_{\eta} L(g^{-1}(\eta)).$$

But the maximum occurs when $g^{-1}(\eta) = \widehat{\theta}$; i.e., take $\widehat{\eta} = g(\widehat{\theta})$.

Suppose g is not one-to-one. For each η in the range of g, define the set (preimage),

$$g^{-1}(\eta) = \{\theta : g(\theta) = \eta\}.$$

The maximum occurs at $\widehat{\theta}$ and the domain of g is Ω which covers $\widehat{\theta}$. Hence, $\widehat{\theta}$ is in one of these preimages and, in fact, it can only be in one preimage. Hence to maximize $L(\eta)$, choose $\widehat{\eta}$ so that $g^{-1}(\widehat{\eta})$ is that unique preimage containing $\widehat{\theta}$. Then $\widehat{\eta} = g(\widehat{\theta})$. ∎

In Example 6.1.5, it might be of interest to estimate $\text{Var}(X) = \theta^2/12$. Hence by Theorem 6.1.2, the mle is $\max\{X_i\}^2/12$. Next, consider Example 6.1.1, where $X_1, \ldots, X_n$ are iid Bernoulli random variables with probability of success p. As shown in the example, $\widehat{p} = \overline{X}$ is the mle of p. Recall that in the large sample confidence interval for p, (5.4.8), an estimate of $\sqrt{p(1-p)}$ is required. By Theorem 6.1.2, the mle of this quantity is $\sqrt{\widehat{p}(1-\widehat{p})}$.

We close this section by showing that maximum likelihood estimators, under regularity conditions, are consistent estimators. Recall that $\mathbf{X}' = (X_1, \ldots, X_n)$.

Theorem 6.1.3. *Assume that $X_1, \ldots, X_n$ satisfy the regularity conditions (R0) - (R2), where θ_0 is the true parameter, and further that $f(x; \theta)$ is differentiable with respect to θ in Ω. Then the likelihood equation,*

$$\frac{\partial}{\partial \theta} L(\theta) = 0$$

or equivalently

$$\frac{\partial}{\partial\theta}l(\theta) = 0$$

has a solution $\widehat{\theta}_n$ such that $\widehat{\theta}_n \overset{P}{\to} \theta_0$.

Proof: Because θ_0 is an interior point in Ω, $(\theta_0 - a, \theta_0 + a) \subset \Omega$, for some $a > 0$. Define S_n to be the event

$$S_n = \{\mathbf{X} : l(\theta_0; \mathbf{X}) > l(\theta_0 - a; \mathbf{X})\} \cap \{\mathbf{X} : l(\theta_0; \mathbf{X}) > l(\theta_0 + a; \mathbf{X})\}.$$

By Theorem 6.1.1, $P(S_n) \to 1$. So we can restrict attention to the event S_n. But on S_n, $l(\theta)$ has a local maximum say, $\widehat{\theta}_n$ such that $\theta_0 - a < \widehat{\theta}_n < \theta_0 + a$ and $l'(\widehat{\theta}_n) = 0$. That is,

$$S_n \subset \left\{\mathbf{X} : |\widehat{\theta}_n(\mathbf{X}) - \theta_0| < a\right\} \cap \left\{\mathbf{X} : l'(\widehat{\theta}_n(\mathbf{X})) = 0\right\}.$$

Therefore,

$$1 = \lim_{n\to\infty} P(S_n) \le \overline{\lim_{n\to\infty}} P\left[\left\{\mathbf{X} : |\widehat{\theta}_n(\mathbf{X}) - \theta_0| < a\right\} \cap \left\{\mathbf{X} : l'(\widehat{\theta}_n(\mathbf{X})) = 0\right\}\right] \le 1;$$

see Remark 4.3.3 for discussion on $\overline{\lim}$. It follows that for the sequence of solutions $\widehat{\theta}_n$, $P[|\widehat{\theta}_n - \theta_0| < a] \to 1$.

The only contentious point in the proof is that the sequence of solutions might depend on a. But we can always choose a solution "closest" to θ_0 in the following way. For each n, the set of all solutions in the interval is bounded, hence the infimum over solutions closest to θ_0 exists. ∎

Note that this theorem is vague in that it discusses solutions of the equation. If, however, we know that the mle is the unique solution of the equation $l'(\theta) = 0$, then it is consistent. We state this as a corollary:

Corollary 6.1.1. *Assume that $X_1, \ldots, X_n$ satisfy the regularity conditions (R0) - (R2), where θ_0 is the true parameter, and that $f(x; \theta)$ is differentiable with respect to θ in Ω. Suppose the likelihood equation has the unique solution $\widehat{\theta}_n$. Then $\widehat{\theta}_n$ is a consistent estimator of θ_0.*

EXERCISES

6.1.1. Let $X_1, X_2, \ldots, X_n$ be a random sample from a $N(\theta, \sigma^2)$ distribution, $-\infty < \theta < \infty$ with σ^2 known. Determine the mle of θ.

6.1.2. Let $X_1, X_2, \ldots, X_n$ be a random sample from a $\Gamma(\alpha = 3, \beta = \theta)$ distribution, $0 < \theta < \infty$. Determine the mle of θ.

6.1.3. Let $X_1, X_2, \ldots, X_n$ represent a random sample from each of the distributions having the following pdfs or pmfs:

(a) $f(x; \theta) = \theta^x e^{-\theta}/x!$, $x = 0, 1, 2, \ldots$, $0 \le \theta < \infty$, zero elsewhere, where $f(0; 0) = 1$.

(b) $f(x; \theta) = \theta x^{\theta-1}$, $0 < x < 1$, $0 < \theta < \infty$, zero elsewhere.

(c) $f(x; \theta) = (1/\theta)e^{-x/\theta}$, $0 < x < \infty$, $0 < \theta < \infty$, zero elsewhere.

(d) $f(x; \theta) = e^{-(x-\theta)}$, $\theta \le x < \infty$, $-\infty < \theta < \infty$, zero elsewhere.

In each case find the mle $\hat{\theta}$ of θ.

6.1.4. Let $Y_1 < Y_2 < \cdots < Y_n$ be the order statistics of a random sample from a distribution with pdf $f(x; \theta) = 1$, $\theta - \frac{1}{2} \le x \le \theta + \frac{1}{2}$, $-\infty < \theta < \infty$, zero elsewhere. Show that every statistic $u(X_1, X_2, \ldots, X_n)$ such that

$$Y_n - \tfrac{1}{2} \le u(X_1, X_2, \ldots, X_n) \le Y_1 + \tfrac{1}{2}$$

is a mle of θ. In particular, $(4Y_1 + 2Y_n + 1)/6$, $(Y_1 + Y_n)/2$, and $(2Y_1 + 4Y_n - 1)/6$ are three such statistics. Thus uniqueness is not, in general, a property of a mle.

6.1.5. Suppose $X_1, \ldots, X_n$ are iid with pdf $f(x; \theta) = 2x/\theta^2$, $0 < x \le \theta$, zero elsewhere, find:

(a) The mle $\hat{\theta}$ for θ.

(b) The constant c so that $E(c\hat{\theta}) = \theta$.

(c) The mle for the median of the distribution.

6.1.6. Suppose $X_1, X_2, \ldots, X_n$ are iid with pdf $f(x; \theta) = (1/\theta)e^{-x/\theta}$, $0 < x < \infty$, zero elsewhere. Find the mle of $P(X \le 2)$.

6.1.7. Let the table

x	0	1	2	3	4	5
Frequency	6	10	14	13	6	1

represent a summary of a sample of size 50 from a binomial distribution having $n = 5$. Find the mle of $P(X \ge 3)$.

6.1.8. Let X_1, X_2, X_3, X_4, X_5 be a random sample from a Cauchy distribution with median θ, that is, with pdf.

$$f(x; \theta) = \frac{1}{\pi} \frac{1}{1 + (x - \theta)^2}, \quad -\infty < x < \infty,$$

where $-\infty < \theta < \infty$. If $x_1 = -1.94$, $x_2 = 0.59$, $x_3 = -5.98$, $x_4 = -0.08$, and $x_5 = -0.77$, find by numerical methods the mle of θ.

6.1.9. Let the table

x	0	1	2	3	4	5
Frequency	7	14	12	13	6	3

represent a summary of a random sample of size 50 from a Poisson distribution. Find the maximum likelihood estimate of $P(X = 2)$.

6.1.10. Let $X_1, X_2, \ldots, X_n$ be a random sample from a Bernoulli distribution with parameter p. If p is restricted so that we know that $\frac{1}{2} \le p \le 1$, find the mle of this parameter.

6.1.11. Let $X_1, X_2, \ldots, X_n$ be random sample from a $N(\theta, \sigma^2)$ distribution, where σ^2 is fixed but $-\infty < \theta < \infty$.

(a) Show that the mle of θ is $\overline{X}$.

(b) If θ is restricted by $0 \le \theta < \infty$, show that the mle of θ is $\widehat{\theta} = \max\{0, \overline{X}\}$.

6.1.12. Let $X_1, X_2, \ldots, X_n$ be a random sample from the Poisson distribution with $0 < \theta \le 2$. Show that the mle of θ is $\widehat{\theta} = \min\{\overline{X}, 2\}$.

6.1.13. Let $X_1, X_2, \ldots, X_n$ ba random sample from a distribution with one of two pdfs. If $\theta = 1$, then $f(x; \theta = 1) = \frac{1}{\sqrt{2\pi}} e^{-x^2/2}$, $-\infty < x < \infty$. If $\theta = 2$, then $f(x; \theta = 2) = 1/[\pi(1 + x^2)]$, $-\infty < x < \infty$. Find the mle of θ.

6.2 Rao-Cramér Lower Bound and Efficiency

In this section we establish a remarkable inequality called the **Rao-Cramér** lower bound which gives a lower bound on the variance of any unbiased estimate. We then show that, under regularity conditions, the variances of the maximum likelihood estimates achieve this lower bound asymptotically.

As in the last section, let X be a random variable with pdf $f(x; \theta)$, $\theta \in \Omega$, where the parameter space Ω is an open interval. In addition to the regularity conditions (6.1.1) of Section 6.1, for the following derivations, we require two more regularity conditions given by

Assumptions 6.2.1. (Additional Regularity Conditions).

(R3): *The pdf $f(x; \theta)$ is twice differentiable as a function of θ.*

(R4): *The integral $\int f(x; \theta)\, dx$ can be differentiated twice under the integral sign as a function of θ.*

Note that conditions (R1)-(R4) mean that the parameter θ does not appear in the endpoints of the interval in which $f(x; \theta) > 0$ and that we can interchange integration and differentiation with respect to θ. Our derivation is for the continuous case but the discrete case can be handled in a similar manner. We begin with the identity

$$1 = \int_{-\infty}^{\infty} f(x; \theta)\, dx.$$

Taking the derivative with respect to θ results in,

$$0 = \int_{-\infty}^{\infty} \frac{\partial f(x; \theta)}{\partial \theta}\, dx.$$

The latter expression can be rewritten as

$$0 = \int_{-\infty}^{\infty} \frac{\partial f(x;\theta)/\partial \theta}{f(x;\theta)} f(x;\theta)\, dx,$$

or, equivalently,

$$0 = \int_{-\infty}^{\infty} \frac{\partial \log f(x;\theta)}{\partial \theta} f(x;\theta)\, dx. \tag{6.2.1}$$

Writing this last equation as an expectation, we have established

$$E\left[\frac{\partial \log f(X;\theta)}{\partial \theta}\right] = 0; \tag{6.2.2}$$

that is, the mean of the random variable $\frac{\partial \log f(X;\theta)}{\partial \theta}$ is 0. If we differentiate (6.2.1) again, it follows that

$$0 = \int_{-\infty}^{\infty} \frac{\partial^2 \log f(x;\theta)}{\partial \theta^2} f(x;\theta)\, dx + \int_{-\infty}^{\infty} \frac{\partial \log f(x;\theta)}{\partial \theta}\frac{\partial \log f(x;\theta)}{\partial \theta} f(x;\theta)\, dx. \tag{6.2.3}$$

The second term of the right-side of this equation can be written as an expectation. We will call this expectation **Fisher information** and denote it by $I(\theta)$; that is,

$$I(\theta) = \int_{-\infty}^{\infty} \frac{\partial \log f(x;\theta)}{\partial \theta}\frac{\partial \log f(x;\theta)}{\partial \theta} f(x;\theta)\, dx = E\left[\left(\frac{\partial \log f(X;\theta)}{\partial \theta}\right)^2\right]. \tag{6.2.4}$$

From Equation (6.2.3), we see that $I(\theta)$ can be computed from

$$I(\theta) = -\int_{-\infty}^{\infty} \frac{\partial^2 \log f(x;\theta)}{\partial \theta^2} f(x;\theta)\, dx. \tag{6.2.5}$$

Using equation (6.2.2), Fisher information is the variance of the random variable $\frac{\partial \log f(X;\theta)}{\partial \theta}$; i.e.,

$$I(\theta) = \mathrm{Var}\left(\frac{\partial \log f(X;\theta)}{\partial \theta}\right). \tag{6.2.6}$$

Usually expression (6.2.5) is easier to compute than expression (6.2.4).

Remark 6.2.1. Note that the information is the weighted mean of either

$$\left[\frac{\partial \log f(x;\theta)}{\partial \theta}\right]^2 \quad \text{or} \quad -\frac{\partial^2 \log f(x;\theta)}{\partial \theta^2},$$

where the weights are given by the pdf $f(x;\theta)$. That is, the greater these derivatives are on the average, the more information that we get about θ. Clearly, if they were equal to zero [so that θ would not be in $\log f(x;\theta)$], there would be zero information about θ. The important function

$$\frac{\partial \log f(x;\theta)}{\partial \theta}$$

is called the **score function**. Recall that it determines the estimating equations for the mle; that is, the mle $\hat{\theta}$ solves

$$\sum_{i=1}^{n} \frac{\partial \log f(x_i; \theta)}{\partial \theta} = 0$$

for θ. ∎

Example 6.2.1 (Information for a Bernoulli Random Variable). Let X be Bernoulli $b(1, \theta)$. Thus

$$
\begin{aligned}
\log f(x; \theta) &= x \log \theta + (1 - x) \log(1 - \theta), \\
\frac{\partial \log f(x; \theta)}{\partial \theta} &= \frac{x}{\theta} - \frac{1 - x}{1 - \theta}, \\
\frac{\partial^2 \log f(x; \theta)}{\partial \theta^2} &= -\frac{x}{\theta^2} - \frac{1 - x}{(1 - \theta)^2}.
\end{aligned}
$$

Clearly,

$$
\begin{aligned}
I(\theta) &= -E\left[\frac{-X}{\theta^2} - \frac{1 - X}{(1 - \theta)^2} \right] \\
&= \frac{\theta}{\theta^2} + \frac{1 - \theta}{(1 - \theta)^2} = \frac{1}{\theta} + \frac{1}{(1 - \theta)} = \frac{1}{\theta(1 - \theta)},
\end{aligned}
$$

which is larger for θ values close to zero or 1. ∎

Example 6.2.2 (Information for a Location Family). Consider a random sample $X_1, \ldots, X_n$ such that

$$X_i = \theta + e_i, \quad i = 1, \ldots, n, \tag{6.2.7}$$

where $e_1, e_2, \ldots, e_n$ are iid with common pdf $f(x)$ and with support $(-\infty, \infty)$. Then the common pdf of X_i is $f_X(x; \theta) = f(x - \theta)$. We call model (6.2.7) a *location model*. Assume that $f(x)$ satisfies the regularity conditions. Then the information is

$$
\begin{aligned}
I(\theta) &= \int_{-\infty}^{\infty} \left(\frac{f'(x - \theta)}{f(x - \theta)} \right)^2 f(x - \theta) \, dx \\
&= \int_{-\infty}^{\infty} \left(\frac{f'(z)}{f(z)} \right)^2 f(z) \, dz, \tag{6.2.8}
\end{aligned}
$$

where the last equality follows from the transformation $z = x - \theta$. Hence, in the location model, the information does not depend on θ.

As an illustration, suppose for $i = 1, \ldots, n$ that X_i has the Laplace pdf, (6.1.6). Then it is easy to see that we can write X_i as,

$$X_i = \theta + e_i, \tag{6.2.9}$$

where $e_1, \ldots, e_n$ are iid with common pdf $f(z) = 2^{-1} \exp\{-|z|\}$, for $-\infty < z < \infty$. As we did in Example 6.1.3, use $\frac{d}{dz}|z| = \text{sgn}(z)$. Then in this case, $f'(z) = -2^{-1}\text{sgn}(z)\exp\{-|z|\}$ and, hence,

$$I(\theta) = \int_{-\infty}^{\infty} \left(\frac{f'(z)}{f(z)}\right)^2 f(z) = \int_{-\infty}^{\infty} f(z)\, dz = 1. \qquad (6.2.10)$$

Note that the Laplace pdf does not satisfy the regularity conditions, but this argument can be made rigorous; see Huber (1981) and, also, Chapter 10. ∎

From (6.2.6), for a sample of size one, say X_1, Fisher information is the variance of the random variable $\frac{\partial \log f(X_1;\theta)}{\partial \theta}$. What about a sample of size n? Let $X_1, X_2, \ldots, X_n$ be a random sample from a distribution having pdf $f(x;\theta)$. As in the last section, let $L(\theta)$, (6.1.1), denote the likelihood function. The function $L(\theta)$ is the pdf of the sample and the random variable whose variance is the information in the sample is given by

$$\frac{\partial \log L(\theta, \mathbf{X})}{\partial \theta} = \sum_{i=1}^{n} \frac{\partial \log f(X_i;\theta)}{\partial \theta}.$$

The summands are iid with common variance $I(\theta)$. Hence, the information in the sample is,

$$\text{Var}\left(\frac{\partial \log L(\theta, \mathbf{X})}{\partial \theta}\right) = nI(\theta). \qquad (6.2.11)$$

Thus the information in a sample of size n is n times the information in a sample of size one. So, in Example 6.2.1, the Fisher information in a random sample of size n from a Bernoulli $b(1, \theta)$ distribution is $n/[\theta(1 - \theta)]$.

We are now ready to obtain the Rao-Cramér lower bound which we state as a theorem.

Theorem 6.2.1 (Rao-Cramér Lower Bound). *Let $X_1, \ldots, X_n$ be iid with common pdf $f(x;\theta)$ for $\theta \in \Omega$. Assume that the regularity conditions (R0)-(R4) hold. Let $Y = u(X_1, X_2, \ldots, X_n)$ be a statistic with mean $E(Y) = E[u(X_1, X_2, \ldots, X_n)] = k(\theta)$. Then*

$$Var(Y) \geq \frac{[k'(\theta)]^2}{nI(\theta)}. \qquad (6.2.12)$$

Proof: The proof is for the continuous case, but the proof for the discrete case is quite similar. Write the mean of Y as

$$k(\theta) = \int_{-\infty}^{\infty} \cdots \int_{-\infty}^{\infty} u(x_1, \ldots, x_n)f(x_1;\theta)\cdots f(x_n;\theta)\, dx_1 \cdots dx_n.$$

Differentiating with respect to θ, we obtain

$$
k'(\theta) = \int_{-\infty}^{\infty} \cdots \int_{-\infty}^{\infty} u(x_1, x_2, \ldots, x_n) \left[\sum_{1}^{n} \frac{1}{f(x_i; \theta)} \frac{\partial f(x_i; \theta)}{\partial \theta} \right]
$$
$$
\times f(x_1; \theta) \cdots f(x_n; \theta) \, dx_1 \cdots dx_n
$$
$$
= \int_{-\infty}^{\infty} \cdots \int_{-\infty}^{\infty} u(x_1, x_2, \ldots, x_n) \left[\sum_{1}^{n} \frac{\partial \log f(x_i; \theta)}{\partial \theta} \right]
$$
$$
\times f(x_1; \theta) \cdots f(x_n; \theta) \, dx_1 \cdots dx_n. \tag{6.2.13}
$$

Define the random variable Z by $Z = \sum_{1}^{n} [\partial \log f(X_i; \theta)/\partial \theta]$. We know from (6.2.2) and (6.2.11) that $E(Z) = 0$ and $\mathrm{Var}(Z) = nI(\theta)$, respectively. Also, equation (6.2.13) can be expressed in terms of expectation as $k'(\theta) = E(YZ)$. Hence, we have

$$
k'(\theta) = E(YZ) = E(Y)E(Z) + \rho \sigma_Y \sqrt{nI(\theta)},
$$

where ρ is the correlation coefficient between Y and Z. Using $E(Z) = 0$, this simplifies to

$$
\rho = \frac{k'(\theta)}{\sigma_Y \sqrt{nI(\theta)}};\cdot
$$

Because $\rho^2 \leq 1$, we have

$$
\frac{[k'(\theta)]^2}{\sigma_Y^2 nI(\theta)} \leq 1,
$$

which, upon rearrangement, is the desired result. ∎

Corollary 6.2.1. *Under the assumptions of Theorem 6.2.1, if $Y = u(X_1, \ldots, X_n)$ is an unbiased estimator of θ, so that $k(\theta) = \theta$, then the Rao-Cramér inequality becomes*

$$
Var(Y) \geq \frac{1}{nI(\theta)}.
$$

Consider the Bernoulli model with probability of success θ which was treated in Example 6.2.1. In the example we showed that $1/nI(\theta) = \theta(1-\theta)/n$. From Example 6.1.1 of Section 6.1, the mle of θ is $\overline{X}$. The mean and variance of a Bernoulli (θ) distribution are θ and $\theta(1 - \theta)$, respectively. Hence the mean and variance of $\overline{X}$ are θ and $\theta(1 - \theta)/n$, respectively. That is, in this case the variance of the mle has attained the Rao-Cramér lower bound.

We now make the following definitions.

Definition 6.2.1. *Let Y be an unbiased estimator of a parameter θ in the case of point estimation. The statistic Y is called an **efficient estimator** of θ if and only if the variance of Y attains the Rao-Cramér lower bound.*

Definition 6.2.2. *In cases in which we can differentiate with respect to a parameter under an integral or summation symbol, the ratio of the Rao-Cramér lower bound to the actual variance of any unbiased estimation of a parameter is called the* **efficiency** *of that estimator.*

Example 6.2.3 (Poisson(θ) Distribution). Let $X_1, X_2, \ldots, X_n$ denote a random sample from a Poisson distribution that has the mean $\theta > 0$. It is known that $\overline{X}$ is an mle of θ; we shall show that it is also an efficient estimator of θ. We have

$$
\begin{aligned}
\frac{\partial \log f(x; \theta)}{\partial \theta} &= \frac{\partial}{\partial \theta}(x \log \theta - \theta - \log x!) \\
&= \frac{x}{\theta} - 1 = \frac{x - \theta}{\theta}.
\end{aligned}
$$

Accordingly,

$$
E\left[\left(\frac{\partial \log f(X; \theta)}{\partial \theta}\right)^2\right] = \frac{E(X - \theta)^2}{\theta^2} = \frac{\sigma^2}{\theta^2} = \frac{\theta}{\theta^2} = \frac{1}{\theta}.
$$

The Rao-Cramér lower bound in this case is $1/[n(1/\theta)] = \theta/n$. But θ/n is the variance of $\overline{X}$. Hence $\overline{X}$ is an efficient estimator of θ. ∎

Example 6.2.4 (Beta($\theta, 1$) Distribution). Let $X_1, X_2, \ldots, X_n$ denote a random sample of size $n > 2$ from a distribution with pdf

$$
f(x; \theta) = \begin{cases} \theta x^{\theta - 1} & \text{for } 0 < x < 1 \\ 0 & \text{elsewhere,} \end{cases} \tag{6.2.14}
$$

where the parameter space is $\Omega = (0, \infty)$. This is the Beta($\theta, 1$) distribution. The derivative of the log of f is

$$
\frac{\partial \log f}{\partial \theta} = \log x + \frac{1}{\theta}. \tag{6.2.15}
$$

From this we have, $\partial^2 \log f / \partial \theta^2 = -\theta^{-2}$. Hence, the information is $I(\theta) = \theta^{-2}$.

Next, we find the mle of θ and investigate its efficiency. The log of the likelihood function is

$$
l(\theta) = \theta \sum_{i=1}^n \log x_i - \sum_{i=1}^n \log x_i + n \log \theta.
$$

The first partial of l is

$$
\frac{\partial \log l}{\partial \theta} = \sum_{i=1}^n \log x_i + \frac{n}{\theta}. \tag{6.2.16}
$$

Setting this to 0 and solving for θ, the mle is $\widehat{\theta} = -n / \sum_{i=1}^n \log X_i$. To obtain the distribution of $\widehat{\theta}$, let $Y_i = -\log X_i$. A straight transformation argument shows that the distribution is $\Gamma(1, 1/\theta)$. Because the X_i's are independent, Theorem 3.3.2 shows that $W = \sum_{i=1}^n Y_i$ is $\Gamma(n, 1/\theta)$. Theorem 3.3.1 shows that,

$$
E[W^k] = \frac{(n + k - 1)!}{\theta^k (n - 1)!}, \tag{6.2.17}
$$

for $k > -n$. So, in particular for $k = -1$, we get

$$E[\widehat{\theta}] = nE[W^{-1}] = \theta \frac{n}{n-1}.$$

Hence, $\widehat{\theta}$ is biased, but the bias vanishes as $n \to \infty$. For $k = -2$, we get

$$E[\widehat{\theta}^2] = n^2 E[W^{-2}] = \theta^2 \frac{n^2}{(n-1)(n-2)},$$

and, hence, after simplifying $E(\widehat{\theta}^2) - [E(\widehat{\theta})]^2$, we obtain

$$\mathrm{Var}(\widehat{\theta}) = \theta^2 \frac{n^2}{(n-1)^2(n-2)}.$$

It follows that this variance is larger than the Rao-Cramér lower bound, and, hence, $\widehat{\theta}$ is not efficient. But notice that its efficiency (as in Definition 6.2.2), converges to 1 as $n \to \infty$. Later in this section, we will say that $\widehat{\theta}$ is asymptotically efficient. ∎

In the above examples, we were able to obtain the mles in closed form along with their distributions, and hence, moments. But this is often not the case. Mles, though, as the next theorem shows, have an asymptotic normal distribution. In fact, mles are asymptotically efficient. Our proof requires the additional regularity condition

Assumptions 6.2.2. (Additional Regularity Condition).

(R5): *The pdf $f(x;\theta)$ is three times differentiable as a function of θ. Further, for all $\theta \in \Omega$, there exists a constant c and a function $M(x)$ such that*

$$\left| \frac{\partial^3}{\partial \theta^3} \log f(x;\theta) \right| \leq M(x),$$

with $E_{\theta_0}[M(X)] < \infty$, for all $\theta_0 - c < \theta < \theta_0 + c$ and all x in the support of X.

Theorem 6.2.2. *Assume $X_1, \ldots, X_n$ are iid with pdf $f(x;\theta_0)$ for $\theta_0 \in \Omega$ such that the regularity conditions (R0) - (R5) are satisfied. Suppose further that Fisher information satisfies $0 < I(\theta_0) < \infty$. Then any consistent sequence of solutions of the mle equations satisfies*

$$\sqrt{n}(\widehat{\theta} - \theta_0) \xrightarrow{D} N\left(0, \frac{1}{I(\theta_0)}\right). \tag{6.2.18}$$

Proof: Expanding the function $l'(\theta)$ into a Taylor series of order two about θ_0 and evaluating it at $\widehat{\theta}_n$, we get

$$l'(\widehat{\theta}_n) = l'(\theta_0) + (\widehat{\theta}_n - \theta_0)l''(\theta_0) + \frac{1}{2}(\widehat{\theta}_n - \theta_0)^2 l'''(\theta_n^*), \tag{6.2.19}$$

where θ_n^* is between θ_0 and $\widehat{\theta}_n$. But $l'(\widehat{\theta}_n) = 0$. Hence, rearranging terms we obtain,

$$\sqrt{n}(\widehat{\theta}_n - \theta_0) = \frac{n^{-1/2}l'(\theta_0)}{-n^{-1}l''(\theta_0) - (2n)^{-1}(\widehat{\theta}_n - \theta_0)l'''(\theta_n^*)}. \tag{6.2.20}$$

By the Central Limit Theorem,

$$\frac{1}{\sqrt{n}}l'(\theta_0) = \frac{1}{\sqrt{n}}\sum_{i=1}^{n}\frac{\partial \log f(X_i; \theta_0)}{\partial \theta} \xrightarrow{D} N(0, I(\theta_0)), \tag{6.2.21}$$

because the summands are iid with $\text{Var}(\partial \log f(X_i; \theta_0)/\partial \theta) = I(\theta_0) < \infty$. Also, by the Law of Large Numbers,

$$-\frac{1}{n}l''(\theta_0) = -\frac{1}{n}\sum_{i=1}^{n}\frac{\partial^2 \log f(X_i; \theta_0)}{\partial \theta^2} \xrightarrow{P} I(\theta_0). \tag{6.2.22}$$

To complete the proof then, we need only show that the second term in the denominator of expression (6.2.20) goes to zero in probability. Because $\widehat{\theta}_n - \theta_0 \to 0$, in probability, by Theorem 4.3.7 this will follow if we can show that $n^{-1}l'''(\theta_n^*)$ is bounded in probability. Let c_0 be the constant defined in Condition (R5). Note that $|\widehat{\theta}_n - \theta_0| < c_0$ implies that $|\theta_n^* - \theta_0| < c_0$, which in turn by Condition (R5) implies the following string of inequalities:

$$\left|-\frac{1}{n}l'''(\theta_n^*)\right| \leq \frac{1}{n}\sum_{i=1}^{n}\left|\frac{\partial^3 \log f(X_i; \theta)}{\partial \theta^3}\right| \leq \frac{1}{n}\sum_{i=1}^{n}M(X_i). \tag{6.2.23}$$

By Condition (R5), $E_{\theta_0}[M(X)] < \infty$; hence, $\frac{1}{n}\sum_{i=1}^{n}M(X_i) \to E_{\theta_0}[M(X)]$, in probability, by the Law of Large Numbers. For the bound, we select $1 + E_{\theta_0}[M(X)]$. Let $\epsilon > 0$ be given. Choose N_1 and N_2 so that

$$n \geq N_1 \quad \Rightarrow \quad P[|\widehat{\theta}_n - \theta_0| < c_0] \geq 1 - \frac{\epsilon}{2} \tag{6.2.24}$$

$$n \geq N_2 \quad \Rightarrow \quad P\left[\left|\frac{1}{n}\sum_{i=1}^{n}M(X_i) - E_{\theta_0}[M(X)]\right| < 1\right] \geq 1 - \frac{\epsilon}{2}. \tag{6.2.25}$$

It follows from (6.2.23), (6.2.24), and (6.2.25) that

$$n \geq \max\{N_1, N_2\} \Rightarrow P\left[\left|-\frac{1}{n}l'''(\theta_n^*)\right| \leq 1 + E_{\theta_0}[M(X)]\right] \geq 1 - \frac{\epsilon}{2};$$

hence, $n^{-1}l'''(\theta_n^*)$ is bounded in probability. $\blacksquare$

We next generalize Definitions 6.2.1 and 6.2.2 concerning efficiency to the asymptotic case.

Definition 6.2.3. *Let $X_1, \ldots, X_n$ be independent and identically distributed with probability density function $f(x; \theta)$. Suppose $\hat{\theta}_{1n} = \hat{\theta}_{1n}(X_1, \ldots, X_n)$ is an estimator of θ_0 such that $\sqrt{n}(\hat{\theta}_{1n} - \theta_0) \xrightarrow{D} N\left(0, \sigma_{\hat{\theta}_{1n}}^2\right)$. Then*

(a) *The* **asymptotic efficiency** *of* $\hat{\theta}_{1n}$ *is defined to be*

$$e(\hat{\theta}_n) = \frac{1/I(\theta_0)}{\sigma^2_{\hat{\theta}_{1n}}}. \qquad (6.2.26)$$

(b) *The estimator* $\hat{\theta}_{1n}$ *is said to be* **asymptotically efficient** *if the ratio in (a) is 1.*

(c) *Let* $\hat{\theta}_{2n}$ *be another estimator such that* $\sqrt{n}(\hat{\theta}_{2n} - \theta_0) \xrightarrow{D} N\left(0, \sigma^2_{\hat{\theta}_{2n}}\right)$. *Then the* **asymptotic relative efficiency**, *(ARE), of* $\hat{\theta}_{1n}$ *to* $\hat{\theta}_{2n}$ *is the reciprocal of the ratio of their respective asymptotic variances; i.e,*

$$e(\hat{\theta}_{1n}, \hat{\theta}_{2n}) = \frac{\sigma^2_{\hat{\theta}_{2n}}}{\sigma^2_{\hat{\theta}_{1n}}}. \qquad (6.2.27)$$

Hence by Theorem 6.2.2, under regularity conditions, maximum likelihood estimators are asymptotically efficient estimators. This is a nice optimality result. Also, if two estimators are asymptotically normal with the same asymptotic mean, then intuitively the estimator with the smaller asymptotic variance would be selected over the other as a better estimator. In this case, the ARE of the selected estimator to the nonselected one is greater than 1.

Example 6.2.5 (ARE of the Sample Median to the Sample Mean). We will obtain this ARE under the Laplace and normal distributions. Consider first the Laplace location model as given in expression (6.2.9); i.e.,

$$X_i = \theta + e_i, \quad i = 1, \ldots n, \qquad (6.2.28)$$

where the e_is are iid with the Laplace pdf (6.1.6). By Example 6.1.3, we know that the mle of θ is the sample median, Q_2. By (6.2.10) the information $I(\theta_0) = 1$, for this distribution; hence, Q_2 is asymptotically normal with mean θ and variance $1/n$. On the other hand, by the Central Limit Theorem, the sample mean $\overline{X}$ is asymptotically normal with mean θ and variance σ^2/n, where $\sigma^2 = \text{Var}(X_i) = \text{Var}(e_i + \theta) = \text{Var}(e_i) = E(e_i^2)$. But

$$E(e_i^2) = \int_{-\infty}^{\infty} z^2 2^{-1} \exp\{-|z|\}\, dz = \int_0^{\infty} z^{3-1} \exp\{-z\}\, dz = \Gamma(3) = 2.$$

Therefore, the $\text{ARE}(Q_2, \overline{X}) = \frac{2}{1} = 2$. Thus if the sample comes from a Laplace distribution, then asymptotically the sample median is twice as efficient as the sample mean.

Next suppose the location model (6.2.28) holds, except now the pdf of e_i is $N(0,1)$. As shown in Chapter 10, under this model Q_2 is asymptotically normal with mean θ and variance $(\pi/2)/n$. Because the variance of $\overline{X}$ is $1/n$, in this case, the $\text{ARE}(Q_2, \overline{X}) = \frac{1}{\pi/2} = 2/\pi = 0.636$. Since $\pi/2 = 1.57$, asymptotically, $\overline{X}$ is 1.57 times more efficient than Q_2 if the sample arises from the normal distribution. ∎

Theorem 6.2.2 is also a practical result for it gives us a way of doing inference. The asymptotic standard deviation of the mle $\widehat{\theta}$ is $[nI(\theta_0)]^{-1/2}$. Because $I(\theta)$ is a continuous function of θ, it follows from Theorems 4.2.4 and 6.1.2 that

$$I(\widehat{\theta}_n) \xrightarrow{P} I(\theta_0).$$

Thus we have a consistent estimate of the asymptotic standard deviation of the mle. Based on this result and the discussion of confidence intervals in Section 5.4, for a specified $0 < \alpha < 1$, the following interval is an approximate $(1 - \alpha)100\%$ confidence interval for θ,

$$\left(\widehat{\theta}_n - z_{\alpha/2} \frac{1}{\sqrt{nI(\widehat{\theta}_n)}}, \widehat{\theta}_n + z_{\alpha/2} \frac{1}{\sqrt{nI(\widehat{\theta}_n)}} \right). \tag{6.2.29}$$

Remark 6.2.2. If we use the asymptotic distributions to construct confidence intervals for θ, the fact that the $ARE(Q_2, \overline{X}) = 2$ when the underlying distribution is the Laplace means that n would need to be twice as large for $\overline{X}$ to get the same length confidence interval as we would if we used Q_2. ∎

A simple corollary to Theorem 6.2.2 yields the asymptotic distribution of a function $g(\widehat{\theta}_n)$ of the mle.

Corollary 6.2.2. *Under the assumptions of Theorem 6.2.2, suppose $g(x)$ is a continuous function of x which is differentiable at θ_0 such that $g'(\theta_0) \neq 0$. Then*

$$\sqrt{n}(g(\widehat{\theta}_n) - g(\theta_0)) \xrightarrow{D} N\left(0, \frac{g'(\theta_0)^2}{I(\theta_0)} \right). \tag{6.2.30}$$

The proof of this corollary follows immediately from the Δ method, Theorem 4.3.9, and Theorem 6.2.2.

The proof of Theorem 6.2.2 contains an asymptotic representation of $\widehat{\theta}$ which will prove useful; hence, we state it as another corollary.

Corollary 6.2.3. *Under the assumptions of Theorem 6.2.2,*

$$\sqrt{n}(\widehat{\theta}_n - \theta_0) = \frac{1}{I(\theta_0)} \frac{1}{\sqrt{n}} \sum_{i=1}^{n} \frac{\partial \log f(X_i; \theta_0)}{\partial \theta} + R_n, \tag{6.2.31}$$

where $R_n \xrightarrow{P} 0$.

The proof is just a rearrangement of equation (6.2.20) and the ensuing results in the proof of Theorem 6.2.2.

Example 6.2.6 (Example 6.2.4 Continued). Let $X_1, \ldots, X_n$ be a random sample having the common pdf (6.2.14). Recall that $I(\theta) = \theta^{-2}$ and that the mle is $\widehat{\theta} = -n/\sum_{i=1}^{n} \log X_i$. Hence, $\widehat{\theta}$ is approximately normally distributed with mean θ

and variance θ^2/n. Based on this, an approximate $(1 - \alpha)100\%$ confidence interval for θ is

$$\widehat{\theta} \pm z_{\alpha/2} \frac{\widehat{\theta}}{\sqrt{n}}.$$

Recall that we were able to obtain the exact distribution of $\widehat{\theta}$ in this case. As Exercise 6.2.12 shows, based on this distribution of $\widehat{\theta}$, an exact confidence interval for θ can be constructed. ■

In obtaining the mle of θ, we are often in the situation of Example 6.1.4; that is, we can verify the existence of the mle but the solution of the equation $l'(\widehat{\theta}) = 0$ cannot be obtained in closed form. In such situations, numerical methods are used. One iterative method that exhibits rapid (quadratic) convergence is Newton's method. The sketch in Figure 6.2.1 helps recall this method. Suppose $\widehat{\theta}^{(0)}$ is an initial guess at the solution. The next guess (one step estimate) is the point $\widehat{\theta}^{(1)}$ which is the horizontal intercept of the tangent line to the curve $l'(\theta)$ at the point $(\widehat{\theta}^{(0)}, l'(\widehat{\theta}^{(0)}))$. A little algebra finds

$$\widehat{\theta}^{(1)} = \widehat{\theta}^{(0)} - \frac{l'(\widehat{\theta}^{(0)})}{l''(\widehat{\theta}^{(0)})}. \tag{6.2.32}$$

We then substitute $\widehat{\theta}^{(1)}$ for $\widehat{\theta}^{(0)}$ and repeat the process. On the figure, trace the second step estimate $\widehat{\theta}^{(2)}$; the process is continued until convergence.

Example 6.2.7 (Example 6.1.4, continued). Recall Example 6.1.4, where the random sample $X_1, \ldots, X_n$ had the common logisitic density

$$f(x; \theta) = \frac{\exp\{-(x - \theta)\}}{(1 + \exp\{-(x - \theta)\})^2}, \quad -\infty < x < \infty, \ -\infty < \theta < \infty. \tag{6.2.33}$$

We showed that likelihood equation had a unique solution but were unable to obtain it in closed form. To use formula (6.2.32) we need the first and second partial derivatives of $l(\theta)$ and an initial guess. Expression (6.1.9) of Example 6.1.4 gives the first partial derivative, from which the second partial is

$$l''(\theta) = -2 \sum_{i=1}^n \frac{\exp\{-(x_i - \theta)\}}{(1 + \exp\{-(x_i - \theta)\})^2}.$$

The logistic distribution is similar to the normal distribution; hence, we can use $\overline{X}$ as our initial guess of θ. The subroutine `mlelogistic` in Appendix B is an R routine which obtains the k-step estimates. ■

We close this section with a remarkable fact. The estimate $\widehat{\theta}^{(1)}$ in equation (6.2.32) is called the **one-step estimator**. As Exercise 6.2.13 shows, this estimator has the same asymptotic distribution as the mle, i.e. (6.2.18), provided that the initial guess $\widehat{\theta}^{(0)}$ is a consistent estimator of θ. That is, the one-step estimate is an asymptotically efficient estimate of θ. This is also true of the other iterative steps.

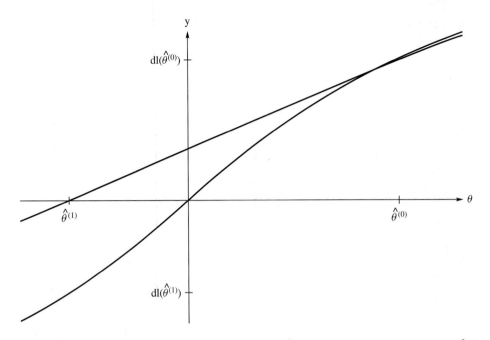

Figure 6.2.1: Beginning with starting value $\widehat{\theta}^{(0)}$, the one step estimate is $\widehat{\theta}^{(1)}$ which is the intersection of the tangent line to the curve $l'(\theta)$ at $\widehat{\theta}^{(0)}$ and the horizontal axis. In the figure, $dl(\theta) = l'(\theta)$.

EXERCISES

6.2.1. Prove that $\overline{X}$, the mean of a random sample of size n from a distribution that is $N(\theta, \sigma^2)$, $-\infty < \theta < \infty$, is, for every known $\sigma^2 > 0$, an efficient estimator of θ.

6.2.2. Given $f(x; \theta) = 1/\theta$, $0 < x < \theta$, zero elsewhere, with $\theta > 0$, formally compute the reciprocal of

$$nE\left\{\left[\frac{\partial \ln f(X:\theta)}{\partial \theta}\right]^2\right\}.$$

Compare this with the variance of $(n+1)Y_n/n$, where Y_n is the largest observation of a random sample of size n from this distribution. Comment.

6.2.3. Given the pdf

$$f(x;\theta) = \frac{1}{\pi[1 + (x-\theta)^2]}, \quad -\infty < x < \infty, \quad -\infty < \theta < \infty.$$

Show that the Rao-Cramér lower bound is $2/n$, where n is the size of a random sample from this Cauchy distribution. What is the asymptotic distribution of $\sqrt{n}(\widehat{\theta}-\theta)$, if $\widehat{\theta}$ is the mle of θ?

6.2.4. Consider Example 6.2.2, where we discussed the location model.

(a) Write the location model when e_i has the logistic pdf given in expression (5.2.8).

(b) Using expression (6.2.8), show that the information, $I(\theta) = 1/3$, for the model in Part (a). *Hint*: In the integral of expression (6.2.8), use the substitution $u = (1 + e^{-z})^{-1}$. Then $du = f(z)dz$, where $f(z)$ is the pdf (5.2.8).

6.2.5. Using the same location model as in Part (a) Exercise 6.2.4, obtain the ARE of the sample median to mle of the model.
Hint: The mle of θ for this model is discussed in Example 6.2.7. Furthermore as shown in Theorem 10.2.3 of Chapter 10, Q_2 is asymptotically normal with asymptotic mean θ and asymptotic variance $1/(4f^2(0)n)$.

6.2.6. Consider a location model (Example 6.2.2) when the error pdf is the contaminated normal (3.4.14) with ϵ proportion of contamination and with σ_c^2 as the variance of the contaminated part. Show that the ARE of the sample median to the sample mean is given by

$$e(Q_2, \overline{X}) = \frac{2[1 + \epsilon(\sigma_c^2 - 1)][1 - \epsilon + (\epsilon/\sigma_c)]^2}{\pi}. \qquad (6.2.34)$$

Use the hint in Exercise 6.2.5 for the median.

(a) If $\sigma_c^2 = 9$, use (6.2.34) to fill in the following table:

ϵ	0	0.05	0.10	0.15
$e(Q_2, \overline{X})$				

(b) Notice from the table that the sample median becomes the "better" estimator when ϵ increases from 0.10 to 0.15. Determine the value for ϵ where this occurs (this involves a third degree polynomial in ϵ, so one way of obtaining the root is to use the Newton algorithm discussed around expression (6.2.32)).

6.2.7. Let X have a gamma distribution with $\alpha = 4$ and $\beta = \theta > 0$.

(a) Find the Fisher information $I(\theta)$.

(b) If $X_1, X_2, \ldots, X_n$ is a random sample from this distribution, show that the mle of θ is an efficient estimator of θ.

(c) What is the asymptotic distribution of $\sqrt{n}(\widehat{\theta} - \theta)$?

6.2.8. Let X be $N(0, \theta)$, $0 < \theta < \infty$.

(a) Find the Fisher information $I(\theta)$.

(b) If $X_1, X_2, \ldots, X_n$ is a random sample from this distribution, show that the mle of θ is an efficient estimator of θ.

(c) What is the asymptotic distribution of $\sqrt{n}(\widehat{\theta} - \theta)$?

6.2.9. If $X_1, X_2, \ldots, X_n$ is a random sample from a distribution with pdf

$$f(x; \theta) = \begin{cases} \frac{3\theta^3}{(x+\theta)^4} & 0 < x < \infty, 0 < \theta < \infty \\ 0 & \text{elsewhere.} \end{cases}$$

Show that $Y = 2\overline{X}$ is an unbiased estimator of θ and determine its efficiency.

6.2.10. Let $X_1, X_2, \ldots, X_n$ be a random sample from a $N(0, \theta)$ distribution. We want to estimate the standard deviation $\sqrt{\theta}$. Find the constant c so that $Y = c \sum_{i=1}^{n} |X_i|$ is an unbiased estimator of $\sqrt{\theta}$ and determine its efficiency.

6.2.11. Let $\overline{X}$ be the mean of a random sample of size n from a $N(\theta, \sigma^2)$ distribution, $-\infty < \theta < \infty, \sigma^2 > 0$. Assume that σ^2 is known. Show that $\overline{X}^2 - \frac{\sigma^2}{n}$ is an unbiased estimator of θ^2 and find its efficiency.

6.2.12. Recall that $\widehat{\theta} = -n/\sum_{i=1}^{n} \log X_i$ is the mle of θ for a Beta$(\theta, 1)$ distribution. Also, $W = -\sum_{i=1}^{n} \log X_i$ has the gamma distribution $\Gamma(n, 1/\theta)$.

(a) Show that $2\theta W$ has a $\chi^2(2n)$ distribution.

(b) Using Part (a), find c_1 and c_2 so that

$$P\left(c_1 < \frac{2\theta n}{\widehat{\theta}} < c_2\right) = 1 - \alpha,$$

for $0 < \alpha < 1$. Next, obtain a $(1 - \alpha)100\%$ confidence interval for θ.

(c) Let $n = 10$ and compare the length of this interval with the length of the interval found in Example 6.2.6.

6.2.13. By using expressions (6.2.21) and (6.2.22) obtain the result for the one-step estimate discussed at the end of this section.

6.2.14. Let S^2 be the sample variance of a random sample of size $n > 1$ from $N(\mu, \theta)$, $0 < \theta < \infty$, where μ is known. We know $E(S^2) = \theta$.

(a) What is the efficiency of S^2?

(b) Under these conditions, what is the mle $\widehat{\theta}$ of θ?

(c) What is the asymptotic distribution of $\sqrt{n}(\widehat{\theta} - \theta)$?

6.3 Maximum Likelihood Tests

The last section presented an inference for pointwise estimation and confidence intervals based on likelihood theory. In this section, we present a corresponding inference for testing hypotheses.

As in the last section, let $X_1, \ldots, X_n$ be iid with pdf $f(x; \theta)$ for $\theta \in \Omega$. In this section, θ is a scalar but in Sections 6.4 and 6.5 extensions to the vector valued case will be discussed. Consider the two-sided hypotheses

$$H_0 : \theta = \theta_0 \text{ versus } H_1 : \theta \neq \theta_0, \tag{6.3.1}$$

where θ_0 is a specified value.

Recall that the likelihood function and its log are given by:

$$L(\theta) = \prod_{i=1}^{n} f(X_i; \theta)$$

$$l(\theta) = \sum_{i=1}^{n} \log f(X_i; \theta).$$

Let $\widehat{\theta}$ denote the maximum likelihood estimate of θ.

To motivate the test, consider Theorem 6.1.1 which says that if θ_0 is the true value of θ then, asymptotically, $L(\theta_0)$ is the maximum value of $L(\theta)$. Consider the ratio of two likelihood functions, namely

$$\Lambda = \frac{L(\theta_0)}{L(\widehat{\theta})}. \tag{6.3.2}$$

Note that $\Lambda \leq 1$, but if H_0 is true Λ should be large (close to 1); while, if H_1 is true, Λ should be smaller. For a specified significance level α, this leads to the intuitive decision rule,

$$\text{Reject } H_0 \text{ in favor of } H_1 \text{ if } \Lambda \leq c, \tag{6.3.3}$$

where c is such that $\alpha = P_{\theta_0}[\Lambda \leq c]$. This test is called the **likelihood ratio test**. Theorem 6.3.1 derives the asymptotic distribution of Λ under H_0, but first we will look at two examples.

Example 6.3.1 (Likelihood Ratio Test for the Exponential Distribution). Suppose $X_1, \ldots, X_n$ are iid with pdf $f(x; \theta) = \theta^{-1} \exp\{-x/\theta\}$, for $x, \theta > 0$. Let the hypotheses be given by (6.3.1). The likelihood function simplifies to

$$L(\theta) = \theta^{-n} \exp\{-(n/\theta)\overline{X}\}.$$

From Example 6.1.2, the mle of θ is $\overline{X}$. After some simplification, the likelihood ratio test statistic simplifies to

$$\Lambda = e^n \left(\frac{\overline{X}}{\theta_0}\right)^n \exp\{-n\overline{X}/\theta_0\}. \tag{6.3.4}$$

The decision rule is to reject H_0 if $\Lambda \le c$. But further simplification of the test is possible. Other than the constant e^n, the test statistic is of the form

$$g(t) = t^n \exp\{-nt\}, \quad t > 0,$$

where $t = \overline{x}/\theta_0$. Using differentiable calculus, it is easy to show that $g(t)$ has a unique critical value at 1, i.e., $g'(1) = 0$, and further that $t = 1$ provides a maximum, because $g''(1) < 0$. As Figure 6.3.1 depicts, $g(t) \le c$ if and only if $t \le c_1$ or $t \ge c_2$. This leads to

$$\Lambda \le c, \text{ if and only if, } \frac{\overline{X}}{\theta_0} \le c_1 \text{ or } \frac{\overline{X}}{\theta_0} \ge c_2.$$

Note that under the null hypothesis, H_0, the statistic $(2/\theta_0)\sum_{i=1}^{n} X_i$ has a χ^2 distribution with $2n$ degrees of freedom. Based on this, the following decision rule results in a level α test:

Reject H_0 if $(2/\theta_0)\sum_{i=1}^{n} X_i \le \chi^2_{1-\alpha/2}(2n)$ or $(2/\theta_0)\sum_{i=1}^{n} X_i \ge \chi^2_{\alpha/2}(2n)$, $\quad$ (6.3.5)

where $\chi^2_{1-\alpha/2}(2n)$ is the lower $\alpha/2$ quantile of a χ^2 distribution with $2n$ degrees of freedom and $\chi^2_{\alpha/2}(2n)$ is the upper $\alpha/2$ quantile of a χ^2 distribution with $2n$ degrees of freedom. Other choices of c_1 and c_2 can be made, but these are usually the choices used in practice. Exercise 6.3.1 investigates the power curve for this test. ∎

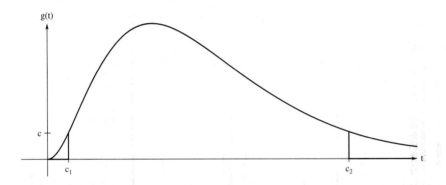

Figure 6.3.1: Plot for Example 6.3.1, showing that the function $g(t) \le c$ if and only if $t \le c_1$ or $t \ge c_2$.

Example 6.3.2 (Likelihood Ratio Test for the Mean of a Normal pdf).
Say $X_1, X_2, \ldots, X_n$ is a random sample from a $N(\theta, \sigma^2)$ distribution where $-\infty < \theta < \infty$ and $\sigma^2 > 0$ is known. Consider the hypotheses

$$H_0 : \theta = \theta_0 \text{ versus } H_1 : \theta \ne \theta_0,$$

where θ_0 is specified. The likelihood function is

$$
L(\theta) = \left(\frac{1}{2\pi\sigma^2}\right)^{n/2} \exp\{-(2\sigma^2)^{-1}\sum_{i=1}^{n}(x_i - \theta)^2\}
$$

$$
= \left(\frac{1}{2\pi\sigma^2}\right)^{n/2} \exp\{-(2\sigma^2)^{-1}\sum_{i=1}^{n}(x_i - \overline{x})^2\} \exp\{-(2\sigma^2)^{-1}n(\overline{x} - \theta)^2\}.
$$

Of course, in $\Omega = \{\theta : -\infty < \theta < \infty\}$, the mle is $\widehat{\theta} = \overline{X}$ and thus,

$$
\Lambda = \frac{L(\theta_0)}{L(\widehat{\theta})} = \exp\{-(2\sigma^2)^{-1}n(\overline{X} - \theta_0)^2\}.
$$

Then $\Lambda \leq c$ is equivalent to $-2\log\Lambda \geq -2\log c$. However,

$$
-2\log\Lambda = \left(\frac{\overline{X} - \theta_0}{\sigma/\sqrt{n}}\right)^2,
$$

which has a $\chi^2(1)$ distribution under H_0. Thus the likelihood ratio test with significance level α states that we reject H_0 and accept H_1 when

$$
-2\log\Lambda = \left(\frac{\overline{X} - \theta_0}{\sigma/\sqrt{n}}\right)^2 \geq \chi_\alpha^2(1). \tag{6.3.6}
$$

In Exercise 6.3.3, the power function of this decision rule is obtained. ■

Other examples are given in the exercises. In the last two examples the likelihood ratio tests simplifies and we are able to get the test in closed form. Often, though, this is impossible. In such cases, similar to Example 6.2.7, we can obtain the mle by iterative routines and, hence, also the test statistic Λ. In Example 6.3.2, $-2\log\Lambda$ had an exact $\chi^2(1)$ null distribution. While not true in general, as the following theorem shows, under regularity conditions, the asymptotic null distribution of $-2\log\Lambda$ is χ^2 with one degree of freedom. Hence, in all cases an asymptotic test can be constructed.

Theorem 6.3.1. *Assume the same regularity conditions as for Theorem 6.2.2. Under the null hypothesis, $H_0 : \theta = \theta_0$,*

$$
-2\log\Lambda \xrightarrow{D} \chi^2(1). \tag{6.3.7}
$$

Proof: Expand the function $l(\theta)$ into a Taylor series about θ_0 of order 1 and evaluate it at the mle, $\widehat{\theta}$. This results in,

$$
l(\widehat{\theta}) = l(\theta_0) + (\widehat{\theta} - \theta_0)l'(\theta_0) + \frac{1}{2}(\widehat{\theta} - \theta_0)^2 l''(\theta_n^*), \tag{6.3.8}
$$

where θ_n^* is between $\widehat{\theta}$ and θ_0. Because $\widehat{\theta} \to \theta_0$, in probability, it follows that $\theta_n^* \to \theta_0$, in probability. This in addition to the fact that the function $l''(\theta)$ is continuous and equation (6.2.22) of Theorem 6.2.2 imply that

$$
-\frac{1}{n}l''(\theta_n^*) \xrightarrow{P} I(\theta_0). \tag{6.3.9}
$$

By Corollary 6.2.3,

$$\frac{1}{\sqrt{n}}l'(\theta_0) = \sqrt{n}(\widehat{\theta} - \theta_0)I(\theta_0) + R_n, \qquad (6.3.10)$$

where $R_n \rightarrow 0$, in probability. If we substitute (6.3.9) and (6.3.10) into expression (6.3.8) and do some simplification, we have

$$-2\log \Lambda = 2(l(\widehat{\theta}) - l(\theta_0)) = \{\sqrt{nI(\theta_0)}(\widehat{\theta} - \theta_0)\}^2 + R_n^*, \qquad (6.3.11)$$

where $R_n^* \rightarrow 0$, in probability. By Theorems 4.3.4 and 6.2.2, the first term on the right side of the above equation converges in distribution to a χ^2 distribution with one degree of freedom. ∎

Define the test statistic $\chi_L^2 = -2\log \Lambda$. For the hypotheses (6.3.1), this theorem suggests the decision rule

$$\text{Reject } H_0 \text{ in favor of } H_1 \text{ if } \chi_L^2 \geq \chi_\alpha^2(1). \qquad (6.3.12)$$

By the last theorem this test has asymptotic level α. If we cannot obtain the test statistic or its distribution in closed form, we can use this asymptotic test.

Besides the likelihood ratio test, in practice two other likelihood related tests are employed. A natural test statistic is based on the asymptotic distribution of $\widehat{\theta}$. Consider the statistic,

$$\chi_W^2 = \left\{ \sqrt{nI(\widehat{\theta})}(\widehat{\theta} - \theta_0) \right\}^2. \qquad (6.3.13)$$

Because $I(\theta)$ is a continuous function, $I(\widehat{\theta}) \rightarrow I(\theta_0)$ in probability under the null hypothesis, (6.3.1). It follows, under H_0, that χ_W^2 has an asymptotic χ^2-distribution with one degree of freedom. This suggests the decision rule,

$$\text{Reject } H_0 \text{ in favor of } H_1, \text{ if } \chi_W^2 \geq \chi_\alpha^2(1). \qquad (6.3.14)$$

As with the test based on χ_L^2, this test has asymptotic level α. Actually the relationship between the two test statistics is strong, because as equation (6.3.11) shows under H_0,

$$\chi_W^2 - \chi_L^2 \xrightarrow{P} 0. \qquad (6.3.15)$$

The test (6.3.14) is often referred to as a **Wald**-type test, after Abraham Wald, who was a prominent statistician of the 20th century.

The third test is called a **scores**-type test, which is often referred to as Rao's score test, after another prominent statistician C. R. Rao. The **scores** are the components of the vector

$$\mathbf{S}(\theta) = \left(\frac{\partial \log f(X_1; \theta)}{\partial \theta}, \ldots, \frac{\partial \log f(X_n; \theta)}{\partial \theta} \right)'. \qquad (6.3.16)$$

In our notation, we have

$$\frac{1}{\sqrt{n}}l'(\theta_0) = \frac{1}{\sqrt{n}} \sum_{i=1}^{n} \frac{\partial \log f(X_i; \theta_0)}{\partial \theta}. \qquad (6.3.17)$$

Define the statistic

$$\chi_R^2 = \left(\frac{l'(\theta_0)}{\sqrt{nI(\theta_0)}} \right)^2. \tag{6.3.18}$$

Under H_0, it follows from expression (6.3.10) that

$$\chi_R^2 = \chi_W^2 + R_{0n}, \tag{6.3.19}$$

where R_{0n} converges to 0 in probability. Hence, the following decision rules defines an asymptotic level α test under H_0,

$$\text{Reject } H_0 \text{ in favor of } H_1, \text{ if } \chi_R^2 \geq \chi_\alpha^2(1). \tag{6.3.20}$$

Example 6.3.3 (Example 6.2.6 Continued). As in Example 6.2.6, let $X_1, \ldots, X_n$ be a random sample having the common Beta$(\theta, 1)$ pdf (6.2.14). We will use this pdf to illustrate the three test statistics discussed above for the hypotheses,

$$H_0 : \theta = 1 \text{ versus } H_1 : \; \theta \neq 1. \tag{6.3.21}$$

Under H_0 $f(x; \theta)$ is the uniform$(0, 1)$ pdf. Recall that $\widehat{\theta} = -n/\sum_{i=1}^n \log X_i$ is the mle of θ. After some simplification, the value of the likelihood function at the mle is

$$L(\widehat{\theta}) = (-\sum_{i=1}^n \log X_i)^{-n} \exp\{-\sum_{i=1}^n \log X_i\} \exp\{n(\log n - 1)\}.$$

Also, $L(1) = 1$. Hence, the likelihood ratio test statistic is $\Lambda = 1/L(\widehat{\theta})$ so that

$$\chi_L^2 = -2 \log \Lambda = 2\{-\sum_{i=1}^n \log X_i - n \log(-\sum_{i=1}^n \log X_i) - n + n \log n\}.$$

Recall that the information for this pdf is $I(\theta) = \theta^{-2}$. For the Wald-type test, we would estimate this consistently by $\widehat{\theta}^{-2}$. The Wald-type test simplifies to

$$\chi_W^2 = \left(\sqrt{\frac{n}{\widehat{\theta}^2}} (\widehat{\theta} - 1) \right)^2 = n\left\{ 1 - \frac{1}{\widehat{\theta}} \right\}^2. \tag{6.3.22}$$

Finally, for the scores-type course, recall from (6.2.15) that the $l'(1)$ is

$$l'(1) = \sum_{i=1}^n \log X_i + n.$$

Hence, the scores type test statistic is

$$\chi_R^2 = \left\{ \frac{\sum_{i=1}^n \log X_i + n}{\sqrt{n}} \right\}^2. \tag{6.3.23}$$

It is easy to show that expressions (6.3.22) and (6.3.23) are the same. From Example 6.2.4, we know the exact distribution of the maximum likelihood estimate. Exercise 6.3.7 uses this distribution to obtain an exact test. ∎

Example 6.3.4 (Likelihood Tests for the Laplace Location Model). Consider the location model

$$X_i = \theta + e_i, \quad i = 1, \ldots, n,$$

where $-\infty < \theta < \infty$ and the random errors e_is are iid each having the Laplace pdf, (6.2.9). Technically, the Laplace distribution does not satisfy all of the regularity conditions (R0) - (R5) but the results below can be derived rigorously; see, for example, Hettmansperger and McKean (1998). Consider testing the hypotheses

$$H_0 : \theta = \theta_0 \text{ versus } H_1 : \theta \neq \theta_0,$$

where θ_0 is specified. Here $\Omega = (-\infty, \infty)$ and $\omega = \{\theta_0\}$. By Example 6.1.3, we know that the mle of θ under Ω is $Q_2 = \text{med}\{X, \ldots, X_n\}$, the sample median. It follows that

$$L(\widehat{\Omega}) = 2^{-n} \exp\{-\sum_{i=1}^{n} |x_i - Q_2|\},$$

while

$$L(\widehat{\omega}) = 2^{-n} \exp\{-\sum_{i=1}^{n} |x_i - \theta_0|\}.$$

Hence, the negative of twice the log of the likelihood ratio test statistic is

$$-2 \log \Lambda = 2 \left[\sum_{i=1}^{n} |x_i - \theta_0| - \sum_{i=1}^{n} |x_i - Q_2| \right]. \tag{6.3.24}$$

Thus the size α asymptotic likelihood ratio test for H_0 versus H_1 rejects H_0 in favor of H_1 if

$$2 \left[\sum_{i=1}^{n} |x_i - \theta_0| - \sum_{i=1}^{n} |x_i - Q_2| \right] \geq \chi_\alpha^2(1).$$

By (6.2.10), the Fisher information for this model is $I(\theta) = 1$. Thus the Wald type test statistic simplifies to

$$\chi_W^2 = [\sqrt{n}(Q_2 - \theta_0)]^2.$$

For the scores test, we have

$$\frac{\partial \log f(x_i - \theta)}{\partial \theta} = \frac{\partial}{\partial \theta} \left[\log \frac{1}{2} - |x_i - \theta| \right] = \text{sgn}(x_i - \theta).$$

Hence, the score vector for this model is $\mathbf{S}(\theta) = (\text{sgn}(X_1 - \theta), \ldots, \text{sgn}(X_n - \theta))'$. From the above discussion, (see equation (6.3.17)), the scores test statistic can be written as

$$\chi_R^2 = (S^*)^2 / n,$$

where

$$S^* = \sum_{i=1}^{n} \text{sgn}(X_i - \theta_0).$$

As Exercise 6.3.4 shows, under H_0, S^* is a linear function of a random variable with a $b(n, 1/2)$ distribution. ■

Which of the three tests should we use? Based on the above discussion, all three tests are asymptotically equivalent under the null hypothesis. Similar to the concept of asymptotic relative efficiency (ARE) we can derive an equivalent concept of efficiency for tests; see Chapter 10 and more advanced books such as Hettmansperger and McKean (1998). However, all three tests have the same asymptotic efficiency. Hence, asymptotic theory offers little help in separating the tests. There have been finite sample comparisons in the literature; but these studies have not selected any of these as a "best" test overall; see Chapter 7 of Lehmann (1999) for more discussion.

EXERCISES

6.3.1. Consider the decision rule (6.3.5) derived in Example 6.3.1. Obtain the distribution of the test statistic under a general alternative and use it to obtain the power function of the test. If computational facilities are available, sketch this power curve for the case when $\theta_0 = 1$, $n = 10$, and $\alpha = 0.05$.

6.3.2. Show that the test with decision rule (6.3.6) is like that of Example 5.6.1 except here σ^2 is known.

6.3.3. Consider the decision rule (6.3.6) derived in Example 6.3.2. Obtain an equivalent test statistic which has a standard normal distribution under H_0. Next obtain the distribution of this test statistic under a general alternative and use it to obtain the power function of the test. If computational facilities are available, sketch this power curve for the case when $\theta_0 = 0$, $n = 10$, $\sigma^2 = 1$, and $\alpha = 0.05$.

6.3.4. Consider Example 6.3.4.

(a) Show that we can write $S^* = 2T - n$ where $T = \#\{X_i > \theta_0\}$.

(b) Show that the scores test for this model is equivalent to rejecting H_0 if $T < c_1$ or $T > c_2$.

(c) Show that under H_0, T has the binomial distribution $b(n, 1/2)$; hence, determine c_1 and c_2 so the test has size α.

(d) Determine the power function for the test based on T as a function of θ.

6.3.5. Let $X_1, X_2, \ldots, X_n$ be a random sample from a $N(\mu_0, \sigma^2 = \theta)$ distribution, where $0 < \theta < \infty$ and μ_0 is known. Show that the likelihood ratio test of $H_0 : \theta = \theta_0$ versus $H_1 : \theta \neq \theta_0$ can be based upon the statistic $W = \sum_{i=1}^{n}(X_i - \mu_0)^2/\theta_0$. Determine the null distribution of W and give, explicitly, the rejection rule for a level α test.

6.3.6. For the test described in Exercise 6.3.5, obtain the distribution of the test statistic under general alternatives. If computational facilities are available, sketch this power curve for the case when $\theta_0 = 1$, $n = 10$, $\mu = 0$, and $\alpha = 0.05$.

6.3.7. Using the results of Example 6.2.4, find an exact size α test for the hypotheses (6.3.21).

6.3.8. Let $X_1, X_2, \ldots, X_n$ be a random sample from a Poisson distribution with mean $\theta > 0$.

(a) Show that the likelihood ratio test of $H_0 : \theta = \theta_0$ versus $H_1 : \theta \neq \theta_0$ is based upon the statistic $Y = \sum_{i=1}^{n} X_i$. Obtain the null distribution of Y.

(b) For $\theta_0 = 2$ and $n = 5$, find the significance level of the test that rejects H_0 if $Y \leq 4$ or $Y \geq 17$.

6.3.9. Let $X_1, X_2, \ldots, X_n$ be a random sample from a Bernoulli $b(1, \theta)$ distribution, where $0 < \theta < 1$.

(a) Show that the likelihood ratio test of $H_0 : \theta = \theta_0$ versus $H_1 : \theta \neq \theta_0$ is based upon the statistic $Y = \sum_{i=1}^{n} X_i$. Obtain the null distribution of Y.

(b) For $n = 100$ and $\theta_0 = 1/2$, find c_1 so that the test rejects H_0 when $Y \leq c_1$ or $Y \geq c_2 = 100 - c_1$ has the approximate significance level of $\alpha = 0.05$. *Hint:* Use the Central Limit Theorem.

6.3.10. Let $X_1, X_2, \ldots, X_n$ be a random sample from a $\Gamma(\alpha = 3, \beta = \theta)$ distribution, where $0 < \theta < \infty$.

(a) Show that the likelihood ratio test of $H_0 : \theta = \theta_0$ versus $H_1 : \theta \neq \theta_0$ is based upon the statistic $W = \sum_{i=1}^{n} X_i$. Obtain the null distribution of $2W/\theta_0$.

(b) For $\theta_0 = 3$ and $n = 5$, find c_1 and c_2 so that the test that rejects H_0 when $W \leq c_1$ or $W \geq c_2$ has significance level 0.05.

6.3.11. Let $X_1, X_2, \ldots, X_n$ be a random sample from a distribution with pdf $f(x; \theta) = \theta \exp\{-|x|^\theta\}/2\Gamma(1/\theta)$, $-\infty < x < \infty$, where $\theta > 0$. Suppose $\Omega = \{\theta : \theta = 1, 2\}$. Consider the hypotheses $H_0 : \theta = 2$ (a normal distribution) versus $H_1 : \theta = 1$ (a double exponential distribution). Show that the likelihood ratio test can be based on the statistic $W = \sum_{i=1}^{n}(X_i^2 - |X_i|)$.

6.3.12. Let $X_1, X_2, \ldots, X_n$ be a random sample from the beta distribution with $\alpha = \beta = \theta$ and $\Omega = \{\theta : \theta = 1, 2\}$. Show that the likelihood ratio test statistic Λ for testing $H_0 : \theta = 1$ versus $H_1 : \theta = 2$ is a function of the statistic $W = \sum_{i=1}^{n} \log X_i + \sum_{i=1}^{n} \log(1 - X_i)$.

6.3.13. Consider a location model

$$X_i = \theta + e_i, \quad i = 1, \ldots, n, \qquad (6.3.25)$$

where $e_1, e_2, \ldots, e_n$ are iid with pdf $f(z)$. There is a nice geometric interpretation for estimating θ. Let $\mathbf{X} = (X_1, \ldots, X_n)'$ and $\mathbf{e} = (e_1, \ldots, e_n)'$ be the vectors of observations and random error, respectively, and let $\boldsymbol{\mu} = \theta \mathbf{1}$ where $\mathbf{1}$ is a vector with all components equal to one. Let V be the subspace of vectors of the form $\boldsymbol{\mu}$; i.e., $V = \{\mathbf{v} : \mathbf{v} = a\mathbf{1}, \text{ for some } a \in R\}$. Then in vector notation we can write the model as

$$\mathbf{X} = \boldsymbol{\mu} + \mathbf{e}, \quad \boldsymbol{\mu} \in V. \qquad (6.3.26)$$

Then we can summarize the model by saying, "Except for the random error vector $\mathbf{e}$, $\mathbf{X}$ would reside in V." Hence, it makes sense intuitively to estimate $\boldsymbol{\mu}$ by a vector in V which is "closest" to $\mathbf{X}$. That is, given a norm $\|\cdot\|$ in R^n choose

$$\widehat{\boldsymbol{\mu}} = \text{Argmin}\|\mathbf{X} - \mathbf{v}\|, \quad \mathbf{v} \in V. \tag{6.3.27}$$

(a) If the error pdf is the Laplace, (6.2.9), show that the minimization in (6.3.27) is equivalent to maximizing the likelihood, when the norm is the l_1 norm given by

$$\|\mathbf{v}\|_1 = \sum_{i=1}^{n} |v_i|. \tag{6.3.28}$$

(b) If the error pdf is the $N(0,1)$, show that the minimization in (6.3.27) is equivalent to maximizing the likelihood, when the norm is given by the square of the l_2 norm

$$\|\mathbf{v}\|_2^2 = \sum_{i=1}^{n} v_i^2. \tag{6.3.29}$$

6.3.14. Continuing with the last exercise, besides estimation there is also a nice geometric interpretation for testing. For the model (6.3.26), consider the hypotheses

$$H_0 : \theta = \theta_0 \text{ versus } H_1 : \theta \neq \theta_0, \tag{6.3.30}$$

where θ_0 is specified. Given a norm $\|\cdot\|$ on R^n, denote by $d(\mathbf{X}, V)$ the distance between $\mathbf{X}$ and the subspace V; i.e., $d(\mathbf{X}, V) = \|\mathbf{X} - \widehat{\boldsymbol{\mu}}\|$, where $\widehat{\boldsymbol{\mu}}$ is defined in equation (6.3.27). If H_0 is true, then $\widehat{\boldsymbol{\mu}}$ should be close to $\boldsymbol{\mu} = \theta_0 \mathbf{1}$ and, hence, $\|\mathbf{X} - \theta_0 \mathbf{1}\|$ should be close to $d(\mathbf{X}, V)$. Denote the difference by

$$RD = \|\mathbf{X} - \theta_0 \mathbf{1}\| - \|\mathbf{X} - \widehat{\boldsymbol{\mu}}\|. \tag{6.3.31}$$

Small values of RD indicate that the null hypothesis is true while large values indicate H_1. So our rejection rule when using RD is

$$\text{Reject } H_0 \text{ in favor of } H_1, \text{ if } RD > c. \tag{6.3.32}$$

(a) If the error pdf is the Laplace, (6.1.6), show that expression (6.3.31) is equivalent to the likelihood ratio test, when the norm is given by (6.3.28).

(b) If the error pdf is the $N(0,1)$, show that expression (6.3.31) is equivalent to the likelihood ratio test when the norm is given by the square of the l_2 norm, (6.3.29).

6.3.15. Let $X_1, X_2, \ldots, X_n$ be a random sample from a distribution with pmf $p(x; \theta) = \theta^x (1 - \theta)^{1-x}$, $x = 0, 1$, where $0 < \theta < 1$. We wish to test $H_0 : \theta = 1/3$ versus $H_1 : \theta \neq 1/3$.

(a) Find Λ and $-2 \log \Lambda$.

(b) Determine the Wald-type test.

(c) What is Rao's score statistic?

6.3.16. Let $X_1, X_2, \ldots, X_n$ be a random sample from a Poisson distribution with mean $\theta > 0$. Test $H_0 : \theta = 2$ against $H_1 : \theta \neq 2$ using

(a) $-2 \log \Lambda$.

(b) a Wald-type statistic.

(c) Rao's score statistic.

6.3.17. Let $X_1, X_2, \ldots, X_n$ be a random sample from a $\Gamma(\alpha, \beta)$-distribution where α is known and $\beta > 0$. Determine the likelihood ratio test for $H_0 : \beta = \beta_0$ against $H_1 : \beta \neq \beta_0$.

6.3.18. Let $Y_1 < Y_2 < \cdots < Y_n$ be the order statistics of a random sample from a uniform distribution on $(0, \theta)$, where $\theta > 0$.

(a) Show that Λ for testing $H_0 : \theta = \theta_0$ against $H_1 : \theta \neq \theta_0$ is $\Lambda = (Y_n/\theta_0)^n$, $Y_n \leq \theta_0$, and $\Lambda = 0$, if $Y_n > \theta_0$.

(b) When H_0 is true, show that $-2 \log \Lambda$ has an exact $\chi^2(2)$ distribution, not $\chi^2(1)$. Note that the regularity conditions are not satisfied.

6.4 Multiparameter Case: Estimation

In this section we discuss the case where $\boldsymbol{\theta}$ is a vector of p parameters. There are analogs to the theorems in the previous sections in which θ is a scalar and we present their results, but for the most part, without proofs. The interested reader can find additional information in more advanced books; see, for instance, Lehmann and Casella (1998) and Rao (1973).

Let $X_1, \ldots, X_n$ be iid with common pdf $f(x; \boldsymbol{\theta})$, where $\boldsymbol{\theta} \in \Omega \subset R^p$. As before, the likelihood function and its log are given by

$$L(\boldsymbol{\theta}) = \prod_{i=1}^{n} f(x_i; \boldsymbol{\theta})$$

$$l(\boldsymbol{\theta}) = \log L(\boldsymbol{\theta}) = \sum_{i=1}^{n} \log f(x_i; \boldsymbol{\theta}), \tag{6.4.1}$$

for $\boldsymbol{\theta} \in \Omega$. The theory requires additional regularity conditions which are listed in the Appendix A, (A.1.1). In keeping with our number scheme in the last two sections, we have labeled these (R6)-(R9). In this section of Chapter 6, when we say under regularity conditions we mean all of the conditions of (6.1.1), (6.2.1), (6.2.2), and (A.1.1) which are relevant to the argument. The discrete case follows in the same way as the continuous case, so in general we will state material in terms of the continuous case.

Note that the proof of Theorem 6.1.1 did not depend on whether the parameter is a scalar or a vector. Therefore, with probability going to one, $L(\boldsymbol{\theta})$ is maximized at the true value of $\boldsymbol{\theta}$. Hence, as an estimate of $\boldsymbol{\theta}$ we will consider the value which maximizes $L(\boldsymbol{\theta})$ or equivalently solves the vector equation $(\partial/\partial\boldsymbol{\theta})l(\boldsymbol{\theta}) = \mathbf{0}$. If it exists this value will be called the **maximum likelihood estimator** (mle) and we will denote it by $\widehat{\boldsymbol{\theta}}$. Often we are interested in a function of $\boldsymbol{\theta}$, say, the parameter $\eta = g(\boldsymbol{\theta})$. Because the second part of the proof of Theorem 6.1.2 remains true for $\boldsymbol{\theta}$ as a vector, $\widehat{\eta} = g(\widehat{\boldsymbol{\theta}})$ is the mle of η.

Example 6.4.1 (Maximum Likelihood Estimates of the Normal pdf). Suppose $X_1, \ldots, X_n$ are iid $N(\mu, \sigma^2)$. In this case, $\boldsymbol{\theta} = (\mu, \sigma^2)'$ and Ω is the product space $(-\infty, \infty) \times (0, \infty)$. The log of the likelihood simplifies to

$$l(\mu, \sigma^2) = -\frac{n}{2} \log 2\pi - n \log \sigma - \frac{1}{2\sigma^2} \sum_{i=1}^{n} (x_i - \mu)^2. \tag{6.4.2}$$

Taking partial derivatives of (6.4.2) with respect to μ and σ and setting them to 0, we get the simultaneous equations,

$$\frac{\partial l}{\partial \mu} = \frac{1}{\sigma^2} \sum_{i=1}^{n} (x_i - \mu) = 0$$

$$\frac{\partial l}{\partial \sigma} = -\frac{n}{\sigma} + \frac{1}{\sigma^3} \sum_{i=1}^{n} (x_i - \mu)^2 = 0.$$

Solving these equations, we obtain $\widehat{\mu} = \overline{X}$ and $\widehat{\sigma} = \sqrt{(1/n) \sum_{i=1}^{n} (X_i - \overline{X})^2}$ as solutions. A check of the second partials shows that these maximize $l(\mu, \sigma^2)$, so these are the mles. Also, by Theorem 6.1.2, $(1/n) \sum_{i=1}^{n} (X_i - \overline{X})^2$ is the mle of σ^2. We know from our discussion in Section 5.4, that these are consistent estimates of μ and σ^2, respectively; that $\widehat{\mu}$ is an unbiased estimate of μ and that $\widehat{\sigma^2}$ is a biased estimate of σ^2 whose bias vanishes as $n \to \infty$. ∎

Example 6.4.2 (General Laplace pdf). Let $X_1, X_2, \ldots, X_n$ be a random sample from the Laplace pdf $f_X(x) = (2b)^{-1} \exp\{-|x - a|/b\}$, $-\infty < x < \infty$, where the parameters (a, b) are in the space $\Omega = \{(a, b) : -\infty < a < \infty, b > 0\}$. Recall in the last sections we looked at the special case where $b = 1$. As we now show, the mle of a is the sample median, regardless of the value of b. The log of the likelihood function is,

$$l(a, b) = -n \log 2 - n \log b - \sum_{i=1}^{n} \left| \frac{x_i - a}{b} \right|.$$

The partial of $l(a, b)$ with respect to a is

$$\frac{\partial l(a, b)}{\partial a} = \frac{1}{b} \sum_{i=1}^{n} \text{sgn} \left\{ \frac{x_i - a}{b} \right\} = \frac{1}{b} \sum_{i=1}^{n} \text{sgn}\{x_i - a\},$$

where the second equality follows because $b > 0$. Setting this partial to 0, we obtain the mle of a to be $Q_2 = \text{med}\{X_1, X_2, \ldots, X_n\}$, just as in Example 6.1.3. Hence, the mle of a is invariant to the parameter b. Taking the partial of $l(a, b)$ with respect to b we obtain

$$\frac{\partial l(a, b)}{\partial b} = -\frac{n}{b} + \frac{1}{b^2} \sum_{i=1}^{n} |x_i - a|.$$

Setting to 0 and solving the two partials simultaneously, we obtain as the mle of b the statistic

$$\widehat{b} = \frac{1}{n} \sum_{i=1}^{n} |X_i - Q_2|. \quad \blacksquare$$

Recall that Fisher information in the scalar case was the variance of the random variable $(\partial/\partial\theta) \log f(X; \theta)$. The analog in the multiparameter case is the variance-covariance matrix of the gradient of $\log f(X; \boldsymbol{\theta})$; that is, the variance-covariance matrix of the random vector given by

$$\bigtriangledown \log f(X; \boldsymbol{\theta}) = \left(\frac{\partial \log f(X; \boldsymbol{\theta})}{\partial \theta_1}, \dots, \frac{\partial \log f(X; \boldsymbol{\theta})}{\partial \theta_p} \right)'. \tag{6.4.3}$$

Fisher information is then defined by the $p \times p$ matrix,

$$\mathbf{I}(\boldsymbol{\theta}) = \text{Cov}\left(\bigtriangledown \log f(X; \boldsymbol{\theta}) \right). \tag{6.4.4}$$

The $(j, k)th$ entry of $\mathbf{I}(\boldsymbol{\theta})$ is given by

$$I_{j,k} = \text{cov}\left(\frac{\partial}{\partial \theta_j} \log f(X; \boldsymbol{\theta}), \frac{\partial}{\partial \theta_k} \log f(X; \boldsymbol{\theta}) \right); \quad j, k = 1, \dots, p. \tag{6.4.5}$$

As in the scalar case, we can simplify this by using the identity $1 = \int f(x; \boldsymbol{\theta}) \, dx$. Under the regularity conditions, as discussed in the second paragraph of this section, the partial derivative of this identity with respect to θ_j results in,

$$0 = \int \frac{\partial}{\partial \theta_j} f(x; \boldsymbol{\theta}) \, dx = \int \left[\frac{\partial}{\partial \theta_j} \log f(x; \boldsymbol{\theta}) \right] f(x; \boldsymbol{\theta}) \, dx$$

$$= E\left[\frac{\partial}{\partial \theta_j} \log f(X; \boldsymbol{\theta}) \right]. \tag{6.4.6}$$

Next on both sides of the first equality above, take the partial derivative with respect to θ_k. After simplification, this results in,

$$0 = \int \left(\frac{\partial^2}{\partial \theta_j \partial \theta_k} \log f(x; \boldsymbol{\theta}) \right) f(x; \boldsymbol{\theta}) \, dx$$

$$+ \int \left(\frac{\partial}{\partial \theta_j} \log f(x; \boldsymbol{\theta}) \frac{\partial}{\partial \theta_k} \log f(x; \boldsymbol{\theta}) \right) f(x; \boldsymbol{\theta}) \, dx;$$

that is,

$$E\left[\frac{\partial}{\partial \theta_j} \log f(X; \boldsymbol{\theta}) \frac{\partial}{\partial \theta_k} \log f(X; \boldsymbol{\theta}) \right] = -E\left[\frac{\partial^2}{\partial \theta_j \partial \theta_k} \log f(X; \boldsymbol{\theta}) \right]. \tag{6.4.7}$$

Using (6.4.6) and (6.4.7) together, we obtain

$$I_{jk} = -E\left[\frac{\partial^2}{\partial\theta_j\partial\theta_k}\log f(X;\boldsymbol{\theta})\right].\tag{6.4.8}$$

Information for a sample follows in the same way as the scalar case. The pdf of the sample is the likelihood function $L(\boldsymbol{\theta};\mathbf{X})$. Replace $f(X;\boldsymbol{\theta})$ by $L(\boldsymbol{\theta};\mathbf{X})$ in the vector given in expression (6.4.3). Because $\log L$ is a sum, this results in the random vector

$$\nabla\log L(\boldsymbol{\theta};\mathbf{X}) = \sum_{i=1}^{n}\nabla\log f(X_i;\boldsymbol{\theta}).\tag{6.4.9}$$

Because the summands are iid with common covariance matrix $\mathbf{I}(\boldsymbol{\theta})$, we have

$$\text{Cov}(\nabla\log L(\boldsymbol{\theta};\mathbf{X})) = n\mathbf{I}(\boldsymbol{\theta}).\tag{6.4.10}$$

As in the scalar case the information in a random sample of size n is n times the information in a sample of size 1.

The diagonal entries of $\mathbf{I}(\boldsymbol{\theta})$ are,

$$I_{ii}(\boldsymbol{\theta}) = \text{Var}\left[\frac{\partial\log f(X;\boldsymbol{\theta})}{\partial\theta_i}\right] = -E\left[\frac{\partial^2}{\partial\theta_i^2}\log f(X_i;\boldsymbol{\theta})\right].$$

This is similar to the case when θ is a scalar, except now $I_{ii}(\boldsymbol{\theta})$ is a function of the vector $\boldsymbol{\theta}$. Recall in the scalar case, that $(nI(\theta))^{-1}$ was the Rao-Cramér lower bound for an unbiased estimate of θ. There is an analog to this in the multiparameter case. In particular, if $Y = u(X_1, \ldots, X_n)$ is an unbiased estimate of θ_j then it can be shown that

$$\text{Var}(Y) \geq \frac{1}{n}\left[\mathbf{I}^{-1}(\boldsymbol{\theta})\right]_{jj};\tag{6.4.11}$$

see, for example, Lehmann (1983). As in the scalar case, we shall call an unbiased estimate **efficient** if its variance attains this lower bound.

Example 6.4.3 (Information Matrix for the Normal pdf). The log of a $N(\mu,\sigma^2)$ pdf is given by

$$\log f(x;\mu,\sigma^2) = -\frac{1}{2}\log 2\pi - \log\sigma - \frac{1}{2\sigma^2}(x-\mu)^2.\tag{6.4.12}$$

The first and second partial derivatives are:

$$\frac{\partial\log f}{\partial\mu} = \frac{1}{\sigma^2}(x-\mu)$$

$$\frac{\partial^2\log f}{\partial\mu^2} = -\frac{1}{\sigma^2}$$

$$\frac{\partial\log f}{\partial\sigma} = -\frac{1}{\sigma} + \frac{1}{\sigma^3}(x-\mu)^2$$

$$\frac{\partial^2\log f}{\partial\sigma^2} = \frac{1}{\sigma^2} - \frac{3}{\sigma^4}(x-\mu)^2$$

$$\frac{\partial^2\log f}{\partial\mu\partial\sigma} = -\frac{2}{\sigma^3}(x-\mu).$$

Upon taking the negative of the expectations of the second partial derivatives, the information matrix for a normal density is

$$\mathbf{I}(\mu, \sigma) = \begin{bmatrix} \frac{1}{\sigma^2} & 0 \\ 0 & \frac{2}{\sigma^2} \end{bmatrix}. \tag{6.4.13}$$

We may want the information matrix for (μ, σ^2). This can be obtained by taking partial derivatives with respect to σ^2 instead of σ; however, in Example 6.4.6, we will obtain it via a transformation. From Example 6.4.1, the maximum likelihood estimates of μ and σ^2 are $\widehat{\mu} = \overline{X}$ and $\widehat{\sigma}^2 = (1/n) \sum_{i=1}^{n} (X_i - \overline{X})^2$, respectively. Based on the information matrix, we note that $\overline{X}$ is an efficient estimate of μ for finite samples. In Example 6.4.6 we will consider the sample variance. ■

Example 6.4.4 (Information Matrix for a Location and Scale Family).
Suppose $X_1, X_2, \ldots, X_n$ is a random sample with common pdf $f_X(x) = b^{-1} f\left(\frac{x-a}{b}\right)$, $-\infty < x < \infty$, where (a, b) is in the space $\Omega = \{(a, b) : -\infty < a < \infty, b > 0\}$ and $f(z)$ is a pdf such that $f(z) > 0$ for $-\infty < z < \infty$. As Exercise 6.4.8 shows, we can model X_i as

$$X_i = a + b e_i, \tag{6.4.14}$$

where the e_is are iid with pdf $f(z)$. This is called a *location and scale model* (LASP). Example 6.4.2 illustrated this model when $f(z)$ had the Laplace pdf. In Exercise 6.4.9, the reader is asked to show that the partial derivatives are:

$$\frac{\partial}{\partial a} \left\{ \log \left[\frac{1}{b} f\left(\frac{x-a}{b}\right) \right] \right\} = -\frac{1}{b} \frac{f'\left(\frac{x-a}{b}\right)}{f\left(\frac{x-a}{b}\right)}$$

$$\frac{\partial}{\partial b} \left\{ \log \left[\frac{1}{b} f\left(\frac{x-a}{b}\right) \right] \right\} = -\frac{1}{b} \left[1 + \frac{\frac{x-a}{b} f'\left(\frac{x-a}{b}\right)}{f\left(\frac{x-a}{b}\right)} \right].$$

Using (6.4.5) and (6.4.6), we then obtain

$$I_{11} = \int_{-\infty}^{\infty} \frac{1}{b^2} \left[\frac{f'\left(\frac{x-a}{b}\right)}{f\left(\frac{x-a}{b}\right)} \right]^2 \frac{1}{b} f\left(\frac{x-a}{b}\right) dx.$$

Now make the substitution $z = (x - a)/b$, $dz = (1/b)dx$. Then we have

$$I_{11} = \frac{1}{b^2} \int_{-\infty}^{\infty} \left[\frac{f'(z)}{f(z)} \right]^2 f(z) \, dz, \tag{6.4.15}$$

hence, information on the location parameter a does not depend on a. As Exercise 6.4.9 shows, upon making this substitution the other entries into the information matrix are:

$$I_{22} = \frac{1}{b^2} \int_{-\infty}^{\infty} \left[1 + \frac{z f'(z)}{f(z)} \right]^2 f(z) \, dz \tag{6.4.16}$$

$$I_{12} = \frac{1}{b^2} \int_{-\infty}^{\infty} z \left[\frac{f'(z)}{f(z)} \right]^2 f(z) \, dz. \tag{6.4.17}$$

Thus, the information matrix can be written as $(1/b)^2$ times a matrix whose entries are free of the parameters a and b. As Exercise 6.4.10 shows, the off-diagonal entries of the information matrix are 0, if the pdf $f(z)$ is symmetric about 0. ∎

Example 6.4.5 (Multinomial Distribution). Consider a random trial which can result in one, and only one, of k outcomes or categories. Let X_j be 1 or 0 depending on whether the jth outcomes occurs or does not, for $j = 1, \ldots, k$. Suppose the probability that outcome j occurs is p_j; hence, $\sum_{j=1}^{k} p_j = 1$. Let $\mathbf{X} = (X_1, \ldots, X_{k-1})'$ and $\mathbf{p} = (p_1, \ldots, p_{k-1})'$. The distribution of $\mathbf{X}$ is multinomial, see Section 3.1. Recall that the pdf is given by

$$f(\mathbf{x}, \mathbf{p}) = \left(\prod_{j=1}^{k-1} p_j^{x_j} \right) (1 - \sum_{j=1}^{k-1} p_j)^{1 - \sum_{j=1}^{k-1} x_j}; \tag{6.4.18}$$

where the parameter space is $\Omega = \{\mathbf{p} : 0 < p_j < 1, j = 1, \ldots, k-1; \sum_{j=1}^{k-1} p_j < 1\}$.

We first obtain the information matrix. The first partial of the log of f with respect to p_i simplifies to

$$\frac{\partial \log f}{\partial p_i} = \frac{x_i}{p_i} - \frac{1 - \sum_{j=1}^{k-1} x_j}{1 - \sum_{j=1}^{k-1} p_j}.$$

The second partial derivatives are given by

$$\frac{\partial^2 \log f}{\partial p_i^2} = -\frac{x_i}{p_i^2} - \frac{1 - \sum_{j=1}^{k-1} x_j}{(1 - \sum_{j=1}^{k-1} p_j)^2}$$

$$\frac{\partial^2 \log f}{\partial p_i \partial p_h} = -\frac{1 - \sum_{j=1}^{k-1} x_j}{(1 - \sum_{j=1}^{k-1} p_j)^2}, \quad i \neq h < k.$$

Recall for this distribution that marginally each random variable X_j has a Bernoulli distribution with mean p_j. Recalling that $p_k = 1 - (p_1 + \cdots + p_{k-1})$, the expectations of the negatives of the second partial derivatives are straightforward and result in the information matrix:

$$\mathbf{I}(\mathbf{p}) = \begin{bmatrix} \frac{1}{p_1} + \frac{1}{p_k} & \frac{1}{p_k} & \cdots & \frac{1}{p_k} \\ \frac{1}{p_k} & \frac{1}{p_2} + \frac{1}{p_k} & \cdots & \frac{1}{p_k} \\ \vdots & \vdots & & \vdots \\ \frac{1}{p_k} & \frac{1}{p_k} & \cdots & \frac{1}{p_{k-1}} + \frac{1}{p_k} \end{bmatrix}. \tag{6.4.19}$$

This is a patterned matrix with inverse (see Page 170 of Graybill, 1969),

$$\mathbf{I}^{-1}(\mathbf{p}) = \begin{bmatrix} p_1(1 - p_1) & -p_1 p_2 & \cdots & -p_1 p_{k-1} \\ -p_1 p_2 & p_2(1 - p_2) & \cdots & -p_2 p_{k-1} \\ \vdots & \vdots & & \vdots \\ -p_1 p_{k-1} & -p_2 p_{k-1} & \cdots & p_{k-1}(1 - p_{k-1}) \end{bmatrix}. \tag{6.4.20}$$

Next, we obtain the mles for a random sample $\mathbf{X}_1, \mathbf{X}_2, \ldots, \mathbf{X}_n$. The likelihood function is given by

$$L(\mathbf{p}) = \prod_{i=1}^{n} \prod_{j=1}^{k-1} p_j^{x_{ji}} \left(1 - \sum_{j=1}^{k-1} p_j \right)^{1 - \sum_{j=1}^{k-1} x_{ji}}. \tag{6.4.21}$$

Let $t_j = \sum_{i=1}^{n} x_{ji}$, for $j = 1, \ldots k - 1$. With simplification the log of L reduces to

$$l(\mathbf{p}) = \sum_{j=1}^{k-1} t_j \log p_j + \left(n - \sum_{j=1}^{k-1} t_j \right) \log \left(1 - \sum_{j=1}^{k-1} p_j \right).$$

The first partial of $l(\mathbf{p})$ with respect to p_h leads to the system of equations:

$$\frac{\partial l(\mathbf{p})}{\partial p_h} = \frac{t_h}{p_h} - \frac{n - \sum_{j=1}^{k-1} t_j}{1 - \sum_{j=1}^{k-1} p_j} = 0, \quad h = 1, \ldots, k-1.$$

It is easily seen that $p_h = t_h/n$ satisfies these equations. Hence, the maximum likelihood estimates are

$$\widehat{p_h} = \frac{\sum_{i=1}^{n} X_{ih}}{n}, \quad h = 1, \ldots, k-1. \tag{6.4.22}$$

Each random variable $\sum_{i=1}^{n} X_{ih}$ is binomial(n, p_h) with variance $np_h(1-p_h)$. Therefore the maximum likelihood estimates are efficient estimates. ∎

As a final note on information, suppose the information matrix is diagonal. Then the lower bound of the variance of the jth estimator (6.4.11) is $1/(n\mathbf{I}_{jj}(\boldsymbol{\theta}))$. Because $\mathbf{I}_{jj}(\boldsymbol{\theta})$ is defined in terms of partial derivatives, see (6.4.5), this is the information in treating all θ_i, except θ_j, as known. For instance in Example 6.4.3, for the normal pdf the information matrix is diagonal; hence, the information for μ could have been obtained by treating σ^2 as known. Example 6.4.4 discusses the information for a general location and scale family. For this general family, of which the normal is a member, the information matrix is diagonal provided the underlying pdf is symmetric.

In the next theorem, we summarize the asymptotic behavior of the maximum likelihood estimator of the vector $\boldsymbol{\theta}$. It shows that the mles are asymptotically efficient estimates.

Theorem 6.4.1. *Let $X_1, \ldots, X_n$ be iid with pdf $f(x; \boldsymbol{\theta})$ for $\boldsymbol{\theta} \in \Omega$. Assume the regularity conditions hold. Then,*

1. *The likelihood equation,*

$$\frac{\partial}{\partial \boldsymbol{\theta}} l(\boldsymbol{\theta}) = \mathbf{0},$$

has a solution $\widehat{\boldsymbol{\theta}}_n$ such that $\widehat{\boldsymbol{\theta}}_n \xrightarrow{P} \boldsymbol{\theta}$.

2. *For any sequence which satisfies (1),*

$$\sqrt{n}(\widehat{\boldsymbol{\theta}}_n - \boldsymbol{\theta}) \xrightarrow{D} N_p(\mathbf{0}, \mathbf{I}^{-1}(\boldsymbol{\theta})).$$

The proof of this theorem can be found in more advanced books; see, for example, Lehmann and Casella (1998). As in the scalar case, the theorem does not assure that the maximum likelihood estimates are unique. But if the sequence of solutions are unique, then they are both consistent and asymptotically normal. In applications, we can often verify uniqueness.

We immediately have the following corollary,

Corollary 6.4.1. *Let $X_1, \ldots, X_n$ be iid with pdf $f(x; \boldsymbol{\theta})$ for $\boldsymbol{\theta} \in \Omega$. Assume the regularity conditions hold. Let $\widehat{\boldsymbol{\theta}}_n$ be a sequence of consistent solutions of the likelihood equation. Then $\widehat{\boldsymbol{\theta}}_n$ are asymptotically efficient estimates; that is, for $j = 1, \ldots, p$,*

$$\sqrt{n}(\widehat{\theta}_{n,j} - \theta_j) \xrightarrow{D} N(0, \mathbf{I}_{jj}^{-1}(\boldsymbol{\theta})).$$

Let $\mathbf{g}$ be a transformation $\mathbf{g}(\boldsymbol{\theta}) = (g_1(\boldsymbol{\theta}), \ldots, g_k(\boldsymbol{\theta}))'$ such that $1 \le k \le p$ and that the $k \times p$ matrix of partial derivatives

$$\mathbf{B} = \left[\frac{\partial g_i}{\partial \theta_j} \right], \quad i = 1, \ldots k; \ j = 1, \ldots, p,$$

has continuous elements and does not vanish in a neighborhood of $\boldsymbol{\theta}$. Let $\widehat{\boldsymbol{\eta}} = \mathbf{g}(\widehat{\boldsymbol{\theta}})$. Then $\widehat{\boldsymbol{\eta}}$ is the mle of $\boldsymbol{\eta} = \mathbf{g}(\boldsymbol{\theta})$. By Theorem 4.5.6,

$$\sqrt{n}(\widehat{\boldsymbol{\eta}} - \boldsymbol{\eta}) \xrightarrow{D} N_k(\mathbf{0}, \mathbf{B}\mathbf{I}^{-1}(\boldsymbol{\theta})\mathbf{B}'). \tag{6.4.23}$$

Hence, the information matrix for $\boldsymbol{\eta}$ is,

$$\mathbf{I}(\boldsymbol{\eta}) = \left(\mathbf{B}\mathbf{I}^{-1}(\boldsymbol{\theta})\mathbf{B}' \right)^{-1}, \tag{6.4.24}$$

provided the inverse exists.

For a simple example of this result, reconsider Example 6.4.3.

Example 6.4.6 (Information for the Variance of a Normal Distribution). Suppose $X_1, \ldots, X_n$ are iid $N(\mu, \sigma^2)$. Recall from Example 6.4.3 that the information matrix was $\mathbf{I}(\mu, \sigma) = \text{diag}\{\sigma^{-2}, 2\sigma^{-2}\}$. Consider the transformation $g(\mu, \sigma) = \sigma^2$. Hence, the matrix of partials $\mathbf{B}$ is the row vector $[0 \ 2\sigma]$. Thus, the information for σ^2 is

$$I(\sigma^2) = \left\{ \begin{bmatrix} 0 & 2\sigma \end{bmatrix} \begin{bmatrix} \frac{1}{\sigma^2} & 0 \\ 0 & \frac{2}{\sigma^2} \end{bmatrix}^{-1} \begin{bmatrix} 0 \\ 2\sigma \end{bmatrix} \right\}^{-1} = \frac{1}{2\sigma^4}.$$

The Rao-Cramér lower bound for the variance of an estimator of σ^2 is $(2\sigma^4)/n$. Recall that the sample variance is unbiased for σ^2 but its variance is $(2\sigma^4)/(n-1)$. Hence, it is not efficient for finite samples but it is asymptotically efficient. ∎

EXERCISES

6.4.1. Let X_1, X_2, and X_3 have a multinomial distribution in which $n = 25$, $k = 4$, and the unknown probabilities are θ_1, θ_2, and θ_3, respectively. Here we can, for convenience, let $X_4 = 25 - X_1 - X_2 - X_3$ and $\theta_4 = 1 - \theta_1 - \theta_2 - \theta_3$. If the observed values of the random variables are $x_1 = 4$, $x_2 = 11$, and $x_3 = 7$, find the maximum likelihood estimates of θ_1, θ_2, and θ_3.

6.4.2. Let $X_1, X_2, \ldots, X_n$ and $Y_1, Y_2, \ldots, Y_m$ be independent random samples from $N(\theta_1, \theta_3)$ and $N(\theta_2, \theta_4)$ distributions, respectively.

(a) If $\Omega \subset R^3$ is defined by

$$\Omega = \{(\theta_1, \theta_2, \theta_3) : -\infty < \theta_i < \infty, i = 1, 2; 0 < \theta_3 = \theta_4 < \infty\},$$

find the mles of $\theta_1, \theta_2, \theta_3$.

(b) If $\Omega \subset R^2$ is defined by

$$\Omega = \{(\theta_1, \theta_3) : -\infty < \theta_1 = \theta_2 < \infty; 0 < \theta_3 = \theta_4 < \infty\},$$

find the mles of θ_1 and θ_3.

6.4.3. Let $X_1, X_2, \ldots, X_n$ be iid, each with the distribution having pdf $f(x; \theta_1, \theta_2) = (1/\theta_2)e^{-(x-\theta_1)/\theta_2}$, $\theta_1 \leq x < \infty$, $-\infty < \theta_2 < \infty$, zero elsewhere. Find the maximum likelihood estimators of θ_1 and θ_2.

6.4.4. The *Pareto distribution* is frequently used a model in study of incomes and has the distribution function

$$F(x; \theta_1, \theta_2) = \begin{cases} 1 - (\theta_1/x)^{\theta_2} & \theta_1 \leq x \\ 0 & \text{elsewhere,} \end{cases}$$

where $\theta_1 > 0$ and $\theta_2 > 0$. If $X_1, X_2, \ldots, X_n$ is a random sample from this distribution, find the maximum likelihood estimators of θ_1 and θ_2.

6.4.5. Let $Y_1 < Y_2 < \cdots < Y_n$ be the order statistics of a random sample of size n from the uniform distribution of the continuous type over the closed interval $[\theta - \rho, \theta + \rho]$. Find the maximum likelihood estimators for θ and ρ. Are these two unbiased estimators?

6.4.6. Let $X_1, X_2, \ldots, X_n$ be a random sample from $N(\mu, \sigma^2)$.

(a) If the constant b is defined by the equation $Pr(X \leq b) = 0.90$, find the mle of b.

(b) If c is given constant, find the mle of $Pr(X \leq c)$.

6.4.7. Consider two Bernoulli distributions with unknown parameters p_1 and p_2. If Y and Z equal the numbers of successes in two independent random samples, each of size n, from the respective distributions, determine the mles of p_1 and p_2 if we know that $0 \leq p_1 \leq p_2 \leq 1$.

6.4.8. Show that if X_i follows the model (6.4.14) then its pdf is $b^{-1}f((x-a)/b)$.

6.4.9. Verify the partial derivatives and the entries of the information matrix for the location and scale family as given in Example 6.4.4.

6.4.10. Suppose the pdf of X is of a location and scale family as defined in Example 6.4.4. Show that if $f(z) = f(-z)$ then the entry I_{12} of the information matrix is 0. Then argue that in this case the mles of a and b are asymptotically independent.

6.4.11. Suppose $X_1, X_2, \ldots, X_n$ are iid $N(\mu, \sigma^2)$. Show that X_i follows a location and scale family as given in Example 6.4.4. Obtain the entries of the information matrix as given in this example and show that they agree with the information matrix determined in Example 6.4.3.

6.5 Multiparameter Case: Testing

In the multiparameter case, hypotheses of interest often specify $\boldsymbol{\theta}$ to be in a sub-region of the space. For example, suppose X has a $N(\mu, \sigma^2)$ distribution. The full space is $\Omega = \{(\mu, \sigma^2) : \sigma^2 > 0, -\infty < \mu < \infty\}$. This is a two-dimensional space. We may be interested though in testing that $\mu = \mu_0$, where μ_0 is a specified value. Here, we are not concerned about the parameter σ^2. Under H_0, the parameter space is the one-dimensional space $\omega = \{(\mu_0, \sigma^2) : \sigma^2 > 0\}$. We say that H_0 is defined in terms of one constraint on the space Ω.

In general, let $X_1, \ldots, X_n$ be iid with pdf $f(x; \boldsymbol{\theta})$ for $\boldsymbol{\theta} \in \Omega \subset R^p$. As in the last section, we will assume that the regularity conditions listed in (6.1.1), (6.2.1), (6.2.2), and (A.1.1) are satisfied. In this section, we will invoke these by the phrase under regularity conditions. The hypotheses of interest are,

$$H_0 : \boldsymbol{\theta} \in \omega \text{ versus } H_1 : \boldsymbol{\theta} \in \Omega \cap \omega^c, \tag{6.5.1}$$

where $\omega \subset \Omega$ is defined in terms of q, $0 < q \leq p$, independent constraints of the form, $g_1(\boldsymbol{\theta}) = a_1, \ldots, g_q(\boldsymbol{\theta}) = a_q$. The functions $g_1, \ldots, g_q$ must be continuously differentiable. This implies ω is a $p - q$ dimensional space. Based on Theorem 6.1.1, the true parameter maximizes the likelihood function, so an intuitive test statistic is given by the likelihood ratio

$$\Lambda = \frac{\max_{\boldsymbol{\theta} \in \omega} L(\boldsymbol{\theta})}{\max_{\boldsymbol{\theta} \in \Omega} L(\boldsymbol{\theta})}. \tag{6.5.2}$$

Large values (close to one) of Λ suggests that H_0 is true, while small values indicate H_1 is true. For a specified level α, $0 < \alpha < 1$, this suggests the decision rule

$$\text{Reject } H_0 \text{ in favor of } H_1, \text{ if } \Lambda \leq c, \tag{6.5.3}$$

where c is such that $\alpha = \max_{\boldsymbol{\theta} \in \omega} P_{\boldsymbol{\theta}}[\Lambda \leq c]$. As in the scalar case, this test often has optimal properties; see Section 6.3. To determine c, we need to determine the distribution of Λ or a function of Λ when H_0 is true.

Let $\widehat{\boldsymbol{\theta}}$ denote the maximum likelihood estimator when the parameter space is the full space Ω and let $\widehat{\boldsymbol{\theta}}_0$ denote the maximum likelihood estimator when the parameter space is the reduced space ω. For convenience, define $L(\widehat{\Omega}) = L\left(\widehat{\boldsymbol{\theta}}\right)$ and $L(\widehat{\omega}) = L\left(\widehat{\boldsymbol{\theta}}_0\right)$. Then we can write the test statistic as

$$\Lambda = \frac{L(\widehat{\omega})}{L(\widehat{\Omega})}. \tag{6.5.4}$$

Example 6.5.1 (LRT for the Mean of a Normal pdf). Let $X_1, \ldots, X_n$ be a random sample from a normal distribution with mean μ and variance σ^2. Suppose we are interested in testing

$$H_0 : \mu = \mu_0 \text{ versus } H_1 : \mu \neq \mu_0, \tag{6.5.5}$$

where μ_0 is specified. Let $\Omega = \{(\mu, \sigma^2) : -\infty < \mu < \infty, \sigma^2 > 0\}$ denote the full model parameter space. The reduced model parameter space is the one dimensional subspace $\omega = \{(\mu_0, \sigma^2) : \sigma^2 > 0\}$. By Example 6.4.1, the mles of μ and σ^2 under Ω are $\widehat{\mu} = \overline{X}$ and $\widehat{\sigma}^2 = (1/n)\sum_{i=1}^{n}(X_i - \overline{X})^2$, respectively. Under Ω, the maximum value of the likelihood function is

$$L(\widehat{\Omega}) = \frac{1}{(2\pi)^{n/2}} \frac{1}{(\widehat{\sigma}^2)^{n/2}} \exp\{-(n/2)\}. \tag{6.5.6}$$

Following Example 6.4.1, it is easy to show that under the reduced parameter space ω, $\widehat{\sigma}_0^2 = (1/n)\sum_{i=1}^{n}(X_i - \mu_0)^2$. Thus the maximum value of the likelihood function under ω is:

$$L(\widehat{\omega}) = \frac{1}{(2\pi)^{n/2}} \frac{1}{(\widehat{\sigma}_0^2)^{n/2}} \exp\{-(n/2)\}. \tag{6.5.7}$$

The likelihood ratio test statistic is the ratio of $L(\widehat{\omega})$ to $L(\widehat{\Omega})$; i.e,

$$\Lambda = \left(\frac{\sum_{i=1}^{n}(X_i - \overline{X})^2}{\sum_{i=1}^{n}(X_i - \mu_0)^2}\right)^{n/2}. \tag{6.5.8}$$

The likelihood ratio test rejects H_0, if $\Lambda \leq c$, but this is equivalent to rejecting H_0, if $\Lambda^{-2/n} \geq c'$. Next consider the identity,

$$\sum_{i=1}^{n}(X_i - \mu_0)^2 = \sum_{i=1}^{n}(X_i - \overline{X})^2 + n(\overline{X} - \mu_0)^2. \tag{6.5.9}$$

Substituting (6.5.9) for $\sum_{i=1}^{n}(X_i - \mu_0)^2$, after simplification, the test becomes reject H_0 if

$$1 + \frac{n(\overline{X} - \mu_0)^2}{\sum_{i=1}^{n}(X_i - \overline{X})^2} \geq c',$$

or equivalently, reject H_0 if

$$\left\{\frac{\sqrt{n}(\overline{X} - \mu_0)}{\sqrt{\sum_{i=1}^{n}(X_i - \overline{X})^2/(n-1)}}\right\}^2 \geq c'' = (c'-1)(n-1).$$

Let T denote the expression within braces on the leftside of this inequality. Then the decision rule is equivalent to,

$$\text{Reject } H_0 \text{ in favor of } H_1, \text{ if } |T| \geq c^*, \tag{6.5.10}$$

where $\alpha = P_{H_0}[|T| \geq c^*]$. Of course this is the two sided version of the t test presented in Example 5.5.4. If we take c to be $t_{\alpha/2, n-1}$, the upper $\alpha/2$ critical value of a t distribution with $n-1$ degrees of freedom, then our test will have exact level α. ∎

Other examples of likelihood ratio tests for normal distributions can be found in the exercises.

We are not always as fortunate as in Example 6.5.1 to obtain the likelihood ratio test in a simple form. Often it is difficult or perhaps impossible to obtain its finite sample distribution. But, as the next theorem shows, we can always obtain an asymptotic test based on it.

Theorem 6.5.1. *Let $X_1, \ldots, X_n$ be iid with pdf $f(x; \boldsymbol{\theta})$ for $\boldsymbol{\theta} \in \Omega \subset R^p$. Assume the regularity conditions hold. Let $\widehat{\boldsymbol{\theta}}_n$ be a sequence of consistent solutions of the likelihood equation when the parameter space is the full space Ω. Let $\widehat{\boldsymbol{\theta}}_{0,n}$ be a sequence of consistent solutions of the likelihood equation when the parameter space is the reduced space ω, which has dimension $p - q$. Let Λ denote the likelihood ratio test statistic given in (6.5.4). Under H_0, (6.5.1),*

$$-2 \log \Lambda \xrightarrow{D} \chi^2(q). \tag{6.5.11}$$

A proof of this theorem can be found in Rao (1973).

There are analogs of the Wald-type and scores-types tests, also. The Wald-type test statistic is formulated in terms of the constraints, which define H_0, evaluated at the mle under Ω. We will not formally state it here, but as the following example shows it is often a straightforward formulation. The interested reader can find a discussion of these tests in Lehmann (1999).

A careful reading of the development of this chapter shows that much of it remains the same if X is a random vector. The next example demonstrates this.

Example 6.5.2 (Application of a Multinomial Distribution). As an example, consider a poll for a presidential race with k candidates. Those polled are asked to select the person for which they would vote if the election were held tomorrow. Assuming that those polled are selected independently of one another and that each can select one and only one candidate, the multinomial model seems appropriate. In this problem, suppose we are interested in comparing how the two "leaders" are doing. In fact, say the null hypothesis of interest is that they are equally favorable. This can be modeled with a multinomial model which has the three categories: (1) and (2) for the two leading candidates and (3) for all other candidates. Our observation is a vector (X_1, X_2) where X_i is 1 or 0 depending on whether Category i is selected or not. If both are 0 then Category (3) has been selected. Let p_i denote the probability that Category i is selected. Then the pdf of (X_1, X_2) is the trinomial density,

$$f(x_1, x_2; p_1, p_2) = p_1^{x_1} p_2^{x_2} (1 - p_1 - p_2)^{1 - x_1 - x_2}, \tag{6.5.12}$$

for $x_i = 0, 1, i = 1, 2; x_1 + x_2 \leq 1$, where the parameter space is $\Omega = \{(p_1, p_2) : 0 < p_i < 1, p_1 + p_2 < 1\}$. Suppose $(X_{11}, X_{21}), \ldots, (X_{1n}, X_{2n})$ are a random sample from this distribution. We shall consider the hypotheses,

$$H_0 : p_1 = p_2 \text{ versus } H_1 : p_1 \neq p_2. \tag{6.5.13}$$

We first derive the likelihood ratio test. Let $T_j = \sum_{i=1}^{n} X_{ji}$ for $j = 1, 2$. From Example 6.4.5, we know that the maximum likelihood estimates are $\widehat{p}_j = T_j/n$, for $j = 1, 2$. The value of the likelihood function (6.4.21) at the mles under Ω is

$$L\left(\widehat{\Omega}\right) = \hat{p}_1^{n\hat{p}_1} \hat{p}_2^{n\hat{p}_2} (1 - \hat{p}_1 - \hat{p}_2)^{n(1 - \hat{p}_1 - \hat{p}_2)}.$$

Under the null hypothesis, let p be the common value of p_1 and p_2. The pdf of (X_1, X_2) is

$$f(x_1, x_2; p) = p^{x_1 + x_2} (1 - 2p)^{1 - x_1 - x_2}; \quad x_1, x_2 = 0, 1; x_1 + x_2 \leq 1, \tag{6.5.14}$$

where the parameter space is $\omega = \{p : 0 < p < 1\}$. The likelihood under ω is

$$L(p) = p^{t_1 + t_2} (1 - 2p)^{n - t_1 - t_2}. \tag{6.5.15}$$

Differentiating the $\log L(p)$, with respect to p and setting the derivative to 0 results in the following maximum likelihood estimate, under ω:

$$\widehat{p}_0 = \frac{t_1 + t_2}{2n} = \frac{\widehat{p}_1 + \widehat{p}_2}{2}, \tag{6.5.16}$$

where $\widehat{p}_1$ and $\widehat{p}_2$ are the mles under Ω. The likelihood function evaluated at the mle under ω simplifies to

$$L\left(\hat{\omega}\right) = \left(\frac{\hat{p}_1 + \hat{p}_2}{2}\right)^{n(\hat{p}_1 + \hat{p}_2)} (1 - \hat{p}_1 - \hat{p}_2)^{n(1 - \hat{p}_1 - \hat{p}_2)}. \tag{6.5.17}$$

The reciprocal of the likelihood ratio test statistic then simplifies to

$$\Lambda^{-1} = \left(\frac{2\widehat{p}_1}{\widehat{p}_1 + \widehat{p}_2}\right)^{n\widehat{p}_1} \left(\frac{2\widehat{p}_2}{\widehat{p}_1 + \widehat{p}_2}\right)^{n\widehat{p}_2}. \tag{6.5.18}$$

Based on Theorem 6.5.11, an asymptotic level α test rejects H_0 if $2 \log \Lambda^{-1} > \chi_\alpha^2(1)$.

This is an example where the Wald's test can easily be formulated. The constraint under H_0, is $p_1 - p_2 = 0$. Hence, the Wald-type statistic is $W = \widehat{p}_1 - \widehat{p}_2$, which can be expressed as $W = [1, -1][\widehat{p}_1 ; \widehat{p}_2]'$. Recall that the information matrix and its inverse were found for k categories in Example 6.4.5. From Theorem 6.4.1 we then have

$$\begin{bmatrix} \widehat{p}_1 \\ \widehat{p}_2 \end{bmatrix} \text{ is approximately } N_2 \left(\begin{pmatrix} p_1 \\ p_2 \end{pmatrix}, \frac{1}{n} \begin{bmatrix} p_1(1 - p_1) & -p_1 p_2 \\ -p_1 p_2 & p_2(1 - p_2) \end{bmatrix} \right). \tag{6.5.19}$$

As shown in Example 6.4.5 the finite sample moments are the same as the asymptotic moments. Hence the variance of W is

$$
\begin{aligned}
\text{Var}(W) &= [1, -1]\frac{1}{n}\left[\begin{array}{cc} p_1(1-p_1) & -p_1 p_2 \\ -p_1 p_2 & p_2(1-p_2) \end{array}\right]\left[\begin{array}{c} 1 \\ -1 \end{array}\right] \\
&= \frac{p_1 + p_2 - (p_1 - p_2)^2}{n}.
\end{aligned}
$$

Because W is asymptotically normal, an asymptotic level α test for the hypotheses (6.5.13) is to reject H_0 if $\chi_W^2 \geq \chi_\alpha(1)$ where

$$
\chi_W^2 = \frac{(\widehat{p}_1 - \widehat{p}_2)^2}{(\widehat{p}_1 + \widehat{p}_2 - (\widehat{p}_1 - \widehat{p}_2)^2)/n}.
$$

It also follows that an asymptotic $(1-\alpha)100\%$ confidence interval for the difference $p_1 - p_2$ is

$$
\widehat{p}_1 - \widehat{p}_2 \pm z_{\alpha/2}\left(\frac{\widehat{p}_1 + \widehat{p}_2 - (\widehat{p}_1 - \widehat{p}_2)^2}{n}\right)^{1/2}.
$$

Returning to the polling situation discussed at the beginning of this example, we would say the race is too close to call if 0 is in this confidence interval. ∎

Example 6.5.3 (Two Sample Binomial Proportions). In Example 6.5.2, we developed tests for $p_1 = p_2$ based on a single sample from a multinomial distribution. Now consider the situation where $X_1, X_2, \ldots, X_{n_1}$ is a random sample from a $b(1, p_1)$ distribution, $Y_1, Y_2, \ldots, Y_{n_2}$ is a random sample from a $b(1, p_2)$ distribution, and the X_is and Y_js are mutually independent. The hypotheses of interest are

$$
H_0 : p_1 = p_2 \text{ versus } H_1 : p_1 \neq p_2. \tag{6.5.20}
$$

This situation occurs in practice when, for instance, we are comparing the president's rating from one month to the next. The full and reduced model parameter spaces are given respectively by $\Omega = \{(p_1, p_2) : 0 < p_i < 1, i = 1, 2\}$ and $\omega = \{(p, p) : 0 < p < 1\}$. The likelihood function for the full model simplifies to

$$
L(p_1, p_2) = p_1^{n_1 \overline{x}}(1 - p_1)^{n_1 - n_1 \overline{x}} p_2^{n_2 \overline{y}}(1 - p_2)^{n_2 - n_2 \overline{y}}. \tag{6.5.21}
$$

It follows immediately that the mles of p_1 and p_2 are $\overline{x}$ and $\overline{y}$, respectively. Note for the reduced model, we can combine the samples into one large sample from a $b(n, p)$ distribution, where $n = n_1 + n_2$ is the combined sample size. Hence, for the reduced model the mle of p is

$$
\widehat{p} = \frac{\sum_{i=1}^{n_1} x_i + \sum_{i=1}^{n_2} y_i}{n_1 + n_2} = \frac{n_1 \overline{x} + n_2 \overline{y}}{n}, \tag{6.5.22}
$$

i.e., a weighted average of the individual sample proportions. Using this, the reader is asked to derive the LRT for the hypotheses (6.5.20) in Exercise 6.5.9. We next derive the Wald-type test. Let $\widehat{p}_1 = \overline{x}$ and $\widehat{p}_2 = \overline{y}$. From the Central Limit Theorem, we have

$$
\frac{\sqrt{n_i}(\widehat{p}_i - p_i)}{\sqrt{p_i(1 - p_i)}} \xrightarrow{D} Z_i, \quad i = 1, 2,
$$

where Z_1 and Z_2 are iid $N(0,1)$ random variables. Assume for $i = 1, 2$ that, as $n \to \infty$, $n_i/n \to \lambda_i$, where $0 < \lambda_i < 1$ and $\lambda_1 + \lambda_2 = 1$. As Exercise 6.5.10 shows

$$\sqrt{n}[(\widehat{p}_1 - \widehat{p}_2) - (p_1 - p_2)] \xrightarrow{D} N\left(0, \frac{1}{\lambda_1}p_1(1 - p_1) + \frac{1}{\lambda_2}p_2(1 - p_2)\right). \qquad (6.5.23)$$

It follows that the random variable

$$Z = \frac{(\widehat{p}_1 - \widehat{p}_2) - (p_1 - p_2)}{\sqrt{\frac{p_1(1-p_1)}{n_1} + \frac{p_2(1-p_2)}{n_2}}} \qquad (6.5.24)$$

has an approximate $N(0,1)$ distribution. Under H_0, $p_1 - p_2 = 0$. We could use Z as a test statistic, provided we replace the parameters $p_1(1 - p_1)$ and $p_2(1 - p_2)$ in its denominator with a consistent estimate. Recall that $\widehat{p}_i \to p_i$, $i = 1, 2$, in probability. Thus under H_0, the statistic

$$Z^* = \frac{\widehat{p}_1 - \widehat{p}_2}{\sqrt{\frac{\widehat{p}_1(1-\widehat{p}_1)}{n_1} + \frac{\widehat{p}_2(1-\widehat{p}_2)}{n_2}}} \qquad (6.5.25)$$

has an approximate $N(0,1)$ distribution. Hence, an approximate level α test is to reject H_0, if $|z^*| \geq z_{\alpha/2}$. Another consistent estimator of the denominator is discussed in Exercise 6.5.11. ∎

EXERCISES

6.5.1. In Example 6.5.1 let $n = 10$, and let the experimental value of the random variables yield $\overline{x} = 0.6$ and $\sum_1^{10}(x_i - \overline{x})^2 = 3.6$. If the test derived in that example is used, do we accept or reject $H_0 : \theta_1 = 0$ at the 5 percent significance level?

6.5.2. Let $X_1, X_2, \ldots, X_n$ be a random sample from the distribution $N(\theta_1, \theta_2)$. Show that the likelihood ratio principle for testing $H_0 : \theta_2 = \theta_2'$ specified, and θ_1 unspecified, against $H_1 : \theta_2 \neq \theta_2'$, θ_1 unspecified, leads to a test that rejects when $\sum_1^n(x_i - \overline{x})^2 \leq c_1$ or $\sum_1^n(x_i - \overline{x})^2 \geq c_2$, where $c_1 < c_2$ are selected appropriately.

6.5.3. Let $X_1, \ldots, X_n$ and $Y_1, \ldots, Y_m$ be independent random samples from the distributions $N(\theta_1, \theta_3)$ and $N(\theta_2, \theta_4)$, respectively.

(a) Show that the likelihood ratio for testing $H_0 : \theta_1 = \theta_2$, $\theta_3 = \theta_4$ against all alternatives is given by

$$\frac{\left[\sum_1^n(x_i - \overline{x})^2/n\right]^{n/2}\left[\sum_1^m(y_i - \overline{y})^2/m\right]^{m/2}}{\left\{\left[\sum_1^n(x_i - u)^2 + \sum_1^m(y_i - u)^2\right]\bigg/(m + n)\right\}^{(n+m)/2}},$$

where $u = (n\overline{x} + m\overline{y})/(n + m)$.

(b) Show that the likelihood ratio test for testing $H_0 : \theta_3 = \theta_4$, θ_1 and θ_2 unspecified, against $H_1 : \theta_3 \neq \theta_4$, θ_1 and θ_2 unspecified, can be based on the random variable

$$F = \frac{\sum_1^n (X_i - \overline{X})^2 / (n-1)}{\sum_1^m (Y_i - \overline{Y})^2 / (m-1)}.$$

6.5.4. Let $X_1, X_2, \ldots, X_n$ and $Y_1, Y_2, \ldots, Y_m$ be independent random samples from the two normal distributions $N(0, \theta_1)$ and $N(0, \theta_2)$.

(a) Find the likelihood ratio Λ for testing the composite hypothesis $H_0 : \theta_1 = \theta_2$ against the composite alternative $H_1 : \theta_1 \neq \theta_2$.

(b) This Λ is a function of what F-statistic that would actually be used in this test?

6.5.5. Let X and Y be two independent random variables with respective pdfs

$$f(x; \theta_i) = \begin{cases} \left(\frac{1}{\theta_i}\right) e^{-x/\theta_i} & 0 < x < \infty, 0 < \theta_i < \infty \\ 0 & \text{elsewhere,} \end{cases}$$

for $i = 1, 2$. To test $H_0 : \theta_1 = \theta_2$ against $H_1 : \theta_1 \neq \theta_2$, two independent samples of sizes n_1 and n_2, respectively, were taken from these distributions. Find the likelihood ratio Λ and show that Λ can be written as a function of a statistic having an F-distribution, under H_0.

6.5.6. Consider the two uniform distributions with respective pdfs

$$f(x; \theta_i) = \begin{cases} \frac{1}{2\theta_i} & -\theta_i < x < \theta_i, -\infty < \theta_i < \infty \\ 0 & \text{elsewhere,} \end{cases}$$

for $i = 1, 2$. The null hypothesis is $H_0 : \theta_1 = \theta_2$ while the alternative is $H_1 : \theta_1 \neq \theta_2$. Let $X_1 < X_2 < \cdots < X_{n_1}$ and $Y_1 < Y_2 < \cdots < Y_{n_2}$ be the order statistics of two independent random samples from the respective distributions. Using the likelihood ratio Λ, find the statistic used to test H_0 against H_1. Find the distribution of $-2 \log \Lambda$ when H_0 is true. Note that in this nonregular case the number of degrees of freedom is two times the difference of the dimension of Ω and ω.

6.5.7. Let $(X_1, Y_1), (X_2, Y_2), \ldots, (X_n, Y_n)$ be a random sample from a bivariate normal distribution with $\mu_1, \mu_2, \sigma_1^2 = \sigma_2^2 = \sigma^2, \rho = \frac{1}{2}$, where μ_1, μ_2, and $\sigma^2 > 0$ are unknown real numbers. Find the likelihood ratio Λ for testing $H_0 : \mu_1 = \mu_2 = 0$, σ^2 unknown against all alternatives. The likelihood ratio Λ is a function of what statistic that has a well-known distribution?

6.5.8. Let n independent trials of an experiment be such that $x_1, x_2, \ldots, x_k$ are the respective numbers of times that the experiment ends in the mutually exclusive and exhaustive events $C_1, C_2, \ldots, C_k$. If $p_i = P(C_i)$ is constant throughout the n trials, then the probability of that particular sequence of trials is $L = p_1^{x_1} p_2^{x_2} \cdots p_k^{x_k}$.

(a) Recalling that $p_1 + p_2 + \cdots + p_k = 1$, show that the likelihood ratio for testing $H_0 : p_i = p_{i0} > 0$, $i = 1, 2, \ldots, k$, against all alternatives is given by

$$\Lambda = \prod_{i=1}^{k} \left(\frac{(p_{i0})^{x_i}}{(x_i/n)^{x_i}} \right).$$

(b) Show that

$$-2 \log \Lambda = \sum_{i=1}^{k} \frac{x_i (x_i - n p_{0i})^2}{(n p_i')^2},$$

where p_i' is between p_{0i} and x_i/n.

Hint: Expand $\log p_{i0}$ in a Taylor's series with the remainder in the term involving $(p_{i0} - x_i/n)^2$.

(c) For large n, argue that $x_i/(n p_i')^2$ is approximated by $1/(n p_{i0})$ and hence

$$-2 \log \Lambda \approx \sum_{i=1}^{k} \frac{(x_i - n p_{0i})^2}{n p_{0i}}, \quad \text{when } H_0 \text{ is true.}$$

Theorem 6.5.1 says that the right-hand member of this last equation defines a statistic that has an approximate chi-square distribution with $k-1$ degrees of freedom. Note that

$$\text{dimension of } \Omega - \text{dimension of } \omega = (k-1) - 0 = k - 1.$$

6.5.9. Finish the derivation of the LRT found in Example 6.5.3. Simplify as much as possible.

6.5.10. Show that expression (6.5.23) of Example 6.5.3 is true.

6.5.11. As discussed in Example 6.5.3, Z, (6.5.25), can be used as a test statistic provided we have a consistent estimator of $p_1(1-p_1)$ and $p_2(1-p_2)$ when H_0 is true. In the example, we discussed an estimator which is consistent under both H_0 and H_1. Under H_0, though, $p_1(1-p_1) = p_2(1-p_2) = p(1-p)$, where $p = p_1 = p_2$. Show that the statistic (6.5.22) is a consistent estimator of p, under H_0. Thus determine another test of H_0.

6.5.12. A machine shop that manufactures toggle levers has both a day and a night shift. A toggle lever is defective if a standard nut cannot be screwed onto the threads. Let p_1 and p_2 be the proportion of defective levers among those manufactured by the day and night shifts, respectively. We shall test the null hypothesis, $H_0 : p_1 = p_2$, against a two-sided alternative hypothesis based on two random samples, each of 1000 levers taken from the production of the respective shifts. Use the test statistic Z^* given in Example 6.5.3.

(a) Sketch a standard normal pdf illustrating the critical region having $\alpha = 0.05$.

(b) If $y_1 = 37$ and $y_2 = 53$ defectives were observed for the day and night shifts, respectively, calculate the value of the test statistic and the approximate p-value (note that this is a two-sided test). Locate the calculated test statistic on your figure in Part (a) and state your conclusion. Obtain the approximate p-value of the test.

6.5.13. For the situation given in Part (b) of Exercise 6.5.12, calculate the tests defined in Exercises 6.5.9 and 6.5.11. Obtain the approximate p-values of all three tests. Discuss the results.

6.6   The EM Algorithm

In practice, we are often in the situation where part of the data is missing. For example, we may be observing lifetimes of mechanical parts which have been put on test and some of these parts are still functioning when the statistical analysis is carried out. In this section, we introduce the EM Algorithm which frequently can be used in these situations to obtain maximum likelihood estimates. Our presentation is brief. For further information, the interested reader can consult the literature in this area including the monograph by McLachlan and Krishnan (1997). Although, for convenience, we will write in terms of continuous random variables; the theory in this section holds for the discrete case as well.

Suppose we consider a sample of n items, where n_1 of the items are observed while $n_2 = n - n_1$ items are not observable. Denote the observed items by $\mathbf{X}' = (X_1, X_2, \ldots, X_{n_1})$ and the unobserved items by $\mathbf{Z}' = (Z_1, Z_2, \ldots, Z_{n_2})$. Assume that the X_is are iid with pdf $f(x|\theta)$, where $\theta \in \Omega$. Assume that Z_js and the X_is are mutually independent. The conditional notation will prove useful here. Let $g(\mathbf{x}|\theta)$ denote the joint pdf of $\mathbf{X}$. Let $h(\mathbf{x}, \mathbf{z}|\theta)$ denote the joint pdf of the observed and the unobserved items. Let $k(\mathbf{z}|\theta, \mathbf{x})$ denote the conditional pdf of the missing data given the observed data. By the definition of a conditional pdf, we have the identity

$$k(\mathbf{z}|\theta, \mathbf{x}) = \frac{h(\mathbf{x}, \mathbf{z}|\theta)}{g(\mathbf{x}|\theta)}. \tag{6.6.1}$$

The **observed likelihood** function is $L(\theta|\mathbf{x}) = g(\mathbf{x}|\theta)$. The **complete likelihood** function is defined by

$$L^c(\theta|\mathbf{x}, \mathbf{z}) = h(\mathbf{x}, \mathbf{z}|\theta). \tag{6.6.2}$$

Our goal is maximize the likelihood function $L(\theta|\mathbf{x})$ by using the complete likelihood $L^c(\theta|\mathbf{x}, \mathbf{z})$ in this process.

Using (6.6.1), we derive the following basic identity for an arbitrary but fixed

$\theta_0 \in \Omega$:

$$
\begin{aligned}
\log L(\theta|\mathbf{x}) &= \int \log L(\theta|\mathbf{x}) k(\mathbf{z}|\theta_0, \mathbf{x}) \, d\mathbf{z} \\
&= \int \log g(\mathbf{x}|\theta) k(\mathbf{z}|\theta_0, \mathbf{x}) \, d\mathbf{z} \\
&= \int [\log h(\mathbf{x}, \mathbf{z}|\theta) - \log k(\mathbf{z}|\theta, \mathbf{x})] k(\mathbf{z}|\theta_0, \mathbf{x}) \, d\mathbf{z} \\
&= \int [\log[h(\mathbf{x}, \mathbf{z}|\theta)] k(\mathbf{z}|\theta_0, \mathbf{x}) \, d\mathbf{z} - \int [\log[k(\mathbf{z}|\theta, \mathbf{x})] k(\mathbf{z}|\theta_0, \mathbf{x}) \, d\mathbf{z} \\
&= E_{\theta_0}[\log L^c(\theta|\mathbf{x}, \mathbf{Z})|\theta_0, \mathbf{x}] - E_{\theta_0}[\log k(\mathbf{Z}|\theta, \mathbf{x})|\theta_0, \mathbf{x}], \qquad (6.6.3)
\end{aligned}
$$

where the expectations are taken under the conditional pdf $k(\mathbf{z}|\theta_0, \mathbf{x})$. Define the first term on the right-side of (6.6.3) to be the function

$$
Q(\theta|\theta_0, \mathbf{x}) = E_{\theta_0}[\log L^c(\theta|\mathbf{x}, \mathbf{Z})|\theta_0, \mathbf{x}]. \qquad (6.6.4)
$$

The expectation which defines the function Q is called the E step of the EM algorithm.

Recall that we want to maximize $\log L(\theta|\mathbf{x})$. As discussed below, we need only maximize $Q(\theta|\theta_0, \mathbf{x})$. This maximization is called the M step of the EM algorithm.

Denote by $\widehat{\theta}^{(0)}$ an initial estimate of θ, perhaps based on the observed likelihood. Let $\widehat{\theta}^{(1)}$ be the argument which maximizes $Q(\theta|\widehat{\theta}^{(0)}, \mathbf{x})$. This is the first-step estimate of θ. Proceeding this way, we obtain a sequence of estimates $\widehat{\theta}^{(m)}$. We formally define this algorithm as follows:

Algorithm 6.6.1 (EM Algorithm). *Let $\widehat{\theta}^{(m)}$ denote the estimate on the mth step. To compute the estimate on the $(m + 1)$st step do:*

1. *Expectation Step: Compute*

$$
Q(\theta|\widehat{\theta}^{(m)}, \mathbf{x}) = E_{\widehat{\theta}^{(m)}}[\log L^c(\theta|\mathbf{x}, \mathbf{Z})|\widehat{\theta}_m, \mathbf{x}], \qquad (6.6.5)
$$

where the expectation is taken under the conditional pdf $k(\mathbf{z}|\widehat{\theta}^{(m)}, \mathbf{x})$.

2. *Maximization Step: Let*

$$
\widehat{\theta}^{(m+1)} = Argmax \, Q(\theta|\widehat{\theta}^{(m)}, \mathbf{x}). \qquad (6.6.6)
$$

Under strong assumptions, it can be shown that $\widehat{\theta}^{(m)}$ converges in probability to the maximum likelihood estimate, as $m \to \infty$. We will not show these results, but as the next theorem shows $\widehat{\theta}^{(m+1)}$ always increases the likelihood over $\widehat{\theta}^{(m)}$.

Theorem 6.6.1. *The sequence of estimates $\widehat{\theta}^{(m)}$, defined by Algorithm 6.6.1, satisfies*

$$
L(\widehat{\theta}^{(m+1)}|\mathbf{x}) \geq L(\widehat{\theta}^{(m)}|\mathbf{x}). \qquad (6.6.7)
$$

Proof: Because $\widehat{\theta}^{(m+1)}$ maximizes $Q(\theta|\widehat{\theta}^{(m)}, \mathbf{x})$, we have

$$Q(\widehat{\theta}^{(m+1)}|\widehat{\theta}^{(m)}, \mathbf{x}) \geq Q(\widehat{\theta}^{(m)}|\widehat{\theta}^{(m)}, \mathbf{x});$$

that is,

$$E_{\widehat{\theta}^{(m)}}[\log L^c(\widehat{\theta}^{(m+1)}|\mathbf{x}, \mathbf{Z})] \geq E_{\widehat{\theta}^{(m)}}[\log L^c(\widehat{\theta}^{(m)}|\mathbf{x}, \mathbf{Z})]; \tag{6.6.8}$$

where the expectation is taken under the pdf $k(\mathbf{z}|\widehat{\theta}^{(m)}, \mathbf{x})$. By expression (6.6.3), we can complete the proof by showing that

$$E_{\widehat{\theta}^{(m)}}[\log k(\mathbf{Z}|\widehat{\theta}^{(m+1)}, \mathbf{x})] \leq E_{\widehat{\theta}^{(m)}}[\log k(\mathbf{Z}|\widehat{\theta}^{(m)}, \mathbf{x})]. \tag{6.6.9}$$

Keep in mind that these expectations are taken under the conditional pdf of $\mathbf{Z}$ given $\widehat{\theta}^{(m)}$ and $\mathbf{x}$. An application of Jensen's inequality, (1.10.5), yields

$$
\begin{aligned}
E_{\widehat{\theta}^{(m)}}\left\{\log\left[\frac{k(\mathbf{Z}|\widehat{\theta}^{(m+1)}, \mathbf{x})}{k(\mathbf{Z}|\widehat{\theta}^{(m)}, \mathbf{x})}\right]\right\} &\leq \log E_{\widehat{\theta}^{(m)}}\left[\frac{k(\mathbf{Z}|\widehat{\theta}^{(m+1)}, \mathbf{x})}{k(\mathbf{Z}|\widehat{\theta}^{(m)}, \mathbf{x})}\right] \\
&= \log\int\frac{k(\mathbf{z}|\widehat{\theta}^{(m+1)}, \mathbf{x})}{k(\mathbf{z}|\widehat{\theta}^{(m)}, \mathbf{x})}k(\mathbf{z}|\widehat{\theta}^{(m)}, \mathbf{x})\, d\mathbf{z} \\
&= \log(1) = 0. \tag{6.6.10}
\end{aligned}
$$

This last result establishes (6.6.9) and, hence, finishes the proof. ∎

As an example, suppose $X_1, X_2, \ldots, X_{n_1}$ are iid with pdf $f(x - \theta)$, for $-\infty < x < \infty$, where $-\infty < \theta < \infty$. Denote the cdf of X_i by $F(x - \theta)$. Let $Z_1, Z_2, \ldots, Z_{n_2}$ denote the censored observations. For these observations, we only know that $Z_j > a$, for some a which is known and that the Z_js are independent of the X_is. Then the observed and complete likelihoods are given by

$$L(\theta|\mathbf{x}) = [1 - F(a - \theta)]^{n_2}\prod_{i=1}^{n_1}f(x_i - \theta), \tag{6.6.11}$$

$$L^c(\theta|\mathbf{x}, \mathbf{z}) = \prod_{i=1}^{n_1}f(x_i - \theta)\prod_{i=1}^{n_2}f(z_i - \theta). \tag{6.6.12}$$

By expression (6.6.1), the conditional distribution $\mathbf{Z}$ given $\mathbf{X}$ is the ratio of (6.6.13) to (6.6.11), that is,

$$
\begin{aligned}
k(\mathbf{z}|\theta, \mathbf{x}) &= \frac{\prod_{i=1}^{n_1}f(x_i - \theta)\prod_{i=1}^{n_2}f(z_i - \theta)}{[1 - F(a - \theta)]^{n_2}\prod_{i=1}^{n_2}f(z_i - \theta)} \\
&= [1 - F(a - \theta)]^{-n_2}\prod_{i=1}^{n_2}f(z_i - \theta), \quad a < z_i, i = 1, \ldots, n_2 \tag{6.6.13}
\end{aligned}
$$

Thus $\mathbf{Z}$ and $\mathbf{X}$ are indedpendent and $Z_1, \ldots, Z_{n_2}$ are iid with the common pdf $f(z - \theta)/[1 - F(a - \theta)]$, for $z > a$. Based on these observations and expression

(6.6.13), we have the following derivation:

$$
\begin{aligned}
Q(\theta|\theta_0, \mathbf{x}) &= E_{\theta_0}[\log L^c(\theta|\mathbf{x}, \mathbf{Z})] \\
&= E_{\theta_0}[\sum_{i=1}^{n_1} \log f(x_i - \theta) + \sum_{i=1}^{n_2} \log f(Z_i - \theta)] \\
&= \sum_{i=1}^{n_1} \log f(x_i - \theta) + n_2 E_{\theta_0}[\log f(Z - \theta)] \\
&= \sum_{i=1}^{n_1} \log f(x_i - \theta) \\
&\quad + n_2 \int_a^\infty \log f(z - \theta) \frac{f(z - \theta_0)}{1 - F(a - \theta_0)} \, dz.
\end{aligned}
\tag{6.6.14}
$$

This last result is the E step of the EM Algorithm. For the M step, we need the partial derivative of $Q(\theta|\theta_0, \mathbf{x})$ with respect to θ. This is easily found to be

$$
\frac{\partial Q}{\partial \theta} = -\left\{ \sum_{i=1}^{n_1} \frac{f'(x_i - \theta)}{f(x_i - \theta)} + n_2 \int_a^\infty \frac{f'(z - \theta)}{f(z - \theta)} \frac{f(z - \theta_0)}{1 - F(a - \theta_0)} \, dz \right\}.
\tag{6.6.15}
$$

Assuming that $\theta_0 = \widehat{\theta}_0$, the first step EM estimate would be the value of θ, say $\widehat{\theta}^{(1)}$, which solves $\frac{\partial Q}{\partial \theta} = 0$. In the next example, we obtain the solution for a normal model.

Example 6.6.1. Assume the censoring model given above, but now assume that X has a $N(\theta, 1)$ distribution. Then $f(x) = \phi(x) = (2\pi)^{-1/2} \exp\{-x^2/2\}$. It is easy to show that $f'(x)/f(x) = -x$. Letting $\Phi(z)$ denote, as usual, the cdf of a standard normal random variable, by (6.6.15) the partial derivative of $Q(\theta|\theta_0, \mathbf{x})$ with respect to θ for this model simplifies to:

$$
\begin{aligned}
\frac{\partial Q}{\partial \theta} &= \sum_{i=1}^{n_1}(x_i - \theta) + n_2 \int_a^\infty (z - \theta) \frac{1}{\sqrt{2\pi}} \frac{\exp\{-(z - \theta_0)^2/2\}}{1 - \Phi(a - \theta_0)} \, dz \\
&= n_1(\bar{x} - \theta) + n_2 \int_a^\infty (z - \theta_0) \frac{1}{\sqrt{2\pi}} \frac{\exp\{-(z - \theta_0)^2/2\}}{1 - \Phi(a - \theta_0)} \, dz - n_2(\theta - \theta_0) \\
&= n_1(\bar{x} - \theta) + \frac{n_2}{1 - \Phi(a - \theta_0)} \phi(a - \theta_0) - n_2(\theta - \theta_0).
\end{aligned}
$$

Solving $\partial Q/\partial \theta = 0$ for θ determines the EM step estimates. In particular, given that $\widehat{\theta}^{(m)}$ is the EM estimate on the mth step, the $(m+1)$st step estimate is

$$
\widehat{\theta}^{(m+1)} = \frac{n_1}{n} \bar{x} + \frac{n_2}{n} \widehat{\theta}^{(m)} + \frac{n_2}{n} \frac{\phi(a - \widehat{\theta}^{(m)})}{1 - \Phi(a - \widehat{\theta}^{(m)})},
\tag{6.6.16}
$$

where $n = n_1 + n_2$. ∎

For our second example, consider a mixture problem involving normal distributions. Suppose Y_1 has a $N(\mu_1, \sigma_1^2)$ distribution and Y_2 has a $N(\mu_2, \sigma_2^2)$ distribution. Let W be a Bernoulli random variable independent of Y_1 and Y_2 and with probability of success $\pi = P(W = 1)$. Suppose the random variable we observe is $X = (1 - W)Y_1 + WY_2$. In this case, the vector of parameters is given by $\boldsymbol{\theta}' = (\mu_1, \mu_2, \sigma_1, \sigma_2, \pi)$. As shown in Section 3.4, the pdf of the mixture random variable X is

$$f(x) = (1 - \pi)f_1(x) + \pi f_2(x), \quad -\infty < x < \infty, \tag{6.6.17}$$

where $f_j(x) = \sigma_j^{-1}\phi((x-\mu_j)/\sigma_j))$, $j = 1, 2$, and $\phi(z)$ is the pdf of a standard normal random variable. Suppose we observe a random sample $\mathbf{X}' = (X_1, X_2, \ldots, X_n)$ from this mixture distribution with pdf $f(x)$. Then the log of the likelihood function is

$$l(\boldsymbol{\theta}|\mathbf{x}) = \sum_{i=1}^{n} \log[(1 - \pi)f_1(x_i) + \pi f_2(x_i)]. \tag{6.6.18}$$

In this mixture problem, the unobserved data are the random variables which identify the distribution membership. For $i = 1, 2, \ldots, n$, define the random variables

$$W_i = \begin{cases} 0 & \text{if } X_i \text{ has pdf } f_1(x) \\ 1 & \text{if } X_i \text{ has pdf } f_2(x). \end{cases}$$

These variables, of course, constitute the random sample on the Bernoulli random variable W. Accordingly, assume that $W_1, W_2, \ldots, W_n$ are iid Bernoulli random variables with probability of success π. The complete likelihood function is

$$L^c(\boldsymbol{\theta}|\mathbf{x}, \mathbf{w}) = \prod_{W_i=0} f_1(x_i) \prod_{W_i=1} f_1(x_i).$$

Hence, the log of the complete likelihood function is

$$\begin{aligned} l^c(\boldsymbol{\theta}|\mathbf{x}, \mathbf{w}) &= \sum_{W_i=0} \log f_1(x_i) + \sum_{W_i=1} \log f_2(x_i) \\ &= \sum_{i=1}^{n}[(1 - w_i)\log f_1(x_i) + w_i \log f_2(x_i)]. \end{aligned} \tag{6.6.19}$$

For the E step of the algorithm, we need the conditional expectation of W_i given $\mathbf{x}$ under $\boldsymbol{\theta}_0$; that is,

$$E_{\boldsymbol{\theta}_0}[W_i|\boldsymbol{\theta}_0, \mathbf{x}] = P[W_i = 1|\boldsymbol{\theta}_0, \mathbf{x}].$$

An estimate of this expectation is the likelihood of x_i being drawn from distribution two, which is given by

$$\gamma_i = \frac{\widehat{\pi} f_{2,0}(x_i)}{(1 - \widehat{\pi})f_{1,0}(x_i) + \widehat{\pi} f_{2,0}(x_i)}, \tag{6.6.20}$$

where the subscript 0 signifies that the parameters at $\boldsymbol{\theta}_0$ are being used. Expression (6.6.20) is intuitively evident; see McLachlan and Krishnan (1997) for more discussion. Replacing w_i by γ_i in expression (6.6.19), the M step of the algorithm is to maximize

$$Q(\boldsymbol{\theta}|\boldsymbol{\theta}_0, \mathbf{x}) = \sum_{i=1}^{n}[(1 - \gamma_i)\log f_1(x_i) + \gamma_i \log f_2(x_i)]. \tag{6.6.21}$$

This maximization is easy to obtain by taking partial derivatives of $Q(\boldsymbol{\theta}|\boldsymbol{\theta}_0, \mathbf{x})$ with respect to the parameters. For example,

$$\frac{\partial Q}{\partial \mu_1} = \sum_{i=1}^{n}(1 - \gamma_i)(-1/2\sigma_1^2)(-2)(x_i - \mu_1).$$

Setting this to 0 and solving for μ_1 yields the estimate of μ_1. The estimates of the other mean and the variances can be obtained similarly. These estimates are:

$$
\begin{aligned}
\widehat{\mu}_1 &= \frac{\sum_{i=1}^{n}(1 - \gamma_i)x_i}{\sum_{i=1}^{n}(1 - \gamma_i)}, \\
\widehat{\sigma}_1^2 &= \frac{\sum_{i=1}^{n}(1 - \gamma_i)(x_i - \widehat{\mu}_1)^2}{\sum_{i=1}^{n}(1 - \gamma_i)}, \\
\widehat{\mu}_2 &= \frac{\sum_{i=1}^{n}\gamma_i x_i}{\sum_{i=1}^{n}\gamma_i}, \\
\widehat{\sigma}_2^2 &= \frac{\sum_{i=1}^{n}\gamma_i(x_i - \widehat{\mu}_2)^2}{\sum_{i=1}^{n}\gamma_i},
\end{aligned}
\tag{6.6.22}
$$

Since γ_i is an estimate of $P[W_i = 1|\boldsymbol{\theta}_0, \mathbf{x}]$, the average $n^{-1}\sum_{i=1}^{n}\gamma_i$ is an estimate of $\pi = P[W_i = 1]$. This average is our estimate of $\widehat{\pi}$.

EXERCISES

6.6.1. Rao (page 368, 1973) considers a problem in the estimation of linkages in genetics. McLachlan and Krishnan (1997) also discuss this problem and we present their model. For our purposes it can be described as a multinomial model with the four categories C_1, C_2, C_3 and C_4. For a sample of size n, let $\mathbf{X} = (X_1, X_2, X_3, X_4)'$ denote the observed frequencies of the four categories. Hence, $n = \sum_{i=1}^{4} X_i$. The probability model is

C_1	C_2	C_3	C_4
$\frac{1}{2} + \frac{1}{4}\theta$	$\frac{1}{4} - \frac{1}{4}\theta$	$\frac{1}{4} - \frac{1}{4}\theta$	$\frac{1}{4}\theta$

where the parameter θ satisfies $0 \leq \theta \leq 1$. In this exercise, we obtain the mle of θ.

  (a) Show that likelihood function is given by

$$L(\theta|\mathbf{x}) = \frac{n!}{x_1!x_2!x_3!x_4!}\left[\frac{1}{2} + \frac{1}{4}\theta\right]^{x_1}\left[\frac{1}{4} - \frac{1}{4}\theta\right]^{x_2+x_3}\left[\frac{1}{4}\theta\right]^{x_4}. \tag{6.6.23}$$

(b) Show that the log of the likelihood function can be expressed as a constant (not involving parameters) plus the term

$$x_1 \log[2 + \theta] + [x_2 + x_3] \log[1 - \theta] + x_4 \log \theta.$$

(c) Obtain the partial of the last expression, set the result to 0, and solve for the mle. (This will result in a quadratic equation which has one positive and one negative root.)

6.6.2. In this exercise, we set up an EM Algorithm to determine the mle for the situation described in Exercise 6.6.1. Split category C_1 into the two subcategories C_{11} and C_{12} with probabilities $1/2$ and $\theta/4$, respectively. Let Z_{11} and Z_{12} denote the respective "frequencies." Then $X_1 = Z_{11} + Z_{12}$. Of course, we cannot observe Z_{11} and Z_{12}. Let $\mathbf{Z} = (Z_{11}, Z_{12})'$.

(a) Obtain the complete likelihood $L^c(\theta|\mathbf{x}, \mathbf{z})$.

(b) Using the last result and (6.6.23), show that the conditional pmf $k(\mathbf{z}|\theta, \mathbf{x})$ is binomial with parameters x_1 and probability of success $\theta/(2 + \theta)$.

(c) Obtain the E step of the EM Algorithm given an initial estimate $\widehat{\theta}^{(0)}$ of θ. That is, obtain

$$Q(\theta|\widehat{\theta}^{(0)}, \mathbf{x}) = E_{\widehat{\theta}^{(0)}}[\log L^c(\theta|\mathbf{x}, \mathbf{Z})|\widehat{\theta}^{(0)}, \mathbf{x}].$$

Recall that this expectation is taken using the conditional pmf $k(\mathbf{z}|\widehat{\theta}^{(0)}, \mathbf{x})$. Keep in mind the next step; i.e., we need only terms that involve θ.

(d) For the M step of the EM Algorithm, solve the equation $\partial Q(\theta|\widehat{\theta}^{(0)}, \mathbf{x})/\partial \theta = 0$. Show that the solution is

$$\widehat{\theta}^{(1)} = \frac{x_1 \widehat{\theta}^{(0)} + 2x_4 + x_4 \widehat{\theta}^{(0)}}{n\widehat{\theta}^{(0)} + 2(x_2 + x_3 + x_4)}. \tag{6.6.24}$$

6.6.3. For the setup of Exercise 6.6.2, show that the following estimator of θ is unbiased.

$$\widetilde{\theta} = n^{-1}(X_1 - X_2 - X_3 + X_4). \tag{6.6.25}$$

6.6.4. Rao (page 368, 1973) presents data for the situation described in Exercise 6.6.1. The observed frequencies are: $\mathbf{x} = (125, 18, 20, 34)'$.

(a) Using computational packages, (eg, R or S-PLUS), with (6.6.25) as the initial estimate, write a program that obtains the stepwise EM estimates $\widehat{\theta}^{(k)}$.

(b) Using the data from Rao, compute the EM estimate of θ with your program. List the sequence of EM estimates, $\{\widehat{\theta}^k\}$, that you obtained. Did your sequence of estimates converge?

(c) Show that the mle using the likelihood approach in Exercise 6.6.1 is the positive root of the equation: $197\theta^2 - 150\theta - 68 = 0$. Compare it with your EM solution. They should be the same within roundoff error.

6.6.5. Suppose $X_1, X_2, \ldots, X_{n_1}$ are a random sample from a $N(\theta, 1)$ distribution. Suppose $Z_1, Z_2, \ldots, Z_{n_2}$ are missing observations. Show that the first step EM estimate is

$$\widehat{\theta}^{(1)} = \frac{n_1 \bar{x} + n_2 \widehat{\theta}^{(0)}}{n},$$

where $\widehat{\theta}^{(0)}$ is an initial estimate of θ and $n = n_1 + n_2$. Note that if $\widehat{\theta}^{(0)} = \bar{x}$, then $\widehat{\theta}^{(k)} = \bar{x}$ for all k.

6.6.6. Consider the situation described in Example 6.6.1. But suppose we have left censoring. That is, if $Z_1, Z_2, \ldots, Z_{n_2}$ are the censored items then all we know is that each $Z_j < a$. Obtain the EM Algorithm estimate of θ.

6.6.7. Suppose the following data follow the model of Example 6.6.1.

2.01	0.74	0.68	1.50+	1.47	1.50+	1.50+	1.52
0.07	-0.04	-0.21	0.05	-0.09	0.67	0.14	

where the superscript + denotes that the observation was censored at 1.50. Write a computer program to obtain the EM Algorithm estimate of θ.

6.6.8. The following data are observations of the random variable $X = (1-W)Y_1 + WY_2$, where W has a Bernoulli distribution with probability of success 0.70; Y_1 has a $N(100, 20^2)$ distribution; Y_2 has a $N(120, 25^2)$ distribution; W and Y_1 are independent; and W and Y_2 are independent.

119.0	96.0	146.2	138.6	143.4	98.2	124.5
114.1	136.2	136.4	184.8	79.8	151.9	114.2
145.7	95.9	97.3	136.4	109.2	103.2	

Program the EM Algorithm for this mixing problem as discussed at the end of the section. Use a dotplot to obtain initial estimates of the parameters. Compute the estimates. How close are they to the true parameters?

Chapter 7

Sufficiency

7.1 Measures of Quality of Estimators

In Chapter 6 we presented procedures for finding point estimates, interval estimates, and tests of statistical hypotheses based on likelihood theory. In this and the next chapter, we present some optimal point estimates and tests for certain situations. We first consider point estimation.

In this chapter, as in Chapter 6, we find it convenient to use the letter f to denote a pmf as well as a pdf. It will be clear from the context whether we are discussing the distributions of discrete or continuous random variables.

Suppose $f(x; \theta)$ for $\theta \in \Omega$ is the pdf (pmf) of a continuous (discrete) random variable X. Consider a point estimator $Y_n = u(X_1, \ldots, X_n)$ based on a sample $X_1, \ldots, X_n$. In Chapter 4, we discussed several properties of point estimators. Recall that Y_n is a consistent estimator (Definition 4.2.2) of θ if Y_n converges to θ in probability; i.e., Y_n is close to θ for large sample sizes. This is definitely a desirable property of a point estimator. Under suitable conditions, Theorem 6.1.3 shows that the maximum likelihood estimator is consistent. Another property was unbiasedness, (Definition 4.1.1), which says that Y_n is an unbiased estimator of θ if $E(Y_n) = \theta$. Recall that maximum likelihood estimators may not be unbiased; although, generally they are asymptotically unbiased, (see Theorem 6.2.2).

If two estimators of θ are unbiased, it would seem that we would choose the one with the smaller variance. This would be especially true if they were both approximately normal because by (5.4.3) the one with the smaller variance would tend to produce shorter asymptotic confidence intervals for θ. This leads to the following definition:

Definition 7.1.1. *For a given positive integer n, $Y = u(X_1, X_2, \ldots, X_n)$ will be called a* **minimum variance unbiased estimator**, *(MVUE), of the parameter θ, if Y is unbiased, that is, $E(Y) = \theta$, and if the variance of Y is less than or equal to the variance of every other unbiased estimator of θ.*

Example 7.1.1. As an illustration, let $X_1, X_2, \ldots, X_9$ denote a random sample from a distribution that is $N(\theta, \sigma^2)$, where $-\infty < \theta < \infty$. Because the statistic

367

$\overline{X} = (X_1 + X_2 + \cdots + X_9)/9$ is $N(\theta, \frac{\sigma^2}{9})$, $\overline{X}$ is an unbiased estimator of θ. The statistic X_1 is $N(\theta, \sigma^2)$, so X_1 is also an unbiased estimator of θ. Although the variance $\frac{\sigma^2}{9}$ of $\overline{X}$ is less than the variance σ^2 of X_1, we cannot say, with $n = 9$, that $\overline{X}$ is the minimum variance unbiased estimator (MVUE) of θ; that definition requires that the comparison be made with every unbiased estimator of θ. To be sure, it is quite impossible to tabulate all other unbiased estimators of this parameter θ, so other methods must be developed for making the comparisons of the variances. A beginning on this problem will be made in this chapter. ∎

Let us now discuss the problem of point estimation of a parameter from a slightly different standpoint. Let $X_1, X_2, \ldots, X_n$ denote a random sample of size n from a distribution that has the pdf $f(x; \theta)$, $\theta \in \Omega$. The distribution may be either of the continuous or the discrete type. Let $Y = u(X_1, X_2, \ldots, X_n)$ be a statistic on which we wish to base a point estimate of the parameter θ. Let $\delta(y)$ be that function of the observed value of the statistic Y which is the point estimate of θ. Thus the function δ *decides* the value of our point estimate of θ and δ is called a *decision function* or a *decision rule*. One value of the decision function, say $\delta(y)$, is called a *decision*. Thus a numerically determined point estimate of a parameter θ is a decision. Now a decision may be correct or it may be wrong. It would be useful to have a measure of the seriousness of the difference, if any, between the true value of θ and the point estimate $\delta(y)$. Accordingly, with each pair, $[\theta, \delta(y)]$, $\theta \in \Omega$, we will associate a nonnegative number $\mathcal{L}[\theta, \delta(y)]$ that reflects this seriousness. We call the function $\mathcal{L}$ the *loss function*. The expected (mean) value of the loss function is called the *risk function*. If $f_Y(y; \theta)$, $\theta \in \Omega$, is the pdf of Y, the risk function $R(\theta, \delta)$ is given by

$$R(\theta, \delta) = E\{\mathcal{L}[\theta, \delta(y)]\} = \int_{-\infty}^{\infty} \mathcal{L}[\theta, \delta(y)] f_Y(y; \theta)\, dy$$

if Y is a random variable of the continuous type. It would be desirable to select a decision function that minimizes the risk $R(\theta, \delta)$ for all values of θ, $\theta \in \Omega$. But this is usually impossible because the decision function δ that minimizes $R(\theta, \delta)$ for one value of θ may not minimize $R(\theta, \delta)$ for another value of θ. Accordingly, we need either to restrict our decision function to a certain class or to consider methods of ordering the risk functions. The following example, while very simple, dramatizes these difficulties.

Example 7.1.2. Let $X_1, X_2, \ldots, X_{25}$ be a random sample from a distribution that is $N(\theta, 1)$, for $-\infty < \theta < \infty$. Let $Y = \overline{X}$, the mean of the random sample, and let $\mathcal{L}[\theta, \delta(y)] = [\theta - \delta(y)]^2$. We shall compare the two decision functions given by $\delta_1(y) = y$ and $\delta_2(y) = 0$ for $-\infty < y < \infty$. The corresponding risk functions are

$$R(\theta, \delta_1) = E[(\theta - Y)^2] = \tfrac{1}{25}$$

and

$$R(\theta, \delta_2) = E[(\theta - 0)^2] = \theta^2.$$

Obviously, if, in fact, $\theta = 0$, then $\delta_2(y) = 0$ is an excellent decision and we have $R(0, \delta_2) = 0$. However, if θ differs from zero by very much, it is equally clear that $\delta_2 = 0$ is a poor decision. For example, if, in fact, $\theta = 2$, $R(2, \delta_2) = 4 > R(2, \delta_1) = \frac{1}{25}$. In general, we see that $R(\theta, \delta_2) < R(\theta, \delta_1)$, provided that $-\frac{1}{5} < \theta < \frac{1}{5}$ and that otherwise $R(\theta, \delta_2) \geq R(\theta, \delta_1)$. That is, one of these decision functions is better than the other for some values of θ and the other decision functions are better for other values of θ. If however, we had restricted our consideration to decision functions δ such that $E[\delta(Y)] = \theta$ for all values of θ, $\theta \in \Omega$, then the decision function $\delta_2(y) = 0$ is not allowed. Under this restriction and with the given $\mathcal{L}[\theta, \delta(y)]$, the risk function is the variance of the unbiased estimator $\delta(Y)$, and we are confronted with the problem of finding the MVUE. Later in this chapter we show that the solution is $\delta(y) = y = \overline{x}$.

Suppose, however, that we do not want to restrict ourselves to decision functions δ, such that $E[\delta(Y)] = \theta$ for all values of θ, $\theta \in \Omega$. Instead, let us say that the decision function that minimizes the maximum of the risk function is the best decision function. Because, in this example, $R(\theta, \delta_2) = \theta^2$ is unbounded, $\delta_2(y) = 0$ is not, in accordance with this criterion, a good decision function. On the other hand, with $-\infty < \theta < \infty$, we have

$$\max_{\theta} R(\theta, \delta_1) = \max_{\theta}(\tfrac{1}{25}) = \tfrac{1}{25}.$$

Accordingly, $\delta_1(y) = y = \overline{x}$ seems to be a very good decision in accordance with this criterion because $\frac{1}{25}$ is small. As a matter of fact, it can be proved that δ_1 is the best decision function, as measured by the *minimax criterion*, when the loss function is $\mathcal{L}[\theta, \delta(y)] = [\theta - \delta(y)]^2$. ∎

In this example we illustrated the following:

1. Without some restriction on the decision function, it is difficult to find a decision function that has a risk function which is uniformly less than the risk function of another decision.

2. One principle of selecting a best decision function is called the *minimax principle*. This principle may be stated as follows: If the decision function given by $\delta_0(y)$ is such that, for all $\theta \in \Omega$,

$$\max_{\theta} R[\theta, \delta_0(y)] \leq \max_{\theta} R[\theta, \delta(y)]$$

for every other decision function $\delta(y)$, then $\delta_0(y)$ is called a *minimax decision function*.

With the restriction $E[\delta(Y)] = \theta$ and the loss function $\mathcal{L}[\theta, \delta(y)] = [\theta - \delta(y)]^2$, the decision function that minimizes the risk function yields an unbiased estimator with minimum variance. If, however, the restriction $E[\delta(Y)] = \theta$ is replaced by some other condition, the decision function $\delta(Y)$, if it exists, which minimizes $E\{[\theta - \delta(Y)]^2\}$ uniformly in θ is sometimes called the *minimum mean-squared-error estimator*. Exercise 7.1.6, 7.1.7, and 7.1.8 provide examples of this type of estimator.

There are two additional observations about decision rules and loss functions that should be made at this point. First, since Y is a statistic, the decision rule $\delta(Y)$ is also a statistic, and we could have started directly with a decision rule based on the observations in a random sample, say $\delta_1(X_1, X_2, \ldots, X_n)$. The risk function is then given by

$$
\begin{aligned}
R(\theta, \delta_1) &= E\{\mathcal{L}[\theta, \delta_1(X_1, \ldots, X_n)]\} \\
&= \int_{-\infty}^{\infty} \cdots \int_{-\infty}^{\infty} \mathcal{L}[\theta, \delta_1(x_1, \ldots, x_n)] f(x_1; \theta) \cdots f(x_n; \theta) \, dx_1 \cdots dx_n
\end{aligned}
$$

if the random sample arises from a continuous-type distribution. We did not do this because, as you will see in this chapter, it is rather easy to find a good statistic, say Y, upon which to base all of the statistical inferences associated with a particular model. Thus we thought it more appropriate to start with a statistic that would be familiar, like the mle $Y = \overline{X}$ in Example 7.1.2. The second decision rule of that example could be written $\delta_2(X_1, X_2, \ldots, X_n) = 0$, a constant no matter what values of $X_1, X_2, \ldots, X_n$ are observed.

The second observation is that we have only used one loss function, namely the *squared-error loss function* $\mathcal{L}(\theta, \delta) = (\theta - \delta)^2$. The *absolute-error loss function* $\mathcal{L}(\theta, \delta) = |\theta - \delta|$ is another popular one. The loss function defined by

$$
\begin{aligned}
\mathcal{L}(\theta, \delta) &= 0, \quad |\theta - \delta| \le a, \\
&= b, \quad |\theta - \delta| > a,
\end{aligned}
$$

where a and b are positive constants, is sometimes referred to as the *goal post loss function*. The reason for this terminology is that football fans recognize that it is similar to kicking a field goal: There is no loss (actually a three-point gain) if within a units of the middle but b units of loss (zero points awarded) if outside that restriction. In addition, loss functions can be asymmetric as well as symmetric as the three previous ones have been. That is, for example, it might be more costly to underestimate the value of θ than to overestimate it. (Many of us think about this type of loss function when estimating the time it takes us to reach an airport to catch a plane.) Some of these loss functions are considered when studying Bayesian estimates in Chapter 11.

Let us close this section with an interesting illustration that raises a question leading to the likelihood principle which many statisticians believe is a quality characteristic that estimators should enjoy. Suppose that two statisticians, A and B, observe 10 independent trials of a random experiment ending in success or failure. Let the probability of success on each trial be θ, where $0 < \theta < 1$. Let us say that each statistician observes one success in these 10 trials. Suppose however, that A had decided to take $n = 10$ such observations in advance and found only one success while B had decided to take as many observations as needed to get the first success, which happened on the $10th$ trial. The model of A is that Y is $b(n = 10, \theta)$ and $y = 1$ is observed. On the other hand, B is considering the random variable Z that has a geometric pdf $g(z) = (1 - \theta)^{z-1}\theta$, $z = 1, 2, 3, \ldots$, and $z = 10$ is observed. In either case, the relative frequency of success is

$$
\frac{y}{n} = \frac{1}{z} = \frac{1}{10},
$$

which could be used as an estimate of θ.

Let us observe, however, that one of the corresponding estimators, Y/n and $1/Z$, is biased. We have

$$E\left(\frac{Y}{10}\right) = \frac{1}{10}E(Y) = \frac{1}{10}(10\theta) = \theta$$

while

$$
\begin{aligned}
E\left(\frac{1}{Z}\right) &= \sum_{z=1}^{\infty} \frac{1}{z}(1-\theta)^{z-1}\theta \\
&= \theta + \tfrac{1}{2}(1-\theta)\theta + \tfrac{1}{3}(1-\theta)^2\theta + \cdots > \theta.
\end{aligned}
$$

That is, $1/Z$ is a biased estimator while $Y/10$ is unbiased. Thus A is using an unbiased estimator while B is not. Should we adjust B's estimator so that it, too, is unbiased?

It is interesting to note that if we maximize the two respective likelihood functions, namely

$$L_1(\theta) = \binom{10}{y}\theta^y(1-\theta)^{10-y}$$

and

$$L_2(\theta) = (1-\theta)^{z-1}\theta,$$

with $n = 10$, $y = 1$, and $z = 10$, we get exactly the same answer, $\hat{\theta} = \frac{1}{10}$. This must be the case, because in each situation we are maximizing $(1-\theta)^9\theta$. Many statisticians believe that this is the way it should be and accordingly adopt the *likelihood principle*:

Suppose two different sets of data from possibly two different random experiments lead to respective likelihood ratios, $L_1(\theta)$ and $L_2(\theta)$, that are proportional to each other. These two data sets provide the same information about the parameter θ and a statistician should obtain the same estimate of θ from either.

In our special illustration, we note that $L_1(\theta) \propto L_2(\theta)$, and the likelihood principle states that statisticians A and B should make the same inference. Thus believers in the likelihood principle would not adjust the second estimator to make it unbiased.

EXERCISES

7.1.1. Show that the mean $\overline{X}$ of a random sample of size n from a distribution having pdf $f(x; \theta) = (1/\theta)e^{-(x/\theta)}$, $0 < x < \infty$, $0 < \theta < \infty$, zero elsewhere, is an unbiased estimator of θ and has variance θ^2/n.

7.1.2. Let $X_1, X_2, \ldots, X_n$ denote a random sample from a normal distribution with mean zero and variance θ, $0 < \theta < \infty$. Show that $\sum_1^n X_i^2/n$ is an unbiased estimator of θ and has variance $2\theta^2/n$.

7.1.3. Let $Y_1 < Y_2 < Y_3$ be the order statistics of a random sample of size 3 from the uniform distribution having pdf $f(x; \theta) = 1/\theta$, $0 < x < \infty$, $0 < \theta < \infty$, zero elsewhere. Show that $4Y_1$, $2Y_2$, and $\frac{4}{3}Y_3$ are all unbiased estimators of θ. Find the variance of each of these unbiased estimators.

7.1.4. Let Y_1 and Y_2 be two independent unbiased estimators of θ. Assume that the variance of Y_1 is twice the variance of Y_2. Find the constants k_1 and k_2 so that $k_1Y_1 + k_2Y_2$ is an unbiased estimator with smallest possible variance for such a linear combination.

7.1.5. In Example 7.1.2 of this section, take $\mathcal{L}[\theta, \delta(y)] = |\theta - \delta(y)|$. Show that $R(\theta, \delta_1) = \frac{1}{5}\sqrt{2/\pi}$ and $R(\theta, \delta_2) = |\theta|$. Of these two decision functions δ_1 and δ_2, which yields the smaller maximum risk?

7.1.6. Let $X_1, X_2, \ldots, X_n$ denote a random sample from a Poisson distribution with parameter θ, $0 < \theta < \infty$. Let $Y = \sum_1^n X_i$ and let $\mathcal{L}[\theta, \delta(y)] = [\theta - \delta(y)]^2$. If we restrict our considerations to decision functions of the form $\delta(y) = b + y/n$, where b does not depend on y, show that $R(\theta, \delta) = b^2 + \theta/n$. What decision function of this form yields a uniformly smaller risk than every other decision function of this form? With this solution, say δ, and $0 < \theta < \infty$, determine $\max_\theta R(\theta, \delta)$ if it exists.

7.1.7. Let $X_1, X_2, \ldots, X_n$ denote a random sample from a distribution that is $N(\mu, \theta)$, $0 < \theta < \infty$, where μ is unknown. Let $Y = \sum_1^n (X_i - \overline{X})^2/n = V$ and let $\mathcal{L}[\theta, \delta(y)] = [\theta - \delta(y)]^2$. If we consider decision functions of the form $\delta(y) = by$, where b does not depend upon y, show that $R(\theta, \delta) = (\theta^2/n^2)[(n^2 - 1)b^2 - 2n(n - 1)b + n^2]$. Show that $b = n/(n+1)$ yields a minimum risk decision functions of this form. Note that $nY/(n + 1)$ is not an unbiased estimator of θ. With $\delta(y) = ny/(n + 1)$ and $0 < \theta < \infty$, determine $\max_\theta R(\theta, \delta)$ if it exists.

7.1.8. Let $X_1, X_2, \ldots, X_n$ denote a random sample from a distribution that is $b(1, \theta)$, $0 \le \theta \le 1$. Let $Y = \sum_1^n X_i$ and let $\mathcal{L}[\theta, \delta(y)] = [\theta - \delta(y)]^2$. Consider decision functions of the form $\delta(y) = by$, where b does not depend upon y. Prove that $R(\theta, \delta) = b^2 n\theta(1 - \theta) + (bn - 1)^2\theta^2$. Show that

$$\max_\theta R(\theta, \delta) = \frac{b^4 n^2}{4[b^2 n - (bn - -1)^2]},$$

provided that the value b is such that $b^2 n \ge 2(bn - 1)^2$. Prove that $b = 1/n$ does not maximize $\max_\theta R(\theta, \delta)$.

7.1.9. Let $X_1, X_2, \ldots, X_n$ be a random sample from a Poisson distribution with mean $\theta > 0$.

(a) Statistician A observes the sample to be the values $x_1, x_2, \ldots, x_n$ with sum $y = \sum x_i$. Find the mle of θ.

(b) Statistician B loses the sample values $x_1, x_2, \ldots, x_n$ but remembers the sum y_1 and the fact that the sample arose from a Poisson distribution. Thus B decides to create some fake observations which he calls $z_1, z_2, \ldots, z_n$ (as he knows they will probably not equal the original x-values) as follows. He notes that the conditional probability of independent Poisson random variables $Z_1, Z_2, \ldots, Z_n$ being equal to $z_1, z_2, \ldots, z_n$, given $\sum z_i = y_1$ is

$$\frac{\frac{\theta^{z_1} e^{-\theta}}{z_1!} \frac{\theta^{z_2} e^{-\theta}}{z_2!} \cdots \frac{\theta^{z_n} e^{-\theta}}{z_n!}}{\frac{(n\theta)^{y_1} e^{n\theta}}{y_1!}} = \frac{y_1!}{z_1! z_2! \cdots z_n!} \left(\frac{1}{n}\right)^{z_1} \left(\frac{1}{n}\right)^{z_2} \cdots \left(\frac{1}{n}\right)^{z_n}$$

since $Y_1 = \sum Z_i$ has a Poisson distribution with mean $n\theta$. The latter distribution is multinomial with y_1 independent trials, each terminating in one of n mutually exclusive and exhaustive ways, each of which has the same probability $1/n$. Accordingly, B runs such a multinomial experiment y_1 independent trials and obtains $z_1, z_2, \ldots, z_n$. Find the likelihood function using these z-values. Is it proportional to that of statistician A?

Hint: Here the likelihood function is the product of this conditional pdf and the pdf of $Y_1 = \sum Z_i$.

7.2   A Sufficient Statistic for a Parameter

Suppose that $X_1, X_2, \ldots, X_n$ is a random sample from a distribution that has pdf $f(x; \theta)$, $\theta \in \Omega$. In Chapters 4 and 6 we constructed statistics to make statistical inferences as illustrated by point and interval estimation and tests of statistical hypotheses. We note that a statistic, for example, $Y = u(X_1, X_2, \ldots, X_n)$, is a form of data reduction. To illustrate, instead of listing all of the individual observations $X_1, X_2, \ldots, X_n$, we might prefer to give only the sample mean $\overline{X}$ or the sample variance S^2. Thus statisticians look for ways of reducing a set of data so that these data can be more easily understood without losing the meaning associated with the entire set of observations.

It is interesting to note that a statistic $Y = u(X_1, X_2, \ldots, X_n)$ really partitions the sample space of $X_1, X_2, \ldots, X_n$. For illustration, suppose we say that the sample was observed and $\overline{x} = 8.32$. There are many points in the sample space which have that same mean of 8.32, and we can consider them as belonging to the set $\{(x_1, x_2, \ldots, x_n) : \overline{x} = 8.32\}$. As a matter of fact, all points on the hyperplane

$$x_1 + x_2 + \cdots + x_n = (8.32)n$$

yield the mean of $\overline{x} = 8.32$, so this hyperplane is the set. However, there are many values that $\overline{X}$ can take and thus there are many such sets. So, in this sense, the sample mean $\overline{X}$, or any statistic $Y = u(X_1, X_2, \ldots, X_n)$, partitions the sample space into a collection of sets.

Often in the study of statistics the parameter θ of the model is unknown; thus we need to make some statistical inference about it. In this section we consider a statistic denoted by $Y_1 = u_1(X_1, X_2, \ldots, X_n)$, which we call a *sufficient statistic* and which we find is good for making those inferences. This sufficient statistic partitions the sample space in such a way that, given

$$(X_1, X_2, \ldots, X_n) \in \{(x_1, x_2, \ldots, x_n) : u_1(x_1, x_2, \ldots, x_n) = y_1\},$$

the conditional probability of $X_1, X_2, \ldots, X_n$ does not depend upon θ. Intuitively, this means that once the set determined by $Y_1 = y_1$ is fixed, the distribution of another statistic, say $Y_2 = u_2(X_1, X_2, \ldots, X_n)$, does not depend upon the parameter θ because the conditional distribution of $X_1, X_2, \ldots, X_n$ does not depend upon θ. Hence it is impossible to use Y_2, given $Y_1 = y_1$, to make a statistical inference about θ. So, in a sense, Y_1 *exhausts* all the information about θ that is contained in the sample. This is why we call $Y_1 = u_1(X_1, X_2, \ldots, X_n)$ a sufficient statistic.

To understand clearly the definition of a sufficient statistic for a parameter θ, we start with an illustration.

Example 7.2.1. Let $X_1, X_2, \ldots, X_n$ denote a random sample from the distribution that has pmf

$$f(x; \theta) = \begin{cases} \theta^x (1 - \theta)^{1-x} & x = 0, 1; \quad 0 < \theta < 1 \\ 0 & \text{elsewhere.} \end{cases}$$

The statistic $Y_1 = X_1 + X_2 + \cdots + X_n$ has the pmf

$$f_{Y_1}(y_1; \theta) = \begin{cases} \binom{n}{y_1} \theta^{y_1} (1 - \theta)^{n-y_1} & y_1 = 0, 1, \ldots, n \\ = 0 & \text{elsewhere.} \end{cases}$$

What is the conditional probability

$$P(X_1 = x_1, \; X_2 = x_2, \ldots, X_n = x_n | Y_1 = y_1) = P(A|B),$$

say, where $y_1 = 0, 1, 2, \ldots, n$? Unless the sum of the integers $x_1, x_2, \ldots, x_n$ (each of which equals zero or 1) is equal to y_1, the conditional probability obviously equals zero because $A \cap B = \phi$. But in the case $y_1 = \sum x_i$, we have that $A \subset B$ so that $A \cap B = A$ and $P(A|B) = P(A)/P(B)$; thus the conditional probability equals

$$\frac{\theta^{x_1}(1-\theta)^{1-x_1}\theta^{x_2}(1-\theta)^{1-x_2}\cdots\theta^{x_n}(1-\theta)^{1-x_n}}{\binom{n}{y_1}\theta^{y_1}(1-\theta)^{n-y_1}} = \frac{\theta^{\sum x_i}(1-\theta)^{n-\sum x_i}}{\binom{n}{\sum x_i}\theta^{\sum x_i}(1-\theta)^{n-\sum x_i}}$$

$$= \frac{1}{\binom{n}{\sum x_i}}.$$

Since $y_1 = x_1 + x_2 + \cdots + x_n$ equals the number of 1's in the n independent trials, this is the conditional probability of selecting a particular arrangement of y_1 ones's and $(n - y_1)$ zeros. Note that this conditional probability does *not* depend upon the value of the parameter θ. ∎

In general, let $f_{Y_1}(y_1; \theta)$ be the pmf of the statistic $Y_1 = u_1(X_1, X_2, \ldots, X_n)$, where $X_1, X_2, \ldots, X_n$ is a random sample arising from a distribution of the discrete type having pmf $f(x; \theta)$, $\theta \in \Omega$. The conditional probability of $X_1 = x_1$, $X_2 = x_2, \ldots, X_n = x_n$, given $Y_1 = y_1$, equals

$$\frac{f(x_1; \theta) f(x_2; \theta) \cdots f(x_n; \theta)}{f_{Y_1}[u_1(x_1, x_2, \ldots, x_n); \theta]},$$

provided that $x_1, x_2, \ldots, x_n$ are such that the fixed $y_1 = u_1(x_1, x_2, \ldots, x_n)$, and equals zero otherwise. We say that $Y_1 = u_1(X_1, X_2, \ldots, X_n)$ is a *sufficient statistic* for θ if and only if this ratio does not depend upon θ. While, with distributions of the continuous type, we cannot use the same argument, we do, in this case, accept the fact that if this ratio does not depend upon θ, then the conditional distribution of $X_1, X_2, \ldots, X_n$, given $Y_1 = y_1$, does not depend upon θ. Thus, in both cases, we use the same definition of a sufficient statistic for θ.

Definition 7.2.1. *Let $X_1, X_2, \ldots, X_n$ denote a random sample of size n from a distribution that has pdf or pmf $f(x; \theta)$, $\theta \in \Omega$. Let $Y_1 = u_1(X_1, X_2, \ldots, X_n)$ be a statistic whose pdf or pmf is $f_{Y_1}(y_1; \theta)$. Then Y_1 is a **sufficient statistic** for θ if and only if*

$$\frac{f(x_1; \theta) f(x_2; \theta) \cdots f(x_n; \theta)}{f_{Y_1}[u_1(x_1, x_2, \ldots, x_n); \theta]} = H(x_1, x_2, \ldots, x_n),$$

where $H(x_1, x_2, \ldots, x_n)$ does not depend upon $\theta \in \Omega$.

Remark 7.2.1. In most cases in this book, $X_1, X_2, \ldots, X_n$ do represent the observations of a random sample; that is, they are iid. It is not necessary, however, in more general situations, that these random variables be independent; as a matter of fact, they do not need to be identically distributed. Thus, more generally, the definition of sufficiency of a statistic $Y_1 = u_1(X_1, X_2, \ldots, X_n)$ would be extended to read that

$$\frac{f(x_1, x_2, \ldots, x_n; \theta)}{f_{Y_1}[u_1(x_1, x_2, \ldots, x_n); \theta)]} = H(x_1, x_2, \ldots, x_n)$$

does not depend upon $\theta \in \Omega$, where $f(x_1, x_2, \ldots, x_n; \theta)$ is the joint pdf or pmf of $X_1, X_2, \ldots, X_n$. There are even a few situations in which we need an extension like this one in this book. ∎

We now give two examples that are illustrative of the definition.

Example 7.2.2. Let $X_1, X_2, \ldots, X_n$ be a random sample from a gamma distribution with $\alpha = 2$ and $\beta = \theta > 0$. Because the mgf associated with this distribution is given by $M(t) = (1 - \theta t)^{-2}$, $t < 1/\theta$, the mgf of $Y_1 = \sum_{i=1}^{n} X_i$ is

$$\begin{aligned} E[e^{t(X_1 + X_2 + \cdots + X_n)}] &= E(e^{tX_1}) E(e^{tX_2}) \cdots E(e^{tX_n}) \\ &= [(1 - \theta t)^{-2}]^n = (1 - \theta t)^{-2n}. \end{aligned}$$

Thus Y_1 has a gamma distribution with $\alpha = 2n$ and $\beta = \theta$, so that its pdf is

$$f_{Y_1}(y_1; \theta) = \begin{cases} \frac{1}{\Gamma(2n)\theta^{2n}} y_1^{2n-1} e^{-y_1/\theta} & 0 < y_1 < \infty \\ 0 & \text{elsewhere.} \end{cases}$$

Thus we have

$$\frac{\left[\dfrac{x_1^{2-1}e^{-x_1/\theta}}{\Gamma(2)\theta^2}\right]\left[\dfrac{x_2^{2-1}e^{-x_2/\theta}}{\Gamma(2)\theta^2}\right]\cdots\left[\dfrac{x_n^{2-1}e^{-x_n/\theta}}{\Gamma(2)\theta^2}\right]}{\dfrac{(x_1+x_2+\cdots+x_n)^{2n-1}e^{-(x_1+x_2+\cdots+x_n)/\theta}}{\Gamma(2n)\theta^{2n}}} = \frac{\Gamma(2n)}{[\Gamma(2)]^n}\frac{x_1x_2\cdots x_n}{(x_1+x_2+\cdots+x_n)^{2n-1}},$$

where $0 < x_i < \infty$, $i = 1, 2, \ldots, n$. Since this ratio does not depend upon θ, the sum Y_1 is a sufficient statistic for θ. ∎

Example 7.2.3. Let $Y_1 < Y_2 < \cdots < Y_n$ denote the order statistics of a random sample of size n from the distribution with pdf

$$f(x;\theta) = e^{-(x-\theta)}I_{(\theta,\infty)}(x).$$

Here we use the indicator function of a set A defined by

$$I_A(x) = \begin{cases} 1 & x \in A \\ 0 & x \notin A. \end{cases}$$

This means, of course, that $f(x;\theta) = e^{-(x-\theta)}$, $\theta < x < \infty$, zero elsewhere. The pdf of $Y_1 = \min(X_i)$ is

$$f_{Y_1}(y_1;\theta) = ne^{-n(y_1-\theta)}I_{(\theta,\infty)}(y_1).$$

Note that $\theta < \min\{x_i\}$ if and only if $\theta < x_i$, for all $i = 1, \ldots, n$. Notationally this can be expressed as $I_{(\theta,\infty)}(\min x_i) = \prod_{i=1}^{n} I_{(\theta,\infty)}(x_i)$. Thus we have that

$$\frac{\prod_{i=1}^{n} e^{-(x_i-\theta)}I_{(\theta,\infty)}(x_i)}{ne^{-n(\min x_i-\theta)}I_{(\theta,\infty)}(\min x_i)} = \frac{e^{-x_1-x_2-\cdots-x_n}}{ne^{-n\min x_i}}.$$

Since this ratio does not depend upon θ, the first order statistic Y_1 is a sufficient statistic for θ. ∎

If we are to show by means of the definition that a certain statistic Y_1 is or is not a sufficient statistic for a parameter θ, we must first of all know the pdf of Y_1, say $f_{Y_1}(y_1;\theta)$. In many instances it may be quite difficult to find this pdf. Fortunately, this problem can be avoided if we will but prove the following *factorization theorem* of Neyman.

Theorem 7.2.1 (Neyman). *Let $X_1, X_2, \ldots, X_n$ denote a random sample from a distribution that has pdf or pmf $f(x;\theta)$, $\theta \in \Omega$. The statistic $Y_1 = u_1(X_1, \ldots, X_n)$ is a sufficient statistic for θ if and only if we can find two nonnegative functions, k_1 and k_2, such that*

$$f(x_1;\theta)f(x_2;\theta)\cdots f(x_n;\theta) = k_1[u_1(x_1, x_2, \ldots, x_n);\theta]k_2(x_1, x_2, \ldots, x_n), \quad (7.2.1)$$

where $k_2(x_1, x_2, \ldots, x_n)$ does not depend upon θ.

Proof. We shall prove the theorem when the random variables are of the continuous type. Assume that the factorization is as stated in the theorem. In our proof we shall make the one-to-one transformation $y_1 = u_1(x_1, x_2, \ldots, x_n)$, $y_2 = u_2(x_1, x_2, \ldots, x_n), \ldots, y_n = u_n(x_1, x_2, \ldots, x_n)$ having the inverse functions $x_1 = w_1(y_1, y_2, \ldots, y_n)$, $x_2 = w_2(y_1, y_2, \ldots, y_n)$, $\ldots, x_n = w_n(y_1, y_2, \ldots, y_n)$ and Jacobian J. The pdf of the statistic $Y_1, Y_2, \ldots, Y_n$ is then given by

$$g(y_1, y_2, \ldots, y_n) = k_1(y_1; \theta)k_2(w_1, w_2, \ldots, w_n)|J|,$$

where $w_i = w_i(y_1, y_2, \ldots, y_n)$, $i = 1, 2, \ldots, n$. The pdf of Y_1, say $f_{Y_1}(y_1; \theta)$, is given by

$$
\begin{aligned}
f_{Y_1}(y_1; \theta) &= \int_{-\infty}^{\infty} \cdots \int_{-\infty}^{\infty} g(y_1, y_2, \ldots, y_n; \theta) \, dy_2 \cdots dy_n \\
&= k_1(y_1; \theta) \int_{-\infty}^{\infty} \cdots \int_{-\infty}^{\infty} |J| k_2(w_1, w_2, \ldots, w_n) \, dy_2 \cdots dy_n.
\end{aligned}
$$

Now the function k_2 does not depend upon θ, nor is θ involved in either the Jacobian J or the limits of integration. Hence the $(n-1)$-fold integral in the right-hand member of the preceding equation is a function of y_1 alone, for example, $m(y_1)$. Thus

$$f_{Y_1}(y_1; \theta) = k_1(y_1; \theta)m(y_1).$$

If $m(y_1) = 0$, then $f_{Y_1}(y_1; \theta) = 0$. If $m(y_1) > 0$, we can write

$$k_1[u_1(x_1, x_2, \ldots, x_n); \theta] = \frac{f_{Y_1}[u_1(x_1, \ldots, x_n); \theta]}{m[u_1(x_1, \ldots, x_n)]},$$

and the assumed factorization becomes

$$f(x_1; \theta) \cdots f(x_n; \theta) = f_{Y_1}[u_1(x_1, \ldots, x_n); \theta] \frac{k_2(x_1, \ldots, x_n)}{m[u_1(x_1, \ldots, x_n)]}.$$

Since neither the function k_2 nor the function m depends upon θ, then in accordance with the definition, Y_1 is a sufficient statistic for the parameter θ.

Conversely, if Y_1 is a sufficient statistic for θ, the factorization can be realized by taking the function k_1 to be the pdf of Y_1, namely the function f_{Y_1}. This completes the proof of the theorem. ∎

This theorem characterizes sufficiency and, as the following examples show, is usually much easier to work with than the definition of sufficiency.

Example 7.2.4. Let $X_1, X_2, \ldots, X_n$ denote a random sample from a distribution that is $N(\theta, \sigma^2)$, $-\infty < \theta < \infty$, where the variance $\sigma^2 > 0$ is known. If $\bar{x} = \sum_{1}^{n} x_i/n$, then

$$\sum_{i=1}^{n}(x_i - \theta)^2 = \sum_{i=1}^{n}[(x_i - \bar{x}) + (\bar{x} - \theta)]^2 = \sum_{i=1}^{n}(x_i - \bar{x})^2 + n(\bar{x} - \theta)^2$$

because

$$2\sum_{i=1}^{n}(x_i - \overline{x})(\overline{x} - \theta) = 2(\overline{x} - \theta)\sum_{i=1}^{n}(x_i - \overline{x}) = 0.$$

Thus the joint pdf of $X_1, X_2, \ldots, X_n$ may be written

$$\left(\frac{1}{\sigma\sqrt{2\pi}}\right)^n \exp\left[-\sum_{i=1}^{n}(x_i - \theta)^2/2\sigma^2\right]$$

$$= \{\exp[-n(\overline{x} - \theta)^2/2\sigma^2]\} \left\{ \frac{\exp\left[-\sum_{i=1}^{n}(x_i - \overline{x})^2/2\sigma^2\right]}{(\sigma\sqrt{2\pi})^n} \right\}$$

Because the first factor of the right-hand member of this equation depends upon $x_1, x_2, \ldots, x_n$ only through $\overline{x}$, and the second factor does not depend upon θ, the factorization theorem implies that the mean $\overline{X}$ of the sample is, for any particular value of σ^2, a sufficient statistic for θ, the mean of the normal distribution. ∎

We could have used the definition in the preceding example because we know that $\overline{X}$ is $N(\theta, \sigma^2/n)$. Let us now consider an example in which the use of the definition is inappropriate.

Example 7.2.5. Let $X_1, X_2, \ldots, X_n$ denote a random sample from a distribution with pdf

$$f(x; \theta) = \begin{cases} \theta x^{\theta-1} & 0 < x < 1 \\ 0 & \text{elsewhere}, \end{cases}$$

where $0 < \theta$. Using the factorization theorem, we will show that the product $u_1(X_1, X_2, \ldots, X_n) = \prod_{i=1}^{n} X_i$ is a sufficient statistic for θ. The joint pdf of $X_1, X_2, \ldots, X_n$ is

$$\theta^n \left(\prod_{i=1}^{n} x_i\right)^{\theta-1} = \left[\theta^n \left(\prod_{i=1}^{n} x_i\right)^{\theta}\right] \left(\frac{1}{\prod_{i=1}^{n} x_i}\right),$$

where $0 < x_i < 1$, $i = 1, 2, \ldots, n$. In the factorization theorem let

$$k_1[u_1(x_1, x_2, \ldots, x_n); \theta] = \theta^n \left(\prod_{i=1}^{n} x_i\right)^{\theta}$$

and

$$k_2(x_1, x_2, \ldots, x_n) = \frac{1}{\prod_{i=1}^{n} x_i}.$$

Since $k_2(x_1, x_2, \ldots, x_n)$ does not depend upon θ, the product $\prod_{i=1}^{n} X_i$ is a sufficient statistic for θ. ∎

There is a tendency for some readers to apply incorrectly the factorization theorem in those instances in which the domain of positive probability density depends upon the parameter θ. This is due to the fact that they do not give proper consideration to the domain of the function $k_2(x_1, x_2, \ldots, x_n)$. This will be illustrated in the next example.

Example 7.2.6. In Example 7.2.3 with $f(x; \theta) = e^{-(x-\theta)}I_{(\theta,\infty)}(x)$, it was found that the first order statistic Y_1 is a sufficient statistic for θ. To illustrate our point about not considering the domain of the function, take $n = 3$ and note that

$$e^{-(x_1-\theta)}e^{-(x_2-\theta)}e^{-(x_3-\theta)} = [e^{-3\max x_i + 3\theta}][e^{-x_1-x_2-x_3+3\max x_i}]$$

or a similar expression. Certainly, in the latter formula, there is no θ in the second factor and it might be assumed that $Y_3 = \max X_i$ is a sufficient statistic for θ. Of course, this is incorrect because we should have written the joint pdf of X_1, X_2, X_3 as

$$\prod_{i=1}^{3}[e^{-(x_i-\theta)}I_{(\theta,\infty)}(x_i)] = [e^{3\theta}I_{(\theta,\infty)}(\min x_i)]\left[\exp\left\{-\sum_{i=1}^{3}x_i\right\}\right]$$

because $I_{(\theta,\infty)}(\min x_i) = I_{(\theta,\infty)}(x_1)I_{(\theta,\infty)}(x_2)I_{(\theta,\infty)}(x_3)$. A similar statement cannot be made with $\max x_i$. Thus $Y_1 = \min X_i$ is the sufficient statistic for θ, not $Y_3 = \max X_i$. ∎

EXERCISES

7.2.1. Let $X_1, X_2, \ldots, X_n$ be iid $N(0, \theta)$, $0 < \theta < \infty$. Show that $\sum_1^n X_i^2$ is a sufficient statistic for θ.

7.2.2. Prove that the sum of the observations of a random sample of size n from a Poisson distribution having parameter θ, $0 < \theta < \infty$, is a sufficient statistic for θ.

7.2.3. Show that the nth order statistic of a random sample of size n from the uniform distribution having pdf $f(x; \theta) = 1/\theta$, $0 < x < \theta$, $0 < \theta < \infty$, zero elsewhere, is a sufficient statistic for θ. Generalize this result by considering the pdf $f(x; \theta) = Q(\theta)M(x)$, $0 < x < \theta$, $0 < \theta < \infty$, zero elsewhere. Here, of course,

$$\int_0^\theta M(x)\,dx = \frac{1}{Q(\theta)}.$$

7.2.4. Let $X_1, X_2, \ldots, X_n$ be a random sample of size n from a geometric distribution that has pmf $f(x; \theta) = (1-\theta)^x\theta$, $x = 0, 1, 2, \ldots$, $0 < \theta < 1$, zero elsewhere. Show that $\sum_1^n X_i$ is a sufficient statistic for θ.

7.2.5. Show that the sum of the observations of a random sample of size n from a gamma distribution that has pdf $f(x; \theta) = (1/\theta)e^{-x/\theta}$, $0 < x < \infty$, $0 < \theta < \infty$, zero elsewhere, is a sufficient statistic for θ.

7.2.6. Let $X_1, X_2, \ldots, X_n$ be a random sample of size n from a beta distribution with parameters $\alpha = \theta$ and $\beta = 2$. Show that the product $X_1 X_2 \cdots X_n$ is a sufficient statistic for θ.

7.2.7. Show that the product of the sample observations is a sufficient statistic for $\theta > 0$ if the random sample is taken from a gamma distribution with parameters $\alpha = \theta$ and $\beta = 6$.

7.2.8. What is the sufficient statistic for θ if the sample arises from a beta distribution in which $\alpha = \beta = \theta > 0$?

7.2.9. We consider a random sample $X_1, X_2, \ldots, X_n$ from a distribution with pdf $f(x; \theta) = (1/\theta) \exp(-x/\theta)$, $0 < x < \infty$, zero elsewhere, where $0 < \theta$. Possibly, in a life testing situation, however, we only observe the first r order statistics $Y_1 < Y_2 < \cdots < Y_r$.

(a) Record the joint pdf of these order statistics and denote it by $L(\theta)$.

(b) Under these conditions, find the mle, $\hat{\theta}$, by maximizing $L(\theta)$.

(c) Find the mgf and pdf of $\hat{\theta}$.

(d) With a slight extension of the definition of sufficiency, is $\hat{\theta}$ a sufficient statistic?

7.3 Properties of a Sufficient Statistic

Suppose $X_1, X_2, \ldots, X_n$ is a random sample on a random variable with pdf or pmf $f(x; \theta)$ where $\theta \in \Omega$. In this section we discuss how sufficiency is used to determine MVUEs. First note that a sufficient estimate is not unique in any sense. For if $Y_1 = u_1(X_1, X_2, \ldots, X_n)$ is a sufficient statistic and $Y_2 = g(Y_1)$ where $g(x)$ is a one-to-one function is a statistic then

$$
\begin{aligned}
f(x_1; \theta) f(x_2; \theta) \cdots f(x_n; \theta) &= k_1[u_1(y_1); \theta] k_2(x_1, x_2, \ldots, x_n) \\
&= k_1[u_1(g^{-1}(y_2)); \theta] k_2(x_1, x_2, \ldots, x_n);
\end{aligned}
$$

hence, by the factorization theorem Y_2 is also sufficient. However, as the theorem below shows, sufficiency can lead to a best point estimate.

We first refer back to Theorem 2.3.1 of Section 2.3: If X_1 and X_2 are random variables such that the variance of X_2 exists, then

$$E[X_2] = E[E(X_2 | X_1)]$$

and

$$\text{var}(X_2) \geq \text{var}[E(X_2 | X_1)].$$

For the adaptation in context of sufficient statistics, we let the sufficient statistic Y_1 be X_1 and Y_2, an unbiased statistic of θ, be X_2. Thus, with $E(Y_2|y_1) = \varphi(y_1)$, we have

$$\theta = E(Y_2) = E[\varphi(Y_1)]$$

and

$$\text{var}(Y_2) \geq \text{var}[\varphi(Y_1)].$$

That is, through this conditioning, the function $\varphi(Y_1)$ of the sufficient statistic Y_1 is an unbiased estimator of θ having a smaller variance than that of the unbiased estimator Y_2. We summarize this discussion more formally in the following theorem, which can be attributed to Rao and Blackwell.

Theorem 7.3.1 (Rao-Blackwell). *Let $X_1, X_2, \ldots, X_n$, n a fixed positive integer, denote a random sample from a distribution (continuous or discrete) that has pdf or pmf $f(x; \theta)$, $\theta \in \Omega$. Let $Y_1 = u_1(X_1, X_2, \ldots, X_n)$ be a sufficient statistic for θ, and let $Y_2 = u_2(X_1, X_2, \ldots, X_n)$, not a function of Y_1 alone, be an unbiased estimator of θ. Then $E(Y_2|y_1) = \varphi(y_1)$ defines a statistic $\varphi(Y_1)$. This statistic $\varphi(Y_1)$ is a function of the sufficient statistic for θ; it is an unbiased estimator of θ; and its variance is less than that of Y_2.*

This theorem tells us that in our search for an MVUE of a parameter, we may, if a sufficient statistic for the parameter exists, restrict that search to functions of the sufficient statistic. For if we begin with an unbiased estimator Y_2 alone, then we can always improve on this by computing $E(Y_2|y_1) = \varphi(y_1)$ so that $\varphi(Y_1)$ is an unbiased estimator with smaller variance than that of Y_2.

After Theorem 7.3.1 many students believe that it is necessary to find first some unbiased estimator Y_2 in their search for $\varphi(Y_1)$, an unbiased estimator of θ based upon the sufficient statistic Y_1. This is not the case at all, and Theorem 7.3.1 simply convinces us that we can restrict our search for a best estimator to functions of Y_1. Furthermore, there is a connection between sufficient statistics and maximum likelihood estimates as the following theorem shows,

Theorem 7.3.2. *Let $X_1, X_2, \ldots, X_n$ denote a random sample from a distribution that has pdf or pmf $f(x; \theta)$, $\theta \in \Omega$. If a sufficient statistic $Y_1 = u_1(X_1, X_2, \ldots, X_n)$ for θ exists and if a maximum likelihood estimator $\hat{\theta}$ of θ also exists uniquely, then $\hat{\theta}$ is a function of $Y_1 = u_1(X_1, X_2, \ldots, X_n)$.*

Proof. Let $f_{Y_1}(y_1; \theta)$ be the pdf or pmf of Y_1. Then by the definition of sufficiency, the likelihood function

$$
\begin{aligned}
L(\theta; x_1, x_2, \ldots, x_n) &= f(x_1; \theta) f(x_2; \theta) \cdots f(x_n; \theta) \\
&= f_{Y_1}[u_1(x_1, x_2, \ldots, x_n); \theta] H(x_1, x_2, \ldots, x_n),
\end{aligned}
$$

where $H(x_1, x_2, \ldots, x_n)$ does not depend upon θ. Thus L and f_{Y_1}, as functions of θ, are maximized simultaneously. Since there is one and only one value of θ

that maximizes L and hence $f_{Y_1}[u_1(x_1, x_2, \ldots, x_n); \theta]$, that value of θ must be a function of $u_1(x_1, x_2, \ldots, x_n)$. Thus the mle $\hat{\theta}$ is a function of the sufficient statistic $Y_1 = u_1(X_1, X_2, \ldots, X_n)$. ∎

We know from Chapter 5 that generally mles are asymptotically unbiased estimators of θ. Hence, one way to proceed is to find a sufficient statistic and then find the mle. Based on this we can often obtained an unbiased estimator which is a function of the sufficient statistic. This process is illustrated in the following example.

Example 7.3.1. Let $X_1, \ldots, X_n$ be iid with pdf

$$f(x; \theta) = \begin{cases} \theta e^{-\theta x} & 0 < x < \infty, \ \theta > 0 \\ 0 & \text{elsewhere.} \end{cases}$$

Suppose we want an MVUE of θ. The joint pdf (likelihood function) is

$$L(\theta; x_1, \ldots, x_n) = \theta^n e^{-\theta \sum_{i=1}^n x_i} \quad \text{for } x_i > 0, \ i = 1, \ldots, n.$$

Hence, by the factorization theorem, the statistic $Y_1 = \sum_{i=1}^n X_i$ is sufficient. The log of the likelihood function is

$$l(\theta) = n \log \theta - \theta \sum_{i=1}^n x_i.$$

Taking the partial with respect to θ of $l(\theta)$ and setting it to 0 results in the mle of θ which is given by

$$Y_2 = \frac{1}{\overline{X}}.$$

The statistic Y_2 is asymptotically unbiased. Hence, as a first step, we shall determine its expectation. In this problem, X_i are iid $\Gamma(1, 1/\theta)$ random variables; hence, $Y_1 = \sum_{i=1}^n X_i$ is $\Gamma(n, 1/\theta)$. Therefore,

$$E\left[\frac{1}{\overline{X}}\right] = nE\left[\frac{1}{\sum_{i=1}^n X_i}\right] = n \int_0^\infty \frac{\theta^n}{\Gamma(n)} x^{-1} x^{n-1} e^{-\theta x} \, dx,$$

making the change of variable $z = \theta x$ and simplifying, results in,

$$E\left[\frac{1}{\overline{X}}\right] = \theta \frac{n}{(n-1)!} \Gamma(n-1) = \theta \frac{n}{n-1}.$$

Thus, the statistic, $\frac{n-1}{\sum_{i=1}^n X_i}$ is a MVUE of θ. ∎

In the next two sections we discover that, in most instances, if there is one function $\varphi(Y_1)$ that is unbiased, $\varphi(Y_1)$ is the only unbiased estimator based on the sufficient statistic Y_1.

Remark 7.3.1. Since the unbiased estimator $\varphi(Y_1)$, where $\varphi(Y_1) = E(Y_2|y_1)$, has variance smaller than that of the unbiased estimator Y_2 of θ, students sometimes reason as follows. Let the function $\Upsilon(y_3) = E[\varphi(Y_1)|Y_3 = y_3]$, where Y_3 is another statistic, which is not sufficient for θ. By the Rao-Blackwell theorem, we have $E[\Upsilon(Y_3)] = \theta$ and $\Upsilon(Y_3)$ has a smaller variance than does $\varphi(Y_1)$. Accordingly, $\Upsilon(Y_3)$ must be better than $\varphi(Y_1)$ as an unbiased estimator of θ. But this is *not* true because Y_3 is not sufficient; thus θ is present in the conditional distribution of Y_1, given $Y_3 = y_3$, and the conditional mean $\Upsilon(y_3)$. So although indeed $E[\Upsilon(Y_3)] = \theta$, $\Upsilon(Y_3)$ is not even a statistic because it involves the unknown parameter θ and hence cannot be used as an estimate. ∎

Example 7.3.2. Let X_1, X_2, X_3 be a random sample from an exponential distribution with mean $\theta > 0$, so that the joint pdf is

$$\left(\frac{1}{\theta}\right)^3 e^{-(x_1+x_2+x_3)/\theta}, \quad 0 < x_i < \infty,$$

$i = 1, 2, 3$, zero elsewhere. From the factorization theorem, we see that $Y_1 = X_1 + X_2 + X_3$ is a sufficient statistic for θ. Of course

$$E(Y_1) = E(X_1 + X_2 + X_3) = 3\theta,$$

and thus $Y_1/3 = \overline{X}$ is a function of the sufficient statistic that is an unbiased estimator of θ.

In addition, let $Y_2 = X_2 + X_3$ and $Y_3 = X_3$. The one-to-one transformation defined by

$$x_1 = y_1 - y_2, \quad x_2 = y_2 - y_3, \quad x_3 = y_3$$

has Jacobian equal to one and the joint pdf of Y_1, Y_2, Y_3 is

$$g(y_1, y_2, y_3; \theta) = \left(\frac{1}{\theta}\right)^3 e^{-y_1/\theta}, \quad 0 < y_3 < y_2 < y_1 < \infty,$$

zero elsewhere. The marginal pdf of Y_1 and Y_3 is found by integrating out y_2 to obtain

$$g_{13}(y_1, y_3; \theta) = \left(\frac{1}{\theta}\right)^3 (y_1 - y_3)e^{-y_1/\theta}, \quad 0 < y_3 < y_1 < \infty,$$

zero elsewhere. The pdf of Y_3 alone is

$$g_3(y_3; \theta) = \frac{1}{\theta}e^{-y_3/\theta}, \quad 0 < y_3 < \infty,$$

zero elsewhere, since $Y_3 = X_3$ is an observation of a random sample from this exponential distribution.

Accordingly, the conditional pdf of Y_1, given $Y_3 = y_3$, is

$$
\begin{aligned}
g_{1|3}(y_1|y_3) &= \frac{g_{13}(y_1, y_3; \theta)}{g_3(y_3; \theta)} \\
&= \left(\frac{1}{\theta}\right)^2 (y_1 - y_3)e^{-(y_1-y_3)/\theta}, \quad 0 < y_3 < y_1 < \infty,
\end{aligned}
$$

zero elsewhere. Thus

$$
\begin{aligned}
E\left(\frac{Y_1}{3}\bigg|y_3\right) &= E\left(\frac{Y_1 - Y_3}{3}\bigg|y_3\right) + E\left(\frac{Y_3}{3}\bigg|y_3\right) \\
&= \left(\frac{1}{3}\right)\int_{y_3}^{\infty}\left(\frac{1}{\theta}\right)^2 (y_1 - y_3)^2 e^{-(y_1 - y_3)/\theta}\,dy_1 + \frac{y_3}{3} \\
&= \left(\frac{1}{3}\right)\frac{\Gamma(3)\theta^3}{\theta^2} + \frac{y_3}{3} = \frac{2\theta}{3} + \frac{y_3}{3} = \Upsilon(y_3).
\end{aligned}
$$

Of course, $E[\Upsilon(Y_3)] = \theta$ and $\text{var}[\Upsilon(Y_3)] \le \text{var}(Y_1/3)$, but $\Upsilon(Y_3)$ is not a statistic as it involves θ and cannot be used as an estimator of θ. This illustrates the preceding remark. ∎

EXERCISES

7.3.1. In each of the Exercises 7.2.1, 7.2.2, 7.2.3, and 7.2.4, show that the mle of θ is a function of the sufficient statistic for θ.

7.3.2. Let $Y_1 < Y_2 < Y_3 < Y_4 < Y_5$ be the order statistics of a random sample of size 5 from the uniform distribution having pdf $f(x;\theta) = 1/\theta$, $0 < x < \theta$, $0 < \theta < \infty$, zero elsewhere. Show that $2Y_3$ is an unbiased estimator of θ. Determine the joint pdf of Y_3 and the sufficient statistic Y_5 for θ. Find the conditional expectation $E(2Y_3|y_5) = \varphi(y_5)$. Compare the variances of $2Y_3$ and $\varphi(Y_5)$.

7.3.3. If X_1, X_2 is a random sample of size 2 from a distribution having pdf $f(x;\theta) = (1/\theta)e^{-x/\theta}$, $0 < x < \infty$, $0 < \theta < \infty$, zero elsewhere, find the joint pdf of the sufficient statistic $Y_1 = X_1 + X_2$ for θ and $Y_2 = X_2$. Show that Y_2 is an unbiased estimator of θ with variance θ^2. Find $E(Y_2|y_1) = \varphi(y_1)$ and the variance of $\varphi(Y_1)$.

7.3.4. Let $f(x, y) = (2/\theta^2)e^{-(x+y)/\theta}$, $0 < x < y < \infty$, zero elsewhere, be the joint pdf of the random variables X and Y.

(a) Show that the mean and the variance of Y are, respectively, $3\theta/2$ and $5\theta^2/4$.

(b) Show that $E(Y|x) = x + \theta$. In accordance with the theory, the expected value of $X + \theta$ is that of Y, namely, $3\theta/2$, and the variance of $X + \theta$ is less than that of Y. Show that the variance of $X + \theta$ is in fact $\theta^2/4$.

7.3.5. In each of Exercises 7.2.1, 7.2.2, and 7.2.3, compute the expected value of the given sufficient statistic and, in each case, determine an unbiased estimator of θ that is a function of that sufficient statistic alone.

7.3.6. Let $X_1, X_2, \ldots, X_n$ be a random sample from a Poisson distribution with mean θ. Find the conditional expectation $E\left(X_1 + 2X_2 + 3X_3 \bigg| \sum_1^n X_i\right)$.

7.4 Completeness and Uniqueness

Let $X_1, X_2, \ldots, X_n$ be a random sample from the Poisson distribution that has pmf

$$f(x; \theta) = \begin{cases} \frac{\theta^x e^{-\theta}}{x!} & x = 0, 1, 2, \ldots; \quad 0 < \theta \\ 0 & \text{elsewhere.} \end{cases}$$

From Exercise 7.2.2, we know that $Y_1 = \sum_{i=1}^{n} X_i$ is a sufficient statistic for θ and its pmf is

$$g_1(y_1; \theta) = \begin{cases} \frac{(n\theta)^{y_1} e^{-n\theta}}{y_1!} & y_1 = 0, 1, 2, \ldots \\ 0 & \text{elsewhere.} \end{cases}$$

Let us consider the family $\{g_1(y_1; \theta) : 0 < \theta\}$ of probability mass functions. Suppose that the function $u(Y_1)$ of Y_1 is such that $E[u(Y_1)] = 0$ for every $\theta > 0$. We shall show that this requires $u(y_1)$ to be zero at every point $y_1 = 0, 1, 2, \ldots$. That is, $E[u(Y_1)] = 0$ for $0 < \theta$ requires

$$0 = u(0) = u(1) = u(2) = u(3) = \cdots.$$

We have for all $\theta > 0$ that

$$\begin{aligned} 0 = E[u(Y_1)] &= \sum_{y_1=0}^{\infty} u(y_1) \frac{(n\theta)^{y_1} e^{-n\theta}}{y_1!} \\ &= e^{-n\theta} \left[u(0) + u(1)\frac{n\theta}{1!} + u(2)\frac{(n\theta)^2}{2!} + \cdots \right]. \end{aligned}$$

Since $e^{-n\theta}$ does not equal zero, we have shown that

$$0 = u(0) + [nu(1)]\theta + \left[\frac{n^2 u(2)}{2} \right] \theta^2 + \cdots.$$

However, if such an infinite (power) series converges to zero for all $\theta > 0$, then each of the coefficients must equal zero. That is,

$$u(0) = 0, \quad nu(1) = 0, \quad \frac{n^2 u(2)}{2} = 0, \ldots$$

and thus $0 = u(0) = u(1) = u(2) = \cdots$, as we wanted to show. Of course, the condition $E[u(Y_1)] = 0$ for all $\theta > 0$ does not place any restriction on $u(y_1)$ when y_1 is not a nonnegative integer. So we see that, in this illustration, $E[u(Y_1)] = 0$ for all $\theta > 0$ requires that $u(y_1)$ equals zero except on a set of points that has probability zero for each pmf $g_1(y_1; \theta)$, $0 < \theta$. From the following definition we observe that the family $\{g_1(y_1; \theta) : 0 < \theta\}$ is complete.

Definition 7.4.1. *Let the random variable Z of either the continuous type or the discrete type have a pdf or pmf that is one member of the family $\{h(z; \theta) : \theta \in \Omega\}$. If the condition $E[u(Z)] = 0$, for every $\theta \in \Omega$, requires that $u(z)$ be zero except on a set of points that has probability zero for each $h(z; \theta)$, $\theta \in \Omega$, then the family $\{h(z; \theta) : \theta \in \Omega\}$ is called a* **complete family** *of probability density or mass functions.*

Remark 7.4.1. In Section 1.8 it was noted that the existence of $E[u(X)]$ implies that the integral (or sum) converges absolutely. This absolute convergence was tacitly assumed in our definition of completeness and it is needed to prove that certain families of probability density functions are complete. ∎

In order to show that certain families of probability density functions of the continuous type are complete, we must appeal to the same type of theorem in analysis that we used when we claimed that the moment generating function uniquely determines a distribution. This is illustrated in the next example.

Example 7.4.1. Consider the family of pdfs $\{h(z;\theta) : 0 < \theta < \infty\}$. Suppose Z has a pdf in this family given by

$$h(z;\theta) = \begin{cases} \frac{1}{\theta}e^{-z/\theta} & 0 < z < \infty \\ 0 & \text{elsewhere.} \end{cases}$$

Let us say that $E[u(Z)] = 0$ for every $\theta > 0$. That is,

$$\frac{1}{\theta}\int_0^\infty u(z)e^{-z/\theta}\,dz = 0, \quad \theta > 0.$$

Readers acquainted with the theory of transformations will recognize the integral in the left-hand member as being essentially the Laplace transform of $u(z)$. In that theory we learn that the only function $u(z)$ transforming to a function of θ which is identically equal to zero is $u(z) = 0$, except (in our terminology) on a set of points that has probability zero for each $h(z;\theta)$, $0 < \theta$. That is, the family $\{h(z;\theta) : 0 < \theta < \infty\}$ is complete. ∎

Let the parameter θ in the pdf or pmf $f(x;\theta)$, $\theta \in \Omega$, have a sufficient statistic $Y_1 = u_1(X_1, X_2, \ldots, X_n)$, where $X_1, X_2, \ldots, X_n$ is a random sample from this distribution. Let the pdf or pmf of Y_1 be $f_{Y_1}(y_1;\theta)$, $\theta \in \Omega$. It has been seen that, if there is any unbiased estimator Y_2 (not a function of Y_1 alone) of θ, then there is at least one function of Y_1 that is an unbiased estimator of θ, and our search for a best estimator of θ may be restricted to functions of Y_1. Suppose it has been verified that a certain function $\varphi(Y_1)$, not a function of θ, is such that $E[\varphi(Y_1)] = \theta$ for all values of θ, $\theta \in \Omega$. Let $\psi(Y_1)$ be another function of the sufficient statistic Y_1 alone, so that we also have $E[\psi(Y_1)] = \theta$ for all value of θ, $\theta \in \Omega$. Hence,

$$E[\varphi(Y_1) - \psi(Y_1)] = 0, \quad \theta \in \Omega.$$

If the family $\{f_{Y_1}(y_1;\theta) : \theta \in \Omega\}$ is complete, the function of $\varphi(y_1) - \psi(y_1) = 0$, except on a set of points that has probability zero. That is, for every other unbiased estimator $\psi(Y_1)$ of θ, we have

$$\varphi(y_1) = \psi(y_1)$$

except possibly at certain special points. Thus, in this sense [namely $\varphi(y_1) = \psi(y_1)$, except on a set of points with probability zero], $\varphi(Y_1)$ is the unique function of Y_1, which is an unbiased estimator of θ. In accordance with the Rao-Blackwell theorem, $\varphi(Y_1)$ has a smaller variance than every other unbiased estimator of θ. That is, the statistic $\varphi(Y_1)$ is the MVUE of θ. This fact is stated in the following theorem of Lehmann and Scheffé.

Theorem 7.4.1 (Lehmann and Scheffé). *Let $X_1, X_2, \ldots, X_n$, n a fixed positive integer, denote a random sample from a distribution that has pdf or pmf $f(x; \theta)$, $\theta \in \Omega$, let $Y_1 = u_1(X_1, X_2, \ldots, X_n)$ be a sufficient statistic for θ, and let the family $\{f_{Y_1}(y_1; \theta) : \theta \in \Omega\}$ be complete. If there is a function of Y_1 that is an unbiased estimator of θ, then this function of Y_1 is the unique MVUE of θ. Here "unique" is used in the sense described in the preceding paragraph.*

The statement that Y_1 is a sufficient statistic for a parameter θ, $\theta \in \Omega$, and that the family $\{f_{Y_1}(y_1; \theta) : \theta \in \Omega\}$ of probability density functions is complete is lengthy and somewhat awkward. We shall adopt the less descriptive, but more convenient, terminology that Y_1 is a *complete sufficient statistic* for θ. In the next section we study a fairly large class of probability density functions for which a complete sufficient statistic Y_1 for θ can be determined by inspection.

EXERCISES

7.4.1. If $az^2 + bz + c = 0$ for more than two values of z, then $a = b = c = 0$. Use this result to show that the family $\{b(2, \theta) : 0 < \theta < 1\}$ is complete.

7.4.2. Show that each of the following families is not complete by finding at least one nonzero function $u(x)$ such that $E[u(X)] = 0$, for all $\theta > 0$.

(a)
$$f(x; \theta) = \begin{cases} \frac{1}{2\theta} & -\theta < x < \theta \quad \text{where } 0 < \theta < \infty \\ 0 & \text{elsewhere.} \end{cases}$$

(b) $N(0, \theta)$, where $0 < \theta < \infty$.

7.4.3. Let $X_1, X_2, \ldots, X_n$ represent a random sample from the discrete distribution having the pmf

$$f(x; \theta) = \begin{cases} \theta^x (1 - \theta)^{1-x} & x = 0, 1, \quad 0 < \theta < 1 \\ 0 & \text{elsewhere.} \end{cases}$$

Show that $Y_1 = \sum_1^n X_i$ is a complete sufficient statistic for θ. Find the unique function of Y_1 that is the MVUE of θ.

Hint: Display $E[u(Y_1)] = 0$, show that the constant term $u(0)$ is equal to zero, divide both members of the equation by $\theta \neq 0$, and repeat the argument.

7.4.4. Consider the family of probability density functions $\{h(z; \theta) : \theta \in \Omega\}$, where $h(z; \theta) = 1/\theta$, $0 < z < \theta$, zero elsewhere.

(a) Show that the family is complete provided that $\Omega = \{\theta : 0 < \theta < \infty\}$.
Hint: For convenience, assume that $u(z)$ is continuous and note that the derivative of $E[u(Z)]$ with respect to θ is equal to zero also.

(b) Show that this family is not complete if $\Omega = \{\theta : 1 < \theta < \infty\}$.
Hint: Concentrate on the interval $0 < z < 1$ and find a nonzero function $u(z)$ on that interval such that $E[u(Z)] = 0$ for all $\theta > 1$.

7.4.5. Show that the first order statistic Y_1 of a random sample of size n from the distribution having pdf $f(x; \theta) = e^{-(x-\theta)}$, $\theta < x < \infty$, $-\infty < \theta < \infty$, zero elsewhere, is a complete sufficient statistic for θ. Find the unique function of this statistic which is the MVUE of θ.

7.4.6. Let a random sample of size n be taken from a distribution of the discrete type with pmf $f(x; \theta) = 1/\theta$, $x = 1, 2, \ldots, \theta$, zero elsewhere, where θ is an unknown positive integer.

(a) Show that the largest observation, say Y, of the sample is a complete sufficient statistic for θ.

(b) Prove that
$$[Y^{n+1} - (Y-1)^{n+1}]/[Y^n - (Y-1)^n]$$
is the unique MVUE of θ.

7.4.7. Let X have the pdf $f_X(x; \theta) = 1/(2\theta)$, for $-\theta < x < \theta$, zero elsewhere, where $\theta > 0$.

(a) Is the statistic $Y = |X|$ a sufficient statistic for θ? Why?

(b) Let $f_Y(y; \theta)$ be the pdf of Y. Is the family $\{f_Y(y; \theta) : \theta > 0\}$ complete? Why?

7.4.8. Let X have the pmf $p(x; \theta) = \frac{1}{2} \binom{n}{x} \theta^{|x|} (1-\theta)^{n-|x|}$, for $x = \pm 1, \pm 2, \ldots, \pm n$, $p(0, \theta) = (1-\theta)^n$, and zero elsewhere, where $0 < \theta < 1$.

(a) Show that this family $\{p(x; \theta) : 0 < \theta < 1\}$ is not complete.

(b) Let $Y = |X|$. Show that Y is a complete and sufficient statistic for θ.

7.4.9. Let $X_1, \ldots, X_n$ be iid with pdf $f(x; \theta) = 1/(3\theta)$, $-\theta < x < 2\theta$, zero elsewhere, where $\theta > 0$.

(a) Find the mle $\widehat{\theta}$ of θ.

(b) Is $\widehat{\theta}$ a sufficient statistic for θ? Why?

(c) Is $(n+1)\widehat{\theta}/n$ the unique MVUE of θ? Why?

7.4.10. Let $Y_1 < Y_2 < \cdots < Y_n$ be the order statistics of a random sample of size n from a distribution with pdf $f(x; \theta) = 1/\theta$, $0 < x < \theta$, zero elsewhere. The statistic Y_n is a complete sufficient statistic for θ and it has pdf
$$g(y_n; \theta) = \frac{n y_n^{n-1}}{\theta^n}, \quad 0 < y_n < \theta,$$

and zero elsewhere.

(a) Find the distribution function $H_n(z; \theta)$ of $Z = n(\theta - Y_n)$.

(b) Find the $\lim_{n \to \infty} H_n(z; \theta)$ and thus the limiting distribution of Z.

7.5   The Exponential Class of Distributions

In this section we discuss an important class of distributions, called the *exponential class*. As we will show, this class possesses complete and sufficient statistics which are readily determined from the distribution.

Consider a family $\{f(x;\theta) : \theta \in \Omega\}$ of probability density or mass functions, where Ω is the interval set $\Omega = \{\theta : \gamma < \theta < \delta\}$, where γ and δ are known constants (they may be $\pm\infty$), and where

$$f(x;\theta) = \begin{cases} \exp[p(\theta)K(x) + S(x) + q(\theta)] & x \in \mathcal{S} \\ 0 & \text{elsewhere,} \end{cases} \tag{7.5.1}$$

where $\mathcal{S}$ is the support of X. In this section we will be concerned with a particular class of the family called the regular exponential class.

Definition 7.5.1 (Regular Exponential Class). *A pdf of the form (7.5.1) is said to be a member of the **regular exponential class** of probability density or mass functions if*

1. *$\mathcal{S}$, the support of X, does not depend upon θ,*

2. *$p(\theta)$ is a nontrivial continuous function of $\theta \in \Omega$,*

3. *Finally,*

    (a) *if X is a continuous random variable then each of $K'(x) \not\equiv 0$ and $S(x)$ is a continuous function of $x \in \mathcal{S}$,*

    (b) *if X is a discrete random variable then $K(x)$ is a nontrivial function of $x \in \mathcal{S}$.*

For example, each member of the family $\{f(x;\theta) : 0 < \theta < \infty\}$, where $f(x;\theta)$ is $N(0,\theta)$, represents a regular case of the exponential class of the continuous type because

$$\begin{aligned} f(x;\theta) &= \frac{1}{\sqrt{2\pi\theta}} e^{-x^2/2\theta} \\ &= \exp\left(-\frac{1}{2\theta}x^2 - \log\sqrt{2\pi\theta}\right), \quad -\infty < x < \infty. \end{aligned}$$

On the other hand, consider the uniform density function given by

$$f(x;\theta) = \begin{cases} \exp\{-\log\theta\} & x \in (0,\theta) \\ 0 & \text{elsewhere.} \end{cases}$$

This can be written in the form (7.5.1) but the support is the interval $(0,\theta)$ which depends on θ. Hence, the uniform family is not a regular exponential family.

Let $X_1, X_2, \ldots, X_n$ denote a random sample from a distribution that represents a regular case of the exponential class. The joint pdf or pmf of $X_1, X_2, \ldots, X_n$ is

$$\exp\left[p(\theta)\sum_1^n K(x_i) + \sum_1^n S(x_i) + nq(\theta)\right]$$

for $x_i \in \mathcal{S}$, $i = 1, 2, \ldots, n$ and zero elsewhere. At points in the $\mathcal{S}$ of X, this joint pdf or pmf may be written as the product of the two nonnegative functions

$$\exp\left[p(\theta)\sum_1^n K(x_i) + nq(\theta)\right]\exp\left[\sum_1^n S(x_i)\right].$$

In accordance with the factorization theorem, Theorem 7.2.1, $Y_1 = \sum_1^n K(X_i)$ is a sufficient statistic for the parameter θ.

Besides the fact that Y_1 is a sufficient statistic, we can obtain the general form of the distribution of Y_1 and its mean and variance. We summarize these results in a theorem. The details of the proof are given in Exercises 7.5.5 and 7.5.8. Exercise 7.5.6 obtains the mgf of Y_1 in the case that $p(\theta) = \theta$.

Theorem 7.5.1. *Let $X_1, X_2, \ldots, X_n$ denote a random sample from a distribution that represents a regular case of the exponential class, with pdf or pmf given by (7.5.1). Consider the statistic $Y_1 = \sum_{i=1}^n K(X_i)$. Then,*

1. *The pdf or pmf of Y_1 has the form,*

$$f_{Y_1}(y_1; \theta) = R(y_1)\exp[p(\theta)y_1 + nq(\theta)] \tag{7.5.2}$$

   for $y_1 \in \mathcal{S}_{Y_1}$ and some function $R(y_1)$. Neither $\mathcal{S}_{Y_1}$ nor $R(y_1)$ depend on θ.

2. *$E(Y_1) = -n\frac{q'(\theta)}{p'(\theta)}$.*

3. *$Var(Y_1) = n\frac{1}{p'(\theta)^3}\{(p''(\theta)q'(\theta) - q''(\theta)p'(\theta)\}$.*

Example 7.5.1. Let X have a Poisson distribution with parameter $\theta \in (0, \infty)$. Then the support of X is the set $\mathcal{S} = \{0, 1, 2, \ldots\}$ which does not depend on θ. Further, the pmf of X on its support is

$$f(x, \theta) = e^{-\theta}\frac{\theta^x}{x!} = \exp\{(\log\theta)x + \log(1/x!) + (-\theta)\}.$$

Hence, the Poisson distribution is a member of the regular exponential class, with $p(\theta) = \log(\theta)$, $q(\theta) = -\theta$, and $K(x) = x$. Therefore if $X_1, X_2, \ldots, X_n$ denotes a random sample on X then the statistic $Y_1 = \sum_{i=1}^n X_i$ is sufficient. But since $p'(\theta) = 1/\theta$ and $q'(\theta) = -1$, Theorem 7.5.1 verifies that the mean of Y_1 is $n\theta$. It is easy to verify that the variance of Y_1 is $n\theta$, also. Finally we can show that the function $R(y_1)$ in Theorem 7.5.1 is given by $R(y_1) = n^{y_1}(1/y_1!)$. ∎

For the regular case of the exponential class, we have shown that the statistic $Y_1 = \sum_1^n K(X_i)$ is sufficient for θ. We now use the form of the pdf of Y_1 given in Theorem 7.5.1 to establish the completness of Y_1.

Theorem 7.5.2. *Let $f(x;\theta)$, $\gamma < \theta < \delta$, be a pdf or pmf of a random variable X whose distribution is a regular case of the exponential class. Then if $X_1, X_2, \ldots, X_n$ (where n is a fixed positive integer) is a random sample from the distribution of X, the statistic $Y_1 = \sum_1^n K(X_i)$ is a sufficient statistic for θ and the family $\{f_{Y_1}(y_1;\theta) : \gamma < \theta < \delta\}$ of probability density functions of Y_1 is complete. That is, Y_1 is a complete sufficient statistic for θ.*

Proof: We have shown above that Y_1 is sufficient. For completeness, suppose that $E[u(Y_1)] = 0$. Expression (7.5.2) of Theorem 7.5.1 gives the pdf of Y_1. Hence, we have the equation

$$\int_{\mathcal{S}_{Y_1}} u(y_1) R(y_1) \exp\{p(\theta)y_1 + nq(\theta)\}\, dy_1 = 0$$

or equivalently since $\exp\{nq(\theta)\} \neq 0$,

$$\int_{\mathcal{S}_{Y_1}} u(y_1) R(y_1) \exp\{p(\theta)y_1\}\, dy_1 = 0$$

for all θ. However, $p(\theta)$ is a nontrivial continuous function of θ, and thus this integral is essentially a type of Laplace transform of $u(y_1)R(y_1)$. The only function of y_1 transforming to the 0 function is the zero function, (except for a set of points with probability zero in our context). That is,

$$u(y_1)R(y_1) \equiv 0.$$

However, $R(y_1) \neq 0$ for all $y_1 \in \mathcal{S}_{Y_1}$ because it is factor in the pdf of Y_1. Hence, $u(y_1) \equiv 0$, (except for a set of points with probability zero). Therefore Y_1 is a complete sufficient statistic for θ. ∎

This theorem has useful implications. In a regular case of form (7.5.1), we can see by inspection that the sufficient statistic is $Y_1 = \sum_1^n K(X_i)$. If we can see how to form a function of Y_1, say $\varphi(Y_1)$, so that $E[\varphi(Y_1)] = \theta$, then the statistic $\varphi(Y_1)$ is unique and is the MVUE of θ.

Example 7.5.2. Let $X_1, X_2, \ldots, X_n$ denote a random sample from a normal distribution that has pdf

$$f(x;\theta) = \frac{1}{\sigma\sqrt{2\pi}} \exp\left[-\frac{(x-\theta)^2}{2\sigma^2}\right], \quad -\infty < x < \infty, \quad -\infty < \theta < \infty,$$

or

$$f(x;\theta) = \exp\left(\frac{\theta}{\sigma^2}x - \frac{x^2}{2\sigma^2} - \log\sqrt{2\pi\sigma^2} - \frac{\theta^2}{2\sigma^2}\right).$$

Here σ^2 is any fixed positive number. This is a regular case of the exponential class with

$$p(\theta) = \frac{\theta}{\sigma^2}, \quad K(x) = x,$$

$$S(x) = -\frac{x^2}{2\sigma^2} - \log\sqrt{2\pi\sigma^2}, \quad q(\theta) = -\frac{\theta^2}{2\sigma^2}.$$

Accordingly, $Y_1 = X_1 + X_2 + \cdots + X_n = n\overline{X}$ is a complete sufficient statistic for the mean θ of a normal distribution for every fixed value of the variance σ^2. Since $E(Y_1) = n\theta$, then $\varphi(Y_1) = Y_1/n = \overline{X}$ is the only function of Y_1 that is an unbiased estimator of θ; and being a function of the sufficient statistic Y_1, it has a minimum variance. That is, $\overline{X}$ is the unique MVUE of θ. Incidentally, since Y_1 is a one-to-one function of $\overline{X}$, $\overline{X}$ itself is also a complete sufficient statistic for θ. ∎

Example 7.5.3 (Example 7.5.1, continued). Reconsider the discussion concerning the Poisson distribution with parameter θ found in Example 7.5.1. Based on this discussion the statistic $Y_1 = \sum_{i=1}^{n} X_i$ was sufficient. It follows from Theorem 7.5.2 that its family of distributions is complete. Since $E(Y_1) = n\theta$, it follows that $\overline{X} = n^{-1}Y_1$ is the unique MVUE of θ. ∎

EXERCISES

7.5.1. Write the pdf

$$f(x;\theta) = \frac{1}{6\theta^4}x^3 e^{-x/\theta}, \quad 0 < x < \infty, \quad 0 < \theta < \infty,$$

zero elsewhere, in the exponential form. If $X_1, X_2, \ldots, X_n$ is a random sample from this distribution, find a complete sufficient statistic Y_1 for θ and the unique function $\varphi(Y_1)$ of this statistic that is the MVUE of θ. Is $\varphi(Y_1)$ itself a complete sufficient statistic?

7.5.2. Let $X_1, X_2, \ldots, X_n$ denote a random sample of size $n > 1$ from a distribution with pdf $f(x;\theta) = \theta e^{-\theta x}$, $0 < x < \infty$, zero elsewhere, and $\theta > 0$. Then $\sum_{1}^{n} X_i$ is a sufficient statistic for θ. Prove that $(n-1)/Y$ is the MVUE of θ.

7.5.3. Let $X_1, X_2, \ldots, X_n$ denote a random sample of size n from a distribution with pdf $f(x;\theta) = \theta x^{\theta-1}$, $0 < x < 1$, zero elsewhere, and $\theta > 0$.

(a) Show that the *geometric mean* $(X_1 X_2 \cdots X_n)^{1/n}$ of the sample is a complete sufficient statistic for θ.

(b) Find the maximum likelihood estimator of θ, and observe that it is a function of this geometric mean.

7.5.4. Let $\overline{X}$ denote the mean of the random sample $X_1, X_2, \ldots, X_n$ from a gamma-type distribution with parameters $\alpha > 0$ and $\beta = \theta > 0$. Compute $E[X_1|\overline{x}]$.
Hint: Can you find directly a function $\psi(\overline{X})$ of $\overline{X}$ such that $E[\psi(\overline{X})] = \theta$? Is $E(X_1|\overline{x}) = \psi(\overline{x})$? Why?

7.5.5. Let X be a random variable with pdf of a regular case of the exponential class. Show that $E[K(X)] = -q'(\theta)/p'(\theta)$, provided these derivatives exist, by differentiating both members of the equality

$$\int_a^b \exp[p(\theta)K(x) + S(x) + q(\theta)]\, dx = 1$$

with respect to θ. By a second differentiation, find the variance of $K(X)$.

7.5.6. Given that $f(x;\theta) = \exp[\theta K(x) + S(x) + q(\theta)]$, $a < x < b$, $\gamma < \theta < \delta$, represents a regular case of the exponential class, show that the moment-generating function $M(t)$ of $Y = K(X)$ is $M(t) = \exp[q(\theta) - q(\theta + t)]$, $\gamma < \theta + t < \delta$.

7.5.7. In the preceding exercise, given that $E(Y) = E[K(X)] = \theta$, prove that Y is $N(\theta, 1)$.
Hint: Consider $M'(0) = \theta$ and solve the resulting differential equation.

7.5.8. If $X_1, X_2, \ldots, X_n$ is a random sample from a distribution that has a pdf which is a regular case of the exponential class, show that the pdf of $Y_1 = \sum_1^n K(X_i)$ is of the form $f_{Y_1}(y_1; \theta) = R(y_1)\exp[p(\theta)y_1 + nq(\theta)]$.
Hint: Let $Y_2 = X_2, \ldots, Y_n = X_n$ be $n - 1$ auxiliary random variables. Find the joint pdf of $Y_1, Y_2, \ldots, Y_n$ and then the marginal pdf of Y_1.

7.5.9. Let Y denote the median and let $\overline{X}$ denote the mean of a random sample of size $n = 2k + 1$ from a distribution that is $N(\mu, \sigma^2)$. Compute $E(Y|\overline{X} = \overline{x})$.
Hint: See Exercise 7.5.4.

7.5.10. Let $X_1, X_2, \ldots, X_n$ be a random sample from a distribution with pdf $f(x;\theta) = \theta^2 x e^{-\theta x}$, $0 < x < \infty$, where $\theta > 0$.

(a) Argue that $Y = \sum_1^n X_i$ is a complete sufficient statistic for θ.

(b) Compute $E(1/Y)$ and find the function of Y which is the unique MVUE of θ.

7.5.11. Let $X_1, X_2, \ldots, X_n$, $n > 2$, be a random sample from the binomial distribution $b(1, \theta)$.

(a) Show that $Y_1 = X_1 + X_2 + \cdots + X_n$ is a complete sufficient statistic for θ.

(b) Find the function $\varphi(Y_1)$ which is the MVUE of θ.

(c) Let $Y_2 = (X_1 + X_2)/2$ and compute $E(Y_2)$.

(d) Determine $E(Y_2|Y_1 = y_1)$.

7.5.12. Let $X_1, X_2, \ldots, X_n$ be a random sample from a distribution with pdf $f(x;\theta) = \theta e^{-\theta x}$, $0 < x < \infty$, zero elsewhere where $0 < \theta$.

(a) What is the complete sufficient statistic, for example Y, for θ?

(b) What function of Y is an unbiased estimator of θ?

7.5.13. Let $X_1, X_2, \ldots, X_n$ be a random sample from a distribution with pdf $f(x; \theta) = \theta^x (1 - \theta)$, $x = 0, 1, 2, \ldots$, zero elsewhere, where $0 \le \theta \le 1$.

(a) Find the mle, $\hat{\theta}$, of θ.

(b) Show that $\sum_{1}^{n} X_i$ is a complete sufficient statistic for θ.

(c) Determine the MVUE of θ.

7.6　Functions of a Parameter

Up to this point we have sought an MVUE of a parameter θ. Not always, however, are we interested in θ but rather in a function of θ. There are several techniques we can use to the find the MVUE. One is by inspection of the expected value of a sufficient statistic. This is how we found the MVUEs in Examples 7.5.2 and 7.5.3 of the last section. In this section and its exercises we offer more examples of the inspection technique. The second technique is based on the conditional expectation of an unbiased estimate given a sufficient statistic. The second example illustrates this technique.

　　Recall in Chapter 5, under regularity conditions we obtained the asymptotic distribution theory for maximum likelihood estimators (mles). This allows certain asymptotic inferences (confidence intervals and tests) for these estimators. Such a simple theory is not available for MVUEs. As Theorem 7.3.2 shows, though, sometimes we can determine the relationship between the mle and the MVUE. In these situations, we can often obtain the asymptotic distribution for the MVUE based on the asymptotic distribution of the mle. We illustrate this for some of the following examples.

Example 7.6.1. Let $X_1, X_2, \ldots, X_n$ denote the observations of a random sample of size $n > 1$ from a distribution that is $b(1, \theta)$, $0 < \theta < 1$. We know that if $Y = \sum_{1}^{n} X_i$, then Y/n is the unique minimum variance unbiased estimator of θ. Now suppose we want to estimate the variance of Y/n which is $\theta(1 - \theta)/n$. Let $\delta = \theta(1 - \theta)$. Because Y is a sufficient statistic for θ, it is known that we can restrict our search to functions of Y. The maximum likelihood estimate of δ which is given by $\tilde{\delta} = (Y/n)(1 - Y/n)$ is a function of the sufficient statistic and seems to be a reasonable starting point. The expectation of this statistic is given by

$$E[\tilde{\delta}] = E\left[\frac{Y}{n}\left(1 - \frac{Y}{n}\right)\right] = \frac{1}{n}E(Y) - \frac{1}{n^2}E(Y^2).$$

Now $E(Y) = n\theta$ and $E(Y^2) = n\theta(1 - \theta) + n^2\theta^2$. Hence

$$E\left[\frac{Y}{n}\left(1 - \frac{Y}{n}\right)\right] = (n - 1)\frac{\theta(1 - \theta)}{n}.$$

If we multiply both members of this equation by $n/(n-1)$, we find that the statistic $\hat{\delta} = (n/(n-1))(Y/n)(1-Y/n) = (n/(n-1))\tilde{\delta}$ is the unique MVUE of δ. Hence, the MVUE of δ/n, the variance of Y/n, is $\hat{\delta}/n$.

It is interesting to compare the mle $\tilde{\delta}$ with $\hat{\delta}$. Recall from Chapter 5 that the mle $\tilde{\delta}$ is a consistent estimate of δ and and that $\sqrt{n}(\tilde{\delta}-\delta)$ is asymptotically normal. Because,

$$\hat{\delta} - \tilde{\delta} = \tilde{\delta}\frac{1}{n-1} \xrightarrow{P} \delta \cdot 0 = 0,$$

it follows that $\hat{\delta}$ is also a consistent estimator of δ. Further,

$$\sqrt{n}(\hat{\delta} - \delta) - \sqrt{n}(\tilde{\delta} - \delta) = \frac{\sqrt{n}}{n-1}\tilde{\delta} \xrightarrow{P} 0. \tag{7.6.1}$$

Hence, $\sqrt{n}(\hat{\delta} - \delta)$ has the same asymptotic distribution as $\sqrt{n}(\tilde{\delta} - \delta)$. Using the Δ-method, Theorem 4.3.9, we can obtain the asymptotic distribution of $\sqrt{n}(\tilde{\delta}-\delta)$. Let $g(\theta) = \theta(1-\theta)$. Then $g'(\theta) = 1 - 2\theta$. Hence, by Theorem 4.3.9, the asymptotic distribution of $\sqrt{n}(\tilde{\delta} - \delta)$, and (7.6.1), we have the asymptotic distribution

$$\sqrt{n}(\hat{\delta} - \delta) \xrightarrow{D} N(0, \theta(1 - \theta)(1 - 2\theta)^2),$$

provided $\theta \neq 1/2$; see Exercise 7.6.10 for the case $\theta = 1/2$. ∎

A somewhat different, but also very important problem in point estimation is considered in the next example. In the example the distribution of a random variable X is described by a pdf $f(x;\theta)$ that depends upon $\theta \in \Omega$. The problem is to estimate the fractional part of the probability for this distribution which is at, or to the left of, a fixed point c. Thus we seek an MVUE of $F(c;\theta)$, where $F(x;\theta)$ is the cdf of X.

Example 7.6.2. Let $X_1, X_2, \ldots, X_n$ be a random sample of size $n > 1$ from a distribution that is $N(\theta, 1)$. Suppose that we wish to find an MVUE of the function of θ defined by

$$P(X \leq c) = \int_{-\infty}^{c} \frac{1}{\sqrt{2\pi}}e^{-(x-\theta)^2/2}\,dx = \Phi(c - \theta),$$

where c is a fixed constant. There are many unbiased estimators of $\Phi(c-\theta)$. We first exhibit one of these, say $u(X_1)$, a function of X_1 alone. We shall then compute the conditional expectation, $E[u(X_1)|\overline{X} = \overline{x}] = \varphi(\overline{x})$, of this unbiased statistic, given the sufficient statistic $\overline{X}$, the mean of the sample. In accordance with the theorems of Rao-Blackwell and Lehmann-Scheffé, $\varphi(\overline{X})$ is the unique MVUE of $\Phi(c - \theta)$.

Consider the function $u(x_1)$, where

$$u(x_1) = \begin{cases} 1 & x_1 \leq c \\ 0 & x_1 > c. \end{cases}$$

The expected value of the random variable $u(X_1)$ is given by

$$E[u(X_1)] = 1 \cdot P[X_1 - \theta \leq c - \theta] = \Phi(c - \theta).$$

That is, $u(X_1)$ is an unbiased estimator of $\Phi(c - \theta)$.

We shall next discuss the joint distribution of X_1 and $\overline{X}$ and the conditional distribution of X_1, given $\overline{X} = \overline{x}$. This conditional distribution enables us to compute the conditional expectation $E[u(X_1)|\overline{X} = \overline{x}] = \varphi(\overline{x})$. In accordance with Exercise 7.6.6 the joint distribution of X_1 and $\overline{X}$ is bivariate normal with mean vector (θ, θ), variances $\sigma_1^2 = 1$ and $\sigma_2^2 = 1/n$, and correlation coefficient $\rho = 1/\sqrt{n}$. Thus the conditional pdf of X_1, given $\overline{X} = \overline{x}$, is normal with linear conditional mean

$$\theta + \frac{\rho\sigma_1}{\sigma_2}(\overline{x} - \theta) = \overline{x}$$

and with variance

$$\sigma_1^2(1 - \rho^2) = \frac{n-1}{n}.$$

The conditional expectation of $u(X_1)$, given $\overline{X} = \overline{x}$, is then

$$
\begin{aligned}
\varphi(\overline{x}) &= \int_{-\infty}^{\infty} u(x_1)\sqrt{\frac{n}{n-1}}\frac{1}{\sqrt{2\pi}}\exp\left[-\frac{n(x_1 - \overline{x})^2}{2(n-1)}\right]dx_1 \\
&= \int_{-\infty}^{c}\sqrt{\frac{n}{n-1}}\frac{1}{\sqrt{2\pi}}\exp\left[-\frac{n(x_1 - \overline{x})^2}{2(n-1)}\right]dx_1.
\end{aligned}
$$

The change of variable $z = \sqrt{n}(x_1 - \overline{x})/\sqrt{n-1}$ enables us to write this conditional expectation as

$$\varphi(\overline{x}) = \int_{-\infty}^{c'}\frac{1}{\sqrt{2\pi}}e^{-z^2/2}\,dz = \Phi(c') = \Phi\left[\frac{\sqrt{n}(c - \overline{x})}{\sqrt{n-1}}\right],$$

where $c' = \sqrt{n}(c - \overline{x})/\sqrt{n-1}$. Thus the unique MVUE of $\Phi(c - \theta)$ is, for every fixed constant c, given by $\varphi(\overline{X}) = \Phi[\sqrt{n}(c - \overline{X})/\sqrt{n-1}]$.

In this example the mle of $\Phi(c - \theta)$ is $\Phi(c - \overline{X})$. These two estimators are close because $\sqrt{n/(n-1)} \to 1$, as $n \to \infty$. ∎

Remark 7.6.1. We should like to draw the attention of the reader to a rather important fact. This has to do with the adoption of a *principle*, such as the principle of unbiasedness and minimum variance. A principle is not a theorem; and seldom does a principle yield satisfactory results in all cases. So far, this principle has provided quite satisfactory results. To see that this is not always the case, let X have a Poisson distribution with parameter θ, $0 < \theta < \infty$. We may look upon X as a random sample of size 1 from this distribution. Thus X is a complete sufficient statistic for θ. We seek the estimator of $e^{-2\theta}$ that is unbiased and has minimum variance. Consider $Y = (-1)^X$. We have

$$E(Y) = E[(-1)^X] = \sum_{x=0}^{\infty}\frac{(-\theta)^x e^{-\theta}}{x!} = e^{-2\theta}.$$

Accordingly, $(-1)^X$ is the MVUE of $e^{-2\theta}$. Here this estimator leaves much to be desired. We are endeavoring to elicit some information about the number $e^{-2\theta}$, where $0 < e^{-2\theta} < 1$; yet our point estimate is either -1 or $+1$, each of which is a very poor estimate of a number between zero and 1. We do not wish to leave the reader with the impression that an MVUE is *bad*. That is not the case at all. We merely wish to point out that if one tries hard enough, one can find instances where such a statistic is *not good*. Incidentally, the maximum likelihood estimator of $e^{-2\theta}$ is, in the case where the sample size equals one, e^{-2X}, which is probably a much better estimator in practice than is the unbiased estimator $(-1)^X$. ∎

EXERCISES

7.6.1. Let $X_1, X_2, \ldots, X_n$ denote a random sample from a distribution that is $N(\theta, 1)$, $-\infty < \theta < \infty$. Find the MVUE of θ^2.
Hint: First determine $E(\overline{X}^2)$.

7.6.2. Let $X_1, X_2, \ldots, X_n$ denote a random sample from a distribution that is $N(0, \theta)$. Then $Y = \sum X_i^2$ is a complete sufficient statistic for θ. Find the MVUE of θ^2.

7.6.3. In the notation of Example 7.6.2 of this section, does $P(-c \le X \le c)$ have an MVUE ? Here $c > 0$.

7.6.4. Let $X_1, X_2, \ldots, X_n$ be a random sample from a Poisson distribution with parameter $\theta > 0$.

(a) Find the MVUE of $P(X \le 1) = (1+\theta)e^{-\theta}$. *Hint:* Let $u(x_1) = 1$, $x_1 \le 1$, zero elsewhere, and find $E[u(X_1)|Y = y]$, where $Y = \sum_{1}^{n} X_i$.

(b) Express the MVUE as a function of the mle.

(c) Determine the asymptotic distribution of the mle.

7.6.5. Let $X_1, X_2, \ldots, X_n$ denote a random sample from a Poisson distribution with parameter $\theta > 0$. From the Remark of this section, we know that $E[(-1)^{X_1}] = e^{-2\theta}$.

(a) Show that $E[(-1)^{X_1}|Y_1 = y_1] = (1 - 2/n)^{y_1}$, where $Y_1 = X_1 + X_2 + \cdots + X_n$.
Hint: First show that the conditional pdf of $X_1, X_2, \ldots, X_{n-1}$, given $Y_1 = y_1$, is multinomial, and hence that of X_1 given $Y_1 = y_1$ is $b(y_1, 1/n)$.

(b) Show that the mle of $e^{-2\theta}$ is $e^{-2\overline{X}}$.

(c) Since $y_1 = n\overline{x}$, show that $(1 - 2/n)^{y_1}$ is approximately equal to $e^{-2\overline{x}}$ when n is large.

7.6.6. As in Example 7.6.2, let $X_1, X_2, \ldots, X_n$ be a random sample of size $n > 1$ from a distribution that is $N(\theta, 1)$. Show that the joint distribution of X_1 and $\overline{X}$ is bivariate normal with mean vector (θ, θ), variances $\sigma_1^2 = 1$ and $\sigma_2^2 = 1/n$, and correlation coefficient $\rho = 1/\sqrt{n}$.

7.6.7. Let a random sample of size n be taken from a distribution that has the pdf $f(x; \theta) = (1/\theta) \exp(-x/\theta) I_{(0,\infty)}(x)$. Find the mle and the MVUE of $P(X \leq 2)$.

7.6.8. Let $X_1, X_2, \ldots, X_n$ be a random sample with the common pdf $f(x) = \theta^{-1} e^{-x/\theta}$, for $x > 0$, zero elsewhere; that is, $f(x)$ is a $\Gamma(1, \theta)$ pdf .

(a) Show that the statistic $\overline{X} = n^{-1} \sum_{i=1}^{n} X_i$ is a complete and sufficient statistic for θ.

(b) Determine the MVUE of θ.

(c) Determine the mle of θ.

(d) Often, though, this pdf is written as $f(x) = \tau e^{-\tau x}$, for $x > 0$, zero elsewhere. Thus $\tau = 1/\theta$. Use Theorem 6.1.2 to determine the mle of τ.

(e) Show that the statistic $\overline{X} = n^{-1} \sum_{i=1}^{n} X_i$ is a complete and sufficient statistic for τ. Show that $(n-1)/(n\overline{X})$ is the MVUE of $\tau = 1/\theta$. Hence, as usual the reciprocal of the mle of θ is the mle of $1/\theta$, but, in this situation, the reciprocal of the MVUE of θ is not the MVUE of $1/\theta$.

(f) Compute the variances of each of the unbiased estimators in Parts (b) and (e).

7.6.9. Consider the situation of the last exercise, but suppose we have the following two independent random samples: (1). $X_1, X_2, \ldots, X_n$ is a random sample with the common pdf $f_X(x) = \theta^{-1} e^{-x/\theta}$, for $x > 0$, zero elsewhere, and (2). $Y_1, Y_2, \ldots, Y_n$ is a random sample with common pdf $f_Y(y) = \tau e^{-\tau y}$, for $y > 0$, zero elsewhere. Assume that $\tau = 1/\theta$.
The last exercise suggests that, for some constant c, $Z = c\overline{X}/\overline{Y}$ might be an unbiased estimator of θ^2. Find this constant c and the variance of Z. *Hint:* Show that $\overline{X}/(\theta^2 \overline{Y})$ has an F-distribution.

7.6.10. Obtain the asymptotic distribution of the MVUE in Example 7.6.1 for the case $\theta = 1/2$.

7.7   The Case of Several Parameters

In many of the interesting problems we encounter, the pdf or pmf may not depend upon a single parameter θ, but perhaps upon two (or more) parameters. In general, our parameter space Ω will be a subset of R^p, but in many of our examples p will be two.

Definition 7.7.1. *Let $X_1, X_2, \ldots, X_n$ denote a random sample from a distribution that has pdf or pmf $f(x; \boldsymbol{\theta})$, where $\boldsymbol{\theta} \in \Omega \subset R^p$. Let S denote the support of X. Let $\mathbf{Y}$ be an m-dimensional random vector of statistics $\mathbf{Y} = (Y_1, \ldots, Y_m)'$, where $Y_i = u_i(X_1, X_2, \ldots, X_n)$, for $i = 1, \ldots, m$. Denote the pdf or pmf of $\mathbf{Y}$ by $f_\mathbf{Y}(\mathbf{y}; \boldsymbol{\theta})$*

for $\mathbf{y} \in R^m$. *The random vector of statistics* $\mathbf{Y}$ *is* **jointly sufficient** *for* $\boldsymbol{\theta}$ *if and only if*

$$\frac{\prod_{i=1}^{n} f(x_i; \boldsymbol{\theta})}{f_{\mathbf{Y}}(\mathbf{y}; \boldsymbol{\theta})} = H(x_1, x_2, \ldots, x_n), \quad \text{for all } x_i \in \mathcal{S},$$

where $H(x_1, x_2, \ldots, x_n)$ *does not depend upon* $\boldsymbol{\theta}$.

In general $m \neq p$, i.e., the number of sufficient statistics does not have to be the same as the number of parameters, but in most of our examples this will be the case.

As may be anticipated, the factorization theorem can be extended. In our notation it can be stated in the following manner. The vector of statistics $\mathbf{Y}$ is jointly sufficient for the parameter $\boldsymbol{\theta} \in \Omega$ if and only if we can find two nonnegative functions k_1 and k_2 such that

$$\prod_{i=1}^{n} f(x_i; \boldsymbol{\theta}) = k_1(\mathbf{y}; \boldsymbol{\theta}) k_2(x_1, \ldots, x_n), \quad \text{for all } x_i \in \mathcal{S}, \tag{7.7.1}$$

where the function $k_2(x_1, x_2, \ldots, x_n)$ does not depend upon $\boldsymbol{\theta}$.

Example 7.7.1. Let $X_1, X_2, \ldots, X_n$ be a random sample from a distribution having pdf

$$f(x; \theta_1, \theta_2) = \begin{cases} \frac{1}{2\theta_2} & \theta_1 - \theta_2 < x < \theta_1 + \theta_2 \\ 0 & \text{elsewhere}, \end{cases}$$

where $-\infty < \theta_1 < \infty$, $0 < \theta_2 < \infty$. Let $Y_1 < Y_2 < \cdots < Y_n$ be the order statistics. The joint pdf of Y_1 and Y_n is given by

$$f_{Y_1,Y_2}(y_1, y_n; \theta_1, \theta_2) = \frac{n(n-1)}{(2\theta_2)^n}(y_n - y_1)^{n-2}, \quad \theta_1 - \theta_2 < y_1 < y_n < \theta_1 + \theta_2,$$

and equals zero elsewhere. Accordingly, the joint pdf of $X_1, X_2, \ldots, X_n$ can be written, for all points in its support (all x_i such that $\theta_1 - \theta_2 < x_i < \theta_1 + \theta_2$),

$$\left(\frac{1}{2\theta_2}\right)^n = \frac{n(n-1)[\max(x_i) - \min(x_i)]^{n-2}}{(2\theta_2)^n} \left(\frac{1}{n(n-1)[\max(x_i) - \min(x_i)]^{n-2}}\right).$$

Since $\min(x_i) \leq x_j \leq \max(x_i)$, $j = 1, 2, \ldots, n$, the last factor does not depend upon the parameters. Either the definition or the factorization theorem assures us that Y_1 and Y_n are joint sufficient statistics for θ_1 and θ_2. ∎

The concept of a complete family of probability density functions is generalized as follows: Let

$$\{f(v_1, v_2, \ldots, v_k; \boldsymbol{\theta}) : \boldsymbol{\theta} \in \Omega\}$$

denote a family of pdfs of k random variables $V_1, V_2, \ldots, V_k$ that depends upon the p-dimensional vector of parameters $\boldsymbol{\theta} \in \Omega$. Let $u(v_1, v_2, \ldots, v_k)$ be a function of $v_1, v_2, \ldots, v_k$ (but not a function of any or all of the parameters). If

$$E[u(V_1, V_2, \ldots, V_k)] = 0$$

for all $\boldsymbol{\theta} \in \Omega$ implies that $u(v_1, v_2, \ldots, v_k) = 0$ at all points $(v_1, v_2, \ldots, v_k)$, except on a set of points that has probability zero for all members of the family of probability density functions, we shall say that the family of probability density functions is a complete family.

In the case where $\boldsymbol{\theta}$ is a vector, we generally consider best estimators of functions of $\boldsymbol{\theta}$; that is, parameters δ where $\delta = g(\boldsymbol{\theta})$ for a specified function g. For example, suppose we are sampling from a $N(\theta_1, \theta_2)$ distribution, where θ_2 is the variance. Let $\boldsymbol{\theta} = (\theta_1, \theta_2)'$ and consider the two parameters $\delta_1 = g_1(\boldsymbol{\theta}) = \theta_1$ and $\delta_2 = g_2(\boldsymbol{\theta}) = \sqrt{\theta_2}$. Hence, we are interested in best estimates of δ_1 and δ_2.

The Rao-Blackwell, Lehmann-Scheffé theory outlined in Sections 7.3 and 7.4 extends naturally to this vector case. Briefly, suppose $\delta = g(\boldsymbol{\theta})$ is the parameter of interest and $\mathbf{Y}$ is a vector of sufficient and complete statistics for $\boldsymbol{\theta}$. Let T be a statistic which is a function of $\mathbf{Y}$, such as, $T = T(\mathbf{Y})$. If $E(T) = \delta$ then T is the unique MVUE of δ.

The remainder of our treatment of the case of several parameters will be restricted to probability density functions that represent what we shall call regular cases of the exponential class. Here, $m = p$.

Definition 7.7.2. *Let X be a random variable with pdf or pmf $f(x; \boldsymbol{\theta})$ where the vector of parameters $\boldsymbol{\theta} \in \Omega \subset R^m$. Let $\mathcal{S}$ denote the support of X. If X is continuous assume that $\mathcal{S} = (a, b)$, where a or b may be $-\infty$ or ∞, respectively. If X is discrete assume that $\mathcal{S} = \{a_1, a_2, \ldots\}$. Suppose $f(x; \boldsymbol{\theta})$ is of the form*

$$f(x; \boldsymbol{\theta}) = \begin{cases} \exp\left[\sum_{j=1}^{m} p_j(\boldsymbol{\theta}) K_j(x) + S(x) + q(\theta_1, \theta_2, \ldots, \theta_m)\right] & \text{for all } x \in \mathcal{S} \\ 0 & \text{elsewhere.} \end{cases}$$

(7.7.2)

*Then we say this pdf or pmf is a member of the **exponential class**. We say it is a **regular case** of the exponential family if, in addition,*

1. *the support does not depend on the vector of parameters $\boldsymbol{\theta}$,*

2. *the space Ω contains a nonempty, m-dimensional open rectangle,*

3. *the $p_j(\boldsymbol{\theta})$, $j = 1, \ldots, m$, are nontrivial, functionally independent, continuous functions of $\boldsymbol{\theta}$,*

4. *and*

   (a) *if X is a continuous random variable, then the m derivatives $K_j'(x)$, for $j = 1, 2, \ldots, m$, are continuous for $a < x < b$ and no one is a linear homogeneous function of the others and $S(x)$ is a continuous function of x, $a < x < b$.*

   (b) *if X is discrete, the $K_j(x)$, $j = 1, 2, \ldots, m$, are nontrivial functions of x on the support $\mathcal{S}$ and no one is a linear homogeneous function of the others.*

Let $X_1, \ldots, X_n$ be a random sample on X where the pdf or the pmf of X is a regular case of the exponential class with the same notation as in Definition 7.7.2. It follows from (7.7.2) that joint pdf or pmf of the sample is given by

$$\prod_{i=1}^{n} f(x_i; \boldsymbol{\theta}) = \exp\left[\sum_{j=1}^{m} p_j(\boldsymbol{\theta}) \sum_{i=1}^{n} K_j(x_i) + nq(\boldsymbol{\theta})\right] \exp\left[\sum_{i=1}^{n} S(x_i)\right], \qquad (7.7.3)$$

for all $x_i \in \mathcal{S}$. In accordance with the factorization theorem, the statistics

$$Y_1 = \sum_{i=1}^{n} K_1(x_i), \quad Y_2 = \sum_{i=1}^{n} K_2(x_i), \ldots, Y_m = \sum_{i=1}^{n} K_m(x_i)$$

are joint sufficient statistics for the m-dimensional vector of parameters $\boldsymbol{\theta}$. It is left as an exercise to prove that the joint pdf of $\mathbf{Y} = (Y_1, \ldots, Y_m)'$ is of the form

$$R(\mathbf{y}) \exp\left[\sum_{j=1}^{m} p_j(\boldsymbol{\theta}) y_j + nq(\boldsymbol{\theta})\right], \qquad (7.7.4)$$

at points of positive probability density. These points of positive probability density and the function $R(\mathbf{y})$ do not depend upon the vector of parameters $\boldsymbol{\theta}$. Moreover, in accordance with a theorem in analysis, it can be asserted that in a regular case of the exponential class, the family of probability density functions of these joint sufficient statistics $Y_1, Y_2, \ldots, Y_m$ is complete when $n > m$. In accordance with a convention previously adopted, we shall refer to $Y_1, Y_2, \ldots, Y_m$ as *joint complete sufficient statistics for the vector of parameters* $\boldsymbol{\theta}$.

Example 7.7.2. Let $X_1, X_2, \ldots, X_n$ denote a random sample from a distribution that is $N(\theta_1, \theta_2)$, $-\infty < \theta_1 < \infty$, $0 < \theta_2 < \infty$. Thus the pdf $f(x; \theta_1, \theta_2)$ of the distribution may be written as

$$f(x; \theta_1, \theta_2) = \exp\left(\frac{-1}{2\theta_2} x^2 + \frac{\theta_1}{\theta_2} x - \frac{\theta_1^2}{2\theta_2} - \ln\sqrt{2\pi\theta_2}\right).$$

Therefore, we can take $K_1(x) = x^2$ and $K_2(x) = x$. Consequently, the statistics

$$Y_1 = \sum_{1}^{n} X_i^2 \quad \text{and} \quad Y_2 = \sum_{1}^{n} X_i$$

are joint complete sufficient statistics for θ_1 and θ_2. Since the relations

$$Z_1 = \frac{Y_2}{n} = \overline{X}, \quad Z_2 = \frac{Y_1 - Y_2^2/n}{n-1} = \frac{\sum (X_i - \overline{X})^2}{n-1}$$

define a one-to-one transformation, Z_1 and Z_2 are also joint complete sufficient statistics for θ_1 and θ_2. Moreover,

$$E(Z_1) = \theta_1 \quad \text{and} \quad E(Z_2) = \theta_2.$$

From completeness, we have that Z_1 and Z_2 are the only functions of Y_1 and Y_2 that are unbiased estimators of θ_1 and θ_2, respectively. Hence, Z_1 and Z_2 are the unique minimum variance estimators of θ_1 and θ_2, respectively. The MVUE of the standard deviation $\sqrt{\theta_2}$ is derived in Exercise 7.7.5. ∎

In this section we have extended the concepts of sufficiency and completeness to the case where $\boldsymbol{\theta}$ is a p-dimensional vector. We now extend these concepts to the case where $\mathbf{X}$ is a k-dimensional random vector. We will only consider the regular exponential case followed by two examples.

Suppose $\mathbf{X}$ is a k-dimensional random vector with pdf or pmf $f(\mathbf{x}; \boldsymbol{\theta})$ where $\boldsymbol{\theta} \in \Omega \subset R^p$. Let $\mathcal{S} \subset R^k$ denote the support of $\mathbf{X}$. Suppose $f(\mathbf{x}; \boldsymbol{\theta})$ is of the form

$$f(\mathbf{x}; \boldsymbol{\theta}) = \begin{cases} \exp\left[\sum_{j=1}^m p_j(\boldsymbol{\theta}) K_j(\mathbf{x}) + S(\mathbf{x}) + q(\boldsymbol{\theta})\right] & \text{for all } \mathbf{x} \in \mathcal{S} \\ 0 & \text{elsewhere.} \end{cases} \tag{7.7.5}$$

Then we say this pdf or pmf is a member of the **exponential class**. If, in addition, $p = m$, the support does not depend on the vector of parameters $\boldsymbol{\theta}$, and conditions similar to those of Definition 7.7.2 hold then we say this pdf is a **regular case** of the exponential class.

Suppose that $\mathbf{X}_1, \ldots, \mathbf{X}_n$ constitute a random sample on $\mathbf{X}$. Then the statistics,

$$Y_j = \sum_{i=1}^n K_j(\mathbf{X}_i), \quad \text{for } j = 1, \ldots, m, \tag{7.7.6}$$

are sufficient and complete statistics for $\boldsymbol{\theta}$. Let $\mathbf{Y} = (Y_1, \ldots, Y_m)'$. Suppose $\delta = g(\boldsymbol{\theta})$ is a parameter of interest. If $T = h(\mathbf{Y})$ for some function h and $E(T) = \delta$ then T is the unique minimum variance unbiased estimator of δ.

Example 7.7.3 (Multinomial). In Example 6.4.5, we consider the mles of the multinomial distribution. In this example we determine the MVUEs of several of the parameters. As in Example 6.4.5, consider a random trial which can result in one, and only one, of k outcomes or categories. Let X_j be 1 or 0 depending on whether the jth outcomes does or does not occur, for $j = 1, \ldots, k$. Suppose the probability that outcome j occurs is p_j; hence, $\sum_{j=1}^k p_j = 1$. Let $\mathbf{X} = (X_1, \ldots, X_{k-1})'$ and $\mathbf{p} = (p_1, \ldots, p_{k-1})'$. The distribution of $\mathbf{X}$ is multinomial and can be found in expression (6.4.18) which can be reexpressed as

$$f(\mathbf{x}, \mathbf{p}) = \exp\left\{\sum_{j=1}^{k-1}\left(\log\left[\frac{p_j}{1 - \sum_{i \neq k} p_i}\right]\right) x_j + \log(1 - \sum_{i \neq k} p_i)\right\}.$$

Because this a regular case of the exponential family, the following statistics, resulting from a random sample $\mathbf{X}_1, \ldots, \mathbf{X}_n$ from the distribution of $\mathbf{X}$, are jointly sufficient and complete for the parameters $\mathbf{p} = (p_1, \ldots, p_{k-1})'$:

$$Y_j = \sum_{i=1}^n X_{ij}, \quad \text{for } j = 1, \ldots, k - 1.$$

Each random variable X_{ij} is Bernoulli with parameter p_j and the variables X_{ij} are independent for $i = 1, \ldots, n$. Hence, the variables Y_j are binomial(n, p_j) for $j = 1, \ldots, k$. Thus, the MVUE of p_j is the statistic $n^{-1} Y_j$.

Next, we shall find the MVUE of $p_j p_l$, for $j \neq l$. Exercise 7.7.8 shows that the mle of $p_j p_l$ is $n^{-2} Y_j Y_l$. Recall from Section 3.1 that the conditional distribution of Y_j given Y_l is $b(n - Y_l), p_j/(1 - p_l))$. As an initial guess at the MVUE, consider the mle, which as shown by Exercise 7.7.8 is $n^{-2} Y_j Y_l$. Hence,

$$
\begin{aligned}
E[n^{-2} Y_j Y_l] &= \frac{1}{n^2} E[E(Y_j Y_l | Y_l)] = \frac{1}{n^2} E[Y_l E(Y_j | Y_l)] \\
&= \frac{1}{n^2} E\left[Y_l (n - Y_l) \frac{p_j}{1 - p_l} \right] = \frac{1}{n^2} \frac{p_j}{1 - p_l} \{E[nY_l] - E[Y_l^2]\} \\
&= \frac{1}{n^2} \frac{p_j}{1 - p_l} \{n^2 p_l - np_l(1 - p_l) - n^2 p_l^2\} \\
&= \frac{1}{n^2} \frac{p_j}{1 - p_l} np_l(n - 1)(1 - p_l) = \frac{(n-1)}{n} p_j p_l.
\end{aligned}
$$

Hence, the MVUE of $p_j p_l$ is $\frac{1}{n(n-1)} Y_j Y_l$. ∎

Example 7.7.4 (Multivariate Normal). Let $\mathbf{X}$ have the multivariate normal distribution $N_k(\boldsymbol{\mu}, \boldsymbol{\Sigma})$, where $\boldsymbol{\Sigma}$ is a positive definite $k \times k$ matrix. The pdf of $\mathbf{X}$ is given in expression (3.5.12). In this case $\boldsymbol{\theta}$ is a $\left(k + \frac{k(k+1)}{2}\right)$ dimensional vector whose first k components consist of the mean vector $\boldsymbol{\mu}$ and whose last $\frac{k(k+1)}{2}$ components consist of the component-wise variances σ_i^2 and the covariances σ_{ij}, for $j \geq i$. The density of $\mathbf{X}$ can be written as,

$$
f_{\mathbf{X}}(\mathbf{x}) = \exp\left\{ -\frac{1}{2} \mathbf{x}' \boldsymbol{\Sigma}^{-1} \mathbf{x} + \boldsymbol{\mu}' \boldsymbol{\Sigma}^{-1} \mathbf{x} - \frac{1}{2} \boldsymbol{\mu}' \boldsymbol{\Sigma}^{-1} \boldsymbol{\mu} - \frac{1}{2} \log |\boldsymbol{\Sigma}| - \frac{k}{2} \log 2\pi \right\},
$$
$$(7.7.7)$$

for $\mathbf{x} \in R^k$. Hence, by (7.7.5) the multivariate normal pdf is a regular case of the exponential class of distributions. We need only identify the functions $K(\mathbf{x})$. The second term in the exponent on the right side of (7.7.7) can be written as $(\boldsymbol{\mu}' \boldsymbol{\Sigma}^{-1}) \mathbf{x}$; hence, $K_1(\mathbf{x}) = \mathbf{x}$. The first term is easily seen to be a linear combination of the products $x_i x_j$, $i, j = 1, 2, \ldots k$ which are the entries of the matrix $\mathbf{x}\mathbf{x}'$. Hence, we can take $K_2(\mathbf{x}) = \mathbf{x}\mathbf{x}'$. Now, let $\mathbf{X}_1, \ldots, \mathbf{X}_n$ be a random sample on $\mathbf{X}$. Based on (7.7.7) then, a set of sufficient and complete statistics is given by,

$$
\mathbf{Y}_1 = \sum_{i=1}^{n} \mathbf{X}_i \text{ and } \mathbf{Y}_2 = \sum_{i=1}^{n} \mathbf{X}_i \mathbf{X}_i'. \tag{7.7.8}
$$

Note that $\mathbf{Y}_1$ is a vector of k statistics and that $\mathbf{Y}_2$ is a $k \times k$ symmetric matrix. Because the matrix is symmetric, we can eliminate the bottom half (elements (i, j) with $i > j$) of the matrix which results in $\left(k + \frac{k(k+1)}{2}\right)$ complete sufficient statistics; i.e., as many complete sufficient statistics as there are parameters.

Based on marginal distributions, it is easy to show that $\overline{X}_j = n^{-1} \sum_{i=1}^{n} X_{ij}$ is the MVUE of μ_j and that $(n - 1)^{-1} \sum_{i=1}^{n} (X_{ij} - \overline{X}_j)^2$ is the MVUE of σ_j^2. The MVUEs of the covariance parameters are obtained in Exerscise 7.7.9. ∎

For our last example, we consider a case where the set of parameters is the cdf.

Example 7.7.5. Let $X_1, X_2, \ldots, X_n$ be a random sample having the common continuous cdf $F(x)$. Let $Y_1 < Y_2 < \cdots < Y_n$ denote the corresponding order statistics. Note that given $Y_1 = y_1, Y_2 = y_2, \ldots, Y_n = y_n$, the conditional distribution of $X_1, X_2, \ldots, X_n$ is discrete with probability $\frac{1}{n!}$ on each of the $n!$ permutations of the vector $(y_1, y_2, \ldots, y_n)$, (because $F(x)$ is continuous we can assume that each of the values $y_1, y_2, \ldots, y_n$ are distinct). That is, the conditional distribution does not depend on $F(x)$. Hence, by the definition of sufficiency the order statistics are sufficient for $F(x)$. Furthermore, while the proof is beyond the scope of this book, it can be shown that the order statistics are also complete; see page 72 of Lehmann and Casella (1998).

Let $T = T(x_1, x_2, \ldots, x_n)$ be any statistic which is *symmetric in its arguments*; i.e., $T(x_1, x_2, \ldots, x_n) = T(x_{i_1}, x_{i_2}, \ldots, x_{i_n})$ for any permutation $(x_{i_1}, x_{i_2}, \ldots, x_{i_n})$ of $(x_1, x_2, \ldots, x_n)$. Then T is a function of the order statistics. This is useful in determining MVUEs for this situation; see Exercises 7.7.12 and 7.7.13. ∎

EXERCISES

7.7.1. Let $Y_1 < Y_2 < Y_3$ be the order statistics of a random sample of size 3 from the distribution with pdf.

$$f(x; \theta_1, \theta_2) = \begin{cases} \frac{1}{\theta_2} \exp\left(-\frac{x-\theta_1}{\theta_2}\right) & \theta_1 < x < \infty, \ -\infty < \theta_1 < \infty, \ 0 < \theta_2 < \infty \\ 0 & \text{elsewhere.} \end{cases}$$

Find the joint pdf of $Z_1 = Y_1$, $Z_2 = Y_2$, and $Z_3 = Y_1 + Y_2 + Y_3$. The corresponding transformation maps the space $\{(y_1, y_2, y_3) : \theta_1 < y_1 < y_2 < y_3 < \infty\}$ onto the space
$$\{(z_1, z_2, z_3) : \theta_1 < z_1 < z_2 < (z_3 - z_1)/2 < \infty\}.$$
Show that Z_1 and Z_3 are joint sufficient statistics for θ_1 and θ_2.

7.7.2. Let $X_1, X_2, \ldots, X_n$ be a random sample from a distribution that has a pdf of form (7.7.2) of this section. Show that $Y_1 = \sum_{i=1}^{n} K_1(X_i), \ldots, Y_m = \sum_{i=1}^{m} K_m(X_i)$ have a joint pdf of form (7.7.4) of this section.

7.7.3. Let $(X_1, Y_1), (X_2, Y_2), \ldots, (X_n, Y_n)$ denote a random sample of size n from a bivariate normal distribution with means μ_1 and μ_2, positive variances σ_1^2 and σ_2^2, and correlation coefficient ρ. Show that $\sum_1^n X_i, \sum_1^n Y_i, \sum_1^n X_i^2, \sum_1^n Y_i^2$, and $\sum_1^n X_i Y_i$ are joint complete sufficient statistics for the five parameters. Are $\overline{X} = \sum_1^n X_i/n, \overline{Y} = \sum_1^n Y_i/n, S_1^2 = \sum_1^n (X_i - \overline{X})^2/(n-1), S_2^2 = \sum_1^n (Y_i - \overline{Y})^2/(n-1)$, and

$\sum_{1}^{n}(X_i - \overline{X})(Y_i - \overline{Y})/(n-1)S_1S_2$ also joint complete sufficient statistics for these parameters?

7.7.4. Let the pdf $f(x; \theta_1, \theta_2)$ be of the form

$$\exp[p_1(\theta_1, \theta_2)K_1(x) + p_2(\theta_1, \theta_2)K_2(x) + S(x) + q_1(\theta_1, \theta_2)], \quad a < x < b,$$

zero elsewhere. Let $K'_1(x) = cK'_2(x)$. Show that $f(x; \theta_1, \theta_2)$ can be written in the form

$$\exp[p_1(\theta_1, \theta_2)K(x) + S(x) + q_1(\theta_1, \theta_2)], \quad a < x < b,$$

zero elsewhere. This is the reason why it is required that no one $K'_j(x)$ be a linear homogeneous function of the others, that is, so that the number of sufficient statistics equals the number of parameters.

7.7.5. In Example 7.2.2, find the MVUE of the standard deviation $\sqrt{\theta_2}$

7.7.6. Let $X_1, X_2, \ldots, X_n$ be a random sample from the uniform distribution with pdf $f(x; \theta_1, \theta_2) = 1/(2\theta_2)$, $\theta_1 - \theta_2 < x < \theta_1 + \theta_2$, where $-\infty < \theta_1 < \infty$ and $\theta_2 > 0$, and the pdf is equal to zero elsewhere.

(a) Show that $Y_1 = \min(X_i)$ and $Y_n = \max(X_i)$, the joint sufficient statistics for θ_1 and θ_2, are complete.

(b) Find the MVUEs of θ_1 and θ_2.

7.7.7. Let $X_1, X_2, \ldots, X_n$ be a random sample from $N(\theta_1, \theta_2)$.

(a) If the constant b is defined by the equation $P(X \leq b) = 0.90$, find the mle and the MVUE of b.

(b) If c is a given constant, find the mle and the MVUE of $P(X \leq c)$.

7.7.8. In the notation of Example 7.7.3, show that the mle of $p_j p_l$ is $n^{-2}Y_j Y_l$.

7.7.9. Refer to Example 7.7.4 on sufficiency for the multivariate normal model.

(a) Determine the MVUE of the covariance parameters σ_{ij}.

(b) Let $h = \sum_{i=1}^{k} a_i \mu_i$, where $a_1, \ldots, a_k$ are specified constants. Find the MVUE for h.

7.7.10. In a personal communication, LeRoy Folks noted that the inverse Gaussian pdf

$$f(x; \theta_1, \theta_2) = \left(\frac{\theta_2}{2\pi x^3}\right)^{1/2} \exp\left[\frac{-\theta_2(x - \theta_1)^2}{2\theta_1^2 x}\right], \quad 0 < x < \infty, \tag{7.7.9}$$

where $\theta_1 > 0$ and $\theta_2 > 0$ is often used to model lifetimes. Find the complete sufficient statistics for (θ_1, θ_2), if $X_1, X_2, \ldots, X_n$ is a random sample from the distribution having this pdf.

7.7.11. Let $X_1, X_2, \ldots, X_n$ be a random sample from a $N(\theta_1, \theta_2)$ distribution.

   (a) Show that $E[(X_1 - \theta_1)^4] = 3\theta_2^2$.

   (b) Find the MVUE of $3\theta_2^2$.

7.7.12. Let $X_1, \ldots, X_n$ be a random sample from a distribution of the continuous type with cdf $F(x)$. Suppose the mean, $\mu = E(X_1)$, exists. Using Example 7.7.5, show that the sample mean, $\overline{X} = n^{-1} \sum_{i=1}^{n} X_i$ is the MVUE of μ.

7.7.13. Let $X_1, \ldots, X_n$ be a random sample from a distribution of the continuous type with cdf $F(x)$. Let $\theta = P(X_1 \leq a) = F(a)$, where a is known. Show that the proportion $n^{-1} \#\{X_i \leq a\}$ is the MVUE of θ.

7.8   Minimal Sufficiency and Ancillary Statistics

In the study of statistics, it is clear that we want to reduce the data contained in the entire sample as much as possible without losing relevant information about the important characteristics of the underlying distribution. That is, a large collection of numbers in the sample is not as meaningful as a few good summary statistics of those data. Sufficient statistics, if they exist, are valuable because we know that the statisticians with those summary measures have as much information as the statistician with the entire sample. Sometimes, however, there are several sets of joint sufficient statistics, and thus we would like to find the simplest one of these sets. For illustration, in a sense, the observations $X_1, X_2, \ldots, X_n$, $n > 2$, of a random sample from $N(\theta_1, \theta_2)$ could be thought of as joint sufficient statistics for θ_1 and θ_2. We know, however, that we can use $\overline{X}$ and S^2 as joint sufficient statistics for those parameters, which is a great simplification over using $X_1, X_2, \ldots, X_n$, particularly if n is large.

In most instances in this chapter, we have been able to find a single sufficient statistic for one parameter or two joint sufficient statistics for two parameters. Possibly the most complicated cases considered so far are given in Example 7.7.3, in which we find $k + k(k+1)/2$ joint sufficient statistics for $k + k(k+1)/2$ parameters; or the multivariate normal distribution given in Example 7.7.4; or the use the order statistics of a random sample for some completely unknown distribution of the continuous type as in Example 7.7.5.

What we would like to do is to change from one set of joint sufficient statistics to another, always reducing the number of statistics involved until we cannot go any further without losing the sufficiency of the resulting statistics. Those statistics that are there at the end of this reduction are called *minimal sufficient statistics*. These are sufficient for the parameters and are functions of every other set of sufficient statistics for those same parameters. Often, if there are k parameters, we can find k joint sufficient statistics that are minimal. In particular, if there is one parameter, we can often find a single sufficient statistic which is minimal. Most of the earlier examples that we have considered illustrate this point, but this is not always the case as shown by the following example.

Example 7.8.1. Let $X_1, X_2, \ldots, X_n$ be a random sample from the uniform distribution over the interval $(\theta - 1, \theta + 1)$ having pdf

$$f(x; \theta) = (\tfrac{1}{2}) I_{(\theta-1,\theta+1)}(x), \quad \text{where} - \infty < \theta < \infty.$$

The joint pdf of $X_1, X_2, \ldots, X_n$ equals the product of $(\tfrac{1}{2})^n$ and certain indicator functions, namely

$$(\tfrac{1}{2})^n \prod_{i=1}^{n} I_{(\theta-1,\theta+1)}(x_i) = (\tfrac{1}{2})^n \{ I_{(\theta-1,\theta+1)}[\min(x_i)] \} \{ I_{(\theta-1,\theta+1)}[\max(x_i)] \},$$

because $\theta - 1 < \min(x_i) \le x_j \le \max(x_i) < \theta + 1$, $j = 1, 2, \ldots, n$. Thus the order statistics $Y_1 = \min(X_i)$ and $Y_n = \max(X_i)$ are the sufficient statistics for θ. These two statistics actually are minimal for this one parameter, as we cannot reduce the number of them to less than two and still have sufficiency. ∎

There is an observation that helps us observe that almost all the sufficient statistics that we have studied thus far are minimal. We have noted that the mle $\hat{\theta}$ of θ is a function of one or more sufficient statistics, when the latter exists. Suppose that this mle $\hat{\theta}$ is also sufficient. Since this sufficient statistic $\hat{\theta}$ is a function of the other sufficient statistics, Theorem 7.3.2, it must be minimal. For example, we have

1. The mle $\hat{\theta} = \overline{X}$ of θ in $N(\theta, \sigma^2)$, σ^2 known, is a minimal sufficient statistic for θ.

2. The mle $\hat{\theta} = \overline{X}$ of θ in a Poisson distribution with mean θ is a minimal sufficient statistic for θ.

3. The mle $\hat{\theta} = Y_n = \max(X_i)$ of θ in the uniform distribution over $(0, \theta)$ is a minimal sufficient statistic for θ.

4. The maximum likelihood estimators $\hat{\theta}_1 = \overline{X}$ and $\hat{\theta}_2 = S^2$ of θ_1 and θ_2 in $N(\theta_1, \theta_2)$ are joint minimal sufficient statistics for θ_1 and θ_2.

From these examples we see that the minimal sufficient statistics do not need to be unique, for any one-to-one transformation of them also provides minimal sufficient statistics. The linkage between minimal sufficient statistics and the mle, however, does not hold in many interesting instances. We illustrate this in the next two examples.

Example 7.8.2. Consider the model given in Example 7.8.1. There we noted that $Y_1 = \min(X_i)$ and $Y_n = \max(X_i)$ are joint sufficient statistics. Also, we have

$$\theta - 1 < Y_1 < Y_n < \theta + 1$$

or, equivalently,

$$Y_n - 1 < \theta < Y_1 + 1.$$

Hence, to maximize the likelihood function so that it equals $(\tfrac{1}{2})^n$, θ can be any value between $Y_n - 1$ and $Y_1 + 1$. For example, many statisticians take the mle to be the mean of these two end points, namely

$$\hat{\theta} = \frac{Y_n - 1 + Y_1 + 1}{2} = \frac{Y_1 + Y_n}{2},$$

which is the midrange. We recognize, however, that this mle is not unique. Some might argue that since $\hat{\theta}$ is an mle of θ and since it is a function of the joint sufficient statistics, Y_1 and Y_n, for θ, it will be a minimal sufficient statistic. This is not the case at all, for $\hat{\theta}$ is not even sufficient. Note that the mle must itself be a sufficient statistic for the parameter before it can be considered the minimal sufficient statistics. ∎

Note that we can model the situation in the last example by,

$$X_i = \theta + W_i, \tag{7.8.1}$$

where $W_1, W_2, \ldots, W_n$ are iid with the common uniform$(-1, 1)$ pdf. Hence, this is an example of a location model. We discuss these models in general next.

Example 7.8.3. Consider a location model given by

$$X_i = \theta + W_i, \tag{7.8.2}$$

where $W_1, W_2, \ldots, W_n$ are iid with the common pdf $f(w)$ and common continuous cdf $F(w)$. From Example 7.7.5, we know that the order statistics $Y_1 < Y_2 < \cdots Y_n$ are a set of complete and sufficient statistics for this situation. Can we obtain a smaller set of minimal sufficient statistics? Consider the following four situations:

(a) Suppose $f(w)$ is the $N(0, 1)$ pdf. Then we know that $\overline{X}$ is both the MVUE and mle of θ. Also, $\overline{X} = n^{-1} \sum_{i=1}^{n} Y_i$; i.e., a function of the order statistics. Hence, $\overline{X}$ is minimal sufficient.

(b) Suppose $f(w) = \exp\{-w\}$, for $w > 0$, zero elsewhere. Then the statistic Y_1 is a sufficient statistic as well as the mle, and thus is minimal sufficient.

(c) Suppose $f(w)$ is the logistic pdf. As discussed in Example 6.1.4, the mle of θ exists and it is easy to compute. As shown on page 38 of Lehmann and Casella (1998), though, the order statistics are minimal sufficient for this situation. That is, no reduction is possible.

(d) Suppose $f(w)$ is the Laplace pdf. It was shown in Example 6.1.3 that the median, Q_2 is the mle of θ, but it is not a sufficient statistic. Further, similar to the logistic pdf, it can be shown that the order statistics are minimal sufficient for this situation. ∎

In general the situation described in Parts (c) and (d), where the mle is obtained rather easily while the set of minimal sufficient statistics is the set of order statistics and no reduction is possible, is the norm for location models.

There is also a relationship between a minimal sufficient statistic and completeness that is explained more fully in Lehmann and Scheffé (1950). Let us say simply and without explanation that for the cases in this book, complete sufficient statistics are minimal sufficient statistics. The converse is not true, however, by noting that in Example 7.8.1 we have

$$E\left[\frac{Y_n - Y_1}{2} - \frac{n-1}{n+1}\right] = 0, \quad \text{for all } \theta.$$

That is, there is a nonzero function of those minimal sufficient statistics, Y_1 and Y_n, whose expectation is zero for all θ.

There are other statistics that almost seem opposites of sufficient statistics. That is, while sufficient statistics contain all the information about the parameters, these other statistics, called *ancillary statistics*, have distributions free of the parameters and seemingly contain no information about those parameters. As an illustration, we know that the variance S^2 of a random sample from $N(\theta, 1)$ has a distribution that does not depend upon θ and hence is an ancillary statistic. Another example is the ratio $Z = X_1/(X_1 + X_2)$, where X_1, X_2 is a random sample from a gamma distribution with known parameter $\alpha > 0$ and unknown parameter $\beta = \theta$, because Z has a beta distribution that is free of θ. There are many examples of ancillary statistics, and we provide some rules that make them rather easy to find with certain models, which we present in the next three examples.

Example 7.8.4 (Location Invariant Statistics). In Example 7.8.3, we introduced the location model. Recall that a random sample $X_1, X_2, \ldots, X_n$ follows this model if

$$X_i = \theta + W_i, \quad i = 1, \ldots, n, \qquad (7.8.3)$$

where $-\infty < \theta < \infty$ is a parameter and $W_1, W_2, \ldots, W_n$ are iid random variables with the pdf $f(w)$ which does not depend on θ. Then the common pdf of X_i is $f(x - \theta)$.

Let $Z = u(X_1, X_2, \ldots, X_n)$ be a statistic such that

$$u(x_1 + d, x_2 + d, \ldots, x_n + d) = u(x_1, x_2, \ldots, x_n),$$

for all real d. Hence,

$$Z = u(W_1 + \theta, W_2 + \theta, \ldots, W_n + \theta) = u(W_1, W_2, \ldots, W_n)$$

is a function of $W_1, W_2, \ldots, W_n$ alone (not of θ). Hence Z must have a distribution that does not depend upon θ. We call $Z = u(X_1, X_2, \ldots, X_n)$ a *location-invariant statistic*.

Assuming a location model, the following are some examples of location-invariant statistics: the sample variance $= S^2$, the sample range $= \max\{X_i\} - \min\{X_i\}$, the mean deviation from the sample median $= (1/n) \sum |X_i - \text{median}(X_i)|$, $X_1 + X_2 - X_3 - X_4$, $X_1 + X_3 - 2X_2$, $(1/n) \sum [X_i - \min(X_i)]$, and so on. ∎

Example 7.8.5 (Scale Invariant Statistics). Next, consider a random sample $X_1, X_2, \ldots, X_n$ which follows a **scale model**; i.e., a model of the form

$$X_i = \theta W_i, \quad i = 1, \ldots, n, \qquad (7.8.4)$$

where $\theta > 0$ and $W_1, W_2, \ldots, W_n$ are iid random variables with pdf $f(w)$ which does not depend on θ. Then the common pdf of X_i is $\theta^{-1} f(x/\theta)$. We call θ a scale parameter. Suppose that $Z = u(X_1, X_2, \ldots, X_n)$ is a statistic such that

$$u(cx_1, cx_2, \ldots, cx_n) = u(x_1, x_2, \ldots, x_n)$$

for all $c > 0$. Then

$$Z = u(X_1, X_2, \ldots, X_n) = u(\theta W_1, \theta W_2, \ldots, \theta W_n) = u(W_1, W_2, \ldots, W_n).$$

Since neither the joint pdf of $W_1, W_2, \ldots, W_n$ nor Z contain θ, the distribution of Z must not depend upon θ. We say that Z is a **scale-invariant statistic**.

The following are some examples of scale-invariant statistics: $X_1/(X_1 + X_2)$, $X_1^2 / \sum_1^n X_i^2$, $\min(X_i)/\max(X_i)$, and so on. ∎

Example 7.8.6 (Location and Scale Invariant Statistics). Finally, consider a random sample $X_1, X_2, \ldots, X_n$ which follows a location and scale model as in Example 7.7.5. That is,

$$X_i = \theta_1 + \theta_2 W_i, \quad i = 1, \ldots, n, \tag{7.8.5}$$

where W_i are iid with the common pdf $f(t)$ which is free of θ_1 and θ_2. In this case the pdf of X_i is $\theta_2^{-1} f((x - \theta_1)/\theta_2)$. Consider the statistic $Z = u(X_1, X_2, \ldots, X_n)$ where

$$u(cx_1 + d, \ldots, cx_n + d) = u(x_1, \ldots, x_n).$$

Then

$$Z = u(X_1, \ldots, X_n) = u(\theta_1 + \theta_2 W_1, \ldots, \theta_1 + \theta_2 W_n) = u(W_1, \ldots, W_n).$$

Since neither the joint pdf of $W_1, \ldots, W_n$ nor Z contains θ_1 and θ_2, the distribution of Z must not depend upon θ_1 nor θ_2. Statistics such as $Z = u(X_1, X_2, \ldots, X_n)$ are called **location and scale invariant statistics**. The following are four examples of such statistics:

(a) $T_1 = [\max(X_i) - \min(X_i)]/S$,

(b) $T_2 = \sum_{i=1}^{n-1}(X_{i+1} - X_i)^2/S^2$,

(c) $T_3 = (X_i - \overline{X})/S$,

(d) $T_4 = |X_i - X_j|/S,, \; ; i \neq j$.

Thus these location invariant, scale invariant, and location and scale invariant statistics provide good illustrations, with the appropriate model for the pdf, of ancillary statistics. Since an ancillary statistic and a complete (minimal) sufficient statistic are such opposites, we might believe that there is, in some sense, no relationship between the two. This is true and in the next section we show that they are independent statistics.

EXERCISES

7.8.1. Let $X_1, X_2, \ldots, X_n$ be a random sample from each of the following distributions involving the parameter θ. In each case find the mle of θ and show that it is a sufficient statistic for θ and hence a minimal sufficient statistic.

(a) $b(1, \theta)$, where $0 \le \theta \le 1$.

(b) Poisson with mean $\theta > 0$.

(c) Gamma with $\alpha = 3$ and $\beta = \theta > 0$.

(d) $N(\theta, 1)$, where $-\infty < \theta < \infty$.

(e) $N(0, \theta)$, where $0 < \theta < \infty$.

7.8.2. Let $Y_1 < Y_2 < \cdots < Y_n$ be the order statistics of a random sample of size n from the uniform distribution over the closed interval $[-\theta, \theta]$ having pdf $f(x; \theta) = (1/2\theta) I_{[-\theta, \theta]}(x)$.

(a) Show that Y_1 and Y_n are joint sufficient statistics for θ.

(b) Argue that the mle of θ is $\hat{\theta} = \max(-Y_1, Y_n)$.

(c) Demonstrate that the mle $\hat{\theta}$ is a sufficient statistic for θ and thus is a minimal sufficient statistic for θ.

7.8.3. Let $Y_1 < Y_2 < \cdots < Y_n$ be the order statistics of a random sample of size n from a distribution with pdf

$$ f(x; \theta_1, \theta_2) = \left(\frac{1}{\theta_2} \right) e^{-(x-\theta_1)/\theta_2} I_{(\theta_1, \infty)}(x), $$

where $-\infty < \theta_1 < \infty$ and $0 < \theta_2 < \infty$. Find the joint minimal sufficient statistics for θ_1 and θ_2.

7.8.4. With random samples from each of the distributions given in Exercises 7.8.1(d), 7.8.2, and 7.8.3, define at least two ancillary statistics that are different from the examples given in the text. These examples illustrate, respectively, location invariant, scale invariant, and location and scale invariant statistics.

7.9 Sufficiency, Completeness and Independence

We have noted that if we have a sufficient statistic Y_1 for a parameter θ, $\theta \in \Omega$, then $h(z|y_1)$, the conditional pdf of another statistic Z, given $Y_1 = y_1$, does not depend upon θ. If, moreover, Y_1 and Z are independent, the pdf $g_2(z)$ of Z is such that $g_2(z) = h(z|y_1)$, and hence $g_2(z)$ must not depend upon θ either. So the independence of a statistic Z and the sufficient statistic Y_1 for a parameter θ means that the distribution of Z does not depend upon $\theta \in \Omega$. That is, Z is an ancillary statistic.

It is interesting to investigate a converse of that property. Suppose that the distribution of an ancillary statistic Z does not depend upon θ; then, are Z and the sufficient statistic Y_1 for θ independent? To begin our search for the answer, we know that the joint pdf of Y_1 and Z is $g_1(y_1; \theta) h(z|y_1)$, where $g_1(y_1; \theta)$ and $h(z|y_1)$

represent the marginal pdf of Y_1 and the conditional pdf of Z given $Y_1 = y_1$, respectively. Thus the marginal pdf of Z is

$$\int_{-\infty}^{\infty} g_1(y_1;\theta)h(z|y_1)\,dy_1 = g_2(z),$$

which, by hypothesis, does not depend upon θ. Because

$$\int_{-\infty}^{\infty} g_2(z)g_1(y_1;\theta)\,dy_1 = g_2(z),$$

if follows, by taking the difference of the last two integrals, that

$$\int_{-\infty}^{\infty} [g_2(z) - h(z|y_1)]g_1(y_1;\theta)\,dy_1 = 0 \qquad (7.9.1)$$

for all $\theta \in \Omega$. Since Y_1 is sufficient statistic for θ, $h(z|y_1)$ does not depend upon θ. By assumption, $g_2(z)$ and hence $g_2(z) - h(z|y_1)$ do not depend upon θ. Now if the family $\{g_1(y_1;\theta) : \theta \in \Omega\}$ is complete, Equation (7.9.1) would require that

$$g_2(z) - h(z|y_1) = 0 \quad \text{or} \quad g_2(z) = h(z|y_1).$$

That is, the joint pdf of Y_1 and Z must be equal to

$$g_1(y_1;\theta)h(z|y_1) = g_1(y_1;\theta)g_2(z).$$

Accordingly, Y_1 and Z are independent, and we have proved the following theorem, which was considered in special cases by Neyman and Hogg and proved in general by Basu.

Theorem 7.9.1. *Let $X_1, X_2, \ldots, X_n$ denote a random sample from a distribution having a pdf $f(x;\theta)$, $\theta \in \Omega$, where Ω is an interval set. Suppose that the statistic Y_1 is a complete and sufficient statistic for θ. Let $Z = u(X_1, X_2, \ldots, X_n)$ be any other statistic (not a function of Y_1 alone). If the distribution of Z does not depend upon θ, then Z is independent of the sufficient statistic Y_1.*

In the discussion above, it is interesting to observe that if Y_1 is a sufficient statistic for θ, then the independence of Y_1 and Z implies that the distribution of Z does not depend upon θ whether $\{g_1(y_1;\theta) : \theta \in \Omega\}$ is or is not complete. Conversely, to prove the independence from the fact that $g_2(z)$ does not depend upon θ, we definitely need the completeness. Accordingly, if we are dealing with situations in which we know that family $\{g_1(y_1;\theta) : \theta \in \Omega\}$ is complete (such as a regular case of the exponential class), we can say that the statistic Z is independent of the sufficient statistic Y_1 if and only if the distribution of Z does not depend upon θ(i.e., Z is an ancillary statistic).

It should be remarked that the theorem (including the special formulation of it for regular cases of the exponential class) extends immediately to probability density functions that involve m parameters for which there exist m joint sufficient statistics. For example, let $X_1, X_2, \ldots, X_n$ be a random sample from a distribution having the pdf $f(x; ; \theta_1, \theta_2)$ that represents a regular case of the exponential class so that there are two joint complete sufficient statistics for θ_1 and θ_2. Then any other statistic $Z = u(X_1, X_2, \ldots, X_n)$ is independent of the joint complete sufficient statistics if and only if the distribution of Z does not depend upon θ_1 or θ_2.

We present an example of the theorem that provides an alternative proof of the independence of $\overline{X}$ and S^2, the mean and the variance of a random sample of size n from a distribution that is $N(\mu, \sigma^2)$. This proof is given as if we were unaware that $(n-1)S^2/\sigma^2$ is $\chi^2(n-1)$, because that fact and the independence were established in Theorem 3.6.1.

Example 7.9.1. Let $X_1, X_2, \ldots, X_n$ denote a random sample of size n from a distribution that is $N(\mu, \sigma^2)$. We know that the mean $\overline{X}$ of the sample is for every known σ^2, a complete sufficient statistic for the parameter μ, $-\infty < \mu < \infty$. Consider the statistic

$$S^2 = \frac{1}{n-1} \sum_{i=1}^{n} (X_i - \overline{X})^2,$$

which is location-invariant. Thus S^2 must have a distribution that does not depend upon μ; and hence, by the theorem, S^2 and $\overline{X}$, the complete sufficient statistic for μ, are independent. ∎

Example 7.9.2. Let $X_1, X_2, \ldots, X_n$ be a random sample of size n from the distribution having pdf

$$f(x; \theta) \quad = \quad e^{-(x-\theta)}, \quad \theta < x < \infty, \quad -\infty < \theta < \infty,$$
$$= \quad 0 \quad \text{elsewhere.}$$

Here the pdf is of the form $f(x-\theta)$, where $f(w) = e^{-w}$, $0 < w < \infty$, zero elsewhere. Moreover, we know (Exercise 7.4.5) that the first order statistic $Y_1 = \min(X_i)$ is a complete sufficient statistic for θ. Hence Y_1 must be independent of each location-invariant statistic $u(X_1, X_2, \ldots, X_n)$, enjoying the property that

$$u(x_1 + d, x_2 + d, \ldots, x_n + d) = u(x_1, x_2, \ldots, x_n)$$

for all real d. Illustrations of such statistics are S^2, the sample range, and

$$\frac{1}{n} \sum_{i=1}^{n} [X_i - \min(X_i)]. \quad ∎$$

Example 7.9.3. Let X_1, X_2 denote a random sample of size $n = 2$ from a distribution with pdf

$$f(x; \theta) \quad = \quad \frac{1}{\theta} e^{-x/\theta}, \quad 0 < x < \infty, \quad 0 < \theta < \infty,$$
$$= \quad 0 \quad \text{elsewhere.}$$

The pdf is of the form $(1/\theta)f(x/\theta)$, where $f(w) = e^{-w}$, $0 < w < \infty$, zero elsewhere. We know that $Y_1 = X_1 + X_2$ is a complete sufficient statistic for θ. Hence Y_1 is independent of every scale-invariant statistic $u(X_1, X_2)$ with the property $u(cx_1, cx_2) = u(x_1, x_2)$. Illustrations of these are X_1/X_2 and $X_1/(X_1 + X_2)$, statistics that have F and beta distributions, respectively. ∎

Example 7.9.4. Let $X_1, X_2, \ldots, X_n$ denote a random sample from a distribution that is $N(\theta_1, \theta_2)$, $-\infty < \theta_1 < \infty$, $0 < \theta_2 < \infty$. In Example 7.7.2 it was proved that the mean $\overline{X}$ and the variance S^2 of the sample are joint complete sufficient statistics for θ_1 and θ_2. Consider the statistic

$$Z = \frac{\sum_{1}^{n-1}(X_{i+1} - X_i)^2}{\sum_{1}^{n}(X_i - \overline{X})^2} = u(X_1, X_2, \ldots, X_n),$$

which satisfies the property that $u(cx_1 + d, \ldots, cx_n + d) = u(x_1, \ldots, x_n)$. That is, the ancillary statistic Z is independent of both $\overline{X}$ and S^2. ∎

In this section we have given several examples in which the complete sufficient statistics are independent of ancillary statistics. Thus in those cases, the ancillary statistics provide no information about the parameters. However, if the sufficient statistics are not complete, the ancillary statistics could provide some information as the following example demonstrates.

Example 7.9.5. We refer back to Examples 7.8.1 and 7.8.2. There the first and nth order statistics, Y_1 and Y_n, were minimal sufficient statistics for θ, where the sample arose from an underlying distribution having pdf $(\frac{1}{2})I_{(\theta-1,\theta+1)}(x)$. Often $T_1 = (Y_1 + Y_n)/2$ is used as an estimator of θ as it is a function of those sufficient statistics which is unbiased. Let us find a relationship between T_1 and the ancillary statistic $T_2 = Y_n - Y_1$.

The joint pdf of Y_1 and Y_n is

$$g(y_1, y_n; \theta) = n(n-1)(y_n - y_1)^{n-2}/2^n, \quad \theta - 1 < y_1 < y_n < \theta + 1,$$

zero elsewhere. Accordingly, the joint pdf of T_1 and T_2 is, since the absolute value of the Jacobian equals 1,

$$h(t_1, t_2; \theta) = n(n-1)t_2^{n-2}/2^n, \quad \theta - 1 + \frac{t_2}{2} < t_1 < \theta + 1 - \frac{t_2}{2}, \quad 0 < t_2 < 2,$$

zero elsewhere. Thus the pdf of T_2 is

$$h_2(t_2; \theta) = n(n-1)t_2^{n-2}(2 - t_2)/2^n, \quad 0 < t_2 < 2,$$

zero elsewhere, which of course is free of θ as T_2 is an ancillary statistic. Thus the conditional pdf of T_1, given $T_2 = t_2$, is

$$h_{1|2}(t_1|t_2; \theta) = \frac{1}{2 - t_2}, \quad \theta - 1 + \frac{t_2}{2} < t_1 < \theta + 1 - \frac{t_2}{2}, \quad 0 < t_2 < 2,$$

zero elsewhere. Note that this is uniform on the interval $(\theta - 1 + t_2/2, \theta + 1 - t_2/2)$; so the conditional mean and variance of T_1 are, respectively,

$$E(T_1|t_2) = \theta \quad \text{and} \quad \text{var}(T_1|t_2) = \frac{(2 - t_2)^2}{12}.$$

Given $T_2 = t_2$, we know something about the conditional variance of T_1. In particular, if that observed value of T_2 is large (close to 2), that variance is small and we can place more reliance on the estimator T_1. On the other hand, a small value of t_2 means that we have less confidence in T_1 as an estimator of θ. It is extremely interesting to note that this conditional variance does not depend upon the sample size n but only on the given value of $T_2 = t_2$. As the sample size increases, T_2 tends to becomes larger and, in those cases, T_1 has smaller conditional variance. ∎

While Example 7.9.5 is a special one demonstrating mathematically that an ancillary statistic can provide some help in point estimation, this does actually happen in practice, too. For illustration, we know that if the sample size is large enough, then

$$T = \frac{\overline{X} - \mu}{S/\sqrt{n}}$$

has an approximate standard normal distribution. Of course, if the sample arises from a normal distribution, $\overline{X}$ and S are independent and T has a t-distribution with $n-1$ degrees of freedom. Even if the sample arises from a symmetric distribution, $\overline{X}$ and S are uncorrelated and T has an approximate t-distribution and certainly an approximate standard normal distribution with sample sizes around 30 or 40. On the other hand, if the sample arises from a highly skewed distribution (say to the right), then $\overline{X}$ and S are highly correlated and the probability $P(-1.96 < T < 1.96)$ is not necessarily close to 0.95 unless the sample size is extremely large (certainly much greater than 30). Intuitively, one can understand why this correlation exists if the underlying distribution is highly skewed to the right. While S has a distribution free of μ (and hence is an ancillary), a large value of S implies a large value of $\overline{X}$, since the underlying pdf is like the one depicted in Figure 7.9.1. Of course, a small value of $\overline{X}$ (say less than the mode) requires a relatively small value of S. This means that unless n is extremely large, it is risky to say that

$$\overline{x} - \frac{1.96s}{\sqrt{n}}, \quad \overline{x} + \frac{1.96s}{\sqrt{n}}$$

provides an approximate 95 percent confidence interval with data from a very skewed distribution. As a matter of fact, the authors have seen situations in which this confidence coefficient is closer to 80 percent, rather than 95 percent, with sample sizes of 30 to 40.

EXERCISES

7.9.1. Let $Y_1 < Y_2 < Y_3 < Y_4$ denote the order statistics of a random sample of size $n = 4$ from a distribution having pdf $f(x; \theta) = 1/\theta$, $0 < x < \theta$, zero elsewhere, where $0 < \theta < \infty$. Argue that the complete sufficient statistic Y_4 for θ is independent of each of the statistics Y_1/Y_4 and $(Y_1 + Y_2)/(Y_3 + Y_4)$.
Hint: Show that the pdf is of the form $(1/\theta)f(x/\theta)$, where $f(w) = 1$, $0 < w < 1$, zero elsewhere.

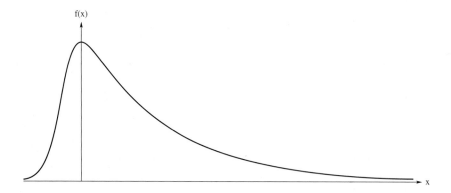

Figure 7.9.1: Graph of a Right Skewed Distribution; see also, Exercise 7.9.14

7.9.2. Let $Y_1 < Y_2 < \cdots < Y_n$ be the order statistics of a random sample from a $N(\theta, \sigma^2)$, $-\infty < \theta < \infty$, distribution. Show that the distribution of $Z = Y_n - \overline{X}$ does not depend upon θ. Thus $\overline{Y} = \sum_1^n Y_i/n$, a complete sufficient statistic for θ is independent of Z.

7.9.3. Let $X_1, X_2, \ldots, X_n$ be iid with the distribution $N(\theta, \sigma^2)$, $-\infty < \theta < \infty$. Prove that a necessary and sufficient condition that the statistics $Z = \sum_1^n a_i X_i$ and $Y = \sum_1^n X_i$, a complete sufficient statistic for θ, are independent is that $\sum_1^n a_i = 0$.

7.9.4. Let X and Y be random variables such that $E(X^k)$ and $E(Y^k) \neq 0$ exist for $k = 1, 2, 3, \ldots$. If the ratio X/Y and its denominator Y are independent, prove that $E[(X/Y)^k] = E(X^k)/E(Y^k), k = 1, 2, 3, \ldots$.
Hint: Write $E(X^k) = E[Y^k(X/Y)^k]$.

7.9.5. Let $Y_1 < Y_2 < \cdots < Y_n$ be the order statistics of a random sample of size n from a distribution that has pdf $f(x; \theta) = (1/\theta)e^{-x/\theta}$, $0 < x < \infty$, $0 < \theta < \infty$, zero elsewhere. Show that the ratio $R = nY_1 \left/ \sum_1^n Y_i \right.$ and its denominator (a complete sufficient statistic for θ) are independent. Use the result of the preceding exercise to determine $E(R^k)$, $k = 1, 2, 3, \ldots$.

7.9.6. Let $X_1, X_2, \ldots, X_5$ be iid with pdf $f(x) = e^{-x}$, $0 < x < \infty$, zero elsewhere. Show that $(X_1 + X_2)/(X_1 + X_2 + \cdots + X_5)$ and its denominator are independent.
Hint: The pdf $f(x)$ is a member of $\{f(x; \theta) : 0 < \theta < \infty\}$, where $f(x; \theta) = (1/\theta)e^{-x/\theta}$, $0 < x < \infty$, zero elsewhere.

7.9.7. Let $Y_1 < Y_2 < \cdots < Y_n$ be the order statistics of a random sample from the normal distribution $N(\theta_1, \theta_2)$, $-\infty < \theta_1 < \infty$, $0 < \theta_2 < \infty$. Show that the joint complete sufficient statistics $\overline{X} = \overline{Y}$ and S^2 for θ_1 and θ_2 are independent of each of $(Y_n - \overline{Y})/S$ and $(Y_n - Y_1)/S$.

7.9.8. Let $Y_1 < Y_2 < \cdots < Y_n$ be the order statistics of a random sample from a distribution with the pdf

$$f(x; \theta_1, \theta_2) = \frac{1}{\theta_2} \exp\left(-\frac{x - \theta_1}{\theta_2}\right),$$

$\theta_1 < x < \infty$, zero elsewhere, where $-\infty < \theta_1 < \infty$, $0 < \theta_2 < \infty$. Show that the joint complete sufficient statistics Y_1 and $\overline{X} = \overline{Y}$ for for the parameters θ_1 and θ_2 are independent of $(Y_2 - Y_1)\bigg/ \displaystyle\sum_1^n (Y_i - Y_1)$.

7.9.9. Let $X_1, X_2, \ldots, X_5$ be a random sample of size $n = 5$ from the normal distribution $N(0, \theta)$.

(a) Argue that the ratio $R = (X_1^2 + X_2^2)/(X_1^2 + \cdots + X_5^2)$ and its denominator $(X_1^2 + \cdots + X_5^2)$ are independent.

(b) Does $5R/2$ have an F-distribution with 2 and 5 degrees of freedom? Explain your answer.

(c) Compute $E(R)$ using Exercise 7.9.4.

7.9.10. Referring to Example 7.9.5 of this section, determine c so that

$$P(-c < T_1 - \theta < c | T_2 = t_2) = 0.95.$$

Use this result to find a 95 percent confidence interval for θ, given $T_2 = t_2$; and note how its length is smaller when the range of t_2 is larger.

7.9.11. Show that $Y = |X|$ is a complete sufficient statistic for $\theta > 0$, where X has the pdf $f_X(x; \theta) = 1/(2\theta)$, for $-\theta < x < \theta$, zero elsewhere. Show that $Y = |X|$ and $Z = \text{sgn}(X)$ are independent.

7.9.12. Let $Y_1 < Y_2 < \cdots < Y_n$ be the order statistics of a random sample from a $N(\theta, \sigma^2)$ distribution, where σ^2 is fixed but arbitrary. Then $\overline{Y} = \overline{X}$ is a complete sufficient statistic for θ. Consider another estimator T of θ, such as $T = (Y_i + Y_{n+1-i})/2$, for $i = 1, 2, \ldots, [n/2]$ or T could be any weighted average of these latter statistics.

(a) Argue that $T - \overline{X}$ and $\overline{X}$ are independent random variables.

(b) Show that $\text{Var}(T) = \text{Var}(\overline{X}) + \text{Var}(T - \overline{X})$.

(c) Since we know the $\text{Var}(\overline{X}) = \sigma^2/n$, it might be more efficient to estimate the $\text{Var}(T)$ by estimating the $\text{Var}(T - \overline{X})$ by Monte Carlo methods rather than doing that with $\text{Var}(T)$ directly, because $\text{Var}(T) \geq \text{Var}(T - \overline{X})$. This is often called the *Monte Carlo Swindle*.

7.9.13. Suppose $X_1, X_2, \ldots, X_n$ is a random sample from a distribution with pdf $f(x; \theta) = (1/2)\theta^3 x^2 e^{-\theta x}$, $0 < x < \infty$, zero elsewhere, where $0 < \theta < \infty$:

(a) Find the mle, $\hat{\theta}$, of θ. Is $\hat{\theta}$ unbiased?

   Hint: Find the pdf of $Y = \sum_1^n X_i$ and then compute $E(\hat{\theta})$.

(b) Argue that Y is a complete sufficient statistic for θ.

(c) Find the MVUE of θ.

(d) Show that X_1/Y and Y are independent.

(e) What is the distribution of X_1/Y?

7.9.14. The pdf depicted in Figure 7.9.1 is given by

$$f_{m_2}(x) = e^x (1 + m_2^{-1} e^x)^{-(m_2+1)}, \quad -\infty < x < \infty, \tag{7.9.2}$$

where $m_2 > 0$, (the pdf graphed is for $m_2 = 0.1$). This is a member of a large family of pdfs, log F-family, which are useful in survival (lifetime) analysis; see Chapter 3 of Hettmansperger and McKean (1998).

(a) Let W be a random variable with pdf (7.9.2). Show that $W = \log Y$, where Y has an F-distribution with 2 and $2m_2$ degrees of freedom.

(b) Show that the pdf becomes the logistic (6.1.8) if $m_2 = 1$.

(c) Consider the location model where

$$X_i = \theta + W_i \quad i = 1, \ldots, n,$$

where $W_1, \ldots, W_n$ are iid with pdf (7.9.2). Similar to the logistic location model, the order statistics are minimal sufficient for this model. Show, similar to Example 6.1.4, that the mle of θ exists.

Chapter 8

Optimal Tests of Hypotheses

8.1 Most Powerful Tests

In Section 5.5 we introduced the concept of hypotheses testing and followed it with the introduction of likelihood ratio tests in Chapter 6. In this chapter we discuss certain best tests.

We are interested in a random variable X which has pdf or pmf $f(x; \theta)$ where $\theta \in \Omega$. We assume that $\theta \in \omega_0$ or $\theta \in \omega_1$ where ω_0 and ω_1 are subsets of Ω and $\omega_0 \cup \omega_1 = \Omega$. We label the hypotheses as

$$H_0 : \theta \in \omega_0 \text{ versus } H_1 : \theta \in \omega_1. \qquad (8.1.1)$$

The hypothesis H_0 is referred to as the **null hypothesis** while H_1 is referred to as the **alternative hypothesis**. The test of H_0 versus H_1 is based on a sample $X_1, \ldots, X_n$ from the distribution of X. In this chapter we will often use the vector $\mathbf{X}' = (X_1, \ldots, X_n)$ to denote the random sample and $\mathbf{x}' = (x_1, \ldots, x_n)$ to denote the values of the sample. Let $\mathcal{S}$ denote the support of the random sample $\mathbf{X}' = (X_1, \ldots, X_n)$.

A **test** of H_0 versus H_1 is based on a subset C of $\mathcal{S}$. This set C is called the **critical region** and its corresponding decision rule is:

$$\begin{aligned} \text{Reject } H_0, \text{ (Accept } H_1), &\qquad \text{if } \mathbf{X} \in C \qquad (8.1.2) \\ \text{Retain } H_0, \text{ (Reject } H_1), &\qquad \text{if } \mathbf{X} \in C^c. \end{aligned}$$

Note that a test is defined by its critical region. Conversely a critical region defines a test.

Recall that the 2×2 decision table, Table 5.5.1, summarizes the results of the hypothesis test in terms of the true state of nature. Besides the correct decisions, two errors can occur. A **Type I** error occurs if H_0 is rejected when it is true while a **Type II** error occurs if H_0 is accepted when H_1 is true. The **size** or **significance level** of the test is the probability of a Type I error, i.e.,

$$\alpha = \max_{\theta \in \omega_0} P_\theta(\mathbf{X} \in C). \qquad (8.1.3)$$

Note that $P_\theta(\mathbf{X} \in C)$ should be read as the probability that $\mathbf{X} \in C$ when θ is the true parameter. Subject to tests having size α, we select tests that minimize Type II error or equivalently maximize the probability of rejecting H_0 when $\theta \in \omega_1$. Recall that the **power function** of a test is given by

$$\gamma_C(\theta) = P_\theta(\mathbf{X} \in C) ; \quad \theta \in \omega_1. \tag{8.1.4}$$

In Chapter 5, we gave examples of tests of hypotheses, while in Sections 6.3 and 6.4, we discussed tests based on maximum likelihood theory. In this chapter we want to construct best tests for certain situations.

We begin with testing a simple hypothesis H_0 against a simple alternative H_1. Let $f(x; \theta)$ denote the pdf or pmf of a random variable X where $\theta \in \Omega = \{\theta', \theta''\}$. Let $\omega_0 = \{\theta'\}$ and $\omega_1 = \{\theta''\}$. Let $\mathbf{X}' = (X_1, \ldots, X_n)$ be a random sample from the distribution of X. We now define a best critical region (and hence a best test) for testing the simple hypothesis H_0 against the alternative simple hypothesis H_1.

Definition 8.1.1. *Let C denote a subset of the sample space. Then we say that C is a **best critical region** of size α for testing the simple hypothesis $H_0 : \theta = \theta'$ against the alternative simple hypothesis $H_1 : \theta = \theta''$ if,*

(a) $P_{\theta'}[\mathbf{X} \in C] = \alpha.$

(b) *and for every subset A of the sample space,*

$$P_{\theta'}[\mathbf{X} \in A] = \alpha \Rightarrow P_{\theta''}[\mathbf{X} \in C] \geq P_{\theta''}[\mathbf{X} \in A].$$

This definition states, in effect, the following: In general, there will be a multiplicity of subsets A of the sample space such that $P_{\theta'}[\mathbf{X} \in A] = \alpha$. Suppose that there is one of these subsets, say C, such that when H_1 is true, the power of the test associated with C is at least as great as the power of the test associated with each other A. Then C is defined as a best critical region of size α for testing H_0 against H_1.

As Theorem 8.1.1 shows, there is a best test for this simple versus simple case. But first, we offer a simple example examining this definition in some detail.

Example 8.1.1. Consider the one random variable X that has a binomial distribution with $n = 5$ and $p = \theta$. Let $f(x; \theta)$ denote the pmf of X and let $H_0 : \theta = \frac{1}{2}$ and $H_1 : \theta = \frac{3}{4}$. The following tabulation gives, at points of positive probability density, the value of $f(x; \frac{1}{2})$, $f(x; \frac{3}{4})$, and the ratio $f(x; \frac{1}{2})/f(x; \frac{3}{4})$.

x	0	1	2
$f(x; 1/2)$	1/32	5/32	10/32
$f(x; 3/4)$	1/1024	15/1024	90/1024
$f(x; 1/2)/f(x; 3/4)$	32/1	32/3	32/9

x	3	4	5
$f(x; 1/2)$	10/32	5/32	1/32
$f(x; 3/4)$	270/1024	405/1024	243/1024
$f(x; 1/2)/f(x; 3/4)$	32/27	32/81	32/243

We shall use one random value of X to test the simple hypothesis $H_0 : \theta = \frac{1}{2}$ against the alternative simple hypothesis $H_1 : \theta = \frac{3}{4}$, and we shall first assign the significance level of the test to be $\alpha = \frac{1}{32}$. We seek a best critical region of size $\alpha = \frac{1}{32}$. If $A_1 = \{x : x = 0\}$ or $A_2 = \{x : x = 5\}$, then $P_{\{\theta=1/2\}}(X \in A_1) = P_{\{\theta=1/2\}}(X \in A_2) = \frac{1}{32}$ and there is no other subset A_3 of the space $\{x : x = 0, 1, 2, 3, 4, 5\}$ such that $P_{\{\theta=1/2\}}(X \in A_3) = \frac{1}{32}$. Then either A_1 or A_2 is the best critical region C of size $\alpha = \frac{1}{32}$ for testing H_0 against H_1. We note that $P_{\{\theta=1/2\}}(X \in A_1) = \frac{1}{32}$ and $P_{\{\theta=3/4\}}(X \in A_1) = \frac{1}{1024}$. Thus, if the set A_1 is used as a critical region of size $\alpha = \frac{1}{32}$, we have the intolerable situation that the probability of rejecting H_0 when H_1 is true (H_0 is false) is much less than the probability of rejecting H_0 when H_0 is true.

On the other hand, if the set A_2 is used as a critical region, then $P_{\{\theta=1/2\}}(X \in A_2) = \frac{1}{32}$ and $P_{\{\theta=3/4\}}(X \in A_2) = \frac{243}{1024}$. That is, the probability of rejecting H_0 when H_1 is true is much greater than the probability of rejecting H_0 when H_0 is true. Certainly, this is a more desirable state of affairs, and actually A_2 is the best critical region of size $\alpha = \frac{1}{32}$. The latter statement follows from the fact that when H_0 is true, there are but two subsets, A_1 and A_2, of the sample space, each of whose probability measure is $\frac{1}{32}$ and the fact that

$$\frac{243}{1024} = P_{\{\theta=3/4\}}(X \in A_2) > P_{\{\theta=3/4\}}(X \in A_1) = \frac{1}{1024}.$$

It should be noted in this problem that the best critical region $C = A_2$ of size $\alpha = \frac{1}{32}$ is found by including in C the point (or points) at which $f(x; \frac{1}{2})$ is *small* in comparison with $f(x; \frac{3}{4})$. This is seen to be true once it is observed that the ratio $f(x; \frac{1}{2})/f(x; \frac{3}{4})$ is a minimum at $x = 5$. Accordingly, the ratio $f(x; \frac{1}{2})/f(x; \frac{3}{4})$, which is given in the last line of the above tabulation, provides us with a precise tool by which to find a best critical region C for certain given values of α. To illustrate this, take $\alpha = \frac{6}{32}$. When H_0 is true, each of the subsets $\{x : x = 0, 1\}$, $\{x : x = 0, 4\}$, $\{x : x = 1, 5\}$, $\{x : x = 4, 5\}$ has probability measure $\frac{6}{32}$. By direct computation it is found that the best critical region of this size is $\{x : x = 4, 5\}$. This reflects the fact that the ratio $f(x; \frac{1}{2})/f(x; \frac{3}{4})$ has its two smallest values for $x = 4$ and $x = 5$. The power of this test, which has $\alpha = \frac{6}{32}$ is

$$P_{\{\theta=3/4\}}(X = 4, 5) = \frac{405}{1024} + \frac{243}{1024} = \frac{648}{1024} \quad \blacksquare$$

The preceding example should make the following theorem, due to Neyman and Pearson, easier to understand. It is an important theorem because it provides a systematic method of determining a best critical region.

Theorem 8.1.1. Neyman-Pearson Theorem. *Let $X_1, X_2, \ldots, X_n$, where n is a fixed positive integer, denote a random sample from a distribution that has pdf or pmf $f(x; \theta)$. Then the likelihood of $X_1, X_2, \ldots, X_n$ is*

$$L(\theta; \mathbf{x}) = \prod_{i=1}^{n} f(x_i; \theta), \quad \text{for } \mathbf{x}' = (x_1, \ldots, x_n).$$

Let θ' and θ'' be distinct fixed values of θ so that $\Omega = \{\theta : \theta = \theta', \theta''\}$, and let k be a positive number. Let C be a subset of the sample space such that:

(a) $\dfrac{L(\theta'; \mathbf{x})}{L(\theta''; \mathbf{x})} \leq k$, *for each point* $\mathbf{x} \in C$.

(b) $\dfrac{L(\theta'; \mathbf{x})}{L(\theta''; \mathbf{x})} \geq k$, *for each point* $\mathbf{x} \in C^c$.

(c) $\alpha = P_{H_0}[\mathbf{X} \in C]$.

Then C is a best critical region of size α for testing the simple hypothesis $H_0 : \theta = \theta'$ against the alternative simple hypothesis $H_1 : \theta = \theta''$.

Proof: We shall give the proof when the random variables are of the continuous type. If C is the only critical region of size α, the theorem is proved. If there is another critical region of size α, denote it by A. For convenience, we shall let $\int \cdots \int_R L(\theta; x_1, \ldots, x_n)\, dx_1 \cdots dx_n$ be denoted by $\int_R L(\theta)$. In this notation we wish to show that

$$\int_C L(\theta'') - \int_A L(\theta'') \geq 0.$$

Since C is the union of the disjoint sets $C \cap A$ and $C \cap A^c$ and A is the union of the disjoint sets $A \cap C$ and $A \cap C^c$, we have

$$\int_C L(\theta'') - \int_A L(\theta'') = \int_{C \cap A} L(\theta'') + \int_{C \cap A^c} L(\theta'') - \int_{A \cap C} L(\theta'') - \int_{A \cap C^c} L(\theta'')$$

$$= \int_{C \cap A^c} L(\theta'') - \int_{A \cap C^c} L(\theta''). \qquad (8.1.5)$$

However, by the hypothesis of the theorem, $L(\theta'') \geq (1/k) L(\theta')$ at each point of C, and hence at each point of $C \cap A^c$; thus

$$\int_{C \cap A^c} L(\theta'') \geq \frac{1}{k} \int_{C \cap A^c} L(\theta').$$

But $L(\theta'') \leq (1/k) L(\theta')$ at each point of C^c, and hence at each point of $A \cap C^c$; accordingly,

$$\int_{A \cap C^c} L(\theta'') \leq \frac{1}{k} \int_{A \cap C^c} L(\theta').$$

These inequalities imply that

$$\int_{C \cap A^c} L(\theta'') - \int_{A \cap C^c} L(\theta'') \geq \frac{1}{k} \int_{C \cap A^c} L(\theta') - \frac{1}{k} \int_{A \cap C^c} L(\theta');$$

and, from Equation (8.1.5), we obtain

$$\int_C L(\theta'') - \int_A L(\theta'') \geq \frac{1}{k} \left[\int_{C \cap A^c} L(\theta') - \int_{A \cap C^c} L(\theta') \right]. \qquad (8.1.6)$$

However,

$$
\int_{C \cap A^c} L(\theta') - \int_{A \cap C^c} L(\theta') = \int_{C \cap A^c} L(\theta') + \int_{C \cap A} L(\theta')
$$
$$
- \int_{A \cap C} L(\theta') - \int_{A \cap C^c} L(\theta')
$$
$$
= \int_C L(\theta') - \int_A L(\theta') = \alpha - \alpha = 0.
$$

If this result is substituted in inequality (8.1.6), we obtain the desired result,

$$
\int_C L(\theta'') - \int_A L(\theta'') \geq 0.
$$

If the random variables are of the discrete type, the proof is the same with integration replaced by summation. ∎

Remark 8.1.1. As stated in the theorem, conditions (a), (b), and (c) are sufficient ones for region C to be a best critical region of size α. However, they are also necessary. We discuss this briefly. Suppose there is a region A of size α that does not satisfy (a) and (b) and that is as powerful at $\theta = \theta''$ as C, which satisfies (a), (b), and (c). Then expression (8.1.5) would be zero, since the power at θ'' using A is equal to that using C. It can be proved that to have expression (8.1.5) equal zero A must be of the same form as C. As a matter of fact in the continuous case, A and C would essentially be the same region; that is, they could differ only by a set having probability zero. However, in the discrete case, if $P_{H_0}[L(\theta') = kL(\theta'')]$ is positive, A and C could be different sets, but each would necessarily enjoy conditions (a), (b), and (c) to be a best critical region of size α. ∎

It would seem that a test should have the property that its power should never fall below its significance level; otherwise, the probability of falsely rejecting H_0 (level) is higher than the probability of correctly rejecting H_0 (power). As the next corollary shows, the most powerful test does satisfy this property. In Section 8.3 we will say the best test is unbiased.

Corollary 8.1.1. *As in Theorem 8.1.1, let C be the critical region of the best test of $H_0 : \theta = \theta'$ versus $H_1 : \theta = \theta''$. Suppose the significance level of the test is α. Let $\gamma_C(\theta'') = P_{\theta''}[\mathbf{X} \in C]$ denote the power of the test. Then $\alpha \leq \gamma_C(\theta'')$.*

Proof: Consider the "unreasonable" test in which the data are ignored, but a Bernoulli trial is performed which has probability α of success. If the trial ends in success, we reject H_0. The level of this test is α. Because the power of a test is the probability of rejecting H_0 when H_1 is true, the power of this unreasonable test is α also. But C is the best critical region of size α and thus has power greater than or equal to the power of the unreasonable test. That is, $\gamma_C(\theta'') \geq \alpha$, which is the desired result. ∎

Another aspect of Theorem 8.1.1 to be emphasized is that if we take C to be the set of all points $\mathbf{x}$ which satisfy

$$\frac{L(\theta'; \mathbf{x})}{L(\theta''; \mathbf{x})} \leq k, \quad k > 0,$$

then, in accordance with the theorem, C will be a best critical region. This inequality can frequently be expressed in one of the forms (where c_1 and c_2 are constants)

$$u_1(\mathbf{x}; \theta', \theta'') \leq c_1,$$

or

$$u_2(\mathbf{x}; \theta', \theta'') \geq c_2.$$

Suppose that it is the first form, $u_1 \leq c_1$. Since θ' and θ'' are given constants, $u_1(\mathbf{X}; \theta', \theta'')$ is a statistic; and if the pdf or pmf of this statistic can be found when H_0 is true, then the significance level of the test of H_0 against H_1 can be determined from this distribution. That is,

$$\alpha = P_{H_0}[u_1(\mathbf{X}; \theta', \theta'') \leq c_1].$$

Moreover, the test may be based on this statistic; for if the observed vector values of $\mathbf{X}$ is $\mathbf{x}$, we reject H_0 (accept H_1) if $u_1(\mathbf{x}) \leq c_1$.

A positive number k determines a best critical region C whose size is $\alpha = P_{H_0}[\mathbf{X} \in C]$ for that particular k. It may be that this value of α is unsuitable for the purpose at hand; that is, it is too large or too small. However, if there is a statistic $u_1(\mathbf{X})$ as in the preceding paragraph, whose pdf or pmf can be determined when H_0 is true, we need not experiment with various values of k to obtain a desirable significance level. For if the distribution of the statistic is known, or can be found, we may determine c_1 such that $P_{H_0}[u_1(\mathbf{X}) \leq c_1]$ is a desirable significance level.

An illustrative example follows.

Example 8.1.2. Let $\mathbf{X}' = (X_1, \ldots, X_n)$ denote a random sample from the distribution that has the pdf

$$f(x; \theta) = \frac{1}{\sqrt{2\pi}} \exp\left(-\frac{(x-\theta)^2}{2}\right), \quad -\infty < x < \infty.$$

It is desired to test the simple hypothesis $H_0 : \theta = \theta' = 0$ against the alternative simple hypothesis $H_1 : \theta = \theta'' = 1$. Now

$$\frac{L(\theta'; \mathbf{x})}{L(\theta''; \mathbf{x})} = \frac{(1/\sqrt{2\pi})^n \exp\left[-\sum_1^n x_i^2 / 2\right]}{(1/\sqrt{2\pi})^n \exp\left[-\sum_1^n (x_i - 1)^2 / 2\right]}$$

$$= \exp\left(-\sum_1^n x_i + \frac{n}{2}\right).$$

If $k > 0$, the set of all points $(x_1, x_2, \ldots, x_n)$ such that

$$\exp\left(-\sum_1^n x_i + \frac{n}{2}\right) \leq k$$

is a best critical region. This inequality holds if and only if

$$-\sum_1^n x_i + \frac{n}{2} \leq \log k$$

or, equivalently,

$$\sum_1^n x_i \geq \frac{n}{2} - \log k = c.$$

In this case, a best critical region is the set $C = \left\{(x_1, x_2, \ldots, x_n) : \sum_1^n x_i \geq c\right\}$, where c is a constant that can be determined so that the size of the critical region is a desired number α. The event $\sum_1^n X_i \geq c$ is equivalent to the event $\overline{X} \geq c/n = c_1$, for example, so the test may be based upon the statistic $\overline{X}$. If H_0 is true, that is, $\theta = \theta' = 0$, then $\overline{X}$ has a distribution that is $N(0, 1/n)$. For a given positive integer n, the size of the sample and a given significance level α, the number c_1 can be found from Table III in Appendix B, so that $P_{H_0}(\overline{X} \geq c_1) = \alpha$. Hence, if the experimental values of $X_1, X_2, \ldots, X_n$ were, respectively, $x_1, x_2, \ldots, x_n$, we would compute $\overline{x} = \sum_1^n x_i/n$. If $\overline{x} \geq c_1$, the simple hypothesis $H_0 : \theta = \theta' = 0$ would be rejected at the significance level α; if $\overline{x} < c_1$, the hypothesis H_0 would be accepted. The probability of rejecting H_0 when H_0 is true is α; the probability of rejecting H_0, when H_0 is false, is the value of the power of the test at $\theta = \theta'' = 1$. That is

$$P_{H_1}(\overline{X} \geq c_1) = \int_{c_1}^{\infty} \frac{1}{\sqrt{2\pi}\sqrt{1/n}} \exp\left[-\frac{(\overline{x}-1)^2}{2(1/n)}\right] d\overline{x}.$$

For example, if $n = 25$ and if α is selected to be 0.05, then from Table III we find $c_1 = 1.645/\sqrt{25} = 0.329$. Thus the power of this best test of H_0 against H_1 is 0.05 when H_0 is true, and is

$$\int_{0.329}^{\infty} \frac{1}{\sqrt{2\pi}\sqrt{\frac{1}{25}}} \exp\left[-\frac{(\overline{x}-1)^2}{2(\frac{1}{25})}\right] d\overline{x} = \int_{-3.355}^{\infty} \frac{1}{\sqrt{2\pi}} e^{-w^2/2}\, dw = 0.999+,$$

when H_1 is true. ∎

There is another aspect of this theorem that warrants special mention. It has to do with the number of parameters that appear in the pdf. Our notation suggests that there is but one parameter. However, a careful review of the proof will reveal that nowhere was this needed or assumed. The pdf or pmf may depend upon any finite number of parameters. What is essential is that the hypothesis H_0 and the alternative hypothesis H_1 be simple, namely that they completely specify the distributions. With this in mind, we see that the simple hypotheses H_0 and H_1 do not need to be hypotheses about the parameters of a distribution, nor, as a matter of fact, do the random variables $X_1, X_2, \ldots, X_n$ need to be independent. That is, if H_0 is the simple hypothesis that the joint pdf or pmf is $g(x_1, x_2, \ldots, x_n)$, and if H_1 is the alternative simple hypothesis that the joint pdf or pmf is $h(x_1, x_2, \ldots, x_n)$, then C is a best critical region of size α for testing H_0 against H_1 if, for $k > 0$:

1. $\dfrac{g(x_1, x_2, \ldots, x_n)}{h(x_1, x_2, \ldots, x_n)} \leq k$ for $(x_1, x_2, \ldots, x_n) \in C$.

2. $\dfrac{g(x_1, x_2, \ldots, x_n)}{h(x_1, x_2, \ldots, x_n)} \geq k$ for $(x_1, x_2, \ldots, x_n) \in C^c$.

3. $\alpha = P_{H_0}[(X_1, X_2, \ldots, X_n) \in C]$.

An illustrative example follows.

Example 8.1.3. Let $X_1, \ldots, X_n$ denote a random sample from a distribution which has a pmf $f(x)$ that is positive on and only on the nonnegative integers. It is desired to test the simple hypothesis

$$H_0 : f(x) = \begin{cases} \frac{e^{-1}}{x!} & x = 0, 1, 2, \ldots \\ 0 & \text{elsewhere,} \end{cases}$$

against the alternative simple hypothesis

$$H_1 : f(x) = \begin{cases} (\frac{1}{2})^{x+1} & x = 0, 1, 2, \ldots \\ 0 & \text{elsewhere.} \end{cases}$$

Here

$$\frac{g(x_1, \ldots, x_n)}{h(x_1, \ldots, x_n)} = \frac{e^{-n}/(x_1! x_2! \cdots x_n!)}{(\frac{1}{2})^n (\frac{1}{2})^{x_1 + x_2 + \cdots + x_n}}$$

$$= \frac{(2e^{-1})^n 2^{\sum x_i}}{\prod_1^n (x_i!)}.$$

If $k > 0$, the set of points $(x_1, x_2, \ldots, x_n)$ such that

$$\left(\sum_1^n x_i \right) \log 2 - \log \left[\prod_1^n (x_i!) \right] \leq \log k - n \log(2e^{-1}) = c$$

is a best critical region C. Consider the case of $k = 1$ and $n = 1$. The preceding inequality may be written $2^{x_1}/x_1! \leq e/2$. This inequality is satisfied by all points in the set $C = \{x_1 : x_1 = 0, 3, 4, 5, \ldots\}$. Thus the power of the test when H_0 is true is

$$P_{H_0}(X_1 \in C) = 1 - P_{H_0}(X_1 = 1, 2) = 0.448,$$

approximately, in accordance with Table I of Appendix B; i.e., the level of this test is 0.448. The power of the test when H_1 is true is given by

$$P_{H_1}(X_1 \in C) = 1 - P_{H_1}(X_1 = 1, 2) = 1 - (\tfrac{1}{4} + \tfrac{1}{8}) = 0.625. \quad \blacksquare$$

Note that these results are consistent with Corollary 8.1.1.

Remark 8.1.2. In the notation of this section, say C is a critical region such that

$$\alpha = \int_C L(\theta') \quad \text{and} \quad \beta = \int_{C^c} L(\theta''),$$

where α and β equal the respective probabilities of the Type I and Type II errors associated with C. Let d_1 and d_2 be two given positive constants. Consider a certain linear function of α and β, namely

$$\begin{aligned}
d_1 \int_C L(\theta') + d_2 \int_{C^c} L(\theta'') &= d_1 \int_C L(\theta') + d_2 \left[1 - \int_C L(\theta'')\right] \\
&= d_2 + \int_C [d_1 L(\theta') - d_2 L(\theta'')].
\end{aligned}$$

If we wished to minimize this expression, we would select C to be the set of all $(x_1, x_2, \ldots, x_n)$ such that

$$d_1 L(\theta') - d_2 L(\theta'') < 0$$

or, equivalently,

$$\frac{L(\theta')}{L(\theta'')} < \frac{d_2}{d_1}, \quad \text{for all } (x_1, x_2, \ldots, x_n) \in C,$$

which according to the Neyman-Pearson theorem provides a best critical region with $k = d_2/d_1$. That is, this critical region C is one that minimizes $d_1\alpha + d_2\beta$. There could be others, including points on which $L(\theta')/L(\theta'') = d_2/d_1$, but these would still be best critical regions according to the Neyman-Pearson theorem. $\quad \blacksquare$

EXERCISES

8.1.1. In Example 8.1.2 of this section, let the simple hypotheses read $H_0 : \theta = \theta' = 0$ and $H_1 : \theta = \theta'' = -1$. Show that the best test of H_0 against H_1 may be carried out by use of the statistic $\overline{X}$, and that if $n = 25$ and $\alpha = 0.05$, the power of the test is 0.999+ when H_1 is true.

8.1.2. Let the random variable X have the pdf $f(x; \theta) = (1/\theta)e^{-x/\theta}$, $0 < x < \infty$, zero elsewhere. Consider the simple hypothesis $H_0 : \theta = \theta' = 2$ and the alternative hypothesis $H_1 : \theta = \theta'' = 4$. Let X_1, X_2 denote a random sample of size 2 from this distribution. Show that the best test of H_0 against H_1 may be carried out by use of the statistic $X_1 + X_2$.

8.1.3. Repeat Exercise 8.1.2 when $H_1 : \theta = \theta'' = 6$. Generalize this for every $\theta'' > 2$.

8.1.4. Let $X_1, X_2, \ldots, X_{10}$ be a random sample of size 10 from a normal distribution $N(0, \sigma^2)$. Find a best critical region of size $\alpha = 0.05$ for testing $H_0 : \sigma^2 = 1$ against $H_1 : \sigma^2 = 2$. Is this a best critical region of size 0.05 for testing $H_0 : \sigma^2 = 1$ against $H_1 : \sigma^2 = 4$? Against $H_1 : \sigma^2 = \sigma_1^2 > 1$?

8.1.5. If $X_1, X_2, \ldots, X_n$ is a random sample from a distribution having pdf of the form $f(x; \theta) = \theta x^{\theta-1}$, $0 < x < 1$, zero elsewhere, show that a best critical region for testing $H_0 : \theta = 1$ against $H_1 : \theta = 2$ is $C = \left\{ (x_1, x_2, \ldots, x_n) : c \leq \prod_{i=1}^{n} x_i \right\}$.

8.1.6. Let $X_1, X_2, \ldots, X_{10}$ be a random sample from a distribution that is $N(\theta_1, \theta_2)$. Find a best test of the simple hypothesis $H_0 : \theta_1 = \theta_1' = 0$, $\theta_2 = \theta_2' = 1$ against the alternative simple hypothesis $H_1 : \theta_1 = \theta_1'' = 1$, $\theta_2 = \theta_2'' = 4$.

8.1.7. Let $X_1, X_2, \ldots, X_n$ denote a random sample from a normal distribution $N(\theta, 100)$. Show that $C = \left\{ (x_1, x_2, \ldots, x_n) : c \leq \overline{x} = \sum_{1}^{n} x_i/n \right\}$ is a best critical region for testing $H_0 : \theta = 75$ against $H_1 : \theta = 78$. Find n and c so that

$$P_{H_0}[(X_1, X_2, \ldots, X_n) \in C] = P_{H_0}(\overline{X} \geq c) = 0.05$$

and

$$P_{H_1}[(X_1, X_2, \ldots, X_n) \in C] = P_{H_1}(\overline{X} \geq c) = 0.90,$$

approximately.

8.1.8. If $X_1, X_2, \ldots, X_n$ is a random sample from a beta distribution with parameters $\alpha = \beta = \theta > 0$, find a best critical region for testing $H_0 : \theta = 1$ against $H_1 : \theta = 2$.

8.1.9. Let $X_1, X_2, \ldots, X_n$ be iid with pmf $f(x; p) = p^x(1-p)^{1-x}$, $x = 0, 1$, zero elsewhere. Show that $C = \left\{ (x_1, \ldots, x_n) : \sum_{1}^{n} x_i \leq c \right\}$ is a best critical region for testing $H_0 : p = \frac{1}{2}$ against $H_1 : p = \frac{1}{3}$. Use the Central Limit Theorem to find n and c so that approximately $P_{H_0}\left(\sum_{1}^{n} X_i \leq c \right) = 0.10$ and $P_{H_1}\left(\sum_{1}^{n} X_i \leq c \right) = 0.80$.

8.1.10. Let $X_1, X_2, \ldots, X_{10}$ denote a random sample of size 10 from a Poisson distribution with mean θ. Show that the critical region C defined by $\sum_{1}^{10} x_i \geq 3$ is a best critical region for testing $H_0 : \theta = 0.1$ against $H_1 : \theta = 0.5$. Determine, for this test, the significance level α and the power at $\theta = 0.5$.

8.2   Uniformly Most Powerful Tests

This section will take up the problem of a test of a simple hypothesis H_0 against an alternative composite hypothesis H_1. We begin with an example.

Example 8.2.1. Consider the pdf

$$f(x; \theta) = \begin{cases} \frac{1}{\theta} e^{-x/\theta} & 0 < x < \infty \\ 0 & \text{elsewhere,} \end{cases}$$

of Exercises 8.1.2 and 8.1.3. It is desired to test the simple hypothesis $H_0 : \theta = 2$ against the alternative composite hypothesis $H_1 : \theta > 2$. Thus $\Omega = \{\theta : \theta \geq 2\}$. A random sample, X_1, X_2, of size $n = 2$ will be used, and the critical region is $C = \{(x_1, x_2) : 9.5 \leq x_1 + x_2 < \infty\}$. It was shown in the example cited that the significance level of the test is approximately 0.05 and the power of the test when $\theta = 4$ is approximately 0.31. The power function $\gamma(\theta)$ of the test for all $\theta \geq 2$ will now be obtained. We have

$$\begin{aligned} \gamma(\theta) &= 1 - \int_0^{9.5} \int_0^{9.5-x_2} \frac{1}{\theta^2} \exp\left(-\frac{x_1 + x_2}{\theta}\right) \, dx_1 dx_2 \\ &= \left(\frac{\theta + 9.5}{\theta}\right) e^{-9.5/\theta}, \quad 2 \leq \theta. \end{aligned}$$

For example, $\gamma(2) = 0.05$, $\gamma(4) = 0.31$, and $\gamma(9.5) = 2/e \doteq 0.74$. It is shown (Exercise 8.1.3) that the set $C = \{(x_1, x_2) : 9.5 \leq x_1 + x_2 < \infty\}$ is a best critical region of size 0.05 for testing the simple hypothesis $H_0 : \theta = 2$ against each simple hypothesis in the composite hypothesis $H_1 : \theta > 2$. ∎

The preceding example affords an illustration of a test of a simple hypothesis H_0 that is a best test of H_0 against every simple hypothesis in the alternative composite hypothesis H_1. We now define a critical region when it exists, which is a best critical region for testing a simple hypothesis H_0 against an alternative composite hypothesis H_1. It seems desirable that this critical region should be a best critical region for testing H_0 against each simple hypothesis in H_1. That is, the power function of the test that corresponds to this critical region should be at least as great as the power function of any other test with the same significance level for every simple hypothesis in H_1.

Definition 8.2.1. *The critical region C is a* **uniformly most powerful (UMP) critical region** *of size α for testing the simple hypothesis H_0 against an alternative composite hypothesis H_1 if the set C is a best critical region of size α for testing H_0 against each simple hypothesis in H_1. A test defined by this critical region C is called a* **uniformly most powerful (UMP) test**, *with significance level α, for testing the simple hypothesis H_0 against the alternative composite hypothesis H_1.*

As will be seen presently, uniformly most powerful tests do not always exist. However, when they do exist, the Neyman-Pearson theorem provides a technique for finding them. Some illustrative examples are given here.

Example 8.2.2. Let $X_1, X_2, \ldots, X_n$ denote a random sample from a distribution that is $N(0, \theta)$, where the variance θ is an unknown positive number. It will be shown that there exists a uniformly most powerful test with significance level α for testing the simple hypothesis $H_0 : \theta = \theta'$, where θ' is a fixed positive number, against the alternative composite hypothesis $H_1 : \theta > \theta'$. Thus $\Omega = \{\theta : \theta \geq \theta'\}$. The joint pdf of $X_1, X_2, \ldots, X_n$ is

$$L(\theta; x_1, x_2, \ldots, x_n) = \left(\frac{1}{2\pi\theta}\right)^{n/2} \exp\left\{-\frac{1}{2\theta}\sum_{i=1}^{n} x_i^2\right\}.$$

Let θ'' represent a number greater than θ', and let k denote a positive number. Let C be the set of points where

$$\frac{L(\theta' : x_1, x_2, \ldots, x_n)}{L(\theta'' : x_1, x_2, \ldots, x_n)} \leq k,$$

that is, the set of points where

$$\left(\frac{\theta''}{\theta'}\right)^{n/2} \exp\left[-\left(\frac{\theta'' - \theta'}{2\theta'\theta''}\right)\sum_{1}^{n} x_i^2\right] \leq k$$

or, equivalently,

$$\sum_{1}^{n} x_i^2 \geq \frac{2\theta'\theta''}{\theta'' - \theta'}\left[\frac{n}{2}\log\left(\frac{\theta''}{\theta'}\right) - \log k\right] = c.$$

The set $C = \left\{(x_1, x_2, \ldots, x_n) : \sum_{1}^{n} x_i^2 \geq c\right\}$ is then a best critical region for testing the simple hypothesis $H_0 : \theta = \theta'$ against the simple hypothesis $\theta = \theta''$. It remains to determine c, so that this critical region has the desired size α. If H_0 is true, the random variable $\sum_{1}^{n} X_i^2/\theta'$ has a chi-square distribution with n degrees of freedom. Since $\alpha = P_{\theta'}\left(\sum_{1}^{n} X_i^2/\theta' \geq c/\theta'\right)$, c/θ' may be read from Table II in Appendix B and c determined. Then $C = \left\{(x_1, x_2, \ldots, x_n) : \sum_{1}^{n} x_i^2 \geq c\right\}$ is a best critical region of size α for testing $H_0 : \theta = \theta'$ against the hypothesis $\theta = \theta''$. Moreover, for each number θ'' greater than θ', the foregoing argument holds. That is, $C = \left\{(x_1, \ldots, x_n) : \sum_{1}^{n} x_i^2 \geq c\right\}$ is a uniformly most powerful critical region of size α for testing $H_0 : \theta = \theta'$ against $H_1 : \theta > \theta'$. If $x_1, x_2, \ldots, x_n$ denote the experimental values of $X_1, X_2, \ldots, X_n$, then $H_0 : \theta = \theta'$ is rejected at the significance level α, and $H_1 : \theta > \theta'$ is accepted if $\sum_{1}^{n} x_i^2 \geq c$; otherwise, $H_0 : \theta = \theta'$ is accepted.

If, in the preceding discussion, we take $n = 15$, $\alpha = 0.05$, and $\theta' = 3$, then here the two hypotheses will be $H_0 : \theta = 3$ and $H_1 : \theta > 3$. From Table II, $c/3 = 25$ and hence $c = 75$. ∎

Example 8.2.3. Let $X_1, X_2, \ldots, X_n$ denote a random sample from a distribution that is $N(\theta, 1)$, where θ is unknown. It will be shown that there is no uniformly most powerful test of the simple hypothesis $H_0 : \theta = \theta'$, where θ' is a fixed number against the alternative composite hypothesis $H_1 : \theta \neq \theta'$. Thus $\Omega = \{\theta : -\infty < \theta < \infty\}$. Let θ'' be a number not equal to θ'. Let k be a positive number and consider

$$\frac{(1/2\pi)^{n/2} \exp\left[-\sum_{1}^{n}(x_i - \theta')^2/2\right]}{(1/2\pi)^{n/2} \exp\left[-\sum_{1}^{n}(x_i - \theta'')^2/2\right]} \leq k.$$

The preceding inequality may be written as

$$\exp\left\{-(\theta'' - \theta')\sum_{1}^{n} x_i + \frac{n}{2}[(\theta'')^2 - (\theta')^2]\right\} \leq k$$

or

$$(\theta'' - \theta')\sum_{1}^{n} x_i \geq \frac{n}{2}[(\theta'')^2 - (\theta')^2] - \log k.$$

This last inequality is equivalent to

$$\sum_{1}^{n} x_i \geq \frac{n}{2}(\theta'' - \theta') - \frac{\log k}{\theta'' - \theta'},$$

provided that $\theta'' > \theta'$, and it is equivalent to

$$\sum_{1}^{n} x_i \leq \frac{n}{2}(\theta'' - \theta') - \frac{\log k}{\theta'' - \theta'}$$

if $\theta'' < \theta'$. The first of these two expressions defines a best critical region for testing $H_0 : \theta = \theta'$ against the hypothesis $\theta = \theta''$ provided that $\theta'' > \theta'$, while the second expression defines a best critical region for testing $H_0 : \theta = \theta'$ against the hypothesis $\theta = \theta''$ provided that $\theta'' < \theta'$. That is, a best critical region for testing the simple hypothesis against an alternative simple hypothesis, say $\theta = \theta' + 1$, will not serve as a best critical region for testing $H_0 : \theta = \theta'$ against the alternative simple hypothesis $\theta = \theta' - 1$. By definition, then, there is no uniformly most powerful test in the case under consideration.

It should be noted that had the alternative composite hypothesis been either $H_1 : \theta > \theta'$ or $H_1 : \theta < \theta'$, a uniformly most powerful test would exist in each instance. ∎

Example 8.2.4. In Exercise 8.1.10 the reader was asked to show that if a random sample of size $n = 10$ is taken from a Poisson distribution with mean θ, the critical region defined by $\sum_1^n x_i \geq 3$ is a best critical region for testing $H_0 : \theta = 0.1$ against $H_1 : \theta = 0.5$. This critical region is also a uniformly most powerful one for testing $H_0 : \theta = 0.1$ against $H_1 : \theta > 0.1$ because, with $\theta'' > 0.1$,

$$\frac{(0.1)^{\sum x_i} e^{-10(0.1)}/(x_1! x_2! \cdots x_n!)}{(\theta'')^{\sum x_i} e^{-10(\theta'')}/(x_1! x_2! \cdots x_n!)} \leq k$$

is equivalent to

$$\left(\frac{0.1}{\theta''}\right)^{\sum x_i} e^{-10(0.1 - \theta'')} \leq k.$$

The preceding inequality may be written as

$$\left(\sum_1^n x_i\right) (\log 0.1 - \log \theta'') \leq \log k + 10(1 - \theta'')$$

or, since $\theta'' > 0.1$, equivalently as

$$\sum_1^n x_i \geq \frac{\log k + 1 - 10\theta''}{\log 0.1 - \log \theta''}.$$

Of course, $\sum_1^n x_i \geq 3$ is of the latter form. ∎

Let us make an important observation, although obvious when pointed out. Let $X_1, X_2, \ldots, X_n$ denote a random sample from a distribution that has pdf $f(x; \theta)$, $\theta \in \Omega$. Suppose that $Y = u(X_1, X_2, \ldots, X_n)$ is a sufficient statistic for θ. In accordance with the factorization theorem, the joint pdf of $X_1, X_2, \ldots, X_n$ may be written

$$L(\theta; x_1, x_2, \ldots, x_n) = k_1[u(x_1, x_2, \ldots, x_n); \theta] k_2(x_1, x_2, \ldots, x_n),$$

where $k_2(x_1, x_2, \ldots, x_n)$ does not depend upon θ. Consequently, the ratio

$$\frac{L(\theta'; x_1, x_2, \ldots, x_n)}{L(\theta''; x_1, x_2, \ldots, x_n)} = \frac{k_1[u(x_1, x_2, \ldots, x_n); \theta']}{k_1[u(x_1, x_2, \ldots, x_n); \theta'']}$$

depends upon $x_1, x_2, \ldots, x_n$ only through $u(x_1, x_2, \ldots, x_n)$. Accordingly, if there is a sufficient statistic $Y = u(X_1, X_2, \ldots, X_n)$ for θ and if a best test or a uniformly most powerful test is desired, there is no need to consider tests which are based upon any statistic other than the sufficient statistic. This result supports the importance of sufficiency.

In the above examples, we have presented uniformly most powerful tests. For some families of pdfs and hypotheses, we can obtain general forms of such tests. We will sketch these results for the general one-sided hypotheses of the form,

$$H_0 : \theta \leq \theta' \text{ versus } H_1 : \theta > \theta'. \tag{8.2.1}$$

The other one-sided hypotheses with the null hypothesis $H_0 : \theta \geq \theta'$, is completely analogous. Note that the null hypothesis of (8.2.1) is a composite hypothesis. Recall from Chapter 4 that the level of a test for the hypotheses (8.2.1) is defined by $\max_{\theta \leq \theta'} \gamma(\theta)$, where $\gamma(\theta)$ is the power function of the test. That is, the significance level is the maximum probability of Type I error.

Let $\mathbf{X}' = (X_1, \ldots, X_n)$ be a random sample with common pdf (or pmf) $f(x; \theta)$, $\theta \in \Omega$ and hence, with the likelihood function

$$L(\theta, \mathbf{x}) = \prod_{i=1}^{n} f(x_i; \theta), \quad \mathbf{x}' = (x_1, \ldots, x_n) \, .$$

We will consider the family of pdfs which has monotone likelihood ratio as defined next.

Definition 8.2.2. *We say that the likelihood $L(\theta, \mathbf{x})$ has* **monotone likelihood ratio** *(mlr) in the statistic $y = u(\mathbf{x})$, if for $\theta_1 < \theta_2$, the ratio*

$$\frac{L(\theta_1, \mathbf{x})}{L(\theta_2, \mathbf{x})} \tag{8.2.2}$$

is a monotone function of $y = u(\mathbf{x})$.

Assume then that our likelihood function $L(\theta, \mathbf{x})$ has a monotone decreasing likelihood ratio in the statistic $y = u(\mathbf{x})$. Then the ratio in (8.2.2) is equal to $g(y)$, where g is a decreasing function. The case where the likelihood function has a monotone increasing likelihood ratio, (g is increasing), follows similarly by changing the sense of the inequalities below. Let α denote the significance level. Then we claim that the following test is UMP level α for the hypotheses (8.2.1):

$$\text{Reject } H_0 \text{ if } Y \geq c_Y, \tag{8.2.3}$$

where c_Y is determined by $\alpha = P_{\theta'}[Y \geq c_Y]$.

To show this claim, first consider the simple null hypothesis $H_0' : \theta = \theta'$. Let $\theta'' > \theta'$ be arbitrary but fixed. Let C denote the most powerful critical region for θ' versus θ''. By the Neyman-Pearson Theorem, C is defined by,

$$\frac{L(\theta', \mathbf{X})}{L(\theta'', \mathbf{X})} \leq k, \quad \text{if and only if } \mathbf{X} \in C,$$

where k is determined by $\alpha = P_{\theta'}[\mathbf{X} \in C]$. But by Definition 8.2.2, because $\theta'' > \theta'$,

$$\frac{L(\theta', \mathbf{X})}{L(\theta'', \mathbf{X})} = g(Y) \leq k \Leftrightarrow Y \geq g^{-1}(k).$$

where $g^{-1}(k)$ satisfies $\alpha = P_{\theta'}[Y \geq g^{-1}(k)]$; i.e., $c_Y = g^{-1}(k)$. Hence, the Neyman-Pearson test is equivalent to the test defined by (8.2.3). Furthermore, the test is UMP for θ' versus $\theta'' > \theta'$ because the test only depends on $\theta'' > \theta'$ and $g^{-1}(k)$ is uniquely determined under θ'.

Let $\gamma_Y(\theta)$ denote the power function of the test (8.2.3). To finish, we need to show that $\max_{\theta \le \theta'} \gamma_Y(\theta) = \alpha$. But this follows immediately if we can show that $\gamma_Y(\theta)$ is a nondecreasing function. To see this, let $\theta_1 < \theta_2$. Note that since $\theta_1 < \theta_2$, the test (8.2.3) is the most powerful test for testing θ_1 versus θ_2 with the level $\gamma_Y(\theta_1)$. By Corollary 8.1.1, the power of the test at θ_2 must not be below the level; i.e., $\gamma_Y(\theta_2) \ge \gamma_Y(\theta_1)$. Hence, $\gamma_Y(\theta)$ is a nondecreasing function.

Example 8.2.5. Let $X_1, X_2, \ldots, X_n$ be a random sample from a Bernoulli distribution with parameter $p = \theta$, where $0 < \theta < 1$. Let $\theta' < \theta''$. Consider the ratio of likelihoods,

$$\frac{L(\theta'; x_1, x_2, \ldots, x_n)}{L(\theta''; x_1, x_2, \ldots, x_n)} = \frac{(\theta')^{\sum x_i}(1 - \theta')^{n - \sum x_i}}{(\theta'')^{\sum x_i}(1 - \theta'')^{n - \sum x_i}} = \left[\frac{\theta'(1 - \theta'')}{\theta''(1 - \theta')}\right]^{\sum x_i} \left(\frac{1 - \theta'}{1 - \theta''}\right)^n.$$

Since $\theta'/\theta'' < 1$ and $(1 - \theta'')/(1 - \theta') < 1$, so that $\theta'(1 - \theta'')/\theta''(1 - \theta') < 1$, the ratio is an decreasing function of $y = \sum x_i$. Thus we have a monotone likelihood ratio in the statistic $Y = \sum X_i$.

Consider the hypotheses

$$H_0 : \theta \le \theta' \text{ versus } H_1 : \theta > \theta'. \tag{8.2.4}$$

By our discussion above, the UMP level α decision rule for testing H_0 versus H_1 is given by

$$\text{Reject } H_0 \text{ if } Y = \sum_{i=1}^n X_i \ge c,$$

where c is such that $\alpha = P_{\theta'}[Y \ge c]$. ∎

In the last example concerning a Bernoulli pmf, we obtained a UMP test by showing that its likelihood posseses mlr. The Bernoulli distribution is a regular case of the exponential family and our argument, under the one assumption below, can be generalized to the entire regular exponential family. To show this, suppose that the random sample $X_1, X_2, \ldots, X_n$ arises from a pdf or pmf representing a regular case of the exponential class, namely

$$f(x; \theta) = \begin{cases} \exp[p(\theta)K(x) + S(x) + q(\theta)] & x \in \mathcal{S} \\ 0 & \text{elsewhere}, \end{cases}$$

where the support of X, $\mathcal{S}$, is free of θ. Further assume that $p(\theta)$ is an increasing function of θ. Then

$$\frac{L(\theta')}{L(\theta'')} = \frac{\exp\left[p(\theta')\sum_1^n K(x_i) + \sum_1^n S(x_i) + nq(\theta')\right]}{\exp\left[p(\theta'')\sum_1^n K(x_i) + \sum_1^n S(x_i) + nq(\theta'')\right]}$$

$$= \exp\left\{[p(\theta') - p(\theta'')]\sum_1^n K(x_i) + n[q(\theta') - q(\theta'')]\right\}.$$

If $\theta' < \theta''$, $p(\theta)$ being an increasing function requires this ratio to be a decreasing function of $y = \sum_1^n K(x_i)$. Thus we have a monotone likelihood ratio in the statistic $Y = \sum_1^n K(X_i)$. Hence, consider the hypotheses,

$$H_0 : \theta \leq \theta' \text{ versus } H_1 : \theta > \theta'. \tag{8.2.5}$$

By our discussion above concerning mlr, the UMP level α decision rule for testing H_0 versus H_1 is given by

$$\text{Reject } H_0 \text{ if } Y = \sum_{i=1}^n K(X_i) \geq c,$$

where c is such that $\alpha = P_{\theta'}[Y \geq c]$. Furthermore, the power function of this test is an increasing function in θ.

For the record, consider the other one-sided alternative hypotheses,

$$H_0 : \theta \geq \theta' \text{ versus } H_1 : \theta < \theta'. \tag{8.2.6}$$

The UMP level α decision rule is, for $p(\theta)$ an increasing function,

$$\text{Reject } H_0 \text{ if } Y = \sum_{i=1}^n K(X_i) \leq c,$$

where c is such that $\alpha = P_{\theta'}[Y \leq c]$.

If in the preceding situation with monotone likelihood ratio we test $H_0 : \theta = \theta'$ against $H_1 : \theta > \theta'$, then $\sum K(x_i) \geq c$ would be a uniformly most powerful critical region. From the likelihood ratios displayed in Examples 8.2.2, 8.2.3, 8.2.4, and 8.2.5 we see immediately that the respective critical regions

$$\sum_{i=1}^n x_i^2 \geq c, \quad \sum_{i=1}^n x_i \geq c, \quad \sum_{i=1}^n x_i \geq c, \quad \sum_{i=1}^n x_i \geq c$$

are uniformly most powerful for testing $H_0 : \theta = \theta'$ against $H_1 : \theta > \theta'$.

There is a final remark that should be made about uniformly most powerful tests. Of course, in Definition 8.2.1, the word *uniformly* is associated with θ; that is, C is a best critical region of size α for testing $H_0 : \theta = \theta_0$ against all θ values given by the composite alternative H_1. However, suppose that the form of such a region is

$$u(x_1, x_2, \ldots, x_n) \leq c.$$

Then this form provides uniformly most powerful critical regions for all attainable α values by, of course, appropriately changing the value of c. That is, there is a certain uniformity property, also associated with α, that is not always noted in statistics texts.

EXERCISES

8.2.1. Let X have the pmf $f(x;\theta) = \theta^x(1-\theta)^{1-x}$, $x = 0,1$, zero elsewhere. We test the simple hypothesis $H_0 : \theta = \frac{1}{4}$ against the alternative composite hypothesis $H_1 : \theta < \frac{1}{4}$ by taking a random sample of size 10 and rejecting $H_0 : \theta = \frac{1}{4}$ if and only if the observed values $x_1, x_2, \ldots, x_{10}$ of the sample observations are such that $\sum_{1}^{10} x_i \leq 1$. Find the power function $\gamma(\theta)$, $0 < \theta \leq \frac{1}{4}$, of this test.

8.2.2. Let X have a pdf of the form $f(x;\theta) = 1/\theta$, $0 < x < \theta$, zero elsewhere. Let $Y_1 < Y_2 < Y_3 < Y_4$ denote the order statistics of a random sample of size 4 from this distribution. Let the observed value of Y_4 be y_4. We reject $H_0 : \theta = 1$ and accept $H_1 : \theta \neq 1$ if either $y_4 \leq \frac{1}{2}$ or $y_4 > 1$. Find the power function $\gamma(\theta)$, $0 < \theta$, of the test.

8.2.3. Consider a normal distribution of the form $N(\theta, 4)$. The simple hypothesis $H_0 : \theta = 0$ is rejected, and the alternative composite hypothesis $H_1 : \theta > 0$ is accepted if and only if the observed mean $\overline{x}$ of a random sample of size 25 is greater than or equal to $\frac{3}{5}$. Find the power function $\gamma(\theta)$, $0 \leq \theta$, of this test.

8.2.4. Consider the distributions $N(\mu_1, 400)$ and $N(\mu_2, 225)$. Let $\theta = \mu_1 - \mu_2$. Let $\overline{x}$ and $\overline{y}$ denote the observed means of two independent random samples, each of size n, from these two distributions. We reject $H_0 : \theta = 0$ and accept $H_1 : \theta > 0$ if and only if $\overline{x} - \overline{y} \geq c$. If $\gamma(\theta)$ is the power function of this test, find n and c so that $\gamma(0) = 0.05$ and $\gamma(10) = 0.90$, approximately.

8.2.5. If in Example 8.2.2 of this section $H_0 : \theta = \theta'$, where θ' is a fixed positive number, and $H_1 : \theta < \theta'$, show that the set $\left\{ (x_1, x_2, \ldots, x_n) : \sum_{1}^{n} x_i^2 \leq c \right\}$ is a uniformly most powerful critical region for testing H_0 against H_1.

8.2.6. If, in Example 8.2.2 of this section, $H_0 : \theta = \theta'$, where θ' is a fixed positive number, and $H_1 : \theta \neq \theta'$, show that there is no uniformly most powerful test for testing H_0 against H_1.

8.2.7. Let $X_1, X_2, \ldots, X_{25}$ denote a random sample of size 25 from a normal distribution $N(\theta, 100)$. Find a uniformly most powerful critical region of size $\alpha = 0.10$ for testing $H_0 : \theta = 75$ against $H_1 : \theta > 75$.

8.2.8. Let $X_1, X_2, \ldots, X_n$ denote a random sample from a normal distribution $N(\theta, 16)$. Find the sample size n and a uniformly most powerful test of $H_0 : \theta = 25$ against $H_1 : \theta < 25$ with power function $\gamma(\theta)$ so that approximately $\gamma(25) = 0.10$ and $\gamma(23) = 0.90$.

8.2.9. Consider a distribution having a pmf of the form $f(x;\theta) = \theta^x(1-\theta)^{1-x}$, $x = 0,1$, zero elsewhere. Let $H_0 : \theta = \frac{1}{20}$ and $H_1 : \theta > \frac{1}{20}$. Use the central limit theorem to determine the sample size n of a random sample so that a uniformly most powerful test of H_0 against H_1 has a power function $\gamma(\theta)$, with approximately $\gamma(\frac{1}{20}) = 0.05$ and $\gamma(\frac{1}{10}) = 0.90$.

8.2.10. Illustrative Example 8.2.1 of this section dealt with a random sample of size $n = 2$ from a gamma distribution with $\alpha = 1$, $\beta = \theta$. Thus the mgf of the distribution is $(1 - \theta t)^{-1}$, $t < 1/\theta$, $\theta \geq 2$. Let $Z = X_1 + X_2$. Show that Z has a gamma distribution with $\alpha = 2$, $\beta = \theta$. Express the power function $\gamma(\theta)$ of Example 8.2.1 in terms of a single integral. Generalize this for a random sample of size n.

8.2.11. Let $X_1, X_2, \ldots, X_n$ be a random sample from a distribution with pdf $f(x; \theta) = \theta x^{\theta - 1}$, $0 < x < 1$, zero elsewhere, where $\theta > 0$. Find a sufficient statistic for θ and show that a uniformly most powerful test of $H_0 : \theta = 6$ against $H_1 : \theta < 6$ is based on this statistic.

8.2.12. Let X have the pdf $f(x; \theta) = \theta^x (1 - \theta)^{1-x}$, $x = 0, 1$, zero elsewhere. We test $H_0 : \theta = \frac{1}{2}$ against $H_1 : \theta < \frac{1}{2}$ by taking a random sample $X_1, X_2, \ldots, X_5$ of size $n = 5$ and rejecting H_0 if $Y = \sum_1^n X_i$ is observed to be less than or equal to a constant c.

(a) Show that this is a uniformly most powerful test.

(b) Find the significance level when $c = 1$.

(c) Find the significance level when $c = 0$.

(d) By using a *randomized test*, as discussed in Example 5.6.4, modify the tests given in Parts (b) and (c) to find a test with significance level $\alpha = \frac{2}{32}$.

8.2.13. Let $X_1, \ldots, X_n$ denote a random sample from a gamma-type distribution with $\alpha = 2$ and $\beta = \theta$. Let $H_0 : \theta = 1$ and $H_1 : \theta > 1$.

(a) Show that there exists a uniformly most powerful test for H_0 against H_1, determine the statistic Y upon which the test may be based, and indicate the nature of the best critical region.

(b) Find the pdf of the statistic Y in Part (a). If we want a significance level of 0.05, write an equation which can be used to determine the critical region. Let $\gamma(\theta)$, $\theta \geq 1$, be the power function of the test. Express the power function as an integral.

8.3   Likelihood Ratio Tests

In the first section of this chapter we presented most powerful tests for simple versus simple hypotheses. In the second section, we extended this theory to uniformly most powerful tests for essentially one-sided alternative hypotheses and families of distributions which have monotone likelihood ratio. What about the general case? That is, suppose the random variable X has pdf or pmf $f(x; \boldsymbol{\theta})$ where $\boldsymbol{\theta}$ is a vector of parameters in Ω. Let $\omega \subset \Omega$ and consider the hypotheses

$$H_0 : \boldsymbol{\theta} \in \omega \text{ versus } H_1 : \boldsymbol{\theta} \in \Omega \cap \omega^c. \tag{8.3.1}$$

There are complications in extending the optimal theory to this general situation which are addressed in more advanced books; see, in particular, Lehmann (1986). We will illustrate some of these complications with an example. Suppose X has a $N(\theta_1, \theta_2)$ distribution and that we want to test $\theta_1 = \theta_1'$, where θ_1' is specified. In the notation of (8.3.1), $\boldsymbol{\theta} = (\theta_1, \theta_2)$, $\Omega = \{\boldsymbol{\theta} : -\infty < \theta_1 < \infty, \theta_2 > 0\}$, and $\omega = \{\boldsymbol{\theta} : \theta_1 = \theta_1', \theta_2 > 0\}$. Notice that $H_0 : \boldsymbol{\theta} \in \omega$ is a composite null hypothesis. Let $X_1, \ldots, X_n$ be a random sample on X.

Assume for the moment that θ_2 is known. Then H_0 becomes the simple hypothesis $\theta_1 = \theta_1'$. This is essentially the situation discussed in Example 8.2.3. There it was shown that no UMP test exists for this situation. If the hypotheses were

$$H_0^* : \theta_1 \le \theta_1' \text{ versus } H_1^* : \theta_1 > \theta_1', \qquad (8.3.2)$$

then the test based on the critical region $C_1 = \left\{ \overline{X} > \theta_1' + \sqrt{\frac{\theta_2}{n}} z_\alpha \right\}$, where $z_\alpha = \Phi^{-1}(1 - \alpha)$ and Φ is the cdf of a standard normal random variable, is a UMP test of size α. Although this test has the most power for detecting $\theta_1 > \theta_1'$, it has very little power to detect $\theta_1 < \theta_1'$. In fact, the power is below the level α in this case. Recall in the discussion before Corollary 8.1.1 that we referred to this as intolerable for a test.

To avoid this situation, the concept of unbiasedness is introduced. A test is said to be **unbiased**, if its power never falls below the significance level. For example, by Corollary 8.1.1 the most powerful test of a simple versus simple hypotheses is unbiased. The tests of one-sided alternatives based on mlr pdfs are unbiased. But the test based on the critical region C_1 of (8.3.2) is biased for the two sided alternative. In the theory of optimal tests, only unbiased tests are considered. Within this class, the uniformly most powerful test is selected. In our illustrative example, the test based on the critical region $C_2 = \left\{ |\overline{X} - \theta_1'| > \sqrt{\frac{\theta_2}{n}} z_{\alpha/2} \right\}$ can be shown to be UMP unbiased.

In practice, though, the variance θ_2 is unknown. In this case, to establish theory for optimal tests involves the concept of what is called conditional tests. We will not pursue this any further in this text, but refer the interested reader to Lehmann (1986).

Recall from Chapter 6 that the likelihood ratio tests (6.3.3) can be used to test general hypotheses such as (8.3.1). There is no guarantee that they will be optimal. However, as are tests based on the Neyman-Pearson Theorem, they are based on a ratio of likelihood functions. In many situations, the likelihood ratio test statistics are optimal. In the example above on testing for the mean of a normal distribution, with known variance, the likelihood ratio test is the same as the UMP unbiased test. When the variance is unknown, the likelihood ratio test results in the one-sample t-test as shown in Example 6.5.1 of Chapter 6. This is the same as the conditional test discussed in Lehmann (1986).

In Chapter 6, we presented likelihood ratio tests for several situations. In the remainder of this section, we present likelihood ratio tests for other situations when sampling from normal distributions. As just cited, the one-sample t-test to test for the mean of a normal distribution with unknown variance was derived in Example 6.5.1. We next derive the two-sample t-test.

Example 8.3.1. Let the independent random variables X and Y have distributions that are $N(\theta_1, \theta_3)$ and $N(\theta_2, \theta_3)$, where the means θ_1 and θ_2 and common variance θ_3 are unknown. Then $\Omega = \{(\theta_1, \theta_2, \theta_3) : -\infty < \theta_1 < \infty, -\infty < \theta_2 < \infty, 0 < \theta_3 < \infty\}$. Let $X_1, X_2, \ldots, X_n$ and $Y_1, Y_2, \ldots, Y_m$ denote independent random samples from these distributions. The hypothesis $H_0 : \theta_1 = \theta_2$, unspecified, and θ_3 unspecified, is to be tested against all alternatives. Then $\omega = \{(\theta_1, \theta_2, \theta_3) : -\infty < \theta_1 = \theta_2 < \infty, 0 < \theta_3 < \infty\}$. Here $X_1, X_2, \ldots, X_n, Y_1, Y_2, \ldots, Y_m$ are $n + m > 2$ mutually independent random variables having the likelihood functions

$$L(\omega) = \left(\frac{1}{2\pi\theta_3}\right)^{(n+m)/2} \exp\left\{-\frac{1}{2\theta_3}\left[\sum_1^n (x_i - \theta_1)^2 + \sum_1^m (y_i - \theta_1)^2\right]\right\}$$

and

$$L(\Omega) = \left(\frac{1}{2\pi\theta_3}\right)^{(n+m)/2} \exp\left\{-\frac{1}{2\theta_3}\left[\sum_1^n (x_i - \theta_1)^2 + \sum_1^m (y_i - \theta_2)^2\right]\right\}.$$

If $\frac{\partial \log L(\omega)}{\partial \theta_1}$ and $\frac{\partial \log L(\omega)}{\partial \theta_3}$ are equated to zero, then (Exercise 8.3.2)

$$\sum_1^n (x_i - \theta_1) + \sum_1^m (y_i - \theta_1) = 0,$$

$$\frac{1}{\theta_3}\left[\sum_1^n (x_i - \theta_1)^2 + \sum_1^m (y_i - \theta_1)^2\right] = n + m. \tag{8.3.3}$$

The solutions for θ_1 and θ_3 are, respectively,

$$u = (n+m)^{-1}\left\{\sum_1^n x_i + \sum_1^m y_i\right\},$$

$$w = (n+m)^{-1}\left\{\sum_1^n (x_i - u)^2 + \sum_1^m (y_i - u)^2\right\}.$$

Further, u and w maximize $L(\omega)$. The maximum is

$$L(\hat{\omega}) = \left(\frac{e^{-1}}{2\pi w}\right)^{(n+m)/2}.$$

In a like manner, if

$$\frac{\partial \log L(\Omega)}{\partial \theta_1}, \quad \frac{\partial \log L(\Omega)}{\partial \theta_2}, \quad \frac{\partial \log L(\Omega)}{\partial \theta_3}$$

are equated to zero, then (Exercise 8.3.3)

$$\sum_1^n (x_i - \theta_1) = 0,$$

$$\sum_1^m (y_i - \theta_2) = 0, \tag{8.3.4}$$

$$-(n+m) + \frac{1}{\theta_3}\left[\sum_1^n (x_i - \theta_1)^2 + \sum_1^m (y_i - \theta_2)^2\right] = 0.$$

The solutions for θ_1, θ_2, and θ_3 are, respectively,

$$u_1 \;=\; n^{-1}\sum_1^n x_i,$$

$$u_2 \;=\; m^{-1}\sum_1^m y_i,$$

$$w' \;=\; (n+m)^{-1}\left[\sum_1^n (x_i - u_1)^2 + \sum_1^m (y_i - u_2)^2\right],$$

and, further, u_1, u_2, and w' maximize $L(\Omega)$. The maximum is

$$L(\hat{\Omega}) = \left(\frac{e^{-1}}{2\pi w'}\right)^{(n+m)/2},$$

so that

$$\Lambda(x_1,\ldots,x_n,y_1,\ldots,y_m) = \Lambda = \frac{L(\hat{\omega})}{L(\hat{\Omega})} = \left(\frac{w'}{w}\right)^{(n+m)/2}.$$

The random variable defined by $\Lambda^{2/(n+m)}$ is

$$\frac{\displaystyle\sum_1^n (X_i - \overline{X})^2 + \sum_1^m (Y_i - \overline{Y})^2}{\displaystyle\sum_1^n \{X_i - [(n\overline{X}+m\overline{Y})/(n+m)]\}^2 + \sum_1^n \{Y_i - [(n\overline{X}+m\overline{Y})/(n+m)]\}^2}.$$

Now

$$\sum_1^n \left(X_i - \frac{n\overline{X}+m\overline{Y}}{n+m}\right)^2 = \sum_1^n \left[(X_i - \overline{X}) + \left(\overline{X} - \frac{n\overline{X}+m\overline{Y}}{n+m}\right)\right]^2$$

$$= \sum_1^n (X_i - \overline{X})^2 + n\left(\overline{X} - \frac{n\overline{X}+m\overline{Y}}{n+m}\right)^2$$

and

$$\sum_1^m \left(Y_i - \frac{n\overline{X} + m\overline{Y}}{n+m}\right)^2 = \sum_1^m \left[(Y_i - \overline{Y}) + \left(\overline{Y} - \frac{n\overline{X} + m\overline{Y}}{n+m}\right)\right]^2$$

$$= \sum_1^m (Y_i - \overline{Y})^2 + m\left(\overline{Y} - \frac{n\overline{X} + m\overline{Y}}{n+m}\right)^2.$$

But

$$n\left(\overline{X} - \frac{n\overline{X} + m\overline{Y}}{n+m}\right)^2 = \frac{m^2 n}{(n+m)^2}(\overline{X} - \overline{Y})^2$$

and

$$m\left(\overline{Y} - \frac{n\overline{X} + m\overline{Y}}{n+m}\right)^2 = \frac{n^2 m}{(n+m)^2}(\overline{X} - \overline{Y})^2.$$

Hence, the random variable defined by $\Lambda^{2/(n+m)}$ may be written

$$\frac{\displaystyle\sum_1^n (X_i - \overline{X})^2 + \sum_1^m (Y_i - \overline{Y})^2}{\displaystyle\sum_1^n (X_i - \overline{X})^2 + \sum_1^m (Y_i - \overline{Y})^2 + [nm/(n+m)](\overline{X} - \overline{Y})^2}$$

$$= \frac{1}{1 + \dfrac{[nm/(n+m)](\overline{X} - \overline{Y})^2}{\displaystyle\sum_1^n (X_i - \overline{X})^2 + \sum_1^m (Y_i - \overline{Y})^2}}.$$

If the hypothesis $H_0 : \theta_1 = \theta_2$ is true, the random variable

$$T = \sqrt{\frac{nm}{n+m}}(\overline{X} - \overline{Y}) / \left\{(n+m-2)^{-1}\left[\sum_1^n (X_i - \overline{X})^2 + \sum_1^m (Y_i - \overline{Y})^2\right]\right\}^{1/2}$$

$$(8.3.5)$$

has, in accordance with Section 3.6, a t-distribution with $n + m - 2$ degrees of freedom. Thus the random variable defined by $\Lambda^{2/(n+m)}$ is

$$\frac{n+m-2}{(n+m-2)+T^2}.$$

The test of H_0 against all alternatives may then be based on a t-distribution with $n + m - 2$ degrees of freedom.

The likelihood ratio principle calls for the rejection of H_0 if and only if $\Lambda \le \lambda_0 < 1$. Thus the significance level of the test is

$$\alpha = P_{H_0}[\Lambda(X_1, \ldots, X_n, Y_1, \ldots, Y_m) \le \lambda_0].$$

However, $\Lambda(X_1, \ldots, X_n, Y_1, \ldots, Y_m) \leq \lambda_0$ is equivalent to $|T| \geq c$, and so

$$\alpha = P(|T| \geq c; H_0).$$

For given values of n and m, the number c is determined from Table IV in Appendix B (with $n + m - 2$ degrees of freedom) to yield a desired α. Then H_0 is rejected at a significance level α if and only if $|t| \geq c$, where t is the observed value of T. If, for instance, $n = 10$, $m = 6$, and $\alpha = 0.05$, then $c = 2.145$. ∎

For this last example as well as the one-sample t-test derived in Example 6.5.1, it was found that the likelihood ratio test could be based on a statistic which, when the hypothesis H_0 is true, has a t-distribution. To help us compute the powers of these tests at parameter points other than those described by the hypothesis H_0, we turn to the following definition.

Definition 8.3.1. *Let the random variable W be $N(\delta, 1)$; let the random variable V be $\chi^2(r)$, and W and V be independent. The quotient*

$$T = \frac{W}{\sqrt{V/r}}$$

is said to have a noncentral t-distribution *with r degrees of freedom and noncentrality parameter δ. If $\delta = 0$, we say that T has a central t-distribution.*

In the light of this definition, let us reexamine the t-statistics of Examples 6.5.1 and 8.3.1. In Example 6.5.1 we had

$$t(X_1, \ldots, X_n) = \frac{\sqrt{n}\overline{X}}{\sqrt{\sum_1^n (X_i - \overline{X})^2/(n-1)}}$$

$$= \frac{\sqrt{n}\overline{X}/\sigma}{\sqrt{\sum_1^n (X_i - \overline{X})^2/[\sigma^2(n-1)]}}.$$

Here, where θ_1 is the mean of the normal distribution, $W_1 = \sqrt{n}\overline{X}/\sigma$ is $N(\sqrt{n}\theta_1/\sigma, 1)$, $V_1 = \sum_1^n (X_i - \overline{X})^2/\sigma^2$ is $\chi^2(n-1)$, and W_1 and V_1 are independent. Thus, if $\theta_1 \neq 0$, we see, in accordance with the definition, that $t(X_1, \ldots, X_n)$ has a noncentral t-distribution with $n-1$ degrees of freedom and noncentrality parameter $\delta_1 = \sqrt{n}\theta_1/\sigma$. In Example 8.3.1 we had

$$T = \frac{W_2}{\sqrt{V_2/(n+m-2)}},$$

where

$$W_2 = \sqrt{\frac{nm}{n+m}}(\overline{X} - \overline{Y})\Big/ \sigma$$

and

$$V_2 = \frac{\sum_1^n (X_i - \overline{X})^2 + \sum_1^m (Y_i - \overline{Y})^2}{\sigma^2}.$$

Here W_2 is $N[\sqrt{nm/(n+m)}(\theta_1 - \theta_2)/\sigma, 1]$, V_2 is $\chi^2(n+m-2)$, and W_2 and V_2 are independent. Accordingly, if $\theta_1 \neq \theta_2$, T has a noncentral t-distribution with $n+m-2$ degrees of freedom and noncentrality parameter $\delta_2 = \sqrt{nm/(n+m)}(\theta_1 - \theta_2)/\sigma$. It is interesting to note that $\delta_1 = \sqrt{n}\theta_1/\sigma$ measures the deviation of θ_1 from $\theta_1 = 0$ in units of the standard deviation $\sigma/\sqrt{n}$ of $\overline{X}$. The noncentrality parameter $\delta_2 = \sqrt{nm/(n+m)}(\theta_1 - \theta_2)/\sigma$ is equal to the deviation of $\theta_1 - \theta_2$ from $\theta_1 - \theta_2 = 0$ in units of the standard deviation $\sigma/\sqrt{(n+m)/mn}$ of $\overline{X} - \overline{Y}$.

The packages R and S-PLUS contain functions which evaluate noncentral t-distributional quantities. For example, to obtain the value $P(T \leq t)$ when T has t-distribution with a degrees of freedom and noncentrality parameter b, use the command pt(t, a, ncp=b). For the value of the associated pdf at t, use the command dt(t, a, ncp=b). There are also various tables of the noncentral t-distribution, but they are much too cumbersome to be included in this book.

Remark 8.3.1. The one and two sample tests for normal means, presented in Examples 6.5.1 and 8.3.1 are the tests for normal means presented in most elementary statistics books. They are based on the assumption of normality. What if the underlying distributions are not normal? In that case, with finite variances, the t-test statistics for these situations are asymptotically correct. For example, consider the one sample t-test. Suppose $X_1, \ldots, X_n$ are iid with a common nonnormal pdf which has mean θ_1 and finite variance σ^2. The hypotheses remain the same, i.e, $H_0: \theta_1 = \theta_1'$ versus $H_1: \theta_1 \neq \theta_1'$. The t-test statistic, T_n, is given by

$$T_n = \frac{\sqrt{n}(\overline{X} - \theta_1')}{S_n},\tag{8.3.6}$$

where S_n is the sample standard deviation. Our critical region is $C_1 = \{|T_n| \geq t_{\alpha/2, n-1}\}$. Recall that $S_n \to \sigma$ in probability. Hence, by the Central Limit Theorem, under H_0

$$T_n = \frac{\sigma}{S_n} \frac{\sqrt{n}(\overline{X} - \theta_1')}{\sigma} \xrightarrow{D} Z,\tag{8.3.7}$$

where Z has a standard normal distribution. Hence, the asymptotic test would use the critical region $C_2 = \{|T_n| \geq z_{\alpha/2}\}$. By (8.3.7) the critical region C_2 would have approximate size α. In practice, we would use C_1. Because t critical values are generally larger that z critical values, the use of C_1 would be conservative; i.e, the size of C_1 would be slightly smaller than that of C_2. In terms of robustness, we would say the t-test possesses **robustness of validity**. But the t-test does not possess **robustness of power**. For nonnormal situations, there are more powerful tests than the t-test; see Chapter 10 for discussion.

As Exercise 8.3.4 shows, the two-sample t-test is also asymptotically correct, provided the underlying distributions have the *same* variance. ∎

In Example 8.3.1, in testing the equality of the means of two normal distributions, it was assumed that the unknown variances of the distributions were equal. Let us now consider the problem of testing the equality of these two unknown variances.

Example 8.3.2. We are given the independent random samples $X_1, \ldots, X_n$ and $Y_1, \ldots, Y_m$ from the distributions, which are $N(\theta_1, \theta_3)$ and $N(\theta_2, \theta_4)$, respectively. We have

$$\Omega = \{(\theta_1, \theta_2, \theta_3, \theta_4) : -\infty < \theta_1, \theta_2 < \infty, 0 < \theta_3, \theta_4 < \infty\}.$$

The hypothesis $H_0 : \theta_3 = \theta_4$, unspecified, with θ_1 and θ_2 also unspecified, is to be tested against all alternatives. Then

$$\omega = \{(\theta_1, \theta_2, \theta_3, \theta_4) : -\infty < \theta_1, \theta_2 < \infty, 0 < \theta_3 = \theta_4 < \infty\}.$$

It is easy to show (see Exercise 8.3.8) that the statistic defined by $\Lambda = L(\hat{\omega})/L(\hat{\Omega})$ is a function of the statistic

$$F = \frac{\sum\limits_{1}^{n}(X_i - \overline{X})^2/(n-1)}{\sum\limits_{1}^{m}(Y_i - \overline{Y})^2/(m-1)}. \tag{8.3.8}$$

If $\theta_3 = \theta_4$, this statistic F has an F-distribution with $n-1$ and $m-1$ degrees of freedom. The hypothesis that $(\theta_1, \theta_2, \theta_3, \theta_4) \in \omega$ is rejected if the computed $F \leq c_1$ or if the computed $F \geq c_2$. The constants c_1 and c_2 are usually selected so that, if $\theta_3 = \theta_4$,

$$P(F \leq c_1) = P(F \geq c_2) = \frac{\alpha_1}{2}$$

where α_1 is the desired significance level of this test. ∎

Example 8.3.3. Let the independent random variables X and Y have distributions that are $N(\theta_1, \theta_3)$ and $N(\theta_2, \theta_4)$. In Example 8.3.1 we derived the likelihood ratio test statistic T of the hypothesis $\theta_1 = \theta_2$ when $\theta_3 = \theta_4$, while in Example 8.3.2 we obtained the likelihood ratio test statistic F of the hypothesis $\theta_3 = \theta_4$. The hypothesis that $\theta_1 = \theta_2$ is rejected if the computed $|T| \geq c$, where the constant c is selected so that $\alpha_2 = P(|T| \geq c; \theta_1 = \theta_2, \theta_3 = \theta_4)$ is the assigned significance level of the test. We shall show that, if $\theta_3 = \theta_4$, the likelihood ratio test statistics for equality of variances and equality of means, respectively F and T, are independent. Among other things, this means that if these two tests based on F and T, respectively, are performed sequentially with significance levels α_1 and α_2, the probability of accepting both these hypotheses, when they are true, is $(1 - \alpha_1)(1 - \alpha_2)$. Thus the significance level of this joint test is $\alpha = 1 - (1 - \alpha_1)(1 - \alpha_2)$.

Independence of F and T, when $\theta_3 = \theta_4$, can be established by an appeal to sufficiency and completeness. The three statistics $\overline{X}$, $\overline{Y}$, and $\sum\limits_{1}^{n}(X_i - \overline{X})^2 +$

$\sum_{1}^{n}(Y_i - \overline{Y})^2$ are joint complete sufficient statistics for the three parameters θ_1, θ_2, and $\theta_3 = \theta_4$. Obviously, the distribution of F does not depend upon θ_1, θ_2, or $\theta_3 = \theta_4$, and hence F is independent of the three joint complete sufficient statistics. However, T is a function of these three joint complete sufficient statistics alone, and, accordingly, T is independent of F. It is important to note that these two statistics are independent whether $\theta_1 = \theta_2$ or $\theta_1 \neq \theta_2$. This permits us to calculate probabilities other than the significance level of the test. For example, if $\theta_3 = \theta_4$ and $\theta_1 \neq \theta_2$, then

$$P(c_1 < F < c_2, |T| \geq c) = P(c_1 < F < c_2)P(|T| \geq c).$$

The second factor in the right-hand member is evaluated by using the probabilities of a noncentral t-distribution. Of course, if $\theta_3 = \theta_4$ and the difference $\theta_1 - \theta_2$ is large, we would want the preceding probability to be close to 1 because the event $\{c_1 < F < c_2, |T| \geq c\}$ leads to a correct decision, namely, accept $\theta_3 = \theta_4$ and reject $\theta_1 = \theta_2$. ■

Remark 8.3.2. We caution the reader on this last test for the equality of two variances. In Remark 8.3.1, we discussed that the one- and two-sample t-tests for means are asymptotically correct. The two sample variance test of the last example is not; see, for example, page 126 of Hettmansperger and McKean (1998). If the underlying distributions are not normal then the F-critical values may be far from valid critical values, (unlike the t-critical values for the means tests as discussed in Remark 8.3.1). In a large simulation study Conover, Johnson and Johnson (1981) showed that instead of having the nominal size of $\alpha = 0.05$, the F-test for variances using the F-critical values could have significance levels as high as 0.80, in certain nonnormal situations. Thus, the two sample F-test for variances does not possess robustness of validity. It should only be used in situations where the assumption of normality can be justified. See Exercise 8.3.14 for an illustrative data set. ■

In the above examples, we were able to determine the null distribution of the test statistic. This is often impossible in practice. As discussed in Chapter 6, though, minus twice the log of the likelihood ratio test statistic is asymptotically χ^2 under H_0. Hence, we can obtain an approximate test in most situations.

EXERCISES

8.3.1. In Example 8.3.1, suppose $n = m = 8$, $\overline{x} = 75.2$, $\overline{y} = 78.6$, $\sum_{1}^{8}(x_i - \overline{x})^2 = 71.2$, $\sum_{1}^{8}(y_i - \overline{y})^2 = 54.8$. If we use the test derived in that example, do we accept or reject $H_0 : \theta_1 = \theta_2$ at the 5 percent significance level? Obtain the p-value, see Remark (5.6.1), of the test.

8.3.2. Verify Equations (8.3.3) of Example 8.3.1 of this section.

8.3.3. Verify Equations (8.3.4) of Example 8.3.1 of this section.

8.3.4. Let $X_1, \ldots, X_n$ and $Y_1, \ldots, Y_m$ follow the location model

$$
\begin{aligned}
X_i &= \theta_1 + Z_i, \quad i = 1, \ldots, n \\
Y_i &= \theta_2 + Z_{n+i}, \quad i = 1, \ldots, m,
\end{aligned}
$$

$$(8.3.9)$$

where $Z_1, \ldots, Z_{n+m}$ are iid random variables with common pdf $f(z)$. Assume that $E(Z_i) = 0$ and $\mathrm{Var}(Z_i) = \theta_3 < \infty$.

(a) Show that $E(X_i) = \theta_1$, $E(Y_i) = \theta_2$, and $\mathrm{Var}(X_i) = \mathrm{Var}(Y_i) = \theta_3$.

(b) Consider the hypotheses of Example 8.3.1; i.e,

$$H_0 : \theta_1 = \theta_2 \text{ versus } H_1 : \theta_1 \neq \theta_2.$$

Show that under H_0, the test statistic T given in expression (8.3.5) has a limiting $N(0, 1)$ distribution.

(c) Using Part (b), determine the corresponding large sample test (decision rule) of H_0 versus H_1. (This shows that the test in Example 8.3.1 is asymptotically correct.)

8.3.5. Show that the likelihood ratio principle leads to the same test when testing a simple hypothesis H_0 against an alternative simple hypothesis H_1, as that given by the Neyman-Pearson theorem. Note that there are only two points in Ω.

8.3.6. Let $X_1, X_2, \ldots, X_n$ be a random sample from the normal distribution $N(\theta, 1)$. Show that the likelihood ratio principle for testing $H_0 : \theta = \theta'$, where θ' is specified, against $H_1 : \theta \neq \theta'$ leads to the inequality $|\bar{x} - \theta'| \geq c$.

(a) Is this a uniformly most powerful test of H_0 against H_1?

(b) Is this a uniformly most powerful unbiased test of H_0 against H_1?

8.3.7. Let $X_1, X_2, \ldots, X_n$ be iid $N(\theta_1, \theta_2)$. Show that the likelihood ratio principle for testing $H_0 : \theta_2 = \theta_2'$ specified, and θ_1 unspecified, against $H_1 : \theta_2 \neq \theta_2'$, θ_1 unspecified, leads to a test that rejects when $\sum_1^n (x_i - \bar{x})^2 \leq c_1$ or $\sum_1^n (x_i - \bar{x})^2 \geq c_2$, where $c_1 < c_2$ are selected appropriately.

8.3.8. Let $X_1, \ldots, X_n$ and $Y_1, \ldots, Y_m$ be independent random samples from the distributions $N(\theta_1, \theta_3)$ and $N(\theta_2, \theta_4)$, respectively.

(a) Show that the likelihood ratio for testing $H_0 : \theta_1 = \theta_2$, $\theta_3 = \theta_4$ against all alternatives is given by

$$
\frac{\left[\sum_1^n (x_i - \bar{x})^2 / n\right]^{n/2} \left[\sum_1^m (y_i - \bar{y})^2 / m\right]^{m/2}}{\left\{\left[\sum_1^n (x_i - u)^2 + \sum_1^m (y_i - u)^2\right] \Big/ (m+n)\right\}^{(n+m)/2}},
$$

where $u = (n\bar{x} + m\bar{y})/(n + m)$.

(b) Show that the likelihood ratio for testing $H_0 : \theta_3 = \theta_4$ with θ_1 and θ_4 unspecified can be based on the test statistic F given in expression (8.3.8).

8.3.9. Let $Y_1 < Y_2 < \cdots < Y_5$ be the order statistics of a random sample of size $n = 5$ from a distribution with pdf $f(x; \theta) = \frac{1}{2}e^{-|x-\theta|}$, $-\infty < x < \infty$, for all real θ. Find the likelihood ratio test Λ for testing $H_0 : \theta = \theta_0$ against $H_1 : \theta \neq \theta_0$.

8.3.10. A random sample $X_1, X_2, \ldots, X_n$ arises from a distribution given by

$$H_0 : f(x; \theta) = \frac{1}{\theta}, \quad 0 < x < \theta, \quad \text{zero elsewhere,}$$

or

$$H_1 : f(x; \theta) = \frac{1}{\theta}e^{-x/\theta}, \quad 0 < x < \infty, \quad \text{zero elsewhere.}$$

Determine the likelihood ratio (Λ) test associated with the test of H_0 against H_1.

8.3.11. Consider a random sample $X_1, X_2, \ldots, X_n$ from a distribution with pdf $f(x; \theta) = \theta(1 - x)^{\theta-1}$, $0 < x < 1$, zero elsewhere, where $\theta > 0$.

(a) Find the form of the uniformly most powerful test of $H_0 : \theta = 1$ against $H_1 : \theta > 1$.

(b) What is the likelihood ratio Λ for testing $H_0 : \theta = 1$ against $H_1 : \theta \neq 1$?

8.3.12. Let $X_1, X_2, \ldots, X_n$ and $Y_1, Y_2, \ldots, Y_n$ be independent random samples from two normal distributions $N(\mu_1, \sigma^2)$ and $N(\mu_2, \sigma^2)$, respectively, where σ^2 is the common but unknown variance.

(a) Find the likelihood ratio Λ for testing $H_0 : \mu_1 = \mu_2 = 0$ against all alternatives.

(b) Rewrite Λ so that it is a function of a statistic Z which has a well-known distribution.

(c) Give the distribution of Z under both null and alternative hypotheses.

8.3.13. Let $(X_1, Y_1), (X_2, Y_2), \ldots, (X_n, Y_n)$ be a random sample from a bivariate normal distribution with $\mu_1, \mu_2, \sigma_1^2 = \sigma_2^2 = \sigma^2, \rho = \frac{1}{2}$, where μ_1, μ_2, and $\sigma^2 > 0$ are unknown real numbers. Find the likelihood ratio Λ for testing $H_0 : \mu_1 = \mu_2 = 0$, σ^2 unknown against all alternatives. The likelihood ratio Λ is a function of what statistic that has a well-known distribution?

8.3.14. Let X be a random variable with pdf $f_X(x) = (2b_X)^{-1}\exp\{-|x|/b_X\}$, for $-\infty < x < \infty$ and $b_X > 0$. First, show that the variance of X is $\sigma_X^2 = 2b_X^2$.

Now let Y, independent of X, have pdf $f_Y(y) = (2b_Y)^{-1}\exp\{-|y|/b_Y\}$, for $-\infty < x < \infty$ and $b_Y > 0$. Consider the hypotheses

$$H_0 : \sigma_X^2 = \sigma_Y^2 \quad \text{versus} \quad H_1 : \sigma_X^2 > \sigma_Y^2.$$

To illustrate Remark 8.3.2 for testing these hypotheses, consider the following data set, (taken from page 122 of Hettmansperger and McKean, 1998). Sample 1 represents the values of a sample drawn on X with $b_X = 1$, while Sample 2 represents the values of a sample drawn on Y with $b_Y = 1$. Hence, in this case H_0 is true.

Sample	-.38982	-2.17746	.81368	-.00072
1	$-.11032$	$-.70976$	.45664	.13583
Sample	.76384	$-.57041$	-2.56511	-1.73311
1	.40363	.77812	$-.11548$	
Sample	-1.06716	$-.57712$	.36138	$-.68037$
2	$-.63445$	$-.99624$	$-.18128$	.23957
Sample	$-.77576$	-1.42159	$-.81898$	.32863
2	.21390	1.42551	$-.16589$	

(a) Obtain *comparison boxplots* of these two samples. Comparison boxplots consist of boxplots of both samples drawn on the same scale. Based on these plots, in particular the interquartile ranges, what do you conclude about H_0?

(b) Obtain the F-test (for a one-sided hypothesis) as discussed in Remark 8.3.2 at level $\alpha = 0.10$. What is your conclusion?

(c) The test in Part (b) is not exact. Why?

8.4 The Sequential Probability Ratio Test

Theorem 8.1.1 provides us with a method for determining a best critical region for testing a simple hypothesis against an alternative simple hypothesis. Recall its statement. Let $X_1, X_2, \ldots, X_n$ be a random sample with fixed sample size n from a distribution that has pdf or pmf $f(x; \theta)$, where $\theta = \{\theta : \theta = \theta', \theta''\}$ and θ' and θ'' are known numbers. For this section, we will denote the likelihood of $X_1, X_2, \ldots, X_n$ by

$$L(\theta; n) = f(x_1; \theta) f(x_2; \theta) \cdots f(x_n; \theta),$$

a notation that reveals both the parameter θ and the sample size n. If we reject $H_0 : \theta = \theta'$ and accept $H_1 : \theta = \theta''$ when and only when

$$\frac{L(\theta'; n)}{L(\theta''; n)} \le k,$$

where $k > 0$, then by Theorem 8.1.1 this is a best test of H_0 against H_1.

Let us now suppose that the sample size n is *not* fixed in advance. In fact, let the sample size be a random variable N with sample space $\{1, 2, , 3, \ldots\}$. An interesting procedure for testing the simple hypothesis $H_0 : \theta = \theta'$ against the simple hypothesis $H_1 : \theta = \theta''$ is the following: Let k_0 and k_1 be two positive constants with $k_0 < k_1$. Observe the independent outcomes $X_1, X_2, X_3, \ldots$ in a sequence, for example, $x_1, x_2, x_3, \ldots$, and compute

$$\frac{L(\theta'; 1)}{L(\theta''; 1)}, \frac{L(\theta'; 2)}{L(\theta''; 2)}, \frac{L(\theta'; 3)}{L(\theta''; 3)}, \ldots$$

The hypothesis $H_0 : \theta = \theta'$ is rejected (and $H_1 : \theta = \theta''$ is accepted) if and only if there exists a positive integer n so that $\mathbf{x}_n = (x_1, x_2, \ldots, x_n)$ belongs to the set

$$C_n = \left\{ \mathbf{x}_n : k_0 < \frac{L(\theta', j)}{L(\theta'', j)} < k_1, \ j = 1, \ldots, n-1, \ \text{and} \ \frac{L(\theta', n)}{L(\theta'', n)} \leq k_0 \right\}. \quad (8.4.1)$$

On the other hand, the hypothesis $H_0 : \theta = \theta'$ is accepted (and $H_1 : \theta = \theta''$ is rejected) if and only if there exists a positive integer n so that $(x_1, x_2, \ldots, x_n)$ belongs to the set

$$B_n = \left\{ \mathbf{x}_n : k_0 < \frac{L(\theta', j)}{L(\theta'', j)} < k_1, \ j = 1, \ldots, n-1, \ \text{and} \ \frac{L(\theta', n)}{L(\theta'', n)} \geq k_1 \right\}. \quad (8.4.2)$$

That is, we continue to observe sample observations as long as

$$k_0 < \frac{L(\theta', n)}{L(\theta'', n)} < k_1. \quad (8.4.3)$$

We stop these observations in one of two ways:

1. With rejection of $H_0 : \theta = \theta'$ as soon as

$$\frac{L(\theta', n)}{L(\theta'', n)} \leq k_0,$$

      or

2. With acceptance of $H_0 : \theta = \theta'$ as soon as

$$\frac{L(\theta', n)}{L(\theta'', n)} \geq k_1,$$

A test of this kind is called Wald's *sequential probability ratio test*. Now, frequently, inequality (8.4.3) can be conveniently expressed in an equivalent form

$$c_0(n) < u(x_1, x_2, \ldots, x_n) < c_1(n), \quad (8.4.4)$$

where $u(X_1, X_2, \ldots, X_n)$ is a statistic and $c_0(n)$ and $c_1(n)$ depend on the constants $k_0, k_1, \theta', \theta''$, and on n. Then the observations are stopped and a decision is reached as soon as

$$u(x_1, x_2, \ldots, x_n) \leq c_0(n) \quad \text{or} \quad u(x_1, x_2, \ldots, x_n) \geq c_1(n).$$

We now give an illustrative example.

Example 8.4.1. Let X have a pmf

$$f(x;\theta) = \begin{cases} \theta^x(1-\theta)^{1-x} & x = 0, 1 \\ 0 & \text{elsewhere.} \end{cases}$$

In the preceding discussion of a sequential probability ratio test, let $H_0 : \theta = \frac{1}{3}$ and $H_1 : \theta = \frac{2}{3}$; then, with $\sum x_i = \sum_1^n x_i$,

$$\frac{L(\frac{1}{3},n)}{L(\frac{2}{3},n)} = \frac{(\frac{1}{3})^{\sum x_i}(\frac{2}{3})^{n-\sum x_i}}{(\frac{2}{3})^{\sum x_i}(\frac{1}{3})^{n-\sum x_i}} = 2^{n-2\sum x_i}.$$

If we take logarithms to the base 2, the inequality

$$k_0 < \frac{L(\frac{1}{3},n)}{L(\frac{2}{3},n)} < k_1,$$

with $0 < k_0 < k_1$, becomes

$$\log_2 k_0 < n - 2\sum_1^n x_i < \log_2 k_1,$$

or, equivalently, in the notation of expression (8.4.4),

$$c_0(n) = \frac{n}{2} - \frac{1}{2}\log_2 k_1 < \sum_1^n x_i < \frac{n}{2} - \frac{1}{2}\log_2 k_0 = c_1(n).$$

Note that $L(\frac{1}{3},n)/L(\frac{2}{3},n) \le k_0$ if and only if $c_1(n) \le \sum_1^n x_i$; and $L(\frac{1}{3},n)/L(\frac{2}{3},n) \ge k_1$ if and only if $c_0(n) \le \sum_1^n x_i$. Thus we continue to observe outcomes as long as $c_0(n) < \sum_1^n x_i < c_1(n)$. The observation of outcomes is discontinued with the first value of n of N for which either $c_1(n) \le \sum_1^n x_i$ or $c_0(n) \ge \sum_1^n x_i$. The inequality $c_1(n) \le \sum_1^n x_i$ leads to rejection of $H_0 : \theta = \frac{1}{3}$ (the acceptance of H_1), and the inequality $c_0(n) \ge \sum_1^n x_i$ leads to the acceptance of $H_0 : \theta = \frac{1}{3}$ (the rejection of H_1). ■

Remark 8.4.1. At this point, the reader undoubtedly sees that there are many questions that should be raised in connection with the sequential probability ratio test. Some of these questions are possibly among the following:

1. What is the probability of the procedure continuing indefinitely?

2. What is the value of the power function of this test at each of the points $\theta = \theta'$ and $\theta = \theta''$?

3. If θ'' is one of several values of θ specified by an alternative composite hypothesis, say $H_1 : \theta > \theta'$, what is the power function at each point $\theta \geq \theta'$?

4. Since the sample size N is a random variable, what are some of the properties of the distribution of N? In particular, what is the expected value $E(N)$ of N?

5. How does this test compare with tests that have a fixed sample size n? ■

A course in sequential analysis would investigate these and many other problems. However, in this book our objective is largely that of acquainting the reader with this kind of test procedure. Accordingly, we assert that the answer to question one is zero. Moreover, it can be proved that if $\theta = \theta'$ or if $\theta = \theta''$, $E(N)$ is smaller, for this sequential procedure than the sample size of a fixed-sample-size test which has the same values of the power function at those points. We now consider question two in some detail.

In this section we shall denote the power of the test when H_0 is true by the symbol α and the power of the test when H_1 is true by the symbol $1 - \beta$. Thus α is the probability of committing a type I error (the rejection of H_0 when H_0 is true), and β is the probability of committing a type II error (the acceptance of H_0 when H_0 is false). With the sets C_n and B_n as previously defined, and with random variables of the continuous type, we then have

$$\alpha = \sum_{n=1}^{\infty} \int_{C_n} L(\theta', n), \quad 1 - \beta = \sum_{n=1}^{\infty} \int_{C_n} L(\theta'', n).$$

Since the probability is one that the procedure will terminate, we also have

$$1 - \alpha = \sum_{n=1}^{\infty} \int_{B_n} L(\theta', n), \quad \beta = \sum_{n=1}^{\infty} \int_{B_n} L(\theta'', n).$$

If $(x_1, x_2, \ldots, x_n) \in C_n$, we have $L(\theta', n) \leq k_0 L(\theta'', n)$; hence it is clear that

$$\alpha = \sum_{n=1}^{\infty} \int_{C_n} L(\theta', n) \leq \sum_{n=1}^{\infty} \int_{C_n} k_0 L(\theta'', n) = k_0(1 - \beta).$$

Because $L(\theta', n) \geq k_1 L(\theta'', n)$ at each point of the set B_n, we have

$$1 - \alpha = \sum_{n=1}^{\infty} \int_{B_n} L(\theta', n) \geq \sum_{n=1}^{\infty} \int_{B_n} k_1 L(\theta'', n) = k_1 \beta.$$

Accordingly, it follows that

$$\frac{\alpha}{1 - \beta} \leq k_0, \quad k_1 \leq \frac{1 - \alpha}{\beta}, \tag{8.4.5}$$

provided that β is not equal to zero or one.

Now let α_a and β_a be preassigned proper fractions; some typical values in the applications are 0.01, 0.05, and 0.10. If we take

$$k_0 = \frac{\alpha_a}{1 - \beta_a}, \quad k_1 = \frac{1 - \alpha_a}{\beta_a},$$

then inequalities (8.4.5) become

$$\frac{\alpha}{1 - \beta} \le \frac{\alpha_a}{1 - \beta_a}, \quad \frac{1 - \alpha_a}{\beta_a} \le \frac{1 - \alpha}{\beta}; \qquad (8.4.6)$$

or, equivalently,

$$\alpha(1 - \beta_a) \le (1 - \beta)\alpha_a, \quad \beta(1 - \alpha_a) \le (1 - \alpha)\beta_a.$$

If we add corresponding members of the immediately preceding inequalities, we find that

$$\alpha + \beta - \alpha\beta_a - \beta\alpha_a \le \alpha_a + \beta_a - \beta\alpha_a - \alpha\beta_a$$

and hence

$$\alpha + \beta \le \alpha_a + \beta_a;$$

that is, the sum $\alpha + \beta$ of the probabilities of the two kinds of errors is bounded above by the sum $\alpha_a + \beta_a$ of the preassigned numbers. Moreover, since α and β are positive proper fractions, inequalities (8.4.6) imply that

$$\alpha \le \frac{\alpha_a}{1 - \beta_a}, \quad \beta \le \frac{\beta_a}{1 - \alpha_a};$$

consequently, we have an upper bound on each of α and β. Various investigations of the sequential probability ratio test seem to indicate that in most practical cases, the values of α and β are quite close to α_a and β_a. This prompts us to approximate the power function at the points $\theta = \theta'$ and $\theta = \theta''$ by α_a and $1 - \beta_a$, respectively.

Example 8.4.2. Let X be $N(\theta, 100)$. To find the sequential probability ratio test for testing $H_0 : \theta = 75$ against $H_1 : \theta = 78$ such that each of α and β is approximately equal to 0.10, take

$$k_0 = \frac{0.10}{1 - 0.10} = \frac{1}{9}, \quad k_1 = \frac{1 - 0.10}{0.10} = 9.$$

Since

$$\frac{L(75, n)}{L(78, n)} = \frac{\exp\left[-\sum(x_i - 75)^2/2(100)\right]}{\exp\left[-\sum(x_i - 78)^2/2(100)\right]} = \exp\left(-\frac{6\sum x_i - 459n}{200}\right),$$

the inequality

$$k_0 = \frac{1}{9} < \frac{L(75, n)}{L(78, n)} < 9 = k_1$$

can be rewritten, by taking logarithms, as

$$-\log 9 < \frac{6\sum x_i - 459n}{200} < \log 9.$$

This inequality is equivalent to the inequality

$$c_0(n) = \frac{153}{2}n - \frac{100}{3}\log 9 < \sum_1^n x_i < \frac{153}{2}n + \frac{100}{3}\log 9 = c_1(n).$$

Moreover, $L(75, n)/L(78, n) \leq k_0$ and $L(75, n)/L(78, n) \geq k_1$ are equivalent to the inequalities $\sum_1^n x_i \geq c_1(n)$ and $\sum_1^n x_i \leq c_0(n)$, respectively. Thus the observation of outcomes is discontinued with the first value of n of N for which either $\sum_1^n x_i \geq c_1(n)$ or $\sum_1^n x_i \leq c_0(n)$. The inequality $\sum_1^n x_i \geq c_1(n)$ leads to the rejection of $H_0 : \theta = 75$, and the inequality $\sum_1^n x_i \leq c_0(n)$ leads to the acceptance of $H_0 : \theta = 75$. The power of the test is approximately 0.10 when H_0 is true, and approximately 0.90 when H_1 is true. ∎

Remark 8.4.2. It is interesting to note that a sequential probability ratio test can be thought of as a *random-walk procedure*. To illustrate, the final inequalities of Examples 8.4.1 and 8.4.2 can be written as

$$-\log_2 k_1 < \sum_1^n 2(x_i - 0.5) < -\log_2 k_0$$

and

$$-\frac{100}{3}\log 9 < \sum_1^n (x_i - 76.5) < \frac{100}{3}\log 9,$$

respectively. In each instance, think of starting at the point zero and taking random steps until one of the boundaries is reached. In the first situation the random steps are $2(X_1 - 0.5), 2(X_2 - 0.5), 2(X_3 - 0.5), \ldots$ which have the same length, one, but with random directions. In the second instance, both the length and the direction of the steps are random variables, $X_1 - 76.5, X_2 - 76.5, X_3 - 76.5, \ldots$. ∎

In recent years, there has been much attention devoted to improving quality of products using statistical methods. One such simple method was developed by Walter Shewhart in which a sample of size n of the items being produced is taken and they are measured, resulting in n values. The mean $\bar{x}$ of these n measurements has an approximate normal distribution with mean μ and variance σ^2/n. In practice, μ

and σ^2 must be estimated, but in this discussion, we assume that they are known. From theory we know that the probability is 0.997 that $\bar{x}$ is between

$$\text{LCL} = \mu - \frac{3\sigma}{\sqrt{n}} \quad \text{and} \quad \text{UCL} = \mu + \frac{3\sigma}{\sqrt{n}}.$$

These two values are called the lower (LCL) and upper (UCL) control limits, respectively. Samples like these are taken periodically resulting in a sequence of means, say $\bar{x}_1, \bar{x}_2, \bar{x}_3, \dots$. These are usually plotted; and if they are between the LCL and UCL, we say that the process is *in control*. If one falls outside the limits, this would suggest that the mean μ has shifted, and the process would be investigated.

It was recognized by some that there could be a shift in the mean, say from μ to $\mu + (\sigma/\sqrt{n})$; and it would still be difficult to detect that shift with a single sample mean for now the probability of a single $\bar{x}$ exceeding UCL is only about 0.023. This means that we would need about $1/0.023 \approx 43$ samples, each of size n, on the average before detecting such a shift. This seems too long; so statisticians recognized that they should be cumulating experience as the sequence $\bar{x}_1, \bar{x}_2, \bar{x}_3, \dots$ is observed in order to help them detect the shift sooner. It is the practice to compute the standardized variable $Z = (\overline{X} - \mu)/(\sigma/\sqrt{n})$; thus we state the problem in these terms and provide the solution given by a sequential probability ratio test.

Here Z is $N(\theta, 1)$, and we wish to test $H_0 : \theta = 0$ against $H_1 : \theta = 1$ using the sequence of iid random variables $Z_1, Z_2, \dots, Z_m, \dots$. We use m rather than n, as the latter is the size of the samples taken periodically. We have

$$\frac{L(0, m)}{L(1, m)} = \frac{\exp\left[-\sum z_i^2/2\right]}{\exp\left[-\sum(z_i - 1)^2/2\right]} = \exp\left[-\sum_{i=1}^{n}(z_i - 0.5)\right].$$

Thus

$$k_0 < \exp\left[-\sum_{i=1}^{n}(z_i - 0.5)\right] < k_1$$

can be written as

$$h = -\log k_0 > \sum_{i=1}^{m}(z_i - 0.5) > -\log k_1 = -h.$$

It is true that $-\log k_0 = \log k_1$ when $\alpha_a = \beta_a$. Often, $h = -\log k_0$ is taken to be about 4 or 5, suggesting that $\alpha_a = \beta_a$ is small, like 0.01. As $\sum(z_i - 0.5)$ is cumulating the sum of $z_i - 0.5$, $i = 1, 2, 3, \dots$, these procedures are often called CUSUMS. If the CUSUM $= \sum(z_i - 0.5)$ exceeds h, we would investigate the process, as it seems that the mean has shifted upward. If this shift is to $\theta = 1$, the theory associated with these procedures shows that we need only eight or nine samples on the average, rather than 43, to detect this shift. For more information about these methods, the reader is referred to one of the many books on quality improvement through statistical methods. What we would like to emphasize here is that through sequential methods (not only the sequential probability ratio test), we should take advantage of all past experience that we can gather in making inferences.

EXERCISES

8.4.1. Let X be $N(0, \theta)$ and, in the notation of this section, let $\theta' = 4$, $\theta'' = 9$, $\alpha_a = 0.05$, and $\beta_a = 0.10$. Show that the sequential probability ratio test can be based upon the statistic $\sum_{1}^{n} X_i^2$. Determine $c_0(n)$ and $c_1(n)$.

8.4.2. Let X have a Poisson distribution with mean θ. Find the sequential probability ratio test for testing $H_0 : \theta = 0.02$ against $H_1 : \theta = 0.07$. Show that this test can be based upon the statistic $\sum_{1}^{n} X_i$. If $\alpha_a = 0.20$ and $\beta_a = 0.10$, find $c_0(n)$ and $c_1(n)$.

8.4.3. Let the independent random variables Y and Z be $N(\mu_1, 1)$ and $N(\mu_2, 1)$, respectively. Let $\theta = \mu_1 - \mu_2$. Let us observe independent observations from each distribution, say $Y_1, Y_2, \ldots$ and $Z_1, Z_2, \ldots$. To test sequentially the hypothesis $H_0 : \theta = 0$ against $H_1 : \theta = \frac{1}{2}$, use the sequence $X_i = Y_i - Z_i$, $i = 1, 2, \ldots$. If $\alpha_a = \beta_a = 0.05$, show that the test can be based upon $\overline{X} = \overline{Y} - \overline{Z}$. Find $c_0(n)$ and $c_1(n)$.

8.4.4. Suppose that a manufacturing process makes about 3 percent defective items, which is considered satisfactory for this particular product. The managers would like to decrease this to about 1 percent and clearly want to guard against a substantial increase, say to 5 percent. To monitor the process, periodically $n = 100$ items are taken and the number X of defectives counted. Assume that X is $b(n = 100, p = \theta)$. Based on a sequence $X_1, X_2, \ldots, X_m, \ldots$, determine a sequential probability ratio test that tests $H_0 : \theta = 0.01$ against $H_1 : \theta = 0.05$. (Note that $\theta = 0.03$, the present level, is in between these two values.) Write this test in the form

$$h_0 > \sum_{i=1}^{m} (x_i - nd) > h_1$$

and determine d, h_0, and h_1 if $\alpha_a = \beta_a = 0.02$.

8.4.5. Let $X_1, X_2, \ldots, X_n$ be a random sample from a distribution with pdf $f(x; \theta) = \theta x^{\theta - 1}$, $0 < x < 1$, zero elsewhere.

(a) Find a complete sufficient statistic for θ.

(b) If $\alpha = \beta = \frac{1}{10}$, find the sequential probability ratio test of $H_0 : \theta = 2$ against $H_1 : \theta = 3$.

8.5 Minimax and Classification Procedures

We have considered several procedures which may be used in problems of point estimation. Among these were decision function procedures (in particular, minimax decisions). In this section, we apply minimax procedures to the problem of testing a simple hypothesis H_0 against an alternative simple hypothesis H_1. It is important to observe that these procedures yield, in accordance with the Neyman-Pearson theorem, a best test of H_0 against H_1. We end this section with a discussion on an application of these procedures to a classification problem.

8.5.1 Minimax Procedures

We first investigate the decision function approach to the problem of testing a simple hypothesis against a simple alternative hypothesis. Let the joint pdf of the n random variables $X_1, X_2, \ldots, X_n$ depend upon the parameter θ. Here n is fixed positive integer. This pdf is denoted by $L(\theta; x_1, x_2, \ldots, x_n)$ or, for brevity, by $L(\theta)$. Let θ' and θ'' be distinct and fixed values of θ. We wish to test the simple hypothesis $H_0 : \theta = \theta'$ against the simple hypothesis $H_1 : \theta = \theta''$. Thus the parameter space is $\Omega = \{\theta : \theta = \theta', \theta''\}$. In accordance with the decision function procedure, we need a function δ of the observed values of $X_1, \ldots, X_n$ (or, of the observed value of a statistic Y) that decides which of the two values of θ, θ' or θ'', to accept. That is, the function δ selects either $H_0 : \theta = \theta'$ or $H_1 : \theta = \theta''$. We denote these decisions by $\delta = \theta'$ and $\delta = \theta''$, respectively. Let $\mathcal{L}(\theta, \delta)$ represent the loss function associated with this decision problem. Because the pairs $(\theta = \theta', \delta = \theta')$ and $(\theta = \theta'', \delta = \theta'')$ represent correct decisions, we shall always take $\mathcal{L}(\theta', \theta') = \mathcal{L}(\theta'', \theta'') = 0$. On the other hand, if either $\delta = \theta''$ when $\theta = \theta'$ or $\delta = \theta'$ when $\theta = \theta''$, then a positive value should be assigned to the loss function; that is, $\mathcal{L}(\theta', \theta'') > 0$ and $\mathcal{L}(\theta'', \theta') > 0$.

It has previously been emphasized that a test of $H_0 : \theta = \theta'$ against $H_1 : \theta = \theta''$ can be described in terms of a critical region in the sample space. We can do the same kind of thing with the decision function. That is, we can choose a subset of C of the sample space and if $(x_1, x_2, \ldots, x_n) \in C$, we can make the decision $\delta = \theta''$; whereas, if $(x_1, x_2, \ldots, x_n) \in C^c$, the complement of C, we make the decision $\delta = \theta'$. Thus a given critical region C determines the decision function. In this sense, we may denote the risk function by $R(\theta, C)$ instead of $R(\theta, \delta)$. That is, in a notation used in Section 7.1,

$$R(\theta, C) = R(\theta, \delta) = \int_{C \cup C^c} \mathcal{L}(\theta, \delta) L(\theta).$$

Since $\delta = \theta''$ if $(x_1, \ldots, x_n) \in C$ and $\delta = \theta'$ if $(x_1, \ldots, x_n) \in C^c$, we have

$$R(\theta, C) = \int_C \mathcal{L}(\theta, \theta'') L(\theta) + \int_{C^c} \mathcal{L}(\theta, \theta') L(\theta). \tag{8.5.1}$$

If, in Equation (8.5.1), we take $\theta = \theta'$, then $\mathcal{L}(\theta', \theta') = 0$ and hence

$$R(\theta', C) = \int_C \mathcal{L}(\theta', \theta'') L(\theta') = \mathcal{L}(\theta', \theta'') \int_C L(\theta').$$

On the other hand, if in Equation (8.5.1) we let $\theta = \theta''$, then $\mathcal{L}(\theta'', \theta'') = 0$ and, accordingly,

$$R(\theta'', C) = \int_{C^c} \mathcal{L}(\theta'', \theta') L(\theta'') = \mathcal{L}(\theta'', \theta') \int_{C^c} L(\theta'').$$

It is enlightening to note that, if $\gamma(\theta)$ is the power function of the test associated with the critical region C, then

$$R(\theta', C) = \mathcal{L}(\theta', \theta'')\gamma(\theta') = \mathcal{L}(\theta', \theta'')\alpha,$$

where $\alpha = \gamma(\theta')$ is the significance level; and

$$R(\theta'', C) = \mathcal{L}(\theta'', \theta')[1 - \gamma(\theta'')] = \mathcal{L}(\theta'', \theta')\beta,$$

where $\beta = 1 - \gamma(\theta'')$ is the probability of the type II error.

Let us now see if we can find a minimax solution to our problem. That is, we want to find a critical region C so that

$$\max[R(\theta', C), R(\theta'', C)]$$

is minimized. We shall show that the solution is the region

$$C = \left\{ (x_1, \ldots, x_n) : \frac{L(\theta'; x_1, \ldots, x_n)}{L(\theta''; x_1, \ldots, x_n)} \le k \right\},$$

provided the positive constant k is selected so that $R(\theta', C) = R(\theta'', C)$. That is, if k is chosen so that

$$\mathcal{L}(\theta', \theta'') \int_C L(\theta') = \mathcal{L}(\theta'', \theta') \int_{C^c} L(\theta''),$$

then the critical region C provides a minimax solution. In the case of random variables of the continuous type, k can always be selected so that $R(\theta', C) = R(\theta'', C)$. However, with random variables of the discrete type, we may need to consider an auxiliary random experiment when $L(\theta')/L(\theta'') = k$ in order to achieve the exact equality $R(\theta', C) = R(\theta'', C)$.

To see that C is the minimax solution, consider every other region A for which $R(\theta', C) \ge R(\theta', A)$. A region A for which $R(\theta', C) < R(\theta', A)$ is not a candidate for a minimax solution, for then $R(\theta', C) = R(\theta'', C) < \max[R(\theta', A), R(\theta'', A)]$. Since $R(\theta', C) \ge R(\theta', A)$ means that

$$\mathcal{L}(\theta', \theta'') \int_C L(\theta') \ge \mathcal{L}(\theta', \theta'') \int_A L(\theta'),$$

we have

$$\alpha = \int_C L(\theta') \ge \int_A L(\theta');$$

that is, the significance level of the test associated with the critical region A is less than or equal to α. But C, in accordance with the Neyman-Pearson theorem, is a best critical region of size α. Thus

$$\int_C L(\theta'') \ge \int_A L(\theta'')$$

and

$$\int_{C^c} L(\theta'') \le \int_{A^c} L(\theta'').$$

Accordingly,

$$\mathcal{L}(\theta'', \theta') \int_{C^c} L(\theta'') \le \mathcal{L}(\theta'', \theta') \int_{A^c} L(\theta''),$$

or, equivalently,

$$R(\theta'', C) \le R(\theta'', A).$$

That is,

$$R(\theta', C) = R(\theta'', C) \le R(\theta'', A).$$

This means that

$$\max[R(\theta', C), R(\theta'', C)] \le R(\theta'', A).$$

Then certainly,

$$\max[R(\theta', C), R(\theta'', C)] \le \max[R(\theta', A), R(\theta'', A)],$$

and the critical region C provides a minimax solution, as we wanted to show.

Example 8.5.1. Let $X_1, X_2, \ldots, X_{100}$ denote a random sample of size 100 from a distribution that is $N(\theta, 100)$. We again consider the problem of testing $H_0 : \theta = 75$ against $H_1 : \theta = 78$. We seek a minimax solution with $\mathcal{L}(75, 78) = 3$ and $\mathcal{L}(78, 75) = 1$. Since $L(75)/L(78) \le k$ is equivalent to $\overline{x} \ge c$, we want to determine c, and thus k, so that

$$3P(\overline{X} \ge c; \theta = 75) = P(\overline{X} < c; \theta = 78). \tag{8.5.2}$$

Because $\overline{X}$ is $N(\theta, 1)$, the preceding equation can be rewritten as

$$3[1 - \Phi(c - 75)] = \Phi(c - 78).$$

As requested in Exercise 8.5.4, the reader can show by using Newton's algorithm that the solution to one place is $c = 76.8$. The significance level of the test is $1 - \Phi(1.8) = 0.036$, approximately, and the power of the test when H_1 is true is $1 - \Phi(-1.2) = 0.885$, approximately. ∎

8.5.2 Classification

The summary above has an interesting application to the problem of *classification*, which can be described as follows. An investigator makes a number of measurements on an item and wants to place it into one of several categories (or classify it). For convenience in our discussion, we assume that only two measurements, say X and Y, are made on the item to be classified. Moreover, let X and Y have a joint pdf $f(x, y; \theta)$, where the parameter θ represents one or more parameters. In our simplification, suppose that there are only two possible joint distributions (categories) for X and Y, which are indexed by the parameter values θ' and θ'', respectively. In this case, the problem then reduces to one of observing $X = x$ and $Y = y$ and then testing the hypothesis $\theta = \theta'$ against the hypothesis $\theta = \theta''$, with the classification of X and Y being in accord with which hypothesis is accepted. From the Neyman-Pearson theorem, we know that a best decision of this sort is of the form: If

$$\frac{f(x, y; \theta')}{f(x, y; \theta'')} \leq k$$

choose the distribution indexed by θ''; that is, we classify (x, y) as coming from the distribution indexed by θ''. Otherwise, choose the distribution indexed by θ'; that is, we classify (x, y) as coming from the distribution indexed by θ'. Some discussion on the choice of k follows in the next remark.

Remark 8.5.1 (On the choice of k). Consider the following probabilities:

$$
\begin{aligned}
\pi' &= P[(X, Y) \text{ is drawn from the distribution with pdf } f(x, y; \theta')], \\
\pi'' &= P[(X, Y) \text{ is drawn from the distribution with pdf } f(x, y; \theta'')].
\end{aligned}
$$

Note that $\pi' + \pi'' = 1$. Then it can be shown that the optimal classification rule is determined by taking $k = \pi''/\pi'$; see, for instance, Seber (1984). Hence, if we have prior information on how likely the item is drawn from the distribution with parameter θ' then we can obtain the classification rule. In practice, often each distribution is equilikely. In which case, $\pi' = \pi'' = 1/2$ and hence, $k = 1$. ∎

Example 8.5.2. Let (x, y) be an observation of the random pair (X, Y), which has a bivariate normal distribution with parameters $\mu_1, \mu_2, \sigma_1^2, \sigma_2^2$, and ρ. In Section 3.5 that joint pdf is given by

$$f(x, y; \mu_1, \mu_2, \sigma_1^2, \sigma_2^2) = \frac{1}{2\pi\sigma_1\sigma_2\sqrt{1 - \rho^2}} e^{-q(x, y; \mu_1, \mu_2)/2},$$

for $-\infty < x < \infty$ and $-\infty < y < \infty$, where $\sigma_1 > 0$, $\sigma_2 > 0$, $-1 < \rho < 1$, and

$$q(x, y; \mu_1, \mu_2) = \frac{1}{1 - \rho^2}\left[\left(\frac{x - \mu_1}{\sigma_1}\right)^2 - 2\rho\left(\frac{x - \mu_1}{\sigma_1}\right)\left(\frac{y - \mu_2}{\sigma_2}\right) + \left(\frac{y - \mu_2}{\sigma_2}\right)^2\right].$$

Assume that σ_1^2, σ_2^2, and ρ are known but that we do not know whether the respective means of (X, Y) are (μ_1', μ_2') or (μ_1'', μ_2''). The inequality

$$\frac{f(x, y; \mu_1', \mu_2', \sigma_1^2, \sigma_2^2, \rho)}{f(x, y; \mu_1'', \mu_2'', \sigma_1^2, \sigma_2^2, \rho)} \leq k$$

is equivalent to

$$\tfrac{1}{2}[q(x, y; \mu_1'', \mu_2'') - q(x, y; \mu_1', \mu_2')] \leq \log k.$$

Moreover, it is clear that the difference in the left-hand member of this inequality does not contain terms involving x^2, xy, and y^2. In particular, this inequality is the same as

$$\frac{1}{1 - \rho^2}\left\{\left[\frac{\mu_1' - \mu_1''}{\sigma_1^2} - \frac{\rho(\mu_2' - \mu_2'')}{\sigma_1\sigma_2}\right]x + \left[\frac{\mu_2' - \mu_2''}{\sigma_2^2} - \frac{\rho(\mu_1' - \mu_1'')}{\sigma_1\sigma_2}\right]y\right\}$$

$$\leq \log k + \tfrac{1}{2}[q(0, 0; \mu_1', \mu_2') - q(0, 0; \mu_1'', \mu_2'')],$$

or, for brevity,

$$ax + by \leq c. \tag{8.5.3}$$

That is, if this linear function of x and y in the left-hand member of inequality (8.5.3) is less than or equal to a constant, we classify (x, y) as coming from the bivariate normal distribution with means μ_1'' and μ_2''. Otherwise, we classify (x, y) as arising from the bivariate normal distribution with means μ_1' and μ_2'. Of course, if the prior probabilities can be assigned as discussed in Remark 8.5.1 then k and thus c can be found easily; see Exercise 8.5.3. ∎

Once the rule for classification is established, the statistician might be interested in the two probabilities of misclassifications using that rule. The first of these two is associated with the classification of (x, y) as arising from the distribution indexed by θ'' if, in fact, it comes from that index by θ'. The second misclassification is similar, but with the interchange of θ' and θ''. In the preceding example, the probabilities of these respective misclassifications are

$$P(aX + bY \leq c; \mu_1', \mu_2') \quad \text{and} \quad P(aX + bY > c; \mu_1'', \mu_2'').$$

The distribution of $Z = aX + bY$ follows easily from Theorem 3.5.1, it follows that the distribution of $Z = aX + bY$ is given by,

$$N(a\mu_1 + b\mu_2, a^2\sigma_1^2 + 2ab\rho\sigma_1\sigma_2 + b^2\sigma_2^2).$$

With this information, it is easy to compute the probabilities of misclassifications; see Exercise 8.5.3.

One final remark must be made with respect to the use of the important classification rule established in Example 8.5.2. In most instances the parameter values μ_1', μ_2' and μ_1'', μ_2'' as well as σ_1^2, σ_2^2, and ρ are unknown. In such cases the statistician has usually observed a random sample (frequently called a *training sample*) from each of the two distributions. Let us say the samples have sizes n' and n'', respectively, with sample characteristics

$$\overline{x}', \overline{y}', (s_x')^2, (s_y')^2, r' \quad \text{and} \quad \overline{x}'', \overline{y}'', (s_x'')^2, (s_y'')^2, r''.$$

Accordingly, if in inequality (8.5.3) the parameters $\mu_1', \mu_2', \mu_1'', \mu_2'', \sigma_1^2, \sigma_2^2$, and $\rho\sigma_1\sigma_2$ are replaced by the unbiased estimates

$$\overline{x}', \overline{y}', \overline{x}'', \overline{y}'', \frac{(n'-1)(s_x')^2 + (n''-1)(s_x'')^2}{n' + n'' - 2}, \frac{(n'-1)(s_y')^2 + (n''-1)(s_y'')^2}{n' + n'' - 2},$$
$$\frac{(n'-1)r's_x's_y' + (n''-1)r''s_x''s_y''}{n' + n'' - 2},$$

the resulting expression in the left-hand member is frequently called Fisher's **linear discriminant function**. Since those parameters have been estimated, the distribution theory associated with $aX + bY$ does provide an approximation.

Although we have considered only bivariate distributions in this section, the results can easily be extended to multivariate normal distributions using the results of Section 3.5; see also, Chapter 6 of Seber (1984).

EXERCISES

8.5.1. Let $X_1, X_2, \ldots, X_{20}$ be a random sample of size 20 from a distribution which is $N(\theta, 5)$. Let $L(\theta)$ represent the joint pdf of $X_1, X_2, \ldots, X_{20}$. The problem is to test $H_0 : \theta = 1$ against $H_1 : \theta = 0$. Thus $\Omega = \{\theta : \theta = 0, 1\}$.

(a) Show that $L(1)/L(0) \leq k$ is equivalent to $\bar{x} \leq c$.

(b) Find c so that the significance level is $\alpha = 0.05$. Compute the power of this test if H_1 is true.

(c) If the loss function is such that $\mathcal{L}(1,1) = \mathcal{L}(0,0) = 0$ and $\mathcal{L}(1,0) = \mathcal{L}(0,1) > 0$, find the minimax test. Evaluate the power function of this test at the points $\theta = 1$ and $\theta = 0$.

8.5.2. Let $X_1, X_2, \ldots, X_{10}$ be a random sample of size 10 from a Poisson distribution with parameter θ. Let $L(\theta)$ be the joint pdf of $X_1, X_2, \ldots, X_{10}$. The problem is to test $H_0 : \theta = \frac{1}{2}$ against $H_1 : \theta = 1$.

(a) Show that $L(\frac{1}{2})/L(1) \leq k$ is equivalent to $y = \displaystyle\sum_1^n x_i \geq c$.

(b) In order to make $\alpha = 0.05$, show that H_0 is rejected if $y > 9$ and, if $y = 9$, reject H_0 with probability $\frac{1}{2}$ (using some auxiliary random experiment).

(c) If the loss function is such that $\mathcal{L}(\frac{1}{2}, \frac{1}{2}) = \mathcal{L}(1, 1) = 0$ and $\mathcal{L}(\frac{1}{2}, 1) = 1$ and $\mathcal{L}(1, \frac{1}{2}) = 2$ show that the minimax procedure is to reject H_0 if $y > 6$ and, if $y = 6$, reject H_0 with probability 0.08 (using some auxiliary random experiment).

8.5.3. In Example 8.5.2 let $\mu_1' = \mu_2' = 0$, $\mu_1'' = \mu_2'' = 1$, $\sigma_1^2 = 1$, $\sigma_2^2 = 1$, and $\rho = \frac{1}{2}$.

(a) Find the distribution of the linear function $aX + bY$.

(b) With $k = 1$, compute $P(aX + bY \leq c; \mu_1' = \mu_2' = 0)$ and $P(aX + bY > c; \mu_1'' = \mu_2'' = 1)$.

8.5.4. Determine Newton's algorithm to find the solution of equation (8.5.2). If software is available, write a program which performs your algorithm and then show that the solution is $c = 76.8$. If software is not available, solve (8.5.2) by "trial and error."

8.5.5. Let X and Y have the joint pdf.

$$f(x, y; \theta_1, \theta_2) = \frac{1}{\theta_1 \theta_2} \exp\left(-\frac{x}{\theta_1} - \frac{y}{\theta_2}\right), \quad 0 < x < \infty, \ 0 < y < \infty,$$

zero elsewhere, where $0 < \theta_1$, $0 < \theta_2$. An observation (x, y) arises from the joint distribution with parameters equal to either $(\theta_1' = 1, \theta_2' = 5)$ or $(\theta_1'' = 3, \theta_2'' = 2)$. Determine the form of the classification rule.

8.5.6. Let X and Y have a joint bivariate normal distribution. An observation (x, y) arises from the joint distribution with parameters equal to either

$$\mu_1' = \mu_2' = 0, \quad (\sigma_1^2)' = (\sigma_2^2)' = 1, \quad \rho' = \tfrac{1}{2}$$

or

$$\mu_1'' = \mu_2'' = 1, \quad (\sigma_1^2)'' = 4, \quad (\sigma_2^2)'' = 9, \quad \rho'' = \tfrac{1}{2}.$$

Show that the classification rule involves a second-degree polynomial in x and y.

8.5.7. Let $\boldsymbol{W}' = (W_1, W_2)$ be an observation from one of two bivariate normal distributions, I and II, each with $\mu_1 = \mu_2 = 0$ but with the respective variance-covariance matrices

$$\boldsymbol{V}_1 = \begin{pmatrix} 1 & 0 \\ 0 & 4 \end{pmatrix} \quad \text{and} \quad \boldsymbol{V}_2 = \begin{pmatrix} 3 & 0 \\ 0 & 12 \end{pmatrix}.$$

How would you classify $\boldsymbol{W}$ into I or II?

Chapter 9

Inferences about Normal Models

9.1   Quadratic Forms

A homogeneous polynomial of degree 2 in n variables is called a **quadratic** form in those variables. If both the variables and the coefficients are real, the form is called a **real quadratic** form. Only real quadratic forms will be considered in this book. To illustrate, the form $X_1^2 + X_1 X_2 + X_2^2$ is a quadratic form in the two variables X_1 and X_2; the form $X_1^2 + X_2^2 + X_3^2 - 2X_1 X_2$ is a quadratic form in the three variables X_1, X_2, and X_3; but the form $(X_1 - 1)^2 + (X_2 - 2)^2 = X_1^2 + X_2^2 - 2X_1 - 4X_2 + 5$ is not quadratic form in X_1 and X_2, although it is a quadratic form in the variables $X_1 - 1$ and $X_2 - 2$.

Let $\overline{X}$ and S^2 denote, respectively, the mean and the variance of a random sample $X_1, X_2, \ldots, X_n$ from an arbitrary distribution. Thus

$$
\begin{aligned}
(n-1)S^2 &= \sum_1^n (X_i - \overline{X})^2 = \sum_1^n \left(X_i - \frac{X_1 + X_2 + \cdots + X_n}{n} \right)^2 \\
&= \frac{n-1}{n}(X_1^2 + X_2^2 + \cdots + X_n^2) \\
&\quad - \frac{2}{n}(X_1 X_2 + \cdots + X_1 X_n + \cdots + X_{n-1} X_n)
\end{aligned}
$$

is a quadratic form in the n variables $X_1, X_2, \ldots, X_n$. If the sample arises from a distribution that is $N(\mu, \sigma^2)$, we know that the random variable $(n-1)S^2/\sigma^2$ is $\chi^2(n-1)$ regardless of the value of μ. This fact proved useful in our search for a confidence interval for σ^2 when μ is unknown.

It has been seen that tests of certain statistical hypotheses require a statistic that is a quadratic form. For instance, Example 8.2.2 made use of the statistic $\sum_1^n X_i^2$, which is a quadratic form in the variables $X_1, X_2, \ldots, X_n$. Later in this chapter, tests of other statistical hypotheses will be investigated, and it will be seen that functions of statistics that are quadratic forms will be needed to carry out the tests in an expeditious manner. But first we shall make a study of the distribution of certain quadratic forms in normal and independent random variables.

The following theorem will be proved in Section 9.9.

Theorem 9.1.1. *Let $Q = Q_1 + Q_2 + \cdots + Q_{k-1} + Q_k$, where $Q, Q_1, \ldots, Q_k$ are $k + 1$ random variables that are real quadratic forms in n independent random variables which are normally distributed with common mean and variance μ and σ^2, respectively. Let $Q/\sigma^2, Q_1/\sigma^2, \ldots, Q_{k-1}/\sigma^2$ have chi-square distributions with degrees of freedom $r, r_1, \ldots, r_{k-1}$, respectively. Let Q_k be nonnegative. Then:*

(a) *$Q_1, \ldots, Q_k$ are independent, and hence*

(b) *Q_k/σ^2 has a chi-square distribution with $r - (r_1 + \cdots + r_{k-1}) = r_k$ degrees of freedom.*

Three examples illustrative of the theorem will follow. Each of these examples will deal with a distribution problem that is based on the remarks made in the subsequent paragraph.

Let the random variable X have a distribution that is $N(\mu, \sigma^2)$. Let a and b denote positive integers greater than 1 and let $n = ab$. Consider a random sample of size $n = ab$ from this normal distribution. The observations of the random sample will be denoted by the symbols

$$
\begin{array}{ccccc}
X_{11}, & X_{12}, & \ldots, & X_{1j}, & \ldots, & X_{1b} \\
X_{21}, & X_{22}, & \ldots, & X_{2j}, & \ldots, & X_{2b} \\
\vdots & & & & & \\
X_{i1}, & X_{i2}, & \ldots, & X_{ij}, & \ldots, & X_{ib} \\
\vdots & & & & & \\
X_{a1}, & X_{a2}, & \ldots, & X_{aj}, & \ldots, & X_{ab}.
\end{array}
$$

By assumption these $n = ab$ random variables are independent, and each has the same normal distribution with mean μ and variance σ^2. Thus, if we wish, we may consider each row as being a random sample of size b from the given distribution; and we may consider each column as being a random sample of size a from the given distribution. We now define $a + b + 1$ statistics. They are

$$
\overline{X}_{..} = \frac{X_{11} + \cdots + X_{1b} + \cdots + X_{a1} + \cdots + X_{ab}}{ab} = \frac{\displaystyle\sum_{i=1}^{a}\sum_{j=1}^{b} X_{ij}}{ab},
$$

$$
\overline{X}_{i.} = \frac{X_{i1} + X_{i2} + \cdots + X_{ib}}{b} = \frac{\displaystyle\sum_{j=1}^{b} X_{ij}}{b}, \quad i = 1, 2, \ldots, a,
$$

and

$$\overline{X}_{\cdot j} = \frac{X_{1j} + X_{2j} + \cdots + X_{aj}}{a} = \frac{\sum\limits_{i=1}^{a} X_{ij}}{a}, \quad j = 1, 2, \ldots, b.$$

Thus the statistic $\overline{X}_{\cdot\cdot}$ is the mean of the random sample of size $n = ab$; the statistics $\overline{X}_{1\cdot}, \overline{X}_{2\cdot}, \ldots, \overline{X}_{a\cdot}$ are, respectively, the means of the rows; and the statistics $\overline{X}_{\cdot 1}, \overline{X}_{\cdot 2}, \ldots, \overline{X}_{\cdot b}$ are, respectively, the means of the columns. Examples illustrative of the theorem follow.

Example 9.1.1. Consider the variance S^2 of the random sample of size $n = ab$. We have the algebraic identity

$$
\begin{aligned}
(ab - 1)S^2 &= \sum_{i=1}^{a} \sum_{j=1}^{b} (X_{ij} - \overline{X}_{\cdot\cdot})^2 \\
&= \sum_{i=1}^{a} \sum_{j=1}^{b} [(X_{ij} - \overline{X}_{i\cdot}) + (\overline{X}_{i\cdot} - \overline{X}_{\cdot\cdot})]^2 \\
&= \sum_{i=1}^{a} \sum_{j=1}^{b} (X_{ij} - \overline{X}_{i\cdot})^2 + \sum_{i=1}^{a} \sum_{j=1}^{b} (\overline{X}_{i\cdot} - \overline{X}_{\cdot\cdot})^2 \\
&\quad + 2 \sum_{i=1}^{a} \sum_{j=1}^{b} (X_{ij} - \overline{X}_{i\cdot})(\overline{X}_{i\cdot} - \overline{X}_{\cdot\cdot}).
\end{aligned}
$$

The last term of the right-hand member of this identity may be written

$$2 \sum_{i=1}^{a} \left[(\overline{X}_{i\cdot} - \overline{X}_{\cdot\cdot}) \sum_{j=1}^{b} (X_{ij} - \overline{X}_{i\cdot}) \right] = 2 \sum_{i=1}^{a} [(\overline{X}_{i\cdot} - \overline{X}_{\cdot\cdot})(b\overline{X}_{i\cdot} - b\overline{X}_{i\cdot})] = 0,$$

and the term

$$\sum_{i=1}^{a} \sum_{j=1}^{b} (\overline{X}_{i\cdot} - \overline{X}_{\cdot\cdot})^2$$

may be written

$$b \sum_{i=1}^{a} (\overline{X}_{i\cdot} - \overline{X}_{\cdot\cdot})^2.$$

Thus

$$(ab - 1)S^2 = \sum_{i=1}^{a} \sum_{j=1}^{b} (X_{ij} - \overline{X}_{i\cdot})^2 + b \sum_{i=1}^{a} (\overline{X}_{i\cdot} - \overline{X}_{\cdot\cdot})^2,$$

or, for brevity,

$$Q = Q_1 + Q_2.$$

We shall use Theorem 9.1.1 with $k = 2$ to show that Q_1 and Q_2 are independent. Since S^2 is the variance of a random variable of size $n = ab$ from the given normal distribution, then $(ab - 1)S^2/\sigma^2$ has a chi-square distribution with $ab - 1$ degrees of freedom. Now

$$\frac{Q_1}{\sigma^2} = \sum_{i=1}^{a} \left[\sum_{j=1}^{b} (X_{ij} - \overline{X}_{i.})^2/\sigma^2 \right].$$

For each fixed value of i, $\sum_{j=1}^{b} (X_{ij} - \overline{X}_{i.})^2$ is the product of $(b - 1)$ and the variance of a random sample of size b from the given normal distribution and accordingly, $\sum_{j=1}^{b} (X_{ij} - \overline{X}_{i.})^2/\sigma^2$ has a chi-square distribution with $b - 1$ degrees of freedom. Because the X_{ij} are independent, Q_1/σ^2 is the sum of a independent random variables, each having a chi-square distribution with $b - 1$ degrees of freedom. Hence Q_1/σ^2 has a chi-square distribution with $a(b - 1)$ degrees of freedom. Now $Q_2 = b\sum_{i=1}^{a} (\overline{X}_{i.} - \overline{X}_{..})^2 \geq 0$. In accordance with the theorem, Q_1 and Q_2 are independent, and Q_2/σ^2 has a chi-square distribution with $ab - 1 - a(b-1) = a - 1$ degrees of freedom. ∎

Example 9.1.2. In $(ab - 1)S^2$ replace $X_{ij} - \overline{X}_{..}$ by $(X_{ij} - \overline{X}_{.j}) + (\overline{X}_{.j} - \overline{X}_{..})$ to obtain

$$(ab - 1)S^2 = \sum_{j=1}^{b} \sum_{i=1}^{a} [(X_{ij} - \overline{X}_{.j}) + (\overline{X}_{.j} - \overline{X}_{..})]^2,$$

or

$$(ab - 1)S^2 = \sum_{j=1}^{b} \sum_{i=1}^{a} (X_{ij} - \overline{X}_{.j})^2 + a \sum_{j=1}^{b} (\overline{X}_{.j} - \overline{X}_{..})^2,$$

or, for brevity,

$$Q = Q_3 + Q_4.$$

It is easy to show (Exercise 9.1.1) that Q_3/σ^2 has a chi-square distribution with $b(a - 1)$ degrees of freedom. Since $Q_4 = a\sum_{j=1}^{b} (\overline{X}_{.j} - \overline{X}_{..})^2 \geq 0$, the theorem enables us to assert that Q_3 and Q_4 are independent and that Q_4/σ^2 has a chi-square distribution with $ab - 1 - b(a - 1) = b - 1$ degrees of freedom. ∎

Example 9.1.3. In $(ab - 1)S^2$ replace $X_{ij} - \overline{X}_{..}$ by $(\overline{X}_{i.} - \overline{X}_{..}) + (\overline{X}_{.j} - \overline{X}_{..}) + (X_{ij} - \overline{X}_{i.} - \overline{X}_{.j} + \overline{X}_{..})$ to obtain (Exercise 9.1.2)

$$(ab - 1)S^2 = b\sum_{i=1}^{a} (\overline{X}_{i.} - \overline{X}_{..})^2 + a\sum_{j=1}^{b} (\overline{X}_{.j} - \overline{X}_{..})^2 + \sum_{j=1}^{b} \sum_{i=1}^{a} (X_{ij} - \overline{X}_{i.} - \overline{X}_{.j} + \overline{X}_{..})^2,$$

or, for brevity,

$$Q = Q_2 + Q_4 + Q_5,$$

where Q_2 and Q_4 are defined in Example 9.1.1 and 9.1.2. From Examples 9.1.1 and 9.1.2, Q/σ^2, Q_2/σ^2, and Q_4/σ^2 have chi-square distributions with $ab - 1$, $a - 1$, and $b - 1$ degrees of freedom, respectively. Since $Q_5 \geq 0$, the theorem asserts that Q_2, Q_4, and Q_5 are independent and that Q_5/σ^2 has a chi-square distribution with $ab - 1 - (a - 1) - (b - 1) = (a - 1)(b - 1)$ degrees of freedom. ∎

Once these quadratic form statistics have been shown to be independent, a multiplicity of F-statistics can be defined. For instance,

$$\frac{Q_4/[\sigma^2(b - 1)]}{Q_3/[\sigma^2 b(a - 1)]} = \frac{Q_4/(b - 1)}{Q_3/[b(a - 1)]}$$

has an F-distribution with $b - 1$ and $b(a - 1)$ degrees of freedom; and

$$\frac{Q_4/[\sigma^2(b - 1)]}{Q_5/[\sigma^2(a - 1)(b - 1)]} = \frac{Q_4/(b - 1)}{Q_5/(a - 1)(b - 1)}$$

has an F-distribution with $b - 1$ and $(a - 1)(b - 1)$ degrees of freedom. In the subsequent sections it will be shown that some likelihood ratio tests of certain statistical hypotheses can be based on these F-statistics.

EXERCISES

9.1.1. In Example 9.1.2 verify that $Q = Q_3 + Q_4$ and that Q_3/σ^2 has a chi-square distribution with $b(a - 1)$ degrees of freedom.

9.1.2. In Example 9.1.3 verify that $Q = Q_2 + Q_4 + Q_5$.

9.1.3. Let $X_1, X_2, \ldots, X_n$ be a random sample from a normal distribution $N(\mu, \sigma^2)$. Show that

$$\sum_{i=1}^{n}(X_i - \overline{X})^2 = \sum_{i=2}^{n}(X_i - \overline{X}')^2 + \frac{n - 1}{n}(X_1 - \overline{X}')^2,$$

where $\overline{X} = \sum_{i=1}^{n} X_i/n$ and $\overline{X}' = \sum_{i=2}^{n} X_i/(n - 1)$.

Hint: Replace $X_i - \overline{X}$ by $(X_i - \overline{X}') - (X_1 - \overline{X}')/n$. Show that $\sum_{i=2}^{n}(X_i - \overline{X}')^2/\sigma^2$ has a chi-square distribution with $n - 2$ degrees of freedom. Prove that the two terms in the right-hand member are independent. What then is the distribution of

$$\frac{[(n - 1)/n](X_1 - \overline{X}')^2}{\sigma^2}?$$

9.1.4. Let $X_{ijk}, i = 1, \ldots, a; j = 1, \ldots, b; k = 1, \ldots, c$, be a random sample of size $n = abc$ from a normal distribution $N(\mu, \sigma^2)$. Let $\overline{X}_{\cdots} = \sum_{k=1}^{c} \sum_{j=1}^{b} \sum_{i=1}^{a} X_{ijk}/n$ and $\overline{X}_{i\cdots} = \sum_{k=1}^{c} \sum_{j=1}^{b} X_{ijk}/bc$. Prove that

$$\sum_{i=1}^{a} \sum_{j=1}^{b} \sum_{k=1}^{c} (X_{ijk} - \overline{X}_{\cdots})^2 = \sum_{i=1}^{a} \sum_{j=1}^{b} \sum_{k=1}^{c} (X_{ijk} - \overline{X}_{i\cdots})^2 + bc \sum_{i=1}^{a} (\overline{X}_{i\cdots} - \overline{X}_{\cdots})^2.$$

Show that $\sum_{i=1}^{a} \sum_{j=1}^{b} \sum_{k=1}^{c} (X_{ijk} - \overline{X}_{i\cdots})^2/\sigma^2$ has a chi-square distribution with $a(bc - 1)$ degrees of freedom. Prove that the two terms in the right-hand member are independent. What, then, is the distribution of $bc \sum_{i=1}^{a} (\overline{X}_{i\cdots} - \overline{X}_{\cdots})^2/\sigma^2$? Furthermore, let $\overline{X}_{\cdot j\cdot} = \sum_{k=1}^{c} \sum_{i=1}^{a} X_{ijk}/ac$ and $\overline{X}_{ij\cdot} = \sum_{k=1}^{c} X_{ijk}/c$. Show that

$$\sum_{i=1}^{a} \sum_{j=1}^{b} \sum_{k=1}^{c} (X_{ijk} - \overline{X}_{\cdots})^2 = \sum_{i=1}^{a} \sum_{j=1}^{b} \sum_{k=1}^{c} (X_{ijk} - \overline{X}_{ij\cdot})^2$$

$$+ bc \sum_{i=1}^{a} (\overline{X}_{i\cdots} - \overline{X}_{\cdots})^2 + ac \sum_{j=1}^{b} (\overline{X}_{\cdot j\cdot} - \overline{X}_{\cdots})^2$$

$$+ c \sum_{i=1}^{a} \sum_{j=1}^{b} (\overline{X}_{ij\cdot} - \overline{X}_{i\cdots} - \overline{X}_{\cdot j\cdot} + \overline{X}_{\cdots}).$$

Prove that the four terms in the right-hand member, when divided by σ^2, are independent chi-square variables with $ab(c - 1)$, $a - 1$, $b - 1$, and $(a - 1)(b - 1)$ degrees of freedom, respectively.

9.1.5. Let X_1, X_2, X_3, X_4 be a random sample of size $n = 4$ from the normal distribution $N(0, 1)$. Show that $\sum_{i=1}^{4} (X_i - \overline{X})^2$ equals

$$\frac{(X_1 - X_2)^2}{2} + \frac{[X_3 - (X_1 + X_2)/2]^2}{3/2} + \frac{[X_4 - (X_1 + X_2 + X_3)/3]^2}{4/3}$$

and argue that these three terms are independent, each with a chi-square distribution with 1 degree of freedom.

9.2   One-way ANOVA

Consider b independent random variables that have normal distributions with unknown means $\mu_1, \mu_2, \ldots, \mu_b$, respectively, and unknown but common variance σ^2.

For each $j = 1, 2, \ldots, b$, let $X_{1j}, X_{2j}, \ldots, X_{aj}$ represent a random sample of size a from the normal distribution with mean μ_j and variance σ^2. The appropriate model for the observations is

$$X_{ij} = \mu_j + e_{ij} ; \quad i = 1, \ldots, a, j = 1, \ldots, b, \tag{9.2.1}$$

where e_{ij} are iid $N(0, \sigma^2)$. Suppose that it is desired to test the composite hypothesis $H_0 : \mu_1 = \mu_2 = \cdots = \mu_b = \mu$, μ unspecified, against all possible alternative hypotheses H_1. A likelihood ratio test will be used.

Such problems often arise in practice. For example, suppose for a certain type of disease there are b drugs which can be used to treat it and we are interested in determining which drug is best in terms of a certain response. Let X_j denote this response when drug j is applied and let $\mu_j = E(X_j)$. If we assume that X_j is $N(\mu_j, \sigma^2)$, then the above null hypothesis says that all the drugs are equally effective. We often summarize this problem by saying that we have one factor at b levels. In this case the factor is the treatment of the disease and each level corresponds to one of the treatment drugs. Model (9.2.1) is called a **one-way** model. As we will see, the likelihood ratio test can be thought of in terms of estimates of variance. Hence, this is an example of an analysis of variance (ANOVA). In short, we say that this example is a one-way ANOVA problem.

Here the total parameter space is

$$\Omega = \{(\mu_1, \mu_2, \ldots, \mu_b, \sigma^2) : -\infty < \mu_j < \infty, \ 0 < \sigma^2 < \infty\}$$

and

$$\omega = \{(\mu_1, \mu_2, \ldots, \mu_b, \sigma^2) : -\infty < \mu_1 = \mu_2 = \cdots = \mu_b = \mu < \infty, \ 0 < \sigma^2 < \infty\}.$$

The likelihood functions, denoted by $L(\omega)$ and $L(\Omega)$ are, respectively,

$$L(\omega) = \left(\frac{1}{2\pi\sigma^2}\right)^{ab/2} \exp\left[-\frac{1}{2\sigma^2}\sum_{j=1}^{b}\sum_{i=1}^{a}(x_{ij} - \mu)^2\right]$$

and

$$L(\Omega) = \left(\frac{1}{2\pi\sigma^2}\right)^{ab/2} \exp\left[-\frac{1}{2\sigma^2}\sum_{j=1}^{b}\sum_{i=1}^{a}(x_{ij} - \mu_j)^2\right].$$

Now

$$\frac{\partial \log L(\omega)}{\partial \mu} = \sigma^{-2}\sum_{j=1}^{b}\sum_{i=1}^{a}(x_{ij} - \mu)$$

and

$$\frac{\partial \log L(\omega)}{\partial(\sigma^2)} = -\frac{ab}{2\sigma^2} + \frac{1}{2\sigma^4}\sum_{j=1}^{b}\sum_{i=1}^{a}(x_{ij} - \mu_j)^2.$$

If we equate these partial derivatives to zero, the solutions for μ and σ^2 are, respectively, in ω,

$$(ab)^{-1} \sum_{j=1}^{b} \sum_{i=1}^{a} x_{ij} = \bar{x}_{..},$$

$$(ab)^{-1} \sum_{j=1}^{b} \sum_{i=1}^{a} (x_{ij} - \bar{x}_{..})^2 = v, \qquad (9.2.2)$$

and these values maximize $L(\omega)$. Furthermore,

$$\frac{\partial \log L(\Omega)}{\partial \mu_j} = \sigma^{-2} \sum_{i=1}^{a} (x_{ij} - \mu_j), \quad j = 1, 2, \ldots, b,$$

and

$$\frac{\partial \log L(\Omega)}{\partial (\sigma^2)} = -\frac{ab}{2\sigma^2} + \frac{1}{2\sigma^4} \sum_{j=1}^{b} \sum_{i=1}^{a} (x_{ij} - \mu_j)^2.$$

If we equate these partial derivatives to zero, the solutions for $\mu_1, \mu_2, \ldots, \mu_b$, and σ^2 are, respectively, in Ω,

$$a^{-1} \sum_{i=1}^{a} x_{ij} = \bar{x}_{.j}, \quad j = 1, 2, \ldots, b,$$

$$(ab)^{-1} \sum_{j=1}^{b} \sum_{i=1}^{a} (x_{ij} - \bar{x}_{.j})^2 = w, \qquad (9.2.3)$$

and these values maximize $L(\Omega)$. These maxima are, respectively,

$$L(\hat{\omega}) = \left[\frac{ab}{2\pi \sum_{j=1}^{b} \sum_{i=1}^{a} (x_{ij} - \bar{x}_{..})^2} \right]^{ab/2} \exp\left[-\frac{ab \sum_{j=1}^{b} \sum_{i=1}^{a} (x_{ij} - \bar{x}_{..})^2}{2 \sum_{j=1}^{b} \sum_{i=1}^{a} (x_{ij} - \bar{x}_{..})^2} \right]$$

$$= \left[\frac{ab}{2\pi \sum_{j=1}^{b} \sum_{i=1}^{a} (x_{ij} - \bar{x}_{..})^2} \right]^{ab/2} e^{-ab/2}$$

and

$$L(\hat{\Omega}) = \left[\frac{ab}{2\pi \sum\limits_{j=1}^{b} \sum\limits_{i=1}^{a} (x_{ij} - \overline{x}_{.j})^2} \right]^{ab/2} e^{-ab/2}.$$

Finally,

$$\Lambda = \frac{L(\hat{\omega})}{L(\hat{\Omega})} = \left[\frac{\sum\limits_{j=1}^{b} \sum\limits_{i=1}^{a} (x_{ij} - \overline{x}_{.j})^2}{\sum\limits_{j=1}^{b} \sum\limits_{i=1}^{a} (x_{ij} - \overline{x}_{..})^2} \right]^{ab/2}.$$

In the notation of Section 9.1, the statistics defined by the functions $\overline{x}_{..}$ and v given by Equations (9.2.2) of this section are

$$\overline{X}_{..} = \frac{1}{ab} \sum_{j=1}^{b} \sum_{i=1}^{a} X_{ij} \quad \text{and} \quad V = \frac{1}{ab} \sum_{j=1}^{b} \sum_{i=1}^{a} (X_{ij} - \overline{X}_{..})^2 = \frac{Q}{ab}; \qquad (9.2.4)$$

while the statistics defined by the functions $\overline{x}_{.1}, \overline{x}_{.2}, \dots, \overline{x}_{.b}$ and w given by Equations (9.2.3) in this section are, respectively, given by the formulas $\overline{X}_{.j} = \sum\limits_{i=1}^{a} X_{ij}/a$, $j = 1, 2, \dots, b$, and $Q_3/ab = \sum\limits_{j=1}^{b} \sum\limits_{i=1}^{a} (X_{ij} - \overline{X}_{.j})^2/ab$. Thus, in the notation of Section 9.1, $\Lambda^{2/ab}$ defines the statistic Q_3/Q.

We reject the hypothesis H_0 if $\Lambda \leq \lambda_0$. To find λ_0 so that we have a desired significance level α, we must assume that the hypothesis H_0 is true. If the hypothesis H_0 is true, the random variables X_{ij} constitute a random sample of size $n = ab$ from a distribution that is normal with mean μ and variance σ^2. Thus, by Example 9.1.2 we have that $Q = Q_3 + Q_4$, where $Q_4 = a \sum\limits_{j=1}^{b} (\overline{X}_{.j} - \overline{X}_{..})^2$; that Q_3 and Q_4 are independent; and that Q_3/σ^2 and Q_4/σ^2 have chi-square distributions with $b(a-1)$ and $b-1$ degrees of freedom, respectively. Thus the statistic defined by $\lambda^{2/ab}$ may be written

$$\frac{Q_3}{Q_3 + Q_4} = \frac{1}{1 + Q_4/Q_3}.$$

The significance level of the test of H_0 is

$$\alpha = P_{H_0} \left[\frac{1}{1 + Q_4/Q_3} \leq \lambda_0^{2/ab} \right]$$

$$= P_{H_0} \left[\frac{Q_4/(b-1)}{Q_3/([b(a-1)]} \geq c \right],$$

where

$$c = \frac{b(a-1)}{b-1}(\lambda_0^{-2/ab} - 1).$$

But

$$F = \frac{Q_4/[\sigma^2(b-1)]}{Q_3/[\sigma^2 b(a-1)]} = \frac{Q_4/(b-1)}{Q_3/[b(a-1)]}$$

has an F-distribution with $b - 1$ and $b(a - 1)$ degrees of freedom. Hence the test of the composite hypothesis $H_0 : \mu_1 = \mu_2 = \cdots = \mu_b = \mu$, μ unspecified, against all possible alternatives may be tested with an F-statistic. The constant c is so selected as to yield the desired value of α.

Remark 9.2.1. It should be pointed out that a test of the equality of the b means μ_j, $j = 1, 2, \ldots, b$, does not require that we take a random sample of size a from each of the b normal distributions. That is, the samples may be of different sizes, for instance, $a_1, a_2, \ldots, a_b$; see Exercise 9.2.1. ∎

Suppose now that we wish to compute the power of the test of H_0 against H_1 when H_0 is false, that is, when we do not have $\mu_1 = \mu_2 = \cdots = \mu_b = \mu$. It will be seen in Section 9.3 that when H_1 is true, no longer is Q_4/σ^2 a random variable that is $\chi^2(b - 1)$. Thus we cannot use an F-statistic to compute the power of the test when H_1 is true. The problem is discussed in Section 9.3.

An observation should be made in connection with maximizing a likelihood function with respect to certain parameters. Sometimes it is easier to avoid the use of the calculus. For example, $L(\Omega)$ of this section can be maximized with respect to μ_j, for every fixed positive σ^2, by minimizing

$$z = \sum_{j=1}^{b}\sum_{i=1}^{a}(x_{ij} - \mu_j)^2$$

with respect to μ_j, $j = 1, 2, \ldots, b$. Now z can be written as

$$z = \sum_{j=1}^{b}\sum_{i=1}^{a}[(x_{ij} - \overline{x}_{.j}) + (\overline{x}_{.j} - \mu_j)]^2$$

$$= \sum_{j=1}^{b}\sum_{i=1}^{a}(x_{ij} - \overline{x}_{.j})^2 + a\sum_{j=1}^{b}(\overline{x}_{.j} - \mu_j)^2.$$

Since each term in the right-hand member of the preceding equation is nonnegative, clearly z is a minimum, with respect to μ_j, if we take $\mu_j = \overline{x}_{.j}$, $j = 1, 2, \ldots, b$.

EXERCISES

9.2.1. Let $X_{1j}, X_{2j}, \ldots, X_{a_jj}$ represent independent random samples of sizes a_j from a normal distribution with means μ_j and variances σ^2, $j = 1, 2, \ldots, b$. Show that

$$\sum_{j=1}^{b}\sum_{i=1}^{a_j}(X_{ij} - \overline{X}_{..})^2 = \sum_{j=1}^{b}\sum_{i=1}^{a_j}(X_{ij} - \overline{X}_{.j})^2 + \sum_{j=1}^{b}a_j(\overline{X}_{.j} - \overline{X}_{..})^2,$$

or $Q' = Q_3' + Q_4'$. Here $\overline{X}_{..} = \sum_{j=1}^{b}\sum_{i=1}^{a_j}X_{ij} / \sum_{j=1}^{b}a_j$ and $\overline{X}_{.j} = \sum_{i=1}^{a_j}X_{ij}/a_j$. If $\mu_1 = \mu_2 = \cdots = \mu_b$, show that Q'/σ^2 and Q_3'/σ^2 have chi-square distributions. Prove that Q_3' and Q_4' are independent, and hence Q_4'/σ^2 also has a chi-square distribution. If the likelihood ratio Λ is used to test $H_0 : \mu_1 = \mu_2 = \cdots = \mu_b = \mu$, μ unspecified and σ^2 unknown against all possible alternatives, show that $\Lambda \leq \lambda_0$ is equivalent to the computed $F \geq c$, where

$$F = \frac{\left(\sum_{j=1}^{b}a_j - b\right)Q_4'}{(b-1)Q_3'}.$$

What is the distribution of F when H_0 is true?

9.2.2. Consider the T-statistic that was derived through a likelihood ratio for testing the equality of the means of two normal distributions having common variance in Example 8.3.1. Show that T^2 is exactly the F-statistic of Exercise 9.2.1 with $a_1 = n$, $a_2 = m$, and $b = 2$. Of course, $X_1, \ldots, X_n, \overline{X}$ are replaced with $X_{11}, \ldots, X_{1n}, \overline{X}_{1.}$ and $Y_1, \ldots, Y_m, \overline{Y}$ by $X_{21} \ldots, X_{2m}, \overline{X}_{2.}$.

9.2.3. In Exercise 9.2.1, show that the linear functions $X_{ij} - \overline{X}_{.j}$ and $\overline{X}_{.j} - \overline{X}_{..}$ are uncorrelated.
Hint: Recall the definition of $\overline{X}_{.j}$ and $\overline{X}_{..}$ and, without loss of generality, we can let $E(X_{ij}) = 0$ for all i, j.

9.2.4. The following are observations associated with independent random samples from three normal distributions having equal variances and respective means μ_1, μ_2, μ_3.

I	II	III
0.5	2.1	3.0
1.3	3.3	5.1
-1.0	0.0	1.9
1.8	2.3	2.4
	2.5	4.2
		4.1

Compute the F-statistic that is used to test $H_0 : \mu_1 = \mu_2 = \mu_3$.

9.2.5. Using the notation of this section, assume that the means satisfy the condition that $\mu = \mu_1 + (b-1)d = \mu_2 - d = \mu_3 - d = \cdots = \mu_b - d$. That is, the last $b-1$ means are equal but differ from the first mean μ_1, provided that $d \neq 0$. Let independent random samples of size a be taken from the b normal distributions with common unknown variance σ^2.

(a) Show that the maximum likelihood estimators of μ and d are $\hat{\mu} = \overline{X}_{\cdot\cdot}$ and

$$\hat{d} = \frac{\sum\limits_{j=2}^{b} \overline{X}_{\cdot j}/(b-1) - \overline{X}_{\cdot 1}}{b}.$$

(b) Using Exercise 9.1.3, find Q_6 and $Q_7 = c\hat{d}^2$ so that, when $d = 0$, Q_7/σ^2 is $\chi^2(1)$ and

$$\sum_{i=1}^{a} \sum_{j=1}^{b} (X_{ij} - \overline{X}_{\cdot\cdot})^2 = Q_3 + Q_6 + Q_7.$$

(c) Argue that the three terms in the right-hand member of Part (b), once divided by σ^2, are independent random variables with chi-square distributions, provided that $d = 0$.

(d) The ratio $Q_7/(Q_3 + Q_6)$ times what constant has an F-distribution, provided that $d = 0$? Note that this F is really the square of the two-sample T used to test the equality of the mean of the first distribution and the common mean of the other distributions, in which the last $b-1$ samples are combined into one.

9.2.6. Let μ_1, μ_2, μ_3 be, respectively, the means of three normal distributions with a common but unknown variance σ^2. In order to test, at the $\alpha = 5$ percent significance level, the hypothesis $H_0 : \mu_1 = \mu_2 = \mu_3$ against all possible alternative hypotheses, we take an independent random sample of size 4 from each of these distributions. Determine whether we accept or reject H_0 if the observed values from these three distributions are, respectively,

$$
\begin{array}{ccccc}
X_1 : & 5 & 9 & 6 & 8 \\
X_2 : & 11 & 13 & 10 & 12 \\
X_3 : & 10 & 6 & 9 & 9
\end{array}
$$

9.2.7. The driver of a diesel-powered automobile decided to test the quality of three types of diesel fuel sold in the area based on mpg. Test the null hypothesis that the three means are equal using the following data. Make the usual assumptions and take $\alpha = 0.05$.

$$
\begin{array}{lccccc}
\text{Brand A:} & 38.7 & 39.2 & 40.1 & 38.9 & \\
\text{Brand B:} & 41.9 & 42.3 & 41.3 & & \\
\text{Brand C:} & 40.8 & 41.2 & 39.5 & 38.9 & 40.3
\end{array}
$$

9.3 Noncentral χ^2 and F Distributions

Let $X_1, X_2, \ldots, X_n$ denote independent random variables that are $N(\mu_i, \sigma^2)$, $i = 1, 2, \ldots, n$, and consider the quadratic form $Y = \sum_1^n X_i^2/\sigma^2$. If each μ_i is zero, we know that Y is $\chi^2(n)$. We shall now investigate the distribution of Y when each μ_i is not zero. The mgf of Y is given by

$$M(t) = E\left[\exp\left(t\sum_{i=1}^n \frac{X_i^2}{\sigma^2}\right)\right]$$

$$= \prod_{i=1}^n E\left[\exp\left(t\frac{X_i^2}{\sigma^2}\right)\right].$$

Consider

$$E\left[\exp\left(\frac{tX_i^2}{\sigma^2}\right)\right] = \int_{-\infty}^{\infty} \frac{1}{\sigma\sqrt{2\pi}}\exp\left[\frac{tx_i^2}{\sigma^2} - \frac{(x_i - \mu_i)^2}{2\sigma^2}\right]\,dx_i.$$

The integral exists if $t < \frac{1}{2}$. To evaluate the integral, note that

$$\frac{tx_i^2}{\sigma^2} - \frac{(x_i - \mu_i)^2}{2\sigma^2} = -\frac{x_i^2(1 - 2t)}{2\sigma^2} + \frac{2\mu_i x_i}{2\sigma^2} - \frac{\mu_i^2}{2\sigma^2}$$

$$= \frac{t\mu_i^2}{\sigma^2(1 - 2t)} - \frac{1 - 2t}{2\sigma^2}\left(x_i - \frac{\mu_i}{1 - 2t}\right)^2.$$

Accordingly, with $t < \frac{1}{2}$, we have

$$E\left[\exp\left(\frac{tX_i^2}{\sigma^2}\right)\right] = \exp\left[\frac{t\mu_i^2}{\sigma^2(1 - 2t)}\right]\int_{-\infty}^{\infty} \frac{1}{\sigma\sqrt{2\pi}}\exp\left[-\frac{1 - 2t}{2\sigma^2}\left(x_i - \frac{\mu_i}{1 - 2t}\right)^2\right]\,dx_i.$$

If we multiply the integrand by $\sqrt{1 - 2t}$, $t < \frac{1}{2}$, we have the integral of a normal pdf with mean $\mu_i/(1 - 2t)$ and variance $\sigma^2/(1 - 2t)$. Thus

$$E\left[\exp\left(\frac{tX_i^2}{\sigma^2}\right)\right] = \frac{1}{\sqrt{1 - 2t}}\exp\left[\frac{t\mu_i^2}{\sigma^2(1 - 2t)}\right],$$

and the mgf of $Y = \sum_1^n X_i^2/\sigma^2$ is given by

$$M(t) = \frac{1}{(1 - 2t)^{n/2}}\exp\left[\frac{t\sum_1^n \mu_i^2}{\sigma^2(1 - 2t)}\right], \quad t < \frac{1}{2}. \tag{9.3.1}$$

A random variable that has an mgf of the functional form

$$M(t) = \frac{1}{(1 - 2t)^{r/2}}e^{t\theta/(1 - 2t)}, \tag{9.3.2}$$

where $t < \frac{1}{2}$, $0 < \theta$, and r is a positive integer, is said to have a **noncentral chi-square distribution** with r degrees of freedom and noncentrality parameter θ. If one sets the noncentrality parameter $\theta = 0$, one has $M(t) = (1 - 2t)^{-r/2}$, which is the mgf of a random variable that is $\chi^2(r)$. Such a random variable can appropriately be called a **central chi-square variable**. We shall use the symbol $\chi^2(r, \theta)$ to denote a noncentral chi-square distribution that has the parameters r and θ; and we shall say that a random variable is $\chi^2(r, \theta)$ when that random variable has this kind of distribution. The symbol $\chi^2(r, 0)$ is equivalent to $\chi^2(r)$. Thus our random variable $Y = \sum_{1}^{n} X_i^2/\sigma^2$ of this section is $\chi^2\left(n, \sum_{1}^{n} \mu_i^2/\sigma^2\right)$. If each μ_i is equal to zero, then Y is $\chi^2(n, 0)$ or, more simply, Y is $\chi^2(n)$.

The noncentral chi-square variables in which we have interest are certain quadratic forms in normally distributed variables divided by a variance σ^2. In our example it is worth noting that the noncentrality parameter of $\sum_{1}^{n} X_i^2/\sigma^2$, which is $\sum_{1}^{n} \mu_i^2/\sigma^2$, may be computed by replacing each X_i in the quadratic form by its mean μ_i, $i = 1, 2, \ldots, n$. This is no fortuitous circumstance; any quadratic form $Q = Q(X_1, \ldots, X_n)$ in normally distributed variables, which is such that Q/σ^2 is $\chi^2(r, \theta)$, has $\theta = Q(\mu_1, \mu_2, \ldots, \mu_n)/\sigma^2$; and if Q/σ^2 is a chi-square variable (central or noncentral) for certain real values of $\mu_1, \mu_2, \ldots, \mu_n$, it is chi-square (central or noncentral) for *all* real values of these means.

It should be pointed out that Theorem 9.1.1, Section 9.1, is valid whether the random variables are central or noncentral chi-square variables.

We next discuss the noncentral F-distribution. If U and V are independent and are, respectively, $\chi^2(r_1)$ and $\chi^2(r_2)$, the random variable F has been defined by $F = r_2 U/r_1 V$. Now suppose, in particular, that U is $\chi^2(r_1, \theta)$, V is $\chi^2(r_2)$, and that U and V are independent. The distribution of the random variable $r_2 U/r_1 V$ is called a **noncentral F-distribution** with r_1 and r_2 degrees of freedom with noncentrality parameter θ. Note that the noncentrality parameter of F is precisely the noncentrality parameter of the random variable U, which is $\chi^2(r_1, \theta)$.

There are R and S-PLUS commands which will compute the cdf of noncentral χ^2 and F random variables. For example, suppose we want to compute $P(Y \leq y)$ where Y has χ^2-distribution with d degrees of freedom and noncentrality parameter b. This probability is returned with the command `pchisq(y,d,b)`. The corresponding value of the pdf at y is computed by the command `dchisq(y,d,b)`. As another example, suppose we want $P(W \geq w)$ where W has an F-distribution with n1 and n2 degrees of freedom and noncentrality parameter b. This is computed by the command `1-pf(w,n1,n2,b)`, while the command `df(w,n1,n2,b)` computes the value of the density of W at w. Tables of the noncentral chi-square and noncentral F-distributions are available in the literature, also.

EXERCISES

9.3.1. Let Y_i, $i = 1, 2, \ldots, n$, denote independent random variables that are, respectively, $\chi^2(r_i, \theta_i)$, $i = 1, 2, \ldots, n$. Prove that $Z = \sum_1^n Y_i$ is $\chi^2\left(\sum_1^n r_i, \sum_1^n \theta_i\right)$.

9.3.2. Compute the mean and the variance of a random variable that is $\chi^2(r, \theta)$.

9.3.3. Compute the mean of a random variable that has a noncentral F-distribution with degrees of freedom r_1 and $r_2 > 2$ and noncentrality parameter θ.

9.3.4. Show that the square of a noncentral T random variable is a noncentral F random variable.

9.3.5. Let X_1 and X_2 be two independent random variables. Let X_1 and $Y = X_1 + X_2$ be $\chi^2(r_1, \theta_1)$ and $\chi^2(r, \theta)$, respectively. Here $r_1 < r$ and $\theta_1 \leq \theta$. Show that X_2 is $\chi^2(r - r_1, \theta - \theta_1)$.

9.3.6. In Exercise 9.2.1, if $\mu_1, \mu_2, \ldots, \mu_b$ are not equal, what are the distributions of Q_3'/σ^2, Q_4'/σ^2, and F?

9.4 Multiple Comparisons

Consider b independent random variables that have normal distributions with unknown means $\mu_1, \mu_2, \ldots, \mu_b$, respectively, and with unknown but common variance σ^2. Let $k_1, \ldots, k_b$ represent b known real constants that are not all zero. We want to find a confidence interval of $\sum_1^b k_j \mu_j$, a linear function of the means $\mu_1, \mu_2, \ldots, \mu_b$. To do this, we take a random sample $X_{1j}, X_{2j}, \ldots, X_{aj}$ of size a from the distribution $N(\mu_j, \sigma^2)$, $j = 1, 2, \ldots, b$. If we denote $\sum_{i=1}^a X_{ij}/a$ by $\overline{X}_{.j}$, then we know that $\overline{X}_{.j}$ is $N(\mu_j, \sigma^2/a)$, that $\sum_1^a (X_{ij} - \overline{X}_{.j})^2/\sigma^2$ is $\chi^2(a - 1)$, and that the two random variables are independent. Since the independent random samples are taken from the b distributions, the $2b$ random variables $\overline{X}_{.j}$, $\sum_1^a (X_{ij} - \overline{X}_{.j})^2/\sigma^2$, $j = 1, 2, \ldots, b$, are independent. Moreover, $\overline{X}_{.1}, \overline{X}_{.2}, \ldots, \overline{X}_{.b}$ and

$$\sum_{j=1}^b \sum_{i=1}^a \frac{(X_{ij} - \overline{X}_{.j})^2}{\sigma^2}$$

are independent and the latter is $\chi^2[b(a-1)]$. Let $Z = \sum_1^b k_j \overline{X}_{\cdot j}$. Then Z is normal with mean $\sum_1^b k_j \mu_j$ and variance $\left(\sum_1^b k_j^2\right) \sigma^2/a$, and Z is independent of

$$V = \frac{1}{b(a-1)} \sum_{j=1}^b \sum_{i=1}^a (X_{ij} - \overline{X}_{\cdot j})^2.$$

Hence the random variable

$$T = \frac{(\sum_1^b k_j \overline{X}_{\cdot j} - \sum_1^b k_j \mu_j)/\sqrt{(\sigma^2/a) \sum_1^b k_i^2}}{\sqrt{V/\sigma^2}} = \frac{\sum_1^b k_j \overline{X}_{\cdot j} - \sum_1^b k_j \mu_j}{\sqrt{(V/a) \sum_1^b k_j^2}}$$

has a t-distribution with $b(a-1)$ degrees of freedom. A positive number c can be found in Table IV in Appendix B, for certain values of α, $0 < \alpha < 1$, such that $P(-c \leq T \leq c) = 1 - \alpha$. It follows that the probability is $1 - \alpha$ that

$$\sum_1^b k_j \overline{X}_{\cdot j} - c\sqrt{\left(\sum_1^b k_j^2\right) \frac{V}{a}} \leq \sum_1^b k_j \mu_j \leq \sum_1^b k_j \overline{X}_{\cdot j} + c\sqrt{\left(\sum_1^b k_j^2\right) \frac{V}{a}}.$$

The observed values of $\overline{X}_{\cdot j}$, $j = 1, 2, \ldots, b$, and V will provide a $100(1 - \alpha)$ percent confidence interval for $\sum_1^b k_j \mu_j$.

It should be observed that the confidence interval for $\sum_1^b k_j \mu_j$ depends upon the particular choice of $k_1, k_2, \ldots, k_b$. It is conceivable that we may be interested in more than one linear function of $\mu_1, \mu_2, \ldots, \mu_b$, such as $\mu_2 - \mu_1$, $\mu_3 - (\mu_1 + \mu_2)/2$, or $\mu_1 + \cdots + \mu_b$. We can, of course, find for each $\sum_1^b k_j \mu_j$ a random interval that has a preassigned probability of including that particular $\sum_1^b k_j \mu_j$. But how can we compute the probability that *simultaneously* these random intervals include their respective linear functions of $\mu_1, \mu_2, \ldots, \mu_b$? The following procedure of **multiple comparisons**, due to Scheffé, is one solution to this problem.

The random variable

$$\frac{\sum_{j=1}^b (\overline{X}_{\cdot j} - \mu_j)^2}{\sigma^2/a}$$

is $\chi^2(b)$ and, because it is a function of $\overline{X}_{\cdot 1}, \ldots, \overline{X}_{\cdot b}$ alone, it is independent of the random variable

$$V = \frac{1}{b(a-1)} \sum_{j=1}^{b} \sum_{i=1}^{a} (X_{ij} - \overline{X}_{.j})^2.$$

Hence the random variable

$$F = \frac{a \sum_{j=1}^{b} (\overline{X}_{.j} - \mu_j)^2 / b}{V}$$

has an F-distribution with b and $b(a-1)$ degrees of freedom. From Table V in Appendix B, for certain values of α, we can find a constant d such that $P(F \leq d) = 1 - \alpha$ or

$$P\left[\sum_{j=1}^{b} (\overline{X}_{.j} - \mu_j)^2 \leq bd\frac{V}{a} \right] = 1 - \alpha.$$

Note that $\sum_{j=1}^{b} (\overline{X}_{.j} - \mu_j)^2$ is the square of the distance, in b-dimensional space, from the point $(\mu_1, \mu_2, \ldots, \mu_b)$ to the random point $(\overline{X}_{.1}, \overline{X}_{.2}, \ldots, \overline{X}_{.b})$. Consider a space of dimension b and let $(t_1, t_2, \ldots, t_b)$ denote the coordinates of a point in that space. An equation of a hyperplane that passes through the point $(\mu_1, \mu_2, \ldots, \mu_b)$ is given by

$$k_1(t_1 - \mu_1) + k_2(t_2 - \mu_2) + \cdots + k_b(t_b - \mu_b) = 0, \tag{9.4.1}$$

where not all the real numbers k_j, $j = 1, 2, \ldots, b$, are equal to zero. The square of the distance from this hyperplane to the point $(t_1 = \overline{X}_{.1}, t_2 = \overline{X}_{.2}, \ldots, t_b = \overline{X}_{.b})$ is

$$\frac{[k_1(\overline{X}_{.1} - \mu_1) + k_2(\overline{X}_{.2} - \mu_2) + \cdots + k_b(\overline{X}_{.b} - \mu_b)]^2}{k_1^2 + k_2^2 + \cdots + k_b^2}. \tag{9.4.2}$$

From the geometry of the situation it follows that $\sum_{1}^{b} (\overline{X}_{.j} - \mu_j)^2$ is equal to the maximum of expression (9.4.2) with respect to $k_1, k_2, \ldots, k_b$. Thus the inequality $\sum_{1}^{b} (\overline{X}_{.j} - \mu_j)^2 \leq (bd)(V/a)$ holds if and only if

$$\frac{\left[\sum_{1}^{b} k_j(\overline{X}_{.j} - \mu_j) \right]^2}{\sum_{j=1}^{b} k_j^2} \leq bd\frac{V}{a}, \tag{9.4.3}$$

for every real $k_1, k_2, \ldots, k_b$, not all zero. Accordingly, these two equivalent events have the same probability, $1 - \alpha$. However, inequality (9.4.3) may be written in the form

$$\left| \sum_1^b k_j \overline{X}_{.j} - \sum_1^b k_j \mu_j \right| \leq \sqrt{bd \left(\sum_1^b k_j^2 \right) \frac{V}{a}}.$$

Thus the probability is $1 - \alpha$ that simultaneously, for *all* real $k_1, k_2, \ldots, k_b$, not all zero,

$$\sum_1^b k_j \overline{X}_{.j} - \sqrt{bd \left(\sum_1^b k_j^2 \right) \frac{V}{a}} \leq \sum_1^b k_j \mu_j \leq \sum_1^b k_j \overline{X}_{.j} + \sqrt{bd \left(\sum_1^b k_j^2 \right) \frac{V}{a}}. \quad (9.4.4)$$

Denote by A the event where inequality (9.4.4) is true for all real $k_1, \ldots, k_b$, and denote by B the event where that inequality is true for a finite number of b-tuples $(k_1, \ldots, k_b)$. If A occurs, then B occurs; hence, $P(A) \leq P(B)$. In the applications, one is often interested only in a finite number of linear functions $\sum_1^b k_j \mu_j$. Once the observed values are available, we obtain from (9.4.4) a confidence interval for each of these linear functions. Since $P(B) \geq P(A) = 1 - \alpha$, we have a confidence coefficient of at least $100(1 - \alpha)$ percent that the linear functions are in these respective confidence intervals.

Remark 9.4.1. If the sample sizes, say $a_1, a_2, \ldots, a_b$, are unequal, inequality (9.4.4) becomes

$$\sum_1^b k_j \overline{X}_{.j} - \sqrt{bd \left(\sum_1^b \frac{k_j^2}{a_j} \right) V} \leq \sum_1^b k_j \mu_j \leq \sum_1^b k_j \overline{X}_{.j} + \sqrt{bd \left(\sum_1^b \frac{k_j^2}{a_j} \right) V}, \quad (9.4.5)$$

where

$$\overline{X}_{.j} = \frac{\sum_{i=1}^{a_j} X_{ij}}{a_j}, \quad V = \frac{\sum_{j=1}^{b} \sum_{i=1}^{a_j} (X_{ij} - \overline{X}_{.j})^2}{\sum_1^b (a_j - 1)},$$

and d is selected from Table V with b and $\sum_1^b (a_j - 1)$ degrees of freedom. Inequality (9.4.5) reduces to inequality (9.4.4) when $a_1 = a_2 = \cdots = a_b$.

Moreover, if we restrict our attention to linear functions of the form $\sum_1^b k_j \mu_j$ with $\sum_1^b k_j = 0$ (such linear functions are called *contrasts*), the radical in inequality (9.4.5) is replaced by

$$\sqrt{d(b - 1) \sum_1^b \frac{k_j^2}{a_j} V},$$

where d is now found in Table V with $b - 1$ and $\sum_1^b (a_j - 1)$ degrees of freedom. ∎

In multiple comparisons based on the Scheffé procedure, one often finds that the length of a confidence interval is much greater than the length of a $100(1 - \alpha)$ percent confidence interval for a particular linear function $\sum_1^b k_j \mu_j$. But this is to be expected because in one case the probability $1 - \alpha$ applies to just one event, and in the other it applies to the simultaneous occurrence of many events. One reasonable way to reduce the length of these intervals is to take a larger value of α, say 0.25, instead of 0.05. After all, it is still a very strong statement to say that the probability is 0.75 that *all* these events occur. There are, however, other multiple comparison procedures which are often used in practice. One of these is the Bonferroni procedure described in Exercise 9.4.2. This procedure can be used for a finite number of confidence intervals and, as Exercise 9.4.3 shows, the concept is easily extended to tests of hypotheses. In the case of the $\binom{b}{2}$ pairwise comparisons of means, i.e. comparisons of the form $\mu_i - \mu_j$, the procedure most often used is the Tukey-Kramer procedure; see Miller (1981) and Hsu (1996) for discussion.

EXERCISES

9.4.1. If $A_1, A_2, \ldots, A_k$ are events, prove, by induction, Boole's inequality

$$P(A_1 \cup A_2 \cup \cdots \cup A_k) \leq \sum_1^k P(A_i).$$

Then show that

$$P(A_1^c \cap A_2^c \cap \cdots \cap A_k^c) \geq 1 - \sum_1^b P(A_i).$$

9.4.2 (Bonferroni Multiple Comparison Procedure). In the notation of this section, let $(k_{i1}, k_{i2}, \ldots, k_{ib})$, $i = 1, 2, \ldots, m$, represent a finite number of b-tuples. The problem is to find simultaneous confidence intervals for $\sum_{j=1}^{b} k_{ij} \mu_j$, $i = 1, 2, \ldots, m$, by a method different from that of Scheffé. Define the random variable T_i by

$$\left(\sum_{j=1}^{b} k_{ij} \overline{X}_{\cdot j} - \sum_{j=1}^{b} k_{ij} \mu_j \right) \Big/ \sqrt{\left(\sum_{j=1}^{b} k_{ij}^2 \right) V/a}, \quad i = 1, 2, \ldots, m.$$

(a) Let the event A_i^c be given by $-c_i \leq T_i \leq c_i$, $i = 1, 2, \ldots, m$. Find the random variables U_i and W_i such that $U_i \leq \sum_1^b k_{ij} \mu_j \leq W_j$ is equivalent to A_i^c.

(b) Select c_i such that $P(A_i^c) = 1 - \alpha/m$; that is, $P(A_i) = \alpha/m$. Use Exercise 9.4.1 to determine a lower bound on the probability that simultaneously the random intervals $(U_1, W_1), \ldots, (U_m, W_m)$ include $\sum_{j=1}^{b} k_{1j} \mu_j, \ldots, \sum_{j=1}^{b} k_{mj} \mu_j$, respectively.

(c) Let $a = 3$, $b = 6$, and $\alpha = 0.05$. Consider the linear functions $\mu_1 - \mu_2$, $\mu_2 - \mu_3$, $\mu_3 - \mu_4$, $\mu_4 - (\mu_5 + \mu_6)/2$, and $(\mu_1 + \mu_2 + \cdots + \mu_6)/6$. Here $m = 5$. Show that the lengths of the confidence intervals given by the results of Part (b) are shorter than the corresponding ones given by the method of Scheffé as described in the text. If m becomes sufficiently large, however, this is not the case.

9.4.3. Extend the Bonferroni procedure described in the last problem to simultaneous testing. That is, suppose we have m hypotheses of interest: H_{0i} versus H_{1i}, $i = 1, \ldots, m$. For testing H_{0i} versus H_{1i}, let $C_{i,\alpha}$ be a critical region of size α and assume H_{0i} is rejected if $\mathbf{X}_i \in C_{i,\alpha}$, for a sample $\mathbf{X}_i$. Determine a rule so that we can simultaneously test these m hypotheses with a Type I error rate less than or equal to α.

9.5   The Analysis of Variance

Recall the one-way analysis of variance (ANOVA) problem considered in Section 9.2 which was concerned with one factor at b levels. In this section, we are concerned with the situation where we have two factors A and B with levels a and b, respectively. Let X_{ij}, $i = 1, 2, \ldots, a$ and $j = 1, 2, \ldots, b$, denote the response for Factor A at level i and Factor B at level j. Denote the total sample size by $n = ab$. We shall assume that the X_{ij}'s are independent normally distributed random variables with common variance σ^2. Denote the mean of X_{ij} by μ_{ij}. The mean μ_{ij} is often referred to as the mean of the $(i, j)th$ cell. For our first model, we will consider the **additive model** where

$$\mu_{ij} = \overline{\mu} + (\overline{\mu}_{i.} - \overline{\mu}) + (\overline{\mu}_{.j} - \overline{\mu}) \; ; \qquad (9.5.1)$$

that is, the mean in the $(i, j)th$ cell is due to additive effects of the levels, i of Factor A and j of Factor B, over the average (constant) $\overline{\mu}$. Let $\alpha_i = \overline{\mu}_{i.} - \overline{\mu}$, $i = 1, \ldots, a$; $\beta_j = \overline{\mu}_{.j} - \overline{\mu}$, $j = 1, \ldots, b$; and $\mu = \overline{\mu}$. Then the model can be written more simply as,

$$\mu_{ij} = \mu + \alpha_i + \beta_j, \qquad (9.5.2)$$

where $\sum_{i=1}^{a} \alpha_i = 0$ and $\sum_{j=1}^{b} \beta_j = 0$. We refer to this model as being a **two-way** ANOVA model.

For example, take $a = 2$, $b = 3$, $\mu = 5$, $\alpha_1 = 1$, $\alpha_2 = -1$, $\beta_1 = 1$, $\beta_2 = 0$, and $\beta_3 = -1$. Then the cell means are:

		Factor B		
		1	2	3
Factor A	1	$\mu_{11} = 7$	$\mu_{12} = 6$	$\mu_{13} = 5$
	2	$\mu_{21} = 5$	$\mu_{22} = 4$	$\mu_{23} = 3$

Note that for each i, the plots of μ_{ij} versus j are parallel. This is true for additive models in general; see Exercise 9.5.8. We will call these plots **mean profile plots**. Had we taken $\beta_1 = \beta_2 = \beta_3 = 0$, then the cell means would be:

		Factor B	
	1	2	3
Factor A 1	$\mu_{11} = 6$	$\mu_{12} = 6$	$\mu_{13} = 6$
2	$\mu_{21} = 4$	$\mu_{22} = 4$	$\mu_{23} = 4$

The hypotheses of interest are,

$$H_{0A}:\ \alpha_1 = \cdots = \alpha_a = 0 \text{ versus } H_{1A}:\ \alpha_i \neq 0, \text{ for some } i, \tag{9.5.3}$$

and

$$H_{0B}:\ \beta_1 = \cdots = \beta_b = 0 \text{ versus } H_{1B}:\ \beta_j \neq 0, \text{ for some } j. \tag{9.5.4}$$

If H_{0A} is true then by (9.5.2) the mean of the $(i, j)th$ cell does not depend on the level of A. The second example above is under H_{0B}. The cell means remain the same from column to column for a specified row. We call these hypotheses **main effect** hypotheses.

Remark 9.5.1. The model just described, and others similar to it, are widely used in statistical applications. Consider a situation in which it is desirable to investigate the effects of two factors that influence an outcome. Thus the variety of a grain and the type of fertilizer used influence the yield; or the teacher and the size of the class may influence the score on a standardized test. Let X_{ij} denote the yield from the use of variety i of a grain and type j of fertilizer. A test of the hypothesis that $\beta_1 = \beta_2 = \cdots = \beta_b = 0$ would then be a test of the hypothesis that the mean yield of each variety of grain is the same regardless of the type of fertilizer used. ∎

To construct a test of the composite hypothesis H_{0B} versus H_{1B}, we could obtain the corresponding likelihood ratio. However, to gain more insight into such a test, let us reconsider the likelihood ratio test of Section 9.2, namely that of the equality of the means of b distributions. There the important quadratic forms are Q, Q_3, and Q_4, which are related through the equation $Q = Q_4 + Q_3$. That is,

$$(ab - 1)S^2 = \sum_{j=1}^{b}\sum_{i=1}^{a}(\overline{X}_{.j} - \overline{X}_{..})^2 + \sum_{j=1}^{b}\sum_{i=1}^{a}(X_{ij} - \overline{X}_{.j})^2,$$

so we see that the total sum of squares, $(ab - 1)S^2$, is decomposed into a sum of squares, Q_4, *among* columns means and a sum of squares, Q_3, *within* columns. The latter sum of squares, divided by $n = ab$, is the mle of σ^2, provided that the parameters are in Ω; and we denote it by $\hat{\sigma}_\Omega^2$. Of course, $(ab - 1)S^2/ab$ is the mle of σ^2 under ω, here denoted by $\hat{\sigma}_\omega^2$. So the likelihood ratio $\Lambda = (\hat{\sigma}_\Omega^2/\hat{\sigma}_\omega^2)^{ab/2}$ is a monotone function of the statistic

$$F = \frac{Q_4/(b - 1)}{Q_3/[b(a - 1)]}$$

upon which the test of the equality of means is based.

To help find a test for H_{0B} versus H_{1B}, (9.5.4), return to the decomposition of Example 9.1.3, Section 9.1, namely $Q = Q_2 + Q_4 + Q_5$. That is

$$(ab-1)S^2 = \sum_{i=1}^{a}\sum_{j=1}^{b}(\overline{X}_{i.} - \overline{X}_{..})^2 + \sum_{i=1}^{a}\sum_{j=1}^{b}(\overline{X}_{.j} - \overline{X}_{..})^2 + \sum_{i=1}^{a}\sum_{j=1}^{b}(X_{ij} - \overline{X}_{i.} - \overline{X}_{.j} + \overline{X}_{..})^2,$$

thus the total sum of squares is decomposed into that among *rows* (Q_2), that among *columns* (Q_4), and that *remaining* (Q_5). It is interesting to observe that $\hat{\sigma}_{\Omega}^2 = Q_5/ab$ is the mle of σ^2 under Ω and

$$\hat{\sigma}_{\omega}^2 = \frac{(Q_4 + Q_5)}{ab} = \sum_{i=1}^{a}\sum_{j=1}^{b} \frac{(X_{ij} - \overline{X}_{i.})^2}{ab}$$

is that estimator under ω. A useful monotone function of the likelihood ratio $\Lambda = (\hat{\sigma}_{\Omega}^2/\hat{\sigma}_{\omega}^2)^{ab/2}$ is

$$F = \frac{Q_4/(b-1)}{Q_5/[(a-1)(b-1)]},$$

which has, under H_0, an F-distribution with $b-1$ and $(a-1)(b-1)$ degrees of freedom. The hypothesis H_0 is rejected if $F \geq c$, where $\alpha = P_{H_0}(F \geq c)$. This is the likelihood ratio test for H_{0B} versus H_{1B}.

If we are to compute the power function of the test, we need the distribution of F when H_{0B} is not true. From Section 9.3 we know, when H_{1B} is true, that Q_4/σ^2 and Q_5/σ^2 are independent (central or noncentral) chi-square variables. We shall compute the noncentrality parameters of Q_4/σ^2 and Q_5/σ^2 when H_1 is true. We have $E(X_{ij}) = \mu + \alpha_i + \beta_j$, $E(\overline{X}_{i.}) = \mu + \alpha_i$, $E(\overline{X}_{.j}) = \mu + \beta_j$ and $E(\overline{X}_{..}) = \mu$. Accordingly, the noncentrality parameter Q_4/σ^2 is

$$\frac{a}{\sigma^2}\sum_{j=1}^{b}(\mu + \beta_j - \mu)^2 = \frac{a}{\sigma^2}\sum_{j=1}^{b}\beta_j^2$$

and that of Q_5/σ^2 is

$$\sigma^{-2}\sum_{j=1}^{b}\sum_{i=1}^{a}(\mu + \alpha_i + \beta_j - \mu - \alpha_i - \mu - \beta_j + \mu)^2 = 0.$$

Thus, if the hypothesis H_{0B} is not true, F has a noncentral F-distribution with $b-1$ and $(a-1)(b-1)$ degrees of freedom and noncentrality parameter $a\sum_{j=1}^{b}\beta_j^2/\sigma^2$. The desired probabilities can then be found in tables of the noncentral F-distribution.

A similar argument can be used to construct the F needed to test the equality of row means; that is, H_{0A} versus H_{1A}, (9.5.3). The F test statistic is essentially the ratio of the sum of squares among rows and Q_5. In particular, this F is defined by

$$F = \frac{Q_2/(a-1)}{Q_5/[(a-1)(b-1)]}$$

and under $H_0 : \alpha_1 = \alpha_2 = \cdots = \alpha_a = 0$, has an F-distribution with $a-1$ and $(a-1)(b-1)$ degrees of freedom.

The analysis-of-variance problem that has just been discussed is usually referred to as a *two-way classification with one observation per cell*. Each combination of i

and j determines a cell; thus there is a total of ab cells in this model. Let us now investigate another two-way classification problem, but in this case we take $c > 1$ independent observations per cell.

Let X_{ijk}, $i = 1, 2, \ldots, a$, $j = 1, 2, \ldots, b$, and $k = 1, 2, \ldots, c$, denote $n = abc$ random variables which are independent and which have normal distributions with common, but unknown, variance σ^2. Denote the the mean of each X_{ijk}, $k = 1, 2, \ldots, c$ by μ_{ij}. Under the additive model, (9.5.1), the mean of each cell depended on its row and column, but often the mean is cell specific. To allow this, consider the parameters,

$$
\begin{aligned}
\gamma_{ij} &= \mu_{ij} - \{\mu + (\overline{\mu}_{i.} - \mu) + (\overline{\mu}_{.j} - \mu)\} \\
&= \mu_{ij} - \overline{\mu}_{i.} - \overline{\mu}_{.j} + \mu,
\end{aligned}
$$

for $i = 1, \ldots a, j = 1, \ldots, b$. Hence γ_{ij} reflects the specific contribution to the cell mean over and above the additive model. These parameters are called **interaction parameters**. Using the second form (9.5.2), we can write the cell means as,

$$
\mu_{ij} = \mu + \alpha_i + \beta_j + \gamma_{ij}, \tag{9.5.5}
$$

where $\sum_{i=1}^{a} \alpha_i = 0$, $\sum_{j=1}^{b} \beta_j = 0$, and $\sum_{i=1}^{a} \gamma_{ij} = \sum_{j=1}^{b} \gamma_{ij} = 0$. This model is called a **two-way** model with interaction.

For example, take $a = 2$, $b = 3$, $\mu = 5$, $\alpha_1 = 1$, $\alpha_2 = -1$, $\beta_1 = 1$, $\beta_2 = 0$, $\beta_3 = -1$, $\gamma_{11} = 1$, $\gamma_{12} = 1$, $\gamma_{13} = -2$, $\gamma_{21} = -1$, $\gamma_{22} = -1$, and $\gamma_{23} = 2$. Then the cell means are

		Factor B		
		1	2	3
Factor A	1	$\mu_{11} = 8$	$\mu_{12} = 7$	$\mu_{13} = 3$
	2	$\mu_{21} = 4$	$\mu_{22} = 3$	$\mu_{23} = 5$

Note that, if each $\gamma_{ij} = 0$, then the cell means are:

		Factor B		
		1	2	3
Factor A	1	$\mu_{11} = 7$	$\mu_{12} = 6$	$\mu_{13} = 5$
	2	$\mu_{21} = 5$	$\mu_{22} = 4$	$\mu_{23} = 3$

Note that the mean profile plots for this second example are parallel but the first (where interaction is present) are not.

The major hypotheses of interest for the interaction model are:

$$
H_{0AB} : \gamma_{ij} = 0 \text{ for all } i, j \text{ versus } H_{1AB} : \gamma_{ij} \neq 0, \text{ for some } i, j. \tag{9.5.6}
$$

From Exercise 9.1.4 of Section 9.1 we have that

$$\sum_{i=1}^{a}\sum_{j=1}^{b}\sum_{k=1}^{c}(X_{ijk} - \overline{X}_{...})^2 = bc\sum_{i=1}^{a}(\overline{X}_{i..} - \overline{X}_{...})^2 + ac\sum_{j=1}^{b}(\overline{X}_{.j.} - \overline{X}_{...})^2$$

$$+ c\sum_{i=1}^{a}\sum_{j=1}^{b}(\overline{X}_{ij.} - \overline{X}_{i..} - \overline{X}_{.j.} + \overline{X}_{...})^2$$

$$+ \sum_{i=1}^{a}\sum_{j=1}^{b}\sum_{k=1}^{c}(X_{ijk} - \overline{X}_{ij.})^2,$$

that is, the total sum of squares is decomposed into that due to *row* differences, that due to *column* differences, that due to *interaction,* and that *within cells.* The test of H_{0AB} versus H_{1AB}, is based upon an F with $(a-1)(b-1)$ and $ab(c-1)$ degrees of freedom.

$$F = \frac{\left[c\sum_{i=1}^{a}\sum_{j=1}^{b}(\overline{X}_{ij.} - \overline{X}_{i..} - \overline{X}_{.j.} + \overline{X}_{...})^2\right] / [(a-1)(b-1)]}{[\sum\sum\sum(X_{ijk} - \overline{X}_{ij.})^2] / [ab(c-1)]}.$$

The reader should verify that the noncentrality parameter of this F-distribution is equal to $c\sum_{j=1}^{b}\sum_{i=1}^{a}\gamma_{ij}^2/\sigma^2$. Thus F is central when $H_0 : \gamma_{ij} = 0$, $i = 1, 2, \ldots, a$, $j = 1, 2, \ldots, b$, is true.

If $H_0 : \gamma_{ij} = 0$ is accepted, then one usually continues to test $\alpha_i = 0$, $i = 1, 2, \ldots, a$, by using the test statistic

$$F = \frac{bc\sum_{i=1}^{a}(\overline{X}_{i..} - \overline{X}_{...})^2/(a-1)}{\sum_{i=1}^{a}\sum_{j=1}^{b}\sum_{k=1}^{c}(X_{ijk} - \overline{X}_{ij.})^2/[ab(c-1)]},$$

which has a null F-distribution with $a-1$ and $ab(c-1)$ degrees of freedom. Similarly the test of $\beta_j = 0$, $j = 1, 2, \ldots, b$, proceeds by using the test statistic

$$F = \frac{ac\sum_{j=1}^{b}(\overline{X}_{.j.} - \overline{X}_{...})^2/(b-1)}{\sum_{i=1}^{a}\sum_{j=1}^{b}\sum_{k=1}^{c}(X_{ijk} - \overline{X}_{ij.})^2/[ab(c-1)]},$$

which has a null F-distribution with $b-1$ and $ab(c-1)$ degrees of freedom.

EXERCISES

9.5.1. Show that

$$\sum_{j=1}^{b}\sum_{i=1}^{a}(X_{ij} - \overline{X}_{i.})^2 = \sum_{j=1}^{b}\sum_{i=1}^{a}(X_{ij} - \overline{X}_{i.} - \overline{X}_{.j} + \overline{X}_{..})^2 + a\sum_{j=1}^{b}(\overline{X}_{.j} - \overline{X}_{..})^2.$$

9.5.2. If at least one $\gamma_{ij} \neq 0$, show that the F, which is used to test that each interaction is equal to zero, has noncentrality parameter equal to $c\sum_{j=1}^{b}\sum_{i=1}^{a}\gamma_{ij}^2/\sigma^2$.

9.5.3. Using the background of the two-way classification with one observation per cell, show that the maximum likelihood estimator of α_i, β_j, and μ are $\hat{\alpha}_i = \overline{X}_{i.} - \overline{X}_{..}$, $\hat{\beta}_j = \overline{X}_{.j} - \overline{X}_{..}$, and $\hat{\mu} = \overline{X}_{..}$, respectively. Show that these are unbiased estimators of their respective parameters and compute $\text{var}(\hat{\alpha}_i)$, $\text{var}(\hat{\beta}_j)$, and $\text{var}(\hat{\mu})$.

9.5.4. Prove that the linear functions $X_{ij} - \overline{X}_{i.} - \overline{X}_{.j} + \overline{X}_{..}$ and $\overline{X}_{.j} - \overline{X}_{..}$ are uncorrelated, under the assumptions of this section.

9.5.5. Given the following observations associated with a two-way classification with $a = 3$ and $b = 4$, compute the F-statistic used to test the equality of the column means $(\beta_1 = \beta_2 = \beta_3 = \beta_4 = 0)$ and the equality of the row means $(\alpha_1 = \alpha_2 = \alpha_3 = 0)$, respectively.

Row/Column	1	2	3	4
1	3.1	4.2	2.7	4.9
2	2.7	2.9	1.8	3.0
3	4.0	4.6	3.0	3.9

9.5.6. With the background of the two-way classification with $c > 1$ observations per cell, show that the maximum likelihood estimators of the parameters are

$$\hat{\alpha}_i = \overline{X}_{i..} - \overline{X}_{...}$$
$$\hat{\beta}_j = \overline{X}_{.j.} - \overline{X}_{...}$$
$$\hat{\gamma}_{ij} = \overline{X}_{ij.} - \overline{X}_{i..} - \overline{X}_{.j.} + \overline{X}_{...}$$
$$\hat{\mu} = \overline{X}_{...}.$$

Show that these are unbiased estimators of the respective parameters. Compute the variance of each estimator.

9.5.7. Given the following observations in a two-way classification with $a = 3$, $b = 4$, and $c = 2$, compute the F-statistics used to test that all interactions are equal to zero $(\gamma_{ij} = 0)$, all column means are equal $(\beta_j = 0)$, and all row means are equal $(\alpha_i = 0)$, respectively.

Row/Column	1	2	3	4
1	3.1	4.2	2.7	4.9
	2.9	4.9	3.2	4.5
2	2.7	2.9	1.8	3.0
	2.9	2.3	2.4	3.7
3	4.0	4.6	3.0	3.9
	4.4	5.0	2.5	4.2

9.5.8. For the additive model (9.5.1) show that the mean profile plots are parallel. The sample mean profile plots are given by, plot $\overline{X}_{ij}$ versus j, for each i. These offer a graphical diagnostic for interaction detection. Obtain these plots for the last exercise.

9.5.9. We wish to compare compressive strengths of concrete corresponding to $a = 3$ different drying methods (treatments). Concrete is mixed in batches that are just large enough to produce three cylinders. Although care is taken to achieve uniformity, we expect some variability among the $b = 5$ batches used to obtain the following compressive strengths. (There is little reason to suspect interaction and hence only one observation is taken in each cell.)

Treatment	Batch				
	B_1	B_2	B_3	B_4	B_5
A_1	52	47	44	51	42
A_2	60	55	49	52	43
A_3	56	48	45	44	38

(a) Use the 5 percent significance level and test $H_A : \alpha_1 = \alpha_2 = \alpha_3 = 0$ against all alternatives.

(b) Use the 5 percent significance level and test $H_B : \beta_1 = \beta_2 = \beta_3 = \beta_4 = \beta_5 = 0$ against all alternatives.

9.5.10. With $a = 3$ and $b = 4$, find μ, α_i, β_j and γ_{ij} if μ_{ij}, for $i = 1, 2, 3$ and $j = 1, 2, 3, 4$, are given by

$$
\begin{array}{cccc}
6 & 7 & 7 & 12 \\
10 & 3 & 11 & 8 \\
8 & 5 & 9 & 10
\end{array}
$$

9.6 A Regression Problem

There is often interest in the relation between two variables, for example, a student's scholastic aptitude test score in mathematics and this same student's grade in calculus. Frequently, one of these variables, say x, is known in advance of the other, and hence there is interest in predicting a future random variable Y. Since Y is a random variable, we cannot predict its future observed value $Y = y$ with certainty. Thus let us first concentrate on the problem of estimating the mean of Y, that is, $E(Y)$. Now $E(Y)$ is usually a function of x; for example, in our illustration with the calculus grade, say Y, we would expect $E(Y)$ to increase with increasing mathematics aptitude score x. Sometimes $E(Y) = \mu(x)$ is assumed to be of a given form, such as a linear or quadratic or exponential function; that is, $\mu(x)$ could be assumed to be equal to $\alpha + \beta x$ or $\alpha + \beta x + \gamma x^2$ or $\alpha e^{\beta x}$. To estimate $E(Y) = \mu(x)$, or equivalently the parameters α, β, and γ, we observe the random variable Y for each of n possible different values of x, say $x_1, x_2, \ldots, x_n$, which are not all equal. Once the n independent experiments have been performed, we have n pairs of known numbers $(x_1, y_1), (x_2, y_2), \ldots, (x_n, y_n)$. These pairs are then used to estimate the mean $E(Y)$. Problems like this are often classified under *regression* because $E(Y) = \mu(x)$ is frequently called a regression curve.

Remark 9.6.1. A model for the mean such as $\alpha + \beta x + \gamma x^2$, is called a **linear model** because it is linear in the parameters α, β, and γ. Thus $\alpha e^{\beta x}$ is not a linear model because it is not linear in α and β. Note that, in Sections 9.1 to 9.4, all the means were linear in the parameters and hence, are linear models. ∎

Let us begin with the case in which $E(Y) = \mu(x)$ is a linear function. Denote by Y_i the response at x_i and consider the model,

$$Y_i = \alpha + \beta(x_i - \overline{x}) + e_i, \quad i = 1, \ldots, n, \tag{9.6.1}$$

where $\overline{x} = n^{-1} \sum_{i=1}^{n} x_i$ and $e_1, \ldots, e_n$ are iid random variables with a common $N(0, \sigma^2)$ distribution. Hence, $E(Y_i) = \alpha + \beta(x_i - \overline{x})$, $\text{Var}(Y_i) = \sigma^2$, and Y_i has $N(\alpha + \beta(x_i - \overline{x}), \sigma^2)$ distribution. The n points are $(x_1, y_1), (x_2, y_2), \ldots, (x_n, y_n)$; so the first problem is that of fitting a straight line to the set of points. Figure 9.6.1 shows a **scatterplot** of sixty observations $(x_1, y_1), \ldots, (x_{60}, y_{60})$ drawn from a linear model of the form (9.6.1).

The joint pdf of $Y_1, \ldots, Y_n$ is the product of the individual probability density functions; that is, the likelihood function equals

$$L(\alpha, \beta, \sigma^2) = \prod_{i=1}^{n} \frac{1}{\sqrt{2\pi\sigma^2}} \exp\left\{ -\frac{[y_i - \alpha - \beta(x_i - \overline{x})]^2}{2\sigma^2} \right\}$$

$$= \left(\frac{1}{2\pi\sigma^2} \right)^{n/2} \exp\left\{ -\frac{1}{2\sigma^2} \sum_{i=1}^{n} [y_i - \alpha - \beta(x_i - \overline{x})]^2 \right\}.$$

To maximize $L(\alpha, \beta, \sigma^2)$, or, equivalently, to minimize

$$-\log L(\alpha, \beta, \sigma^2) = \frac{n}{2} \log(2\pi\sigma^2) + \frac{\sum_{i=1}^{n} [y_i - \alpha - \beta(x_i - \overline{x})]^2}{2\sigma^2},$$

we must select α and β to minimize

$$H(\alpha, \beta) = \sum_{i=1}^{n} [y_i - \alpha - \beta(x_i - \overline{x})]^2.$$

Since $|y_i - \alpha - \beta(x_i - \overline{x})| = |y_i - \mu(x_i)|$ is the vertical distance from the point (x_i, y_i) to the line $y = \mu(x)$, (see the dashed-line segment in Figure 9.6.1), we note that $H(\alpha, \beta)$ represents the sum of the squares of those distances. Thus selecting α and

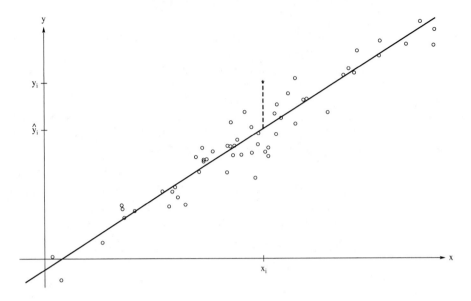

Figure 9.6.1: The plot shows the Least Squares fitted line (solid line) to a set of data. The dashed-line segment from $(x_i, \hat{y}_i)$ to (x_i, y_i) shows the deviation of (x_i, y_i) from its fit.

β so that the sum of the squares is minimized means that we are fitting the straight line to the data by the **method of least squares** (LS).

To minimize $H(\alpha, \beta)$, we find the two first partial derivatives

$$\frac{\partial H(\alpha, \beta)}{\partial \alpha} = 2 \sum_{i=1}^{n} [y_i - \alpha - \beta(x_i - \overline{x})](-1)$$

and

$$\frac{\partial H(\alpha, \beta)}{\partial \beta} = 2 \sum_{i=1}^{n} [y_i - \alpha - \beta(x_i - \overline{x})][-(x_i - \overline{x})].$$

Setting $\partial H(\alpha, \beta)/\partial \alpha = 0$, we obtain

$$\sum_{i=1}^{n} y_i - n\alpha - \beta \sum_{i=1}^{n} (x_i - \overline{x}) = 0. \tag{9.6.2}$$

Since

$$\sum_{i=1}^{n} (x_i - \overline{x}) = 0,$$

we have that

$$\sum_{i=1}^{n} y_i - n\alpha = 0$$

and thus

$$\hat{\alpha} = \overline{Y}.$$

The equation $\partial H(\alpha, \beta)/\partial \beta = 0$ yields, with α replaced by $\overline{y}$,

$$\sum_{i=1}^{n}(y_i - \overline{y})(x_i - \overline{x}) - \beta \sum_{i=1}^{n}(x_i - \overline{x})^2 = 0 \qquad (9.6.3)$$

or, equivalently,

$$\hat{\beta} = \frac{\sum_{i=1}^{n}(Y_i - \overline{Y})(x_i - \overline{x})}{\sum_{i=1}^{n}(x_i - \overline{x})^2} = \frac{\sum_{i=1}^{n} Y_i(x_i - \overline{x})}{\sum_{i=1}^{n}(x_i - \overline{x})^2}.$$

Equations (9.6.2) and (9.6.3) are the estimating equations for the LS solutions for this simple linear model. The **fitted value** at the point (x_i, y_i) is given by

$$\hat{y}_i = \hat{\alpha} + \hat{\beta}(x_i - \overline{x}), \qquad (9.6.4)$$

which is shown on Figure 9.6.1. The fitted value $\hat{y}_i$ is also called the **predicted value** of y_i at x_i. The **residual** at the point (x_i, y_i) is given by

$$\hat{e}_i = y_i - \hat{y}_i, \qquad (9.6.5)$$

which is also shown on Figure 9.6.1. Residual means "what is left" and the residual in regression is exactly that, i.e., what is left over after the fit. The relationship between the fitted values and the residuals is explored in Exercise 9.6.11.

To find the maximum likelihood estimator of σ^2, consider the partial derivative

$$\frac{\partial[-\log L(\alpha, \beta, \sigma^2)]}{\partial(\sigma^2)} = \frac{n}{2\sigma^2} - \frac{\sum_{i=1}^{n}[y_i - \alpha - \beta(x_i - \overline{x})]^2}{2(\sigma^2)^2}.$$

Setting this equal to zero and replacing α and β by their solutions $\hat{\alpha}$ and $\hat{\beta}$, we obtain

$$\hat{\sigma}^2 = \frac{1}{n}\sum_{i=1}^{n}[Y_i - \hat{\alpha} - \hat{\beta}(x_i - \overline{x})]^2.$$

Of course, due to invariance of mles, $\hat{\sigma} = \sqrt{\hat{\sigma}^2}$. Note that in terms of the residuals, $\hat{\sigma}^2 = n^{-1}\sum_{i=1}^{n} \hat{e}_i^2$. As shown in Exercise 9.6.11, the average of the residuals is 0.

Since $\hat{\alpha}$ is linear function of independent and normally distributed random variables, $\hat{\alpha}$ has a normal distribution with mean

$$E(\hat{\alpha}) = E\left(\frac{1}{n}\sum_{i=1}^{n} Y_i\right) = \frac{1}{n}\sum_{i=1}^{n} E(Y_i)$$

$$= \frac{1}{n}\sum_{i=1}^{n}[\alpha + \beta(x_i - \overline{x})] = \alpha,$$

and variance

$$\text{var}(\hat{\alpha}) = \sum_{i=1}^{n} \left(\frac{1}{n}\right)^2 \text{var}(Y_i) = \frac{\sigma^2}{n}.$$

The estimator $\hat{\beta}$ is also a linear function of $Y_1, Y_2, \ldots, Y_n$ and hence has a normal distribution with mean

$$E(\hat{\beta}) = \frac{\sum_{i=1}^{n}(x_i - \overline{x})[\alpha + \beta(x_i - \overline{x})]}{\sum_{i=1}^{n}(x_i - \overline{x})^2}$$

$$= \frac{\alpha \sum_{i=1}^{n}(x_i - \overline{x}) + \beta \sum_{i=1}^{n}(x_i - \overline{x})^2}{\sum_{i=1}^{n}(x_i - \overline{x})^2} = \beta$$

and variance

$$\text{var}(\hat{\beta}) = \sum_{i=1}^{n} \left[\frac{x_i - \overline{x}}{\sum_{i=1}^{n}(x_i - \overline{x})^2}\right]^2 \text{var}(Y_i)$$

$$= \frac{\sum_{i=1}^{n}(x_i - \overline{x})^2}{[\sum_{i=1}^{n}(x_i - \overline{x})^2]^2} \sigma^2 = \frac{\sigma^2}{\sum_{i=1}^{n}(x_i - \overline{x})^2}.$$

In summary, the estimators $\hat{\alpha}$ and $\hat{\beta}$ are linear functions of the independent normal random variables $Y_1, \ldots, Y_n$. In Exercise 9.6.10 it is further shown that covariance between $\hat{\alpha}$ and $\hat{\beta}$ is zero. It follows that $\hat{\alpha}$ and $\hat{\beta}$ are independent random variables with a bivariate normal distribution; that is,

$$\begin{pmatrix} \hat{\alpha} \\ \hat{\beta} \end{pmatrix} \text{ has a } N_2 \left(\begin{pmatrix} \alpha \\ \beta \end{pmatrix}, \sigma^2 \begin{bmatrix} \frac{1}{n} & 0 \\ 0 & \frac{1}{\sum_{i=1}^{n}(x_i - \overline{x})^2} \end{bmatrix} \right) \text{ distribution.} \qquad (9.6.6)$$

Next, we consider the estimator of σ^2. It can be shown (Exercise 9.6.6) that

$$\sum_{i=1}^{n}[Y_i - \alpha - \beta(x_i - \overline{x})]^2 = \sum_{i=1}^{n}\{(\hat{\alpha} - \alpha) + (\hat{\beta} - \beta)(x_i - \overline{x})$$

$$+ [Y_i - \hat{\alpha} - \hat{\beta}(x_i - \overline{x})]\}^2$$

$$= n(\hat{\alpha} - \alpha)^2 + (\hat{\beta} - \beta)^2 \sum_{i=1}^{n}(x_i - \overline{x})^2 + n\hat{\sigma}^2.$$

or for brevity,

$$Q = Q_1 + Q_2 + Q_3.$$

Here $Q, Q_1, Q_2,$ and Q_3 are real quadratic forms in the variables

$$Y_i - \alpha - \beta(x_i - \overline{x}), \quad i = 1, 2, \ldots, n.$$

In this equation, Q represents the sum of the squares of n independent random variables that have normal distributions with means zero and variances σ^2. Thus Q/σ^2 has a χ^2 distribution with n degrees of freedom. Each of the random variables $\sqrt{n}(\hat{\alpha} - \alpha)/\sigma$ and $\sqrt{\sum_{i=1}^{n}(x_i - \overline{x})^2}(\hat{\beta} - \beta)/\sigma$ has a normal distribution with zero mean and unit variance; thus each of Q_1/σ^2 and Q_2/σ^2 has a χ^2 distribution with 1 degree of freedom. Since Q_3 is nonnegative, we have, in accordance with Theorem 9.1.1, that Q_1, Q_2, and Q_3 are independent, so that Q_3/σ^2 has a χ^2 distribution with $n - 1 - 1 = n - 2$ degrees of freedom. That is, $n\hat{\sigma}^2/\sigma^2$ has a χ^2 distribution with $n - 2$ degrees of freedom.

We now extend this discussion to obtain inference for the parameters α and β. It follows from the above derivations that each of the random variables

$$T_1 = \frac{[\sqrt{n}(\hat{\alpha} - \alpha)]/\sigma}{\sqrt{Q_3/[\sigma^2(n-2)]}} = \frac{\hat{\alpha} - \alpha}{\sqrt{\hat{\sigma}^2/(n-2)}}$$

and

$$T_2 = \frac{\left[\sqrt{\sum_{i=1}^n (x_i - \bar{x})^2}(\hat{\beta} - \beta)\right]/\sigma}{\sqrt{Q_3/[\sigma^2(n-2)]}} = \frac{\hat{\beta} - \beta}{\sqrt{n\hat{\sigma}^2/[(n-2)\sum_1^n (x_i - \bar{x})^2]}}$$

has a t-distribution with $n - 2$ degrees of freedom. These facts enable us to obtain confidence intervals for α and β; see Exercise 9.6.3. The fact that $n\hat{\sigma}^2/\sigma^2$ has a χ^2 distribution with $n - 2$ degrees of freedom provides a means of determining a confidence interval for σ^2. These are some of the statistical inferences about the parameters to which reference was made in the introductory remarks of this section.

Remark 9.6.2. The more discerning reader should quite properly question our construction of T_1 and T_2 immediately above. We know that the *squares* of the linear forms are independent of $Q_3 = n\hat{\sigma}^2$, but we do not know, at this time, that the linear forms themselves enjoy this independence. A more general problem is solved in Chapter 12 and the present case is a special instance. ∎

Example 9.6.1 (Geometry of the Least Squares Fit). In the modern literature linear models are usually expressed in terms of matrices and vectors, which we briefly introduce in this example. Furthermore, this allows us to discuss the simple geometry behind the least squares fit. Consider then Model (9.6.1). Write the vectors $\mathbf{Y} = (Y_1, \ldots, Y_n)'$, $\mathbf{e} = (e_1, \ldots, e_n)'$, and $\mathbf{x}_c = (x_1 - \bar{x}, \ldots, x_n - \bar{x})'$. Let $\mathbf{1}$ denote the $n \times 1$ vector whose components are all one. Then Model (9.6.1) can be expressed equivalently as

$$\begin{aligned} \mathbf{Y} &= \alpha\mathbf{1} + \beta\mathbf{x}_c + \mathbf{e} \\ &= [\mathbf{1} \ \mathbf{x}_c]\begin{pmatrix} \alpha \\ \beta \end{pmatrix} + \mathbf{e} \\ &= \mathbf{X}\boldsymbol{\beta} + \mathbf{e}, \end{aligned} \tag{9.6.7}$$

where $\mathbf{X}$ is the $n \times 2$ matrix with columns $\mathbf{1}$ and $\mathbf{x}_c$ and $\boldsymbol{\beta} = (\alpha, \beta)'$. Next, let $\boldsymbol{\theta} = E(\mathbf{Y}) = \mathbf{X}\boldsymbol{\beta}$. Finally, let V be the 2-dimensional subspace of R^n spanned by the columns of $\mathbf{X}$; i.e., V is the range of the matrix $\mathbf{X}$. Hence, we can also express the model succinctly as

$$\mathbf{Y} = \boldsymbol{\theta} + \mathbf{e}, \quad \boldsymbol{\theta} \in V. \tag{9.6.8}$$

Hence, except for random error vector $\mathbf{e}$, $\mathbf{Y}$ would lie in V. It makes sense intuitively then, as suggested by Figure 9.6.2, to estimate $\boldsymbol{\theta}$ by the vector in V which is "closest" (in Euclidean distance) to $\mathbf{Y}$; that is, by $\hat{\boldsymbol{\theta}}$ where

$$\hat{\boldsymbol{\theta}} = \text{Argmin}_{\boldsymbol{\theta} \in V} \|\mathbf{Y} - \boldsymbol{\theta}\|^2, \tag{9.6.9}$$

where the square of the **Euclidean norm** is given by $\|\mathbf{u}\|^2 = \sum_{i=1}^{n} u_i^2$, for $\mathbf{u} \in R^n$. As shown in Exercise 9.6.11 and depicted on the plot in Figure 9.6.2, $\hat{\boldsymbol{\theta}} = \hat{\alpha}\mathbf{1} + \hat{\beta}\mathbf{x}_c$, where $\hat{\alpha}$ and $\hat{\beta}$ are the least squares estimates given above. Also, the vector $\hat{\mathbf{e}} = \mathbf{Y} - \hat{\boldsymbol{\theta}}$ is the vector of residuals and $n\hat{\sigma}^2 = \|\hat{\mathbf{e}}\|^2$. Also, just as depicted in Figure 9.6.2 the angle between the vectors $\hat{\boldsymbol{\theta}}$ and $\hat{\mathbf{e}}$ is a right angle. In linear models, we say that $\hat{\boldsymbol{\theta}}$ is the projection of $\mathbf{Y}$ onto the subspace V. ∎

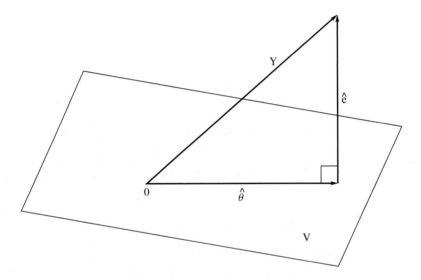

Figure 9.6.2: The sketch shows the geometry of least squares. The vector of responses is $\mathbf{Y}$, the fit is $\hat{\boldsymbol{\theta}}$, and the vector of residuals is $\hat{\mathbf{e}}$

EXERCISES

9.6.1. Student's scores on the mathematics portion of the ACT examination, x, and on the final examination in the first-semester calculus (200 points possible), y, are given.

(a) Calculate the least squares regression line for these data.

(b) Plot the points and the least squares regression line on the same graph.

(c) Find point estimates for α, β, and σ^2.

(d) Find 95 percent confidence intervals for α and β under the usual assumptions.

x	y	x	y
25	138	20	100
20	84	25	143
26	104	26	141
26	112	28	161
28	88	25	124
28	132	31	118
29	90	30	168
32	183		

9.6.2 (Telephone Data). Consider the data presented below. The responses (y) for this data set are the numbers of telephone calls (tens of millions) made in Belgium for the years 1950 through 1973. Time, the years, serves as the predictor variable (x). The data are discussed on page 151 of Hettmansperger and McKean (1998).

Year	50	51	52	53	54	55
No. Calls	0.44	0.47	0.47	0.59	0.66	0.73
Year	56	57	58	59	60	61
No. Calls	0.81	0.88	1.06	1.20	1.35	1.49
Year	62	63	64	65	66	67
No. Calls	1.61	2.12	11.90	12.40	14.20	15.90
Year	68	69	70	71	72	73
No. Calls	18.20	21.20	4.30	2.40	2.70	2.90

(a) Calculate the least squares regression line for these data.

(b) Plot the points and the least squares regression line on the same graph.

(c) What is the reason for the poor least squares fit?

9.6.3. Find $(1-\alpha)100\%$ confidence intervals for the parameters α and β in Model 9.6.1.

9.6.4. Consider Model (9.6.1). Let $\eta_0 = E(Y|x = x_0 - \bar{x})$. The least squares estimator of η_0 is $\hat{\eta}_0 = \hat{\alpha} + \hat{\beta}(x_0 - \bar{x})$.

(a) Using (9.6.6), determine the distribution of $\hat{\eta}_0$.

(b) Obtain a $(1-\alpha)100\%$ confidence interval for η_0.

9.6.5. Assume that the sample $(x_1, Y_1), \ldots, (x_n, Y_n)$ follows the linear model (9.6.1). Suppose Y_0 is a future observation at $x = x_0 - \bar{x}$ and we want to determine a predictive interval for it. Assume that the model (9.6.1) holds for Y_0; i.e., Y_0 has a $N(\alpha + \beta(x_0 - \bar{x}), \sigma^2)$ distribution. We will use $\hat{\eta}_0$ of Exercise 9.6.4 as our prediction of Y_0.

(a) Obtain the distribution of $Y_0 - \hat{\eta}_0$. Use the fact that the future observation Y_0 is independent of the sample $(x_1, Y_1), \ldots, (x_n, Y_n)$.

(b) Determine a t-statistic with numerator $Y_0 - \hat{\eta}_0$.

(c) Now beginning with $1 - \alpha = P[-t_{\alpha/2,n-2} < t < t_{\alpha/2,n-2}]$, where $0 < \alpha < 1$, determine a $(1 - \alpha)100\%$ predictive interval for Y_0.

(d) Compare this predictive interval with the confidence interval obtained in Exercise 9.6.4. Intuitively, why is the predictive interval larger?

9.6.6. Show that

$$\sum_{i=1}^{n}[Y_i - \alpha - \beta(x_i - \bar{x})]^2 = n(\hat{\alpha} - \alpha)^2 + (\hat{\beta} - \beta)^2 \sum_{i=1}^{n}(x_i - \bar{x})^2 + \sum_{i=1}^{n}[Y_i - \hat{\alpha} - \hat{\beta}(x_i - \bar{x})]^2.$$

9.6.7. Let the independent random variables $Y_1, Y_2, \ldots, Y_n$ have, respectively, the probability density functions $N(\beta x_i, \gamma^2 x_i^2)$, $i = 1, 2, \ldots, n$, where the given numbers $x_1, x_2, \ldots, x_n$ are not all equal and no one is zero. Find the maximum likelihood estimators of β and γ^2.

9.6.8. Let the independent random variables $Y_1, \ldots, Y_n$ have the joint pdf.

$$L(\alpha, \beta, \sigma^2) = \left(\frac{1}{2\pi\sigma^2}\right)^{n/2} \exp\left\{-\frac{1}{2\sigma^2}\sum_{1}^{n}[y_i - \alpha - \beta(x_i - \bar{x})]^2\right\},$$

where the given numbers $x_1, x_2, \ldots, x_n$ are not all equal. Let $H_0 : \beta = 0$ (α and σ^2 unspecified). It is desired to use a likelihood ratio test to test H_0 against all possible alternatives. Find Λ and see whether the test can be based on a familiar statistic.

Hint: In the notation of this section show that

$$\sum_{1}^{n}(Y_i - \hat{\alpha})^2 = Q_3 + \hat{\beta}^2 \sum_{1}^{n}(x_i - \bar{x})^2.$$

9.6.9. Using the notation of Section 9.2, assume that the means μ_j satisfy a linear function of j, namely $\mu_j = c + d[j - (b+1)/2]$. Let independent random samples of size a be taken from the b normal distributions having means $\mu_1, \mu_2, \ldots, \mu_b$, respectively, and common unknown variance σ^2.

(a) Show that the maximum likelihood estimators of c and d are, respectively, $\hat{c} = \overline{X}_{..}$ and

$$\hat{d} = \frac{\sum_{j=1}^{b}[j - (b-1)/2](\overline{X}_{.j} - \overline{X}_{..})}{\sum_{j=1}^{b}[j - (b+1)/2]^2}.$$

(b) Show that

$$\sum_{i=1}^{a}\sum_{j=1}^{b}(X_{ij} - \overline{X}_{..})^2 = \sum_{i=1}^{a}\sum_{j=1}^{b}\left[X_{ij} - \overline{X}_{..} - \hat{d}\left(j - \frac{b+1}{2}\right)\right]^2$$

$$+ \hat{d}^2 \sum_{j=1}^{b}a\left(j - \frac{b+1}{2}\right)^2.$$

(c) Argue that the two terms in the right-hand member of Part (b), once divided by σ^2, are independent random variables with χ^2 distributions provided that $d = 0$.

(d) What F-statistic would be used to test the equality of the means, that is, $H_0 : d = 0$?

9.6.10. Show that the covariance between $\hat{\alpha}$ and $\hat{\beta}$ is zero.

9.6.11. Reconsider Example 9.6.1.

(a) Show that $\hat{\boldsymbol{\theta}} = \hat{\alpha}\mathbf{1} + \hat{\beta}\mathbf{x}_c$, where $\hat{\alpha}$ and $\hat{\beta}$ are the least squares estimators derived in this section.

(b) Show that the vector $\hat{\mathbf{e}} = \mathbf{Y} - \hat{\boldsymbol{\theta}}$ is the vector of residuals; i.e., its ith entry is $\hat{e}_i$, (9.6.5).

(c) As depicted in Figure 9.6.2, show that the angle between the vectors $\hat{\boldsymbol{\theta}}$ and $\hat{\mathbf{e}}$ is a right angle.

(d) Show that the residuals sum to zero; i.e., $\mathbf{1}'\hat{\mathbf{e}} = 0$.

9.6.12. Fit $y = a + x$ to the data

x	0	1	2
y	1	3	4

by the method of least squares.

9.6.13. Fit by the method of least squares the plane $z = a + bx + cy$ to the five points $(x, y, z) : (-1, -2, 5), (0, -2, 4), (0, 0, 4), (1, 0, 2), (2, 1, 0)$.

9.6.14. Let the 4×1 matrix $\mathbf{Y}$ be multivariate normal $N(\mathbf{X}\boldsymbol{\beta}, \sigma^2 \mathbf{I})$, where the 4×3 matrix $\mathbf{X}$ equals

$$\mathbf{X} = \begin{bmatrix} 1 & 1 & 2 \\ 1 & -1 & 2 \\ 1 & 0 & -3 \\ 1 & 0 & -1 \end{bmatrix}$$

and $\boldsymbol{\beta}$ is the 3×1 regression coeffient matrix.

(a) Find the mean matrix and the covariance matrix of $\hat{\boldsymbol{\beta}} = (\mathbf{X}'\mathbf{X})^{-1}\mathbf{X}'\mathbf{Y}$.

(b) If we observe $\mathbf{Y}'$ to be equal to $(6, 1, 11, 3)$, compute $\hat{\boldsymbol{\beta}}$.

9.6.15. Suppose $\mathbf{Y}$ is an $n \times 1$ random vector, $\mathbf{X}$ is an $n \times p$ matrix of known constants of rank p, and $\boldsymbol{\beta}$ is a $p \times 1$ vector of regression coefficients. Let $\mathbf{Y}$ have a $N(\mathbf{X}\boldsymbol{\beta}, \sigma^2 \mathbf{I})$ distribution. Discuss the joint pdf of $\hat{\boldsymbol{\beta}} = (\mathbf{X}'\mathbf{X})^{-1}\mathbf{X}'\mathbf{Y}$ and $\mathbf{Y}'[\mathbf{I} - \mathbf{X}(\mathbf{X}'\mathbf{X})^{-1}\mathbf{X}']\mathbf{Y}/\sigma^2$.

9.6.16. Let the independent normal random variables $Y_1, Y_2, \ldots, Y_n$ have, respectively, the probability density functions $N(\mu, \gamma^2 x_i^2)$, $i = 1, 2, \ldots, n$, where the given $x_1, x_2, \ldots, x_n$ are not all equal and no one of which is zero. Discuss the test of the hypothesis $H_0 : \gamma = 1$, μ unspecified, against all alternatives $H_1 : \gamma \neq 1$, μ unspecified.

9.6.17. Let $Y_1, Y_2, \ldots, Y_n$ be n independent normal variables with common unknown variance σ^2. Let Y_i have mean βx_i, $i = 1, 2, \ldots, n$, where $x_1, x_2, \ldots, x_n$ are known but not all the same and β is an unknown constant. Find the likelihood ratio test for $H_0 : \beta = 0$ against all alternatives. Show that this likelihood ratio test can be based on a statistic that has a well-known distribution.

9.7   A Test of Independence

Let X and Y have a bivariate normal distribution with means μ_1 and μ_2, positive variances σ_1^2 and σ_2^2, and correlation coefficient ρ. We wish to test the hypothesis that X and Y are independent. Because two jointly normally distributed random variables are independent if and only if $\rho = 0$, we test the hypothesis $H_0 : \rho = 0$ against the hypothesis $H_1 : \rho \neq 0$. A likelihood ratio test will be used. Let $(X_1, Y_1), (X_2, Y_2), \ldots, (X_n, Y_n)$ denote a random sample of size $n > 2$ from the bivariate normal distribution; that is, the joint pdf of these $2n$ random variables is given by

$$f(x_1, y_1)f(x_2, y_2) \cdots f(x_n, y_n).$$

Although it is fairly difficult to show, the statistic that is defined by the likelihood ratio Λ is a function of the statistic, which is the mle of ρ, namely

$$R = \frac{\sum_{i=1}^{n}(X_i - \overline{X})(Y_i - \overline{Y})}{\sqrt{\sum_{i=1}^{n}(X_i - \overline{X})^2 \sum_{i=1}^{n}(Y_i - \overline{Y})^2}}. \tag{9.7.1}$$

This statistic R is called the sample **correlation coefficient** of the random sample. Following the discussion after expression (4.5.5), the statistic R is a consistent estimate of ρ; see Exercise 9.7.5. The likelihood ratio principle, which calls for the rejection of H_0 if $\Lambda \leq \lambda_0$, is equivalent to the computed value of $|R| \geq c$. That is, if the absolute value of the correlation coefficient of the sample is too large, we reject the hypothesis that the correlation coefficient of the distribution is equal to zero. To determine a value of c for a satisfactory significance level, it will be necessary to obtain the distribution of R, or a function of R, when H_0 is true. This will now be done.

Let $X_1 = x_1, X_2 = x_2, \ldots, X_n = x_n$, $n > 2$, where $x_1, x_2, \ldots, x_n$ and $\overline{x} = \sum_1^n x_i/n$ are fixed numbers such that $\sum_1^n (x_i - \overline{x})^2 > 0$. Consider the conditional pdf of $Y_1, Y_2, \ldots, Y_n$ given that $X_1 = x_1, X_2 = x_2, \ldots, X_n = x_n$. Because $Y_1, Y_2, \ldots, Y_n$ are independent and, with $\rho = 0$, are also independent of $X_1, X_2, \ldots, X_n$, this conditional pdf is given by

$$\left(\frac{1}{\sqrt{2\pi}\sigma_2}\right)^n \exp\left\{-\frac{1}{2\sigma^2}\sum_1^n (y_i - \mu_2)^2\right\}.$$

Let R_c be the correlation coefficient, given $X_1 = x_1, X_2 = x_2, \ldots, X_n = x_n$, so that

$$\frac{R_c\sqrt{\sum_{i=1}^n (Y_i - \overline{Y})^2}}{\sqrt{\sum_{i=1}^n (x_i - \overline{x})^2}} = \frac{\sum_{i=1}^n (x_i - \overline{x})(Y_i - \overline{Y})}{\sum_{i=1}^n (x_i - \overline{x})^2} = \frac{\sum_{i=1}^n (x_i - \overline{x})Y_i}{\sum_{i=1}^n (x_i - \overline{x})^2}$$

is like $\hat{\beta}$ of Section 9.6 and has mean zero when $\rho = 0$. Thus, referring to T_2 of Section 9.6, we see that

$$\frac{R_c\sqrt{\sum (Y_i - \overline{Y})^2}/\sqrt{\sum (x_i - \overline{x})^2)}}{\sqrt{\frac{\sum_{i=1}^n \left\{Y_i - \overline{Y} - \left[R_c\sqrt{\sum_{j=1}^n (Y_j - \overline{Y})^2}/\sqrt{\sum_{j=1}^n (x_j - \overline{x})^2}\right](x_i - \overline{x})\right\}^2}{(n-2)\sum_{j=1}^n (x_j - \overline{x})^2}}} = \frac{R_c\sqrt{n-2}}{\sqrt{1 - R_c^2}} \qquad (9.7.2)$$

has, given $X_1 = x_1, \ldots, X_n = x_n$, a conditional t-distribution with $n - 2$ degrees of freedom. Note that the pdf, say $g(t)$, of this t-distribution does not depend upon $x_1, x_2, \ldots, x_n$. Now the joint pdf of $X_1, X_2, \ldots, X_n$ and $R\sqrt{n-2}/\sqrt{1 - R^2}$, where

$$R = \frac{\sum_1^n (X_i - \overline{X})(Y_i - \overline{Y})}{\sqrt{\sum_1^n (X_i - \overline{X})^2 \sum_1^n (Y_i - \overline{Y})^2}},$$

is the product of $g(t)$ and the joint pdf of $X_1, \ldots, X_n$. Integration on $x_1, x_2, \ldots, x_n$ yields the marginal pdf of $R\sqrt{n-2}/\sqrt{1 - R^2}$, because $g(t)$ does not depend upon $x_1, x_2, \ldots, x_n$ it is obvious that this marginal pdf is $g(t)$, the conditional pdf of $R\sqrt{n-2}/\sqrt{1 - R^2}$. The change-of-variable technique can now be used to find the pdf of R.

Remark 9.7.1. Since R has, when $\rho = 0$, a conditional distribution that does not depend upon $x_1, x_2, \ldots, x_n$ (and hence that conditional distribution is, in fact, the marginal distribution of R), we have the remarkable fact that R is independent of $X_1, X_2, \ldots, X_n$. It follows that R is independent of *every function* of $X_1, X_2, \ldots, X_n$ alone, that is, a function that does not depend upon any Y_i. In like manner, R is independent of every function of $Y_1, Y_2, \ldots, Y_n$ alone. Moreover, a careful review of the argument reveals that nowhere did we use the fact that X has a normal marginal distribution. Thus, if X and Y are independent, and if Y has a normal distribution, then R has the same conditional distribution whatever is the distribution of X, subject to the condition $\sum_1^n (x_i - \bar{x})^2 > 0$. Moreover, if $P\left[\sum_1^n (X_i - \overline{X})^2 > 0\right] = 1$, then R has the same marginal distribution whatever is the distribution of X. ∎

If we write $T = R\sqrt{n-2}/\sqrt{1-R^2}$, where T has a t-distribution with $n - 2 > 0$ degrees of freedom, it is easy to show by the change-of-variable technique (Exercise 9.7.4), that the pdf of R is given by

$$g(t) = \begin{cases} \frac{\Gamma[(n-1)/2]}{\Gamma(\frac{1}{2})\Gamma[(n-2)/2]}(1 - r^2)^{(n-4)/2} & -1 < r < 1 \\ 0 & \text{elsewhere.} \end{cases} \qquad (9.7.3)$$

We have now solved the problem of the distribution of R, when $\rho = 0$ and $n > 2$, or perhaps more conveniently, that of $R\sqrt{n-2}/\sqrt{1-R^2}$. The likelihood ratio test of the hypothesis $H_0 : \rho = 0$ against all alternatives $H_1 : \rho \neq 0$ may be based either on the statistic R or on the statistic $R\sqrt{n-2}/\sqrt{1-R^2} = T$, although the latter is easier to use. In either case the significance level of the test is

$$\alpha = P_{H_0}(|R| \geq c_1) = P_{H_0}(|T| \geq c_2),$$

where the constants c_1 and c_2 are chosen so as to give the desired value of α.

Remark 9.7.2. It is possible to obtain an approximate test of size α by using the fact that

$$W = \frac{1}{2} \log \left(\frac{1+R}{1-R} \right)$$

has an approximate normal distribution with mean $\frac{1}{2} \log[(1 + \rho)/(1 - \rho)]$ and with variance $1/(n - 3)$. We accept this statement without proof. Thus a test of $H_0 : \rho = 0$ can be based on the statistic

$$Z = \frac{\frac{1}{2} \log[(1 + R)/(1 - R)] - \frac{1}{2} \log[(1 + \rho)/(1 - \rho)]}{\sqrt{1/(n - 3)}},$$

with $\rho = 0$ so that $\frac{1}{2} \log[(1 + \rho)/(1 - \rho)] = 0$. However, using W, we can also test an hypothesis like $H_0 : \rho = \rho_0$ against $H_1 : \rho \neq \rho_0$, where ρ_0 is not necessarily zero. In that case the hypothesized mean of W is

$$\frac{1}{2} \log \left(\frac{1 + \rho_0}{1 - \rho_0} \right). \quad ∎$$

EXERCISES

9.7.1. Show that

$$R = \frac{\sum_{1}^{n}(X_i - \overline{X})(Y_i - \overline{Y})}{\sqrt{\sum_{1}^{n}(X_i - \overline{X})^2 \sum_{1}^{n}(Y_i - \overline{Y})^2}} = \frac{\sum_{1}^{n}X_iY_i - n\overline{X}\,\overline{Y}}{\sqrt{\left(\sum_{1}^{n}X_i^2 - n\overline{X}^2\right)\left(\sum_{1}^{n}Y_i^2 - n\overline{Y}^2\right)}}$$

9.7.2. A random sample of size $n = 6$ from a bivariate normal distribution yields a value of the correlation coefficient of 0.89. Would we accept or reject, at the 5 percent significance level, the hypothesis that $\rho = 0$.

9.7.3. Verify Equation (9.7.2) of this section.

9.7.4. Verify the pdf (9.7.3) of this section.

9.7.5. Using the results of Section 4.5, show that R, (9.7.1), is a consistent estimate of ρ.

9.7.6. Two experiments gave the following results:

n	$\overline{x}$	$\overline{y}$	s_x	s_y	r
100	10	20	5	8	0.70
200	12	22	6	10	0.80

Calculate r for the combined sample.

9.8 The Distributions of Certain Quadratic Forms

Remark 9.8.1. It is essential that the reader have the background of the multivariate normal distribution as given in Section 3.5 to understand Sections 9.8 and 9.9. ∎

Remark 9.8.2. We will make use of the **trace** of a square matrix. If $\mathbf{A} = [a_{ij}]$ is an $n \times n$ matrix, then we define the trace of $\mathbf{A}$, $(\text{tr }\mathbf{A})$, to be the sum of its diagonal entries, i.e.,

$$\text{tr }\mathbf{A} = \sum_{i=1}^{n} a_{ii}. \tag{9.8.1}$$

The trace of a matrix has several interesting properties. One is that it is a linear operator, that is,

$$\text{tr}\,(a\mathbf{A} + b\mathbf{B}) = a\,\text{tr }\mathbf{A} + b\,\text{tr }\mathbf{B}. \tag{9.8.2}$$

A second useful property is: If $\mathbf{A}$ is an $n \times m$ matrix, $\mathbf{B}$ is an $m \times k$ matrix, and $\mathbf{C}$ is a $k \times n$ matrix, then

$$\text{tr}\,(\mathbf{ABC}) = \text{tr}\,(\mathbf{BCA}) = \text{tr}\,(\mathbf{CAB}). \tag{9.8.3}$$

The reader is asked to prove these facts in Exercise 9.8.7. Finally, a simple but useful property is that $\text{tr }a = a$, for any scalar a. ∎

We begin this section with a more formal but equivalent definition of a quadratic form. Let $\mathbf{X} = (X_1, \ldots, X_n)$ be an n-dimensional random vector and let $\mathbf{A}$ be a real $n \times n$ symmetric matrix. Then the random variable $Q = \mathbf{X}'\mathbf{AX}$ is called a **quadratic form** in $\mathbf{X}$. Due to the symmetry of $\mathbf{A}$ there are several ways we can write Q:

$$Q = \mathbf{X}'\mathbf{AX} = \sum_{i=1}^{n}\sum_{j=1}^{n} a_{ij}X_iX_j = \sum_{i=1}^{n} a_{ii}X_i^2 + \sum\sum_{i \neq j} a_{ij}X_iX_j \quad (9.8.4)$$

$$= \sum_{i=1}^{n} a_{ii}X_i^2 + 2\sum\sum_{i<j} a_{ij}X_iX_j. \quad (9.8.5)$$

These are very useful random variables in analysis of variance models. As the following theorem shows, the mean of a quadratic form is easily obtained.

Theorem 9.8.1. *Suppose the n-dimensional random vector $\mathbf{X}$ has mean $\boldsymbol{\mu}$ and variance-covariance matrix $\boldsymbol{\Sigma}$. Let $Q = \mathbf{X}'\mathbf{AX}$, where $\mathbf{A}$ is a real $n \times n$ symmetric matrix. Then*

$$E(Q) = tr\mathbf{A}\boldsymbol{\Sigma} + \boldsymbol{\mu}'\mathbf{A}\boldsymbol{\mu}. \quad (9.8.6)$$

Proof: Using the trace operator and property (9.8.3), we have

$$\begin{aligned} E(Q) = E(tr\mathbf{X}'\mathbf{AX}) &= E(tr\mathbf{A}\mathbf{X}\mathbf{X}') \\ &= tr\mathbf{A}E(\mathbf{X}\mathbf{X}') \\ &= tr\mathbf{A}(\boldsymbol{\Sigma} + \boldsymbol{\mu}\boldsymbol{\mu}') \\ &= tr\mathbf{A}\boldsymbol{\Sigma} + \boldsymbol{\mu}'\mathbf{A}\boldsymbol{\mu}, \end{aligned}$$

where the third line follows from Theorem 2.6.2. ∎

Example 9.8.1. (Sample Variance) Let $\mathbf{X}' = (X_1, \ldots, X_n)$ be an n-dimesional vector of random variables. Let $\mathbf{1}' = (1, \ldots, 1)$ be the n-dimensional vector whose components are one. Consider the quadratic form $Q = \mathbf{X}'(\mathbf{I} - \frac{1}{n}\mathbf{J})\mathbf{X}$, where $\mathbf{J} = \mathbf{1}\mathbf{1}'$; hence, $\mathbf{J}$ is an $n \times n$ matrix with all entries equal to 1. Note that the off diagonal entries of $(\mathbf{I} - \frac{1}{n}\mathbf{J})$ are $-n^{-1}$ while the diagonal entries are $1 - n^{-1}$; hence by (9.8.4) Q simplifies to,

$$\begin{aligned} Q &= \sum_{i=1}^{n} X_i^2\left(1 - \frac{1}{n}\right) + \sum\sum_{i \neq j}\left(-\frac{1}{n}\right)X_iX_j \\ &= \sum_{i=1}^{n} X_i^2\left(1 - \frac{1}{n}\right) - \frac{1}{n}\sum_{i=1}^{n} X_i\sum_{j=1}^{n} X_j + \frac{1}{n}\sum_{i=1}^{n} X_i^2 \\ &= \sum_{i=1}^{n} X_i^2 - n\overline{X}^2 = (n-1)S^2, \quad (9.8.7) \end{aligned}$$

where $\overline{X}$ and S^2 denote the sample mean and variance of $X_1, \ldots, X_n$.

Suppose we further assume that $X_1, \ldots, X_n$ are iid random variables with common mean μ and variance σ^2. Using Theorem 9.8.1 we can obtain yet another proof that S^2 is an unbiased estimate of σ^2. Note that the mean of the random vector $\mathbf{X}$ is $\mu\mathbf{1}$ and that its variance-covariance matrix is $\sigma^2\mathbf{I}$. Based on Theorem 9.8.1 we find immediately that,

$$E(S^2) = \frac{1}{n-1}\left\{\mathrm{tr}(\mathbf{I} - \frac{1}{n}\mathbf{J})\sigma^2\mathbf{I} + \mu^2(\mathbf{1'1} - \frac{1}{n}\mathbf{1'11'1})\right\} = \sigma^2 . \quad \blacksquare$$

The spectral decomposition of symmetric matrices will prove quite useful in this part of the chapter. As discussed around expression (3.5.4), a real symmetric matrix $\mathbf{A}$ can be diagonalized as

$$\mathbf{A} = \mathbf{\Gamma'\Lambda\Gamma}, \tag{9.8.8}$$

where $\mathbf{\Lambda}$ is the diagonal matrix $\mathbf{\Lambda} = \mathrm{diag}(\lambda_1, \ldots, \lambda_n)$, $\lambda_1 \geq \cdots \geq \lambda_n$ are the eigenvalues of $\mathbf{A}$, and the columns of $\mathbf{\Gamma'} = [\mathbf{v}_1 \cdots \mathbf{v}_n]$ are the corresponding orthonormal eigenvectors, (i.e., $\mathbf{\Gamma}$ is an orthogonal matrix). Recall from linear algebra that the rank of $\mathbf{A}$ is the number of nonzero eigenvalues. Further, because $\mathbf{\Lambda}$ is diagonal, we can write this expression as

$$\mathbf{A} = \sum_{i=1}^{n} \lambda_i \mathbf{v}_i \mathbf{v}_n'. \tag{9.8.9}$$

For normal random variables, we can use this last equation to obtain the mgf of the quadratic form Q.

Theorem 9.8.2. *Let $\mathbf{X'} = (X_1, \ldots, X_n)$ where $X_1, \ldots, X_n$ are iid $N(0, \sigma^2)$. Consider the quadratic form $Q = \sigma^{-2}\mathbf{X'AX}$ for a symmetric matrix $\mathbf{A}$ of rank $r \leq n$. Then Q has the moment generating function*

$$M(t) = \prod_{i=1}^{r}(1 - 2t\lambda_i)^{-1/2} = |\mathbf{I} - 2t\mathbf{A}|^{-1/2}, \tag{9.8.10}$$

where $\lambda_1, \ldots \lambda_r$ are the nonzero eigenvalues of $\mathbf{A}$, $|t| < 1/(2\lambda^*)$, and the value of λ^* is given by $\lambda^* = \max_{1 \leq i \leq r}|\lambda_i|$.

Proof: Write the spectral decomposition of $\mathbf{A}$ as in expression (9.8.9). Since the rank of $\mathbf{A}$ is r, exactly r of the eigenvalues are not 0. Denote the nonzero eigenvalues by $\lambda_1, \ldots, \lambda_r$. Then we can write Q as

$$Q = \sum_{i=1}^{r} \lambda_i(\sigma^{-1}\mathbf{v}_i'\mathbf{X})^2. \tag{9.8.11}$$

Let $\mathbf{\Gamma_1'} = [\mathbf{v}_1 \cdots \mathbf{v}_r]$ and define the r-dimensional random vector $\mathbf{W}$ by $\mathbf{W} = \sigma^{-1}\mathbf{\Gamma_1 X}$. Since $\mathbf{X}$ is $N_n(\mathbf{0}, \sigma^2\mathbf{I}_n)$ and $\mathbf{\Gamma_1'\Gamma_1} = \mathbf{I}_r$, Theorem 3.5.1 shows that $\mathbf{W}$ has a $N_r(\mathbf{0}, \mathbf{I}_r)$ distribution. In terms of the W_i we can write (9.8.11) as

$$Q = \sum_{i=1}^{r} \lambda_i W_i^2. \tag{9.8.12}$$

Because $W_1, \ldots, W_r$ are independent $N(0,1)$ random variables, $W_1^2, \ldots, W_r^2$ are independent $\chi^2(1)$ random variables. Thus the mgf of Q is,

$$
\begin{aligned}
E[\exp\{tQ\}] &= E\left[\exp\left\{\sum_{i=1}^{r} t\lambda_i W_i^2\right\}\right] \\
&= \prod_{i=1}^{r} E[\exp\{t\lambda_i W_i^2\}] = \prod_{i=1}^{r}(1 - 2t\lambda_i)^{-1/2}. \quad (9.8.13)
\end{aligned}
$$

The last equality holds if we assume that $|t| < 1/(2\lambda^*)$, where $\lambda^* = \max_{1 \le i \le r}|\lambda_i|$; see Exercise 9.8.6 To obtain the second form in (9.8.10), recall that the determinant of an orthogonal matrix is one. The result then follows from

$$
\begin{aligned}
|\mathbf{I} - 2t\mathbf{A}| = |\boldsymbol{\Gamma}'\boldsymbol{\Gamma} - 2t\boldsymbol{\Gamma}'\boldsymbol{\Lambda}\boldsymbol{\Gamma}| &= |\boldsymbol{\Gamma}'(\mathbf{I} - 2t\boldsymbol{\Lambda})\boldsymbol{\Gamma}| \\
&= |\mathbf{I} - 2t\boldsymbol{\Lambda}| = \left\{\prod_{i=1}^{r}(1 - 2t\lambda_i)^{-1/2}\right\}^{-2}. \quad \blacksquare
\end{aligned}
$$

Example 9.8.2. To illustrate this theorem, suppose X_i, $i = 1, 2, \ldots, n$ are independent random variables with X_i distributed as $N(\mu_i, \sigma_i^2)$, $i = 1, 2, \ldots, n$, respectively. Let $Z_i = (X_i - \mu_i)/\sigma_i$. We know that $\sum_{i=1}^{n} Z_i^2$ has χ^2 distribution with n degrees of freedom. To illustrate Theorem 9.8.2, let $\mathbf{Z}' = (Z_1, \ldots, Z_n)$. Let $Q = \mathbf{Z}'\mathbf{I}\mathbf{Z}$. Hence, the symmetric matrix associated with Q is the identity matrix $\mathbf{I}$, which has n eigenvalues, all of value 1; i.e., $\lambda_i \equiv 1$. By Theorem 9.8.2, the mgf of Q is $(1 - 2t)^{-n/2}$; i.e, Q is distributed χ^2 with n degrees of freedom. $\blacksquare$

In general, from Theorem 9.8.2, note how close the mgf of the quadratic form Q is the mgf of a χ^2 distribution. The next two theorems give conditions where this is true.

Theorem 9.8.3. *Let $\mathbf{X}' = (X_1, X_2, \ldots, X_n)$ have a $N_n(\boldsymbol{\mu}, \boldsymbol{\Sigma})$ distribution where $\boldsymbol{\Sigma}$ is positive definite. Then $Q = (\mathbf{X} - \boldsymbol{\mu})'\boldsymbol{\Sigma}^{-1}(\mathbf{X} - \boldsymbol{\mu})$ has a $\chi^2(n)$ distribution.*

Proof: Write the spectral decomposition of $\boldsymbol{\Sigma}$ as $\boldsymbol{\Sigma} = \boldsymbol{\Gamma}'\boldsymbol{\Lambda}\boldsymbol{\Gamma}$, where $\boldsymbol{\Gamma}$ is an orthogonal matrix and $\boldsymbol{\Lambda} = \mathrm{diag}\{\lambda_1, \ldots, \lambda_n\}$ is a diagonal matrix whose diagonal entries are the eigenvalues of $\boldsymbol{\Sigma}$. Becuase $\boldsymbol{\Sigma}$ is positive definite all $\lambda_i > 0$. Hence we can write,

$$
\boldsymbol{\Sigma}^{-1} = \boldsymbol{\Gamma}'\boldsymbol{\Lambda}^{-1}\boldsymbol{\Gamma} = \boldsymbol{\Gamma}'\boldsymbol{\Lambda}^{-1/2}\boldsymbol{\Gamma}\boldsymbol{\Gamma}'\boldsymbol{\Lambda}^{-1/2}\boldsymbol{\Gamma},
$$

where $\boldsymbol{\Lambda}^{-1/2} = \mathrm{diag}\{\lambda_1^{-1/2}, \ldots, \lambda_n^{-1/2}\}$. Thus, we have

$$
Q = \left\{\boldsymbol{\Lambda}^{-1/2}\boldsymbol{\Gamma}(\mathbf{X} - \boldsymbol{\mu})\right\}' \mathbf{I} \left\{\boldsymbol{\Lambda}^{-1/2}\boldsymbol{\Gamma}(\mathbf{X} - \boldsymbol{\mu})\right\}.
$$

But by Theorem 3.5.1, it is easy to show that the random vector $\boldsymbol{\Lambda}^{-1/2}\boldsymbol{\Gamma}(\mathbf{X} - \boldsymbol{\mu})$ has a $N_n(\mathbf{0}, \mathbf{I})$ distribution; hence, Q has $\chi^2(n)$ distribution. $\blacksquare$

The remarkable fact that the random variable Q in the last theorem is $\chi^2(n)$ stimulates a number of questions about quadratic forms in normally distributed variables. We would like to treat this problem generally, but limitations of space forbid this, and we find it necessary to restrict ourselves to some special cases; see, for instance, Stapleton (1995) for discussion.

Recall from linear algebra that a symmetric matrix $\mathbf{A}$ is **idempotent** if $\mathbf{A}^2 = \mathbf{A}$. In Section 9.1, we have already met some idempotent matrices. For example, the matrix $\mathbf{I} - \frac{1}{n}\mathbf{J}$ of Example 9.8.1 is idempotent. Idempotent matrices possess some important characteristics. Suppose λ is an eigenvalue of an idempotent matrix $\mathbf{A}$ with corresponding eigenvector $\mathbf{v}$. Then the following identity is true,

$$\lambda\mathbf{v} = \mathbf{A}\mathbf{v} = \mathbf{A}^2\mathbf{v} = \lambda\mathbf{A}\mathbf{v} = \lambda^2\mathbf{v}.$$

Hence, $\lambda(\lambda - 1)\mathbf{v} = \mathbf{0}$. Since $\mathbf{v} \neq \mathbf{0}$, $\lambda = 0$ or 1. Conversely, if the eigenvalues of a real symmetric matrix are only 0's and 1's then it is idempotent; see Exercise 9.8.10. Thus the rank of an idempotent matrix $\mathbf{A}$ is the number of its eigenvalues which are 1. Denote the spectral decomposition of $\mathbf{A}$ by $\mathbf{A} = \mathbf{\Gamma}'\mathbf{\Lambda}\mathbf{\Gamma}$, where $\mathbf{\Lambda}$ is a diagonal matrix of eigenvalues and $\mathbf{\Gamma}$ is an orthogonal matrix whose columns are the corresponding orthonormal eigenvectors. Because the diagonal entries of $\mathbf{\Lambda}$ are 0 or 1 and $\mathbf{\Gamma}$ is orthogonal, we have

$$\operatorname{tr}\mathbf{A} = \operatorname{tr}\mathbf{\Lambda}\mathbf{\Gamma}\mathbf{\Gamma}' = \operatorname{tr}\mathbf{\Lambda} = \operatorname{rank}(\mathbf{A}),$$

i.e, the rank of an idempotent matrix is equal to its trace.

Theorem 9.8.4. *Let $\mathbf{X}' = (X_1, \ldots, X_n)$, where $X_1, \ldots, X_n$ are iid $N(0, \sigma^2)$. Let $Q = \sigma^{-2}\mathbf{X}'\mathbf{A}\mathbf{X}$ for a symmetric matrix $\mathbf{A}$ with rank r. Then Q has $\chi^2(r)$ distribution if and only if $\mathbf{A}$ is idempotent.*

Proof: By Theorem 9.8.2, the mgf of Q is,

$$M_Q(t) = \prod_{i=1}^{r}(1 - 2t\lambda_i)^{-1/2}, \tag{9.8.14}$$

where $\lambda_1, \ldots, \lambda_r$ are the r nonzero eigenvalues of $\mathbf{A}$. Suppose, first, that $\mathbf{A}$ is idempotent. Then $\lambda_1 = \cdots = \lambda_r = 1$ and the mgf of Q is $M_Q(t) = (1 - 2t)^{-r/2}$; i.e., Q has $\chi^2(r)$ distribution. Next, suppose Q has $\chi^2(r)$ distribution. Then for t in a neighborhood of 0, we have the identity

$$\prod_{i=1}^{r}(1 - 2t\lambda_i)^{-1/2} = (1 - 2t)^{-r/2},$$

which upon squaring both sides leads to

$$\prod_{i=1}^{r}(1 - 2t\lambda_i) = (1 - 2t)^{r},$$

By the uniqueness of the factorization of polynomials, $\lambda_1 = \cdots = \lambda_r = 1$. Hence, $\mathbf{A}$ is idempotent. ∎

Example 9.8.3. Based on this last theorem, we can obtain quickly the distribution of the sample variance when sampling from a normal distribution. Suppose $X_1, X_2, \ldots, X_n$ are iid $N(\mu, \sigma^2)$. Let $\mathbf{X} = (X_1, X_2, \ldots, X_n)'$. Then $\mathbf{X}$ has a $N_n(\mu\mathbf{1}, \sigma^2\mathbf{I})$ distribution, where $\mathbf{1}$ denotes a $n \times 1$ vector with all components equal to 1. Let $S^2 = (n-1)^{-1} \sum_{i=1}^{n} (X_i - \overline{X})^2$. Then by Example 9.8.1, we can write

$$\frac{(n-1)S^2}{\sigma^2} = \sigma^{-2}\mathbf{X}' \left(\mathbf{I} - \frac{1}{n}\mathbf{J}\right) \mathbf{X} = \sigma^{-2}(\mathbf{X} - \mu\mathbf{1})' \left(\mathbf{I} - \frac{1}{n}\mathbf{J}\right) (\mathbf{X} - \mu\mathbf{1}),$$

where the last equality holds because $\left(\mathbf{I} - \frac{1}{n}\mathbf{J}\right) \mathbf{1} = \mathbf{0}$. Because the matrix $\mathbf{I} - \frac{1}{n}\mathbf{J}$ is idempotent, $\operatorname{tr}\left(\mathbf{I} - \frac{1}{n}\mathbf{J}\right) = n-1$, and $\mathbf{X} - \mu\mathbf{1}$ is $N_n(\mathbf{0}, \sigma^2\mathbf{I})$, it follows from Theorem 9.8.4 that $(n-1)S^2/\sigma^2$ has $\chi^2(n-1)$ distribution. ∎

Remark 9.8.3. If the normal distribution in Theorem 9.8.4 is $N_n(\boldsymbol{\mu}, \sigma^2\mathbf{I})$, the condition $\mathbf{A}^2 = \mathbf{A}$ remains a necessary and sufficient condition that Q/σ^2 have a chi-square distribution. In general, however, Q/σ^2 is not central $\chi^2(r)$ but instead, Q/σ^2 has a noncentral chi-square distribution if $\mathbf{A}^2 = \mathbf{A}$. The number of degrees of freedom is r, the rank of $\mathbf{A}$, and the noncentrality parameter is $\boldsymbol{\mu}'\mathbf{A}\boldsymbol{\mu}/\sigma^2$. If $\boldsymbol{\mu} = \mu\mathbf{1}$, then $\boldsymbol{\mu}'\mathbf{A}\boldsymbol{\mu} = \mu^2 \sum_{i,j} a_{ij}$, where $\mathbf{A} = [a_{ij}]$. Then, if $\mu \neq 0$, the conditions $\mathbf{A}^2 = \mathbf{A}$ and $\sum_{i,j} a_{ij} = 0$ are necessary and sufficient conditions that Q/σ^2 be central $\chi^2(r)$. Moreover, the theorem may be extended to a quadratic form in random variables which have a multivariate normal distribution with positive definite covariance matrix $\boldsymbol{\Sigma}$; here the necessary and sufficient condition that Q have a chi-square distribution is $\mathbf{A}\boldsymbol{\Sigma}\mathbf{A} = \mathbf{A}$. See Exercise 9.8.9. ∎

EXERCISES

9.8.1. Let $Q = X_1 X_2 - X_3 X_4$, where X_1, X_2, X_3, X_4 is a random sample of size 4 from a distribution which is $N(0, \sigma^2)$. Show that Q/σ^2 does not have a chi-square distribution. Find the mgf of Q/σ^2.

9.8.2. Let $\mathbf{X}' = [X_1, X_2]$ be bivariate normal with matrix of means $\boldsymbol{\mu}' = [\mu_1, \mu_2]$ and positive definite covariance matrix $\boldsymbol{\Sigma}$. Let

$$Q_1 = \frac{X_1^2}{\sigma_1^2(1-\rho^2)} - 2\rho\frac{X_1 X_2}{\sigma_1\sigma_2(1-\rho^2)} + \frac{X_2^2}{\sigma_2^2(1-\rho^2)}.$$

Show that Q_1 is $\chi^2(r, \theta)$ and find r and θ. When and only when does Q_1 have a central chi-square distribution?

9.8.3. Let $\mathbf{X}' = [X_1, X_2, X_3]$ denote a random sample of size 3 from a distribution that is $N(4, 8)$ and let

$$\mathbf{A} = \begin{pmatrix} \frac{1}{2} & 0 & \frac{1}{2} \\ 0 & 1 & 0 \\ \frac{1}{2} & 0 & \frac{1}{2} \end{pmatrix}.$$

Let $Q = \mathbf{X}'\mathbf{A}\mathbf{X}/\sigma^2$.

(a) Use Theorem 9.8.1 to find the $E(Q)$.

(b) Justify the assertion that Q is $\chi^2(2, 6)$.

9.8.4. Suppose $X_1, \ldots, X_n$ are independent random variables with the common mean μ but with unequal variances $\sigma_i^2 = \text{Var}(X_i)$.

(a) Determine the variance of $\overline{X}$.

(b) Determine the constant K so that $Q = K \sum_{i=1}^n (X_i - \overline{X})^2$ is an unbiased estimate of the variance of $\overline{X}$. (*Hint:* Proceed as in Example 9.8.3.)

9.8.5. Suppose $X_1, \ldots, X_n$ are correlated random variables, with common mean μ and variance σ^2 but with correlations ρ (all correlations are the same).

(a) Determine the variance of $\overline{X}$.

(b) Determine the constant K so that $Q = K \sum_{i=1}^n (X_i - \overline{X})^2$ is an unbiased estimate of the variance of $\overline{X}$. (*Hint:* Proceed as in Example 9.8.3.)

9.8.6. Fill in the details for expression (9.8.13).

9.8.7. For the trace operator defined in expression (9.8.1), prove the following properties are true.

(a) If $\mathbf{A}$ and $\mathbf{B}$ are $n \times n$ matrices and a and b are scalars, then

$$\text{tr}\,(a\mathbf{A} + b\mathbf{B}) = a\,\text{tr}\,\mathbf{A} + b\,\text{tr}\,\mathbf{B}.$$

(b) If $\mathbf{A}$ is an $n \times m$ matrix, $\mathbf{B}$ is an $m \times k$ matrix, and $\mathbf{C}$ is a $k \times n$ matrix, then

$$\text{tr}\,(\mathbf{ABC}) = \text{tr}\,(\mathbf{BCA}) = \text{tr}\,(\mathbf{CAB}).$$

(c) If $\mathbf{A}$ is a square matrix and if $\mathbf{\Gamma}$ is an orthogonal matrix, use the result of Part (a) to show that $\text{tr}(\mathbf{\Gamma}'\mathbf{A}\mathbf{\Gamma}) = \text{tr}\mathbf{A}$.

(d) $\mathbf{A}$ is a real symmetric idempotent matrix, use the result of Part (b) to prove that the rank of $\mathbf{A}$ is equal to $\text{tr}\mathbf{A}$.

9.8.8. Let $\mathbf{A} = [a_{ij}]$ be a real symmetric matrix. Prove that $\sum_i \sum_j a_{ij}^2$ is equal to the sum of the squares of the eigenvalues of $\mathbf{A}$.

Hint: If $\mathbf{\Gamma}$ is an orthogonal matrix, show that $\sum_j \sum_i a_{ij}^2 = \text{tr}(\mathbf{A}^2) = \text{tr}(\mathbf{\Gamma}'\mathbf{A}^2\mathbf{\Gamma}) = \text{tr}[(\mathbf{\Gamma}'\mathbf{A}\mathbf{\Gamma})(\mathbf{\Gamma}'\mathbf{A}\mathbf{\Gamma})]$.

9.8.9. Suppose $\mathbf{X}$ has a $N_n(\mathbf{0}, \mathbf{\Sigma})$, where $\mathbf{\Sigma}$ is positive definite. Let $Q = \mathbf{X}'\mathbf{A}\mathbf{X}$ for a symmetric matrix $\mathbf{A}$ with rank r. Prove Q has $\chi^2(r)$ distribution if and only if $\mathbf{A}\mathbf{\Sigma}\mathbf{A} = \mathbf{A}$.

Hint: Write Q as

$$Q = (\mathbf{\Sigma}^{-1/2}\mathbf{X})'\mathbf{\Sigma}^{1/2}\mathbf{A}\mathbf{\Sigma}^{1/2}(\mathbf{\Sigma}^{-1/2}\mathbf{X}),$$

where $\mathbf{\Sigma}^{1/2} = \mathbf{\Gamma}'\mathbf{\Lambda}^{1/2}\mathbf{\Gamma}$ and $\mathbf{\Sigma} = \mathbf{\Gamma}'\mathbf{\Lambda}\mathbf{\Gamma}$ is the spectral decomposition of $\mathbf{\Sigma}$. Then use Theorem 9.8.4.

9.8.10. Suppose $\mathbf{A}$ is a real symmetric matrix. If the eigenvalues of $\mathbf{A}$ are only 0's and 1's then prove that $\mathbf{A}$ is idempotent.

9.9   The Independence of Certain Quadratic Forms

We have previously investigated the independence of linear functions of normally distributed variables. In this section we shall prove some theorems about the independence of quadratic forms. We shall confine our attention to normally distributed variables that constitute a random sample of size n from a distribution that is $N(0, \sigma^2)$.

Remark 9.9.1. In the proof of the next theorem, we use the fact that if $\mathbf{A}$ is an $m \times n$ matrix of rank n (i.e., $\mathbf{A}$ has full column rank), then the matrix $\mathbf{A}'\mathbf{A}$ is nonsingular. A proof of this linear algebra fact is sketched in Exercises 12.3.2 and 12.3.3 of Chapter 12. ∎

Theorem 9.9.1 (Craig). *Let $\mathbf{X}' = (X_1, \ldots, X_n)$ where $X_1, \ldots, X_n$ are iid $N(0, \sigma^2)$ random variables. For real symmetric matrices $\mathbf{A}$ and $\mathbf{B}$, let $Q_1 = \sigma^{-2}\mathbf{X}'\mathbf{A}\mathbf{X}$ and $Q_2 = \sigma^{-2}\mathbf{X}'\mathbf{B}\mathbf{X}$ denote quadratic forms in $\mathbf{X}$. The random variables Q_1 and Q_2 are independent if and only if $\mathbf{A}\mathbf{B} = \mathbf{0}$.*

Proof: First, we obtain some preliminary results. Based on these results, the proof follows immediately. Assume the ranks of the matrices $\mathbf{A}$ and $\mathbf{B}$ are r and s, respectively. Let $\mathbf{\Gamma}_1'\mathbf{\Lambda}_1\mathbf{\Gamma}_1$ denote the spectral decomposition of $\mathbf{A}$. Denote the r nonzero eigenvalues of $\mathbf{A}$ by $\lambda_1, \ldots, \lambda_r$. Without loss of generality, assume that these nonzero eigenvalues of $\mathbf{A}$ are the first r elements on the main diagonal of $\mathbf{\Lambda}_1$ and let $\mathbf{\Gamma}_{11}'$ be the $n \times r$ matrix whose columns are the corresponding eigenvectors. Finally let $\mathbf{\Lambda}_{11} = \text{diag}\{\lambda_1, \ldots, \lambda_r\}$. Then we can write the spectral decomposition of $\mathbf{A}$ in either of the two ways:

$$\mathbf{A} = \mathbf{\Gamma}_1'\mathbf{\Lambda}_1\mathbf{\Gamma}_1 = \mathbf{\Gamma}_{11}'\mathbf{\Lambda}_{11}\mathbf{\Gamma}_{11}. \tag{9.9.1}$$

Note that we can write Q_1 as,

$$Q_1 = \sigma^{-2}\mathbf{X}'\mathbf{\Gamma}_{11}'\mathbf{\Lambda}_{11}\mathbf{\Gamma}_{11}\mathbf{X} = \sigma^{-2}(\mathbf{\Gamma}_{11}\mathbf{X})'\mathbf{\Lambda}_{11}(\mathbf{\Gamma}_{11}\mathbf{X}) = \mathbf{W}_1'\mathbf{\Lambda}_{11}\mathbf{W}_1, \tag{9.9.2}$$

where $\mathbf{W}_1 = \sigma^{-1}\mathbf{\Gamma}_{11}\mathbf{X}$. Next, obtain a similar representation based on the s nonzero eigenvalues $\gamma_1, \ldots, \gamma_s$ of $\mathbf{B}$. Let $\mathbf{\Lambda}_{22} = \text{diag}\{\gamma_1, \ldots, \gamma_s\}$ denote the $s \times s$ diagonal matrix of nonzero eigenvalues and form the $n \times s$ matrix $\mathbf{\Gamma}_{21}' = [\mathbf{u}_1 \cdots \mathbf{u}_s]$ of corresponding eigenvectors. Then we can write the spectral decomposition of $\mathbf{B}$ as

$$\mathbf{B} = \mathbf{\Gamma}_{21}'\mathbf{\Lambda}_{22}\mathbf{\Gamma}_{21}. \tag{9.9.3}$$

Also, we can write Q_2 as,

$$Q_2 = \mathbf{W}_2'\mathbf{\Lambda}_{22}\mathbf{W}_2, \tag{9.9.4}$$

where $\mathbf{W}_2 = \sigma^{-1}\mathbf{\Gamma}_{21}\mathbf{X}$. Letting $\mathbf{W}' = (\mathbf{W}_1', \mathbf{W}_2')$, we have

$$\mathbf{W} = \sigma^{-1}\begin{bmatrix} \mathbf{\Gamma}_{11} \\ \mathbf{\Gamma}_{21} \end{bmatrix}\mathbf{X}.$$

Because $\mathbf{X}$ has a $N_n(\mathbf{0}, \sigma^2\mathbf{I})$ distribution, Theorem 3.5.1 shows that $\mathbf{W}$ has an $(r+s)$ dimensional multivariate normal distribution with mean $\mathbf{0}$ and variance-covariance matrix

$$\text{Var}(\mathbf{W}) = \begin{bmatrix} \mathbf{I}_r & \mathbf{\Gamma}_{11}\mathbf{\Gamma}_{21}' \\ \mathbf{\Gamma}_{21}\mathbf{\Gamma}_{11}' & \mathbf{I}_s \end{bmatrix}. \tag{9.9.5}$$

Finally using (9.9.1) and (9.9.3), we have the identity

$$\mathbf{AB} = \{\mathbf{\Gamma}_{11}'\mathbf{\Lambda}_{11}\}\mathbf{\Gamma}_{11}\mathbf{\Gamma}_{21}'\{\mathbf{\Lambda}_{22}\mathbf{\Gamma}_{21}\}. \tag{9.9.6}$$

Let $\mathbf{U}$ denote the matrix in the first set of braces. Note that $\mathbf{U}$ has full column rank so in particular $(\mathbf{U}'\mathbf{U})^{-1}$ exists. Let $\mathbf{V}$ denote the matrix in the second set of braces. Note that $\mathbf{V}$ has full row rank so in particular $(\mathbf{V}\mathbf{V}')^{-1}$ exists. Hence, we can express this identity as,

$$(\mathbf{U}'\mathbf{U})^{-1}\mathbf{AB}(\mathbf{V}\mathbf{V}')^{-1} = \mathbf{\Gamma}_{11}\mathbf{\Gamma}_{21}'. \tag{9.9.7}$$

For the proof then, suppose $\mathbf{AB} = \mathbf{0}$. Then, by (9.9.7), $\mathbf{\Gamma}_{11}\mathbf{\Gamma}_{21}' = \mathbf{0}$ and, hence, by (9.9.5), the random vectors $\mathbf{W}_1$ and $\mathbf{W}_2$ are independent. Therefore, by (9.9.2) and (9.9.4) Q_1 and Q_2 are independent.

Conversely, if Q_1 and Q_2 are independent then

$$E^{-2}[\exp\{t_1Q_1 + t_2Q_2\}] = E^{-2}[\exp\{t_1Q_1\}]E^{-2}[\exp\{t_2Q_2\}], \tag{9.9.8}$$

for (t_1, t_2) in an open neighborhood of $(0,0)$. Note that $t_1Q_1 + t_2Q_2$ is a quadratic form in $\mathbf{X}$ with symmetric matrix $t_1\mathbf{A}+t_2\mathbf{B}$. Recall that the matrix $\mathbf{\Gamma}_1$ is orthogonal and, hence, has determinant one. Using this and Theorem 9.8.2, we can write the left side of (9.9.8) as

$$
\begin{aligned}
E^{-2}[\exp\{t_1Q_1 + t_2Q_2\}] &= |\mathbf{I}_n - 2t_1\mathbf{A} - 2t_2\mathbf{B}| \\
&= |\mathbf{\Gamma}_1'\mathbf{\Gamma}_1 - 2t_1\mathbf{\Gamma}_1'\mathbf{\Lambda}_1\mathbf{\Gamma}_1 - 2t_2\mathbf{\Gamma}_1'(\mathbf{\Gamma}_1\mathbf{B}\mathbf{\Gamma}_1')\mathbf{\Gamma}_1| \\
&= |\mathbf{I}_n - 2t_1\mathbf{\Lambda}_1 - 2t_2\mathbf{D}|, \tag{9.9.9}
\end{aligned}
$$

where the matrix $\mathbf{D}$ is given by

$$\mathbf{D} = \mathbf{\Gamma}_1\mathbf{B}\mathbf{\Gamma}_1' = \begin{bmatrix} \mathbf{D}_{11} & \mathbf{D}_{12} \\ \mathbf{D}_{21} & \mathbf{D}_{22} \end{bmatrix}, \tag{9.9.10}$$

and $\mathbf{D}_{11}$ is $r \times r$. By (9.9.2), (9.9.3), and Theorem 9.8.2, the right side of (9.9.8) can be written as,

$$E^{-2}[\exp\{t_1Q_1\}]E^{-2}[\exp\{t_2Q_2\}] = \left\{\prod_{i=1}^{r}(1 - 2t_1\lambda_i)\right\}|\mathbf{I}_n - 2t_2\mathbf{D}|. \tag{9.9.11}$$

This leads to the identity,

$$|\mathbf{I}_n - 2t_1\mathbf{\Lambda}_1 - 2t_2\mathbf{D}| = \left\{\prod_{i=1}^{r}(1 - 2t_1\lambda_i)\right\}|\mathbf{I}_n - 2t_2\mathbf{D}|, \tag{9.9.12}$$

for (t_1, t_2) in an open neighborhood of $(0,0)$.

The coefficient of $(-2t_1)^r$ on the right side of (9.9.12) is $\lambda_1 \cdots \lambda_r|\mathbf{I} - 2t_2\mathbf{D}|$. It is not so easy to find the coefficient of $(-2t_1)^r$ in the left side of the equation (9.9.12). Conceive of expanding this determinant in terms of minors of order r formed from the first r columns. One term in this expansion is the product of the minor of order

r in the upper left-hand corner, namely, $|\mathbf{I}_r - 2t_1\mathbf{\Lambda}_{11} - 2t_2\mathbf{D}_{11}|$, and the minor of order $n - r$ in the lower right-hand corner, namely, $|\mathbf{I}_{n-r} - 2t_2\mathbf{D}_{22}|$. Moreover, this product is the only term in the expansion of the determinant that involves $(-2t_1)^r$. Thus the coefficient of $(-2t_1)^r$ in the left-hand member of Equation (9.9.12) is $\lambda_1 \cdots \lambda_r |\mathbf{I}_{n-r} - 2t_2\mathbf{D}_{22}|$. If we equate these coefficients of $(-2t_1)^r$, we have,

$$|\mathbf{I} - 2t_2\mathbf{D}| = |\mathbf{I}_{n-r} - 2t_2\mathbf{D}_{22}|, \tag{9.9.13}$$

for t_2 in an open neighborhood of 0. Equation (9.9.13) implies that the nonzero eigenvalues of the matrices $\mathbf{D}$ and $\mathbf{D}_{22}$ are the same (see Exercise 9.9.8). Recall that the sum of the squares of the eigenvalues of a symmetric matrix is equal to the sum of the squares of the elements of that matrix (see Exercise 9.8.8). Thus the sum of the squares of the elements of matrix $\mathbf{D}$ is equal to the sum of the squares of the elements of $\mathbf{D}_{22}$. Since the elements of the matrix $\mathbf{D}$ are real, it follows that each of the elements of $\mathbf{D}_{11}, \mathbf{D}_{12}$, and $\mathbf{D}_{21}$ are zero. Hence, we can write

$$0 = \mathbf{\Lambda}_1\mathbf{D} = \mathbf{\Gamma}_1\mathbf{A}\mathbf{\Gamma}_1'\mathbf{\Gamma}_1\mathbf{B}\mathbf{\Gamma}_1',$$

because $\mathbf{\Gamma}_1$ is an orthogonal matrix, $\mathbf{AB} = \mathbf{0}$. ∎

Remark 9.9.2. Theorem 9.9.1 remains valid if the random sample is from a distribution which is $N(\mu, \sigma^2)$, whatever is the real value of μ. Moreover, Theorem 9.9.1 may be extended to quadratic forms in random variables that have a joint multivariate normal distribution with a positive definite covariance matrix $\mathbf{\Sigma}$. The necessary and sufficient condition for the independence of two such quadratic forms with symmetric matrices $\mathbf{A}$ and $\mathbf{B}$ then becomes $\mathbf{A}\mathbf{\Sigma}\mathbf{B} = \mathbf{0}$. In our Theorem 9.9.1, we have $\mathbf{\Sigma} = \sigma^2\mathbf{I}$, so that $\mathbf{A}\mathbf{\Sigma}\mathbf{B} = \mathbf{A}\sigma^2\mathbf{I}\mathbf{B} = \sigma^2\mathbf{A}\mathbf{B} = \mathbf{0}$. ∎

The following theorem is from Hogg and Craig (1958).

Theorem 9.9.2 (Hogg and Craig). *Let* $Q = Q_1 + \cdots + Q_{k-1} + Q_k$, *where* $Q, Q_1, \ldots, Q_{k-1}, Q_k$ *are* $k + 1$ *random variables that are quadratic forms in the observations of a random sample of size* n *from a distribution which is* $N(0, \sigma^2)$. *Let* Q/σ^2 *be* $\chi^2(r)$, *let* Q_i/σ^2 *be* $\chi^2(r_i)$, $i = 1, 2, \ldots, k - 1$, *and let* Q_k *be nonnegative. Then the random variables* $Q_1, Q_2, \ldots, Q_k$ *are independent and hence,* Q_k/σ^2 *is* $\chi^2(r_k = r - r_1 - \cdots - r_{k-1})$.

Proof: Take first the case of $k = 2$ and let the real symmetric matrices Q, Q_1, and Q_2 be denoted, respectively, by $\mathbf{A}, \mathbf{A}_1, \mathbf{A}_2$. We are given that $Q = Q_1 + Q_2$ or, equivalently, that $\mathbf{A} = \mathbf{A}_1 + \mathbf{A}_2$. We are also given that Q/σ^2 is $\chi^2(r)$ and that Q_1/σ^2 is $\chi^2(r_1)$. In accordance with Theorem 9.8.4, we have $\mathbf{A}^2 = \mathbf{A}$ and $\mathbf{A}_1^2 = \mathbf{A}$. Since $Q_2 \geq 0$, each of the matrices $\mathbf{A}, \mathbf{A}_1$, and $\mathbf{A}_2$ is positive semidefinite. Because $\mathbf{A}^2 = \mathbf{A}$, we can find an orthogonal matrix $\mathbf{\Gamma}$ such that

$$\mathbf{\Gamma}'\mathbf{A}\mathbf{\Gamma} = \begin{bmatrix} \mathbf{I}_r & \mathbf{0} \\ \mathbf{0} & \mathbf{0} \end{bmatrix}.$$

If we multiply both members of $\mathbf{A} = \mathbf{A}_1 + \mathbf{A}_2$ on the left by $\mathbf{\Gamma}'$ and on the right by $\mathbf{\Gamma}$, we have

$$\begin{bmatrix} \mathbf{I}_r & \mathbf{0} \\ \mathbf{0} & \mathbf{0} \end{bmatrix} = \mathbf{\Gamma}'\mathbf{A}_1\mathbf{\Gamma} + \mathbf{\Gamma}'\mathbf{A}_2\mathbf{\Gamma}.$$

Now each of A_1 and A_2, and hence each of $\Gamma' A_1 \Gamma$ and $\Gamma' A_2 \Gamma$ is positive semidefinite. Recall that, if a real symmetric matrix is positive semidefinite, each element on the principal diagonal is positive or zero. Moreover, if an element on the principal diagonal is zero, then all elements in that row and all elements in that column are zero. Thus $\Gamma' A \Gamma = \Gamma' A_1 \Gamma + \Gamma' A_2 \Gamma$ can be written as

$$
\begin{bmatrix} I_r & 0 \\ 0 & 0 \end{bmatrix} = \begin{bmatrix} G_r & 0 \\ 0 & 0 \end{bmatrix} + \begin{bmatrix} H_r & 0 \\ 0 & 0 \end{bmatrix}. \tag{9.9.14}
$$

Since $A_1^2 = A_1$, we have

$$
(\Gamma' A_1 \Gamma)^2 = \Gamma' A_1 \Gamma = \begin{bmatrix} G_r & 0 \\ 0 & 0 \end{bmatrix}.
$$

If we multiply both members of Equation (9.9.14) on the left by the matrix $\Gamma' A_1 \Gamma$, we see that

$$
\begin{bmatrix} G_r & 0 \\ 0 & 0 \end{bmatrix} = \begin{bmatrix} G_r & 0 \\ 0 & 0 \end{bmatrix} + \begin{bmatrix} G_r H_r & 0 \\ 0 & 0 \end{bmatrix}
$$

or, equivalently, $\Gamma' A_1 \Gamma = \Gamma' A_1 \Gamma + (\Gamma' A_1 \Gamma)(\Gamma' A_2 \Gamma)$. Thus $(\Gamma' A_1 \Gamma) \times (\Gamma' A_2 \Gamma) = 0$ and $A_1 A_2 = 0$. In accordance with Theorem 9.9.1, Q_1 and Q_2 are independent. This independence immediately implies that Q_2/σ^2 is $\chi^2(r_2 = r - r_1)$. This completes the proof when $k = 2$. For $k > 2$, the proof may be made by induction. We shall merely indicate how this can be done by using $k = 3$. Take $A = A_1 + A_2 + A_3$, where $A^2 = A$, $A_1^2 = A_1$, $A_2^2 = A_2$, and A_3 is positive semidefinite. Write $A = A_1 + (A_2 + A_3) = A_1 + B_1$, say. Now $A^2 = A$, $A_1^2 = A_1$, and B_1 is positive semidefinite. In accordance with the case of $k = 2$, we have $A_1 B_1 = 0$, so that $B_1^2 = B_1$. With $B_1 = A_2 + A_3$, where $B_1^2 = B_1$, $A_2^2 = A_2$, it follows from the case of $k = 2$ that $A_2 A_3 = 0$ and $A_3^2 = A_3$. If we regroup by writing $A = A_2 + (A_1 + A_3)$, we obtain $A_1 A_3 = 0$, and so on. ∎

Remark 9.9.3. In our statement of Theorem 9.9.2 we took $X_1, X_2, \ldots, X_n$ to be observations of a random sample from a distribution which is $N(0, \sigma^2)$. We did this because our proof of Theorem 9.9.1 was restricted to that case. In fact, if $Q', Q_1', \ldots, Q_k'$ are quadratic forms in any normal variables (including multivariate normal variables), if $Q' = Q_1' + \cdots + Q_k'$, if $Q', Q_1', \ldots, Q_{k-1}'$ are central or noncentral chi-square, and if Q_k' is nonnegative, then $Q_1', \ldots, Q_k'$ are independent and Q_k' is either central or noncentral chi-square. ∎

This section will conclude with a proof of a frequently quoted theorem due to Cochran.

Theorem 9.9.3 (Cochran). *Let $X_1, X_2, \ldots, X_n$ denote a random sample from a distribution which is $N(0, \sigma^2)$. Let the sum of the squares of these observations be written in the form*

$$
\sum_1^n X_i^2 = Q_1 + Q_2 + \cdots + Q_k,
$$

where Q_j is a quadratic form in $X_1, X_2, \ldots, X_n$, with matrix $\boldsymbol{A}_j$ which has rank r_j, $j = 1, 2, \ldots, k$. The random variables $Q_1, Q_2, \ldots, Q_k$ are independent and Q_j/σ^2 is $\chi^2(r_j)$, $j = 1, 2, \ldots, k$, if and only if $\sum_1^n r_j = n$.

Proof. First assume the two conditions $\sum_1^n r_j = n$ and $\sum_1^n X_i^2 = \sum_1^n Q_j$ to be satisfied. The latter equation implies that $\boldsymbol{I} = \boldsymbol{A}_1 + \boldsymbol{A}_2 + \cdots + \boldsymbol{A}_k$. Let $\boldsymbol{B}_i = \boldsymbol{I} - \boldsymbol{A}_i$; that is, $\boldsymbol{B}_i$ is the sum of the matrices $\boldsymbol{A}_1, \ldots, \boldsymbol{A}_k$ exclusive of $\boldsymbol{A}_i$. Let R_i denote the rank of $\boldsymbol{B}_i$. Since the rank of the sum of several matrices is less than or equal to the sum of the ranks, we have $R_i \leq \sum_1^n r_j - r_i = n - r_i$. However, $\boldsymbol{I} = \boldsymbol{A}_i + \boldsymbol{B}_i$, so that $n \leq r_i + R_i$ and $n - r_i \leq R_i$. Hence, $R_i = n - r_i$. The eigenvalues of $\boldsymbol{B}_i$ are the roots of the equation $|\boldsymbol{B}_i - \lambda \boldsymbol{I}| = 0$. Since $\boldsymbol{B}_i = \boldsymbol{I} - \boldsymbol{A}_i$, this equation can be written as $|\boldsymbol{I} - \boldsymbol{A}_i - \lambda \boldsymbol{I}| = 0$. Thus we have $|\boldsymbol{A}_i - (1 - \lambda)\boldsymbol{I}| = 0$. But each root of the last equation is one minus an eigenvalue of $\boldsymbol{A}_i$. Since $\boldsymbol{B}_i$ has exactly $n - R_i = r_i$ eigenvalues that are zero, then $\boldsymbol{A}_i$ has exactly r_i eigenvalues that are equal to one. However, r_i is the rank of $\boldsymbol{A}_i$. Thus each of the r_i nonzero eigenvalues of $\boldsymbol{A}_i$ is one. That is, $\boldsymbol{A}_i^2 = \boldsymbol{A}_i$ and thus $Q_i/\sigma^2(r_i)$, $i = 1, 2, \ldots, k$. In accordance with Theorem 9.9.2, the random variables $Q_1, Q_2, \ldots, Q_k$ are independent.

To complete the proof of Theorem 9.9.3, take

$$\sum_1^n X_i^2 = Q_1 + Q_2 + \cdots + Q_k,$$

let $Q_1, Q_2, \ldots, Q_k$ be independent, and let Q_j/σ^2 be $\chi^2(r_j)$, $j = 1, 2, \ldots, k$. Then $\sum_1^n Q_j/\sigma^2$ is $\chi^2\left(\sum_1^n r_j\right)$. But $\sum_1^n Q_j/\sigma^2 = \sum_1^n X_i^2/\sigma^2$ is $\chi^2(n)$. Thus $\sum_1^n r_j = n$ and the proof is complete. ∎

EXERCISES

9.9.1. Let X_1, X_2, X_3 be a random sample from the normal distribution $N(0, \sigma^2)$. Are the quadratic forms $X_1^2 + 3X_1X_2 + X_2^2 + X_1X_3 + X_3^2$ and $X_1^2 - 2X_1X_2 + \frac{2}{3}X_2^2 - 2X_1X_2 - X_3^2$ independent or dependent?

9.9.2. Let $X_1, X_2, \ldots, X_n$ denote a random sample of size n from a distribution which is $N(0, \sigma^2)$. Prove that $\sum_1^n X_i^2$ and every quadratic form, which is nonidentically zero in $X_1, X_2, \ldots, X_n$, are dependent.

9.9.3. Let X_1, X_2, X_3, X_4 denote a random sample of size 4 from a distribution which is $N(0, \sigma^2)$. Let $Y = \sum_1^4 a_i X_i$, where a_1, a_2, a_3, and a_4 are real constants. If Y^2 and $Q = X_1X_2 - X_3X_4$ are independent, determine a_1, a_2, a_3, and a_4.

9.9.4. Let A be the real symmetric matrix of a quadratic form Q in the observations of a random sample of size n from a distribution which is $N(0, \sigma^2)$. Given that Q and the mean $\overline{X}$ of the sample are independent, what can be said of the elements of each row (column) of A?

Hint: Are Q and $\overline{X}^2$ independent?

9.9.5. Let $A_1, A_2, \ldots, A_k$ be the matrices of $k > 2$ quadratic forms $Q_1, Q_2, \ldots, Q_k$ in the observations of a random sample of size n from a distribution which is $N(0, \sigma^2)$. Prove that the pairwise independence of these forms implies that they are mutually independent.

Hint: Show that $A_i A_j = 0$, $i \neq j$, permits $E[\exp(t_1 Q_1 + t_2 Q_2 + \cdots t_k Q_k)]$ to be written as a product of the mgfs of $Q_1, Q_2, \ldots, Q_k$.

9.9.6. Let $X' = [X_1, X_2, \ldots, X_n]$, where $X_1, X_2, \ldots, X_n$ are observations of a random sample from a distribution which is $N(0, \sigma^2)$. Let $b' = [b_1, b_2, \ldots, b_n]$ be a real nonzero vector, and let A be a real symmetric matrix of order n. Prove that the linear form $b'X$ and the quadratic form $X'AX$ are independent if and only if $b'A = 0$. Use this fact to prove that $b'X$ and $X'AX$ are independent if and only if the two quadratic forms, $(b'X)^2 = X'bb'X$ and $X'AX$ are independent.

9.9.7. Let Q_1 and Q_2 be two nonnegative quadratic forms in the observations of a random sample from a distribution which is $N(0, \sigma^2)$. Show that another quadratic form Q is independent of $Q_1 + Q_2$ if and only if Q is independent of each of Q_1 and Q_2.

Hint: Consider the orthogonal transformation that diagonalizes the matrix of $Q_1 + Q_2$. After this transformation, what are the forms of the matrices Q, Q_1 and Q_2 if Q and $Q_1 + Q_2$ are independent?

9.9.8. Prove that Equation (9.9.13) of this section implies that the nonzero eigenvalues of the matrices D and D_{22} are the same.

Hint: Let $\lambda = 1/(2t_2)$, $t_2 \neq 0$, and show that Equation (9.9.13) is equivalent to $|D - \lambda I| = (-\lambda)^r |D_{22} - \lambda I_{n-r}|$.

9.9.9. Here Q_1 and Q_2 are quadratic forms in observations of a random sample from $N(0, 1)$. If Q_1 and Q_2 are independent and if $Q_1 + Q_2$ has a chi-square distribution, prove that Q_1 and Q_2 are chi-square variables.

9.9.10. Often in regression the mean of the random variable Y is a linear function of p-values $x_1, x_2, \ldots, x_p$, say $\beta_1 x_1 + \beta_2 x_2 + \cdots + \beta_p x_p$, where $\beta' = (\beta_1, \beta_2, \ldots, \beta_p)$ are the *regression coefficients*. Suppose that n values, $Y' = (Y_1, Y_2, \ldots, Y_n)$ are observed for the x-values in $X = [x_{ij}]$, where X is an $n \times p$ *design matrix* and its *i*th row is associated with Y_i, $i = 1, 2, \ldots, n$. Assume that Y is multivariate normal with mean $X\beta$ and variance-covariance matrix $\sigma^2 I$, where I is the $n \times n$ identity matrix.

(a) Note that $Y_1, Y_2, \ldots, Y_n$ are independent. Why?

(b) Since Y should approximately equal its mean $X\beta$, we estimate β by solving the *normal equations* $X'Y = X'X\beta$ for β. Assuming that $X'X$ is non-singular, solve the equations to get $\hat{\beta} = (X'X)^{-1}X'Y$. Show that $\hat{\beta}$ has a multivariate normal distribution with mean $\boldsymbol{\beta}$ and variance-covariance matrix $\sigma^2(\boldsymbol{X}'\boldsymbol{X})^{-1}$.

(c) Show that

$$(\boldsymbol{Y} - \boldsymbol{X}\boldsymbol{\beta})'(\boldsymbol{Y} - \boldsymbol{X}\boldsymbol{\beta}) = (\hat{\boldsymbol{\beta}} - \boldsymbol{\beta})'(\boldsymbol{X}'\boldsymbol{X})(\hat{\boldsymbol{\beta}} - \boldsymbol{\beta}) + (\boldsymbol{Y} - \boldsymbol{X}\hat{\boldsymbol{\beta}})'(\boldsymbol{Y} - \boldsymbol{X}\hat{\boldsymbol{\beta}}),$$

say $Q = Q_1 + Q_2$ for convenience.

(d) Show that Q_1/σ^2 is $\chi^2(p)$.

(e) Show that Q_1 and Q_2 are independent.

(f) Argue that Q_2/σ^2 is $\chi^2(n-p)$.

(g) Find c so that cQ_1/Q_2 has an F-distribution.

(h) The fact that a value d can be found so that $P(cQ_1/Q_2 \le d) = 1 - \alpha$ could be used to find a $100(1 - \alpha)$ percent confidence ellipsoid for $\boldsymbol{\beta}$. Explain.

9.9.11. Say that G.P.A. (Y) is thought to be a linear function of a "coded" high school rank (x_2) and a "coded" American College Testing score (x_3), namely, $\beta_1 + \beta_2 x_2 + \beta_3 x_3$. Note that all x_1 values equal 1. We observe the following five points:

x_1	x_2	x_3	Y
1	1	2	3
1	4	3	6
1	2	2	4
1	4	2	4
1	3	2	4

(a) Compute $\boldsymbol{X}'\boldsymbol{X}$ and $\hat{\boldsymbol{\beta}} = (\boldsymbol{X}'\boldsymbol{X})^{-1}\boldsymbol{X}'\boldsymbol{Y}$.

(b) Compute a 95 percent confidence ellipsoid for $\boldsymbol{\beta}' = (\beta_1, \beta_2, \beta_3)$. See (h) of Exercise 9.9.10.

Chapter 10

Nonparametric Statistics

10.1 Location Models

In this chapter, we present some nonparametric procedures for the simple location problems. As we shall show, the test procedures associated with these methods are distribution-free under null hypotheses. We also obtain point estimators and confidence intervals associated with these tests. The distributions of the estimators are not distribution-free; hence, we use the term **rank-based** to refer collectively to these procedures. The asymptotic relative efficiencies of these procedures are easily obtained, thus facilitating comparisons among them and procedures that we have discussed in earlier chapters. We also obtain estimators that are asymptotically efficient; that is, they achieve asymptotically the Rao-Cramér bound.

Our purpose is not a rigorous development of these concepts and at times we will simply sketch the theory. A rigorous treatment can be found in several advanced texts, such as Randles and Wolfe (1979) or Hettmansperger and McKean (1998).

In this and the following section, we consider the one-sample problem. For the most part, we consider continuous random variables X with cdf and pdf $F_X(x)$ and $f_X(x)$, respectively. In this and the succeeding chapters we want to identify classes of parameters. Think of a parameter as a function of the cdf (or pdf) of a given random variable. For example, consider the mean μ of X. We can write it as, $\mu_X = T(F_X)$ if T is defined as

$$T(F_X) = \int_{-\infty}^{\infty} x f_X(x)\, dx.$$

As another example, recall that the median of a random variable X is a parameter ξ such that $F_X(\xi) = 1/2$; i.e., $\xi = F_X^{-1}(1/2)$. Hence, in this notation, we say that the parameter ξ is defined by the function $T(F_X) = F_X^{-1}(1/2)$. Note that these Ts are functions of the cdfs (or pdfs). We shall call them **functionals**.

Functionals induce nonparametric estimators naturally. Let $X_1, X_2, \ldots, X_n$ denote a random sample from some distribution with cdf $F(x)$ and let $T(F)$ be a functional. Recall the empirical distribution function of the sample is given by

$$\widehat{F}_n(x) = n^{-1}[\#\{X_i \le x\}], \quad -\infty < x < \infty. \qquad (10.1.1)$$

Because $\widehat{F}_n(x)$ is a cdf, $T(\widehat{F}_n)$ is well defined. Furthermore, $T(\widehat{F}_n)$ depends only on the sample; hence, it is a statistic. We call $T(\widehat{F}_n)$ the *induced estimator* of $T(F)$. For example, if $T(F)$ is the mean of the distribution then it is easy to see that $T(\widehat{F}_n) = \overline{X}$; see Exercise 10.1.3. Likewise, if $T(F) = F^{-1}(1/2)$ is the median of the distribution then $T(\widehat{F}_n) = Q_2$, the sample median.

We begin with the definition of a location functional.

Definition 10.1.1. *Let X be a continuous random variable with cdf $F_X(x)$ and pdf $f_X(x)$. We say that $T(F_X)$ is a **location functional** if it satisfies:*

$$\text{If } Y = X + a, \text{ then } T(F_Y) = T(F_X) + a, \text{ for all } a \in R \qquad (10.1.2)$$
$$\text{If } Y = aX, \text{ then } T(F_Y) = aT(F_X), \text{ for all } a \ne 0. \qquad (10.1.3)$$

For example, suppose T is the mean functional; i.e., $T(F_X) = E(X)$. Let $Y = X + a$ then $E(Y) = E(X + a) = E(X) + a$. Secondly, if $Y = aX$ then $E(Y) = aE(X)$. Hence the mean is a location functional. The next example shows that the median is a location functional.

Example 10.1.1. Let $F(x)$ be the cdf of X and let $T(F_X) = F_X^{-1}(1/2)$ be the median functional of X. Note another way to state this is: $F_X(T(F_X)) = 1/2$. Let $Y = X + a$. It then follows that the cdf of Y is $F_Y(y) = F_X(y - a)$. The following identity shows that $T(F_Y) = T(F_X) + a$:

$$F_Y(T(F_X) + a) = F_X(T(F_X) + a - a) = F_X(T(F_X)) = 1/2.$$

Next, suppose $Y = aX$. If $a > 0$, then $F_Y(y) = F_X(y/a)$ and, hence,

$$F_Y(aT(F_X)) = F_X(aT(F_X)/a) = F_X(T(F_X)) = 1/2.$$

Thus, $T(F_Y) = aT(F_X)$, when $a > 0$.

On the other hand, if $a < 0$, then $F_Y(y) = 1 - F_X(y/a)$. Hence,

$$F_Y(aT(F_X)) = 1 - F_X(aT(F_X)/a) = 1 - F_X(T(F_X)) = 1 - 1/2 = 1/2.$$

Therefore, (10.1.3) holds for all $a \ne 0$. Thus the median is a location parameter.

Recall that the median is a percentile, namely the $50th$ percentile of a distribution. As Exercise 10.1.1 shows, the median is the only percentile which is a location functional. ∎

We will often continue to use parameter notation to denote functionals. For example, $\theta_X = T(F_X)$.

In Chapters 5 and 6, we wrote the location model for specified pdfs. In this chapter, we write it for a general pdf in terms of a specified location functional. Let X be a random variable with cdf $F_X(x)$ and pdf $f_X(x)$. Let $\theta_X = T(F_X)$ be a location functional. Define the random variable ε to be $\varepsilon = X - T(F_X)$. Then by (10.1.2), $T(F_\varepsilon) = 0$; i.e, ε has location 0, according to T. Further, the pdf of X can be written as $f_X(x) = f(x - T(F_X))$ where $f(x)$ is the pdf of ε.

Definition 10.1.2 (Location Model). *Let $\theta_X = T(F_X)$ be a location functional. We say that the observations $X_1, X_2, \ldots, X_n$ follow a **location model** with functional $\theta_X = T(F_X)$, if*

$$X_i = \theta_X + \varepsilon_i, \qquad (10.1.4)$$

where $\varepsilon_1, \varepsilon_2, \ldots, \varepsilon_n$, are iid random variables with pdf $f(x)$ and $T(F_\varepsilon) = 0$. Hence, from the above discussion, $X_1, X_2, \ldots, X_n$ are iid with pdf $f_X(x) = f(x - T(F_X))$.

Example 10.1.2. Let ε be a random variable with cdf $F(x)$, such that $F(0) = 1/2$. Assume that $\varepsilon_1, \varepsilon_2, \ldots, \varepsilon_n$, are iid with cdf $F(x)$. Let $\theta \in R$ and define:

$$X_i = \theta + \varepsilon_i, \quad i = 1, 2, \ldots, n.$$

Then $X_1, X_2, \ldots, X_n$ follow the location model with the locational functional θ, which is the median of X_i. ■

Note that the location model very much depends on the functional. It forces one to state clearly which location functional is being used in order to write the model statement. For the class of symmetric densities, though, all location functionals are the same.

Theorem 10.1.1. *Let X be a random variable with cdf $F_X(x)$ and pdf $f_X(x)$ such that the distribution of X is symmetric about a. Let $T(F_X)$ be any location functional. Then $T(F_X) = a$.*

Proof: By (10.1.2), we have

$$T(F_{X-a}) = T(F_X) - a. \qquad (10.1.5)$$

Since the distribution of X is symmetric about a, it is easy to show that $X - a$ and $-(X - a)$ have the same distribution; see Exercise 10.1.2. Hence, using (10.1.2) and (10.1.3), we have,

$$T(F_{X-a}) = T(F_{-(X-a)}) = -(T(F_X) - a) = -T(F_X) + a. \qquad (10.1.6)$$

Putting (10.1.5) and (10.1.6) together gives the result. ■

Hence, the assumption of symmetry is very appealing. For under it, the concept of "center" is unique.

EXERCISES

10.1.1. Let X be a continuous random variable with cdf $F(x)$. For $0 < p < 1$, let ξ_p be the *p*th quantile, i.e., $F(\xi_p) = p$. If $p \neq 1/2$, show that while property (10.1.2) holds, property (10.1.3) does not. Thus, ξ_p is not a location parameter.

10.1.2. Let X be a continuous random variable with pdf $f(x)$. Suppose $f(x)$ is symmetric about a; i.e., $f(x - a) = f(-(x - a))$. Show that the random variables $X - a$ and $-(X - a)$ have the same pdf.

10.1.3. Let $\widehat{F}_n(x)$ denote the empirical cdf of the sample $X_1, X_2, \ldots, X_n$. The distribution of $\widehat{F}_n(x)$ puts mass $1/n$ at each sample item X_i. Show that its mean is $\overline{X}$. If $T(F) = F^{-1}(1/2)$ is the median, show that $T(\widehat{F}_n) = Q_2$, the sample median.

10.1.4. Let X be a random variable with cdf $F(x)$ and let $T(F)$ be a functional. We say that $T(F)$ is a **scale functional** if it satisfies the three properties

$$
\begin{array}{llll}
(i) & T(F_{aX}) & = & aT(F_X), \quad \text{for } a > 0 \\
(ii) & T(F_{X+b}) & = & T(F_X), \quad\quad \text{for all } b \ ` \\
(iii) & T(F_{-X}) & = & T(F_X).
\end{array}
$$

Show that the following functionals are scale functionals.

(a) The standard deviation, $T(F_X) = (\text{Var}(X))^{1/2}$.

(b) The interquartile range, $T(F_X) = F_X^{-1}(3/4) - F_X^{-1}(1/4)$.

10.2   Sample Median and Sign Test

In this section, we shall consider inference for the median of a distribution using the sample median. Fundamental to this discussion is the sign test statistic which we discuss first.

Let $X_1, X_2, \ldots, X_n$ be a random sample which follows the location model

$$ X_i = \theta + \varepsilon_i, \tag{10.2.1} $$

where $\varepsilon_1, \varepsilon_2, \ldots, \varepsilon_n$ are iid with cdf $F(x)$, pdf $f(x)$, and median 0. Note that in terms of Section 10.1, the location functional is the median and hence, θ is the median of X_i. We begin with a test for the one-sided hypotheses,

$$ H_0 : \theta = \theta_0 \text{ versus } H_1 : \theta > \theta_0. \tag{10.2.2} $$

Consider the statistic,

$$ S = S(\theta_0) = \#\{X_i > \theta_0\}, \tag{10.2.3} $$

which is called the **sign statistic** because it counts the number of positive signs in the differences $X_i - \theta_0$, $i = 1, 2, \ldots, n$. If we define $I(x > a)$ to be 1 or 0 depending on whether $x > a$ or $x \leq a$ then we can express S as

$$ S = \sum_{i=1}^{n} I(X_i > \theta_0). \tag{10.2.4} $$

Note that if H_0 is true, then we expect about half of the observations to exceed θ_0, while if H_1 is true we expect more than one-half the observations to exceed θ_0. Consider then the test of the hypotheses (10.2.2) given by

$$ \text{Reject } H_0 \text{ in favor of } H_1 \text{ if } S \geq c. \tag{10.2.5} $$

Under the null hypothesis, the random variables $I(X_i > \theta_0)$ are iid Bernoulli $b(1, 1/2)$ random variables. Hence, the null distribution of S is $b(n, 1/2)$ with mean $n/2$ and variance $n/4$. Note that under H_0, the sign test does not depend on the distribution of X_i. We call such a test, a **distribution free** test.

For a level α test, select c to be c_α where c_α is the upper α critical point of a binomial $b(n, 1/2)$ distribution. Then $P_{H_0}(S \geq c_\alpha) = \alpha$. The test statistic, though, has a discrete distribution so for an exact test there are only a finite number of levels α available. The values of c_α can be found in tables; see, for instance, Hollander and Wolfe (1999). If the computer packages R or S-PLUS are available then these critical values are easily obtained. For instance, the command `pbinom(0:15,15,.5)` will return the cdf of a binomial distribution with $n = 15$ and $p = 0.5$ from which all possible levels can be seen.

For a given data set, the p-value associated with the sign test is given by $\widehat{p} = P_{H_0}(S \geq s)$ where s is the realized value of S based on the sample. Tables are available to find these p-values. If the reader has access to either the R or S-PLUS statistical packages, then the command `1 - pbinom(s-1,n,.5)` computes $\widehat{p} = P_{H_0}(S \geq s)$.

It is convenient at times to use a large sample test based on the asymptotic distribution of the test statistic. By the Central Limit Theorem, under H_0, the standardized statistic $[S(\theta_0) - (n/2)]/\sqrt{n}/2$ is asymptotically normal, $N(0, 1)$. Hence, the large sample test rejects H_0 if

$$[S(\theta_0) - (n/2)]/\sqrt{n}/2 \geq z_\alpha; \tag{10.2.6}$$

see Exercise 10.2.2.

We briefly touch on the two-sided hypotheses given by

$$H_0 : \theta = \theta_0 \text{ versus } H_1 : \theta \neq \theta_0. \tag{10.2.7}$$

The following symmetric decision rule seems appropriate:

$$\text{Reject } H_0 \text{ in favor of } H_1 \text{ if } S \leq c_1 \text{ or if } S \geq n - c_1. \tag{10.2.8}$$

For a level α test, c_1 would be chosen such that $\alpha/2 = P_{H_0}(S \leq c_1)$. The critical point could be found by a statistics package or tables. Recall that the p-value is given by $\widehat{p} = 2\min\{P_{H_0}(S \leq s), P_{H_0}(S \geq s)\}$, where s is the realized value of S based on the sample.

Example 10.2.1 (Shoshoni Rectangles). A golden rectangle is a rectangle in which the ratio of the width (w) to length (l) is the golden ratio, which is approximately 0.618. It can be characterized in various ways. For example, $w/l = l/(w+l)$ characterizes the golden rectangle. It is considered to be an aesthetic standard in Western civilization and appears in art and architecture going back to the ancient Greeks. It now appears in such items as credit and business cards. In a cultural anthropology study, DuBois (1960) reports on a study of the Shoshoni beaded baskets. These baskets contain beaded rectangles and the question was whether the Shoshonis use the same aesthetic standard as the West. Let X denote the ratio of the width to length of a Shoshoni beaded basket. Let θ be the median of X. The hypotheses of interest are

$$H_0 : \theta = 0.618 \text{ versus } H_A : \theta \neq 0.618.$$

This is a two-sided hypotheses. It follows from the above discussion that the sign test rejects H_0 in favor of H_1 if $S(0.618) \leq c$ or $S(0.618) \geq n - c$.

A sample of 20 width to length (ordered) ratios from Shoshoni baskets resulted in the data

Width to Length Ratios of Rectangles

0.553	0.570	0.576	0.601	0.606	0.606	0.609	0.611	0.615	0.628
0.654	0.662	0.668	0.670	0.672	0.690	0.693	0.749	0.844	0.933

For these data, $S(0.618) = 9$ with $2P(b(20, .5) \leq 9) = 0.8238$ as the p-value. Thus, there is no evidence to reject H_0 based on these data.

A boxplot and a normal $q-q$ plot of the data are given in Figure 10.2.1. Notice that the data contain two, possibly three, potential outliers. The data do not appear to be drawn from a normal distribution. ■

We next obtain several useful results concerning the power function of the sign test for the hypotheses (10.2.2). Because we can always subtract θ_0 from each X_i, we can assume without loss of generality that $\theta_0 = 0$. The following function will prove useful here and in the associated estimation and confidence intervals described below. Define

$$S(\theta) = \#\{X_i > \theta\}. \tag{10.2.9}$$

The sign test statistic is given by $S(\theta_0)$. We can easily describe the function $S(\theta)$. First note that we can write it in terms of the order statistics $Y_1 < \cdots < Y_n$ of $X_1, \ldots, X_n$ because $\#\{Y_i > \theta\} = \#\{X_i > \theta\}$. Now if $\theta < Y_1$ then all the Y_is are larger than θ hence $S(\theta) = n$. Next, if $Y_1 \leq \theta < Y_2$ then $S(\theta) = n - 1$. Continuing this way, we see that $S(\theta)$ is is a decreasing step function of θ, which steps down one unit at each order statistic Y_i, attaining its maximum and minimum values n and 0 at Y_1 and Y_n, respectively. Figure 10.2.2 depicts this function. We need the following translation property.

Lemma 10.2.1. *For every* k,

$$P_\theta[S(0) \geq k] = P_0[S(-\theta) \geq k]. \tag{10.2.10}$$

Proof: Note that the left side of equation (10.2.10) concerns the probability of the event $\#\{X_i > 0\}$, where X_i has median θ. The right side concerns the probability of the event $\#\{(X_i + \theta) > 0\}$, where the random variable $X_i + \theta$ has median θ, (because under $\theta = 0$, X_i has median 0). Hence, the left and right sides give the same probability. ■

Based on this lemma, it is easy to show that the power function sign test is monotone for one-sided tests.

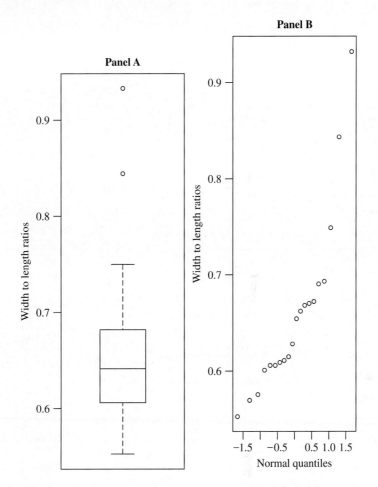

Figure 10.2.1: Boxplot (Panel A) and normal q–q plot (Panel B) of the Shoshoni data.

Theorem 10.2.1. *Suppose Model (10.2.1) is true. Let $\gamma(\theta)$ be the power function of the sign test of level α for the hypotheses (10.2.2). Then $\gamma(\theta)$ is a nondecreasing function of θ.*

Proof: Let c_α denote the $b(n, 1/2)$ upper critical value as given above. Without loss of generality, assume that $\theta_0 = 0$. The power function of the sign test is

$$\gamma(\theta) = P_\theta[S(0) \geq c_\alpha], \quad \text{for } -\infty < \theta < \infty.$$

Suppose $\theta_1 < \theta_2$. Then $-\theta_1 > -\theta_2$ and hence, since, $S(\theta)$ is nonincreasing,

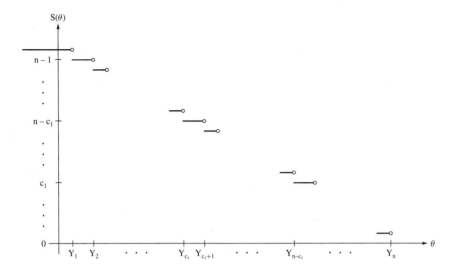

Figure 10.2.2: The sketch shows the graph of the decreasing step function $S(\theta)$. The function drops one unit at each order statistic Y_i.

$S(-\theta_1) \leq S(-\theta_2)$. This and Lemma 10.2.1 yield

$$
\begin{aligned}
\gamma(\theta_1) &= P_{\theta_1}[S(0) \geq c_\alpha] \\
&= P_0[S(-\theta_1) \geq c_\alpha] \\
&\leq P_0[S(-\theta_2) \geq c_\alpha] \\
&= P_{\theta_2}[S(0) \geq c_\alpha] \\
&= \gamma(\theta_2),
\end{aligned}
$$

which is the desired result. ∎

This is a very desirable property for any test. Because the monotonicity of the power function of the sign test holds for all $\theta \in R$, we can extend the simple null hypothesis of (10.2.2) to the composite null hypothesis

$$H_0 : \theta \leq \theta_0 \text{ versus } H_1 : \theta > \theta_0. \tag{10.2.11}$$

Recall from Definition 5.5.4 of Chapter 5 that the size of the test for a composite null hypothesis is given by $\max_{\theta \leq \theta_0} \gamma(\theta)$. Because $\gamma(\theta)$ is nondecreasing, the size of the sign test is α for this extended null hypothesis. As a second result, it follows immediately that the sign test is an unbiased test; see Section 8.3. As Exercise 10.2.7 shows, the power function of the sign test for the other one-sided alternative, $H_1 : \theta < \theta_0$, is nonincreasing.

Under an alternative, say $\theta = \theta_1$, the test statistic S has the binomial distribution $b(n, p_1)$, where p_1 is given by,

$$p_1 = P_{\theta_1}(X > 0) = 1 - F(-\theta_1), \tag{10.2.12}$$

where $F(x)$ is the cdf of ε in Model (10.2.1). Hence, S is not distribution free under alternative hypotheses. As in Exercise 10.2.3, we can determine the power of the test for specified θ_1 and $F(x)$. We want to compare the power of the sign test to other size α tests, in particular, the test based on the sample mean. However, for these comparison purposes, we need more general results. Some of which are obtained in the next subsection.

10.2.1   Asymptotic Relative Efficiency

One solution to this problem is to consider the behavior of a test under a sequence of local alternatives. In this section, we will often take $\theta_0 = 0$ in hypotheses (10.2.2). As noted before expression (10.2.9), this is without loss of generality. For the hypotheses (10.2.2), consider the sequence of alternatives

$$H_0 : \ \theta = 0 \text{ versus } H_{1n} : \ \theta_n = \frac{\delta}{\sqrt{n}}, \tag{10.2.13}$$

where $\delta > 0$. Note that this sequence of alternatives converges to the null hypothesis as $n \to \infty$. We often call such a sequence of alternatives **local alternatives**. The idea is to consider how the power function of a test behaves relative to the power functions of other tests under this sequence of alternatives. We will only sketch this development. For more details the reader can consult more advanced books cited in Section 10.1. As a first step in that direction, we obtain the asymptotic power lemma for the sign test.

Consider the large sample size α test given by (10.2.6). Under the alternative θ_n, we can approximate the mean of this test as follows:

$$
\begin{aligned}
E_{\theta_n}\left[\frac{1}{\sqrt{n}}\left(S(0) - \frac{n}{2}\right)\right] &= E_0\left[\frac{1}{\sqrt{n}}\left(S(-\theta_n) - \frac{n}{2}\right)\right] \\
&= \frac{1}{\sqrt{n}}\sum_{i=1}^{n} E_0[I(X_i > -\theta_n)] - \frac{\sqrt{n}}{2} \\
&= \frac{1}{\sqrt{n}}\sum_{i=1}^{n} P_0(X_i > -\theta_n) - \frac{\sqrt{n}}{2} \\
&= \sqrt{n}\left(1 - F(-\theta_n) - \frac{1}{2}\right) \\
&= \sqrt{n}\left(\frac{1}{2} - F(-\theta_n)\right) \\
&\doteq \sqrt{n}\theta_n f(0) = \delta f(0), \tag{10.2.14}
\end{aligned}
$$

where the step to the last line is due to the mean value theorem. It can be shown in more advanced texts that the variance of $\frac{n^{-1/2}(S(0)-(n/2))}{1/2}$ converges to 1 under θ_n, just as under H_0, and that, futhermore, $[S(0) - (n/2) - \sqrt{n}\delta f(0)]/(\sqrt{n}/2)$ has a limiting standard normal distribution. This leads to the *asymptotic power lemma* which we state in the form of a theorem.

Theorem 10.2.2 (Asymptotic Power Lemma). *Consider the sequence of hypotheses (10.2.13). The limit of the power function of the large sample, size α, sign test is*

$$\lim_{n \to \infty} \gamma(\theta_n) = 1 - \Phi(z_\alpha - \delta \tau_S^{-1}), \tag{10.2.15}$$

where $\tau_S = 1/[2f(0)]$ and $\Phi(z)$ is the cdf of a standard normal random variable.

Proof: Using expression (10.2.14) and the discussion which followed its derivation, we have

$$
\begin{aligned}
\gamma(\theta_n) &= P_{\theta_n}\left[\frac{n^{-1/2}[S(0) - (n/2)]}{1/2} \geq z_\alpha\right] \\
&= P_{\theta_n}\left[\frac{n^{-1/2}[S(0) - (n/2) - \sqrt{n}\delta f(0)]}{1/2} \geq z_\alpha - \delta 2 f(0)\right] \\
&\to 1 - \Phi(z_\alpha - \delta 2 f(0)),
\end{aligned}
$$

which was to be shown. ■

As shown in Exercise 10.2.4, the parameter τ_S is a scale parameter (functional) as defined in Exercise 10.1.4 of the last section. We will later show that $\tau_S/\sqrt{n}$ is the asymptotic standard deviation of the sample median.

There were several approximations used in the proof of Theorem 10.2.2. A rigorous proof can be found in more advanced texts; such as those cited in Section 10.1. It will be quite helpful for the next sections to reconsider the approximation of the mean given in (10.2.14) in terms of another concept called efficacy. Consider another standardization of the test statistic given by

$$\overline{S}(0) = \frac{1}{n}\sum_{i=1}^{n} I(X_i > 0), \tag{10.2.16}$$

where the bar notation is used to signify that $\overline{S}(0)$ is an average and, in this case under H_0, converges in probability to $\frac{1}{2}$. Let $\mu(\theta) = E_\theta(\overline{S}(0) - \frac{1}{2})$. Then by expression (10.2.14) we have,

$$\mu(\theta_n) = E_\theta\left(\overline{S}(0) - \frac{1}{2}\right) = \frac{1}{2} - F(-\theta_n). \tag{10.2.17}$$

Let $\sigma_{\overline{S}}^2 = \mathrm{Var}(\overline{S}(0)) = \frac{1}{4n}$. Finally define the **efficacy** of the sign test to be:

$$c_S = \lim_{n \to \infty} \frac{\mu'(0)}{\sqrt{n}\sigma_{\overline{S}}}. \tag{10.2.18}$$

That is, the efficacy is the rate of change of the mean of the test statistic at the null divided by the product of $\sqrt{n}$ and the standard deviation of the test statistic at the null. So the efficacy increases with an increase in this rate as it should. We use this formulation of efficacy throughout this chapter.

Hence by expression (10.2.14) the efficacy of the sign test is

$$c_S = \frac{f(0)}{1/2} = 2f(0) = \tau_S^{-1}, \tag{10.2.19}$$

the reciprocal of the scale parameter τ_S. In terms of efficacy we can write the conclusion of the Asymptotic Power Lemma as,

$$\lim_{n \to \infty} \gamma(\theta_n) = 1 - \Phi(z_\alpha - \delta c_S). \tag{10.2.20}$$

This is not a coincidence and it will be true for the procedures we consider in the next section.

Remark 10.2.1. In this chapter, we compare nonparametric procedures with traditional parametric procedures. For instance, we compare the sign test with the test based on the sample mean. Traditionally, tests based on sample means are referred to as t-tests. Even though our comparisons are asymptotic and we could use the terminology of z-tests we will instead use the traditional terminology of t-tests. ∎

As a second illustration of efficacy, we determine the efficacy of the t-test for the mean. Assume that the random variables ε_i in Model (10.2.1) are symmetrically distributed about 0 and their mean exists. Hence, the parameter θ is the location parameter. In particular, $\theta = E(X_i) = \text{med}(X_i)$. Denote the variance of X_i by σ^2. This allows us to easily compare the sign and t-tests. Recall for hypotheses (10.2.2) that the t-test rejects H_0 in favor of H_1 if $\overline{X} \geq c$. The form of the test statistic is then $\overline{X}$. Furthermore, we have

$$\mu_{\overline{X}}(\theta) = E_\theta(\overline{X}) = \theta, \tag{10.2.21}$$

and

$$\sigma_{\overline{X}}^2(0) = V_0(\overline{X}) = \frac{\sigma^2}{n}. \tag{10.2.22}$$

Thus by (10.2.21) and (10.2.22), the efficacy of the t-test is

$$c_t = \lim_{n \to \infty} \frac{\mu'_{\overline{X}}(0)}{\sqrt{n}(\sigma/\sqrt{n})} = \frac{1}{\sigma}. \tag{10.2.23}$$

As confirmed in Exercise 10.2.8, the asymptotic power of the large sample level α, t-test under the sequence of alternatives (10.2.13) is $1 - \Phi(z_\alpha - \delta c_t)$. Thus we can compare the sign and t-tests by comparing their efficacies. We do this from the perspective of sample size determination.

Assume without loss of generality that $H_0 : \theta = 0$. Now suppose we want to determine the sample size so that a level α sign test can detect the alternative $\theta^* > 0$ with (approximate) probability γ^*. That is, find n so that

$$\gamma^* = \gamma(\theta^*) = P_{\theta^*}\left[\frac{S(0) - (n/2)}{\sqrt{n}/2} \geq z_\alpha\right]. \tag{10.2.24}$$

Write $\theta^* = \sqrt{n}\theta^*/\sqrt{n}$. Then using the asymptotic power lemma we have

$$\gamma^* = \gamma(\sqrt{n}\theta^*/\sqrt{n}) \doteq 1 - \Phi(z_\alpha - \sqrt{n}\theta^* \tau_S^{-1}).$$

Now denote z_{γ^*} to be the upper $1 - \gamma^*$ quantile of the standard normal distribution. Then from this last equation we have

$$z_{\gamma^*} = z_\alpha - \sqrt{n}\theta^* \tau_S^{-1}.$$

Solving for n, we get

$$n_S = \left(\frac{(z_\alpha - z_{\gamma^*})\tau_S}{\theta^*} \right)^2. \tag{10.2.25}$$

As outlined in Exercise 10.2.8, for this situation the sample size determination for the test based on the sample mean is

$$n_{\overline{X}} = \left(\frac{(z_\alpha - z_{\gamma^*})\sigma}{\theta^*} \right)^2, \tag{10.2.26}$$

where $\sigma^2 = \text{Var}(\varepsilon)$.

Suppose we have two tests of the same level for which the asymptotic power lemma holds and for each we determine the sample size necessary to achieve power γ^* at the alternative θ^*. Then the ratio of these sample sizes is called the **asymptotic relative efficiency** (ARE) between the tests. We show later that this is the same as the ARE defined in Chapter 5 between estimators. Hence, the ARE of the sign test to the t-test is

$$ARE(S, t) = \frac{n_{\overline{X}}}{n_S} = \frac{\sigma^2}{\tau_S^2} = \frac{c_S^2}{c_t^2}. \tag{10.2.27}$$

Note that this is the same relative efficiency that was discussed in Example 6.2.5 when the sample median was compared to the sample mean. In the next two examples we revisit this discussion by examining the AREs when X_i has a normal distribution or a Laplace (double exponential) distribution.

Example 10.2.2 ($ARE(S, t)$: normal distribution). Suppose $X_1, X_2, \ldots, X_n$ follow the location model (10.1.4) where $f(x)$ is a $N(0, \sigma^2)$ pdf. Then $\tau_S = (2f(0))^{-1} = \sqrt{\pi/2}\sigma$. Hence, the $ARE(S, t)$ is given by,

$$ARE(S, t) = \frac{\sigma^2}{\tau_S^2} = \frac{\sigma^2}{(\pi/2)\sigma^2} = \frac{2}{\pi} \doteq 0.637. \tag{10.2.28}$$

Hence, at the normal distribution the sign test is only 64% as efficient as the t-test. In terms of sample size at the normal distribution, the t-test requires a smaller sample, $.64n_s$, where n_s is the sample size of the sign test, to achieve the same power as the sign test. A cautionary note is needed here because this is asymptotic efficiency. There have been ample empirical (simulation) studies which give credence to these numbers. ∎

Example 10.2.3 ($ARE(S,t)$ at the Laplace distribution). For this example, consider Model (10.1.4) where $f(x)$ is the Laplace pdf $f(x) = (2b)^{-1}\exp\{-|x|/b\}$ for $-\infty < x < \infty$ and $b > 0$. Then $\tau_S = (2f(0))^{-1} = b$ while $\sigma^2 = E(X^2) = 2b^2$. Hence, the $ARE(S,t)$ is given by,

$$ARE(S,t) = \frac{\sigma^2}{\tau_S^2} = \frac{2b^2}{b^2} = 2. \tag{10.2.29}$$

So, at the Laplace distribution, the sign test is (asymptotically) twice as efficient as the t-test. In terms of sample size at the Laplace distribution, the t-test requires twice as large a sample as the sign test to achieve the same asymptotic power as the sign test. ∎

The normal distribution has much lighter tails than the Laplace distribution, because the two pdfs are proportional to $\exp\{-t^2/2\sigma^2\}$ and $\exp\{-|t|/b\}$, respectively. Based on the last two examples, it seems that the t-test is more efficient for light-tailed distributions while the sign test is more efficient for heavier-tailed distributions. This is true in general and we illustrate this in the next example where we can easily vary the tail weight from light to heavy.

Example 10.2.4 ($ARE(S,t)$ at a family of contaminated normals). Consider the location Model (10.1.4) where the cdf of ε_i is the contaminated normal cdf given in expression (3.4.16). Assume that $\theta_0 = 0$. Recall for this distribution $(1-\epsilon)$ proportion of the time the sample is drawn from a $N(0, b^2)$ distribution while ϵ proportion of the time the sample is drawn from a $N(0, b^2\sigma_c^2)$ distribution. The corresponding pdf is given by

$$f(x) = \frac{1-\epsilon}{b}\phi\left(\frac{x}{b}\right) + \frac{\epsilon}{b\sigma_c}\phi\left(\frac{x}{b\sigma_c}\right), \tag{10.2.30}$$

where $\phi(z)$ is the the pdf of a standard normal random variable. As shown in Section 3.4, the variance of ε_i is $b^2(1 + \epsilon(\sigma_c^2 - 1))$. Also, $\tau_s = b\sqrt{\pi/2}/[1 - \epsilon + (\epsilon/\sigma_c)]$. Thus, the ARE is

$$ARE(S,t) = \frac{2}{\pi}[(1 + \epsilon(\sigma_c^2 - 1))][1 - \epsilon + (\epsilon/\sigma_c)]^2. \tag{10.2.31}$$

If either of ϵ or σ_c increases then the contamination effect becomes larger, generally resulting in a heavier-tailed distribution. For example, the following table (see Exercise 6.2.6) shows the AREs for various values of ϵ when σ_c set at 3.0:

ϵ	0	0.01	0.02	0.03	0.05	0.10	0.15	0.25
$ARE(S,t)$	0.636	0.678	0.718	0.758	0.832	0.998	1.134	1.326

Notice that for this range of ϵ and for $\sigma_c = 3$, the ARE increases as the proportion of contamination increases. It takes, however, over 10% contamination before the sign test becomes more efficient than the t-test. ∎

10.2.2   Estimating Equations Based on Sign Test

In practice, we often want to estimate θ, the median of X_i, in Model 10.2.1. The associated point estimate based on the sign test can be described with a simple geometry, which is analogous to the geometry of the sample mean. As Exercise 10.2.5 shows, the sample mean $\overline{X}$ is such that

$$\overline{X} = \text{Argmin} \sqrt{\sum_{i=1}^{n}(X_i - \theta)^2}. \tag{10.2.32}$$

The quantity $\sqrt{\sum_{i=1}^{n}(X_i - \theta)^2}$ is the Euclidean distance between the vector of observations $\mathbf{X} = (X_1, X_2, \ldots, X_n)'$ and the vector $\theta\mathbf{1}$. If we simply interchange the square root and the summation symbols, we go from the Euclidean distance to the L_1 distance. Let

$$\widehat{\theta} = \text{Argmin} \sum_{i=1}^{n} |X_i - \theta|. \tag{10.2.33}$$

To determine $\widehat{\theta}$, simply differentiate the quantity on the right side with respect to θ. We then obtain,

$$\frac{\partial}{\partial \theta} \sum_{i=1}^{n} |X_i - \theta| = -\sum_{i=1}^{n} \text{sgn}(X_i - \theta).$$

Setting this to 0, we obtain the estimating equations (EE)

$$\sum_{i=1}^{n} \text{sgn}(X_i - \theta) = 0, \tag{10.2.34}$$

whose solution is the sample median Q_2.

Because our observations are continuous random variables, we have the identity

$$\sum_{i=1}^{n} \text{sgn}(X_i - \theta) = 2S(\theta) - n.$$

Hence, the sample median also solves $S(\theta) \dot{=} n/2$. Consider again Figure 10.2.2. Imagine $n/2$ on the vertical axis. This is halfway in the total drop of $S(\theta)$, from n to 0. The order statistic on the horizontal axis corresponding to $n/2$ is essentially the sample median (middle order statistic). In terms of testing, this last equation says that, based on the data, the sample median is the "most acceptable" hypothesis, because $n/2$ is the null expected value of the test statistic. We often think of this as estimation by the inversion of a test.

We now sketch the asymptotic distribution of the sample median. Assume without loss of generality that the true median of X_i is 0. Let $x \in R$. Using the fact that $S(\theta)$ is nonincreasing and the identity $S(\theta) \dot{=} n/2$, we have the following equivalencies:

$$\{\sqrt{n}Q_2 \leq x\} \Leftrightarrow \left\{Q_2 \leq \frac{x}{\sqrt{n}}\right\} \Leftrightarrow \left\{S\left(\frac{x}{\sqrt{n}}\right) \leq \frac{n}{2}\right\}.$$

Hence, we have

$$
\begin{aligned}
P_0(\sqrt{n}Q_2 \le x) &= P_0\left[S\left(\frac{x}{\sqrt{n}}\right) \le \frac{n}{2}\right] \\
&= P_{-x/\sqrt{n}}\left[S(0) \le \frac{n}{2})\right] \\
&= P_{-x/\sqrt{n}}\left[\frac{S(0) - (n/2)}{\sqrt{n}/2} \le 0\right] \\
&\rightarrow \Phi(0 - x\tau_S^{-1}) = P(\tau_S Z \le x),
\end{aligned}
$$

where Z has a standard normal distribution, Notice that the limit was obtained by invoking the Asymptotic Power Lemma with $\alpha = 0.5$ and hence, $z_\alpha = 0$. Rearranging the last term above, we obtain the asymptotic distribution of the sample median which we state as a theorem:

Theorem 10.2.3. *For the random sample $X_1, X_2, \ldots, X_n$, assume that Model (10.2.1) holds. Suppose that $f(0) > 0$. Then*

$$\sqrt{n}(Q_2 - \theta) \rightarrow N(0, \tau_S^2), \tag{10.2.35}$$

where $\tau_S = (2f(0))^{-1}$.

In Section 6.2 we defined the ARE between two estimators to be the reciprocal of their asymptotic variances. For the sample median and mean, this is the same ratio as that based on sample size determinations of their respective tests given above in expression (10.2.27).

10.2.3   Confidence Interval for the Median

Suppose the random sample $X_1, X_2, \ldots, X_n$ follows the location model (10.2.1). In this subsection, we develop a confidence for the median θ of X_i. Assuming that θ is the true median, the random variable $S(\theta)$, (10.2.9), has a binomial $b(n, 1/2)$ distribution. For $0 < \alpha < 1$, select c_1 so that $P_\theta[S(\theta) \le c_1] = \alpha/2$. Hence, we have

$$1 - \alpha = P_\theta[c_1 < S(\theta) < n - c_1]. \tag{10.2.36}$$

Recall in our derivation for the t-confidence interval for the mean in Chapter 3, we began with such a statement and then "inverted" the pivot random variable $t = \sqrt{n}(\overline{X} - \mu)/S$ (S in this expression is the sample standard deviation) to obtain an equivalent inequality with μ isolated in the middle. In this case, the function $S(\theta)$ does not have an inverse but it is a decreasing step function of θ and the inversion can still be performed. As depicted in Figure 10.2.2, $c_1 < S(\theta) < n - c_1$ if and only if $Y_{c_1+1} \le \theta < Y_{n-c_1}$, where $Y_1 < Y_2 < \cdots < Y_n$ are the order statistics of the sample $X_1, X_2, \ldots, X_n$. Therefore, the interval $[Y_{c_1+1}, Y_{n-c_1})$ is a $(1 - \alpha)100\%$ confidence interval for the median θ. Because the order statistics are continuous random variables, the interval (Y_{c_1+1}, Y_{n-c_1}) is an equivalent confidence interval.

If n is large, then there is a large sample approximation to c_1. We know from the Central Limit Theorem that $S(\theta)$ is approximately normal with mean $n/2$ and variance $n/4$. Then, using the continuity correction, we obtain the approximation

$$c_1 \doteq \frac{n}{2} - \frac{\sqrt{n}z_{\alpha/2}}{2} - \frac{1}{2}, \tag{10.2.37}$$

where $\Phi(-z_{\alpha/2}) = \alpha/2$; see Exercise 10.2.6.

Example 10.2.5 (Example 10.2.1, Continued). There are 20 data points in the Shoshoni basket data. The sample median is $0.5(0.628 + 0.654) = 0.641$, width to length. Because $0.021 = P_{H_0}(S \leq 5)$, a 95.8% confidence interval for θ is the interval $(y_6, y_{15}) = (0.606, 0.672)$ which includes 0.618, the ratio of the width to length which characterizes the golden rectangle. ∎

EXERCISES

10.2.1. Sketch the Figure 10.2.2 for the Shoshoni basket data found in Example 10.2.1. Show the values of the test statistic, point estimate, and the 95.8% confidence interval of Example 10.2.5 on the sketch.

10.2.2. Show that the test given by (10.2.6) has asymptotically level α; that is, show that under H_0

$$[S(\theta_0) - (n/2)]/(\sqrt{n}/2) \xrightarrow{D} Z,$$

where Z has a $N(0, 1)$ distribution.

10.2.3. Let θ denote the median of a random variable X. Consider tseting

$$H_0 : \theta = 0 \text{ versus } H_A : \theta > 0 .$$

Suppose we have a sample of size $n = 25$.

(a) Let S denote the sign test statistic. Determine the level of the test: reject H_0 if $S \geq 16$.

(b) Determine the power of the test in Part (a), if X has $N(0.5, 1)$ distribution.

(c) Assuming X has finite mean $\mu = \theta$, consider the asymptotic test, reject H_0 if $\overline{X}/(\sigma/\sqrt{n}) \geq k$. Assuming that $\sigma = 1$, determine k so the asymptotic test has the same level as the test in Part (a). Then determine the power of this test for the situation in Part (b).

10.2.4. Recall the definition of a scale functional given in Exercise 10.1.4. Show that the parameter τ_S defined in Theorem 10.2.2 is a scale functional.

10.2.5. Show that the sample mean solves the equation (10.2.32).

10.2.6. Derive the approximation (10.2.37).

10.2.7. Show that the power function of the sign test is nonincreasing for the hypotheses

$$H_0 : \ \theta = \theta_0 \text{ versus } H_1 : \ \theta < \theta_0. \qquad (10.2.38)$$

10.2.8. Let $X_1, X_2, \ldots, X_n$ be a random sample which follows the location model (10.2.1). In this exercise we want to compare the sign tests and t-test of the hypotheses (10.2.2); so we will assume the random errors, ε_i are symmetrically distributed about 0. Let $\sigma^2 = \text{Var}(\varepsilon_i)$. Hence, the mean and the median are the same for this location model. Assume, also, that $\theta_0 = 0$. Consider the large sample version of the t-test, which rejects H_0 in favor of H_1 if $\overline{X}/(\sigma/\sqrt{n}) > z_\alpha$.

(a) Obtain the power function, $\gamma_t(\theta)$, of the large sample version of the t-test.

(b) Show that $\gamma_t(\theta)$ is nondecreasing in θ.

(c) Show that $\gamma_t(\theta_n) \to 1 - \Phi(z_\alpha - \sigma\theta^*)$, under the sequence of local alternatives (10.2.13).

(d) Based on Part (c), obtain the sample size determination for the t-test to detect θ^* with approximate power γ^*.

(e) Derive the $ARE(S,t)$ given in (10.2.27).

10.3 Signed-Rank Wilcoxon

Let $X_1, X_2, \ldots, X_n$ be a random sample that follows Model 10.2.1. Inference for θ based on the sign test is simple and requires few assumptions about the underlying distribution of X_i. On the other hand, sign procedures have the low efficiency of 0.64 relative to procedures based on the t-test given an underlying normal distribution. In this section, we discuss a nonparametric procedure that does attain high efficiency relative to the t-test. We make the additional assumption that the pdf $f(x)$ of ε_i in Model 10.2.1 is symmetric; i.e., $f(x) = f(-x)$, for all $x \in R$. Hence, X_i is symmetrically distributed about θ. By Theorem 10.1.1, all location parameters are identical.

First consider the one-sided hypotheses given by (10.2.2). As in Section 10.2, without loss of generality we can assume that $\theta_0 = 0$; otherwise, we would consider the sample $X_1 - \theta_0, \ldots, X_n - \theta_0$. Under a symmetric pdf, observations X_i which are the same distance from 0 are equilikely and hence, should receive the same weight. A test statistic which does this is the **signed-rank Wilcoxon** given by

$$T = \sum_{i=1}^{n} \text{sgn}(X_i)R|X_i|, \qquad (10.3.1)$$

where $R|X_i|$ denotes the rank of X_i among $|X_1|, \ldots, |X_n|$, where the rankings are from low to high. Intuitively, under the null hypothesis, we expect half the X_is to be positive and half to be negative. Further the ranks should be uniformly distributed on the integers $\{1, 2, \ldots, n\}$. Hence values of T around 0 are indicative of H_0. On the other hand, if H_1 is true then we expect more than half X_is to be positive and further, the positive observations are more likely to receive the higher ranks. Thus an appropriate decision rule is

$$\text{Reject } H_0 \text{ in favor of } H_1 \text{ if } T \geq c, \tag{10.3.2}$$

where c is determined by the level α of the test.

Given α, we need the null distribution of T to determine the critical point c. The set of integers $\{-n(n+1)/2, -[n(n+1)/2]+2, \ldots n(n+1)/2\}$ form the support of T. Also, from Section 10.2, we know that the signs are iid with support $\{-1, 1\}$ and pmf

$$p(-1) = p(1) = \frac{1}{2}. \tag{10.3.3}$$

A key result is the following lemma:

Lemma 10.3.1. *Under H_0 and symmetry about 0 for the pdf, $|X_1|, \ldots, |X_n|$ are independent of $sgn(X_1), \ldots, sgn(X_n)$.*

Proof: Because $X_1, \ldots, X_n$ is a random sample from the cdf $F(x)$, it suffices to show that $P[|X_i| \leq x, \text{sgn}(X_i) = 1] = P[|X_i| \leq x]P[\text{sgn}(X_i) = 1]$. But due to H_0 and the symmetry of $f(x)$ this follows from the following string of equalities:

$$
\begin{aligned}
P[|X_i| \leq x, \text{sgn}(X_i) = 1] &= P[0 < X_i \leq x] = F(x) - \frac{1}{2} \\
&= [2F(x) - 1]\frac{1}{2} = P[|X_i| \leq x]P[\text{sgn}(X_i) = 1]. \quad \blacksquare
\end{aligned}
$$

Based on this lemma, the ranks of the X_is are independent of the signs of the X_is. Note that the ranks are a permutation of the integers $1, 2, \ldots, n$. By the lemma this independence is true for any permutation. In particular, suppose we use the permutation that orders the absolute values. For example, suppose the observations are $-6.1, 4.3, 7.2, 8.0, -2.1$. Then the permutation $5, 2, 1, 3, 4$ orders the absolute values; that is, the fifth observation is smallest in absolute value, second observation is the next smallest, etc. This is called the permutation of the *anti-ranks*, which we denote generally by by $i_1, i_2, \ldots, i_n$. Using the anti-ranks, we can write T as

$$T = \sum_{j=1}^{n} j\,\text{sgn}(X_{i_j}), \tag{10.3.4}$$

where by Lemma 10.3.1 $\text{sgn}(X_{i_j})$ are iid with support $\{-1, 1\}$ and pmf (10.3.3).

Based on this observation, for $s \in R$, the mgf of T is

$$
\begin{aligned}
E[\exp\{sT\}] &= E\left[\exp\left\{\sum_{j=1}^{n} sj\operatorname{sgn}(X_{i_j})\right\}\right] \\
&= \prod_{j=1}^{n} E[\exp\{sj\operatorname{sgn}(X_{i_j})\}] \\
&= \prod_{j=1}^{n} \left(\frac{1}{2}e^{-sj} + \frac{1}{2}e^{+sj}\right) \\
&= \frac{1}{2^n} \prod_{j=1}^{n} \left(e^{-sj} + e^{+sj}\right).
\end{aligned}
\tag{10.3.5}
$$

Because the mgf does not depend on the underlying symmetric pdf $f(x)$, the test statistic T is distribution free under H_0. Although the pmf of T cannot be obtained in closed form, this mgf can be used to generate the pmf for a specified n; see Exercise 10.3.1.

Because the $\operatorname{sgn}(X_{i_j})$s are mutually independent with mean zero, it follows that $E_{H_0}[T] = 0$. Further because variance of $\operatorname{sgn}(X_{i_j})$ is one, we have

$$
\operatorname{Var}_{H_0}(T) = \sum_{j=1}^{n} \operatorname{Var}_{H_0}(j\operatorname{sgn}(X_{i_j})) = \sum_{j=1}^{n} j^2 = n(n+1)(2n+1)/6
$$

We summarize these results in the following theorem:

Theorem 10.3.1. *Assume that Model (10.2.1) is true for the random sample $X_1, \ldots, X_n$. Assume also that the pdf $f(x)$ is symmetric about 0. Then under H_0,*

$$
T \text{ is distribution free with a symmetric pmf} \tag{10.3.6}
$$
$$
E_{H_0}[T] = 0 \tag{10.3.7}
$$
$$
Var_{H_0}(T) = \frac{n(n+1)(2n+1)}{6} \tag{10.3.8}
$$
$$
\frac{T}{\sqrt{Var_{H_0}(T)}} \text{ has an asymptotically } N(0,1) \text{ distribution.} \tag{10.3.9}
$$

Proof: The first part of (10.3.6) and the expressions (10.3.7) and (10.3.8) were derived above. The asymptotic distribution of T certainly is plausible and can be found in more advanced books. To obtain the second part of (10.3.6), we need to show that the distribution of T is symmetric about 0. But by the mgf of Y, (10.3.5), we have

$$
E[\exp\{s(-T)\}] = E[\exp\{(-s)T\}] = E[\exp\{sT\}].
$$

Hence, T and $-T$ have the same distribution, so T is symmetrically distributed about 0. ∎

Critical values for the decision rule (10.3.2) can be obtained for the exact distribution of T. Tables of the exact distribution can be found in applied nonparametric books such as Hollander and Wolfe (1999). Discussion on the use of R and S-PLUS is given in the next paragraph. But note that the support of T is much denser than that of the sign test, so the normal approximation is good even for a sample size of 10.

There is another formulation of T which will be convenient. Let T^+ denote the sum of the ranks of the positive X_is. Then, because the sum of all ranks is $n(n+1)/2$, we have

$$
\begin{aligned}
T &= \sum_{i=1}^{n} \operatorname{sgn}(X_i) R|X_i| = \sum_{X_i > 0} R|X_i| - \sum_{X_i < 0} R|X_i| \\
&= 2 \sum_{X_i > 0} R|X_i| - \frac{n(n+1)}{2} \\
&= 2T^+ - \frac{n(n+1)}{2}.
\end{aligned} \tag{10.3.10}
$$

Hence, T^+ is a linear function of T and hence, is an equivalent formulation of the signed-rank tests statistic T. For the record we note the null mean and variance of T^+:

$$
E_{H_0}(T^+) = \frac{n(n+1)}{4} \quad \text{and} \quad \operatorname{Var}_{H_0}(T^+) = \frac{n(n+1)(2n+1)}{24}. \tag{10.3.11}
$$

If the reader has the computer language R or S-PLUS at hand then the function **psignrank** evaluates the cdf of T^+. For example, for sample size n, the probability $P(T^+ \leq t)$ is computed by the command **psirank(t,n)**.

Let $X_i > 0$ and consider all X_j such that $-X_i < X_j < X_i$. Thus all the averages $(X_i + X_j)/2$, under these restrictions, are positive, including $(X_i + X_i)/2$. From the restriction though, the number of these positive averages is simply the $R|X_i|$. Doing this for all $X_i > 0$, we obtain

$$
T^+ = \#_{i \leq j}\{(X_j + X_i)/2 > 0\}. \tag{10.3.12}
$$

The pairwise averages $(X_j + X_i)/2$ are often called the *Walsh averages*. Hence the signed rank Wilcoxon can be obtained by counting up the number of positive Walsh averages.

Example 10.3.1 (Zea Mays Data of Darwin). Reconsider the data set discussed in Example 5.5.1. Recall that W_i is the difference in heights of the cross-fertilized plant minus the self-fertilized plant in pot i, for $i = 1, \ldots, 15$. Let θ be the location parameter and consider the one-sided hypotheses

$$
H_0 : \theta = 0 \text{ versus } H_1 : \theta > 0. \tag{10.3.13}
$$

Table 10.3.1 displays the data and the signed ranks.

Adding up the ranks of the positive items in column 5 of Table 10.3.1, we obtain $T^+ = 96$. Using the exact distribution the R command is 1-psignrank(95,15)),

Table 10.3.1: Signed-Ranks for Darwin Data, Example 10.3.1.

Pot	Cross-fertilized	Self-fertilized	Difference	Signed-Rank
1	23.500	17.375	6.125	11
2	12.000	20.375	-8.375	-14
3	21.000	20.000	1.000	2
4	22.000	20.000	2.000	4
5	19.125	18.375	0.750	1
6	21.550	18.625	2.925	5
7	22.125	18.625	3.500	7
8	20.375	15.250	5.125	9
9	18.250	16.500	1.750	3
10	21.625	18.000	3.625	8
11	23.250	16.250	7.000	12
12	21.000	18.000	3.000	6
13	22.125	12.750	9.375	15
14	23.000	15.500	7.500	13
15	12.000	18.000	-6.000	-10

we obtain the p-value, $\widehat{p} = P_{H_0}(T^+ \geq 96) = 0.021$. For comparison, the asymptotic p-value, using the continuity correction is:

$$P_{H_0}(T^+ \geq 96) = P_{H_0}(T^+ \geq 95.5) \doteq P\left(Z \geq \frac{95.5 - 60}{\sqrt{15 \cdot 16 \cdot 31/24}}\right)$$

$$= P(Z \geq 2.016) = 0.022,$$

which is quite close to the exact value of 0.021. ∎

Based on the identity (10.3.12), we obtain a useful process. Let

$$T^+(\theta) = \#_{i \leq j}\{[(X_j - \theta) + (X_i - \theta)]/2 > 0\} = \#_{i \leq j}\{(X_j + X_i)/2 > \theta\}. \quad (10.3.14)$$

The process associated with $T^+(\theta)$ is much like the sign process, (10.2.9). Let $W_1 < W_2 < \cdots < W_{n(n+1)/2}$ denote the $n(n+1)/2$ ordered Walsh averages. Then a graph of $T^+(\theta)$ would appear as in Figure 10.2.2, except the ordered Walsh averages would be on the horizontal axis and the largest value on the vertical would be $n(n+1)/2$. Hence, the function $T^+(\theta)$ is a decreasing step function of θ which steps down one unit at each Walsh average. This observation greatly simplifies the discussion on the properties of the signed-rank Wilcoxon.

Let c_α denote the critical value of a level α test of the hypotheses (10.2.2) based on the signed-rank test statistic T^+; i.e., $\alpha = P_{H_0}(T^+ \geq c_\alpha)$. Let $\gamma_{SW}(\theta) = P_\theta(T^+ \geq c_\alpha)$, for $\theta \geq \theta_0$, denote the power function of the test. The translation property, Lemma 10.2.1, holds for the signed-rank Wilcoxon. Hence, as in Theorem 10.2.1, the power function is a nondecreasing function of θ. In particular, the signed-rank Wilcoxon test is an unbiased test.

10.3.1   Asymptotic Relative Efficiency

We will investigate the efficiency of the signed-rank Wilcoxon by first determining its efficacy. Without loss of generality we can assume that $\theta_0 = 0$. Consider the same sequence of local alternatives discussed in the last section, i.e.,

$$H_0 : \theta = 0 \text{ versus } H_{1n} : \theta_n = \frac{\delta}{\sqrt{n}}, \tag{10.3.15}$$

where $\delta > 0$. Contemplate the modified statistic, which is the average of $T^+(\theta)$,

$$\overline{T}^+(\theta) = \frac{2}{n(n+1)}T^+(\theta). \tag{10.3.16}$$

Then by (10.3.11),

$$E_0[\overline{T}^+(0)] = \frac{2}{n(n+1)}\frac{n(n+1)}{4} = \frac{1}{2} \text{ and } \sigma^2_{\overline{T}^+}(0) = \text{Var}_0[\overline{T}^+(0)] = \frac{2n+1}{6n(n+1)}. \tag{10.3.17}$$

Let $a_n = 2/n(n+1)$. Note that we can decompose $\overline{T}^+(\theta_n)$ into two parts as

$$\overline{T}^+(\theta_n) = a_n S(\theta_n) + a_n \sum_{i<j} I(X_i + X_j > 2\theta_n) = a_n S(\theta_n) + a_n T^*(\theta_n), \tag{10.3.18}$$

where $S(\theta)$ is the sign process (10.2.9) and

$$T^*(\theta_n) = \sum_{i<j} I(X_i + X_j > 2\theta_n). \tag{10.3.19}$$

To obtain the efficacy, we require the mean

$$\mu_{\overline{T}^+}(\theta_n) = E_{\theta_n}[\overline{T}^+(0)] = E_0[\overline{T}^+(-\theta_n)]. \tag{10.3.20}$$

But by (10.2.14), $a_n E_0(S(-\theta_n)) = a_n n(2^{-1} - F(-\theta_n)) \to 0$. Hence, we need only be concerned with the second term in (10.3.18). But note that the Walsh averages in $T^*(\theta)$ are identically distributed. Thus

$$a_n E_0(T^*(-\theta_n)) = a_n \binom{n}{2} P_0(X_1 + X_2 > -2\theta_n). \tag{10.3.21}$$

This latter probability can be expressed as follows

$$
\begin{aligned}
P_0(X_1 + X_2 > -2\theta_n) &= E_0[P_0(X_1 > -2\theta_n - X_2 | X_2)] = E_0[1 - F(-2\theta_n - X_2)] \\
&= \int_{-\infty}^{\infty} [1 - F(-2\theta_n - x)]f(x)\,dx \\
&= \int_{-\infty}^{\infty} F(2\theta_n + x)f(x)\,dx \\
&\doteq \int_{-\infty}^{\infty} [F(x) + 2\theta_n f(x)]f(x)\,dx \\
&= \frac{1}{2} + 2\theta_n \int_{-\infty}^{\infty} f^2(x)\,dx;
\end{aligned} \tag{10.3.22}
$$

where we have used the facts that X_1 and X_2 are iid, symmetrically distributed about 0, and the mean value function. Hence

$$\mu_{\overline{T}^+}(\theta_n) \doteq a_n \binom{n}{2} \left(\frac{1}{2} + 2\theta_n \int_{-\infty}^{\infty} f^2(x)\, dx \right). \qquad (10.3.23)$$

Putting (10.3.17) and (10.3.23) together, we have the efficacy

$$c_{T^+} = \lim_{n \to \infty} \frac{\mu'_{\overline{T}^+}(0)}{\sqrt{n}\sigma_{\overline{T}^+}(0)} = \sqrt{12} \int_{-\infty}^{\infty} f^2(x)\, dx. \qquad (10.3.24)$$

In a more advanced text, this development can be made into a rigorous argument for the following asymptotic power lemma.

Theorem 10.3.2 (Asymptotic Power Lemma). *Consider the sequence of hypotheses (10.3.15). The limit of the power function of the large sample, size α, signed-rank Wilcoxon test is given by,*

$$\lim_{n \to \infty} \gamma_{SR}(\theta_n) = 1 - \Phi(z_\alpha - \delta \tau_W^{-1}), \qquad (10.3.25)$$

where $\tau_W = 1/\sqrt{12} \int_{-\infty}^{\infty} f^2(x)\, dx$ is the reciprocal of the efficacy c_{T^+} and $\Phi(z)$ is the cdf of a standard normal random variable.

As shown in Exercise 10.3.7, the parameter τ_W is a scale functional.

The arguments used in the determination of the sample size in Section 10.2 for the sign test were based on the asymptotic power lemma; hence, these arguments follow almost verbatim for the signed-rank Wilcoxon. In particular, the sample size needed so that a level α signed-rank Wilcoxon test of the hypotheses (10.2.2) can detect the alternative $\theta = \theta_0 + \theta^*$ with approximate probability γ^* is

$$n_W = \left(\frac{(z_\alpha - z_{\gamma^*})\tau_W}{\theta^*} \right)^2. \qquad (10.3.26)$$

Using (10.2.26), the ARE between the signed-rank Wilcoxon test and the t-test based on the sample mean is

$$ARE(T, t) = \frac{n_t}{n_T} = \frac{\sigma^2}{\tau_W^2}. \qquad (10.3.27)$$

We now derive some AREs between the Wilcoxon and the t-test. As noted above, the parameter τ_W is a scale functional and, hence, varies directly with scale transformations of the form aX, for $a > 0$. Likewise the standard deviation σ is also a scale functional. Therefore, because the AREs are ratios of scale functionals, they are scale invariant. Hence, for derivations of AREs, we can select a pdf with convenient choice of scale. For example, if we are considering an ARE at the normal distribution, we can work with a $N(0, 1)$ pdf.

Table 10.3.2: AREs among the sign, signed-rank Wilcoxon, and the t-tests for contaminated normals with $\sigma_c = 3$ and proportion of contamination ϵ.

ϵ	0.00	0.01	0.02	0.03	0.05	0.10	0.15	0.25
$ARE(W,t)$	0.955	1.009	1.060	1.108	1.196	1.373	1.497	1.616
$ARE(S,t)$	0.637	0.678	0.719	0.758	0.833	0.998	1.134	1.326
$ARE(W,S)$	1.500	1.487	1.474	1.461	1.436	1.376	1.319	1.218

Example 10.3.2 ($ARE(W,t)$ at the normal distribution). If $f(x)$ is a $N(0,1)$ pdf, then

$$
\tau_W^{-1} = \sqrt{12} \int_{-\infty}^{\infty} \left(\frac{1}{\sqrt{2\pi}} \exp\{-x^2/2\} \right)^2 dx
$$

$$
= \frac{\sqrt{12}}{\sqrt{2}\sqrt{2\pi}} \int_{-\infty}^{\infty} \frac{1}{\sqrt{2\pi}(1/\sqrt{2})} \exp\{-2^{-1}(x/(1/\sqrt{2}))^2\} \, dx = \sqrt{\frac{3}{\pi}}
$$

Hence $\tau_W^2 = \pi/3$. Since $\sigma = 1$, we have

$$
ARE(W,t) = \frac{\sigma^2}{\tau_W^2} = \frac{3}{\pi} = 0.955. \tag{10.3.28}
$$

As discussed above this ARE holds for all normal distributions. Hence, at the normal distribution the Wilcoxon signed-rank test is 95.5% efficient as the t-test. The Wilcoxon is called a *highly efficient* procedure. ∎

Example 10.3.3 ($ARE(W,t)$ at a family of contaminated normals). For this example, suppose that $f(x)$ is the pdf of a contaminated normal distribution. For convenience, we will use the standardized pdf given in expression (10.2.30) with $b = 1$. Recall for this distribution $(1-\epsilon)$ proportion of the time the sample is drawn from a $N(0,1)$ distribution while ϵ proportion of the time the sample is drawn from a $N(0,\sigma_c^2)$ distribution. Recall that the variance is $\sigma^2 = 1 + \epsilon(\sigma_c^2 - 1)$. Note that the formula for the pdf $f(x)$ is given in expression (3.4.14). In Exercise 10.3.2 it is shown that

$$
\int_{-\infty}^{\infty} f^2(x) \, dx = \frac{(1-\epsilon)^2}{2\sqrt{\pi}} + \frac{\epsilon^2}{6\sqrt{\pi}} + \frac{\epsilon(1-\epsilon)}{2\sqrt{\pi}}. \tag{10.3.29}
$$

Based on this, an expression for the ARE can be obtained; see Exercise 10.3.2. We used this expression to determine the AREs between the Wilcoxon and the t-tests for the situations with $\sigma_c = 3$ and ϵ varying from 0.01 to 0.25 displaying them in Table 10.3.2. For convenience, we have also displayed the AREs between the sign test and these two tests.

Note that the signed-rank Wilcoxon is more efficient than the t-test even at 1% contamination and increases to 150% efficiency for 15% contamination. ∎

10.3.2 Estimating Equations Based on Signed-rank Wilcoxon

For the sign procedure, the estimation of θ was based on minimizing the L_1 norm. The estimator associated with the signed-rank test minimizes another norm which is discussed in Exercises 10.3.4 and 10.3.5. Recall that we also show that location estimator based on the sign test could be obtained by inverting the test. Considering this for the Wilcoxon, the estimator $\widehat{\theta}_W$ solves

$$T^+(\widehat{\theta}_W) = \frac{n(n+1)}{4}. \qquad (10.3.30)$$

Using the description of the function $T^+(\theta)$ after its definition, (10.3.14), it is easily seen that $\widehat{\theta}_W = \text{median}\left\{\frac{X_i + X_j}{2}\right\}$; i.e., the median of the Walsh averages. This is often called the Hodges-Lehmann estimator because of several seminal articles by Hodges and Lehmann on the properties of this estimator; see Hodges and Lehmann (1963).

Several computer packages obtain the Hodges-Lehmann estimate. For example, the minitab (1991) command wint will return it. Also, this estimate is computed at the web site www.stat.wmich.edu/slab/RGLM.

Once again, we can use practically the same argument that we used for the sign process to obtain the asymptotic distribution of the Hodges-Lehmann estimator. We summarize the result in the theorem

Theorem 10.3.3. *Consider a random sample $X_1, X_2, X_3, \ldots, X_n$ which follows Model (10.2.1). Suppose that $f(x)$ is symmetric about 0. Then*

$$\sqrt{n}(\widehat{\theta}_W - \theta) \rightarrow N(0, \tau_W^2), \qquad (10.3.31)$$

where $\tau_W = \left(\sqrt{12} \int_{-\infty}^{\infty} f^2(x)\, dx\right)^{-1}$.

Using this theorem, the AREs based on asymptotic variances for the signed-rank Wilcoxon are the same as those defined above.

10.3.3 Confidence Interval for the Median

Because of the similarity between the processes $S(\theta)$ and $T^+(\theta)$, confidence intervals for θ based on the signed-rank Wilcoxon follow the same way as those based on $S(\theta)$. For a given level α, let c_{W1} denote the critical point of the signed-rank Wilcoxon distribution such that $P_\theta[T^+(\theta) \le c_{W1}] = \alpha/2$. As in Section 10.2.3, we then have that

$$\begin{aligned}
1 - \alpha &= P_\theta[c_{W1} < T^+(\theta) < n - c_{W1}] \\
&= P_\theta[W_{c_{W1}+1} \le \theta < W_{m-c_{W1}}], \qquad (10.3.32)
\end{aligned}$$

where $m = n(n+1)/2$ denotes the number of Walsh averages. Therefore, the interval $[W_{c_{W1}+1}, W_{m-c_{W1}})$ is a $(1-\alpha)100\%$ confidence interval for θ.

We can use the asymptotic null distribution of T^+, (10.3.9), to obtain the following approximation to c_{W1}. As shown in Exercise 10.3.3,

$$c_{W1} \doteq \frac{n(n+1)}{4} - z_{\alpha/2}\sqrt{\frac{n(n+1)(2n+1)}{24}} - \frac{1}{2}, \tag{10.3.33}$$

where $\Phi(-z_{\alpha/2}) = \alpha/2$.

Besides computing the Hodges-Lehmann estimate, the minitab command `wint` will obtain the confidence interval $[W_{c_{W1}+1}, W_{m-c_{W1}})$. This can also be obtained at the web site `www.stat.wmich.edu/slab/RGLM`.

Example 10.3.4 (Zea Mays Data of Darwin, Continued). Reconsider the data set discussed in Example 10.3.1, we used the signed-rank Wilcoxon to test the hypotheses the the effect θ was 0. We now obtain the estimate of θ and a 95% confidence interval for it based on the signed-rank Wilcoxon. Recall that $n = 15$; hence, there are 120 Walsh averages. Using a computer package, we sorted these Walsh averages. The point estimate of the effect is the median of these averages which is 3.14. Hence, we estimate that cross-fertilized Zea Mays will grow 3.14 inches taller than self-fertilized Zea Mays. The approximate cutoff point for the confidence interval given by expression (10.3.33) is $c_{W1} = 25$. Hence, a 95% confidence interval for θ is $[W_{26}, W_{95}) = [0.500, 5.250)$; that is, we are 95% confident that the true effect is between 0.500 to 5.250 inches. ∎

EXERCISE

10.3.1. (a) For $n = 3$, expand the mgf (10.3.5) to show that the distribution of the signed-rank Wilcoxon is given by

j	-6	-4	-2	0	2	4	6
$P(T = j)$	$\frac{1}{8}$	$\frac{1}{8}$	$\frac{1}{8}$	$\frac{2}{8}$	$\frac{1}{8}$	$\frac{1}{8}$	$\frac{1}{8}$

(b) Obtain the distribution of the signed-rank Wilcoxon for $n = 4$.

10.3.2. Assume that $f(x)$ has the contaminated normal pdf given in expression (3.4.14). Derive expression (10.3.29) and use it to obtain $ARE(W, t)$ for this pdf.

10.3.3. Use the asymptotic null distribution of T^+, (10.3.9), to obtain the approximation (10.3.33) to c_{W1}.

10.3.4. For a vector $\mathbf{v} \in R^n$, define the function

$$\|\mathbf{v}\| = \sum_{i=1}^{n} R(|v_i|)|v_i|. \tag{10.3.34}$$

Show that this function is a norm on R^n; that is, it satisfies the properties:

1. $\|\mathbf{v}\| \geq 0$ and $\|\mathbf{v}\| = 0$ if and only if $\mathbf{v} = \mathbf{0}$.

2. $\|a\mathbf{v}\| = |a|\|\mathbf{v}\|$, for all $a \in R$.

3. $\|\mathbf{u} + \mathbf{v}\| \leq \|\mathbf{u}\| + \|\mathbf{v}\|$, for all $\mathbf{u}, \mathbf{v} \in R^n$.

For the triangle inequality use the antirank version, that is

$$\|\mathbf{v}\| = \sum_{j=1}^{n} j |v_{i_j}|. \tag{10.3.35}$$

Then use the following fact: If we have two sets of n numbers, for example, $\{t_1, t_2, \ldots, t_n\}$ and $\{s_1, s_2, \ldots, s_n\}$, then the largest sum of pairwise products, one from each set, is given by $\sum_{j=1}^{n} t_{i_j} s_{k_j}$, where $\{i_j\}$ and $\{k_j\}$ are the antiranks for the t_i and s_i, respectively; i.e., $t_{i_1} \leq t_{i_2} \leq \cdots \leq t_{i_n}$ and $s_{k_1} \leq s_{k_2} \leq \cdots \leq s_{k_n}$.

10.3.5. Consider the norm given in Exercise 10.3.4. For a location model, define the estimate of θ to be

$$\widehat{\theta} = \mathrm{Argmin}_\theta \|X_i - \theta\|. \tag{10.3.36}$$

Show that $\widehat{\theta}$ is the Hodges-Lehmann estimate; i.e., satisfies (10.4.25). *Hint*: Use the antirank version (10.3.35) of the norm when differentiating with respect to θ.

10.3.6. Prove that a pdf (or pmf) $f(x)$ is symmetric about 0 if and only if its mgf is symmetric about 0, provided the mgf exists.

10.3.7. In Exercise 10.1.4, we defined the term scale functional. Show that the parameter τ_W, (10.3.25), is a scale functional.

10.4   Mann-Whitney-Wilcoxon Procedure

Suppose $X_1, X_2, \ldots, X_{n_1}$ is a random sample from a distribution with a continuous cdf $F(x)$ and pdf $f(x)$ and $Y_1, Y_2, \ldots, Y_{n_2}$ is a random sample from a distribution with a continuous cdf $G(x)$ and pdf $g(x)$. For this situation there is a natural null hypothesis given by $H_0 : F(x) = G(x)$ for all x; i.e., the samples are from the same distribution. What about alternative hypotheses besides the general alternative not H_0? An interesting alternative is that X is *stochastically larger* than Y, which is defined by $G(x) \geq F(x)$, for all x, with strict inequality for at least one x. This alternative hypothesis is discussed in the exercises.

For the most part in his section, we will consider the location model. In this case, $G(x) = F(x - \Delta)$ for some value of Δ. Hence, the null hypothesis becomes $H_0 : \Delta = 0$. The parameter Δ is often called the *shift* between the distributions and, in this case, the distribution of Y is the same as the distribution of $X + \Delta$; that is,

$$P(Y \leq y) = F(y - \Delta) = P(X + \Delta \leq y). \tag{10.4.1}$$

If $\Delta > 0$, then Y is stochastically larger than X; see Exercise 10.4.5.

In the shift case, the parameter Δ is independent of what location functional is used. To see this, suppose we select an arbitrary location functional for X, say, $T(F_X)$. Then we can write X_i as

$$X_i = T(F) + \varepsilon_i,$$

where $\varepsilon_1, \ldots, \varepsilon_{n_1}$ are iid with $T(F_\varepsilon) = 0$. By (10.4.1) it follows that

$$Y_j = T(F_X) + \Delta + \varepsilon_j, \quad j = 1, 2, \ldots, n_2.$$

Hence, $T(F_Y) = T(F_X) + \Delta$. Therefore, $\Delta = T(F_Y) - T(F_X)$ for any location functional; i.e, Δ is the same no matter what functional is chosen to model location.

Assume then that the shift model holds for the two samples. Alternatives of interest are the usual one- and two-sided alternatives. For convenience we pick on the one-sided hypotheses given by,

$$H_0 : \Delta = 0 \text{ versus } H_1 : \Delta > 0 . \tag{10.4.2}$$

The exercises consider the other hypotheses. Under H_0, the distributions of X and Y are the same and we can combine the samples to have one large sample of $n = n_1 + n_2$ observations. Suppose we rank the combined samples from 1 to n and consider the statistic

$$W = \sum_{j=1}^{n_2} R(Y_j), \tag{10.4.3}$$

where $R(Y_j)$ denotes the rank of Y_j in the combined sample of n items. This statistic is often called the **Mann-Whitney-Wilcoxon** (MWW) statistic. Under H_0 the ranks should be uniformly distributed between the X_is and the Y_js; however, under $H_1 : \Delta > 0$, the Y_js should get most of the large ranks. Hence, an intuitive rejection rule is given by,

$$\text{Reject } H_0 \text{ in favor of } H_1 \text{ if } W \geq c. \tag{10.4.4}$$

We now discuss the null distribution of W which will enable us to select c for the decision rule based on a specified level α. Under H_0, the ranks of the Y_js are equilikely to be any subset of size n_2 from a set of n elements. Recall that there are $\binom{n}{n_2}$ such subsets; therefore, if $\{r_1, \ldots, r_{n_2}\}$ is a subset of size n_2 from $\{1, \ldots, n\}$ then

$$P[R(Y_1) = r_1, \ldots, R(Y_{n_2}) = r_{n_2}] = \binom{n}{n_2}^{-1}. \tag{10.4.5}$$

This implies that the statistic W is distribution free under H_0. Although, the null distribution of W cannot be obtained in closed form, there are recursive algorithms which obtain this distribution; see Chapter 2 of the text by Hettmansperger and McKean (1998). In the same way, the distribution of a single rank $R(Y_j)$ is uniformly distributed on the integers $\{1, \ldots, n\}$, under H_0. Hence, we immediately have

$$E_{H_0}(W) = \sum_{j=1}^{n_2} E_{H_0}(R(Y_j)) = \sum_{j=1}^{n_2} \sum_{i=1}^{n} i \frac{1}{n} = \sum_{j=1}^{n_2} \frac{n(n+1)}{2n} = \frac{n_2(n+1)}{2}.$$

The variance is displayed below (10.4.8) and a derivation of a more general case is given in Section 10.5. It also can be shown that W is asymptotically normal. We summarize these items in the theorem below.

Theorem 10.4.1. *Suppose $X_1, X_2, \ldots, X_{n_1}$ is a random sample from a distribution with a continuous cdf $F(x)$ and $Y_1, Y_2, \ldots, Y_{n_2}$ is a random sample from a distribution with a continuous cdf $G(x)$. Suppose $H_0 : F(x) = G(x)$, for all x. If H_0 is true, then*

$$W \text{ is distribution free with a symmetric pmf} \tag{10.4.6}$$

$$E_{H_0}[W] = \frac{n_2(n+1)}{2} \tag{10.4.7}$$

$$Var_{H_0}(W) = \frac{n_1 n_2(n+1)}{12} \tag{10.4.8}$$

$$\frac{W - n_2(n+1)/2}{\sqrt{Var_{H_0}(W)}} \text{ has an asymptotically } N(0,1) \text{ distribution.} \tag{10.4.9}$$

The only item of the theorem not discussed above is the symmetry of the null distribution, which we will show later after considering an example:

Example 10.4.1 (Water Wheel Data Set). In an experiment discussed in Abebe et al. (2001), mice were placed in a wheel that is partially submerged in water. If they keep the wheel moving, they will avoid the water. The response is the number of wheel revolutions per minute. Group 1 is a placebo group while Group 2 consists of mice that are under the influence of a drug. The data are:

Group 1 X	2.3	0.3	5.2	3.1	1.1	0.9	2.0	0.7	1.4	0.3
Group 2 Y	0.8	2.8	4.0	2.4	1.2	0.0	6.2	1.5	28.8	0.7

Comparison dotplots of the data (asked for in Exercise 10.4.6) show that the two data sets are similar except for the large outlier in the treatment group. A two-sided hypothesis seems appropriate in this case. Notice that a few of the data points in the data set have the same value (are tied). This happens in real data sets. We will follow the usual practice and use the average of the ranks involved to break ties. For example, the observations $x_2 = x_{10} = 0.3$ are tied and the ranks involved for the combined data are 2 and 3. Hence, we use 2.5 for the ranks of each of these observations. Continuing in this way, the Wilcoxon test statistic is $w = \sum_{j=1}^{10} R(y_j) = 116.50$. The null mean and variance of W are 105 and 175, respectively. The asymptotic test statistic is $z = (116.5 - 105)/\sqrt{175} = 0.869$ with p-value $2(1 - \Phi(0.869)) = 0.38$; (see below for a discussion on exact p-values). Hence, H_0 would not be rejected. The test confirms the comparison dotplots of the data. The t-test based on the difference in means is discussed in Exercise 10.4.6. ∎

We next want to derive some properties of the test statistic and then use these properties to discuss point estimation and confidence intervals for Δ. As in the last section, another way of writing W will prove helpful in these regards. Without loss of generality, assume that the Y_js are in order. Recall the distributions of X_i and Y_j are continuous; hence, we treat the observations as distinct. Thus,

$R(Y_j) = \#_i\{X_i < Y_j\} + \#_i\{Y_i \le Y_j\}$. This leads to

$$W = \sum_{j=1}^{n_2} R(Y_j) \;=\; \sum_{j=1}^{n_2} \#_i\{X_i < Y_j\} + \sum_{j=1}^{n_2} \#_i\{Y_i \le Y_j\}$$

$$\qquad\qquad = \;\; \#_{i,j}\{Y_j > X_i\} + \frac{n_2(n_2+1)}{2}. \qquad (10.4.10)$$

Let $U = \#_{i,j}\{Y_j > X_i\}$, then we have $W = U + n_2(n_2+1)/2$. Hence, an equivalent test for the hypotheses (10.4.2) is to reject H_0 if $U \ge c_2$. It follows immediately from Theorem 10.4.1 that, under H_0, U is distribution free with mean $n_1 n_2/2$ and variance (10.4.8) and that it has an asymptotic normal distribution. The symmetry of the null distribution of either U or W can now be easily obtained. Under H_0, both X_i and Y_j have the same distribution, so the distributions of U and $U' = \#_{i,j}\{X_i > Y_j\}$ must be the same. Furthermore, $U + U' = n_1 n_2$. This leads to,

$$P_{H_0}\left(U - \frac{n_1 n_2}{2} = u\right) \;=\; P_{H_0}\left(n_1 n_2 - U' - \frac{n_1 n_2}{2} = u\right)$$

$$\qquad\qquad = \; P_{H_0}\left(U' - \frac{n_1 n_2}{2} = -u\right)$$

$$\qquad\qquad = \; P_{H_0}\left(U - \frac{n_1 n_2}{2} = -u\right),$$

which yields the desired symmetry result in Theorem 10.4.1.

Tables for the distribution of U can be found in the literature; see, for instance Hollander and Wolfe (1999). Many computer packages also return p-values and critical values. If the reader has access to R or S-PLUS, the command `pwilcox(u,n1,n2)` will compute $P(U \le u)$, where `n1` and `n2` denote the sample sizes.

Note that if $G(x) = F(x - \Delta)$, then $Y_j - \Delta$ has the same distribution as X_i. So the process of interest here is

$$U(\Delta) = \#_{i,j}\{(Y_j - \Delta) > X_i\} = \#_{i,j}\{Y_j - X_i > \Delta)\}. \qquad (10.4.11)$$

Hence, $U(\Delta)$ is counting the number of differences $Y_j - X_i$ which exceed Δ. Let $D_1 < D_2 < \cdots < D_{n_1 n_2}$ denote the $n_1 n_2$ ordered differences of $Y_j - X_i$. Then the graph of $U(\Delta)$ is the same as that in Figure 10.2.2, except the D_is are on the horizontal axis and the n on the vertical axis is replaced by $n_1 n_2$; that is, $U(\Delta)$ is a decreasing step function of Δ which steps down one unit at each difference D_i with the maximum value of $n_1 n_2$.

We can then proceed as in the last two sections to obtain properties of inference based on the Wilcoxon. Let c_α denote the critical value of a level α test of the hypotheses (10.2.2) based on the statistic U; i.e., $\alpha = P_{H_0}(U \ge c_\alpha)$. Let $\gamma_U(\Delta) = P_\Delta(U \ge c_\alpha)$, for $\Delta \ge 0$, denote the power function of the test. The translation property, Lemma 10.2.1, holds for the process $U(\Delta)$. Hence, as in Theorem 10.2.1, the power function is a nondecreasing function of Δ. In particular, the Wilcoxon test is an unbiased test.

10.4.1 Asymptotic Relative Efficiency

The asymptotic relative efficiency (ARE) of the Wilcoxon follows along similar lines as for the sign test statistic in Section 10.2.1. Here, consider the sequence of local alternatives given by

$$H_0 : \Delta = 0 \text{ versus } H_{1n} : \Delta_n = \frac{\delta}{\sqrt{n}}, \qquad (10.4.12)$$

where $\delta > 0$. We also assume that

$$\frac{n_1}{n} \to \lambda_1, \ \frac{n_2}{n} \to \lambda_2 \text{ where } \lambda_1 + \lambda_2 = 1. \qquad (10.4.13)$$

This assumption implies that $n_1/n_2 \to \lambda_1/\lambda_2$; i.e, the sample sizes maintain the same ratio asymptotically.

To determine the efficacy of the MWW, consider the modified statistic

$$\overline{U}(\Delta) = \frac{1}{n_1 n_2} U(\Delta). \qquad (10.4.14)$$

It follows immediately that

$$\mu_U(0) = E_0(\overline{U}(0)) = \tfrac{1}{2} \quad \text{and} \quad \overline{\sigma}_U^2(0) = \tfrac{n+1}{12 n_1 n_2}. \qquad (10.4.15)$$

Because the pairs (X_i, Y_j)s are iid we have,

$$
\begin{aligned}
\mu_U(\Delta_n) = E_{\Delta_n}(\overline{U}(0)) = E_0(\overline{U}(-\Delta_n)) &= \frac{n_1 n_2}{n_1 n_2} P_0(Y - X > -\Delta_n) \\
&= P_0(Y - X > -\Delta_n). \quad (10.4.16)
\end{aligned}
$$

The independence of X and Y and the fact $\int_{-\infty}^{\infty} F(x)f(x)\,dx = 1/2$ gives

$$
\begin{aligned}
P_0(Y - X > -\Delta_n) &= E_0(P_0[Y > X - \Delta_n | X]) \\
&= E_0(1 - F(X - \Delta_n)) \\
&= 1 - \int_{-\infty}^{\infty} F(x - \Delta_n)f(x)\,dx \\
&= \frac{1}{2} + \int_{-\infty}^{\infty} (F(x) - F(x - \Delta_n))f(x)\,dx \\
&\doteq \frac{1}{2} + \Delta_n \int_{-\infty}^{\infty} f^2(x)\,dx, \qquad (10.4.17)
\end{aligned}
$$

where we have applied the mean value theorem to obtain the last line. Putting together (10.4.15) and (10.4.17), we have the efficacy

$$c_U = \lim_{n \to \infty} \frac{\mu_{\overline{U}}'(0)}{\sqrt{n}\,\sigma_{\overline{U}}(0)} = \sqrt{12}\sqrt{\lambda_1 \lambda_2} \int_{-\infty}^{\infty} f^2(x)\,dx. \qquad (10.4.18)$$

This argument can be made into a rigorous argument for the following theorem

Theorem 10.4.2 (Asymptotic Power Lemma). *Consider the sequence of hypotheses (10.4.12). The limit of the power function of the size α Mann-Whitney-Wilcoxon test is given by*

$$\lim_{n\to\infty} \gamma_U(\Delta_n) = 1 - \Phi(z_\alpha - \sqrt{\lambda_1\lambda_2}\delta\tau_W^{-1}), \qquad (10.4.19)$$

where $\tau_W = 1/\sqrt{12}\int_{-\infty}^{\infty} f^2(x)\,dx$ is the reciprocal of the efficacy c_U and $\Phi(z)$ is the cdf of a standard normal random variable.

As in the last two sections we can use this theorem to establish a relative measure of efficiency by considering sample size determination. Consider the hypotheses (10.4.2). Suppose we want to determine the sample size $n = n_1 + n_2$ for a level α MWW test to detect the alternative Δ^* with approximate power γ^*. By Theorem 10.4.2, we have the equation

$$\gamma^* = \gamma_U(\sqrt{n}\Delta^*/\sqrt{n}) \doteq 1 - \Phi(z_\alpha - \sqrt{\lambda_1\lambda_2}\sqrt{n}\Delta^*\tau_W^{-1}). \qquad (10.4.20)$$

This leads to the equation

$$z_{\gamma^*} = z_\alpha - \sqrt{\lambda_1\lambda_2}\delta\tau_W^{-1}, \qquad (10.4.21)$$

where $\Phi(z_{\gamma^*}) = 1 - \gamma^*$. Solving for n, we obtain

$$n_U \doteq \left(\frac{(z_\alpha - z_{\gamma^*})\tau_W}{\Delta^*\sqrt{\lambda_1\lambda_2}}\right)^2. \qquad (10.4.22)$$

To use this in applications, the sample size proportions $\lambda_1 = n_1/n$ and $\lambda_2 = n_2/n$ must be given. As Exercise 10.4.2 points out, the most powerful two-sample designs have sample size proportions of $1/2$;, i.e., equal sample sizes.

To use this to obtain the asymptotic relative efficiency between the MWW and the two-sample pooled t-test, Exercise 10.4.2 shows that the sample size needed for the two-sample t-tests to attain approximate power γ^* to detect Δ^* is given by

$$n_{LS} \doteq \left(\frac{(z_\alpha - z_{\gamma^*})\sigma}{\Delta^*\sqrt{\lambda_1\lambda_2}}\right)^2, \qquad (10.4.23)$$

where σ is the variance of e_i. Hence, as in the last section the asymptotic relative efficiency between the Wilcoxon test (MWW) and the t-test is the ratio of the sample sizes (10.4.22) and (10.4.23) which is

$$\text{ARE(MWW, LS)} = \frac{\sigma^2}{\tau_W^2}. \qquad (10.4.24)$$

Note that this is the same ARE as derived in the last section between the signed rank Wilcoxon and the t-test. If $f(x)$ is a normal pdf, then the MWW has efficiency 95.5% relative to the pooled t-test. Thus the MWW tests lose little efficiency at the normal. On the other hand, it is much more efficient than the pooled t-test at the family of contaminated normals (with $\epsilon > 0$) as in Example 10.3.3.

10.4.2 Estimating Equations Based on the Mann-Whitney-Wilcoxon

As with the signed-rank Wilcoxon procedure in the last section, we will invert the test statistic to obtain an estimate of Δ. As discussed in the next section this estimate can be defined in terms of minimizing a norm. The estimator $\widehat{\theta}_W$ solves the estimating equations

$$U(\Delta) = E_{H_0}(U) = \frac{n_1 n_2}{2}. \qquad (10.4.25)$$

Recalling the description of the process $U(\Delta)$ described above, it is clear that the Hodges-Lehmann estimator is given by

$$\widehat{\Delta}_U = \mathrm{med}_{i,j}\{Y_j - X_i\}. \qquad (10.4.26)$$

The asymptotic distribution of the estimate follows in the same way as in the last section based on the process $U(\Delta)$ and the asymptotic power lemma, Theorem 10.4.2. We avoid sketching the proof and will simply state the result as a theorem

Theorem 10.4.3. *Assume that the random variables $X_1, X_2, \ldots, X_{n_1}$ are iid with pdf $f(x)$ and that the random variables $Y_1, Y_2, \ldots, Y_{n_2}$ are iid with pdf $f(x - \Delta)$. Then*

$$\widehat{\Delta}_U \text{ has an approximate } N\left(\Delta, \tau_W^2\left(\frac{1}{n_1} + \frac{1}{n_2}\right)\right) \text{ distribution}, \qquad (10.4.27)$$

where $\tau_W = \left(\sqrt{12} \int_{-\infty}^{\infty} f^2(x)\, dx\right)^{-1}$.

As Exercise 10.4.3 shows, provided the $\mathrm{Var}(\varepsilon_i) = \sigma^2 < \infty$, the LS estimate $\overline{Y} - \overline{X}$ of Δ has the following approximate distribution:

$$\overline{Y} - \overline{X} \text{ has an approximate } N\left(\Delta, \sigma^2\left(\frac{1}{n_1} + \frac{1}{n_2}\right)\right) \text{ distribution.} \qquad (10.4.28)$$

Note that the ratio of the asymptotic variances of $\widehat{\Delta}_U$ is given by the ratio (10.4.24). Hence, the ARE of the tests agrees with the ARE of the corresponding estimates.

10.4.3 Confidence Interval for the Shift Parameter Δ

The confidence interval for Δ corresponding to the MWW estimate is derived the same way as the Hodges-Lehmann estimate in the last section. For a given level α, let c denote the critical point of the MWW distribution such that $P_\Delta[U(\Delta) \le c] = \alpha/2$. As in Section 10.2.3, we then have

$$\begin{aligned}
1 - \alpha &= P_\Delta[c < U(\Delta) < n_1 n_2 - c] \\
&= P_\Delta[D_{c+1} \le \Delta < D_{n_1 n_2 - c}], \qquad (10.4.29)
\end{aligned}$$

where $D_1 < D_2 < \cdots < D_{n_1 n_2}$ denote the order differences $Y_j - X_i$. Therefore, the interval $[D_{c+1}, D_{n_1 n_2 - c})$ is a $(1 - \alpha)100\%$ confidence interval for Δ. Using the null asymptotic distribution of the MWW test statistic U, we have the following approximation for c

$$c \doteq \frac{n_1 n_2}{2} - z_{\alpha/2} \sqrt{\frac{n_1 n_2 (n+1)}{12}} - \frac{1}{2}, \qquad (10.4.30)$$

where $\Phi(-z_{\alpha/2}) = \alpha/2$; see Exercise 10.4.4.

Example 10.4.2 (Example 10.4.1, Continued). Returning to Example 10.4.1, the MWW estimate of Δ is $\widehat{\Delta} = 1.15$. The asymptotic rule for the selection of the differences which enter a 95% confidence interval gives the value for $c \doteq 24$. Hence, the confidence interval is (D_{25}, D_{76}) which for this data set has the value $(-0.7, 2.6)$. Hence, as with the test statistic, the confidence interval covers the null hypothesis of $\Delta = 0$. ∎

There are statistical computing packages that obtain the Hodges-Lehmann estimate and confidence intervals. For example, Minitab has the command `mann`. These statistics can also be obtained at the web site `www.stat.wmich.edu/slab/RGLM`.

EXERCISES

10.4.1. By considering the asymptotic power lemma, Theorem 10.4.2, show that the equal sample size situation $n_1 = n_2$ is the most powerful design among designs with $n_1 + n_2 = n$, n fixed, when level and alternatives are also fixed. *Hint:* Show that this problem is equivalent to maximizing the function

$$g(n_1) = \frac{n_1(n - n_1)}{n^2},$$

and then obtain the result.

10.4.2. Consider the asymptotic version of the t test for the hypotheses (10.4.2) which is discussed in Section 5.6.

(a) Using the set up of Lemma 10.4.2, derive the corresponding asymptotic power lemma for this test.

(b) Use your result in Part (a) to obtain expression (10.4.23).

10.4.3. Use the Central Limit Theorem to show that expression (10.4.28) is true.

10.4.4. For the cutoff index c of the confidence interval (10.4.29) for Δ, derive the approximation given in expression (10.4.30).

10.4.5. Let X be a continuous random variable with cdf $F(x)$. Suppose $Y = X + \Delta$, where $\Delta > 0$. Show that Y is stochastically larger than X.

10.4.6. Consider the data given in Example 10.4.1.

(a) Obtain comparison dotplots of the data.

(b) Show that the difference in sample means is 3.11 which is much larger than the MWW estimate of shift. What accounts for this discrepancy?

(c) Show that the 95% confidence interval for Δ using t is given by $(-2.7, 8.92)$. Why is this interval so much larger than the corresponding MWW interval?

(d) Show that the value of the t-test statistic, discussed in Section 5.6, for this data set is 1.12 with p-value 0.28. Although, as with the MWW results, this would be considered insignificant based on the comparison dotplots; this seems more significant than warranted.

10.5 General Rank Scores

Suppose we are interested in estimating the center of a symmetric distribution using an estimator which corresponds to a distribution-free procedure. Presently our choice would be either the sign test or the signed-rank Wilcoxon test. If the sample is drawn from a normal distribution, then of the two we would choose the signed-rank Wilcoxon because it is much more efficient than the sign test at the normal distribution. But the Wilcoxon is not fully efficient. This raises the question: Is there is a distribution-free procedure which is fully efficient at the normal distribution, i.e., has efficiency of 100% relative to the t-test at the normal? More generally, suppose we specify a distribution. Is there a distribution-free procedure which has 100% efficiency relative to the mle at that distribution? In general, the answer to both of these questions is yes. In this section, we explore these questions for the two-sample location problem since this problem generalizes immediately to the regression problem of Section 10.7. A similar theory can be developed for the one-sample problem; see Chapter 1 of Hettmansperger and McKean (1998).

As in the last section, let $X_1, X_2, \ldots, X_{n_1}$ be a random sample from the continuous distribution with cdf and pdf, respectively, $F(x)$ and $f(x)$. Let $Y_1, Y_2, \ldots, Y_{n_2}$ be a random sample from the continuous distribution with cdf and pdf, respectively, $F(x - \Delta)$ and $f(x - \Delta)$, where Δ is the shift in location. Let $n = n_1 + n_2$ denote the combined sample sizes. Consider the hypotheses

$$H_0 : \Delta = 0 \text{ versus } H_1 : \Delta > 0. \tag{10.5.1}$$

We first define a general class of rank scores. Let $\varphi(u)$ be a nondecreasing function defined on the interval $(0, 1)$, such that $\int_0^1 \varphi^2(u)\, du < \infty$. We call $\varphi(u)$ a **score** function. Without loss of generality, we will standardize this function so that $\int_0^1 \varphi(u)\, du = 0$ and $\int_0^1 \varphi^2(u)\, du = 1$. Next, define the scores $a_\varphi(i) = \varphi[i/(n+1)]$, for $i = 1, \ldots, n$. Then $a_\varphi(1) \le a_\varphi(2) \le \cdots \le a_\varphi(n)$ and assume that $\sum_{i=1}^n a(i) = 0$, (this essentially follows from $\int \varphi(u)\, du = 0$, see Exercise 10.5.11). Consider the test statistic

$$W_\varphi = \sum_{j=1}^{n_2} a_\varphi(R(Y_j)), \tag{10.5.2}$$

where $R(Y_j)$ denotes the rank of Y_j in the combined sample of n observations. Since the scores are nondecreasing, a natural rejection rule is given by

$$\text{Reject } H_0 \text{ in favor of } H_1 \text{ if } W_\varphi \geq c. \tag{10.5.3}$$

Note if we use the linear score function $\varphi(u) = \sqrt{12}(u - (1/2))$ then

$$W_\varphi = \sum_{j=1}^{n_2} \sqrt{12} \left(\frac{R(Y_j)}{n+1} - \frac{1}{2} \right) = \frac{\sqrt{12}}{n+1} \sum_{j=1}^{n_2} \left(R(Y_j) - \frac{n+1}{2} \right)$$

$$= \frac{\sqrt{12}}{n+1} W - \frac{\sqrt{12}n}{2} \tag{10.5.4}$$

where W is the MWW test statistic, (10.4.3). Hence, the special case of a linear score function results in the MWW test statistic.

To complete the decision rule (10.5.2), we need the null distribution of the test statistic W_φ. But many of its properties follow along the same lines as that of the MWW test. First, W_φ is distribution free because, under the null hypothesis, every subset of ranks for the Y_js are equilikely. In general, the distribution of W_φ cannot be obtained in closed form but it can be generated recursively similar to the distribution of the MWW-test statistic. Next, to obtain the null mean of W_φ use the fact that $R(Y_j)$ is uniform on the integers $1, 2, \ldots, n$. Because $\sum_{i=1}^{n} a_\varphi(i) = 0$ we then have

$$E_{H_0}(W_\varphi) = \sum_{j=1}^{n_2} E_{H_0}(a_\varphi(R(Y_j))) = \sum_{j=1}^{n_2} \sum_{i=1}^{n} a_\varphi(i) \frac{1}{n} = 0. \tag{10.5.5}$$

To determine the null variance, first define the quantity s_a^2 by the equation

$$E_{H_0}(a_\varphi^2(R(Y_j))) = \sum_{i=1}^{n} a_\varphi^2(i) \frac{1}{n} = \frac{1}{n} \sum_{i=1}^{n} a_\varphi^2(i) = \frac{1}{n} s_a^2;. \tag{10.5.6}$$

As Exercise 10.5.3 shows, $s_a^2/n \doteq 1$. Since $E_{H_0}(W_\varphi) = 0$, we have

$$\begin{aligned}
\text{Var}_{H_0}(W_\varphi) &= E_{H_0}(W_\varphi^2) = \sum_{j=1}^{n_2} \sum_{j'=1}^{n_2} E_{H_0}[a_\varphi(R(Y_j))a_\varphi(R(Y_{j'}))] \\
&= \sum_{j=1}^{n_2} E_{H_0}[a_\varphi^2(R(Y_j))] + \sum\sum_{j \neq j'} E_{H_0}[a_\varphi(R(Y_j))a_\varphi(R(Y_{j'}))] \\
&= \frac{n_2}{n} s_a^2 - \frac{n_2(n_2-1)}{n(n-1)} s_a^2 \tag{10.5.7} \\
&= \frac{n_1 n_2}{n(n-1)} s_a^2; \tag{10.5.8}
\end{aligned}$$

see Exercise 10.5.1 for the derivation of the second term in expression (10.5.7). In more advanced books it is shown that W_φ is asymptotically normal under H_0. Hence, the corresponding asymptotic decision rule of level α is

$$\text{Reject } H_0 \text{ in favor of } H_1 \text{ if } z = \frac{W_\varphi}{\sqrt{\text{Var}_{H_0}(W_\varphi)}} \geq z_\alpha. \tag{10.5.9}$$

To answer the questions posed in the first paragraph of this section, the efficacy of the test statistic W_φ is needed. To proceed along the lines of the last section, define the process

$$W_\varphi(\Delta) = \sum_{j=1}^{n_2} a_\varphi(R(Y_j - \Delta)), \qquad (10.5.10)$$

where $R(Y_j - \Delta)$ denotes the rank of $Y_j - \Delta$ among $X_1, \ldots, X_{n_1}, Y_1 - \Delta, \ldots, Y_{n_2} - \Delta$. In the last section, the process for the MWW statistic was also written in terms of counts of the differences $Y_j - X_i$. We are not as fortunate here; but as the next theorem shows, this general process is a simple decreasing step function of Δ.

Theorem 10.5.1. *The process $W_\varphi(\Delta)$ is a decreasing step function of Δ which steps down at each difference $Y_j - X_i$, $i = 1, \ldots, n_1$ and $j = 1, \ldots, n_2$. Its maximum and minimum values are $\sum_{j=n_1+1}^{n} a_\varphi(j) \geq 0$ and $\sum_{j=1}^{n_2} a_\varphi(j) \leq 0$, respectively.*

Proof: Suppose $\Delta_1 < \Delta_2$ and $W_\varphi(\Delta_1) \neq W_\varphi(\Delta_2)$. Hence, the assignment of the ranks among the X_i and $Y_j - \Delta$ must differ at Δ_1 and Δ_2; that is, then there must be a j and an i such that $Y_j - \Delta_2 < X_i$ and $Y_j - \Delta_1 > X_i$. This implies that $\Delta_1 < Y_j - X_i < \Delta_2$. Thus $W_\varphi(\Delta)$ changes values at the differences $Y_j - X_i$. To show it is decreasing, suppose $\Delta_1 < Y_j - X_i < \Delta_2$ and there are no other differences between Δ_1 and Δ_2. Then $Y_j - \Delta_1$ and X_i must have adjacent ranks; otherwise, there would be more than one difference between Δ_1 and Δ_2. Since $Y_j - \Delta_1 > X_i$ and $Y_j - \Delta_2 < X_i$, we have

$$R(Y_j - \Delta_1) = R(X_i) + 1 \text{ and } R(Y_j - \Delta_2) = R(X_i) - 1 .$$

Also, in the expression for $W_\varphi(\Delta)$, only the rank of the Y_j term has changed in the interval $[\Delta_1, \Delta_2]$. Therefore, since the scores are nondecreasing,

$$
\begin{aligned}
W_\varphi(\Delta_1) - W_\varphi(\Delta_2) &= \sum_{k \neq j} a_\varphi(R(Y_k - \Delta_1)) + a_\varphi(R(Y_j - \Delta_1)) \\
&\quad - \left[\sum_{k \neq j} a_\varphi(R(Y_k - \Delta_2)) + a_\varphi(R(Y_j - \Delta_2)) \right] \\
&= a_\varphi(R(X_i) + 1)) - a_\varphi(R(X_i) - 1)) \geq 0.
\end{aligned}
$$

Because $W_\varphi(\Delta)$ is a decreasing step function and steps only at the differences $Y_j - X_i$, its maximum value occurs when $\Delta < Y_j - X_i$, for all i, j; i.e., when $X_i < Y_j - \Delta$, for all i, j. Hence in this case, the variables $Y_j - \Delta$ must get all the high ranks, so

$$\max_{\Delta} W_\varphi(\Delta) = \sum_{j=n_1+1}^{n} a_\varphi(j).$$

Note that this maximum value must be nonnegative. For suppose it was strictly negative, then at least one $a_\varphi(j) < 0$ for $j = n_1 + 1, \ldots, n$. Because the scores are nondecreasing then $a_\varphi(i) < 0$ for all $i = 1, \ldots, n_1$. This leads to the contradiction

$$0 > \sum_{j=n_1+1}^{n} a_\varphi(j) \geq \sum_{j=n_1+1}^{n} a_\varphi(j) + \sum_{j=1}^{n_1} a_\varphi(j) = 0.$$

The results for the minimum value are obtained in the same way; see Exercise 10.5.5. ∎

As Exercise 10.5.6 shows, the translation property, Lemma 10.2.1, holds for the process $W_\varphi(\Delta)$. Using this result and the last theorem, we can show the power function of the test statistic W_φ for the hypotheses (10.5.1) is nondecreasing. Hence, the test is unbiased.

10.5.1   Efficacy

We next sketch the derivation the efficacy of the test based on W_φ. Our arguments can be made rigorous; see advanced texts. Consider the statistic

$$\overline{W}_\varphi(0) = \frac{1}{n}W_\varphi(0). \tag{10.5.11}$$

Based on (10.5.5) and (10.5.8), we have

$$\mu_\varphi(0) = E_0(\overline{W}_\varphi(0)) = 0 \quad \text{and} \quad \sigma_\varphi^2 = \text{Var}_0(\overline{W}_\varphi(0)) = \frac{n_1 n_2}{n(n-1)}n^{-2}s_a^2. \tag{10.5.12}$$

Notice from Exercise 10.5.3 that the variance of $\overline{W}_\varphi(0)$ is of order $O(n^{-1})$. We have

$$\mu_\varphi(\Delta) = E_\Delta[\overline{W}_\varphi(0)] = E_0[\overline{W}_\varphi(-\Delta)] = \frac{1}{n}\sum_{j=1}^{n_2} E_0[a_\varphi(R(Y_j + \Delta))]. \tag{10.5.13}$$

Suppose that $\widehat{F}_{n_1}$ and $\widehat{F}_{n_2}$ are the empirical cdfs of the random samples $X_1, \ldots, X_{n_1}$ and $Y_1, \ldots, Y_{n_2}$, respectively. The relationship between the ranks and empirical cdfs follows as

$$
\begin{aligned}
R(Y_j + \Delta) &= \#_k\{Y_k + \Delta \le Y_j + \Delta\} + \#_i\{X_i \le Y_j + \Delta\} \\
&= \#_k\{Y_k \le Y_j\} + \#_i\{X_i \le Y_j + \Delta\} \\
&= n_2\widehat{F}_{n_2}(Y_j) + n_1\widehat{F}_{n_1}(Y_j + \Delta).
\end{aligned} \tag{10.5.14}
$$

Substituting this last expression into expression (10.5.13), we get

$$
\begin{aligned}
\mu_\varphi(\Delta) &= \frac{1}{n}\sum_{j=1}^{n_2} E_0\left\{\varphi\left[\frac{n_2}{n+1}\widehat{F}_{n_2}(Y_j) + \frac{n_1}{n+1}\widehat{F}_{n_1}(Y_j + \Delta)\right]\right\} &(10.5.15) \\
&\to \lambda_2 E_0\left\{\varphi\left[\lambda_2 F(Y) + \lambda_1 F(Y + \Delta)\right]\right\} &(10.5.16) \\
&= \lambda_2 \int_{-\infty}^{\infty} \varphi\left[\lambda_2 F(Y) + \lambda_1 F(Y + \Delta)\right] f(y)\,dy. &(10.5.17)
\end{aligned}
$$

The limit in expression (10.5.16) is actually a double limit which follows from $\widehat{F}_{n_i}(x) \to F(x)$, $i = 1, 2$, under H_0, and the observation that upon substituting F for the empirical cdfs in expression (10.5.15), the sum contains identically distributed random variables and thus, the same expectation. These approximations can be made rigorous. It follows immediately that

$$\frac{d}{d\Delta}\mu_\varphi(\Delta) = \lambda_2 \int_{-\infty}^{\infty} \varphi'\left[\lambda_2 F(Y) + \lambda_1 F(Y + \Delta)\right] \lambda_1 f(y + \Delta) f(y)\, dy.$$

Hence,

$$\mu_\varphi'(0) = \lambda_1 \lambda_2 \int_{-\infty}^{\infty} \varphi'[F(y)] f^2(y)\, dy. \tag{10.5.18}$$

From (10.5.12),

$$\sqrt{n}\sigma_\varphi = \sqrt{n}\sqrt{\frac{n_1 n_2}{n(n-1)}} \frac{1}{\sqrt{n}} \sqrt{\frac{1}{n}s_a^2} \to \sqrt{\lambda_1 \lambda_2}. \tag{10.5.19}$$

Based on (10.5.18) and (10.5.19), the efficacy of W_φ is given by

$$c_\varphi = \lim_{n \to \infty} \frac{\mu_\varphi'(0)}{\sqrt{n}\sigma_\varphi} = \sqrt{\lambda_1 \lambda_2} \int_{-\infty}^{\infty} \varphi'[F(y)] f^2(y)\, dy. \tag{10.5.20}$$

Using the efficacy, the asymptotic power can be derived for the test statistic W_φ. Consider the sequence of local alternatives given by (10.4.12) and the level α asymptotic test based on W_φ. Denote the power function of the test by $\gamma_\varphi(\Delta_n)$. Then it can be shown that

$$\lim_{n \to \infty} \gamma_\varphi(\Delta_n) = 1 - \Phi(z_\alpha - c_\varphi \delta), \tag{10.5.21}$$

where $\Phi(z)$ is the cdf of a standard normal random variable. Sample size determination based on the test statistic W_φ proceeds as in the last few sections; see Exercise 10.5.7.

10.5.2   Estimating Equations Based on General Scores

Suppose we are using the scores $a_\varphi(i) = \varphi(i/(n+1))$ discussed in Section 10.5.1. Recall that the mean of the test statistic W_φ is 0. Hence, the corresponding estimator of Δ solves the estimating equations

$$W_\varphi(\widehat{\Delta}) \doteq 0. \tag{10.5.22}$$

By Theorem 10.5.1, $W_\varphi(\widehat{\Delta})$ is a decreasing step function of Δ. Furthermore the maximum value is positive and the minimum value is negative, (only degenerate cases would result in one or both of these as 0); hence, the solution to equation (10.5.22) exists. Because $W_\varphi(\widehat{\Delta})$ is a step function it may not be unique. When it is not unique, though, as with Wilcoxon and median procedures, there is an interval of solutions so the midpoint of the interval can be chosen. Numerically, this is an easy equation to solve because simple iterative techniques such as the bisection method or the method of false position can be used; see the discussion on page 186 of Hettmansperger and McKean (1998). The asymptotic distribution of the estimator can be derived using the asymptotic power lemma and is given by,

$$\widehat{\Delta}_\varphi \text{ has an approximate } N\left(\Delta, \tau_\varphi^2 \left(\frac{1}{n_1} + \frac{1}{n_2}\right)\right) \text{ distribution,} \tag{10.5.23}$$

where

$$\tau_\varphi = \left[\int_{-\infty}^{\infty} \varphi'[F(y)]f^2(y)\,dy \right]^{-1}. \tag{10.5.24}$$

Hence, the efficacy can be expressed as $c_\varphi = \sqrt{\lambda_1 \lambda_2}\tau_\varphi^{-1}$. As Exercise 10.5.8 shows, the parameter τ_φ is a scale parameter. Since the efficacy is $c_\varphi = \sqrt{\lambda_1 \lambda_2}\tau_\varphi^{-1}$, the efficacy varies inversely with scale. This observation will be helpful in the next subsection.

10.5.3 Optimization: Best Estimates

We can now answer the questions posed in the first paragraph. For a given pdf $f(x)$, we show that in general we can select a score function which maximizes the power of the test and which minimizes the asymptotic variance of the estimator. Under certain conditions we show that estimators based on this optimal score function have the same efficiency as maximum likelihood estimators (mles); i.e, they obtain the Rao-Cramér Lower Bound.

As above let $X_1, \ldots, X_{n_1}$ be a random sample from the continuous cdf $F(x)$ with pdf $f(x)$. Let $Y_1, \ldots, Y_{n_2}$ be a random sample from the continuous cdf $F(x - \Delta)$ with pdf $f(x - \Delta)$. The problem is to choose φ to maximize the efficacy c_φ given in expression (10.5.20). Note that maximizing the efficacy is equivalent to minimizing the asymptotic variance of the corresponding estimator of Δ.

For a general score function $\varphi(u)$, consider its efficacy given by expression (10.5.20). Without loss of generality, the relative sample sizes in this expression can be ignored, so we consider $c_\varphi^* = (\sqrt{\lambda_1 \lambda_2})^{-1}c_\varphi$. If we make the change of variables $u = F(y)$ and then integrate by parts, we get

$$\begin{aligned}
c_\varphi^* &= \int_{-\infty}^{\infty} \varphi'[F(y)]f^2(y)\,dy \\
&= \int_0^1 \varphi'(u)f(F^{-1}(u))\,du \\
&= \int_0^1 \varphi(u)\left[-\frac{f'(F^{-1}(u))}{f(F^{-1}(u))} \right]du. \tag{10.5.25}
\end{aligned}$$

Recall that the score function $\int \varphi^2(u)\,du = 1$. Thus, we can state the problem as

$$\begin{aligned}
\max_{\varphi} c_\varphi^{*2} &= \max_{\varphi} \left\{ \int_0^1 \varphi(u)\left[-\frac{f'(F^{-1}(u))}{f(F^{-1}(u))} \right]du \right\}^2 \\
&= \left\{ \max_{\varphi} \frac{\left\{ \int_0^1 \varphi(u)\left[-\frac{f'(F^{-1}(u))}{f(F^{-1}(u))} \right]du \right\}^2}{\int_0^1 \varphi^2(u)\,du \int_0^1 \left[\frac{f'(F^{-1}(u))}{f(F^{-1}(u))} \right]^2 du} \right\} \int_0^1 \left[\frac{f'(F^{-1}(u))}{f(F^{-1}(u))} \right]^2 du.
\end{aligned}$$

The quantity that we are maximizing in this last expression, however, is the square of a correlation coefficient which achieves its maximum value 1. Therefore by choosing the score function $\varphi(u) = \varphi_f(u)$ where

$$\varphi_f(u) = -\kappa \frac{f'(F^{-1}(u))}{f(F^{-1}(u))}, \qquad (10.5.26)$$

and κ is a constant chosen so that $\int \varphi_f^2(u)\, du = 1$, then the correlation coefficient is 1 and the maximum value is

$$I(f) = \int_0^1 \left[\frac{f'(F^{-1}(u))}{f(F^{-1}(u))} \right]^2 du, \qquad (10.5.27)$$

which is Fisher information for the location model. We call the score function given by (10.5.26) the **optimal score function**.

In terms of estimation, if $\widehat{\Delta}$ is the corresponding estimator then, according to (10.5.24), it has the asymptotic variance

$$\tau_\varphi^2 = \left[\frac{1}{I(f)} \right] \left(\frac{1}{n_1} + \frac{1}{n_2} \right). \qquad (10.5.28)$$

Thus, the estimator $\widehat{\Delta}$ achieves asymptotically the Rao-Cramér lower bound; that is, $\widehat{\Delta}$ is an asymptotically efficient estimator of Δ. In terms of asymptotic relative efficiency, the ARE between the estimator $\widehat{\Delta}$ and the mle of Δ is 1. Thus we have answered the second question of the first paragraph of this section.

Now we look at some examples. The initial example assumes that the distribution of ε_i is normal, which will answer the leading question at the beginning of this section. First though, note an invariance which simplifies matters. Suppose Z is a scale and location transformation of a random variable X; i.e., $Z = a(X - b)$, where $a > 0$ and $-\infty < b < \infty$. Because the efficacy varies indirectly with scale, we have $c_{fz}^2 = a^{-2} c_{fx}^2$. Furthermore, as Exercise 10.5.8 shows, the efficacy is invariant to location and also that $I(f_Z) = a^{-2} I(f_X)$. Hence, the quantity maximized above is invariant to changes in location and scale. In particular, in the derivation of optimal scores only the form of the density is important.

Example 10.5.1 (Normal Scores). Suppose the error random variable ε_i has a normal distribution. Based on the discussion in the last paragraph, we can take the pdf of a $N(0, 1)$ distribution as the form of the density. So consider, $f_Z(z) = \phi(z) = (2\pi)^{-1/2} \exp\{-2^{-1} z^2\}$. Then $-\phi'(z) = z\phi(z)$. Let $\Phi(z)$ denote the cdf of Z. Hence, the optimal score function is

$$\varphi_N(u) = -\kappa \frac{\varphi'(\Phi^{-1}(u))}{\varphi(\Phi^{-1}(u))} = \Phi^{-1}(u); \qquad (10.5.29)$$

see Exercise 10.5.4 which shows that $\kappa = 1$, as well as $\int \varphi_N(u)\, du = 0$. The corresponding scores, $a_N(i) = \Phi^{-1}(i/(n+1))$, are often called the **normal scores**. Denote the process by

$$W_N(\Delta) = \sum_{j=1}^{n_2} \Phi^{-1}[R(Y_j - \Delta)/(n+1)]. \qquad (10.5.30)$$

The associated test statistic for the hypotheses (10.5.1) is the statistic $W_N = W_N(0)$. The estimator of Δ solves the estimating equations

$$W_N(\widehat{\Delta}_N) \doteq 0. \tag{10.5.31}$$

Although the estimate cannot be obtained in closed form, this equation is relatively easy to solve numerically. From the above discussion, $\mathrm{ARE}(\widehat{\Delta}_N, \overline{Y} - \overline{X}) = 1$ at the normal distribution. Hence, normal score procedures are fully efficient at the normal distribution. Actually, a much more powerful result can be obtained for symmetric distributions. It can be shown that $\mathrm{ARE}(\widehat{\Delta}_N, \overline{Y} - \overline{X}) \geq 1$ at all symmetric distributions. ∎

Example 10.5.2 (Wilcoxon Scores). If the random errors, ε_i, $i = 1, 2, \ldots, n$, have a logistic distribution with pdf $f_Z(z) = \exp\{-z\}/(1 + \exp\{-z\})^2$. The cdf is $F_Z(z) = (1 + \exp\{-z\})^{-1}$. As Execise 10.5.10 shows

$$-\frac{f_Z'(z)}{f_Z(z)} = F_Z(z)(1 - \exp\{-z\}) \quad \text{and} \quad F_Z^{-1}(u) = \log\frac{u}{1-u}. \tag{10.5.32}$$

Upon standardization, this leads to the optimal score function,

$$\varphi_W(u) = \sqrt{12}(u - (1/2)), \tag{10.5.33}$$

that is, the Wilcoxon scores. The properties of the inference based on Wilcoxon scores are discussed in Section 10.4. Let $\widehat{\Delta}_W = \mathrm{med}\,\{Y_j - X_i\}$ denote the corresponding estimate. Recall that the $\mathrm{ARE}(\widehat{\Delta}_W, \overline{Y} - \overline{X}) = 0.955$ at the normal. Hodges and Lehmann (1956) showed that $\mathrm{ARE}(\widehat{\Delta}_W, \overline{Y} - \overline{X}) \geq 0.864$ over all symmetric distributions. ∎

Example 10.5.3. As a numerical illustration we consider some generated normal observations. The first sample, labeled X, was generated from a $N(48, 10^2)$ distribution while the second sample, Y, was generated from a $N(58, 10^2)$ distribution. There are 15 observations in each sample. The data are displayed in Table 10.5.1, and along with the data, the ranks and the normal scores are exhibited. We will consider tests of the two-sided hypotheses $H_0 : \Delta = 0$ versus $H_1 : \Delta \neq 0$ for the Wilcoxon, normal scores, and Student t procedures.

As the following comparison dotplots show, the second sample observations appear to be larger than those from the first sample.

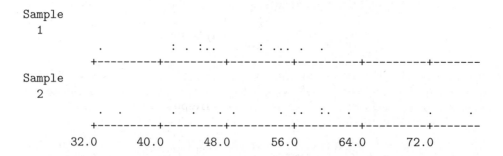

Table 10.5.1: Data for Example 10.5.3

Sample 1 (X)			Sample 2 (Y)		
Data	Ranks	Normal Scores	Data	Ranks	Normal Scores
51.9	15	−0.04044	59.2	24	0.75273
56.9	23	0.64932	49.1	14	−0.12159
45.2	11	−0.37229	54.4	19	0.28689
52.3	16	0.04044	47.0	13	−0.20354
59.5	26	0.98917	55.9	21	0.46049
41.4	4	−1.13098	34.9	3	−1.30015
46.4	12	−0.28689	62.2	28	1.30015
45.1	10	−0.46049	41.6	6	−0.86489
53.9	17	0.12159	59.3	25	0.86489
42.9	7	−0.75273	32.7	1	−1.84860
41.5	5	−0.98917	72.1	29	1.51793
55.2	20	0.37229	43.8	8	−0.64932
32.9	2	−1.51793	56.8	22	0.55244
54.0	18	0.20354	76.7	30	1.84860
45.0	9	−0.55244	60.3	27	1.13098

The test statistics along with their standardized versions and p-values are:

Method	Test Statistic	Standardized	p-value	Estimate of Δ
Student t	$\overline{Y} - \overline{X} = 5.43$	1.47	0.15	5.43
Wilcoxon	$W = 270$	1.56	0.12	5.20
Normal Scores	$W_N = 3.73$	1.48	0.14	5.10

Notice that the standardized tests statistics and their corresponding p-values are quite similar and all would result in the same decision regarding the hypotheses. As shown in the table, the corresponding point estimates of Δ are also alike. The estimates were obtained at the web site www.stat.wmich.edu/slab/RGLM. ∎

Example 10.5.4 (Sign Scores). For our final example, suppose that the random errors, $\varepsilon_1, \varepsilon_2, \ldots, \varepsilon_n$, have a Laplace distribution. Consider the convenient form, $f_Z(z) = 2^{-1} \exp\{-|z|\}$. Then $f'_Z(z) = -2^{-1} \text{sgn}(z) \exp\{-|z|\}$ and hence, $-f'_Z(F_Z^{-1}(u))/f_Z(F_Z^{-1}(u)) = \text{sgn}(z)$. But $F_Z^{-1}(u) > 0$ if and only if $u > 1/2$. The optimal score function is

$$\varphi_S(u) = \text{sgn}\left(u - \frac{1}{2}\right), \tag{10.5.34}$$

which is easily shown to be standardized. The corresponding process is

$$W_S(\Delta) = \sum_{j=1}^{n_2} \text{sgn}\left[R(Y_j - \Delta) - \frac{n+1}{2}\right]. \tag{10.5.35}$$

Because of the signs, this test statistic can be written in a simpler form which is often called *Mood's* test; see Exercise 10.5.12.

We can also obtain the associated estimator in closed form. The estimator solves the equation

$$\sum_{j=1}^{n_2} \text{sgn}\left[R(Y_j - \Delta) - \frac{n+1}{2}\right] = 0. \qquad (10.5.36)$$

For this equation, we rank the variables

$$\{X_1, \ldots, X_{n_1}, Y_1 - \Delta, \ldots, Y_{n_2} - \Delta\}.$$

Because ranks, though, are invariant to a constant shift we obtain the same ranks if we rank the variables

$$\{X_1 - \text{med}\{X_i\}, \ldots, X_{n_1} - \text{med}\{X_i\}, Y_1 - \Delta - \text{med}\{X_i\}, \ldots, Y_{n_2} - \Delta - \text{med}\{X_i\}.$$

Therefore, the solution to equation (10.5.36) is easily seen to be

$$\widehat{\Delta}_S = \text{med}\{Y_j\} - \text{med}\{X_i\}. \quad \blacksquare \qquad (10.5.37)$$

Other examples are given in the exercises.

EXERCISES

10.5.1. Complete the derivation of the null variance of the test statistic W_φ by showing the second term in expression (10.5.7) is true. *Hint:* Use the fact that under H_0, for $j \neq j'$, the pair $(a_\varphi(R(Y_j)), a_\varphi(R(Y_{j'})))$ is unifomly distributed on the pairs of integers (i, i'), $i, i' = 1, 2, \ldots, n$, $i \neq i'$.

10.5.2. For the Wilcoxon score function $\varphi(u) = \sqrt{12}(u - (1/2))$, obtain the value of s_a. Then show that the $V_{H_0}(W_\varphi)$ given in expression (10.5.8) is the same (except for standardization) as the variance of the MWW statistic of Section 10.4.

10.5.3. Recall that the scores have been standardized so that $\int_{-\infty}^{\infty} \varphi^2(u)\, du = 1$. Use this and a Riemann sum to show that $n^{-1}s_a^2 \to 1$, where s_a^2 is defined in expression (10.5.6).

10.5.4. Show that the normal scores, (10.5.29), derived in Example 10.5.1 are standardized; that is, $\int_0^1 \varphi_N(u)\, du = 0$ and $\int_0^1 \varphi_N^2(u)\, du = 1$.

10.5.5. In Theorem 10.5.1, show that the minimum value of $W_\varphi(\Delta)$ is given by $\sum_{j=1}^{n_2} a_\varphi(j)$ and that it is nonpositive.

10.5.6. Show that $E_\Delta[W_\varphi(0)] = E_0[W_\varphi(-\Delta)]$.

10.5.7. Consider the hypotheses (10.4.2). Suppose we select the score function $\varphi(u)$ and the corresponding test based on W_φ. Suppose we want to determine the sample size $n = n_1 + n_2$ for this test of significance level α to detect the alternative Δ^* with approximate power γ^*. Assuming that the sample sizes n_1 and n_2 are the same, show that

$$n \doteq \left(\frac{(z_\alpha - z_{\gamma^*})2\tau_\varphi}{\Delta^*}\right)^2. \qquad (10.5.38)$$

10.5.8. In the context of this section, show the following invariances:

(a) Show that the parameter τ_φ, (10.5.24), is a scale functional as defined in Exercise 10.1.4.

(b) Show Part (a) implies that the efficacy, (10.5.20), is invariant to the location and varies indirectly with scale.

(c) Suppose Z is a scale and location transformation of a random variable X; i.e., $Z = a(X - b)$, where $a > 0$ and $-\infty < b < \infty$. Show that $I(f_Z) = a^{-2}I(f_X)$.

10.5.9. Consider the scale parameter τ_φ, (10.5.24), when normal scores are used; i.e., $\varphi(u) = \Phi^{-1}(u)$. Suppose we are sampling from a $N(\mu, \sigma^2)$ distribution. Show that $\tau_\varphi = \sigma$.

10.5.10. In the context of Example 10.5.2 obtain the results in expression (10.5.32).

10.5.11. Let the scores $a(i)$ be generated by $a_\varphi(i) = \varphi(i/(n+1))$, for $i = 1, \ldots, n$. where $\int_0^1 \varphi(u)\, du = 0$ and $\int_0^1 \varphi^2(u)\, du = 1$. Using Riemann sums, with subintervals of equal length, of the integrals $\int_0^1 \varphi(u)\, du$ and $\int_0^1 \varphi^2(u)\, du$, show that $\sum_{i=1}^n a(i) \doteq 0$ and $\sum_{i=1}^n a^2(i) \doteq n$.

10.5.12. Consider the sign scores test procedure discussed in Example 10.5.4.

(a) Show that $W_S = 2W_S^* - n_2$, where $W_S^* = \#_j\left\{R(Y_j) > \frac{n+1}{2}\right\}$. Hence, W_S^* is an equivalent test statistic. Find the null mean and variance of W_S.

(b) Show that $W_S^* = \#_j\{Y_j > \theta^*\}$, where θ^* is the combined sample median.

(c) Suppose n is even. Letting $W_{XS}^* = \#_i\{X_i > \theta^*\}$, show that we can table W_S^* in the following 2×2 contingency table with all margins fixed:

	Y	X	
Num. items $> \theta^*$	W_S^*	W_{XS}^*	$\frac{n}{2}$
Num. items $< \theta^*$	$n_2 - W_S^*$	$n_1 - W_{XS}^*$	$\frac{n}{2}$
	n_2	n_1	n

Show that the usual χ^2 goodness-of-fit is the same as Z_S^2 where Z_S is the standardized z-test based on W_S. This is often called **Mood's Median Test**.

10.5.13. Recall the data discussed in Example 10.5.3.

(a) Obtain the contingency table described in Exercise 10.5.12.

(b) Obtain the χ^2 goodness-of-fit test statistic associated with the table and use it to test at level 0.05 the hypotheses $H_0 : \Delta = 0$ versus $H_1 : \Delta \neq 0$.

(c) Obtain the point estimate of Δ given in expression (10.5.37).

10.5.14. Optimal signed-rank based methods exist for the one sample problem, also. In this exercise, we briefly discuss these methods. Let $X_1, X_2, \ldots, X_n$ follow the location model

$$X_i = \theta + e_i, \tag{10.5.39}$$

where $e_1, e_2, \ldots, e_n$ are iid with pdf $f(x)$ which is symmetric about 0; i.e., $f(-x) = f(x)$.

(a) Show that under symmetry the optimal two sample score function (10.5.26) satisfies

$$\varphi_f(1 - u) = -\varphi_f(u), \quad 0 < u < 1, \tag{10.5.40}$$

that is, $\varphi_f(u)$ is an odd function about $\frac{1}{2}$. Show that a function satisfying (10.5.40) is 0 at $u = \frac{1}{2}$.

(b) For a two sample score function $\varphi(u)$ which is odd about $\frac{1}{2}$, define the function $\varphi^+(u) = \varphi[(u+1)/2]$; i.e, the top half of $\varphi(u)$. Show that $\varphi^+(u) \geq 0$, provided $\varphi(u)$ is nondecreasing.

(c) Assume for the remainder of the problem that $\varphi^+(u)$ is nonnegative and non-decreasing on the interval $(0, 1)$. Define the scores $a^+(i) = \varphi^+[i/(n + 1)]$, $i = 1, 2, \ldots, n$, and the corresponding statistic

$$W_{\varphi^+} = \sum_{i=1}^{n} \text{sgn}(X_i) a^+(R|X_i|). \tag{10.5.41}$$

Show that W_{φ^+} reduces to a linear function of the signed-rank test statistic (10.3.1), if $\varphi(u) = 2u - 1$.

(d) Show that W_{φ^+} reduces to a linear function of the sign test statistic (10.2.3), if $\varphi(u) = \text{sgn}(2u - 1)$.

Note: Suppose Model (10.5.39) is true and we take $\varphi(u) = \varphi_f(u)$ where $\varphi_f(u)$ is given by (10.5.26). If we choose $\varphi^+(u) = \varphi[(u+1)/2]$ to generate the signed-rank scores then it can be shown that the corresponding test statistic W_{φ^+} is optimal, among all signed-rank tests.

(e) Consider the hypotheses

$$H_0 : \theta = 0 \text{ versus } H_1 : \theta > 0.$$

Our decision rule for the statistic W_{φ^+} is to reject H_0 in favor of H_1, if $W_{\varphi^+} \geq k$, for some k. Write W_{φ^+} in terms of the anti-ranks, (10.3.4). Show that W_{φ^+} is distribution-free under H_0.

(f) Determine the mean and variance of W_{φ^+} under H_0.

(g) Assuming that when properly standardized the null distribution is asymptotically normal, determine the asymptotic test.

10.6 Adaptive Procedures

In the last section, we presented fully efficient rank-based procedures for testing and estimation. As with mle methods, though, the underlying form of the distribution must be known in order to select the optimal rank score function. In practice, often the underlying distribution is not known. In this case, we could select a score function, such as the Wilcoxon which is fairly efficient for moderate to heavy-tailed error distributions. Or perhaps the distribution of the errors is thought to be quite close to a normal distribution then the normal scores would be a proper choice. Suppose we use a technique which bases the score selection on the data. These techniques are called **adaptive** procedures. Such a procedure could attempt to estimate the score function; see, for example, Naranjo and McKean (1997). However, large data sets are often needed for these. There are other adaptive procedures which attempt to select a score from a finite class of scores based on some criteria. In this section, we look at an adaptive testing procedure for testing which retains the distribution-free property.

Frequently, an investigator is tempted to evaluate several test statistics associated with a single hypothesis and then use the one statistic that best supports his or her position, usually rejection. Obviously, this type of procedure changes the actual significance level of the test from the nominal α that is used. However, there is a way in which the investigator can first look at the data and then select a test statistic without changing this significance level. For illustration, suppose there are three possible test statistics W_1, W_2, W_3 of the hypothesis H_0 with respective critical regions C_1, C_2, C_3 such that $P(W_i \in C_i; H_0) = \alpha$, $i = 1, 2, 3$. Moreover, suppose that a statistic Q, based upon the same data, selects one and only one of the statistics W_1, W_2, W_3, and that W is then used to test H_0. For example, we choose to use the test statistic W_i if $Q \in D_i$, $i = 1, 2, 3$, where the events defined by D_1, D_2, and D_3 are mutually exclusive and exhaustive. Now if Q and each W_i are independent when H_0 is true, then the probability of rejection, using the entire procedure (selecting and testing), is, under H_0,

$$P_{H_0}(Q \in D_1, W_1 \in C_1) + P_{H_0}(Q \in D_2, W_2 \in C_2) + P_{H_0}(Q \in D_3, W_3 \in C_3)$$
$$= P_{H_0}(Q \in D_1)P_{H_0}(W_1 \in C_1) + P_{H_0}(Q \in D_2)P_{H_0}(W_2 \in C_2)$$
$$+ P_{H_0}(Q \in D_3)P_{H_0}(W_3 \in C_3)$$
$$= \alpha[P_{H_0}(Q \in D_1) + P_{H_0}(Q \in D_2) + P_{H_0}(Q \in D_3)] = \alpha.$$

That is, the procedure of selecting W_i using an independent statistic Q and then constructing a test of significance level α with the statistic W_i has overall significance level α.

Of course, the important element in this procedure is the ability to be able to find a selector Q that is independent of each test statistic W. This can frequently be done by using the fact that the complete sufficient statistic for the parameters, given by H_0, are independent of every statistic whose distribution is free of those parameters. For illustration, if independent random samples of sizes n_1 and n_2 arise from two normal distributions with respective means μ_1 and μ_2 and common variance σ^2, then the complete sufficient statistics $\overline{X}, \overline{Y}$, and

$$V = \sum_{1}^{n_1} (X_i - \overline{X})^2 + \sum_{1}^{n_2} (Y_i - \overline{Y})^2$$

for μ_1, μ_2, and σ^2 are independent of every statistic whose distribution is free of μ_1, μ_2, and σ^2 such as the statistics

$$\frac{\displaystyle\sum_{1}^{n_1} (X_i - \overline{X})^2}{\displaystyle\sum_{1}^{n_2} (Y_i - \overline{Y})^2}, \quad \frac{\displaystyle\sum_{1}^{n_1} |X_i - \text{median}(X_i)|}{\displaystyle\sum_{1}^{n_2} |Y_i - \text{median}(Y_i)|}, \quad \frac{\text{range}(X_1, X_2, \ldots, X_{n_1})}{\text{range}(Y_1, Y_2, \ldots, Y_{n_2})}.$$

Thus, in general, we would hope to be able to find a selector Q that is a function of the complete sufficient statistics for the parameters, under H_0, so that it is independent of the test statistic.

It is particularly interesting to note that it is relatively easy to use this technique in *nonparametric* methods by using the independence result based upon complete sufficient statistics for *parameters*. For the situations here, we must find complete sufficient statistics for a cdf, F, of the continuous type. In Chapter 6, it is shown that the order statistics $Y_1 < Y_2 < \cdots < Y_n$ of a random sample of size n from a distribution of the continuous type with pdf $F'(x) = f(x)$ are sufficient statistics for the "parameter" f (or F). Moreover, if the family of distributions contains all probability density functions of the continuous type, the family of joint probability density functions of $Y_1, Y_2, \ldots, Y_n$ is also complete. That is, the order statistics $Y_1, Y_2, \ldots, Y_n$ are complete sufficient statistics for the parameters f (or F).

Accordingly, our selector Q will be based upon those complete sufficient statistics, the order statistics under H_0. This allows us to independently choose a distribution-free test appropriate for this type of underlying distribution, and thus optimize (maximize the power) our inference.

A statistical test that maintains the significance level close to a desired significance level α for a wide variety of underlying distributions with good (not necessarily the best for any one type of distribution) power for all these distributions is described as being *robust*. As an illustration, the pooled t-test (Student's t) used to test the equality of the means of two normal distributions is quite robust *provided* that the underlying distributions are rather close to normal ones with common variance. However, if the class of distributions includes those that are not too close to normal ones, such as contaminated normal distributions, the test based upon t is *not* robust; the significance level is not maintained and the power of the t-test can be quite low for heavy-tailed distributions. As a matter of fact, the test based on the Mann-Whitney-Wilcoxon statistic (Section 10.4) is a much more robust test than that based upon t if the class of distributions includes those with heavy tails.

In the following example, we illustrate a robust adaptive distribution-free procedure in the setting of the two sample problem.

Example 10.6.1. Let $X_1, X_2, \ldots, X_{n_1}$ be a random sample from a continuous-type distribution with cdf $F(x)$ and let $Y_1, Y_2, \ldots, Y_{n_2}$ be a random sample from a distribution with the cdf $F(x - \Delta)$. Let $n = n_1 + n_2$ denote the combined sample size. We test

$$H_0 : \; \Delta = 0 \text{ versus } H_1 : \; \Delta > 0,$$

by using one of four distribution-free statistics, the one being the Wilcoxon and the other three being modifications of the Wilcoxon. In particular, the test statistics are

$$W_i = \sum_{j=1}^{n_2} a_i[R(Y_j)], \quad i = 1, 2, 3, 4, \tag{10.6.1}$$

where

$$a_i(j) = \varphi_i[j/(n+1)],$$

and the four functions are displayed in Figure 10.6.1. The score function $\varphi_1(u)$ is the Wilcoxon. The score function $\varphi_2(u)$ is the sign score function. The score function $\varphi_3(u)$ is good for short-tailed distributions, and $\varphi_4(u)$ is good for long, right-skewed distributions with shift alternatives.

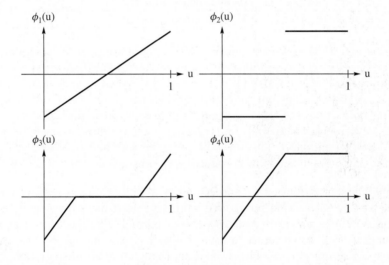

Figure 10.6.1: Plots of the score functions $\varphi_1(u)$, $\varphi_2(u)$, $\varphi_3(u)$ and $\varphi_4(u)$.

We combine the two samples into one denoting the order statistics of the combined sample by $V_1 < V_2 < \cdots < V_n$. These are complete sufficient statistics for $F(x)$ under the null hypothesis. For $i = 1, \ldots, 4$, the test statistic W_i is distribution free under H_0 and, in particular, the distribution of W_i does not depend on $F(x)$. Therefore, each W_i is independent of $V_1, V_2, \ldots, V_n$. We use a pair of selector statistics (Q_1, Q_2) which are functions of $V_1, V_2, \ldots, V_n$ and hence, are also independent of each W_i. The first is

$$Q_1 = \frac{\overline{U}_{.05} - \overline{M}_{.5}}{\overline{M}_{.5} - \overline{L}_{.05}}, \tag{10.6.2}$$

where $\overline{U}_{.05}$, $\overline{M}_{.5}$, and $\overline{L}_{.05}$ are the averages of the largest 5% of the Vs, the middle 50% of the Vs, and the smallest 5% of the Vs, respectively. If Q_1 is large (say 2 or more) then the right tail of the distribution seems longer than the left tail; that is, there is an indication that the distribution is skewed to the right. On the other hand, if $Q_1 < \frac{1}{2}$, the sample indicates that the distribution may be skewed to the left. The second selector statistic is

$$Q_2 = \frac{\overline{U}_{.05} - \overline{L}_{.05}}{\overline{U}_{.5} - \overline{L}_{.5}}. \tag{10.6.3}$$

Large values of Q_2 indicate that the distribution is heavy tailed while small values indicate that the distribution is light tailed. Rules are needed for score selection and here, we make use of the benchmarks proposed in an article by Hogg et al. (1975). These rules are tabulated below, along with their benchmarks:

Benchmark	Distribution Indicated	Score Selected
$Q_2 > 7$	Heavy-tailed symmetric	φ_2
$Q_1 > 2$ and $Q_2 < 7$	Right-skewed	φ_4
$Q_1 \leq 2$ and $Q_2 \leq 2$	Light-tailed symmetric	φ_3
Elsewhere	Moderate heavy-tailed	φ_1

Hogg et al. (1975) performed a Monte Carlo power study of this adaptive procedure over a number of distributions with different kurtosis and skewness coefficients. In the study, both the adaptive procedure and the Wilcoxon test maintain their α level over the distributions but the Student t does not. Moreover the Wilcoxon test has better power than the t test as the distribution deviates much from the normal (kurtosis $= 3$ and skewness $= 0$), but the adaptive procedure is much better than the Wilcoxon for the short tailed distributions, the very heavy tailed distributions, and the highly skewed distributions which were considered in the study. ∎

The adaptive distribution-free procedure that we have discussed is for testing. Suppose we have a location model and were interested in estimating the shift in locations Δ. For example, if the true F is a normal cdf then a good choice for the estimator of Δ would be the estimator based on the normal scores procedure discussed in Example 10.5.1. The estimators, though, are not distribution free and hence, the above reasoning will not hold. Also, the combined sample observations $X_1, \ldots, X_{n_1}, Y_1, \ldots, Y_{n_2}$ are not identically distributed. There are adaptive procedures based on residuals, $X_1, \ldots, X_{n_1}, Y_1 - \widehat{\Delta}, \ldots, Y_{n_2} - \widehat{\Delta}$ where $\widehat{\Delta}$ is an initial estimator of Δ; see page 212 of Hettmansperger and McKean (1998) for discussion.

EXERCISES

10.6.1. Consider the data in Example 10.5.3.

(a) Set up the adaptive procedure of Example 10.6.1 for this data set, using the four score functions $\varphi_1(u)$, $\varphi_2(u)$, $\varphi_3(u)$, and $\varphi_4(u)$ graphed in Figure 10.6.1.

(b) Obtain the actual scores $a_j(i) = \varphi_j(i/(n+1))$ for $j = 1, 2, 3,$ and 4 and $n = 30$, if the score functions are given by

$$
\begin{aligned}
\varphi_1(u) &= \quad 2u - 1 \quad 0 < u < 1 \\
\varphi_2(u) &= \quad \mathrm{sgn}(2u - 1) \quad 0 < u < 1 \\
\varphi_3(u) &= \left\{
\begin{array}{ll}
4u - 1 & 0 < u \leq \frac{1}{4} \\
0 & \frac{1}{4} < u \leq \frac{3}{4} \\
4u - 3 & \frac{3}{4} < u < 1
\end{array}
\right. \\
\varphi_4(u) &= \left\{
\begin{array}{ll}
4u - 1 & 0 < u \leq \frac{1}{2} \\
1 & \frac{1}{2} < u < 1
\end{array}
\right.
\end{aligned}
$$

(c) Standardize the scores so that they sum to 0.

(d) Obtain the asymptotic test statistics for each of these score functions.

(e) Use the adaptive procedure to test

$$H_0 : \Delta = 0 \text{ versus } H_1 : \Delta > 0,$$

where $\Delta = \mu_Y - \mu_X$. Obtain the p-value of the test.

10.6.2. Use the adaptive procedure of the last exercise on the data of Example 10.4.1.

10.6.3. Let $F(x)$ be a distribution function of a distribution of the continuous type which is symmetric about its median θ. We wish to test $H_0 : \theta = 0$ against $H_1 : \theta > 0$. Use the fact that the $2n$ values, X_i and $-X_i$, $i = 1, 2, \ldots, n$, after ordering, are complete sufficient statistics for F, provided that H_0 is true.

(a) As in Exercise 10.5.14, determine the one-sample signed-rank test statistics corresponding to the two sample score functions $\varphi_1(u)$, $\varphi_2(u)$, and $\varphi_3(u)$ defined in the last exercise. Use the asymptotic test statistics. Note that these scores functions are odd about $\frac{1}{2}$, hence, their top-halves serve as score functions for signed-rank statistics.

(b) We are assuming symmetric distributions in this problem; hence, we will only choose Q_2 as our score selector. If $Q_2 \geq 7$ then select $\varphi_2(u)$; if $2 < Q_2 < 7$ then select $\varphi_1(u)$; and finally if $Q_2 \leq 2$ then select $\varphi_3(u)$. Construct this adaptive distribution-free test.

(c) Use your adaptive procedure on the *Zea Mays* Data of Darwin, Example 10.3.1. Obtain the p-value.

10.7 Simple Linear Model

In this section, we consider the simple linear model and briefly develop the rank-based procedures for it.

Consider the simple linear model given by

$$Y_i = \alpha + \beta(x_i - \overline{x}) + \varepsilon_i, \quad i = 1, 2, \ldots, n, \tag{10.7.1}$$

where $\varepsilon_1, \varepsilon_2, \ldots, \varepsilon_n$ are iid with continuous cdf $F(x)$ and pdf $f(x)$. In this model, the variables $x_1, x_2, \ldots, x_n$ are considered fixed. The parameter β is the slope parameter. It is the expected change (provided expectations exist) when x increases by one unit. A natural null hypothesis is

$$H_0: \ \beta = 0 \text{ versus } H_1: \ \beta \neq 0. \tag{10.7.2}$$

Under H_0, the distribution of Y is free of x.

We will present a geometry for rank-based procedures for linear models in Chapter 12. Here, it is easier to present a development which parallels the preceding sections. Hence, we introduce a rank test of H_0 and then invert the test to estimate β. Before doing this, though, we present an example which shows that the two sample location problem of Section 10.4 is a regression problem.

Example 10.7.1. As in Section 10.4, let $X_1, X_2, \ldots, X_{n_1}$ be a random sample from a distribution with a continuous cdf $F(x - \alpha)$, where α is a location parameter. Let $Y_1, Y_2, \ldots, Y_{n_2}$ be a random sample from with cdf $F(x - \alpha - \Delta)$. Hence, Δ is the shift between the cdfs of X_i and Y_j. Redefine the observations as $Z_i = X_i$, for $i = 1, \ldots, n_1$, and $Z_{n_1+i} = Y_i$, for $i = n_1 + 1, \ldots, n$, where $n = n_1 + n_2$. Let c_i be 0 or 1 depending on whether $1 \leq i \leq n_1$ or $n_1 + 1 \leq i \leq n$. Then we can write the two sample location models as

$$Z_i = \alpha + \Delta c_i + \varepsilon_i, \tag{10.7.3}$$

where $\varepsilon_1, \varepsilon_2, \ldots, \varepsilon_n$ are iid with cdf $F(x)$. Hence, the shift in locations is the slope parameter from this viewpoint. ∎

Suppose the regression model, (12.2.2), holds and, further, that H_0 is true. Then we would expect that Y_i and $x_i - \overline{x}$ are not related and, in particular, that they are uncorrelated. Hence, as a test statistic one could consider $\sum_{i=1}^{n}(x_i - \overline{x})Y_i$. As Exercise 10.7.2 shows, if we additionally assume that the random errors ε_i are normally distributed, this test statistic properly standardized is the likelihood ratio test statistic. Reasoning in the same way, for a specified score function we would expect that $a_\varphi(R(Y_i))$ and $x_i - \overline{x}$ are uncorrelated, under H_0. Therefore, consider the test statistic

$$T_\varphi = \sum_{i=1}^{n}(x_i - \overline{x})a_\varphi(R(Y_i)), \tag{10.7.4}$$

where $R(Y_i)$ denotes the rank of Y_i among $Y_1, \ldots, Y_n$ and $a_\varphi(i) = \varphi(i/(n+1))$ for a nondecreasing score function $\varphi(u)$ which is standardized so that $\int \varphi(u)\,du = 0$ and $\int \varphi^2(u)\,du = 1$. Values of T_φ close to 0 indicate H_0.

Assume H_0 is true. Then $Y_1, \ldots, Y_n$ are iid random variables. Hence, any permutation of the integers $\{1, 2, \ldots, n\}$ are equilikely to be the ranks of $Y_1, \ldots, Y_n$. So the distribution of T_φ is free of $F(x)$. Note that the distribution will depend on the $x_1, x_2, \ldots, x_n$. Thus tables of the distribution are not available; although, with high speed computing this distribution can be generated. Because $R(Y_i)$ is uniformly distributed on the integers $\{1, 2, \ldots, n\}$, it is easy to show that the null expectation of T_φ is zero. The null variance follows as that of W_φ of Section 10.5, so we have left the details for Exercise 10.7.4. To summarize, the null moments are given by

$$E_{H_0}(T_\varphi) = 0 \quad \text{and} \quad \text{Var}_{H_0}(T_\varphi) = \tfrac{1}{n-1} s_a^2 \sum_{i=1}^{n} (x_i - \overline{x})^2, \tag{10.7.5}$$

where s_a^2 is the sum of the squares of the scores (10.5.6). Also, it can be shown that the test statistic is asymptotically normal. Therefore, an asymptotic level α decision rule for the hypotheses (10.7.2) is given by

$$\text{Reject } H_0 \text{ in favor of } H_1 \text{ if } |z| = \left| \frac{T_\varphi}{\sqrt{\text{Var}_{H_0}(T_\varphi)}} \right| \geq z_{\alpha/2}. \tag{10.7.6}$$

The associated process is given by

$$T_\varphi(\beta) = \sum_{i=1}^{n} (x_i - \overline{x}) a_\varphi(R(Y_i - x_i\beta)). \tag{10.7.7}$$

Hence, the corresponding estimate of β is given by $\widehat{\beta}_\varphi$ which solves the estimating equations

$$T_\varphi(\widehat{\beta}_\varphi) \dot{=} 0. \tag{10.7.8}$$

Similar to the Theorem 10.5.1, it can be shown that $T_\varphi(\beta)$ is a decreasing step function of β which steps down at each sample slope $(Y_j - Y_i)/(x_j - x_i)$; see Exercise 10.7.3. Thus, the estimate exists. It cannot be obtained in closed form but simple iterative techniques can be used to find the solution. In the regression problem, though, prediction of Y is often of interest which requires an estimate of α, also. Notice that such an estimate can be obtained as a location estimate based on residuals. This is discussed in some detail in Section 3.5.2 of Hettmansperger and McKean (1998). For our purposes we consider the median of the residuals; that is, we estimate α as

$$\widehat{\alpha} = \text{med}\{Y_i - \widehat{\beta}_\varphi(x_i - \overline{x})\}. \tag{10.7.9}$$

The Wilcoxon estimates of slope and intercept are computed by several packages. The minitab command `rregr` will obtain the Wilcoxon fit. Terpstra and McKean (2004) have written a collection of R and S-PLUS functions which obtain this fit. It can also be obtained at the web site `www.stat.wmich.edu/slab/RGLM`. This web site was used for the following example.

Example 10.7.2 (Telephone Data). Consider the regression data discussed in Exercise 9.6.2. Recall that the responses (y) for this data set are the numbers of telephone calls (tens of millions) made in Belgium for the years 1950 through 1973, while time in years serves as the predictor variable (x). The data are plotted in Figure 10.7.1. For this example, we used Wilcoxon scores to fit the model (12.2.2).

This resulted in the estimates $\widehat{\beta}_W = 0.145$ and $\widehat{\alpha} = -7.13$. The Wilcoxon fitted value $\widehat{Y}_{\varphi,i} = -7.13 + 0.145x_i$ is plotted in Figure 10.7.1. The least squares fit $\widehat{Y}_{LS,i} = -26.0 + 0.504x_i$, found in Exercise 9.6.2, is also plotted. Note that the Wilcoxon fit is much less sensitive to the outliers than the least squares fit.

The outliers in this data set were recording errors; see page 25 of Rousseeuw and Leroy (1987) for more discussion. ∎

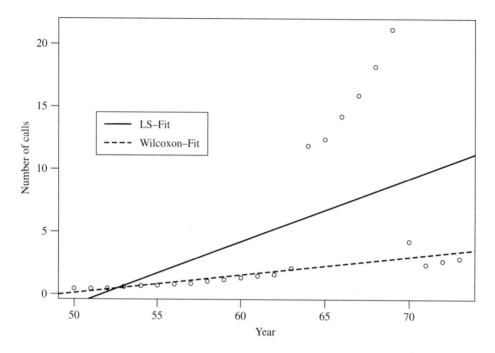

Figure 10.7.1: Plot of Telephone Data, Example 10.7.2, Overlaid with Wilcoxon and LS Fits.

Similar to Lemma 10.2.1, a translation property holds for the process $T(\beta)$ given by

$$E_\beta[T(0)] = E_0[T(-\beta)]; \tag{10.7.10}$$

see Exercise 10.7.1. Further, as Exercise 10.7.5 shows, this property implies that the power curve for the one-sided tests of $H_0 : \beta = 0$ are monotone, assuring the unbiasedness of the tests based on T_φ.

We can now derive the efficacy of the process. Let $\mu_T(\beta) = E_\beta[T(0)]$ and $\sigma_T^2(0) = \text{Var}_0[T(0)]$. Expression (10.7.5) gives the result for $\sigma_T^2(0)$. Recall that for the mean $\mu_T(\beta)$ we need its derivative at 0. We freely use the relationship between rankings and the empirical cdf and then approximate this empirical cdf with the true cdf. Hence,

$$
\begin{aligned}
\mu_T(\beta) = E_\beta[T(0)] \ &= \ E_0[T(-\beta)] = \sum_{i=1}^n (x_i - \overline{x}) E_0[a_\varphi(R(Y_i + x_i\beta))] \\
&= \ \sum_{i=1}^n (x_i - \overline{x}) E_0\left[\varphi\left(\frac{n\widehat{F}_n(Y_i + x_i\beta)}{n+1}\right)\right] \\
&\doteq \ \sum_{i=1}^n (x_i - \overline{x}) E_0[\varphi(F(Y_i + x_i\beta))] \\
&= \ \sum_{i=1}^n (x_i - \overline{x}) \int_{-\infty}^\infty \varphi(F(y + x_i\beta)) f(y)\, dy. \qquad (10.7.11)
\end{aligned}
$$

Differentiating this last expression we have

$$
\mu_T'(\beta) = \sum_{i=1}^n (x_i - \overline{x}) x_i \int_{-\infty}^\infty \varphi'(F(y + x_i\beta)) f(y + x_i\beta) f(y)\, dy,
$$

which yields

$$
\mu_T'(0) = \sum_{i=1}^n (x_i - \overline{x})^2 \int_{-\infty}^\infty \varphi'(F(y)) f^2(y)\, dy. \qquad (10.7.12)
$$

We need one assumption on the $x_1, x_2, \ldots, x_n$; namely, $n^{-1}\sum_{i=1}^n (x_i - \overline{x})^2 \to \sigma_x^2$, where $0 < \sigma_x^2 < \infty$. Recall that $(n-1)^{-1} s_a^2 \to 1$. Therefore, the efficacy of the process $T(\beta)$ is given by

$$
\begin{aligned}
c_T \ &= \ \lim_{n\to\infty} \frac{\mu_T'(0)}{\sqrt{n}\,\sigma_T(0)} = \lim_{n\to\infty} \frac{\sum_{i=1}^n (x_i - \overline{x})^2 \int_{-\infty}^\infty \varphi'(F(y)) f^2(y)\, dy}{\sqrt{n}\,\sqrt{(n-1)^{-1} s_a^2}\, \sqrt{\sum_{i=1}^n (x_i - \overline{x})^2}} \\
&= \ \sigma_x \int_{-\infty}^\infty \varphi'(F(y)) f^2(y)\, dy. \qquad (10.7.13)
\end{aligned}
$$

Using this, an asymptotic power lemma can be derived for the test based on T_φ; see expression (10.7.17) of Exercise 10.7.6. Based on this, it can be shown that the asymptotic distribution of the estimator $\widehat{\beta}_\varphi$ is given by

$$
\widehat{\beta}_\varphi \text{ has an approximate } N\left(\beta, \tau_\varphi^2\left(\sum_{i=1}^n (x_i - \overline{x})^2\right)^{-1}\right) \text{ distribution.} \qquad (10.7.14)
$$

Define the scale parameter τ_φ by $\tau_\varphi = \left(\int_{-\infty}^\infty \varphi'(F(y)) f^2(y)\, dy\right)^{-1}$.

Remark 10.7.1. The least squares (LS) estimates for Model (12.2.2) were discussed in Section 9.6 in the case that the random errors $\varepsilon_1, \varepsilon_2, \ldots, \varepsilon_n$ are iid with a $N(0, \sigma^2)$ distribution. In general if the random errors are not necessairly normally distributed, the asymptotic distribution of the LS estimates is given in Theorem 12.4.1 under certain assumptions. In particular for Model (12.2.2) under these conditions, the LS estimator of β, say $\widehat{\beta}_{LS}$ is

$$
\widehat{\beta}_{LS} \text{ has an approximate } N\left(\beta, \sigma^2\left(\sum_{i=1}^n (x_i - \overline{x})^2\right)^{-1}\right) \text{ distribution,} \qquad (10.7.15)
$$

where σ^2 is the variance of ε_i. Based on (10.7.14) and (10.7.15), it follows that the ARE between the rank-based and LS estimators is given by

$$ARE(\widehat{\beta}_\varphi, \widehat{\beta}_{LS}) = \frac{\sigma^2}{\tau_\varphi^2}. \tag{10.7.16}$$

Hence, if Wilcoxon scores are used this ARE is the same as the ARE between the Wilcoxon and t-procedures in the one- and two-sample location models. ∎

EXERCISES

10.7.1. Establish expression (10.7.10). To do this, note first that the expression is the same as

$$E_\beta\left[\sum_{i=1}^n (x_i - \overline{x})a_\varphi(R(Y_i))\right] = E_0\left[\sum_{i=1}^n (x_i - \overline{x})a_\varphi(R(Y_i + x_i\beta))\right].$$

Show that the cdfs of Y_i (under β) and $Y_i + (x_i - \overline{x})\beta$ (under 0) are the same.

10.7.2. Assume that Model 12.2.2 is true and that, further, $\varepsilon_1, \ldots, \varepsilon_n$ are iid $N(0, \sigma^2)$ random variables. Show that the test statistic $\sum_{i=1}^n (x_i - \overline{x})Y_i$, properly standardized, is the likelihood ratio test of the hypotheses (10.7.2).

10.7.3. Suppose we have a two-sample model given by (10.7.3). Assuming Wilcoxon scores, show that the test statistic (10.7.4) is equivalent to the Wilcoxon test statistic found in expression (10.4.3).

10.7.4. Show that the null variance of the test statistic T_φ is the value given in (10.7.5).

10.7.5. Show that the translation property (10.7.10) implies that the power curve for the one-sided tests based on the test statistic T_φ of $H_0 : \beta = 0$ is monotone.

10.7.6. Consider the sequence of local alternatives given by the hypotheses

$$H_0 : \beta = 0 \text{ versus } H_{1n} : \beta = \beta_n = \frac{\beta_1}{\sqrt{n}},$$

where $\beta_1 > 0$. Let $\gamma(\beta)$ be the power function discussed in Exercise 10.7.5 for an asymptotic level α test based on the test statistic T_φ. Using the mean value theorem to approximate $\mu_T(\beta_n)$, sketch a proof of the limit

$$\lim_{n \to \infty} \gamma(\beta_n) = 1 - \Phi(z_\alpha - c_T\beta_1). \tag{10.7.17}$$

10.8 Measures of Association

In the last section, we discussed the simple linear regression model in which the random variables, Ys, were the responses or dependent variables while the xs were the independent variables and were thought of as fixed. Regression models occur in several ways. In an experimental design, the values of the independent variables are prespecified and the responses are observed. Bioassays (dose-response experiments) are examples. The doses are fixed and the responses are observed. If the experimental design is performed in a controlled environment (for example all other variables are controlled) it may be possible to establish cause and effect between x and Y. On the other hand, in observational studies both the xs and Ys are observed. In the regression setting, we are still interested in predicting Y in terms of x, but usually cause and effect between x and Y are precluded in such studies (other variables besides x may be changing).

In this section, we focus on observational studies but are interested in the strength of the association between Y and x. So both X and Y are treated as random variables in this section and the underlying distribution of interest is the bivariate distribution of the pair (X, Y). We will assume that this bivariate distribution is continuous with cdf $F(x, y)$ and pdf $f(x, y)$.

Hence, let (X, Y) be a pair of random variables. A natural null model (baseline model) is that there is no relationship between X and Y; that is, the null hypothesis is given by $H_0 : X$ and Y are independent. Alternatives, though, depend on which measure of association is of interest. For example, if we are interested in the correlation between X and Y we would use the correlation coefficient ρ, (Section 9.7), as our measure of the association. A two-sided alternative in this case is $H_1 : \rho \neq 0$. Recall that independence between X and Y imply that $\rho = 0$, but that the converse is not true. However, the contrapositive is true, that is, $\rho \neq 0$ implies that X and Y are dependent. So in rejecting H_0 we would conclude that X and Y are dependent. Furthermore, the size of ρ indicates the strength of the correlation between X and Y.

10.8.1 Kendall's τ

The first measure of association that we consider in this section is a measure of the *monotonicity* between X and Y. Monotonicity is an easily understood association between X and Y. Let (X_1, Y_1) and (X_2, Y_2) be independent pairs with the same bivariate distribution, (discrete or continuous). We say these pairs are **concordant** if sgn $\{(X_1 - X_2)(Y_1 - Y_2)\} = 1$ and are **discordant** if sgn $\{(X_1 - X_2)(Y_1 - Y_2)\} = -1$. The variables X and Y have an increasing relationship if the pairs tend to be concordant and a decreasing relationship if the pairs tend to be discordant. A measure of this is given by **Kendall's** τ,

$$\tau = P[\text{sgn} \{(X_1 - X_2)(Y_1 - Y_2)\} = 1] - P[\text{sgn} \{(X_1 - X_2)(Y_1 - Y_2)\} = -1]. \quad (10.8.1)$$

As Exercise 10.8.1 shows, $-1 \leq \tau \leq 1$. Positive values of τ indicate increasing monotonicity, negatives values indicate decreasing monotonicity, and $\tau = 0$ reflects neither. Furthermore, as the following theorem shows, if X and Y are independent then $\tau = 0$.

Theorem 10.8.1. *Let (X_1, Y_1) and (X_2, Y_2) be independent pairs of observations of (X, Y) which has a continuous bivariate distribution. If X and Y are independent, then $\tau = 0$.*

Proof: Let (X_1, Y_1) and (X_2, Y_2) be independent pairs of observations with the same continuous bivariate distribution as (X, Y). Because the cdf is continuous, the sign function is either -1 or 1. By independence, we have

$$
\begin{aligned}
P[\mathrm{sgn}(X_1 - X_2)(Y_1 - Y_2) = 1] &= P[\{X_1 > X_2\} \cap \{Y_1 > Y_2\}] \\
&\quad + P[\{X_1 < X_2\} \cap \{Y_1 < Y_2\}] \\
&= P[X_1 > X_2]P[Y_1 > Y_2] \\
&\quad + P[X_1 < X_2]P[Y_1 < Y_2] \\
&= \left(\frac{1}{2}\right)^2 + \left(\frac{1}{2}\right)^2 = \frac{1}{2}.
\end{aligned}
$$

Likewise, $P[\mathrm{sgn}(X_1 - X_2)(Y_1 - Y_2) = -1] = \frac{1}{2}$; hence, $\tau = 0$. ∎

Relative to Kendall's τ as the measure of association, the two-sided hypotheses of interest here is

$$H_0 : \; X \text{ and } Y \text{ are independent versus } \tau \neq 0. \tag{10.8.2}$$

As Exercise 10.8.1 shows, the converse of Theorem 10.8.1 is false. However, the contrapositive is true; i.e., $\tau \neq 0$ implies that X and Y are dependent. As with the correlation coefficient, in rejecting H_0, we can conclude that X and Y are dependent.

Kendall's τ has a simple unbiased estimator. Let $(X_1, Y_1), (X_2, Y_2), \ldots, (X_n, Y_n)$ be a random sample of the cdf $F(x, y)$. Define the statistic

$$K = \binom{n}{2}^{-1} \sum_{i<j} \mathrm{sgn}\{(X_i - X_j)(Y_i - Y_j)\}; . \tag{10.8.3}$$

Note that for all $i \neq j$, pairs (X_i, Y_i) and (X_j, Y_j) are identically distributed. Thus, $E(K) = \binom{n}{2}^{-1} \binom{n}{2} E[\mathrm{sgn}\{(X_1 - X_2)(Y_1 - Y_2)\}] = \tau$.

In order to use K as a test statistic of the hypotheses (10.8.2), we need its distribution under the null hypothesis. Under H_0, $\tau = 0$ so the $E_{H_0}(K) = 0$. The null variance of K is given by expression (10.8.6); see, for instance, page 205 of Hettmansperger (1984). If all pairs $(X_i, Y_i), (X_j, Y_j)$ of the sample are concordant then $K = 1$ indicating a strictly increasing monotone relationship. On the other hand, if all pairs are discordant then $K = -1$. Thus the range of K is contained in the interval $[-1, 1]$. Also, the summands in expression (10.8.3) are either ± 1. From the proof of Theorem 10.8.1, the probability that a summand is one is $1/2$, which does not depend on the underlying distribution. Hence, the statistic K is distribution-free under H_0. The null distribution of K is symmetric about 0. This is easily seen from the fact that for each concordant pair there is an obvious discordant pair (just reverse an inequality on the Ys) and the fact that concordant and discordant pairs are equilikely under H_0. For tables of the null distribution of K see Hollander and Wolfe (1999). Also, it can be shown that K is asymptotically normal under H_0. We summarize these results in a theorem.

Table 10.8.1: Data for Example 10.8.1.

Year	1500 m	Marathon*	Year	1500 m	Marathon
1896	373.2	3530	1936	227.8	1759
1900	246	3585	1948	229.8	2092
1904	245.4	5333	1952	225.2	1383
1906	252	3084	1956	221.2	1500
1908	243.4	3318	1960	215.6	916
1912	236.8	2215	1964	218.1	731
1920	241.8	1956	1968	214.9	1226
1924	233.6	2483	1972	216.3	740
1928	233.2	1977	1976	219.2	595
1932	231.2	1896	1980	218.4	663

* Actual marathon times are 2 hours + entry

Theorem 10.8.2. *Let $(X_1, Y_1), (X_2, Y_2), \ldots, (X_n, Y_n)$ be a random sample on the bivariate random vector (X, Y) with continuous cdf $F(x, y)$. Under the null hypothesis of independence between X and Y, $F(x, y) = F_X(x) F_Y(y)$, for all (x, y) in the support of (X, Y), the test statistic K satisfies the following properties:*

$$K \text{ is distribution free with a symmetric pmf} \qquad (10.8.4)$$

$$E_{H_0}[K] = 0 \qquad (10.8.5)$$

$$Var_{H_0}(K) = \frac{2}{9} \frac{2n+1}{n(n-1)} \qquad (10.8.6)$$

$$\frac{K}{\sqrt{Var_{H_0}(K)}} \text{ has an asymptotically } N(0,1) \text{ distribution.} \qquad (10.8.7)$$

Based on the asymptotic test, a large sample level α test for the hypotheses (10.8.2) is to reject H_0 if $Z_K > z_{\alpha/2}$ where

$$Z_K = \frac{K}{\sqrt{2(2n+1)/9n(n-1)}}. \qquad (10.8.8)$$

We illustrate this test in the next example.

Example 10.8.1 (Olympic Race Times). Table 10.8.1 displays the winning times for two races in the Olympics beginning with the 1896 Olympics through the 1980 Olympics. The data were taken from Hettmansperger (1984). The times in seconds are for the 1500 m and the marathon. The entries in the table for the marathon race is the actual time minus 2 hours. In Exercise 10.8.2 the reader is asked to scatterplot the times for the two races. The plot shows a strong increasing monotone trend with one obvious outlier (1986 Olympics). An easy calculation shows that $K = 0.695$. The corresponding asymptotic test statistic is $Z_K = 6.27$ with p-value 0.000 which shows strong evidence to reject H_0. ∎

10.8.2   Spearman's Rho

As above, assume that $(X_1, Y_1), (X_2, Y_2), \ldots, (X_n, Y_n)$ is a random sample from a bivariate continuous cdf $F(x, y)$. The population correlation coefficient ρ is a measure of linearity between X and Y. The usual estimate is the sample correlation coefficient given by

$$r = \frac{\sum_{i=1}^{n}(X_i - \overline{X})(Y_i - \overline{Y})}{\sqrt{\sum_{i=1}^{n}(X_i - \overline{X})^2}\sqrt{\sum_{i=1}^{n}(Y_i - \overline{Y})^2}}. \tag{10.8.9}$$

A simple rank analogue is to replace X_i by $R(X_i)$, where $R(X_i)$ denotes the rank of X_i among $X_1, \ldots, X_n$, and likewise Y_i by $R(Y_i)$, where $R(Y_i)$ denotes the rank of Y_i among $Y_1, \ldots, Y_n$. Upon making this substitution, the denominator of the above ratio is a constant. This results in the statistic,

$$r_S = \frac{\sum_{i=1}^{n}(R(X_i) - \frac{n+1}{2})(R(Y_i) - \frac{n+1}{2})}{n(n^2 - 1)/12}, \tag{10.8.10}$$

which is called **Spearman's rho**. The statistic r_S is a correlation coefficient, so the inequality $-1 \leq r_S \leq 1$ is true. Further, as the following theorem shows, independence implies the functional (parameter) corresponding to r_S is 0.

Theorem 10.8.3. *Suppose* $(X_1, Y_1), (X_2, Y_2), \ldots, (X_n, Y_n)$ *is a sample on* (X, Y) *where* (X, Y) *has the continuous cdf* $F(x, y)$. *If* X *and* Y *are independent then* $E(r_S) = 0$.

Proof: Under independence, X_i and Y_j are independent for all i and j; hence, in particular, $R(X_i)$ is independent of $R(Y_i)$. Furthermore, the $R(X_i)$ is uniformly distributed on the integers $\{1, 2, \ldots, n\}$. Therefore, $E(R(X_i)) = (n + 1)/2$, which leads to the result. ∎

Thus the measure of association r_S can be used to test the null hypothesis of independence similar to Kendall's K. Under independence, because the X_is are a random sample, the random vector $(R(X_1), \ldots, R(X_n))$ is equilikely to assume any permutation of the integers $\{1, 2, \ldots, n\}$ and, likewise, the vector of the ranks of the Y_is. Furthermore, under independence the random vector $[R(X_1), \ldots, R(X_n), R(Y_1), \ldots, R(Y_n)]$ is equilikely to assume any of the $(n!)^2$ vectors $(i_1, i_2, \ldots, i_n, j_1, j_2, \ldots, j_n)$, where $(i_1, i_2, \ldots, i_n)$ and $(j_1, j_2, \ldots, j_n)$ are permutations of the integers $\{1, 2, \ldots, n\}$. Hence, under independence the statistic r_S is distribution-free. The distribution is discrete and tables of it can be found, for instance, in Hollander and Wolfe (1999). Similar to Kendall's statistic K, the distribution is symmetric about zero and it has an asymptotic normal distribution with asymptotic variance $1/\sqrt{n-1}$; see Exercise 10.8.6 for a proof of the null variance of r_s. A large sample level α test is to reject independence between X and Y, if $|z_S| > z_{\alpha/2}$, where $z_S = \sqrt{n-1}r_s$. We record these results in a theorem.

Theorem 10.8.4. *Let* $(X_1, Y_1), (X_2, Y_2), \ldots, (X_n, Y_n)$ *be a random sample on the bivariate random vector* (X, Y) *with continuous cdf* $F(x, y)$. *Under the null hypothesis of independence between* X *and* Y, $F(x, y) = F_X(x)F_Y(y)$, *for all* (x, y) *in the* *support of* (X, Y), *the test statistic* r_S *satisfies the following properties:*

$$r_S \text{ is distribution-free, symmetrically distributed about } 0 \quad (10.8.11)$$

$$E_{H_0}[r_S] = 0 \quad (10.8.12)$$

$$Var_{H_0}(r_S) = \frac{1}{\sqrt{n-1}}, \quad (10.8.13)$$

$$\frac{r_S}{\sqrt{Var_{H_0}(r_S)}} \text{ is asymptotically } N(0,1). \quad (10.8.14)$$

The statistic r_S is easy to compute in practice. Simply replace the Xs and Ys by their ranks and use any routine which returns a correlation coefficient. For the data in Example 10.8.1, we used the package minitab to first rank the data with the command `rank` and then used the command `corr` to obtain the correlation coefficient. With R and S-PLUS, r_s is computed by the command `cor(rank(x),rank(y))`, provided the vectors x and y contain the observations.

Example 10.8.2 (Example 10.8.1, Continued). For the data in Example 10.8.1, the value of r_s is 0.905. Therefore, the value of the asymptotic test statistic is $Z_S = 0.905\sqrt{19} = 3.94$. The p-value for a two-sided test is 0.00008; hence, there is strong evidence to reject H_0. ■

If the samples have a strictly increasing monotone relationship, then it is easy to see that $r_S = 1$; while if they have a strictly decreasing monotone relationship then $r_S = -1$. Like Kendall's K statistic, r_S is an estimate of a population parameter, but, except for when X and Y are independent, it is a more complicated expression than τ. It can be shown (see Kendall, 1962) that

$$E(r_S) = \frac{3}{n+1}[\tau + (n-2)(2\gamma - 1)], \quad (10.8.15)$$

where $\gamma = P[(X_2 - X_1)(Y_3 - Y_1) > 0]$. For large n, $E(r_S) \doteq 6(\gamma - 1/2)$, which is a harder parameter to interpret than the measure of concordance τ.

Spearman's rho is based on Wilcoxon scores and hence, can easily be extended to other rank score functions. Some of these measures are discussed in the exercises.

EXERCISES

10.8.1. Show that Kendall's τ satisfies the inequality $-1 \leq \tau \leq 1$.

10.8.2. Consider Example 10.8.1. Let Y = Winning times of the 1500 m race for a particular year and let X = Winning times of the marathon for that year. Obtain a scatterplot of Y versus X and determine the outlying point.

10.8.3. Consider the last exercise as a regression problem. Suppose we are interested in predicting the 1500 m winning time based on the marathon winning time. Assume a simple linear model and obtain the least squares and Wilcoxon (Section 10.7) fits of the data. Overlay the fits on the scatterplot obtained in Exercise 10.8.2. Comment on the fits. What does the slope parameter mean in this problem?

10.8.4. With regards to Exercise 10.8.3, a more interesting predicting problem is the prediction of winning time of either race based on year.

(a) Scatterplot the winning 1500 m race times versus year. Assume a simple linear model (does the assumption make sense) and obtain the least squares and Wilcoxon (Section 10.7) fits of the data. Overlay the fits on the scatterplot. Comment on the fits. What does the slope parameter mean in this problem? Predict the winning time for 1984. How close was your prediction to the true winning time?

(b) Same as Part (a) with the winning times of the marathon for that year.

10.8.5. Spearman's rho is a rank correlation coefficient based on Wilcoxon scores. In this exercise we consider a rank correlation coefficient based on a general score function. Let $(X_1, Y_1), (X_2, Y_2), \ldots, (X_n, Y_n)$ be a random sample from a bivariate continuous cdf $F(x, y)$. Let $a(i) = \varphi(i/(n+1))$ where $\sum_{i=1}^n a(i) = 0$. In particular, $\bar{a} = 0$. As in expression (10.5.6), let $s_a^2 = \sum_{i=1}^n a^2(i)$. Consider the rank correlation coefficient,

$$r_a = \frac{1}{s_a^2} \sum_{i=1}^n a(R(X_i))a(R(Y_i)). \qquad (10.8.16)$$

(a) Show that r_a is a correlation coefficient on the sample

$$\{(a[R(X_1)], a[R(Y_1)]), (a[R(X_2)], a[R(Y_2)]), \ldots, (a[R(X_n)], a[R(Y_n)])\}.$$

(b) For the score function $\varphi(u) = \sqrt{12}(u - (1/2))$, show that $r_a = r_S$, Spearman's rho.

(c) Obtain r_a for the sign score function $\varphi(u) = \text{sgn}(u - (1/2))$. Call this rank correlation coefficient r_{qc}, (the subscript qc will be obvious from Exercise 10.8.7).

10.8.6. Consider the general score rank correlation coefficient r_a defined in Exercise 10.8.5. Consider the null hypothesis $H_0 : X$ and Y are independent.

(a) Show that $E_{H_0}(r_a) = 0$.

(b) Based on Part (a) and H_0, as a first step in obtaining the null variance show that the following expression is true:

$$\text{Var}_{H_0}(r_a) = \frac{1}{s_a^4} \sum_{i=1}^n \sum_{j=1}^n E_{H_0}[a(R(X_i))a(R(X_j))]E_{H_0}[a(R(Y_i))a(R(Y_j))].$$

(c) To determine the expectation in the last expression, consider the two cases: $i = j$ and $i \neq j$. Then using uniformity of the distribution of the ranks, show that

$$\text{Var}_{H_0}(r_a) = \frac{1}{s_a^4} \frac{1}{n-1} s_a^4 = \frac{1}{n-1}. \qquad (10.8.17)$$

10.8.7. Consider the rank correlation coefficient given by r_{qc} in Part (c) of Exercise 10.8.5. Let Q_{2X} and Q_{2Y} denote the medians of the samples $X_1, \ldots, X_n$ and $Y_1, \ldots, Y_n$, respectively. Now consider the four quadrants:

$$
\begin{aligned}
I &= \{(x, y) : x > Q_{2X}, y > Q_{2Y}\}, \\
II &= \{(x, y) : x < Q_{2X}, y > Q_{2Y}\}, \\
III &= \{(x, y) : x < Q_{2X}, y < Q_{2Y}\}, \\
IV &= \{(x, y) : x > Q_{2X}, y < Q_{2Y}\}.
\end{aligned}
$$

Show essentially that

$$
r_{qc} = \frac{1}{n}\{\#(X_i, Y_i) \in I + \#(X_i, Y_i) \in III - \#(X_i, Y_i) \in II - \#(X_i, Y_i) \in IV\}.
$$
$$(10.8.18)$$

Hence, r_{qc} is referred to as the *quadrant count* correlation coefficient.

10.8.8. Set up the asymptotic test of independence for r_{qc} of the last exercise. Then use it to test for independence between the 1500 m race times and the marathon race times of the data in Example 10.8.1.

10.8.9. Obtain the rank correlation coefficient when normal scores are used; that is, the scores are $a(i) = \Phi^{-1}(i/(n+1))$, $i = 1, \ldots n$. Call it r_N. Set up the asymptotic test of independence for r_N of the last exercise. Then use it to test for independence between the 1500 m race times and the marathon race times of the data in Example 10.8.1.

10.8.10. Suppose that the hypothesis H_0 concerns the independence of two random variables X and Y. That is, we wish to test $H_0 : F(x, y) = F_1(x)F_2(y)$, where F, F_1, and F_3 are the respective joint and marginal distribution functions of the continuous type, against all alternatives. Let $(X_1, Y_1), (X_2, Y_2), \ldots, (X_n, Y_n)$ be a random sample from the joint distribution. Under H_0, the order statistics of $X_1, X_2, \ldots, X_n$ and the order statistics of $Y_1, Y_2, \ldots, Y_n$ are, respectively, complete sufficient statistics for F_1 and F_2, under H_0. Use r_S, r_{qc}, and r_N it to create an adaptive distribution-free test of H_0.

Remark 10.8.1. It is interesting to note that in an adaptive procedure it would be possible to use different score functions for the Xs and Ys. That is, the order statistics of the X values might suggest one score function and those of the Ys another score function. Under the null hypothesis of independence, the resulting procedure would produce an α level test. ∎

Chapter 11

Bayesian Statistics

11.1 Subjective Probability

Subjective probability is the foundation of Bayesian methods. So in this section we briefly discuss it. Suppose a person has assigned $P(C) = \frac{2}{5}$ to some event C. Then the *odds* against C would be

$$O(C) = \frac{1 - P(C)}{P(C)} = \frac{1 - \frac{2}{5}}{\frac{2}{5}} = \frac{3}{2}.$$

Moreover, if that person is willing to bet, he or she is willing to accept either side of the bet: (1) win 3 units if C occurs and lose 2 if it does not occur or (2) win 2 units if C does not occur and lose 3 if it does. If that is not the case, then that person should review his or her subjective probability of event C.

This is really much like two children dividing a candy bar as equal as possible: One divides it and the other gets to choose which of the two parts seems most desirable; that is, the larger. Accordingly, the child dividing the candy bar tries extremely hard to cut it as equal as possible. Clearly, this is exactly what the person selecting the subjective probability does as he or she must be willing to take either side of the bet with the odds established.

Let us now say the reader is willing to accept that the subjective probability $P(C)$ as the fair price for event C, given that you will win one unit in case C occurs and, of course, lose $P(C)$ if it does not occur. Then it turns out, all rules (definitions and theorems) on probability found in Chapter 1 follow for subjective probabilities. We do not give proofs of them all but only two of them, and some of the others are left as exercises. These proofs were given to us in a personal communication from George Woodworth, University of Iowa.

Theorem 11.1.1. *If C_1 and C_2 are mutually exclusive, then*

$$P(C_1 \cup C_2) = P(C_1) + P(C_2).$$

Proof: Suppose a person thinks a fair price for C_1 is $p_1 = P(C_1)$ and that for C_2 is $p_2 = P(C_2)$. However, that person believes the fair price for $C_1 \cup C_2$ is p_3 which differs from $p_1 + p_2$. Say, $p_3 < p_1 + p_2$ and let the difference be $d = (p_1 + p_2) - p_3$. A gambler offers this person the price $p_3 + \frac{d}{4}$ for $C_1 \cup C_2$. That person takes the offer because it is better than p_3. The gambler sells C_1 at a discount price of $p_1 - \frac{d}{4}$ and sells C_2 at a discount price of $p_2 - \frac{d}{4}$ to that person. Being a rational person with those given prices of p_1, p_2, and p_3, all three of these deals seem very satisfactory. However, that person received $p_3 + \frac{d}{4}$ and paid $p_1 + p_2 - \frac{d}{2}$. Thus before any bets are paid off, that person has

$$p_3 + \frac{d}{4} - \left(p_1 + p_2 - \frac{d}{2} \right) = p_3 - p_1 - p_2 + \frac{3d}{4} = -\frac{d}{4}.$$

That is, the person is down $\frac{d}{4}$ before any bets are settled.

- Suppose C_1 happens: the gambler has $C_1 \cup C_2$ and the person has C_1; so they exchange units and the person is still down $\frac{d}{4}$ The same thing occurs if C_2 happens.

- Suppose neither C_1 or C_2 happens, then the gambler and that person receive zero, and the person is still down $\frac{d}{4}$.

- Of course, C_1 and C_2 can not occur together since they are mutually exclusive.

Thus we see that it is bad for that person to assign

$$p_3 = P(C_1 \cup C_2) < p_1 + p_2 = P(C_1) + P(C_2),$$

because the gambler can put that person in a position to lose $(p_1 + p_2 - p_3)/4$ no matter what happens. This is sometimes referred to as a **Dutch book**.

The argument when $p_3 > p_1 + p_2$ is similar and can also lead to a Dutch book; it is left as an exercise. Thus p_3 must equal $p_1 + p_2$ to avoid a Dutch book; that is, $P(C_1 \cup C_2) = P(C_1) + P(C_2)$. ■

Let us prove another one.

Theorem 11.1.2. *If $C_1 \subset C_2$, then $P(C_1) \leq P(C_2)$.*

Proof: Say the person believes the fair prices are such that $p_1 = P(C_1) > p_2 = P(C_2)$. Then if $d = p_1 - p_2$, the gambler sells C_1 to that person for $p_1 - \frac{d}{4}$ and buys C_2 from that person for $p_2 + \frac{d}{4}$. If the person truly believes $p_1 > p_2$, both of these are good deals. Yet before any bets are settled, that person has

$$p_2 + \frac{d}{4} - \left(p_1 - \frac{d}{4} \right) = p_2 - p_1 + \frac{d}{2} = -d + \frac{d}{2} = -\frac{d}{2};$$

that is, that person is down $\frac{d}{2}$.

- If C_1 is true, then C_2 is true, and both receive one unit from each other and that person is still down $\frac{d}{2}$.

- If C_2 happens, but C_1 does not, then the gambler receives one unit from that person and the latter is down $1 + \frac{d}{2}$.

- If neither C_1 nor C_2 happens, neither the gambler nor the person receives anything and the person is still down $\frac{d}{2}$.

The person loses no matter what happens; that is, we have a Dutch book when $p_1 > p_2$. So $p_1 > p_2$ is unfavorable, and thus it must be that the fair prices are $p_1 \leq p_2$. ∎

In the exercises, we give hints how to show that:

$P(\mathcal{C}) = 1$ (Exercise 11.1.3),

$P(C^c) = 1 - P(C)$ (Exercise 11.1.4),

If $C_1 \subset C_2$ and $C_2 \subset C_1$ (that is, $C_1 \equiv C_2$), then $P(C_1) = P(C_2)$ (Exercise 11.1.5),

If C_1, C_2, and C_3 are mutually exclusive, then $P(C_1 \cup C_2 \cup C_3) = PC_1) + P(C_2) + p(C_3)$ (Exercise 11.1.6),

$P(C_1 \cup C_2) = P(C_1) + P(C_2) - P(C_1 \cap C_2)$ (Exercise 11.1.7).

The Bayesian continues to consider subjective conditional probabilities, such as $P(C_1|C_2)$, which is the fair price of C_1 only if C_2 is true. If C_2 is not true, the bet is off. Of course, $P(C_1|C_2)$ could differ from $P(C_1)$. To illustrate, say C_2 is the event that "it will rain today" and C_1 is the event that "a certain person who will be outside on that day will catch a cold." Most of us would probably assign the fair prices so that

$$P(C_1) < P(C_1|C_2).$$

Consequently, a person has a better chance of getting a cold on a rainy day.

The Bayesian can go on to argue that

$$P(C_1 \cap C_2) = P(C_2)P(C_1|C_2),$$

recalling that the bet $P(C_1|C_2)$ is called off if C_2 does not happen by creating a Dutch book situation. However, we will not consider that argument here, and simply state that all the rules of subjective probabilities are the same as those of Chapter 1 using this subjective approach to probability.

EXERCISES

11.1.1. The following amounts are bet on horses A, B, C, D, E to win.

Horse	Amount
A	$600,000
B	$200,000
C	$100,000
D	$75,000
E	$25,000
Total	$1,000,000

Suppose the track wants to take 20% off the top, namely $200,000. Determine the payoff for winning with a two dollar bet on each of the five horses. (In this exercise, we do not concern ourselves with "place" and "show.")
Hint: Figure out what would be a fair payoff so that the track does not take any money, (that is, the track's take is zero), and then compute 80% of those payoffs.

11.1.2. In the proof of 11.1.1, we considered the case in which $p_3 < p_1 + p_2$. Now, say the person believes that $p_3 > p_1 + p_2$ and create a Dutch book for him.
Hint: Let $d = p_3 - (p_1 + p_2)$. The gambler buys from the person C_1 at a premium price of $p_1 + d/4$ and C_2 for $p_2 + d/4$. Then the gambler sells $C_1 \cup C_2$ to that person at a discount of $p_3 - d/4$. All those are good deals for that person who believes that p_1, p_2, p_3 are correct with $p_3 > p_1 + p_2$. Show that the person has a Dutch book.

11.1.3. Show that $P(\mathcal{C}) = 1$.
Hint: Suppose a person thinks $P(\mathcal{C}) = p \neq 1$. Consider two cases: $p > 1$ and $p < 1$. In the first case, say $d = p - 1$ and the gambler sells the person $\mathcal{C}$ at a discount price $1 + d/2$. Of course, Ω happens and the gambler pays the person one unit, but he is down $1 + d/2 - 1 = d/2$; therefore, he has a Dutch book. Proceed with the other case.

11.1.4. Show that $P(C^c) = 1 - P(C)$.
Hint: $C^c \cup C = \mathcal{C}$ and use result (11.1.1) and Exercise 11.1.3.

11.1.5. Show that if $C_1 \subset C_2$ and $C_2 \subset C_1$ (that is, $C_1 \equiv C_2$), then $P(C_1) = P(C_2)$.
Hint: Use result (11.1.2) twice.

11.1.6. Show that if C_1, C_2, and C_3 are mutually exclusive, then $P(C_1 \cup C_2 \cup C_3) = P(C_1) + P(C_2) + P(C_3)$.
Hint: Write $C_1 \cup C_2 \cup C_3 = C_1 \cup (C_2 \cup C_3)$ and use result (11.1.1) twice.

11.1.7. Show that $P(C_1 \cup C_2) = P(C_1) + P(C_2) - P(C_1 \cap C_2)$.
Hint: $C_1 \cup C_2 \equiv C_1 \cup (C_1^c \cap C_2)$ and $C_2 \equiv (C_1 \cap C_2) \cup (C_1^c \cap C_2)$. Use result (11.1.1) twice and a little algebra.

11.2   Bayesian Procedures

To understand the Bayesian inference, let us review Bayes Theorem, (1.4.1), in a situation in which we are trying to determine something about a parameter of a distribution. Suppose we have a Poisson distribution with parameter $\theta > 0$, and we know that the parameter is either equal to $\theta = 2$ or $\theta = 3$. In Bayesian inference, the parameter is treated as a random variable Θ. Suppose for this example, we assign subjective **prior** probabilities of $P(\Theta = 2) = \frac{1}{3}$ and $P(\Theta = 3) = \frac{2}{3}$ to the two possible values. These subjective probabilities are based upon past experiences, and it might be unrealistic that Θ can only take one of two values, instead of a continuous $\theta > 0$ (we address this immediately after this introductory illustration). Now suppose a random sample of size $n = 2$ results in the observations $x_1 = 2$,

$x_2 = 4$. Given these data, what is the **posterior** probabilites of $\Theta = 2$ and $\Theta = 3$? By Bayes Theorem, we have

$$P(\Theta = 2 | X_1 = 2, X_2 = 4) = \frac{P(\Theta = 2 \text{ and } X_1 = 2, X_2 = 4)}{P(X_1 = 2, X_2 = 4)}$$

$$= \frac{\left(\frac{1}{3}\right)\frac{e^{-2}2^2}{2!}\frac{e^{-2}2^4}{4!}}{\left(\frac{1}{3}\right)\frac{e^{-2}2^2}{2!}\frac{e^{-2}2^4}{4!} + \left(\frac{2}{3}\right)\frac{e^{-3}3^2}{2!}\frac{e^{-3}3^4}{4!}}$$

$$= 0.245.$$

Similarly,

$$P(\Theta = 3 | X_1 = 2, X_2 = 4) = 1 - 0.245 = 0.755.$$

That is, with the observations $x_1 = 2, x_2 = 4$, the posterior probability of $\Theta = 2$ was smaller than the prior probability of $\Theta = 2$. Similarly, the posterior probability of $\Theta = 3$ was greater than the corresponding prior. That is, the observations $x_1 = 2, x_2 = 4$ seemed to favor $\Theta = 3$ more than $\Theta = 2$; and that seems to agree with our intuition as $\bar{x} = 3$. Now let us address in general a more realistic situation in which we place a prior pdf $h(\theta)$ on a support which is a continuium.

11.2.1 Prior and Posterior Distributions

We shall now describe the Bayesian approach to the problem of estimation. This approach takes into account any prior knowledge of the experiment that the statistician has and it is one application of a principle of statistical inference that may be called **Bayesian statistics**. Consider a random variable X that has a distribution of probability that depends upon the symbol θ, where θ is an element of a well-defined set Ω. For example, if the symbol θ is the mean of a normal distribution, Ω may be the real line. We have previously looked upon θ as being a parameter, albeit an unknown parameter. Let us now introduce a random variable Θ that has a distribution of probability over the set Ω; and just as we look upon x as a possible value of the random variable X, we now look upon θ as a possible value of the random variable Θ. Thus the distribution of X depends upon θ, an experimental value of the random variable Θ. We shall denote the pdf of Θ by $h(\theta)$ and we take $h(\theta) = 0$ when θ is not an element of Ω. The pdf $h(\theta)$ is called the **prior** pdf of Θ. Moreover, we now denote the pdf of X by $f(x|\theta)$ since we think of it as a conditional pdf of X, given $\Theta = \theta$. For clarity in this chapter, we will use the following summary of this model:

$$\begin{aligned} X|\theta &\sim f(x|\theta) \\ \Theta &\sim h(\theta). \end{aligned} \tag{11.2.1}$$

Suppose that $X_1, X_2, \ldots, X_n$ is a random sample from the conditional distribution of X given $\Theta = \theta$ with pdf $f(x|\theta)$. Vector notation will be convenient in this chapter. Let $\mathbf{X}' = (X_1, X_2, \ldots, X_n)$ and $\mathbf{x}' = (x_1, x_2, \ldots, x_n)$. Thus we can write the joint conditional pdf of $\mathbf{X}$, given $\Theta = \theta$, as

$$L(\mathbf{x}\,|\,\theta) = f(x_1|\theta)f(x_2|\theta)\cdots f(x_n|\theta). \tag{11.2.2}$$

Thus the joint pdf of $\mathbf{X}$ and Θ is

$$g(\mathbf{x}, \theta) = L(\mathbf{x}\,|\,\theta)h(\theta). \tag{11.2.3}$$

If Θ is a random variable of the continuous type, the joint marginal pdf of $\mathbf{X}$ is given by

$$g_1(\mathbf{x}) = \int_{-\infty}^{\infty} g(\mathbf{x}, \theta)\, d\theta. \tag{11.2.4}$$

If Θ is a random variable of the discrete type, integration would be replaced by summation. In either case the conditional pdf of Θ, given the sample $\mathbf{X}$, is

$$k(\theta|\mathbf{x}) = \frac{g(\mathbf{x}, \theta)}{g_1(\mathbf{x})} = \frac{L(\mathbf{x}\,|\,\theta)h(\theta)}{g_1(\mathbf{x})}. \tag{11.2.5}$$

The distribution defined by this conditional pdf is called the **posterior distribution** and (11.2.5) is called the **posterior pdf**. The prior distribution reflects the subjective belief of Θ before the sample is drawn while the posterior distribution is the conditional distribution of Θ after the sample is drawn. Further discussion on these distributions follows an illustrative example.

Example 11.2.1. Consider the model

$$X_i|\theta \quad \sim \quad \text{iid Poisson}(\theta)$$
$$\Theta \quad \sim \quad \Gamma(\alpha, \beta), \alpha \text{ and } \beta \text{ are known.}$$

Hence, the random sample is drawn from a Poisson distribution with mean θ and the prior distribution is a $\Gamma(\alpha, \beta)$ distribution. Let $\mathbf{X}' = (X_1, X_2, \ldots, X_n)$. Thus in this case, the joint conditional pdf of $\mathbf{X}$, given $\Theta = \theta$, (11.2.2), is

$$L(\mathbf{x}\,|\,\theta) = \frac{\theta^{x_1}e^{-\theta}}{x_1!} \cdots \frac{\theta^{x_n}e^{-\theta}}{x_n!}, \quad x_i = 0, 1, 2, \ldots, i = 1, 2, \ldots, n,$$

and the prior pdf is

$$h(\theta) = \frac{\theta^{\alpha-1}e^{-\theta/\beta}}{\Gamma(\alpha)\beta^{\alpha}}, \quad 0 < \theta < \infty.$$

Hence, the joint mixed continuous discrete pdf is given by

$$g(\mathbf{x}, \theta) = L(\mathbf{x}\,|\,\theta)h(\theta) = \left[\frac{\theta^{x_1}e^{-\theta}}{x_1!} \cdots \frac{\theta^{x_n}e^{-\theta}}{x_n!}\right]\left[\frac{\theta^{\alpha-1}e^{-\theta/\beta}}{\Gamma(\alpha)\beta^{\alpha}}\right],$$

provided that $x_i = 0, 1, 2, 3, \ldots, i = 1, 2, \ldots, n$ and $0 < \theta < \infty$, and is equal to zero elsewhere. Then the marginal distribution of the sample, (11.2.4), is

$$g_1(\mathbf{x}) = \int_0^{\infty} \frac{\theta^{\sum x_i + \alpha - 1}e^{-(n+1/\beta)\theta}}{x_1!\cdots x_n!\Gamma(\alpha)\beta^{\alpha}}\, d\theta = \frac{\Gamma\left(\sum_1^n x_i + \alpha\right)}{x_1!\cdots x_n!\Gamma(\alpha)\beta^{\alpha}(n+1/\beta)^{\sum x_i + \alpha}}. \tag{11.2.6}$$

Finally, the posterior pdf of Θ, given $\mathbf{X} = \mathbf{x}$, (11.2.5), is

$$k(\theta|\mathbf{x}) = \frac{L(\mathbf{x}\,|\,\theta)h(\theta)}{g_1(\mathbf{x})} = \frac{\theta^{\sum x_i + \alpha - 1} e^{-\theta/[\beta/(n\beta+1)]}}{\Gamma\left(\sum x_i + \alpha\right)[\beta/(n\beta+1)]^{\sum x_i + \alpha}}, \tag{11.2.7}$$

provided that $0 < \theta < \infty$, and is equal to zero elsewhere. This conditional pdf is one of the gamma type with parameters $\alpha^* = \sum_{i=1}^{n} x_i + \alpha$ and $\beta^* = \beta/(n\beta+1)$. Notice that the posterior pdf reflects both prior information (α, β) and sample information $(\sum_{i=1}^{n} x_i)$. ■

In Example 11.2.1 notice that it is not really necessary to determine the marginal pdf $g_1(\mathbf{x})$ to find the posterior pdf $k(\theta|\mathbf{x})$. If we divide $L(\mathbf{x}\,|\,\theta)h(\theta)$ by $g_1(\mathbf{x})$, we must get the product of a factor, which depends upon $\mathbf{x}$ but does *not* depend upon θ, say $c(\mathbf{x})$, and

$$\theta^{\sum x_i + \alpha - 1} e^{-\theta/[\beta/(n\beta+)]}.$$

That is,

$$k(\theta|\mathbf{x}) = c(\mathbf{x})\theta^{\sum x_i + \alpha - 1} e^{-\theta/[\beta/(n\beta+)]},$$

provided that $0 < \theta < \infty$ and $x_i = 0, 1, 2, \ldots,\ i = 1, 2, \ldots, n$. However, $c(\mathbf{x})$ must be that "constant" needed to make $k(\theta|\mathbf{x})$ a pdf, namely

$$c(\mathbf{x}) = \frac{1}{\Gamma\left(\sum x_i + \alpha\right)[\beta/(n\beta+1)]^{\sum x_i + \alpha}}.$$

Accordingly, we frequently write that $k(\theta|\mathbf{x})$ is proportional to $L(\mathbf{x}\,|\,\theta)h(\theta)$; that is, the posterior pdf can be written as

$$k(\theta|\mathbf{x}) \propto L(\mathbf{x}\,|\,\theta)h(\theta). \tag{11.2.8}$$

Note that in the right-hand member of this expression all factors involving constants and $\mathbf{x}$ alone (not θ) can be dropped. For illustration, in solving the problem presented in Example 11.2.1, we simply write

$$k(\theta|\mathbf{x}) \propto \theta^{\sum x_i} e^{-n\theta} \theta^{\alpha-1} e^{-\theta/\beta}$$

or, equivalently,

$$k(\theta|\mathbf{x}) \propto \theta^{\sum x_i + \alpha - 1} e^{-\theta/[\beta/(n\beta+1)]},$$

$0 < \theta < \infty$ and is equal to zero elsewhere. Clearly, $k(\theta|\mathbf{x})$ must be gamma pdf with parameters $\alpha^* = \sum x_i + \alpha$ and $\beta^* = \beta/(n\beta+1)$.

There is another observation that can be made at this point. Suppose that there exists a sufficient statistic $Y = u(\mathbf{X})$ for the parameter so that

$$L(\mathbf{x}\,|\,\theta) = g[u(\mathbf{x})|\theta]H(\mathbf{x}),$$

where now $g(y|\theta)$ is the pdf of Y, given $\Theta = \theta$. Then we note that

$$k(\theta|\mathbf{x}) \propto g[u(\mathbf{x})|\theta]h(\theta);$$

because the factor $H(\mathbf{x})$ that does not depend upon θ can be dropped. Thus if a sufficient statistic Y for the parameter exists, we can begin with the pdf of Y if we wish and write

$$k(\theta|y) \propto g(y|\theta)h(\theta), \tag{11.2.9}$$

where now $k(\theta|y)$ is the conditional pdf of Θ given the sufficient statistic $Y = y$. In the case of a sufficient statistic Y, we will also use $g_1(y)$ to denote the marginal pdf of Y; that is, in the continuous case,

$$g_1(y) = \int_{-\infty}^{\infty} g(y|\theta)h(\theta)\,d\theta.$$

11.2.2   Bayesian Point Estimation

Suppose we want a point estimator of θ. From the Bayesian viewpoint, this really amounts to selecting a decision function δ, so that $\delta(\mathbf{x})$ is a predicted value of θ (an experimental value of the random variable Θ) when both the computed value $\mathbf{x}$ and the conditional pdf $k(\theta|\mathbf{x})$ are known. Now, in general, how would we predict an experimental value of any random variable, say W, if we want our prediction to be "reasonably close" to the value to be observed? Many statisticians would predict the mean, $E(W)$, of the distribution of W; others would predict a median (perhaps unique) of the distribution of W; and some would have other predictions. However, it seems desirable that the choice of the decision function should depend upon a loss function $\mathcal{L}[\theta, \delta(\mathbf{x})]$. One way in which this dependence upon the loss function can be reflected is to select the decision function δ in such a way that the conditional expectation of the loss is a minimum. A **Bayes' estimate** is a decision function δ that minimizes

$$E\{\mathcal{L}[\Theta, \delta(\mathbf{x})]|\mathbf{X} = \mathbf{x}\} = \int_{-\infty}^{\infty} \mathcal{L}[\theta, \delta(\mathbf{x})]k(\theta|\mathbf{x})\,d\theta,$$

if Θ is a random variable of the continuous type. That is,

$$\delta(\mathbf{x}) = \text{Argmin} \int_{-\infty}^{\infty} \mathcal{L}[\theta, \delta(\mathbf{x})]k(\theta|\mathbf{x})\,d\theta. \tag{11.2.10}$$

The associated random variable $\delta(\mathbf{X})$ is called a **Bayes' estimator** of θ. The usual modification of the right-hand member of this equation is made for random variables of the discrete type. If the loss function is given by $\mathcal{L}[\theta, \delta(\mathbf{x})] = [\theta - \delta(\mathbf{x})]^2$, then the Bayes' estimate is $\delta(\mathbf{x}) = E(\Theta|\mathbf{x})$, the mean of the conditional distribution of Θ, given $\mathbf{X} = \mathbf{x}$. This follows from the fact that $E[(W - b)^2]$, if it exists, is a minimum when $b = E(W)$. If the loss function is given by $\mathcal{L}[\theta, \delta(\mathbf{x})] = |\theta - \delta(\mathbf{x})|$, then a median of the conditional distribution of Θ, given $\mathbf{X} = \mathbf{x}$, is the Bayes' solution. This follows from the fact that $E(|W - b|)$, if it exists, is a minimum when b is equal to any median of the distribution of W.

    It is easy to generalize this to estimate a function of θ, for instance, $l(\theta)$ for a specified function $l(\theta)$. For the loss function $\mathcal{L}[\theta, \delta(\mathbf{x})]$, a **Bayes estimate** of $l(\theta)$

is a decision function δ that minimizes

$$E\{\mathcal{L}[l(\Theta), \delta(\mathbf{x})]|\mathbf{X} = \mathbf{x}\} = \int_{-\infty}^{\infty} \mathcal{L}[l(\theta), \delta(\mathbf{x})]k(\theta|\mathbf{x})\,d\theta.$$

The random variable $\delta(\mathbf{X})$ is called a **Bayes' estimator** of $l(\theta)$.

The conditional expectation of the loss, given $\mathbf{X} = \mathbf{x}$, defines a random variable that is a function of the sample $\mathbf{X}$. The expected value of that function of $\mathbf{X}$, in the notation of this section, is given by

$$\int_{-\infty}^{\infty} \left\{ \int_{-\infty}^{\infty} \mathcal{L}[\theta, \delta(\mathbf{x})]k(\theta|\mathbf{x})\,d\theta \right\} g_1(\mathbf{x})\,d\mathbf{x} = \int_{-\infty}^{\infty} \left\{ \int_{-\infty}^{\infty} \mathcal{L}[\theta, \delta(\mathbf{x})]L(\mathbf{x}|\theta)\,d\mathbf{x} \right\} h(\theta)d\theta,$$

in the continuous case. The integral within the braces in the latter expression is, for every given $\theta \in \Theta$, the **risk function** $R(\theta, \delta)$; accordingly, the latter expression is the mean value of the risk, or the expected risk. Because a Bayes' estimate $\delta(\mathbf{x})$ minimizes

$$\int_{-\infty}^{\infty} \mathcal{L}[\theta, \delta(\mathbf{x})]k(\theta|\mathbf{x})\,d\theta$$

for every $\mathbf{x}$ for which $g(\mathbf{x}) > 0$, it is evident that a Bayes' estimate $\delta(\mathbf{x})$ minimizes this mean value of the risk. We now give two illustrative examples.

Example 11.2.2. Consider the model

$$
\begin{aligned}
X_i|\theta &\sim \text{ iid binomial, } b(1, \theta) \\
\Theta &\sim \text{ beta}(\alpha, \beta), \ \alpha \text{ and } \beta \text{ are known,}
\end{aligned}
$$

that is, the prior pdf is

$$h(\theta) = \begin{cases} \frac{\Gamma(\alpha+\beta)}{\Gamma(\alpha)\Gamma(\beta)}\theta^{\alpha-1}(1-\theta)^{\beta-1} & 0 < \theta < 1 \\ 0 & \text{elsewhere.} \end{cases}$$

where α and β are assigned positive constants. We seek a decision function δ that is a Bayes' solution. The sufficient statistic is $Y = \sum_1^n X_i$, which has a $b(n, \theta)$ distribution. Thus the conditional pdf of Y given $\Theta = \theta$ is

$$g(y|\theta) = \begin{cases} \binom{n}{y}\theta^y(1-\theta)^{n-y} & y = 0, 1, \ldots, n \\ 0 & \text{elsewhere.} \end{cases}$$

Thus by (11.2.9), the conditional pdf of Θ, given $Y = y$ at points of positive probability density, is

$$k(\theta|y) \propto \theta^y(1-\theta)^{n-y}\theta^{\alpha-1}(1-\theta)^{\beta-1}, \quad 0 < \theta < 1.$$

That is,

$$k(\theta|y) = \frac{\Gamma(n+\alpha+\beta)}{\Gamma(\alpha+y)\Gamma(n+\beta-y)}\theta^{\alpha+y-1}(1-\theta)^{\beta+n-y-1}, \quad 0 < \theta < 1,$$

and $y = 0, 1, \ldots, n$. Hence, the posterior pdf is a beta density function with parameters $(\alpha + y, \beta + n - y)$. We take squared error loss, i.e., $\mathcal{L}[\theta, \delta(y)] = [\theta - \delta(y)]^2$, as the loss function. Then, the Bayesian point estimate of θ is the mean of this beta pdf which is

$$\delta(y) = \frac{\alpha + y}{\alpha + \beta + n}.$$

It is very instructive to note that this Bayes' estimator can be written as

$$\delta(y) = \left(\frac{n}{\alpha + \beta + n}\right)\frac{y}{n} + \left(\frac{\alpha + \beta}{\alpha + \beta + n}\right)\frac{\alpha}{\alpha + \beta}$$

which is a weighted average of the maximum likelihood estimate y/n of θ and the mean $\alpha/(\alpha + \beta)$ of the prior pdf of the parameter. Moreover, the respective weights are $n/(\alpha+\beta+n)$ and $(\alpha+\beta)/(\alpha+\beta+n)$. Note that for large n the Bayes' estimate is close to the maximum likelihood estimate of θ and that, furthermore, $\delta(Y)$ is a consistent estimator of θ. Thus we see that α and β should be selected so that not only is $\alpha/(\alpha + \beta)$ the desired prior mean, but the sum $\alpha + \beta$ indicates the worth of the prior opinion relative to a sample of size n. That is, if we want our prior opinion to have as much weight as a sample size of 20, we would take $\alpha + \beta = 20$. So if our prior mean is $\frac{3}{4}$, we have that α and β are selected so that $\alpha = 15$ and $\beta = 5$. ∎

Example 11.2.3. For this example, we have the normal model,

$$X_i|\theta \quad \sim \quad \text{iid } N(\theta, \sigma^2), \text{where } \sigma^2 \text{ is known}$$
$$\Theta \quad \sim \quad N(\theta_0, \sigma_0^2), \text{where } \theta_0 \text{ and } \sigma_0^2 \text{ are known.}$$

Then $Y = \overline{X}$ is a sufficient statistic. Hence, an equivalent formulation of the model is

$$Y|\theta \quad \sim \quad N(\theta, \sigma^2/n), \text{where } \sigma^2 \text{ is known}$$
$$\Theta \quad \sim \quad N(\theta_0, \sigma_0^2), \text{where } \theta_0 \text{ and } \sigma_0^2 \text{ are known.}$$

Then for the posterior pdf we have

$$k(\theta|y) \propto \frac{1}{\sqrt{2\pi}\sigma/\sqrt{n}} \frac{1}{\sqrt{2\pi}\sigma_0} \exp\left[-\frac{(y - \theta)^2}{2(\sigma^2/n)} - \frac{(\theta - \theta_0)^2}{2\sigma_0^2}\right].$$

If we eliminate all constant factors (including factors involving only y), we have

$$k(\theta|y) \propto \exp\left[-\frac{(\sigma_0^2 + \sigma^2/n)\theta^2 - 2(y\sigma_0^2 + \theta_0\sigma^2/n)\theta}{2(\sigma^2/n)\sigma_0^2}\right].$$

This can be simplified, by completing the square to read (after eliminating factors not involving θ)

$$k(\theta|y) \propto \exp\left[-\frac{\left(\theta - \dfrac{y\sigma_0^2 + \theta_0\sigma^2/n}{\sigma_0^2 + \sigma^2/n}\right)^2}{\dfrac{2(\sigma^2/n)\sigma_0^2}{(\sigma_0^2 + \sigma^2/n)}}\right].$$

That is, the posterior pdf of the parameter is obviously normal with mean

$$\frac{y\sigma_0^2 + \theta_0\sigma^2/n}{\sigma_0^2 + \sigma^2/n} = \left(\frac{\sigma_0^2}{\sigma_0^2 + \sigma^2/n}\right) y + \left(\frac{\sigma^2/n}{\sigma_0^2 + \sigma^2/n}\right) \theta_0 \qquad (11.2.11)$$

and variance $(\sigma^2/n)\sigma_0^2/(\sigma_0^2 + \sigma^2/n)$. If the square-error loss function is used, this posterior mean is the Bayes' estimator. Again, note that it is a weighted average of the maximum likelihood estimate $y = \overline{x}$ and the prior mean θ_0. As in the last example, for large n the Bayes' estimator is close to the maximum likelihood estimator and $\delta(Y)$ is a consistent estimator of θ. Thus the Bayesian procedures permit the decision maker to enter his or her prior opinions into the solution in a very formal way such that the influences of these prior notions will be less and less as n increases. ∎

In Bayesian statistics all the information is contained in the posterior pdf $k(\theta|y)$. In Examples 11.2.2 and 11.2.3 we found Bayesian point estimates using the square-error loss function. It should be noted that if $\mathcal{L}[\delta(y), \theta] = |\delta(y) - \theta|$, the absolute value of the error, then the Bayes' solution would be the median of the posterior distribution of the parameter, which is given by $k(\theta|y)$. Hence, the Bayes' estimator changes, *as it should*, with different loss functions.

11.2.3   Bayesian Interval Estimation

If an interval estimate of θ is desired, we can find two functions $u(\mathbf{x})$ and $v(\mathbf{x})$ so that the conditional probability

$$P[u(\mathbf{x}) < \Theta < v(\mathbf{x})|\mathbf{X} = \mathbf{x}] = \int_{u(\mathbf{x})}^{v(\mathbf{x})} k(\theta|\mathbf{x})\, d\theta$$

is large, for example, 0.95. Then the interval $u(\mathbf{x})$ to $v(\mathbf{x})$ is an interval estimate of θ in the sense that the conditional probability of Θ belonging to that interval is equal to 0.95. These intervals are often called **credible** or **probability intervals**, so as not to confuse them with confidence intervals.

Example 11.2.4. As an illustration, consider Example 11.2.3 where $X_1, X_2, \ldots, X_n$ is a random sample from a $N(\theta, \sigma^2)$ distribution, where σ^2 is known, and the prior distribution is a normal $N(\theta_0, \sigma_0^2)$ distribution. The statistic $Y = \overline{X}$ is sufficient. Recall that the posterior pdf of Θ given $Y = y$ was normal with mean and variance given near expression (11.2.11). Hence, a credible interval is found by taking the mean of the posterior distribution and adding and subtracting 1.96 of its standard deviation; that is, the interval

$$\frac{y\sigma_0^2 + \theta_0\sigma^2/n}{\sigma_0^2 + \sigma^2/n} \pm 1.96\sqrt{\frac{(\sigma^2/n)\sigma_0^2}{\sigma_0^2 + \sigma^2/n}}$$

forms a credible interval of probability 0.95 for θ. ∎

Example 11.2.5. Recall Example 11.2.1 where $\mathbf{X}' = (X_1, X_2, \ldots, X_n)$ is a random sample from a Poisson distribution with mean θ and a $\Gamma(\alpha, \beta)$ prior, with α and β known, is considered. As given by expression (11.2.7) the posterior pdf is a $\Gamma(y + \alpha, \beta/(n\beta + 1))$ pdf, where $y = \sum_{i=1}^{n} x_i$. Hence, if we use the squared error loss function, the Bayes' point estimate of θ is the mean of the posterior

$$\delta(y) = \frac{\beta(y + \alpha)}{n\beta + 1} = \frac{n\beta}{n\beta + 1} \frac{y}{n} + \frac{\alpha\beta}{n\beta + 1}.$$

As with the other Bayes' estimates we have discussed in this section, for large n this estimate is close to the maximum likelihood estimate and the statistic $\delta(Y)$ is a consistent estimate of θ. To obtain a credible interval, note that the posterior distribution of $\frac{2(n\beta+1)}{\beta}\Theta$ is $\chi^2(2(\sum_{i=1}^{n} x_i + \alpha))$. Based on this, the following interval is a $(1 - \alpha)100\%$ credible interval for θ:

$$\left(\frac{\beta}{2(n\beta + 1)} \chi^2_{1-(\alpha/2)}(2(\sum_{i=1}^{n} x_i + \alpha)), \frac{\beta}{2(n\beta + 1)} \chi^2_{\alpha/2}(2(\sum_{i=1}^{n} x_i + \alpha)) \right), \quad (11.2.12)$$

where $\chi^2_{1-(\alpha/2)}(2(\sum_{i=1}^{n} x_i + \alpha))$ and $\chi^2_{\alpha/2}(2(\sum_{i=1}^{n} x_i + \alpha))$ are the lower and upper χ^2 quantiles for a χ^2 distribution with $2(\sum_{i=1}^{n} x_i + \alpha)$ degrees of freedom. ∎

11.2.4 Bayesian Testing Procedures

As above, let X be a random variable with pdf (pmf) $f(x|\theta)$, $\theta \in \Omega$. Suppose we are interested in testing the hypotheses

$$H_0 : \theta \in \omega_0 \text{ versus } H_1 : \theta \in \omega_1,$$

where $\omega_0 \cup \omega_1 = \Omega$ and $\omega_0 \cap \omega_1 = \phi$. A simple Bayesian procedure to test these hypotheses proceeds as follows. Let $h(\theta)$ denote the prior distribution of the prior random variable Θ; let $\mathbf{X}' = (X_1, X_2, \ldots, X_n)$ denote a random sample on X; and denote the posterior pdf or pmf by $k(\theta|\mathbf{x})$. We use the posterior distribution to compute the following conditional probabilities

$$P(\Theta \in \omega_0|\mathbf{x}) \text{ and } P(\Theta \in \omega_1|\mathbf{x}).$$

In the Bayesian framework, these conditional probabilities represent the truth of H_0 and H_1, respectively. A simple rule is to

$$\text{Accept } H_0 \text{ if } P(\Theta \in \omega_0|\mathbf{x}) \geq P(\Theta \in \omega_1|\mathbf{x}),$$

otherwise, accept H_1; that is, accept the hypothesis which has the greater conditional probability. Note that, the condition $\omega_0 \cap \omega_1 = \phi$ is required but $\omega_0 \cup \omega_1 = \Omega$ is not necessary. More than two hypotheses may be tested at the same time, in which case a simple rule would be to accept the hypothesis with the greater conditional probability. We finish this subsection with a numerical example.

Example 11.2.6. Referring again to Example 11.2.1 where $\mathbf{X}' = (X_1, X_2, \ldots, X_n)$ is a random sample from a Poisson distribution with mean θ. Suppose we are interested in testing

$$H_0 : \theta \leq 10 \text{ versus } H_1 : \theta > 10. \qquad (11.2.13)$$

Suppose we think θ is about 12 but are not quite sure. Hence, we choose the $\Gamma(10, 1.2)$ pdf as our prior which is shown in the left panel of Figure 11.2.1. The mean of the prior is 12 but as the plot shows there is some variability, (the variance of the prior distribution is 14.4). The data for the problem are

$$\begin{array}{cccccccccc} 11 & 7 & 11 & 6 & 5 & 9 & 14 & 10 & 9 & 5 \\ 8 & 10 & 8 & 10 & 12 & 9 & 3 & 12 & 14 & 4 \end{array}$$

(these are the values of a random sample of size $n = 20$ taken from a Poisson distribution with mean 8; of course, in practice we would not know the mean is 8). The value of sufficient statistic is $y = \sum_{i=1}^{20} x_i = 177$. Hence, from Example 11.2.1, the posterior distribution is a $\Gamma\left(177 + 10, \frac{1.2}{20(1.2)+1}\right) = \Gamma(187, 0.048)$ distribution, which is shown in the right panel of Figure 11.2.1. Note that the data have moved the mean to the left of 12 to $187(0.048) = 8.976$, which is the Bayes estimate (under square error loss) of θ. Using a statistical computing package (we used the `pgamma` command in R), we compute the posterior probability of H_0 as

$$P[\Theta \leq 10 | y = 177] = P[\Gamma(187, 0.048) \leq 10] = 0.9368.$$

Thus $P[\Theta > 10 | y = 177] = 1 - 0.9368 = 0.0632$; consequently, our rule would accept H_0.

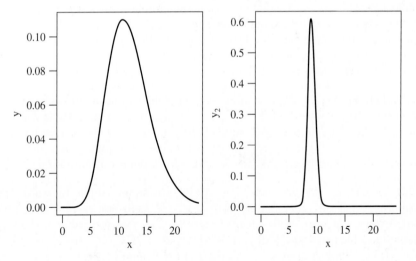

Figure 11.2.1: Prior (left panel) and posterior (right panel) pdfs of Example 11.2.6

The 95% credible interval, (11.2.12), is $(7.77, 10.31)$, which does contain 10; see Exercise 11.2.7 for details. ∎

11.2.5   Bayesian Sequential Procedures

Finally, we should observe what a Bayesian would do if additional data were collected beyond $x_1, x_2, \ldots, x_n$. In such a situation, the posterior distribution found with the observations $x_1, x_2, \ldots, x_n$ becomes the new prior distribution, and the additional observations give a new posterior distribution and inferences would be made from that second posterior. Of course, this can continue with even more observations. That is, the second posterior becomes the new prior, and the next set of observations give the next posterior from which the inferences can be made. Clearly, this gives the Bayesian an excellent way of handling sequential analysis. They can continue taking data, always updating the previous posterior, which has become a new prior distribution. Everything a Bayesian needs for inferences is in that final posterior distribution obtained by this sequential procedure.

EXERCISES

11.2.1. Let Y have a binomial distribution in which $n = 20$ and $p = \theta$. The prior probabilities on θ are $P(\theta = 0.3) = 2/3$ and $P(\theta = 0.5) = 1/3$. If $y = 9$, what are the posterior probabilities for $\theta = 0.3$ and $\theta = 0.5$?

11.2.2. Let $X_1, X_2, \ldots, X_n$ be a random sample from a distribution that is $b(1, \theta)$. Let the prior of Θ be a beta one with parameters α and β. Show that the posterior pdf $k(\theta | x_1, x_2, \ldots, x_n)$ is exactly the same as $k(\theta | y)$ given in Example 11.2.2.

11.2.3. Let $X_1, X_2, \ldots, X_n$ denote a random sample from a distribution that is $N(\theta, \sigma^2)$, where $-\infty < \theta < \infty$ and σ^2 is a given positive number. Let $Y = \overline{X}$ denote the mean of the random sample. Take the loss function to be $\mathcal{L}[\theta, \delta(y)] = |\theta - \delta(y)|$. If θ is an observed value of the random variable Θ, that is, $N(\mu, \tau^2)$, where $\tau^2 > 0$ and μ are known numbers, find the Bayes' solution $\delta(y)$ for a point estimate θ.

11.2.4. Let $X_1, X_2, \ldots, X_n$ denote a random sample from a Poisson distribution with mean θ, $0 < \theta < \infty$. Let $Y = \sum_1^n X_i$. Use the loss function to be $\mathcal{L}[\theta, \delta(y)] = [\theta - \delta(y)]^2$. Let θ be an observed value of the random variable Θ. If Θ has the pdf $h(\theta) = \theta^{\alpha-1} e^{-\theta/\beta} / \Gamma(\alpha)\beta^\alpha$, for $0 < \theta < \infty$, zero elsewhere, where $\alpha > 0$, $\beta > 0$ are known numbers, find the Bayes' solution $\delta(y)$ for a point estimate for θ.

11.2.5. Let Y_n be the nth order statistic of a random sample of size n from a distribution with pdf $f(x|\theta) = 1/\theta$, $0 < x < \theta$, zero elsewhere. Take the loss function to be $\mathcal{L}[\theta, \delta(y)] = [\theta - \delta(y_n)]^2$. Let θ be an observed value of the random variable Θ, which has pdf $h(\theta) = \beta\alpha^\beta/\theta^{\beta+1}$, $\alpha < \theta < \infty$, zero elsewhere, with $\alpha > 0$, $\beta > 0$. Find the Bayes' solution $\delta(y_n)$ for a point estimate of θ.

11.2.6. Let Y_1 and Y_2 be statistics that have trinomial distribution with parameters n, θ_1, and θ_2. Here θ_1 and θ_2 are observed values of the random variables Θ_1 and Θ_2, which have a Dirichlet distribution with known parameters α_1, α_2, and α_3; see expression (3.3.6). Show that the conditional distribution of Θ_1 and Θ_2 is Dirichlet and determine the conditional means $E(\Theta_1 | y_1, y_2)$ and $E(\Theta_2 | y_1, y_2)$.

11.2.7. For Example 11.2.6, obtain the 95% credible interval for θ. Next obtain the value of the mle for θ and the 95% confidence interval for θ discussed in Chapter 6.

11.2.8. In Example 11.2.2 let $n = 30$, $\alpha = 10$, and $\beta = 5$ so that $\delta(y) = (10+y)/45$ is the Bayes' estimate of θ.

(a) If Y has a binomial distribution $b(30, \theta)$, compute the risk $E\{[\theta - \delta(Y)]^2\}$.

(b) Find values of θ for which the risk of Part (a) is less than $\theta(1-\theta)/30$, the risk associated with the maximum likelihood estimator Y/n of θ.

11.2.9. Let Y_4 be the largest order statistic of a sample of size $n = 4$ from a distribution with uniform pdf $f(x; \theta) = 1/\theta$, $0 < x < \theta$, zero elsewhere. If the prior pdf of the parameter $g(\theta) = 2/\theta^3$, $1 < \theta < \infty$, zero elsewhere, find the Bayesian estimator $\delta(Y_4)$ of θ, based upon the sufficient statistic Y_4, using the loss function $|\delta(y_4) - \theta|$.

11.2.10. Refer to Example 11.2.3, suppose we select $\sigma_0^2 = d\sigma^2$, where σ^2 was known in that example. What value do we assign to d so that the variance of posterior is two-thirds the variance of $Y = \overline{X}$, namely σ^2/n?

11.3 More Bayesian Terminology and Ideas

Suppose $\mathbf{X}' = (X_1, X_2, \ldots, X_n)$ represents a random sample with likelihood $L(\mathbf{x}|\theta)$ and we assume a prior pdf $h(\theta)$. The joint marginal pdf of $\mathbf{X}$ is given by

$$g_1(\mathbf{x}) = \int_{-\infty}^{\infty} L(\mathbf{x}|\theta)h(\theta)d\theta.$$

This is often called the pdf of the **predictive distribution** of $\mathbf{X}$ because it provides the best description of the probabilities about $\mathbf{X}$ given the likelihood and the prior. An illustration of this is provided in expression (11.2.6) of Example 11.2.1. Again note that this predictive distribution is highly dependent on the probability models for X and Θ.

In this section, we will consider two classes of prior distributions. The first class is the class of conjugate priors defined by

Definition 11.3.1. *A class of prior pdfs for the family of distributions with pdfs $f(x|\theta)$, $\theta \in \Omega$ is said to define a **conjugate family of distributions** if the posterior pdf of the parameter is in the same family of distributions as the prior.*

As an illustration consider Example 11.2.5 where the pmf of X_i given θ was Poisson with mean θ. In this example, we selected a gamma prior and the resulting posterior distribution was of the gamma family also. Hence, the gamma pdf forms a conjugate class of priors for this Poisson model. This was true also for Example 11.2.2 where the conjugate family was beta and the model was a binomial and for Example 11.2.3 where both the model and the prior were normal.

To motivate our second class of priors, consider the binomial model, $b(1, \theta)$, presented in Example 11.2.2. Thomas Bayes (1763) took as a prior the beta distribution with $\alpha = \beta = 1$, namely $h(\theta) = 1, 0 < \theta < 1$, zero elsewhere, because he argued that he did not have much prior knowledge about θ. However, we note that this leads to the estimate of

$$\left(\frac{n}{n+2}\right)\left(\frac{y}{n}\right) + \left(\frac{2}{n+2}\right)\left(\frac{1}{2}\right).$$

We often call this a *shrinkage* estimate because the estimate $\frac{y}{n}$ is pulled a little towards the prior mean of $\frac{1}{2}$, although Bayes tried to avoid having the prior influence the inference.

Haldane (1948) did note, however, that if a prior beta pdf with $\alpha = \beta = 0$, then the shrinkage estimate would reduce to the mle y/n. Of course, a beta pdf with $\alpha = \beta = 0$ is not a pdf at all for it would be such that

$$h(\theta) \propto \frac{1}{\theta(1-\theta)}, \quad 0 < \theta < 1,$$

zero elsewhere, and

$$\int_0^1 \frac{c}{\theta(1-\theta)} d\theta$$

does not exist. However, such priors are used if when combined with the likelihood, we obtain a posterior pdf which is a proper pdf. By **proper**, we mean that it integrates to a positive constant. Here, in this example we obtain the posterior pdf of

$$f(\theta|y) \propto \theta^{y-1}(1-\theta)^{n-y-1},$$

which is proper provided $y > 1$ and $n - y > 1$. Of course, the posterior mean is y/n.

Definition 11.3.2. *Let* $\mathbf{X}' = (X_1, X_2, \ldots, X_n)$ *be a random sample from the distribution with pdf* $f(x|\theta)$. *A prior* $h(\theta) \geq 0$ *for this family is said to be* **improper** *if it is not a pdf but the function* $k(\theta|\mathbf{x}) \propto L(\mathbf{x}|\theta)h(\theta)$ *can be made proper.*

A **noninformative prior** is a prior which treats all values of θ the same, that is, uniformly. Continuous noninformative priors are often improper. As an example, suppose we have a normal distribution $N(\theta_1, \theta_2)$ in which both θ_1 and $\theta_2 > 0$ are unknown. A noninformative prior for θ_1 is $h_1(\theta_1) = 1, -\infty < \theta < \infty$. Clearly, this is not a pdf. An improper prior for θ_2 is $h_2(\theta_2) = c_2/\theta_2, 0 < \theta_2 < \infty$, zero elsewhere. Note that $\log \theta_2$ is uniformly distributed between $-\infty < \log \theta_2 < \infty$. Hence, in this way it is a noninformative prior. In addition, assume the parameters are independent. Then the joint prior, which is improper, is

$$h_1(\theta_1)h_2(\theta_2) \propto 1/\theta_2, \quad -\infty < \theta_1 < \infty, 0 < \theta_2 < \infty. \qquad (11.3.1)$$

Using this prior, we present the Bayes solution for θ_1 in the next example.

Example 11.3.1. Let $X_1, X_2, \ldots, X_n$ be a random sample from a $N(\theta_1, \theta_2)$ distribution. Recall that $\overline{X}$ and $S^2 = (n-1) \sum_{i=1}^{n} (X_i - \overline{X})^2$ are sufficient statistics. Suppose we use the improper prior given by (11.3.1). Then the posterior distribution is given by

$$k_{12}(\theta_1, \theta_2 | \overline{x}, s^2) \propto \left(\frac{1}{\theta_2} \right) \left(\frac{1}{\sqrt{2\pi\theta_2}} \right)^n \exp\left[-\frac{1}{2} \left\{ (n-1)s^2 + n(\overline{x} - \theta_1)^2 \right\} / \theta_2 \right]$$

$$\propto \left(\frac{1}{\theta_2} \right)^{\frac{n}{2}+1} \exp\left[-\frac{1}{2} \left\{ (n-1)s^2 + n(\overline{x} - \theta_1)^2 \right\} / \theta_2 \right].$$

To get the conditional pdf of θ_1, given $\overline{x}$, s^2, we integrate out θ_2

$$k_1(\theta_1 | \overline{x}, s^2) = \int_0^{\infty} k_{12}(\theta_1, \theta_2 | \overline{x}, s^2) d\theta_2.$$

To carry this out, let us change variables $z = 1/\theta_2$ and $\theta_2 = 1/z$, with Jacobian $-1/z^2$. Thus

$$k_1(\theta_1 | \overline{x}, s^2) \propto \int_0^{\infty} \frac{z^{\frac{n}{2}+1}}{z^2} \exp\left[-\left\{ \frac{(n-1)s^2 + n(\overline{x} - \theta_1)^2}{2} \right\} z \right] dz.$$

Referring to a gamma distribution with $\alpha = n/2$ and $\beta = 2/\{(n-1)s^2 + n(\overline{x} - \theta_1)^2\}$, this result is proportional to

$$k_1(\theta_1 | \overline{x}, s^2) \propto \{(n-1)s^2 + n(\overline{x} - \theta_1)^2\}^{-n/2}.$$

Let us change variables to get more familar results, namely let

$$t = \frac{\theta_1 - \overline{x}}{s/\sqrt{n}} \text{ and } \theta_1 = \overline{x} + ts/\sqrt{n}$$

with Jacobian $s/\sqrt{n}$. This conditional pdf of t, given $\overline{x}$ and s^2, is then

$$k(t | \overline{x}, s^2) \propto \{(n-1)s^2 + (st)^2\}^{-n/2}$$

$$\propto \frac{1}{[1 + t^2/(n-1)]^{[(n-1)+1]/2}}.$$

That is, the conditional pdf of $t = (\theta_1 - \overline{x})/(s/n)$, given $\overline{x}$ and s^2, is a Student t with $n - 1$ degrees of freedom. Since the mean of this pdf is 0 (assuming that $n > 2$), it follows that the Bayes estimator of θ_1 under squared error loss is $\overline{X}$, which is also the mle.

Of course, from $k_1(\theta_1 | \overline{x}, s^2)$ or $k(t | \overline{x}, s^2)$, we can find a credible interval for θ_1. One way of doing this is to select the *highest density region* (HDR) of the pdf θ_1 or that of t. The former is symmetric and unimodal about θ_1 and the latter about zero, but the latter's critical values are tabulated; so we use the HDR of that t-distribution. Thus if we want an interval having probability $1 - \alpha$, we take

$$-t_{\alpha/2} < \frac{\theta_1 - \overline{x}}{s/\sqrt{n}} < t_{\alpha/2}$$

or, equivalently,

$$\bar{x} - t_{\alpha/2}s/\sqrt{n} < \theta_1 < \bar{x} + t_{\alpha/2}s/\sqrt{n}.$$

This interval is the same as the confidence interval for θ_1; see Example 5.4.3. Hence, in this case, the improper prior (11.3.1) leads to the same inference as the traditional analysis. ∎

Usually in a Bayesian analysis, noninformative priors are not used if prior information exists. Let us consider the same situation as in Example 11.3.1, where the model was a $N(\theta_1, \theta_2)$ distribution. Suppose now we consider the **precision** $\theta_3 = 1/\theta_2$ instead of variance θ_2. The likelihood becomes

$$\left(\frac{\theta_3}{2\pi}\right)^{n/2} \exp\left[-\frac{1}{2}\left\{(n-1)s^2 + n(\bar{x}-\theta_1)^2\right\}\theta_3\right]$$

so that it is clear that a conjugate prior for θ_3 is $\Gamma(\alpha, \beta)$. Further, given θ_3, a reasonable prior on θ_1 is $N(\theta_0, \frac{1}{n_0\theta_3})$, where n_0 is selected in some way to reflect how many observations the prior is worth. Thus the joint prior of θ_1 and θ_3 is

$$h(\theta_1, \theta_3) \propto \theta_3^{\alpha-1}e^{-\theta_3/\beta}(n_0\theta_3)^{1/2}e^{-(\theta_1-\theta_0)^2\theta_3 n_0/2}.$$

If this is multiplied by the likelihood function, we obtain the posterior joint pdf of θ_1 and θ_3, namely

$$k(\theta_1, \theta_3|\bar{x}, s^2) \propto \theta_3^{\alpha+n/2+1/2-1} \exp\left[-\frac{1}{2}Q(\theta_1)\theta_3\right],$$

where

$$Q(\theta_1) = \frac{2}{\beta} + n_0(\theta_1 - \theta_0)^2 + [(n-1)s^2 + n(\bar{x}-\theta_1)^2]$$

$$= (n_0 + n_1)\left[\left(\theta_1 - \frac{n_0\theta_0 + n\bar{x}}{n_0 + n}\right)^2\right] + D,$$

with

$$D = \frac{2}{\beta} + (n-1)s^2 + (n_0^{-1} + n^{-1})^{-1}(\theta_0 - \bar{x})^2.$$

If we integrate out θ_3 we obtain

$$k_1(\theta_1|\bar{x}, s^2) \propto \int_0^\infty k(\theta_1, \theta_3|\bar{x}, s^2)d\theta_3$$

$$\propto \frac{1}{[Q(\theta_1)]^{[(n-1)+\alpha]/2}}.$$

To get this in a more familar form, change variables by letting

$$t = \frac{\theta_1 - \frac{n_0\theta_0 + n\bar{x}}{n_0 + n}}{\sqrt{D/[(n_0 + n)(n-1+\alpha-1)]}}$$

with Jacobian $\sqrt{D/[(n_0 + n)(n - 1 + \alpha - 1)]}$. Thus

$$k_2(t|\overline{x}, s^2) \propto \frac{1}{\left[1 + \frac{t^2}{n-1+\alpha-1}\right]^{(n-1+\alpha-1+1)/2}},$$

which is a Student t distribution with $n - 1 + \alpha - 1$ degrees of freedom. The Bayes estimate (under squared error loss) in this case is

$$\frac{n_0\theta_0 + n\overline{x}}{n_0 + n}.$$

It is interesting to note that if we define "new" sample characteristics as

$$n_k = n_0 + n,$$
$$\overline{x}_k = \frac{n_0\theta_0 + n\overline{x}}{n_0 + n},$$
$$s_k^2 = D/(n - 1 + \alpha - 1),$$

then

$$t = \frac{\theta_1 - \overline{x}_k}{s_k/\sqrt{n_k}}$$

has a t-distribution with $n - 1 + \alpha - 1$ degrees of freedom. Of course, using these degrees of freedom, we can find $t_{\gamma/2}$ so that

$$\overline{x}_k \pm t_{\gamma/2}s_k/\sqrt{n_k}$$

is an HDR credible interval estimate for θ_1 with probability $1 - \gamma$. Naturally, it falls upon the Bayesian to assign appropriate values to α, β, n_0, and θ_0. Small values of α and n_0 with a large value of β would create a prior so that this interval estimate would differ very little from the usual one.

Finally, it should be noted that when dealing with symmetric, unimodal posterior distributions it was extremely easy to find the HDR interval estimate. If, however, that posterior distribution is not symmetric, it is more difficult and often the Bayesian would find the interval that has equal probabilities on each tail.

EXERCISES

11.3.1. Let X_1, X_2 be a random sample from a Cauchy distribution with pdf

$$f(x; \theta_1, \theta_2) = \left(\frac{1}{\pi}\right) \frac{\theta_2}{\theta_2^2 + (x - \theta_1)^2}, \quad -\infty < x < \infty,$$

where $-\infty < \theta_1 < \infty$, $0 < \theta_2$. Use the noninformative prior $h(\theta_1, \theta_2) \propto 1$.

(a) Find the posterior pdf of θ_1, θ_2, other than the constant of proportionality.

(b) Evaluate this posterior pdf if $x_1 = 1, x_2 = 4$ for $\theta_1 = 1, 2, 3, 4$ and $\theta_2 = 0.5, 1.0, 1.5, 2.0$.

(c) From the 16 values in Part (b) where does the maximum of the posterior pdf seem to be?

(d) Do you know a computer program that can find the point (θ_1, θ_2) of maximum?

11.3.2. Let $X_1, X_2, \ldots, X_{10}$ be a random sample of size $n = 10$ from a gamma distribution with $\alpha = 3$ and $\beta = 1/\theta$. Suppose we believe that θ has a gamma distribution with $\alpha = 10$ and $\beta = 2$.

(a) Find the posterior distribution of θ.

(b) If the observed $\bar{x} = 18.2$, what is the Bayes point estimate associated with square error loss function.

(c) What is the Bayes point estimate using the mode of the posterior distribution?

(d) Comment on an HDR interval estimate for θ. Would it be easier to find one having equal tail probabilities? *Hint:* Can the posterior distribution be related to a chi-square distribution?

11.3.3. If the normal example of this section had a prior given by $\alpha = 4$, $\beta = 0.5$, $n_0 = 5$, $\theta_0 = 75$, and the observed sample of size $n = 50$ resulted in $\bar{x} = 77.02$, $s^2 = 8.2$.

(a) Find the Bayes point estimate of the mean θ_1

(b) Determine an HDR interval estimate with $1 - \gamma = 0.90$.

11.3.4. Let $f(x|\theta)$, $\theta \in \Omega$ be a pdf with Fisher information, (6.2.4), $I(\theta)$. Consider the Bayes model

$$X|\theta \quad \sim \quad f(x|\theta) , \quad \theta \in \Omega$$
$$\Theta \quad \sim \quad h(\theta) \propto \sqrt{I(\theta)}. \tag{11.3.2}$$

(a) Suppose we are interested in a parameter $\tau = u(\theta)$. Use the chain rule to prove that

$$\sqrt{I(\tau)} = \sqrt{I(\theta)} \left| \frac{\partial \theta}{\partial \tau} \right| . \tag{11.3.3}$$

(b) Show that for the Bayes model (11.3.2), that the prior pdf for τ is proportional to $\sqrt{I(\tau)}$.

The class of priors given by expression (11.3.2) is often called the class of **Jeffreys** priors; see Jeffreys (1961). This exercise shows that Jeffreys priors exhibit an invariance, in that the prior of a parameter τ which is a function of θ is also proportional to the square root of the information for τ.

11.3.5. Consider the Bayes model

$$X_i|\theta \,, i = 1, 2, \ldots n \quad \sim \quad \text{iid with distribution } \Gamma(1,\theta)\,, \theta > 0$$
$$\Theta \quad \sim \quad h(\theta) \propto \frac{1}{\theta}.$$

(a) Show that $h(\theta)$ is in the class of Jeffrys priors.

(b) Show that the posterior pdf is

$$h(\theta|y) \propto \left(\frac{1}{\theta}\right)^{n+2-1} e^{-y/\theta},$$

where $y = \sum_{i=1}^{n} x_i$.

(c) Show that if $\tau = \theta^{-1}$ then the posterior $k(\tau|y)$ is the pdf of a $\Gamma(n, 1/y)$ distribution.

(d) Determine the the posterior pdf of $2y\tau$. Use it to obtain a $(1-\alpha)100\%$ credible interval for θ.

(e) Use the posterior pdf in Part (d) to determine a Bayesian test for the hypotheses $H_0 : \theta \geq \theta_0$ versus $H_1 : \theta < \theta_0$, where θ_0 is specified.

11.3.6. Consider the Bayes model

$$X_i|\theta \,, i = 1, 2, \ldots n \quad \sim \quad \text{iid with distribution Poisson } (\theta)\,, \theta > 0$$
$$\Theta \quad \sim \quad h(\theta) \propto \theta^{-1/2}.$$

(a) Show that $h(\theta)$ is in the class of Jeffrys priors.

(b) Show that the posterior pdf of $2n\theta$ is the pdf of a $\chi^2(2y + 1)$ distribution, where $y = \sum_{i=1}^{n} x_i$.

(c) Use the posterior pdf of Part (b) to obtain a $(1 - \alpha)100\%$ credible interval for θ.

(d) Use the posterior pdf in Part (d) to determine a Bayesian test for the hypotheses $H_0 : \theta \geq \theta_0$ versus $H_1 : \theta < \theta_0$, where θ_0 is specified.

11.3.7. Consider the Bayes model

$$X_i|\theta \,, i = 1, 2, \ldots n \sim \text{iid with distribution } b(1,\theta), \ 0 < \theta < 1.$$

(a) Obtain Jeffrys prior for this model.

(b) Assume squared error loss and obtain the Bayes estimate of θ.

11.3.8. Consider the Bayes model

$$X_i|\theta \,, i = 1, 2, \ldots n \quad \sim \quad \text{iid with distribution } b(1,\theta)\,, 0 < \theta < 1$$
$$\Theta \quad \sim \quad h(\theta) = 1.$$

(a) Obtain the posterior pdf.

(b) Assume squared error loss and obtain the Bayes estimate of θ.

11.4   Gibbs Sampler

From the preceding sections, it is clear that integration techniques play a significant role in Bayesian inference. Hence, we now touch on some of the Monte Carlo techniques used for integration in Bayesian inference.

The Monte Carlo techniques discussed in Chapter 5 can often be used to obtain Bayesian estimates. For example, suppose a random sample is drawn from a $N(\theta, \sigma^2)$, where σ^2 is known. Then $Y = \overline{X}$ is a sufficient statistic. Consider the Bayes model

$$
\begin{aligned}
Y|\theta &\sim N(\theta, \sigma^2/n) \\
\Theta &\sim h(\theta) \propto \exp\{-(\theta - a)/b\}/(1 + \exp\{-[(\theta - a)/b]^2\}), -\infty < \theta < \infty, \\
&\quad a \text{ and } b > 0 \text{ are known,} \tag{11.4.1}
\end{aligned}
$$

i.e., the prior is a logistic distribution. Thus the posterior pdf is

$$
k(\theta|y) = \frac{\frac{1}{\sqrt{2\pi}\sigma/\sqrt{n}} \exp\left\{-\frac{1}{2}\frac{(y-\theta)^2}{\sigma^2/n}\right\} b^{-1}e^{-(\theta-a)/b}/(1 + e^{-[(\theta-a)/b]^2})}{\int_{-\infty}^{\infty} \frac{1}{\sqrt{2\pi}\sigma/\sqrt{n}} \exp\left\{-\frac{1}{2}\frac{(y-\theta)^2}{\sigma^2/n}\right\} b^{-1}e^{-(\theta-a)/b}/(1 + e^{-[(\theta-a)/b]^2})\,d\theta}.
$$

Assuming square error loss, the Bayes estimate is the mean of this posterior distribution. Its computation involves two integrals which cannot be obtained in closed form. We can, however, think of the integration in the following way. Consider the likelihood $f(y|\theta)$ as a function of θ; that is, consider the function

$$
w(\theta) = f(y|\theta) = \frac{1}{\sqrt{2\pi}\sigma/\sqrt{n}} \exp\left\{-\frac{1}{2}\frac{(y-\theta)^2}{\sigma^2/n}\right\}.
$$

We can then write the Bayes estimate as

$$
\begin{aligned}
\delta(y) &= \frac{\int_{-\infty}^{\infty} \theta w(\theta) b^{-1}e^{-(\theta-a)/b}/(1 + e^{-[(\theta-a)/b]^2})\,d\theta}{\int_{-\infty}^{\infty} w(\theta) b^{-1}e^{-(\theta-a)/b}/(1 + e^{-[(\theta-a)/b]^2})\,d\theta} \\
&= \frac{E[\Theta w(\Theta)]}{E[w(\Theta)]}, \tag{11.4.2}
\end{aligned}
$$

where the expectation is taken with Θ having the logistic prior distribution.

The estimation can be carried out by simple Monte Carlo. Independently, generate $\Theta_1, \Theta_2, \ldots, \Theta_m$ from the logistic distribution with pdf as in (11.4.1). This generation is easily computed because the inverse of the logistic cdf is given by $a + b\log\{u/(1-u)\}$, for $0 < u < 1$. Then form the random variable,

$$
T_m = \frac{m^{-1}\sum_{i=1}^{m} \Theta_i w(\Theta_i)}{m^{-1}\sum_{i=1}^{m} w(\Theta_i)}. \tag{11.4.3}
$$

By the Weak Law of Large Numbers, (Theorem 4.2.1), and Slutsky's Theorem 4.3.4, $T_m \to \delta(y)$, in probability. The value of m can be quite large. Thus, simple Monte Carlo techniques enable us to compute this Bayes estimate. Note that we can bootstrap this sample to obtain a confidence interval for $E[\Theta w(\Theta)]/E[w(\Theta)]$; see Exercise 11.4.2.

Besides simple Monte Carlo methods, there are other more complicated Monte Carlo procedures which are useful in Bayesian inference. For motivation, consider the case in which we want to generate an observation which has pdf $f_X(x)$ but this generation is somewhat difficult. Suppose, however, that it is easy to both generate Y, with pdf $f_Y(y)$, and generate an observation from the conditional pdf $f_{X|Y}(x|y)$. As the following theorem shows if we do these sequentially then we can easily generate from $f_X(x)$.

Theorem 11.4.1. *Suppose we generate random variables by the following algorithm:*

$$(1). \quad \text{Generate } Y \sim f_Y(y)$$
$$(2). \quad \text{Generate } X \sim f_{X|Y}(x|Y)$$

Then X has pdf $f_X(x)$.

Proof: To avoid confusion, let W be the random variable generated by the algorithm. We need to show that W has pdf $f_X(x)$. Probabilities of events concerning W are conditional on Y and are taken with respect to the cdf $F_{X|Y}$. Recall that probabilities can always be written as expectations of indicator functions and, hence, for events concerning W, are conditional expectations. In particular for any $t \in R$,

$$
\begin{aligned}
P[W \leq t] &= E[F_{X|Y}(t)] \\
&= \int_{-\infty}^{\infty} \left[\int_{-\infty}^{t} f_{X|Y}(x|y)\, dx \right] f_Y(y)\, dy \\
&= \int_{-\infty}^{t} \left[\int_{-\infty}^{\infty} f_{X|Y}(x|y) f_Y(y)\, dy \right] dx \\
&= \int_{-\infty}^{t} \left[\int_{-\infty}^{\infty} f_{X,Y}(x,y)\, dy \right] dx \\
&= \int_{-\infty}^{t} f_X(x)\, dx.
\end{aligned}
$$

Hence, the random variable generated by the algorithm has pdf $f_X(x)$, as was to be shown. ∎

In the situation of this theorem, suppose we want to determine $E[W(X)]$, for some function $W(x)$, where $E[W^2(X)] < \infty$. Using the algorithm of the theorem, generate independently the sequence $(Y_1, X_1), (Y_2, X_2), \ldots, (Y_m, X_m)$, for a specified value of m, where Y_i is drawn from the pdf $f_Y(y)$ and X_i is generated from the pdf $f_{X|Y}(x|Y)$. Then by the Weak Law of Large Numbers,

$$
\overline{W} = \frac{1}{m} \sum_{i=1}^{m} W(X_i) \xrightarrow{P} \int_{-\infty}^{\infty} W(x) f_X(x)\, dx = E[W(X)].
$$

Furthermore, by the Central Limit Theorem, $\sqrt{m}(\overline{W} - E[W(X)])$ converges in distribution to a $N(0, \sigma_W^2)$ distribution, where $\sigma_W^2 = \text{Var}(W(X))$. If $w_1,, w_2, \ldots, w_m$ is a realization of such a random sample, then an approximate (large sample) $(1 - \alpha)100\%$ confidence interval for $E[W(X)]$ is

$$\overline{w} \pm z_{\alpha/2} \frac{s_W}{\sqrt{m}}, \tag{11.4.4}$$

where $s_W^2 = (m-1)^{-1} \sum_{i=1}^m (w_i - \overline{w})^2$.

To set ideas, we present the following simple example.

Example 11.4.1. Suppose the random variable X has pdf

$$f_X(x) = \begin{cases} 2e^{-x}(1 - e^{-x}) & 0 < x < \infty \\ 0 & \text{elsewhere.} \end{cases} \tag{11.4.5}$$

Suppose Y and $X|Y$ have the respective pdfs

$$f_Y(y) = \begin{cases} 2e^{-2y} & 0 < x < \infty \\ 0 & \text{elsewhere,} \end{cases} \tag{11.4.6}$$

$$f_{X|Y}(x|y) = \begin{cases} e^{-(x-y)} & y < x < \infty \\ 0 & \text{elsewhere.} \end{cases} \tag{11.4.7}$$

Suppose we generate random variables by the following algorithm:

    (1). Generate $Y \sim f_Y(y)$ as in expression (11.4.6)

    (2). Generate $X \sim f_{X|Y}(x|Y)$ as in expression (11.4.7)

Then, by Theorem 11.4.1, X has the pdf (11.4.5). Furthermore, it is easy to generate from the pdfs (11.4.6) and (11.4.7) because the inverses of the respective cdfs are given by $F_Y^{-1}(u) = -2^{-1}\log(1-u)$ and $F_{X|Y}^{-1}(u) = -\log(1-u) + Y$.

As a numerical illustration, the R function condsim1, found in Appendix C, uses this algorithm to generate observations from the pdf (11.4.5). Using this function, we performed $m = 10,000$ simulations of the algorithm. The sample mean and standard deviation were: $\overline{x} = 1.495$ and $s = 1.112$. Hence, a 95% confidence interval for $E(X)$ is $(1.473, 1.517)$ which traps the true value $E(X) = 1.5$; see Exercise 11.4.4. ∎

For the last example, Exercise 11.4.3 establishes the joint distribution of (X, Y) and shows that the marginal pdf of X is given by (11.4.5). Furthermore, as shown in this exercise it is easy to generate from the distribution of X directly. In Bayesian inference, though, we are often dealing with conditional pdfs and theorems such as Theorem 11.4.1 are quite useful.

The main purpose of presenting this algorithm is to motivate another algorithm, called the **Gibbs Sampler**, which is useful in Bayes methodology. We describe it in terms of two random variables. Suppose (X, Y) has pdf $f(x, y)$. Our goal is to generate two streams of iid random variables, one on X and the other on Y. The Gibbs sampler algorithm is:

Algorithm 11.4.1 (Gibbs Sampler). *Let m be a positive integer and X_0 an initial value be given. Then for $i = 1, 2, 3, \ldots, m$,*

(1).   Generate $Y_i | X_{i-1} \sim f(y|x)$

(2).   Generate $X_i | Y_i \sim f(x|y)$.

Note that before entering the *ith* step of the algorithm we have generated X_{i-1}. Let x_{i-1} denote the observed value of X_{i-1}. Then using this value generate sequentially the new Y_i from the pdf $f(y|x_{i-1})$ and then draw (the new) X_i from the pdf $f(x|y_i)$ where y_i is the observed value of Y_i. In advanced texts, it is shown that

$$Y_i \xrightarrow{D} f_Y(y)$$
$$X_i \xrightarrow{D} f_X(x),$$

as $i \to \infty$. Hence, using the Law of Large Numbers, arithmetic averages such as,

$$\frac{1}{m} \sum_{i=1}^{m} W(X_i) \xrightarrow{P} E[W(X)], \text{ as } m \to \infty. \tag{11.4.8}$$

Note that the Gibbs sampler is similar but not quite the same as the algorithm given by Theorem 11.4.1. Consider the sequence of generated pairs,

$$(X_1, Y_1), (X_2, Y_2), \ldots, (X_k, Y_k), (X_{k+1}, Y_{k+1}).$$

Note that to compute (X_{k+1}, Y_{k+1}) we need only the pair (X_k, Y_k) and none of the previous pairs from 1 to $k - 1$. That is, given the present state of the sequence, the future of the sequence is independent of the past. In stochastic processes such a sequence is called a **Markov chain**. Under general conditions, the distribution of Markov chains stablizes (reaches an equilibrium or steady state distribution) as the length of the chain increases. For the Gibbs sampler, the equilibrium distributions are the limiting distributions in the expression (11.4.8) as $i \to \infty$. How large should i be? In practice, usually the chain is allowed to run to some large value i before recording the observations. Furthermore several recordings are run with this value of i and the resulting empirical distributions of the generated random observations are examined for their similarity. Also, the starting value for X_0 is needed; see Casella and George (1992) for discussion. The theory behind the convergences given in the expression (11.4.8), is beyond the scope of this text. There are many excellent references on this theory. Discussion from an elementary level can be found in Casella and George (1992). A good overview can be found in Chapter 7 of Robert and Casella (1999); see, also Lehmann and Casella (1998). We next provide a simple example.

Example 11.4.2. Suppose (X, Y) has the mixed discrete-continuous pdf given by

$$f(x, y) = \begin{cases} \frac{1}{\Gamma(\alpha)} \frac{1}{x!} y^{\alpha + x - 1} e^{-2y} & y > 0 \, ; x = 0, 1, 2, \ldots \\ 0 & \text{elsewhere.} \end{cases} \tag{11.4.9}$$

for $\alpha > 0$. Exercise 11.4.5 shows that this is a pdf and obtains the marginal pdfs. The conditional pdfs, however, are given by

$$f(y|x) \propto y^{\alpha+x-1}e^{-2y} \tag{11.4.10}$$

and

$$f(x|y) \propto e^{-y}\frac{y^x}{x!}. \tag{11.4.11}$$

Hence, the conditional densities are $\Gamma(\alpha+x, 1/2)$ and Poisson (y), respectively. Thus the Gibbs sampler algorithm is: for $i = 1, 2, \ldots, m$

(1).   Generate $Y_i|X_{i-1} \sim \Gamma(\alpha + X_{i-1}, 1/2)$
(2).   Generate $X_i|Y_i \sim \text{Poisson}(Y_i)$.

In particular, for large m and $n > m$,

$$\overline{Y} = (n-m)^{-1}\sum_{i=m+1}^{n} Y_i \xrightarrow{P} E(Y) \tag{11.4.12}$$

$$\overline{X} = (n-m)^{-1}\sum_{i=m+1}^{n} X_i \xrightarrow{P} E(X). \tag{11.4.13}$$

In this case, it can be shown (see Exercise 11.4.5) that both expectations are equal to α. The R routine `gibbser2.s` computes this Gibbs sampler. Using this routine, the authors obtained the following results upon setting $\alpha = 10$, $m = 3000$, and $n = 6000$:

Parameter	Estimate	Sample Estimate	Sample Variance	Approximate 95% Confidence Interval
$E(Y) = \alpha = 10$	$\overline{y}$	10.027	10.775	$(9.910, 10.145)$
$E(X) = \alpha = 10$	$\overline{x}$	10.061	21.191	$(9.896, 10.225)$

where the estimates $\overline{y}$ and $\overline{x}$ are the observed values of the estimators in expressions (11.4.12) and (11.4.13), respectively. The confidence intervals for α are the large sample confidence intervals for means given by expression (5.8.3), using the sample variances found in the fourth column of the table. Note that both confidence intervals trapped $\alpha = 10$. ∎

EXERCISES

11.4.1. Suppose Y has a $\Gamma(1, 1)$ distribution while X given Y has the conditional pdf

$$f(x|y) = \begin{cases} e^{-(x-y)} & 0 < y < x < \infty \\ 0 & \text{elsewhere.} \end{cases}$$

Note that both the pdf of Y and the conditional pdf are easy to simulate.

(a) Set up the algorithm of Theorem 11.4.1 to generate a stream of iid observations with pdf $f_X(x)$.

(b) State how to estimate $E(X)$.

(c) If computational facilities are available, write a computer program to estimate $E(X)$ using your algorithm found in Part (a).

(d) Using your program, obtain a stream of 2000 simulations. Compute your estimate of $E(X)$ and find an approximate 95% confidence interval.

(e) Show that X has a $\Gamma(2,1)$ distribution. Did your confidence interval trap the true value 2?

11.4.2. Carefully write down the algorithm to obtain a bootstrap percentile confidence interval for $E[\Theta w(\Theta)]/E[w(\Theta)]$, using the sample $\Theta_1, \Theta_2, \ldots, \Theta_m$ and the estimator given in expression (11.4.3). If computation facilities are at hand, obtain code for this bootstrap.

11.4.3. Consider Example 11.4.1.

(a) Show that $E(X) = 1.5$.

(b) Obtain the inverse of the cdf of X and use it to show how to generate X directly.

11.4.4. If computation facilities are at hand, obtain another 10,000 simulations similar to that discussed at the end of Example 11.4.1. Use your simulations to obtain a confidence interval for $E(X)$.

11.4.5. Consider Example 11.4.2.

(a) Show that the function given in expression (11.4.9) is a joint, mixed discrete continuous pdf.

(b) Show that the random variable Y has a $\Gamma(\alpha, 1)$ distribution.

(c) Show that the random variable X has a negative binomial distribution with pmf

$$p(x) = \begin{cases} \frac{(\alpha+x-1)\cdots\alpha}{x!} 2^{-(\alpha+x)} & x = 0, 1, 2, \ldots \\ 0 & \text{elsewhere.} \end{cases}$$

(d) Show that $E(X) = \alpha$.

11.4.6. If computation facilities are available write a program (or use `gibbser2.s` of Appendix C) for the Gibbs sampler discussed in Example 11.4.2. Run your program for $\alpha = 10$, $m = 3000$, and $n = 6000$. Compare your results with those of the authors tabled in the example.

11.4.7. Consider the following mixed discrete-continuous pdf for a random vector (X, Y), (discussed in Casella and George, 1992):

$$f(x, y) \propto \begin{cases} \binom{n}{x} y^{x+\alpha-1}(1-y)^{n-x+\beta-1} & x = 0, 1, \ldots, n, \; 0 < y < 1 \\ 0 & \text{elsewhere,} \end{cases}$$

for $\alpha > 0$ and $\beta > 0$.

  (a) Show that this function is indeed a joint, mixed discrete continuous pdf by finding the proper constant of proportionality.

  (b) Determine the conditional pdfs $f(x|y)$ and $f(y|x)$.

  (c) Write the Gibbs sampler algorithm to generate random samples on X and Y.

  (d) Determine the marginal distributions of X and Y.

11.4.8. If computation facilities are available write a program for the Gibbs sampler of Exercise 11.4.7. Run your program for $\alpha = 10$, $\beta = 4$, and $m = 3000$ and $n = 6000$. Obtain estimates (and confidence intervals) of $E(X)$ and $E(Y)$ and compare them with the true parameters.

11.5   Modern Bayesian Methods

The prior pdf has an important influence in Bayesian inference. We need only consider the different Bayes estimators for the normal model based on different priors as shown in Examples 11.2.3 and 11.3.1. One way of having more control over the prior is to model the prior in terms of another random variable. This is called the **hierarchical Bayes** model and it is of the form

$$\begin{aligned} X|\theta &\sim f(x|\theta) \\ \Theta|\gamma &\sim h(\theta|\gamma) \\ \Gamma &\sim \psi(\gamma). \end{aligned} \tag{11.5.1}$$

With this model we can exert control over the prior $h(\theta|\gamma)$ by modifying the pdf of the random variable Γ. A second methodology, **empirical Bayes**, obtains an estimate of γ and plugs it into the posterior pdf. We offer the reader a brief introduction of these procedures in this chapter. There are several good books on Bayesian methods. In particular, Chapter 4 of Lehmann and Casella (1998) discuss these procedures in some detail.

    Consider first the hierarchical Bayes model given by (11.5.1). The parameter γ can be thought of a nuisance parameter. It is often called a *hyperparameter*. As with regular Bayes, the inference focuses on the parameter θ; hence the posterior pdf of interest remains the conditional pdf $k(\theta|\mathbf{x})$.

    In this discussion we will be using several pdfs; hence, we will frequently use g as a generic pdf. It will always be clear from its arguments what distribution it represents. Keep in mind that the conditional pdf $f(\mathbf{x}|\theta)$ does not depend on γ; hence,

$$
\begin{aligned}
g(\theta, \gamma | \mathbf{x}) &= \frac{g(\mathbf{x}, \theta, \gamma)}{g(\mathbf{x})} \\
&= \frac{g(\mathbf{x}|\theta, \gamma) g(\theta, \gamma)}{g(\mathbf{x})} \\
&= \frac{f(\mathbf{x}|\theta) h(\theta|\gamma) \psi(\gamma)}{g(\mathbf{x})}.
\end{aligned}
$$

Therefore, the posterior pdf is given by

$$
k(\theta|\mathbf{x}) = \frac{\int_{-\infty}^{\infty} f(\mathbf{x}|\theta) h(\theta|\gamma) \psi(\gamma) \, d\gamma}{\int_{-\infty}^{\infty} \int_{-\infty}^{\infty} f(\mathbf{x}|\theta) h(\theta|\gamma) \psi(\gamma) \, d\gamma d\theta}. \tag{11.5.2}
$$

Furthermore, assuming square error loss, the Bayes estimate of $W(\theta)$ is

$$
\delta_W(\mathbf{x}) = \frac{\int_{-\infty}^{\infty} \int_{-\infty}^{\infty} W(\theta) f(\mathbf{x}|\theta) h(\theta|\gamma) \psi(\gamma) \, d\gamma d\theta}{\int_{-\infty}^{\infty} \int_{-\infty}^{\infty} f(\mathbf{x}|\theta) h(\theta|\gamma) \psi(\gamma) \, d\gamma d\theta}. \tag{11.5.3}
$$

Recall that we defined the Gibbs sampler in Section 11.4. Here we describe it to obtain the Bayes estimate of $W(\theta)$. For $i = 1, 2, \ldots, m$, where m is specified, the *ith* step of the algorithm is

$$
\begin{aligned}
\Theta_i | \mathbf{x}, \gamma_{i-1} &\sim g(\theta | \mathbf{x}, \gamma_{i-1}) \\
\Gamma_i | \mathbf{x}, \theta_i &\sim g(\gamma | \mathbf{x}, \theta_i).
\end{aligned}
$$

Recall from our discussion in Section 11.4 that

$$
\begin{aligned}
\Theta_i &\xrightarrow{D} k(\theta | \mathbf{x}) \\
\Gamma_i &\xrightarrow{D} g(\gamma | \mathbf{x}),
\end{aligned}
$$

as $i \to \infty$. Furthermore, the arithmetic average,

$$
\frac{1}{m} \sum_{i=1}^{m} W(\Theta_i) \xrightarrow{P} E[W(\Theta) | \mathbf{x}] = \delta_W(\mathbf{x}), \text{ as } m \to \infty. \tag{11.5.4}
$$

In practice, to obtain the Bayes estimate of $W(\theta)$ by the Gibbs sampler, we generate by Monte Carlo the stream of of values $(\theta_1, \gamma_1), (\theta_2, , \gamma_2) \ldots.$ Then choosing large values of m and $n > m$, our estimate of $W(\theta)$ is the average,

$$
\frac{1}{n-m} \sum_{i=m+1}^{n} W(\theta_i). \tag{11.5.5}
$$

Because of the Monte Carlo generation these procedures are often called **MCMC**, for **Markov Chain Monte Carlo** procedures. We next provide two examples.

Example 11.5.1. Reconsider the conjugate family of normal distributions discussed in Example 11.2.3, with $\theta_0 = 0$. Here, we use the model

$$\overline{X}|\Theta \quad \sim \quad N\left(\theta, \frac{\sigma^2}{n}\right), \ \sigma^2 \text{ is known}$$

$$\Theta|\tau^2 \quad \sim \quad N(0, \tau^2)$$

$$\frac{1}{\tau^2} \quad \sim \quad \Gamma(a, b), \ a \text{ and } b \text{ are known}. \qquad (11.5.6)$$

To set up the Gibbs sampler for this hierarchical Bayes model, we need the conditional pdfs $g(\theta|\overline{x}, \tau^2)$ and $g(\tau^2|\overline{x}, \theta)$. For the first we have,

$$g(\theta|\overline{x}, \tau^2) \propto f(\overline{x}|\theta)h(\theta|\tau^2)\psi(\tau^{-2}).$$

As we have been doing, we can ignore standardizing constants; hence, we need only consider the product $f(\overline{x}|\theta)h(\theta|\tau^2)$. But this is a product of two normal pdfs which we obtained in Example 11.2.3. Based on those results, $g(\theta|\overline{x}, \tau^2)$ is the pdf of a $N\left(\frac{\tau^2}{(\sigma^2/n)+\tau^2}\overline{x}, \frac{\tau^2\sigma^2}{\sigma^2+n\tau^2}\right)$. For the second pdf, by ignoring standardizing constants and simplifying, we obtain

$$g\left(\frac{1}{\tau^2}|\overline{x}, \theta\right) \quad \propto \quad f(\overline{x}|\theta)g(\theta|\tau^2)\psi(1/\tau^2)$$

$$\propto \quad \frac{1}{\tau}\exp\left\{-\frac{1}{2}\frac{\theta^2}{\tau^2}\right\}\left(\frac{1}{\tau^2}\right)^{a-1}\exp\left\{-\frac{1}{\tau^2}\frac{1}{b}\right\}$$

$$\propto \quad \left(\frac{1}{\tau^2}\right)^{a+(1/2)-1}\exp\left\{-\frac{1}{\tau^2}\left[\frac{\theta^2}{2}+\frac{1}{b}\right]\right\}, \qquad (11.5.7)$$

which is the pdf of a $\Gamma\left(a+\frac{1}{2}, \left(\frac{\theta^2}{2}+\frac{1}{b}\right)^{-1}\right)$ distribution. Thus the Gibbs sampler for this model is given by: for $i = 1, 2, \ldots, m$,

$$\Theta_i|\overline{x}, \tau_{i-1}^2 \quad \sim \quad N\left(\frac{\tau_{i-1}^2}{(\sigma^2/n)+\tau_{i-1}^2}\overline{x}, \frac{\tau_{i-1}^2\sigma^2}{\sigma^2+n\tau_{i-1}^2}\right)$$

$$\frac{1}{\tau_i^2}|\overline{x}, \Theta_i \quad \sim \quad \Gamma\left(a+\frac{1}{2}, \left(\frac{\theta_i^2}{2}+\frac{1}{b}\right)^{-1}\right). \qquad (11.5.8)$$

As discussed above, for a specified values of large m and $n^* > m$, we collect the chain's values $((\Theta_m, \tau_m), (\Theta_{m+1}, \tau_{m+1}), \ldots, (\Theta_{n^*}, \tau_{n^*}))$ and then obtain the Bayes estimate of θ, (assuming square error loss)

$$\widehat{\theta} = \frac{1}{n^*-m}\sum_{i=m+1}^{n^*}\Theta_i. \qquad (11.5.9)$$

The conditional distribution of Θ given $\overline{x}$ and τ, though, suggest the second estimate given by

$$\widehat{\theta}^* = \frac{1}{n^*-m}\sum_{i=m+1}^{n^*}\frac{\tau_i^2}{\tau_i^2+(\sigma^2/n)}\overline{x}. \quad \blacksquare \qquad (11.5.10)$$

Example 11.5.2. Lehmann and Casella (1998, p. 257) presented the following hierarchical Bayes model

$$
\begin{aligned}
X|\lambda &\sim \text{Poisson}(\lambda) \\
\Lambda|b &\sim \Gamma(1, b) \\
B &\sim g(b) = \tau^{-1}b^{-2}\exp\{-1/b\tau\}, \quad b > 0, \tau > 0.
\end{aligned}
$$

For the Gibbs sampler, we need the two conditional pdfs, $g(\lambda|x, b)$ and $g(b|x, \lambda)$. The joint pdf is

$$
g(x, \lambda, b) = f(x|\lambda)h(\lambda|b)\psi(b). \tag{11.5.11}
$$

Based on the pdfs of the model, (11.5.11), for the first conditional pdf we have

$$
\begin{aligned}
g(\lambda|x, b) &\propto e^{-\lambda}\frac{\lambda^x}{x!}\frac{1}{b}e^{-\lambda/b} \\
&\propto \lambda^{x+1}e^{-\lambda[1+(1/b)]}, \tag{11.5.12}
\end{aligned}
$$

which is the pdf of a $\Gamma(x + 1, b/[b + 1])$ distribution.

For the second conditional pdf, we have

$$
\begin{aligned}
g(b|x, \lambda) &\propto \frac{1}{b}e^{-\lambda/b}\tau^{-1}b^{-2}e^{-1/(b\tau)} \\
&\propto b^{-3}\exp\left\{-\frac{1}{b}\left[\frac{1}{\tau} + \lambda\right]\right\}.
\end{aligned}
$$

In this last expression, making the change of variable $y = 1/b$ which has the Jacobian $db/dy = -y^{-2}$, we obtain

$$
\begin{aligned}
g(b|x, \lambda) &\propto y^3\exp\left\{-y\left[\frac{1}{\tau} + \lambda\right]\right\}y^{-2} \\
&\propto y^{2-1}\exp\left\{-y\left[\frac{1 + \lambda\tau}{\tau}\right]\right\},
\end{aligned}
$$

which is easily seen to be the pdf of the $\Gamma(2, \tau/[\lambda\tau + 1])$ distribution. Therefore, the Gibbs sampler is: for $i = 1, 2, \ldots, m$, where m is specified,

$$
\Lambda_i|x, b_{i-1} \sim \Gamma(x + 1, b_{i-1}/[1 + b_{i-1}])
$$

$$
B_i = Y_i^{-1} \text{ where } Y_i|x, \lambda_i \sim \Gamma(2, \tau/[\lambda_i\tau + 1]). \quad \blacksquare
$$

As a numerical illustration of the last example, suppose we set $\tau = .05$ and observe $x = 6$. The R program `hierarch1.s` in Appendix C performs the Gibbs sampler given in the example. It requires specification of the value of i at which the Gibbs sample commences and the length of the chain beyond this point. We set these values at $m = 1000$ and $n = 4000$, respectively, i.e., the length of the chain used in the estimate is 3000. To see the effect of varying τ has on the Bayes estimator, we performed five Gibbs samplers with the results:

τ	0.040	0.045	0.050	0.055	0.060
$\widehat{\delta}$	6.60	6.49	6.53	6.50	6.44

There is some variation. As discussed in Lehmann and Casella (1998), in general there is less affect on the Bayes estimator due to variability of the hyperparameter than in regular Bayes due to the variance of the prior.

11.5.1   Empirical Bayes

The empirical Bayes model consists of the first two lines of the hierarchical Bayes model, i.e.,

$$\mathbf{X}|\theta \quad \sim \quad f(\mathbf{x}|\theta)$$
$$\Theta|\gamma \quad \sim \quad h(\theta|\gamma).$$

Instead of attempting to model the parameter γ with a pdf as in hierarchical Bayes, empirical Bayes methodology estimates γ based on the data as follows. Recall that

$$
\begin{aligned}
g(\mathbf{x}, \theta|\gamma) &= \frac{g(\mathbf{x}, \theta, \gamma)}{\psi(\gamma)} \\
&= \frac{f(\mathbf{x}|\theta)h(\theta|\gamma)\psi(\gamma)}{\psi(\gamma)} \\
&= f(\mathbf{x}|\theta)h(\theta|\gamma).
\end{aligned}
$$

Consider, then, the likelihood function

$$m(\mathbf{x}|\gamma) = \int_{-\infty}^{\infty} f(\mathbf{x}|\theta)h(\theta|\gamma)\, d\theta. \tag{11.5.13}$$

Using the pdf $m(\mathbf{x}|\gamma)$, we obtain an estimate $\widehat{\gamma} = \widehat{\gamma}(\mathbf{x})$, usually by the method of maximum likelihood. For inference on the parameter θ, the empirical Bayes procedure uses the posterior pdf $k(\theta|\mathbf{x}, \widehat{\gamma})$.

We illustrate the empirical Bayes procedure with the following example.

Example 11.5.3. Consider the same situation discussed in Example 11.5.2, except assume that we have a random sample on X, i.e., consider the model

$$X_i|\lambda, \; i = 1, 2, \ldots, n \quad \sim \quad \text{iid Poisson}(\lambda)$$
$$\Lambda|b \quad \sim \quad \Gamma(1, b).$$

Let $\mathbf{X} = (X_1, X_2, \ldots, X_n)'$. Hence,

$$g(\mathbf{x}|\lambda) = \frac{\lambda^{n\bar{x}}}{x_1! \cdots x_n!} e^{-n\lambda},$$

where $\bar{x} = n^{-1}\sum_{i=1}^{n} x_i$. Thus the pdf we need to maximize is

$$
\begin{aligned}
m(\mathbf{x}|b) &= \int_0^{\infty} g(\mathbf{x}|\lambda)h(\lambda|b)\, d\lambda \\
&= \int_0^{\infty} \frac{1}{x_1! \cdots x_n!} \lambda^{n\bar{x}+1-1} e^{-n\lambda} \frac{1}{b} e^{-\lambda/b}\, d\lambda \\
&= \frac{\Gamma(n\bar{x}+1)[b/(nb+1)]^{n\bar{x}+1}}{x_1! \cdots x_n! b}.
\end{aligned}
$$

Taking the partial derivative of $m(\mathbf{x}|b)$ with respect to b, we obtain

$$\frac{\partial \log m(\mathbf{x}|b)}{\partial b} = -\frac{1}{b} + (n\bar{x} + 1)\frac{1}{b(bn+1)}.$$

Setting this equal to 0 and solving for b, we obtain the solution

$$\widehat{b} = \bar{x}. \tag{11.5.14}$$

To obtain the empirical Bayes estimate of λ we need to compute the posterior pdf with $\widehat{b}$ substituted for b. The posterior pdf is

$$
\begin{aligned}
k(\lambda|\mathbf{x},\widehat{b}) &\propto g(\mathbf{x}|\lambda)h(\lambda|\widehat{b}) \\
&\propto \lambda^{n\bar{x}+1-1}e^{-\lambda[n+(1/\widehat{b})]}, \tag{11.5.15}
\end{aligned}
$$

which is the pdf of a $\Gamma(n\bar{x}+1, \widehat{b}/[n\widehat{b}+1])$ distribution. Therefore, the empirical Bayes estimator under squared error loss is the mean of this distribution, i.e.,

$$\widehat{\lambda} = [n\bar{x}+1]\frac{\widehat{b}}{n\widehat{b}+1} = \bar{x}. \tag{11.5.16}$$

Thus, for the above prior, the empirical Bayes estimate agrees with the mle. ∎

We can use our solution of this last example to obtain the empirical Bayes estimate for Example 11.5.2, also. For in this earlier example, the sample size is one. Thus the empirical Bayes estimate for λ is x. In particular, for the numerical case given at the end of Example 11.5.2, the empirical Bayes estimate has the value 6.

EXERCISES

11.5.1. Consider the Bayes model

$$
\begin{aligned}
X_i|\theta &\sim \text{ iid } \Gamma\left(1, \frac{1}{\theta}\right) \\
\Theta|\beta &\sim \Gamma(2, \beta).
\end{aligned}
$$

By performing the following steps obtain the empirical Bayes estimate of θ.

(a) Obtain the likelihood function

$$m(\mathbf{x}|\beta) = \int_0^\infty f(\mathbf{x}|\theta)h(\theta|\beta)\, d\theta.$$

(b) Obtain the mle $\widehat{\beta}$ of β for the likelihood $m(\mathbf{x}|\beta)$.

(c) Show that the posterior distribution of Θ given $\mathbf{x}$ and $\widehat{\beta}$ is a gamma distribution.

(d) Assuming square error loss, obtain the empirical Bayes estimator.

11.5.2. Consider the hierarchical Bayes model

$$
\begin{aligned}
Y &\sim b(n,p); \quad 0 < p < 1 \\
p|\theta &\sim h(p|\theta) = \theta p^{\theta-1}; \quad \theta > 0 \\
\theta &\sim \Gamma(1,a); \quad a > 0 \text{ is specified.}
\end{aligned}
\tag{11.5.17}
$$

(a) Assuming square error loss, write the Bayes estimate of p as in expression (11.5.3). Integrate relative to θ first. Show that both numerator and denominator are expectations of a beta distribution with parameters $y + 1$ and $n - y + 1$.

(b) Recall the discussion around expression (11.4.2). Write an explicit Monte Carlo algorithm to obtain the Bayes estimate in Part (a).

11.5.3. Reconsider the hierarchical Bayes model (11.5.17) of Exercise 11.5.2.

(a) Show that the conditional pdf $g(p|y,\theta)$ is the pdf of a beta distribution with parameters $y + \theta$ and $n - y + 1$.

(b) Show that the conditional pdf $g(\theta|y,p)$ is the pdf of a gamma distribution with parameters 2 and $\left[\frac{1}{a} - \log p\right]^{-1}$.

(c) Using Parts (a) and (b) and assuming square error loss, write the Gibbs sampler algorithm to obtain the Bayes estimator of p.

11.5.4. For the hierarchical Bayes model of Exercise 11.5.2, set $n = 50$ and $a = 2$. Now draw a θ at random from a $\Gamma(1,2)$ distribution and label it θ^*. Next draw a p at random from the distribution with pdf $\theta^* p^{\theta^*-1}$ and label it p^*. Finally draw a y at random from a $b(n,p^*)$ distribution.

(a) Setting m at 3000, obtain an estimate of θ^* using your Monte Carlo algorithm of Exercise 11.5.2.

(b) Setting m at 3000 and n at 6000, obtain an estimate of θ^* using your Gibbs sampler algorithm of Exercise 11.5.3. Let $p_{3001}, p_{3002}, \ldots, p_{6000}$ denote the stream of values drawn. Recall that these values are (asymptotically) simulated values from the posterior pdf $g(p|y)$. Use this stream of values to obtain a 95% credible interval.

11.5.5. Write the Bayes model of Exercise 11.5.2 as

$$
\begin{aligned}
Y &\sim b(n,p); \quad 0 < p < 1 \\
p|\theta &\sim h(p|\theta) = \theta p^{\theta-1}; \quad \theta > 0.
\end{aligned}
$$

Set up the estimating equations for mle of $g(y|\theta)$; i.e., the first step to obtain the empirical Bayes estimator of p. Simplify as much as possible.

11.5.6. Example 11.5.1 dealt with a hierarchical Bayes model for a conjugate family of normal distributions. Express that model as

$$\overline{X}|\Theta \quad \sim \quad N\left(\theta, \frac{\sigma^2}{n}\right), \; \sigma^2 \text{ is known}$$
$$\Theta|\tau^2 \quad \sim \quad N(0, \tau^2).$$

Obtain the empirical Bayes estimator of θ.

Chapter 12

Linear Models

In this chapter we present an introduction to linear models. These models encompass regression models and analysis of variance models and are the most widely used models in practice. The most popular way to fit linear models is to use the method of least squares (LS). The LS fit, though, is quite sensitive to outliers and influential points. In this chapter, we present two robust concepts, the influence function and breakdown. The influence function is easily shown to predict the LS sensitivity to outliers and influential points. We also present linear models and LS estimation through their geometry. This not only gives a succinct presentation but allows us to easily introduce robust alternatives to LS fitting. We simply change the Euclidean norm used in LS fitting to another norm. We begin with the simple location model and proceed to simple and multiple regression models.

12.1   Robust Concepts

In a few words, we say an estimator is *robust* if it is not sensitive to outliers in the data. In this section, we make this more precise for the location model. Suppose then that $Y_1, Y_2, \ldots Y_n$ is a random sample which follows the location model

$$Y_i = \theta + \varepsilon_i, \quad i = 1, 2, \ldots, n, \tag{12.1.1}$$

where ε_i has cdf $F(t)$ and pdf $f(t)$. Let $F_Y(t)$ and $f_Y(t)$ denote the cdf and pdf of Y, respectively. Then $F_Y(t) = F(t - \theta)$. Note that we can write this model in a matrix formulation, by letting $\mathbf{Y}' = (Y_1, \ldots, Y_n)$, $\boldsymbol{\varepsilon}' = (\varepsilon_1, \ldots, \varepsilon_n)$, and writing

$$\mathbf{Y} = \mathbf{1}\theta + \boldsymbol{\varepsilon}, \tag{12.1.2}$$

where $\mathbf{1}$ is an $n \times 1$ vector whose elements are all one. If we let V denote the subspace of R^n spanned by $\mathbf{1}$, then a simple way of writing Model (12.1.2) is

$$\mathbf{Y} = \boldsymbol{\eta} + \boldsymbol{\varepsilon}, \quad \boldsymbol{\eta} \in V. \tag{12.1.3}$$

If the random error was $\mathbf{0}$, then Y would lie in V. Hence, a simple way to estimate $\boldsymbol{\eta}$ is to determine the vector in V which is "closest" to $\mathbf{Y}$. Here "closest" means in term of a specified norm. In this section we consider two norms and their corresponding estimates.

12.1.1 Norms and Estimating Equations

For our first norm, we take the Euclidean norm which, in this chapter, is labeled as the least squares (LS) norm. The square of this norm is, of course, given by

$$\|\mathbf{v}\|_{LS}^2 = \sum_{1=1}^n v_i^2, \quad \mathbf{v} \in R^n. \tag{12.1.4}$$

Then our LS estimate of θ is $\widehat{\theta}_{LS}$ where

$$\widehat{\theta}_{LS} = \text{Argmin}_{\theta \in R} \|\mathbf{Y} - \mathbf{1}\theta\|_{LS}^2. \tag{12.1.5}$$

Instead of using θ, we could phrase this in terms of estimating $\boldsymbol{\eta}$, but in this section we only use θ; see Exercise 12.1.1.

Equation (12.1.4) is in terms of a *norm*, in this case the LS norm. We are, however, minimizing a function of θ. Hence it is often easier to compute the partial derivative with respect to θ and then solve for the critical point. This is easily seen to result in the **estimating equations**

$$\sum_{i=1}^n (Y_i - \theta) = 0. \tag{12.1.6}$$

The solution to this equation is $\widehat{\theta}_{LS} = \overline{Y}$; i.e., the *sample mean*. Under certain additional assumptions, the LS estimate of θ has optimal properties. For example, if we assume that ε_i has $N(0, \sigma^2)$ distribution, then by Chapter 7 the LS estimate is the MVUE of θ.

For our second norm, we take the least absolute values (L_1) norm given by

$$\|\mathbf{v}\|_{L_1} = \sum_{1=1}^n |v_i|, \quad \mathbf{v} \in R^n. \tag{12.1.7}$$

Then our L_1 estimate of θ is $\widehat{\theta}_{L_1}$ where

$$\widehat{\theta}_{L_1} = \text{Argmin}_{\theta \in R} \|\mathbf{Y} - \mathbf{1}\theta\|_{L_1}. \tag{12.1.8}$$

The corresponding estimating equation is

$$\sum_{i=1}^n \text{sgn}(Y_i - \theta) = 0. \tag{12.1.9}$$

The solution to this equation is seen to be the median of the $Y_1, Y_2, \ldots, Y_n$ which we denote by $\widehat{\theta}_{L_1} = \text{median}\{Y_1, \ldots, Y_n\}$.

To motivate the subsection below on influence functions, we compare these two estimates in terms of their sensitivity functions. We begin our discussion for a general estimator, say, $\widehat{\theta} = \widehat{\theta}(\mathbf{Y})$. What happens to $\widehat{\theta}$ if an outlier is added to the sample? To investigate this, let $\mathbf{y}_n = (y_1, y_2, \ldots y_n)$ be a realization of the sample, let y be the additional point, and denote the augmented sample by $\mathbf{y}'_{n+1} = (\mathbf{y}'_n, y)$. Then a simple measure is the rate of change in the estimate due to y relative to the mass of y, $(1/(n+1))$, i.e.,

$$S(y; \widehat{\theta}) = \frac{\widehat{\theta}(\mathbf{y}_{n+1}) - \widehat{\theta}(\mathbf{y}_n)}{1/(n+1)}. \qquad (12.1.10)$$

This is called the **sensitivity curve** of the estimate $\widehat{\theta}$.

As examples, consider the sample mean and median. For the sample mean, it is easy to see that

$$S(y; \overline{Y}) = \frac{\overline{y}_{n+1} - \overline{y}_n}{1/(n+1)} = y - \overline{y}_n. \qquad (12.1.11)$$

Hence the relative change in the sample mean is a linear function of y. Thus if y is large then the change in sample mean is also large. Actually, the change is unbounded in y. Thus the sample mean is quite sensitive to the size of the outlier. In contrast, consider the sample median in which the sample size n is odd. In this case, the sample median is $\widehat{\theta}_{L_1,n} = y_{(r)}$ where $r = (n+1)/2$. When the additional point y is added, the sample size becomes even and the sample median $\widehat{\theta}_{L_1,n+1}$ is the average of the middle two order statistics. If y varies between these two order statistics then there is some change between the $\widehat{\theta}_{L_1,n}$ and $\widehat{\theta}_{L_1,n+1}$. But once y moves beyond these middle two order statistics, there is no change. Hence, $S(y; \widehat{\theta}_{L_1,n})$ is a bounded function of y. Therefore, $\widehat{\theta}_{L_1,n}$ is much less sensitive to an outlier than the sample mean; see Exercise 12.1.2 for a numerical example.

12.1.2 Influence Functions

One problem with the sensitivity curve is its dependence on the sample. In earlier chapters, we compared estimators in terms of their variances which are functions of the underlying distribution. This is the type of comparison we want to make here.

Recall that the location model (12.1.1) is the model of interest, where $F_Y(t) = F(t - \theta)$ is the cdf of Y and $F(t)$ is the cdf of ε. As discussed in Section 1 of Chapter 10, the parameter θ is a function of the cdf $F_Y(y)$. It is convenient, then, to use function notation such as $\theta = T(F_Y)$. For example, if θ is the mean then $T(F_Y)$ is defined as

$$T(F_Y) = \int_{-\infty}^{\infty} y \, dF_Y(y) = \int_{-\infty}^{\infty} y f_Y(y) \, dy, \qquad (12.1.12)$$

while, if θ is the median then $T(F_Y)$ is defined as

$$T(F_Y) = F_Y^{-1}\left(\frac{1}{2}\right). \qquad (12.1.13)$$

We call $T(F_Y)$ a **functional**. Recall that functionals were also introduced for location models in Chapter 10. It was shown there that $T(F_Y) = T(F) + \theta$. In

Exercise 12.1.3 the reader is asked to show that this is true for the mean and median functionals defined above.

Estimating equations such as those defined in expressions (12.1.6) and (12.1.14) are often quite intuitive, for example based on likelihood equations or methods such as least squares. On the other hand, functionals are more of an abstract concept. But often the estimating equations naturally lead to the functionals. We outline this next for the mean and median functionals.

Let F_n be the empirical distribution function of the realized sample $y_1, y_2, \ldots, y_n$. That is, F_n is the cdf of the distribution which puts mass n^{-1} on each y_i; see (10.1.1). Note that we can write the estimating equation (12.1.6), which defines the sample mean as

$$\sum_{i=1}^{n} (y_i - \theta) \frac{1}{n} = 0. \tag{12.1.14}$$

This is an expectation using the empirical distribution. Since $F_n \to F_Y$ in probability, it would seem that this expectation converges to

$$\int_{-\infty}^{\infty} [y - T(F_Y)] f_Y(y) \, dy = 0. \tag{12.1.15}$$

The solution to the above equation is, of course, $T(F_Y) = E(Y)$.

Likewise, we can write the EE, (12.1.14), which defines the sample median, as

$$\sum_{i=1}^{n} \operatorname{sgn}(Y_i - \theta) \frac{1}{n} = 0. \tag{12.1.16}$$

The corresponding equation for the functional $\theta = T(F_Y)$ is the solution of the equation

$$\int_{-\infty}^{\infty} \operatorname{sgn}[y - T(F_Y)] f_Y(y) \, dy = 0. \tag{12.1.17}$$

Note that this can be written as

$$0 = -\int_{-\infty}^{T(F_Y)} f_Y(y) \, dy + \int_{T(F_Y)}^{\infty} f_Y(y) \, dy = -F_Y[T(F_Y)] + 1 - F_Y[T(F_Y)].$$

Hence, $F_Y[T(F_Y)] = 1/2$ or $T(F_Y) = F_Y^{-1}(1/2)$. Thus $T(F_Y)$ is the median of the distribution of Y.

Now we want to consider how a given functional $T(F_Y)$ changes relative to some perturbation. The analogue of adding an outlier to $F(t)$ is to consider a point-mass contamination of the cdf $F_Y(t)$ at a point y. That is, for $\epsilon > 0$, let

$$F_{y,\epsilon}(t) = (1 - \epsilon) F_Y(t) + \epsilon \Delta_y(t), \tag{12.1.18}$$

where $\Delta_y(t)$ is the cdf with all its mass at y, i.e.,

$$\Delta_y(t) = \begin{cases} 0 & t < y \\ 1 & t \geq y. \end{cases} \tag{12.1.19}$$

The cdf $F_{y,\epsilon}(t)$ is a mixture of two distributions. When sampling from it, $(1-\epsilon)100\%$ of the time an observation is drawn from $F_Y(t)$ while $\epsilon 100\%$ of the time y (an outlier) is drawn. So y has the flavor of the outlier in the sensitivity curve. As Exercise 12.1.4 shows, $F_{y,\epsilon}(t)$ is in the ϵ neighborhood of $F_Y(t)$, based on the sup norm, for all y. That is, for all y, $|F_{y,\epsilon}(t) - F_Y(t)| \leq \epsilon$. Hence, the functional at $F_{y,\epsilon}(t)$ should also be close to $T(F_Y)$. The concept for functionals, corresponding to the sensitivity curve, is the function

$$IF(y; \widehat{\theta}) = \lim_{\epsilon \to 0} \frac{T(F_{y,\epsilon}) - T(F_Y)}{\epsilon}, \qquad (12.1.20)$$

provided the limit exists. The function $IF(y; \widehat{\theta})$ is called the **influence function** of the estimator $\widehat{\theta}$ at y. As the notation suggests, it can be thought of as a derivative of the functional $T(F_{y\epsilon})$ with respect to ϵ evaluated at 0 and we will often determine it this way. Note that for ϵ small,

$$T(F_{y,\epsilon}) \doteq T(F_Y) + \epsilon IF(y; \widehat{\theta});$$

hence, the change of the functional due to point-mass contamination is approximately directly proportional to the influence function. We want estimators, whose influence functions are not sensitive to outliers. Further, as mention above, for any y, $F_{y,\epsilon}(t)$ is close to $F_Y(t)$. Hence, at least, the influence function should be a bounded function of y.

Definition 12.1.1. *The estimator $\widehat{\theta}$ is said to be* **robust** *if $|IF(y; \widehat{\theta})|$ is bounded for all y.*

Hampel (1974) proposed the influence function and discussed its important properties, a few of which we list below. First, however, we determine the influence functions of the sample mean and median.

For the sample mean, recall Section 3.4.1 on mixture distributions. The function $F_{y,\epsilon}(t)$ is the cdf of the random variable $U = I_{1-\epsilon}Y + [1 - I_{1-\epsilon}]W$, where Y, $I_{1-\epsilon}$, and W are independent random variables, Y has cdf $F_Y(t)$, W has cdf $\Delta_y(t)$, and $I_{1-\epsilon}$ is $b(1, 1 - \epsilon)$. Hence,

$$E(U) = (1 - \epsilon)E(Y) + \epsilon E(W) = (1 - \epsilon)E(Y) + \epsilon y.$$

Denote the mean functional by $T_\mu(F_Y) = E(Y)$. In terms of $T_\mu(F)$ we have just shown that

$$T_\mu(F_{y,\epsilon}) = (1 - \epsilon)T_\mu(F_Y) + \epsilon y.$$

Therefore,

$$\frac{\partial T_\mu(F_{y,\epsilon})}{\partial \epsilon} = -T_\mu(F) + y.$$

Hence, the influence function of the sample mean is

$$IF(y; \overline{Y}) = y - \mu, \qquad (12.1.21)$$

where $\mu = E(Y)$. The influence function of the sample mean is linear in y and, hence, is an unbounded function of y. Therefore the sample mean is not a robust estimator. Another way to derive the influence function is to differentiate implicitly equation (12.1.15) when this equation is defined for $F_{y,\epsilon}(t)$; see Exercise 12.1.6. We will obtain influence functions in this way in later sections.

Example 12.1.1 (Influence Function of the Sample Median). In this example, we derive the influence function of the sample median, $\widehat{\theta}_{L_1}$. In this case, the functional is $T_\theta(F) = F^{-1}(1/2)$, i.e., the median of F. To determine the influence function, we first need to determine the functional at the contaminated cdf $F_{y,\epsilon}(t)$; i.e, determine $F_{y,\epsilon}^{-1}(1/2)$. As shown in Exercise 12.1.8, the inverse of the cdf $F_{y,\epsilon}(t)$ is given by

$$F_{y,\epsilon}^{-1}(u) = \begin{cases} F^{-1}\left(\frac{u}{1-\epsilon}\right) & u < F(y) \\ F^{-1}\left(\frac{u-\epsilon}{1-\epsilon}\right) & u \ge F(y), \end{cases} \tag{12.1.22}$$

for $0 < u < 1$. Hence, letting $u = 1/2$, we get

$$T(F_{y,\epsilon}) = F_{y,\epsilon}^{-1}(1/2) = \begin{cases} F_Y^{-1}\left(\frac{1/2}{1-\epsilon}\right) & F_Y^{-1}\left(\frac{1}{2}\right) < y \\ F_Y^{-1}\left(\frac{(1/2)-\epsilon}{1-\epsilon}\right) & F_Y^{-1}\left(\frac{1}{2}\right) > y. \end{cases} \tag{12.1.23}$$

Based on (12.1.23) the partial derivative of $F_{y,\epsilon}^{-1}(1/2)$ with respect to ϵ is seen to be

$$\frac{\partial T_\theta(F_{y,\epsilon})}{\partial \epsilon} = \begin{cases} \frac{(1/2)(1-\epsilon)^{-2}}{f_Y[F_Y^{-1}((1/2)/(1-\epsilon))]} & F_Y^{-1}\left(\frac{1}{2}\right) < y \\ \frac{(-1/2)(1-\epsilon)^{-2}}{f_Y[F_Y^{-1}(\{(1/2)-\epsilon\}/\{1-\epsilon\})]} & F_Y^{-1}\left(\frac{1}{2}\right) > y. \end{cases} \tag{12.1.24}$$

Evaluating this partial derivative at $\epsilon = 0$, we arrive at the influence function of the median

$$IF(y; \widehat{\theta}_{L_1}) = \begin{cases} \frac{1}{2f_Y(\theta)} & \theta < y \\ \frac{-1}{2f_Y(\theta)} & \theta > y \end{cases} = \frac{\text{sgn}(y-\theta)}{2f(\theta)}, \tag{12.1.25}$$

where θ is the median of F_Y. Because this influence function is bounded, the sample median is a robust estimator. ∎

We now list three useful properties of the influence function of an estimator. Note that for the sample mean, $E[IF(Y; \overline{Y})] = E[Y] - \mu = 0$. As Exercise 12.1.9 shows, this is true for the influence function of the sample median, also. Actually, it is true in general. Let $IF(y) = IF(y; \widehat{\theta})$ denote the influence function of the estimator $\widehat{\theta}$ with functional $\theta = T(F_Y)$. Then

$$E[IF(Y)] = 0, \tag{12.1.26}$$

provided expectations exist; see Huber (1981) for discussion. Hence, we have

$$\text{Var}[IF(Y)] = E[IF^2(Y)], \tag{12.1.27}$$

provided the squared expectation exists. A third property of the influence function is the asymptotic result

$$\sqrt{n}[\widehat{\theta} - \theta] = \frac{1}{\sqrt{n}} \sum_{i=1}^n IF(Y_i) + o_p(1). \tag{12.1.28}$$

Assume that the variance (12.1.27) exists, then because $IF(Y_1), \ldots, IF(Y_n)$ are iid with finite variance, the simple Central Limit Theorem and (12.1.28) imply that

$$\sqrt{n}[\widehat{\theta} - \theta] \overset{D}{\to} N(0, E[IF^2(Y)]). \tag{12.1.29}$$

Thus we can obtain the asymptotic distribution of the estimator from its influence function. Under general conditions, expression (12.1.28) holds, but often the verification of the conditions is difficult and the asymptotic distribution can be obtained more easily in another way; see Huber (1981) for discussion. In this chapter, though, we will use (12.1.28) to obtain asymptotic distributions of estimators. Suppose (12.1.28) holds for the estimators $\widehat{\theta}_1$ and $\widehat{\theta}_2$, which are both estimators of the same functional, say, θ. Then letting IF_i denote the influence function of $\widehat{\theta}_i$, $i = 1, 2$, we can express the asymptotic relative efficiency between the two estimators as

$$\mathrm{ARE}(\widehat{\theta}_1, \widehat{\theta}_2) = \frac{E[IF_2^2(Y)]}{E[IF_1^2(Y)]}. \tag{12.1.30}$$

We illustrate these ideas with the sample median.

Example 12.1.2 (Asymptotic Distribution of the Sample Median). The influence function for the sample median $\widehat{\theta}_{L_1}$ is given by (12.1.25). Since $E[\mathrm{sgn}^2(Y - \theta)] = 1$, by expression (12.1.29) the asymptotic distribution of the sample median is

$$\sqrt{n}[\widehat{\theta} - \theta] \overset{D}{\to} N\left(0, [2f_Y(\theta)]^{-2}\right),$$

where θ is the median of the pdf $f_Y(t)$. This agrees with the result given in Chapter 10. ∎

12.1.3 Breakdown Point of an Estimator

The influence function of an estimator measures the sensitivity of an estimator to single outlier, sometimes called the *local sensitivity* of the estimator. We next discuss a measure of *global sensitivity* of an estimator, that is, what proportion of outliers can an estimator tolerate without breaking down.

To make this precise, let $\mathbf{y}' = (y_1, y_2, \ldots, y_n)$ be a realization of a sample. Suppose we corrupt m data points of the sample; i.e., we replace $y_1, \ldots, y_m$ by $y_1^*, \ldots, y_m^*$ where these points are large outliers. Let $\mathbf{y}_m = (y_1^*, \ldots, y_m^*, y_{m+1}, \ldots, y_n)$ denote the corrupted sample. Define the bias of the estimator upon corrupting m data points to be

$$\mathrm{bias}(m, \mathbf{y}_n, \widehat{\theta}) = \sup |\widehat{\theta}(\mathbf{y}_m) - \widehat{\theta}(\mathbf{y}_n)|, \tag{12.1.31}$$

where the sup is taken over all possible corrupted samples $\mathbf{y}_m$. If this bias is infinite, we say that the estimator has **broken down**. The smallest proportion of corruption an estimator can tolerate until its breakdown is called its *finite sample breakdown point*. More precisely, if

$$\epsilon_n^* = \min_m \{m/n : \mathrm{bias}(m, \mathbf{y}_n, \widehat{\theta}) = \infty\}, \tag{12.1.32}$$

then ϵ_n^* is called the **finite sample breakdown point** of $\hat{\theta}$. If the limit

$$\epsilon_n^* \rightarrow \epsilon^*, \tag{12.1.33}$$

exists, we call ϵ^* the **breakdown point** of $\hat{\theta}$.

To determine the breakdown point of the sample mean, suppose we corrupt one data point, say, without loss of generality, the first data point. The corrupted sample is then, $\mathbf{y}' = (y_1^*, y_2, \ldots, y_n)$. Denote the sample mean of the corrupted sample by $\bar{y}^*$. Then it is easy to see that

$$\bar{y}^* - \bar{y} = \frac{1}{n}(y_1^* - y_1).$$

Hence, bias$(1, \mathbf{y}_n, \bar{y})$ is a linear function of y_1^* and can be made as large (in absolute value) as desired by taking y_1^* large (in absolute value). Therefore the finite sample breakdown of the sample mean is $1/n$. Because this goes to 0 as $n \rightarrow \infty$, the breakdown point of the sample mean is 0.

Example 12.1.3 (Breakdown Value of the Sample Median). Next consider the sample median. Let $\mathbf{y}_n = (y_1, y_2, \ldots, y_n)$ be a realization of a random sample. If the sample size is $n = 2k$ then it is easy to see that in a corrupted sample $\mathbf{y}_n$ when $y_{(k)}$ tends to $-\infty$, the median also tends to $-\infty$. Hence, the breakdown value of the sample median is k/n which tends to 0.5. By a similar argument, when the sample size is $n = 2k + 1$, the breakdown value is $(k + 1)/n$ and it also tends to 0.5 as the sample size increases. Hence, we say that the sample median is a 50% breakdown estimate. For a location model, 50% breakdown is the highest possible breakdown point for an estimate. Thus the median achieves the highest possible breakdown point. ∎

EXERCISES

12.1.1. Consider the location model as defined in expression (12.1.3). Let

$$\hat{\boldsymbol{\eta}} = \text{Argmin}_{\boldsymbol{\eta}} \|\mathbf{Y} - \boldsymbol{\eta}\|_{LS}^2.$$

Show that $\hat{\boldsymbol{\eta}} = \bar{y}\mathbf{1}$.

12.1.2. Obtain the sensitivity curves for the sample mean and median for the following data set. Evaluate the curves at the values -300 to 300 in increments of 10 and graph the curves on the same plot. Compare the sensitivity curves.

-9	58	12	-1	-37	0	11	21
18	-24	-4	-53	-9	9	8	

12.1.3. Assume the location model, (12.1.1), is true.

(a) Consider the mean functional defined implicitly in equation (12.1.15). Show that $T(F_Y) = T(F) + \theta$.

(b) Repeat (a) for the median functional defined by equation (12.1.16).

12.1.4. Let $F_{y,\epsilon}(t)$ be the point-mass contaminated cdf given in expression (12.1.18). Show that

$$|F_{y,\epsilon}(t) - F_Y(t)| \le \epsilon,$$

for all t.

12.1.5. Suppose Y is a random variable with mean 0 and variance σ^2. Recall that the function $F_{y,\epsilon}(t)$ is the cdf of the random variable $U = I_{1-\epsilon}Y + [1 - I_{1-\epsilon}]W$ where Y, $I_{1-\epsilon}$, and W are independent random variables, Y has cdf $F_Y(t)$, W has cdf $\Delta_y(t)$, and $I_{1-\epsilon}$ is $b(1, 1-\epsilon)$. Define the functional $V(F_Y) = \text{Var}(Y) = \sigma^2$. Note that the functional at the contaminated cdf $F_{y,\epsilon}(t)$ is the variance of the random variable $U = I_{1-\epsilon}Y + [1 - I_{1-\epsilon}]W$. To derive the influence function of the variance perform the following steps:

(a) Show that $E(U) = \epsilon y$.

(b) Show that $\text{Var}(U) = (1 - \epsilon)\sigma^2 + \epsilon y^2 - \epsilon^2 y^2$.

(c) Obtain the partial derivative of the right side of this last equation with respect to ϵ. This is the influence function.

Hint: Because $I_{1-\epsilon}$ is a Bernoulli random variable, $I_{1-\epsilon}^2 = I_{1-\epsilon}$. Why?

12.1.6. Often influence functions are derived by differentiating implicitly the defining equation for the functional at the contaminated cdf $F_{y,\epsilon}(t)$, (12.1.18). Consider the mean functional with the defining equation (12.1.15). Using the linearity of the differential, first show that the defining equation at the cdf $F_{y,\epsilon}(t)$ can be expressed as

$$0 = \int_{-\infty}^{\infty} [t - T(F_{y,\epsilon})]dF_{y,\epsilon}(t) \quad = \quad (1 - \epsilon) \int_{-\infty}^{\infty} [t - T(F_{y,\epsilon})]f_Y(t)\, dt$$

$$+\epsilon \int_{-\infty}^{\infty} [t - T(F_{y,\epsilon})]\, d_\Delta(t). \quad (12.1.34)$$

Recall that we want $\partial T(F_{y,\epsilon})/\partial \epsilon$. Obtain this by differentiating implicitly the above equation with respect to ϵ.

12.1.7. In Exercise 12.1.5, the influence function of the variance functional was derived directly. Assuming that the mean of Y is 0, note that the variance functional, $V(F_Y)$, also solves the equation

$$0 = \int_{-\infty}^{\infty} [t^2 - V(F_Y)]f_Y(t)\, dt.$$

(a) Determine the natural estimator of the variance by writing the defining equation at the empirical cdf $F_n(t)$, for $Y_1 - \overline{Y}, \ldots Y_n - \overline{Y}$ iid with cdf $F_Y(t)$, and solving for $V(F_n)$.

(b) As in Exercise 12.1.6, write the defining equation for the variance functional at the contaminated cdf $F_{y,\epsilon}(t)$.

(c) Then derive the influence function by implicit differentiation of the defining equation in Part (b).

12.1.8. Show that the inverse of the cdf $F_{y,\epsilon}(t)$ given in expression (12.1.22) is correct.

12.1.9. Let $IF(y)$ be the influence function of the sample median given by (12.1.25). Determine $E[IF(Y)]$ and $\text{Var}[IF(Y)]$.

12.1.10. Let $y_1, y_2, \ldots, y_n$ be a realization of a random sample. Recall that the Hodges-Lehmann estimate of location, (Chapter 10), is the median of the Walsh averages; that is,

$$\widehat{\theta} = \text{med}_{i \leq j} \left\{ \frac{y_i + y_j}{2} \right\}. \tag{12.1.35}$$

Show that the breakdown point of this estimate is 0.29. *Hint:* Suppose we corrupt m data points. We need to determine the value of m which results in corruption of one-half of the Walsh averages. Show that the corruption of m data points leads to

$$p(m) = m + \binom{m}{2} + m(n - m)$$

corrupted Walsh averages. Hence, the finite sample breakdown point is the "correct" solution of the quadratic equation $p(m) = n(n+1)/4$.

12.2   LS and Wilcoxon Estimators of Slope

In Sections 9.6 and 10.7 respectively, we presented the least squares (LS) procedure and a rank-based (Wilcoxon) procedure for fitting simple linear models. In this section, we briefly compare these procedures in terms of their geometries and their robustness properties. We follow the format of the last section.

Recall that the simple linear model is given by

$$Y_i = \alpha + \beta x_{ci} + \varepsilon_i, \quad i = 1, 2, \ldots, n, \tag{12.2.1}$$

where $\varepsilon_1, \varepsilon_2, \ldots, \varepsilon_n$ are continuous random variable which are iid. In this model, we have centered the regression variables; that is, $x_{ci} = x_i - \overline{x}$, where $x_1, x_2, \ldots, x_n$ are considered fixed. The parameter of interest in this section is the slope parameter β, the expected change (provided expectations exist) when the regression variable increases by one unit. The centering of the xs allows us to consider the slope parameter by itself. The results we present are invariant to the intercept parameter α. Estimates of α will be discussed in Subsection 12.2.3. With this in mind, define the random variable e_i to be $\varepsilon_i + \alpha$. Then we can write the model as

$$Y_i = \beta x_{ci} + e_i, \quad i = 1, 2, \ldots, n, \tag{12.2.2}$$

where $e_1, e_2, \ldots, e_n$ are iid with continuous cdf $F(x)$ and pdf $f(x)$. We often refer to the support of Y as the Y-**space**. Likewise we refer to the range of X as the X-**space**. The X-**space** is often referred to as **factor space**.

In Section 9.6 we also wrote the linear model in its matrix formulation. Here, let $\mathbf{Y} = (Y_1, \ldots, Y_n)'$ denote the response vector, let $\mathbf{e} = (e_1, \ldots, e_n)'$ denote the vector of random errors, and let $\mathbf{x}_c = (x_{c1}, \ldots, x_{cn})'$ denote the vector of regression variables. We can then write the linear model in the form

$$\mathbf{Y} = \mathbf{x}_c \beta + \mathbf{e}. \tag{12.2.3}$$

12.2.1 Norms and Estimating Equations

The first procedure is *least squares* (LS). Recall from Example 9.6.1 that the LS estimator of β minimizes the square of the Euclidean distance between $\mathbf{Y}$ and the vector $\mathbf{x}_c \beta$, i.e.,

$$\widehat{\beta}_{LS} = \text{Argmin} \|\mathbf{Y} - \mathbf{x}_c \beta\|_{LS}^2, \tag{12.2.4}$$

where the Euclidean squared-norm is $\|\mathbf{v}\|_{LS}^2 = \sum_{i=1}^n v_i^2$, for $\mathbf{v} \in R^n$. Equivalently taking the partial derivative of $\|\mathbf{Y} - \mathbf{x}_c \beta\|_{LS}^2$ with respect to β and setting the result to 0, the LS estimator solves the equation

$$\sum_{i=1}^n (Y_i - x_{ci} \beta) x_{ci} = 0. \tag{12.2.5}$$

This is the estimating equation (EE) for the LS estimator of β and is often called the **normal equation**. It is easy to see that the LS estimator is

$$\widehat{\beta}_{LS} = \frac{\sum_{i=1}^n x_{ci} Y_i}{\sum_{i=1}^n x_{ci}^2}, \tag{12.2.6}$$

which agrees with expression (9.6) of Chapter 9.

In terms of the matrix model (12.2.3), $\widehat{\beta}_{LS} = (\mathbf{x}_c' \mathbf{x}_c)^{-1} \mathbf{x}_c' \mathbf{Y}$; see Exercise 12.2.1. The fitted or predicted value of $\mathbf{Y}$ is

$$\widehat{\mathbf{Y}}_{LS} = \mathbf{x}_c \widehat{\beta}_{LS} = \mathbf{x}_c (\mathbf{x}_c' \mathbf{x}_c)^{-1} \mathbf{x}_c' \mathbf{Y} = \mathbf{P}_c \mathbf{Y}, \tag{12.2.7}$$

where $\mathbf{P}_c$ is the projection matrix onto the vector space spanned by $\mathbf{x}_c$. Denote the LS residuals by $\widehat{\mathbf{e}}_{LS} = \mathbf{Y} - \widehat{\mathbf{Y}}_{LS}$. Then the vectors $\widehat{\mathbf{Y}}_{LS}$ and $\widehat{\mathbf{e}}_{LS}$ are orthogonal as shown in Figure 9.6.2. See Exercise 12.2.1 for details.

For the *Wilcoxon* estimator, we simply use a different norm. Exercise 12.2.2 shows that the following function is a norm on R^n,

$$\|\mathbf{v}\|_W = \sum_{i=1}^n a(R(v_i)) v_i, \quad \mathbf{v} \in R^n, \tag{12.2.8}$$

where $R(v_i)$ denotes the rank of v_i among $v_1, \ldots, v_n$, $a(i) = \varphi(i/(n+1))$, and $\varphi(u) = \sqrt{12}[u - (1/2)]$. Hence, similar to LS, we define an estimate of β by

$$\widehat{\beta}_W = \text{Argmin} \|\mathbf{Y} - \mathbf{x}_c \beta\|_W, \tag{12.2.9}$$

where the subscript W denotes Wilcoxon because the linear score function $\varphi(u)$ is the Wilcoxon score function that was first used in Section 10.4. If we differentiate $\|\mathbf{Y} - \mathbf{x}_c\beta\|_W$ with respect to β, we see that $\widehat{\beta}_W$ solves the equation

$$\sum_{i=1}^{n} a(R(Y_i - x_{ci}\beta))x_{ci} = 0; \tag{12.2.10}$$

see Exercise 12.2.3. This is the estimating equation for the Wilcoxon estimator of β. Note that this is the same estimator which was discussed in Section 10.7 for the Wilcoxon score function. This equation is the analogue of the LS normal equation.

As with LS, denote the Wilcoxon fitted value and residual vector by

$$\widehat{\mathbf{Y}}_W = \mathbf{x}_c\widehat{\beta}_W \text{ and } \widehat{\mathbf{e}}_W = \mathbf{Y} - \widehat{\mathbf{Y}}_W,$$

respectively. We can display these as in Figure 9.6.2, but the angle between them is not necessarily a right angle. See Exercise 12.2.4 for discussion.

12.2.2   Influence Functions

To determine the robustness properties of these procedures, first consider a probability model corresponding to Model (12.2.2) in which X in addition to Y is a random variable. Assume that the random vector (X, Y) has joint cdf and pdf, $H(x, y)$ and $h(x, y)$, respectively, and satisfies

$$Y = \beta X + e, \tag{12.2.11}$$

where the random variable e has cdf and pdf $F(t)$ and $f(t)$, respectively, and e and X are independent. Since we have centered the xs, we will also assume that $E(X) = 0$. As Exercise 12.2.5 shows

$$P(Y \leq t|X = x) = F(t - \beta x), \tag{12.2.12}$$

and, hence, Y and X are independent if and only if $\beta = 0$.

The functional for the LS estimator easily follows from the LS normal equation (12.2.5). Let H_n denote the empirical cdf of the pairs $(x_1, Y_1), (x_2, Y_2), \ldots, (x_n, Y_n)$; that is, H_n is the cdf corresponding to the discrete distribution which puts probability (mass) of $1/n$ on each point (x_i, Y_i). Then the LS estimating equation, (12.2.5), can be expressed as an expectation with respect to this distribution as

$$\sum_{i=1}^{n}(Y_i - x_{ci}\beta)x_{ci}\frac{1}{n} = 0. \tag{12.2.13}$$

For the probability model, (12.2.11), it follows that the functional $T_{LS}(H)$ corresponding to the LS estimate is is the solution to the equation

$$\int_{-\infty}^{\infty} \int_{-\infty}^{\infty} [y - T_{LS}(H)x]xh(x, y)\,dxdy = 0. \tag{12.2.14}$$

To obtain the functional corresponding to the Wilcoxon estimate, recall the association between the ranks and the empirical cdf; see (10.5.14). We have

$$a(R(Y_i - x_{ci}\beta)) = \varphi \left[\frac{n}{n+1} F_n(Y_i - x_{ci}\beta) \right]. \qquad (12.2.15)$$

Based on the Wilcoxon estimating equations, (12.2.10), and (12.2.15), the functional $T_W(H)$ corresponding to the Wilcoxon estimate satisfies the equation

$$\int_{-\infty}^{\infty} \int_{-\infty}^{\infty} \varphi\{F[y - T_W(H)x]\}xh(x,y)\,dxdy = 0. \qquad (12.2.16)$$

We next derive the influence functions of the LS and Wilcoxon estimators of β. In regression models, we are concerned about the influence of outliers in both the Y and X spaces. Consider then a point-mass distribution with all its mass at the point (x_0, y_0) and let $\Delta_{(x_0,y_0)}(x, y)$ denote the corresponding cdf. Let ϵ denote the probability of sampling from this contaminated distribution, where $0 < \epsilon < 1$. Hence, consider the contaminated distribution with cdf

$$H_\epsilon(x,y) = (1 - \epsilon)H(x,y) + \epsilon\Delta_{(x_0,y_0)}(x,y). \qquad (12.2.17)$$

Because the differential is a linear operator we have

$$dH_\epsilon(x,y) = (1 - \epsilon)dH(x,y) + \epsilon d\Delta_{(x_0,y_0)}(x,y), \qquad (12.2.18)$$

where $dH(x,y) = h(x,y)\,dxdy$; that is, d corresponds to the second mixed partial $d^2/dx\,dy$.

By (12.2.14), the LS functional T_ϵ at the cdf $H_\epsilon(x, y)$ satisfies the equation

$$0 = (1 - \epsilon) \int_{-\infty}^{\infty} \int_{-\infty}^{\infty} x(y - xT_\epsilon)h(x,y)\,dxdy + \epsilon \int_{-\infty}^{\infty} \int_{-\infty}^{\infty} x(y - xT_\epsilon)\,d\Delta_{(x_0,y_0)}(x,y).$$

$$(12.2.19)$$

To find the partial derivative of T_ϵ with respect to ϵ, we simply implicitly differentiate expression (12.2.19) with respect to ϵ which yields

$$\begin{aligned}
0 = & -\int_{-\infty}^{\infty} \int_{-\infty}^{\infty} x(y - T_\epsilon x)h(x,y)\,dxdy \\
& +(1 - \epsilon) \int_{-\infty}^{\infty} \int_{-\infty}^{\infty} x(-x)\frac{\partial T_\epsilon}{\partial \epsilon}h(x,y)\,dxdy \\
& + \int_{-\infty}^{\infty} \int_{-\infty}^{\infty} x(y - xT_\epsilon)\,d\Delta_{(x_0,y_0)}(x,y) + \epsilon B, \qquad (12.2.20)
\end{aligned}$$

where the expression for B is not needed since we are evaluating this partial at $\epsilon = 0$. Notice that at $\epsilon = 0$, $y - T_\epsilon x = y - Tx = y - \beta x$. Hence, at $\epsilon = 0$, the first expression on the rightside of (12.2.20) is 0, while the second expression becomes $-E(X^2)(\partial T/\partial \epsilon)$, where the partial is evaluated at 0. Finally, the third expression becomes $x_0(y_0 - \beta x_0)$. Therefore, solving for the partial $\partial T_\epsilon / \partial \epsilon$ and evaluating at $\epsilon = 0$, we see that the influence function of the LS estimator is given by

$$IF(x_0, y_0; \widehat{\beta}_{LS}) = \frac{(y_0 - \beta x_0)x_0}{E(X^2)}. \tag{12.2.21}$$

Note that the influence function is unbounded in both the Y and X spaces. Hence, the LS estimator will be unduly sensitive to outliers in both spaces. It is not robust.

Based on expression (12.2.16), the Wilcoxon functional at the contaminated distribution satisfies the equation

$$
\begin{aligned}
0 &= (1-\epsilon) \int_{-\infty}^{\infty} \int_{-\infty}^{\infty} x\varphi[F(y - xT_\epsilon)]h(x,y)\,dxdy \\
&\quad + \epsilon \int_{-\infty}^{\infty} \int_{-\infty}^{\infty} x\varphi[F(y - xT_\epsilon)]\,d\Delta_{(x_0,y_0)}(x,y);
\end{aligned}
\tag{12.2.22}
$$

(technically the cdf F should be repaced by the actual cdf of the residual, but the result is the same, see HM p. 426). Proceeding to implicitly differentiate this expression with respect to ϵ, we obtain

$$
\begin{aligned}
0 &= -\int_{-\infty}^{\infty} \int_{-\infty}^{\infty} x\varphi[F(y - xT_\epsilon)]h(x,y)\,dxdy \\
&\quad + (1-\epsilon) \int_{-\infty}^{\infty} \int_{-\infty}^{\infty} x\varphi'[F(y - T_\epsilon x)]f(y - T_\epsilon x)(-x)\frac{\partial T_\epsilon}{\partial \epsilon}h(x,y)\,dxdy \\
&\quad + \int_{-\infty}^{\infty} \int_{-\infty}^{\infty} x\varphi[F(y - xT_\epsilon)]\,d\Delta_{(x_0,y_0)}(x,y) + \epsilon B,
\end{aligned}
\tag{12.2.23}
$$

where the expression for B is not needed since we are evaluating this partial at $\epsilon = 0$. When $\epsilon = 0$ then $Y - TX = e$ and the random variables e and X are independent. Hence, upon setting $\epsilon = 0$, expression (12.2.23) simplifies to

$$0 = -E[\varphi'(F(e))f(e)]E(X^2)\frac{\partial T_\epsilon}{\partial \epsilon}\bigg|_{\epsilon=0} + \varphi[F(y_0 - x_0\beta)]x_0. \tag{12.2.24}$$

Since $\varphi'(u) = \sqrt{12}$, we finally obtain, as the influence function of the Wilcoxon estimator,

$$IF(x_0, y_0; \widehat{\beta}_W) = \frac{\tau\varphi[F(y_0 - \beta x_0)]x_0}{E(X^2)}, \tag{12.2.25}$$

where $\tau = 1/[\sqrt{12} \int f^2(e)\,de]$. Note that the influence function is bounded in the Y space but unbounded in the x space. Thus, unlike the LS estimator, the Wilcoxon estimator is robust against outliers in the Y space, but like the LS estimator it is sensitive to outliers in the X space.

Breakdown for the regression model is based on the corruption of the sample in Model (12.2.2); that is, the sample $(x_{c1}, Y_1), \ldots, (x_{cn}, Y_n)$. Based on the influence functions for both the LS and Wilcoxon estimators, it is clear that corrupting one x_i will breakdown both estimators. This is shown in Exercise 12.2.6. Hence, the breakdown point of both estimators is 0. Weighted versions, though, of the Wilcoxon estimators have bounded influence in both spaces and can achieve 50% breakdown; see Remark 12.4.1 for references. In Section 12.4, we discuss the asymptotic distributions of the LS and Wilcoxon estimators.

12.2.3 Intercept

In practice, the linear model usually contains an intercept parameter; that is, the model is given by (12.2.1) with intercept parameter α. Notice that α is a location parameter of the random variables $Y_i - \beta X_i$. This suggests an estimate of location on the residuals $Y_i - \widehat{\beta} X_i$. For LS, we take the mean of the residuals; i.e.,

$$\widehat{\alpha}_{LS} = n^{-1} \sum_{i=1}^{n} (Y_i - \widehat{\beta}_{LS} X_i) = \overline{Y}, \qquad (12.2.26)$$

because the X_is are centered. Exercise 12.3.13 discusses this estimate in terms of *least squares*. For the Wilcoxon fit, several choices seem appropriate. We will use the median of the Wilcoxon residuals. That is, let

$$\widehat{\alpha}_W = \text{med}_{1 \leq i \leq n} \{Y_i - \widehat{\beta}_W X_i\}. \qquad (12.2.27)$$

Remark 12.2.1 (Computation). There are resources available to compute the Wilcoxon estimates. For example, the web site `www.stat.wmich.edu/slab/RGLM` offers commands which obtains both the Wilcoxon and LS fits. At this site, click on `Simple Regression`. The minitab computer package has the command `rregr` which computes both the Wilcoxon and LS fits. Terpstra et al. (2004) have developed a collection of R and S-PLUS routines which return the Wilcoxon fit. Go to the site `www.stat.wmich.edu/mckean/HMC/Rcode` to download these routines. ∎

EXERCISES

12.2.1. In terms of the matrix model, (12.2.3),

(a) Show that the LS estimate is given by $\widehat{\beta}_{LS} = (\mathbf{x}_c' \mathbf{x}_c)^{-1} \mathbf{x}_c' \mathbf{Y}$.

(b) Show that the vectors $\widehat{\mathbf{Y}}_{LS}$ and $\widehat{\mathbf{e}}_{LS}$ are orthogonal.

12.2.2. Consider the function (12.2.8). By using the correspondence between order statistics and ranks show that

$$\|\mathbf{v}\|_W = \sum_{i=1}^{n} a(R(v_i)) v_i = \sum_{i=1}^{n} a(i) v_{(i)},$$

where $v_{(1)} \leq \cdots \leq v_{(n)}$ are the ordered values of $v_1, \ldots, v_n$. Then by establishing the following properties, show that the function (12.2.8) is a **pseudo-norm** on R^n.

(a) $\|\mathbf{v}\|_W \geq 0$ and $\|\mathbf{v}\|_W = 0$ if and only if $v_1 = v_2 = \cdots = v_n$.

Hint: First, because the scores $a(i)$ sum to 0, show that

$$\sum_{i=1}^{n} a(i)v_{(i)} = \sum_{i<j} a(i)[v_{(i)} - v_{(j)}] + \sum_{i>j} a(i)[v_{(i)} - v_{(j)}],$$

where j is the largest integer in the set $\{1, 2, \ldots, n\}$ such that $a(j) < 0$.

(b) $\|c\mathbf{v}\|_W = |c|\|\mathbf{v}\|_W, \quad$ for all $c \in R$.

(c) $\|\mathbf{v} + \mathbf{w}\|_W \leq \|\mathbf{v}\|_W + \|\mathbf{w}\|_W,$ for all $\mathbf{v}, \mathbf{w} \in R^n$.

Hint: Determine the permutations, say, i_k and j_k of the integers $\{1, 2, \ldots, n\}$ which maximize the $\sum_{k=1}^{n} c_{i_k} d_{j_k}$ for two sets of numbers $\{c_1, \ldots, c_n\}$ and $\{d_1, \ldots, d_n\}$.

12.2.3. Derive the Wilcoxon normal equations, (12.2.9).
Hint: Use the identity

$$\|\mathbf{Y} - \mathbf{x}_C \beta\|_W = \sum_{i=1}^{n} a(i)(Y - x_c\beta)_{(i)}.$$

12.2.4. For the Wilcoxon regression procedure, which vector is orthogonal to $\widehat{\mathbf{Y}}_W$?

12.2.5. For Model (12.2.11), show that equation (12.2.12) holds. Then show that Y and X are independent if and only if $\beta = 0$. Hence, independence is based on a parameter. This is a case where normality is not necessary to have this independence property.

12.2.6. Consider the Telephone Data discussed in Example 10.7.2. It is easily seen in the Figure 10.7.1 that there are seven outliers in the Y space. Based on the estimates discussed in this example, the Wilcoxon estimate of slope is robust to these outliers while the LS estimate is highly sensitive to them.

(a) For this data set change the last value of x from 73 to 173. Notice the drastic change in the LS fit.

(b) Obtain the Wilcoxon estimate for the changed data in Part (a). Notice that it has a drastic change also. To obtain the Wilcoxon fit, click on RGLM at the web site: **www.stat.wmich.edu/slab**.

(c) Using the Wilcoxon estimates of Example 10.7.2, change the the value of Y at $x = 173$ to the predicted value of Y based on the Wilcoxon estimates of Example 10.7.2. Note this point is a "good" point at the outlying x; that is, it fits the model. Now determine the Wilcoxon and LS estimates. Comment on them.

12.2.7. Establish the identity

$$\|\mathbf{v}\|_W = \frac{\sqrt{3}}{2(n+1)} \sum_{i=1}^{n} \sum_{j=1}^{n} |v_i - v_j|, \tag{12.2.28}$$

for all $\mathbf{v} \in R^n$. Thus we have shown that

$$\widehat{\beta}_W = \text{Argmin} \sum_{i=1}^{n} \sum_{j=1}^{n} |(y_i - y_j) - \beta(x_{ci} - x_{cj})|. \tag{12.2.29}$$

Note that the formulation of $\widehat{\beta}_W$ given in expression (12.2.29) allows an easy way to compute the Wilcoxon estimate of slope by using an L_1, (least absolute deviations), routine. This was used in the cited article by Terpstra, et al. for their R or S-PLUS functions which compute the Wilcoxon fit.

12.2.8. This exercise assumes that the package S-PLUS is available or a package which computes an L_1 fit. Consider the simple data set

x	1	2	3	4	5	6
Y	18.1	6.5	10.1	14.9	21.9	22.9

(a) Form the $\binom{6}{2}$ pairwise differences of the xs and the corresponding differences of the Ys.

(b) Use the S-PLUS command `l1fit` to compute the Wilcoxon fit as described in Exercise 12.2.7.

(c) Scatterplot the data and overlay the Wilcoxon and LS fits. Comment on the fits.

12.2.9. Suppose the random variable e has cdf $F(t)$. Let $\varphi(u) = \sqrt{12}[u - (1/2)]$, $0 < u < 1$, denote the Wilcoxon score function.

(a) Show that the random variable $\varphi[F(e_i)]$ has mean 0 and variance 1.

(b) Investigate the mean and variance of $\varphi[F(e_i)]$ for any score function $\varphi(u)$ which satisfies $\int_0^1 \varphi(u)\, du = 0$ and $\int_0^1 \varphi^2(u)\, du = 1$.

12.3 LS Estimation for Linear Models

Recall the motivating example for simple regression presented in Section 9.6. In this example, we are interested in predicting the calculus grade (Y) of a student in terms of his score on an aptitude test (x_1). Other predictors besides x_1 are probably available for this problem. For example, predictors such as the high school grade point average of the student, her high school percentile, and her grade in a precalculus course would seem to contain useful information for the prediction of Y.

In this section, we consider the problem in which we have p predictors $x_1, \ldots, x_p$ and a response variable Y. We could, then, entertain models of the form

$$Y = h(x_1, x_2, \ldots, x_p) + \varepsilon,$$

where ε is a random variable, often called a **random error**, and h is a specified function. In this chapter, we will restrict our attention to the case where h is linear in the β-coefficients. Our data consist of n vectors of the form $(Y_i, x_{i1}, x_{i2}, \ldots, x_{ip})$, for $i = 1, 2, \ldots, n$. We will center the xs, i.e., $x_{cij} = x_{ij} - \overline{x}_j$ where $\overline{x}_j = n^{-1} \sum_{i=1}^{n} x_{ij}$. The linear model is

$$Y_i = \alpha + x_{ci1}\beta_1 + x_{ci2}\beta_2 + \cdots + x_{cip}\beta_p + \varepsilon_i, \quad i = 1, 2, \ldots, n, \qquad (12.3.1)$$

where $\alpha, \beta_1, \ldots, \beta_p$ are unknown parameters, often called **regression coefficients**.

The major assumption for this model is that the random errors $\varepsilon_1, \varepsilon_2, \ldots, \varepsilon_n$ are iid. We say the model is *true* when this assumption holds. In particular, this means that the distribution of the errors does not depend the xs. Under this assumption, a plot of ε_i versus $\alpha + x_{ci1}\beta_1 + x_{ci2}\beta_2 + \cdots + x_{cip}\beta_p$ would be a random scatter. Of course we do not observe the random errors or the regression coefficients, so this plot is impossible to obtain. But once we have estimated the regression coefficients we can obtain an estimate of this plot, called a **residual plot**. Model confirmation is indicated by a random scatter in this plot.

The matrix formulation of the model is much easier to use. We can write Model (12.3.1) as

$$
\begin{bmatrix} Y_1 \\ Y_2 \\ \vdots \\ \vdots \\ Y_n \end{bmatrix}
=
\begin{bmatrix} 1 \\ 1 \\ \vdots \\ \vdots \\ 1 \end{bmatrix} \alpha
+
\begin{bmatrix} x_{c11} & x_{c12} & \cdots & x_{c1p} \\ x_{c21} & x_{c22} & \cdots & x_{c2p} \\ \vdots & \vdots & & \vdots \\ \vdots & \vdots & & \vdots \\ x_{cn1} & x_{cn2} & \cdots & x_{cnp} \end{bmatrix}
\begin{bmatrix} \beta_1 \\ \beta_2 \\ \vdots \\ \beta_p \end{bmatrix}
+
\begin{bmatrix} \varepsilon_1 \\ \varepsilon_2 \\ \vdots \\ \vdots \\ \varepsilon_n \end{bmatrix}
$$

or equivalently

$$\mathbf{Y} = \mathbf{1}_n \alpha + \mathbf{X}_c \boldsymbol{\beta} + \boldsymbol{\varepsilon}. \qquad (12.3.2)$$

A more compact form will be convenient. Let $\mathbf{X} = [\mathbf{1} \ \mathbf{X}_c]$ and let $\mathbf{b} = (\alpha, \boldsymbol{\beta}')'$. Then we can write the model as

$$\mathbf{Y} = \mathbf{Xb} + \boldsymbol{\varepsilon}. \qquad (12.3.3)$$

We will assume that the $n \times (p + 1)$ matrix $\mathbf{X}$ has full column rank $p + 1$. Note further, that the vector $\mathbf{Xb}$ is a linear combination of the columns of the matrix $\mathbf{X}$. Let V be the **column space** of $\mathbf{X}$; that is, the space spanned by the columns of $\mathbf{X}$. Then V is $(p + 1)$-dimensional vector space of R^n. Letting $\boldsymbol{\eta} = \mathbf{Xb}$, the most succinct way of writing Model (12.3.3) is

$$\mathbf{Y} = \boldsymbol{\eta} + \boldsymbol{\varepsilon}, \quad \text{for } \boldsymbol{\eta} \in V. \qquad (12.3.4)$$

There is a nice way to read this last formulation. Except for random error, $\mathbf{Y}$ would lie in the subspace V. Hence, to estimate $\boldsymbol{\eta}$, find a vector in V which lies "closest" (in terms of a given norm) to $\mathbf{Y}$.

12.3.1 Least Squares

The LS estimator of η minimizes the square of the Euclidean distance between the vector $\mathbf{Y}$ and the subspace V; i.e., it is the $\widehat{\eta}$ such that

$$\widehat{\eta} = \text{Argmin}_{\eta \in V} \|\mathbf{Y} - \eta\|^2, \qquad (12.3.5)$$

where $\|\mathbf{v}\|^2 = \sum_{i=1}^{n} v_i^2$ is the squared Euclidean norm. As Exercise 12.3.1 shows, we could proceed as in the last section and find the normal equations. Instead, we introduce the projection of a vector onto a subspace, which facilitates this and ensuing discussions. Let $V^{\perp}$ be the subspace which consists of all vectors in R^n which are orthogonal to all vectors in V; that is,

$$V^{\perp} = \{\mathbf{w} \in R^n : \mathbf{w}'\mathbf{v} = 0, \text{for all } \mathbf{v} \in V\}. \qquad (12.3.6)$$

It is easy to see that $V^{\perp}$ has dimension $n - (p+1)$, where $(p+1)$ is the dimension of V.

Definition 12.3.1. *Let $\mathbf{v}$ be a vector in R^n and let V be a subspace of R^n. We say $\widehat{\mathbf{v}}$ is the projection of v onto V if*

$$\widehat{\mathbf{v}} \ \in \ V \qquad (12.3.7)$$
$$\mathbf{v} - \widehat{\mathbf{v}} \ \in \ V^{\perp}. \qquad (12.3.8)$$

We next give several theorems on projections. The first concerns uniqueness and the second gives an easy way to obtain projections (existence).

Theorem 12.3.1. *Projections are unique.*

Proof: Let $\widehat{\mathbf{v}}_1$ and $\widehat{\mathbf{v}}_2$ be projections of $\mathbf{v}$ onto V. Since V is a subspace, by (12.3.7), $\widehat{\mathbf{v}}_1 - \widehat{\mathbf{v}}_2 \in V$. But $V^{\perp}$ is also a subspace. Hence, $\widehat{\mathbf{v}}_1 - \widehat{\mathbf{v}}_2 = (\mathbf{v} - \widehat{\mathbf{v}}_2) - (\mathbf{v} - \widehat{\mathbf{v}}_1) \in V^{\perp}$. Thus, $\|\widehat{\mathbf{v}}_1 - \widehat{\mathbf{v}}_2\|^2 = 0$, which implies that $\widehat{\mathbf{v}}_1 = \widehat{\mathbf{v}}_2$. ∎

The columns of the matrix $\mathbf{X}$ form a basis for the subspace V; hence, we say that $\mathbf{X}$ is a **basis** matrix for V. We also say that $\mathbf{X}$ has **full column rank**. As Exercise 12.3.3 shows, $\mathbf{X}$ having full column rank implies that $(\mathbf{X}'\mathbf{X})^{-1}$ exists.

Theorem 12.3.2. *Let $\mathbf{X}$ be a basis matrix for a subspace V, let $\mathbf{H} = \mathbf{X}(\mathbf{X}'\mathbf{X})^{-1}\mathbf{X}'$, and let $\mathbf{v}$ be a vector in R^n. Then the projection of $\mathbf{v}$ onto V is $\mathbf{H}\mathbf{v}$.*

Proof: We need only verify the two conditions of Definition 12.3.1. Condition (12.3.7) follows immediately from writing $\mathbf{H}\mathbf{v}$ as

$$\mathbf{H}\mathbf{v} = \mathbf{X}\left\{(\mathbf{X}'\mathbf{X})^{-1}\mathbf{X}'\mathbf{v}\right\},$$

which is clearly a vector in V, the column space of $\mathbf{X}$. Let $\mathbf{u}$ be any vector in V. Then because $\mathbf{X}$ is a basis matrix for V, we can write $\mathbf{u} = \mathbf{X}\mathbf{c}$, where $\mathbf{c} \in R^p$. Then

$$(\mathbf{v} - \mathbf{H}\mathbf{v})'\mathbf{u} = \mathbf{v}'(\mathbf{I} - \mathbf{X}(\mathbf{X}'\mathbf{X})^{-1}\mathbf{X}')\mathbf{X}\mathbf{c} = \mathbf{v}'(\mathbf{X} - \mathbf{X})\mathbf{c} = 0,$$

and, thus, condition (12.3.8) holds. ∎

One immediate result of this theorem is that the projection matrix $\mathbf{H}$ is idempotent (i.e., $\mathbf{H}^2 = \mathbf{H}$) and symmetric. Hence, from Chapter 9, all the eigenvalues of $\mathbf{H}$ are either 0 or 1 and the rank of $\mathbf{H}$ is equal to its trace. Also, as Exercise 12.3.4 shows, the matrix $\mathbf{I} - \mathbf{H}$ is the projection matrix onto $V^{\perp}$.

Based on the last two theorems, to form the projection matrix, we only need a basis matrix, and any basis matrix will do. For example let the columns of $\mathbf{U}$ be an orthonormal (o.n.) basis matrix for V. Then $\mathbf{U}'\mathbf{U} = \mathbf{I}_p$. Hence, by uniqueness, $\mathbf{H} = \mathbf{U}\mathbf{U}' = \mathbf{X}(\mathbf{X}'\mathbf{X})^{-1}\mathbf{X}'$. The formulation with an o.n. basis matrix is often convenient. We next show that the projection of $\mathbf{Y}$ is the LS solution.

Theorem 12.3.3. *Consider Model (12.3.4). Let $\mathbf{H}$ be the projection matrix onto V. Let $\widehat{\boldsymbol{\eta}} = \mathbf{H}\mathbf{Y} = \mathbf{X}(\mathbf{X}'\mathbf{X})^{-1}\mathbf{X}'\mathbf{Y}$. Then $\widehat{\boldsymbol{\eta}}$ is the LS solution, i.e., $\widehat{\boldsymbol{\eta}}$ satisfies equation (12.3.5).*

Proof: Let $\boldsymbol{\eta} \in V$. Then $\mathbf{H}\mathbf{Y} - \boldsymbol{\eta} \in V$. But $(\mathbf{I} - \mathbf{H})\mathbf{Y} \in V^{\perp}$. This leads to

$$
\begin{aligned}
\|\mathbf{Y} - \boldsymbol{\eta}\|^2 &= \|\mathbf{Y} - \mathbf{H}\mathbf{Y} + \mathbf{H}\mathbf{Y} - \boldsymbol{\eta}\|^2 \\
&= \|(\mathbf{I} - \mathbf{H})\mathbf{Y} + (\mathbf{H}\mathbf{Y} - \boldsymbol{\eta})\|^2 \\
&= \|(\mathbf{I} - \mathbf{H})\mathbf{Y}\|^2 + \|(\mathbf{H}\mathbf{Y} - \boldsymbol{\eta})\|^2, \quad (12.3.9)
\end{aligned}
$$

where the last equality follows because $\mathbf{H}\mathbf{Y} - \boldsymbol{\eta} \in V$ and $(\mathbf{I} - \mathbf{H})\mathbf{Y} \in V^{\perp}$. Because $\boldsymbol{\eta}$ does not appear in the first term of the right-side of equation (12.3.9), we can minimize the left-side by taking $\boldsymbol{\eta} = \mathbf{H}\mathbf{Y}$. Hence, the LS solution is the projection $\mathbf{H}\mathbf{Y}$. The uniqueness of the LS solution is obtained in Exercise 12.3.5. ∎

The LS estimate $\widehat{\mathbf{b}}$ of $\mathbf{b}$ must then satisfy

$$\mathbf{X}\widehat{\mathbf{b}} = \mathbf{H}\mathbf{Y} = \mathbf{X}(\mathbf{X}'\mathbf{X})^{-1}\mathbf{X}'\mathbf{Y}. \quad (12.3.10)$$

Upon multiplying both sides of this equation by $\mathbf{X}'$, we get

$$\mathbf{X}'\mathbf{X}\widehat{\mathbf{b}} = \mathbf{X}'\mathbf{Y}. \quad (12.3.11)$$

These are the estimating equations, so called **normal equations**, for the multiple regression model. If we multiply expression (12.3.11) by $(\mathbf{X}'\mathbf{X})^{-1}$, we obtain the LS estimate of $\mathbf{b}$ which is

$$\widehat{\mathbf{b}} = (\mathbf{X}'\mathbf{X})^{-1}\mathbf{X}'\mathbf{Y}. \quad (12.3.12)$$

The estimate $\widehat{\mathbf{Y}} = \mathbf{X}\widehat{\mathbf{b}}$ is called the **fitted** or **predicted** value of $\mathbf{Y}$. Thus the **residual** or estimate of the error vector is given by $\widehat{\boldsymbol{\varepsilon}} = \mathbf{Y} - \widehat{\mathbf{Y}}$. Note that $\widehat{\mathbf{Y}} \in V$ and $\widehat{\boldsymbol{\varepsilon}} \in V^{\perp}$. Hence, $\widehat{\mathbf{Y}} \perp \widehat{\boldsymbol{\varepsilon}}$, a fact that we will use repeatedly.

Recall that the major assumption for Model (12.3.1) is that the distribution of the random errors does not depend on the $\mathbf{x}$ part of the model. A check for this assumption is easily performed by plotting the residuals $\widehat{\boldsymbol{\varepsilon}} = (\widehat{\varepsilon}_1, \widehat{\varepsilon}_2, \ldots, \widehat{\varepsilon}_n)'$ versus the fitted values $\widehat{\mathbf{Y}} = (\widehat{Y}_1, \widehat{Y}_2, \ldots, \widehat{Y}_n)'$. This plot is called the **residual plot**. Randomness in the plot is used as confirmation of the major assumption of the model, while patterns in this plot indicate that the major assumption may be false. Exercise 12.3.8 shows how the patterns in residual plots are helpful in determining more appropriate models.

We present the distributions of the LS estimators in the next two subsections. But as the next theorem shows, under the additional assumptions that $E(\varepsilon_i) = 0$ and $\text{Var}(\varepsilon_i) = \sigma^2 < \infty$, we show that the LS estimators are unbiased and we, further, determine their second moments. Note that under this last assumption, inference with LS estimators requires an estimator of σ^2. Recall in the simple location problem that the estimator of variance is proportional to the sum of the squared deviations from the mean. The deviations in the regression problem are the residuals. Hence, we will take as our estimate of σ^2,

$$\widehat{\sigma}^2 = \frac{1}{n - p - 1} \sum_{i=1}^{n} \widehat{\varepsilon}_i^2. \tag{12.3.13}$$

As the next theorem shows, $\widehat{\sigma}^2$ is an unbiased estimate of σ^2.

Theorem 12.3.4. *Assume that Model (12.3.3) is true and that $E(\varepsilon_i) = 0$ and $Var(\varepsilon_i) = \sigma^2 < \infty$. Then*

(a) $E(\widehat{\mathbf{b}}) = \mathbf{b}$ *and* $Cov(\widehat{\mathbf{b}}) = \sigma^2 (\mathbf{X}'\mathbf{X})^{-1}$,

(b) $E(\widehat{\mathbf{Y}}) = \mathbf{X}\mathbf{b}$ *and* $Cov(\widehat{\mathbf{Y}}) = \sigma^2 \mathbf{H}$,

(c) $E(\widehat{\boldsymbol{\eta}}) = \mathbf{0}$ *and* $Cov(\widehat{\boldsymbol{\eta}}) = \sigma^2 (\mathbf{I} - \mathbf{H})$,

(d) $E(\widehat{\sigma}^2) = \sigma^2$.

Proof: Under the assumptions of Model (12.3.3), note that

$$\begin{aligned} \widehat{\mathbf{b}} &= \mathbf{b} + (\mathbf{X}'\mathbf{X})^{-1}\mathbf{X}'\varepsilon, \\ \widehat{\mathbf{Y}} &= \mathbf{X}\mathbf{b} + \mathbf{H}\varepsilon, \\ \widehat{\varepsilon} &= (\mathbf{I} - \mathbf{H})\varepsilon. \end{aligned} \tag{12.3.14}$$

By the hypotheses, $E(\boldsymbol{\varepsilon}) = \mathbf{0}$ and $\text{Cov}(\boldsymbol{\varepsilon}) = \sigma^2\mathbf{I}$. Using this, the results in Section 3.5, and the facts that $\mathbf{H}$ and $\mathbf{I} - \mathbf{H}$ are symmetric and idempotent, Parts (a) - (c) follow immediately; see Exercise 12.3.9. For Part (d), note first that $(n - p - 1)\widehat{\sigma}^2 = \varepsilon'(\mathbf{I} - \mathbf{H})\varepsilon$. Hence

$$\begin{aligned} E[(n - p - 1)\widehat{\sigma}^2] &= E[\varepsilon'(\mathbf{I} - \mathbf{H})\varepsilon] \\ &= E[\text{tr}(\mathbf{I} - \mathbf{H})\varepsilon\varepsilon'] \\ &= \text{tr}(\mathbf{I} - \mathbf{H})E[\varepsilon\varepsilon'] \\ &= \text{tr}(\mathbf{I} - \mathbf{H})\sigma^2\mathbf{I} = (n - p - 1)\sigma^2. \end{aligned}$$

Thus Part (d) is also true. ∎

12.3.2 Basics of LS Inference under Normal Errors

In this subsection, we obtain the distribution of the LS estimators under the assumption that the random errors of Model 12.3.1 are normally distributed. Exercises 12.3.10 and 12.3.11 show that the LS statistics are sufficient and are mles for the regression parameters.

Theorem 12.3.5. *Suppose Model 12.3.1 is true and that the error random vector ε has a $N_n(\mathbf{0}, \sigma^2\mathbf{I})$ distribution. Then the LS estimators satisfy the following:*

(a) $\widehat{\mathbf{b}}$ *has a $N(\mathbf{b}, \sigma^2(\mathbf{X}'\mathbf{X})^{-1})$ distribution.*

(b) $\widehat{\mathbf{Y}}$ *has a $N(\mathbf{Xb}, \sigma^2\mathbf{H})$ distribution.*

(c) $\widehat{\varepsilon}$ *has a $N(\mathbf{0}, \sigma^2(\mathbf{I} - \mathbf{H}))$ distribution.*

(d) *$(n - p - 1)\widehat{\sigma}^2/\sigma^2$ has a $\chi^2(n - p - 1)$ distribution.*

(e) $\widehat{\mathbf{Y}}$ *and $\widehat{\varepsilon}$ are independent.*

(f) $\widehat{\mathbf{b}}$ *and $\widehat{\sigma}^2$ are independent.*

Proof: From Theorem 12.3.4, we have the above moments of the random vectors. For normality, note that $\mathbf{Y}$ has a $N_n(\mathbf{X}\beta, \sigma^2\mathbf{I})$ distribution. The normality of $\widehat{\mathbf{b}}$, $\widehat{\mathbf{Y}}$, and $\widehat{\varepsilon}$ follow because each of these random vectors is a linear function of $\mathbf{Y}$; see Theorem 3.5.1. Hence, Parts (a) - (c) are true. For Part (d), as in Theorem 12.3.4, we can write

$$\frac{(n - p - 1)\widehat{\sigma}^2}{\sigma^2} = \sigma^{-2}\varepsilon'(\mathbf{I} - \mathbf{H})\varepsilon.$$

Then, because $\mathbf{I} - \mathbf{H}$ is idempotent of rank $n - p - 1$, Theorem 9.8.4 implies Part (d). For the independence between $\widehat{\mathbf{Y}}$ and $\widehat{\varepsilon}$, write them jointly as

$$\left[\begin{array}{c} \widehat{\mathbf{Y}} \\ \widehat{\varepsilon} \end{array}\right] = \left[\begin{array}{c} \mathbf{H} \\ \mathbf{I} - \mathbf{H} \end{array}\right]\mathbf{Y}.$$

Thus, $\widehat{\mathbf{Y}}$ and $\widehat{\varepsilon}$ have a jointly normal distribution and, further, their covariance matrix is

$$\left[\begin{array}{c} \mathbf{H} \\ \mathbf{I} - \mathbf{H} \end{array}\right]\sigma^2\mathbf{I}\left[\begin{array}{c} \mathbf{H} \\ \mathbf{I} - \mathbf{H} \end{array}\right]' = \sigma^2\left[\begin{array}{cc} \mathbf{H} & \mathbf{O} \\ \mathbf{O} & \mathbf{I} - \mathbf{H} \end{array}\right],$$

because $\mathbf{H}$ is idempotent. Thus $\widehat{\mathbf{Y}}$ and $\widehat{\varepsilon}$ have a joint normal distribution and are uncorrelated, and, hence, they are independent random vectors. Because $\widehat{\mathbf{b}}$ and $\widehat{\sigma}^2$ are functions of $\widehat{\mathbf{Y}}$ and $\widehat{\varepsilon}$, respectively, it follows immediately that $\widehat{\mathbf{b}}$ and $\widehat{\sigma}^2$ are independent. ∎

Based on this theorem, we have the following corollary concerning pivot t-random variables. This serves as the basis for the t-confidence intervals and t-tests used in regression. We state it for the regression parameters $\beta_1, \ldots, \beta_p$. The corresponding result for the intercept parameter is given in Exercise 12.3.13. The proof of the corollary is left as an exercise; see Exercise 12.3.12

Corollary 12.3.1. *Under the assumptions of the last theorem, for $j = 1, 2, \ldots, p$ the following random variables have t-distributions with $n - p - 1$ degrees of freedom:*

$$t_j = \frac{\widehat{\beta}_j - \beta}{\widehat{\sigma}\sqrt{(\mathbf{X}_c'\mathbf{X}_c)_{jj}^{-1}}}; \tag{12.3.15}$$

where $(\mathbf{X}_c'\mathbf{X}_c)_{jj}^{-1}$ is the jth diagonal entry of $(\mathbf{X}_c'\mathbf{X}_c)^{-1}$ and $\mathbf{X}_c$ is the centered design matrix as given in expression (12.3.2).

Based on this corollary a $(1 - \alpha)100\%$ confidence interval for β_j is given by

$$\widehat{\beta}_j \pm t_{\alpha/2,n-p-1}\widehat{\sigma}\sqrt{(\mathbf{X}_c'\mathbf{X}_c)_{jj}^{-1}}. \tag{12.3.16}$$

Also, a level α test for the hypotheses $H_0 : \beta_j = 0$ versus $H_0 : \beta_j \neq 0$, $j = 1, 2, \ldots, p$, is given by

$$\text{Reject } H_0 \text{ if } |t| = \frac{|\widehat{\beta}_j|}{\widehat{\sigma}\sqrt{(\mathbf{X}_c'\mathbf{X}_c)_{jj}^{-1}}} > t_{\alpha/2,n-p-1}. \tag{12.3.17}$$

The test statistics are often called *t-ratios*. The noncentrality parameters of these tests are asked for in Exercise 12.3.14.

EXERCISES

12.3.1. Write the linear model as in (12.3.3), i.e., $\mathbf{Y} = \mathbf{Xb} + \boldsymbol{\varepsilon}$. Then the LS estimator of $\mathbf{b}$ satisfies
$$\widehat{\mathbf{b}} = \text{Argmin}\|\mathbf{Y} - \mathbf{Xb}\|^2.$$

(a) Show that
$$\|\mathbf{Y} - \mathbf{Xb}\|^2 = \mathbf{Y}'\mathbf{Y} - 2(\mathbf{X}'\mathbf{Y})'\mathbf{b} + \mathbf{b}'\mathbf{X}'\mathbf{Xb}.$$

(b) Take the partial derivative of this last expression with respect to $\mathbf{b}$ and, hence, derive the normal equations.

12.3.2. Assume that $\mathbf{X}$ is an $n \times p$ matrix. Then the kernel of $\mathbf{X}$ is defined to be the space $\ker(\mathbf{X}) = \{\mathbf{b} : \mathbf{Xb} = \mathbf{0}\}$.

(a) Show that $\ker(\mathbf{X})$ is a subspace of R^p.

(b) The dimension of $\ker(\mathbf{X})$ is called the **nullity** of $\mathbf{X}$ and is denoted by $\nu(\mathbf{X})$. Let $\rho(\mathbf{X})$ denote the rank of $\mathbf{X}$. A fundamental theorem of linear algebra says that $\rho(\mathbf{X}) + \nu(\mathbf{X}) = p$. Use this to show that if $\mathbf{X}$ has full column rank, then $\ker(\mathbf{X}) = \{\mathbf{0}\}$.

12.3.3. Suppose $\mathbf{X}$ is an $n \times p$ matrix with rank p.

(a) Show that $\ker(\mathbf{X}'\mathbf{X}) = \ker(\mathbf{X})$.

(b) Use Part (a) and the last exercise to show that if $\mathbf{X}$ has full column rank then $\mathbf{X}'\mathbf{X}$ is nonsingular.

12.3.4. Write $(\mathbf{I} - \mathbf{H})\mathbf{v} = \mathbf{v} - \widehat{\mathbf{v}}$ and use the definition of projection, (12.3.1), to establish that $\mathbf{I} - \mathbf{H}$ is the projection matrix onto $V^\perp$.

12.3.5. To complete the proof of Theorem 12.3.3, we need to show that the LS estimator $\widehat{\boldsymbol{\eta}} = \mathbf{HY}$ is unique. Suppose that $\widehat{\boldsymbol{\eta}}_2 \in V$ is also a LS solution.

(a) Show that

$$\|\widehat{\boldsymbol{\eta}} - \widehat{\boldsymbol{\eta}}_2\|^2 = 2\|(\mathbf{I} - \mathbf{H})\mathbf{Y}\|^2 - 2\mathbf{Y}'(\mathbf{I} - \mathbf{H})(\mathbf{Y} - \widehat{\boldsymbol{\eta}}_2).$$

(b) Now show that the right side of this last expression is 0 and, hence, establish the uniqueness.

12.3.6. Consider the linear model (12.3.2) with the design matrix given by $\mathbf{X} = [\mathbf{1}\,\mathbf{c}_1 \cdots \mathbf{c}_p]$. Assume that the columns of the design matrix $\mathbf{X}$ are orthogonal. Show that the LS estimate of $\mathbf{b}$ is given by $\widehat{\mathbf{b}}' = (\overline{Y}, \widehat{b}_1, \dots, \widehat{b}_p)$ where $\widehat{b}_j$ is the LS estimate of the simple regression model $Y_i = b_j c_{ij} + \varepsilon_i$, for $j = 1, \dots, p$. That is, the LS multiple regression estimator in this case is found by the individual simple LS regression estimators.

12.3.7. By Exercise 12.3.6, if the design matrix is orthogonal then the LS estimates for multiple regression are the same as the individual simple regressions. In this exercise, consider the data

	X		**Y**
	$\mathbf{c}_1$	$\mathbf{c}_2$	
1	-3	-2.52857	45.6150
1	-2	-2.02857	44.8358
1	-1	-0.62857	57.5003
1	0	-0.12857	41.1391
1	1	1.07143	52.9030
1	2	1.37143	48.2027
1	3	2.87143	56.5706

and the model, $Y_i = \alpha + c_{i1}\beta_1 + c_{i2}\beta_2 + e_i$.

(a) Obtain by LS the individual fits of the two simple models $Y_i = \alpha + c_{ij}\beta_j + e_i$, for $j = 1, 2$.

(b) Obtain the LS fit of the model $Y_i = \alpha + c_{i1}\beta_1 + c_{i2}\beta_2 + e_i$.

(c) Compare the fits.

(d) Plot c_{i1} versus c_{i2}, $i = 1, \dots, n$.

Notice how different the fits are. From an algebraic point-of-view, the reason for this large difference is the high **colinearity** between $\mathbf{c}_1$ and $\mathbf{c}_2$, as shown in the plot.

12.3.8. The data below are generated from the model $Y_i = 0 + 5i + i^2 + \varepsilon_i$, for $i = 1, \dots, 10$, and ε_i iid $N(0, 4^2)$.

i	1	2	3	4	5	6	7	8	9	10
Y_i	3.1	20.1	20.4	31.6	57.0	61.7	86.9	107.5	125.7	148.0

(a) Fit the misspecified model $Y_i = \alpha + \beta_1 i + \varepsilon_i$ by LS and obtain the residual plot. Comment on the plot (Is it random? If not, does it suggest another model to try?).

(b) Same as Part (a) for the fit of the model $Y_i = \alpha + \beta_1 i + \beta_2 i^2 + \varepsilon_i$ by LS.

12.3.9. Fill in the details of the proof of Theorem 12.3.4.

12.3.10. Assume the hypotheses of Theorem 12.3.4 hold. Let $\mathbf{x}'_i$ denote the *ith* row of $\mathbf{X}$.

(a) Show that the likelihood function can be written as

$$L(\mathbf{b}, \sigma^2) = (2\pi)^{-n/2}(\sigma^2)^{-n/2} \exp\left\{-\frac{1}{2\sigma^2}(\mathbf{y}'\mathbf{y} - 2\mathbf{b}'\mathbf{X}'\mathbf{y} + \mathbf{b}'\mathbf{X}'\mathbf{X}\mathbf{b})\right\}.$$

(b) Reexpress this formulation of the likelihood as the regular case of the exponential family and show that the following statistics are sufficient:

$$\mathbf{X}'\mathbf{y} = \sum_{i=1}^{n} y_i \mathbf{x}_i \text{ and } \mathbf{y}'\mathbf{y} = \sum_{i=1}^{n} y_i^2.$$

(c) Obtain the MVUEs of $\mathbf{h}'\mathbf{b}$, (where $\mathbf{h}$ is specified), and σ^2.

12.3.11. Assume the hypotheses of Theorem 12.3.4 hold. Let $\mathbf{x}'_i$ denote the *ith* row of $\mathbf{X}$.

(a) Show that the likelihood function can be written as

$$L(\mathbf{b}, \sigma^2) = (2\pi)^{-n/2}(\sigma^2)^{-n/2} \exp\{\sigma^{-2}\sum_{i=1}^{n}(y_i - \mathbf{x}'\mathbf{b})^2\}.$$

(b) Let σ^2 be arbitrary but fixed. Without differentiating, show the mle of $\mathbf{b}$ is the LS estimator.

(c) Next, obtain the mle of σ^2. Is it unbiased?

12.3.12. Prove Corollary 12.3.1.

12.3.13. For convenience we have centered the independent variables in Model 12.3.2. Suppose, instead, we consider the model

$$\mathbf{Y} = \mathbf{1}_n \alpha^* + \mathbf{X}^* \boldsymbol{\beta} + \boldsymbol{\varepsilon},$$

where the $n \times p$ matrix $\mathbf{X}^*$ is not necessarily centered. Let $\mathbf{H}_1 = n^{-1}\mathbf{1}_n\mathbf{1}'_n$ be the projection matrix onto the space spanned by $\mathbf{1}_n$ and let $\overline{\mathbf{x}}'$ denote the row vector of column averages of $\mathbf{X}^*$.

(a) To show that the LS estimate of $\boldsymbol{\beta}$ remains the same in this model as in (12.3.2) and to determine the LS estimate of α^*, give the reasons for each step in the following derivation:

$$
\begin{aligned}
\|\mathbf{Y} - \alpha^* \mathbf{1}_n - \mathbf{X}^* \boldsymbol{\beta}\|^2 &= \|\mathbf{Y} - \alpha^* \mathbf{1}_n - (\mathbf{H}_1 + \mathbf{I} - \mathbf{H}_1)\mathbf{X}^* \boldsymbol{\beta}\|^2 \\
&= \|\mathbf{Y} - \alpha^* \mathbf{1}_n - \overline{\mathbf{x}}' \boldsymbol{\beta} \mathbf{1} - \mathbf{X}_c \boldsymbol{\beta}\|^2 \\
&= \|\mathbf{Y} - \alpha \mathbf{1}_n - \mathbf{X}_c \boldsymbol{\beta}\|^2,
\end{aligned}
$$

$$(12.3.18)$$

where $\alpha = \alpha^* + \overline{\mathbf{x}}' \boldsymbol{\beta}$. The estimators that minimize (12.3.18) are the LS estimators of α and $\boldsymbol{\beta}$ in Model (12.3.2), (Why?). Hence, the LS estimator of α^* is $\widehat{\alpha}^* = \widehat{\alpha} - \overline{\mathbf{x}}' \widehat{\boldsymbol{\beta}}$, (Why?).

(b) Under the assumptions of Theorem 12.3.5, obtain the distribution of $\widehat{\alpha}^*$ by first showing that the following is true:

$$
\begin{aligned}
\widehat{\alpha}^* &= \overline{Y} - \overline{x}' \widehat{\boldsymbol{\beta}} \\
&= \left[\frac{1}{n} \mathbf{1}' - \overline{x}' (\mathbf{X}_c' \mathbf{X}_c)^{-1} \mathbf{X}_c' \right] \mathbf{Y} \\
&= \left[\frac{1}{n} \mathbf{1}' - \overline{x}' (\mathbf{X}_c' \mathbf{X}_c)^{-1} \mathbf{X}_c' \right] [\mathbf{1}(\alpha^* + \overline{x}' \boldsymbol{\beta}) + \mathbf{X}_c \boldsymbol{\beta} + \mathbf{e}] \\
&= \alpha^* + \left[\frac{1}{n} - \overline{x}' (\mathbf{X}_c' \mathbf{X}_c)^{-1} \mathbf{X}_c' \right] \mathbf{e}.
\end{aligned}
$$

From this last expression, the distribution of $\widehat{\alpha}^*$ can readily be found.

(c) Under the assumptions of Theorem 12.3.5, obtain a confidence interval, at confidence $1 - \gamma$, for α^*.

(d) Under the assumptions of Theorem 12.3.5, obtain a level γ t-test for the hypothesis $H_0 : \alpha^* = 0$ versus $H_1 : \alpha^* \neq 0$.

(e) Is your test in Part (d) practical if there are no data around $\mathbf{0}$? Discuss.

12.3.14. Determine the non-centrality parameters of the tests defined in Corollary 12.3.1.

12.4   Wilcoxon Estimation for Linear Models

The LS inference discussed in Section 12.3.2 is under the assumption of normally distributed errors. For other error distributions, we must appeal to asymptotics. It is, though, a simple extension of the asymptotics developed in Section 12.2 for the simple linear regression model. Also, the development of the robust Wilcoxon estimator in that section is easily extended to the multiple regression model of this section.

As in Section 12.3.2, we are concerned with the regression parameters, $\beta_1, \ldots, \beta_p$. We will discuss the intercept parameter later. Hence, we shall write Model 12.3.2

$$\mathbf{Y} = \mathbf{X}_c\boldsymbol{\beta} + \mathbf{e}, \tag{12.4.1}$$

where $\mathbf{e} = \mathbf{1}\alpha + \boldsymbol{\varepsilon}$. Let V_c be the column space of the matrix $\mathbf{X}_c$. The model is true if the errors $e_1, e_2, \ldots, e_n$ are iid with cdf $F(x)$ and pdf $f(x)$. Except for random error, $\mathbf{Y}$ would lie in V_c. Hence, given a norm on R^n, a simple estimate of $\mathbf{X}_c\boldsymbol{\beta}$ is the vector in V_c which lies closest to $\mathbf{Y}$.

12.4.1 Norms and Estimating Equations

Using the squared Euclidean norm, the LS estimate is given by

$$\widehat{\boldsymbol{\beta}}_{LS} = \operatorname*{Argmin}_{\boldsymbol{\beta} \in R^p} \|\mathbf{Y} - \mathbf{X}_c\boldsymbol{\beta}\|^2, \tag{12.4.2}$$

where $\| \cdot \|_{LS}^2$ is the squared Euclidean norm. As Exercise 12.4.1 shows, $\widehat{\boldsymbol{\beta}}_{LS}$ is the same estimator of the regression coefficients as in Section 12.3. Equivalently, the LS estimator solves the estimating equations

$$\sum_{i=1}^{n} \mathbf{x}_{ci}(Y_i - \mathbf{x}_{ci}'\boldsymbol{\beta}) = \mathbf{0}; \tag{12.4.3}$$

where $\mathbf{x}_{ci}'$ is the ith row of $\mathbf{X}_c$.

The Wilcoxon estimator of $\boldsymbol{\beta}$ in Model 12.4.1, is a vector $\widehat{\boldsymbol{\beta}}_W$ such that

$$\widehat{\boldsymbol{\beta}}_W = \operatorname*{Argmin}_{\boldsymbol{\beta} \in R^p} \|\mathbf{Y} - \mathbf{X}_c\boldsymbol{\beta}\|_W, \tag{12.4.4}$$

where $\| \cdot \|_W$ is the Wilcoxon norm given in expression (12.2.8). If $p = 1$, then this is the Wilcoxon estimate of slope discussed in Section 12.2. Equivalently, the Wilcoxon estimator solves the estimating equations

$$\sum_{i=1}^{n} \mathbf{x}_{ci} a[R(Y_i - \mathbf{x}_{ci}'\boldsymbol{\beta})] = \mathbf{0}, \tag{12.4.5}$$

where $\mathbf{x}_{ci}'$ is the ith row of $\mathbf{X}_c$ and the rank scores are the same as in Section 12.2; i.e., $a(i) = \varphi[i/(n+1)]$ and $\varphi(u) = \sqrt{12}[u - (1/2)]$.

12.4.2 Influence Functions

Because the derivation of the influence functions in the multiple regression model for the LS and Wilcoxon estimators are quite similar to their derivation in the simple linear model, we shall be brief. As in the simple regression model, we now assume that $\mathbf{x}$ is a random vector. The extension of the probability model is given by

$$Y = \mathbf{x}'\boldsymbol{\beta} + e, \tag{12.4.6}$$

where the $(p + 1) \times 1$ random vector $(\mathbf{x}', y)'$ has the joint cdf $H(\mathbf{x}, y)$ and pdf $h(\mathbf{x}, y)$, the random variable e has cdf $F(t)$ and pdf $f(t)$, e and $\mathbf{x}$ are independent, and $E(\mathbf{x}) = \mathbf{0}$.

Influence Function for LS Estimator

Let $\mathbf{T}_{LS}(H)$ denote the functional corresponding to the LS estimator. Based on (12.4.3), it is easy to see that $\mathbf{T}_{LS}(H)$ solves the equation

$$\int_{-\infty}^{\infty} \cdots \int_{-\infty}^{\infty} \mathbf{x}[y - \mathbf{x}'\mathbf{T}_{LS}(H)]h(\mathbf{x}, y)\, d\mathbf{x} dy = \mathbf{0}. \qquad (12.4.7)$$

As in Section 12.2, consider the contaminated cdf $H_\epsilon(\mathbf{x}, y)$ which is based on $\epsilon 100\%$ contamination by the point-mass cdf $\Delta_{(\mathbf{x}_0, y_0)}$; i.e.,

$$H_\epsilon(\mathbf{x}, y) = (1 - \epsilon)H(\mathbf{x}, y) + \epsilon \Delta_{(\mathbf{x}_0, y_0)}(\mathbf{x}, y). \qquad (12.4.8)$$

Analogous to (12.4.7), $\mathbf{T}_\epsilon = \mathbf{T}_{LS}(H_\epsilon)$ satisfies

$$\begin{aligned}
\mathbf{0} = {}& (1 - \epsilon) \int_{-\infty}^{\infty} \cdots \int_{-\infty}^{\infty} \mathbf{x}(y - \mathbf{x}'\mathbf{T}_\epsilon)h(\mathbf{x}, y)\, d\mathbf{x} dy \\
& + \epsilon \int_{-\infty}^{\infty} \cdots \int_{-\infty}^{\infty} \mathbf{x}(y - \mathbf{x}'\mathbf{T}_\epsilon)\, d\Delta_{(\mathbf{x}_0, y_0)}(\mathbf{x}, y). \qquad (12.4.9)
\end{aligned}$$

The only differences, however, between this equation and equation (12.2.19) are that the functional and x are now vectors. We can then implicitly differentiate this equation with respect to ϵ, taking the partial derivative of T_j with respect to ϵ, for $j = 1, \ldots, p$, and entering the results into a row vector. This becomes

$$\begin{aligned}
\mathbf{0} = {}& -\int_{-\infty}^{\infty} \cdots \int_{-\infty}^{\infty} \mathbf{x}(y - \mathbf{x}'\mathbf{T}_\epsilon)h(\mathbf{x}, y)\, d\mathbf{x} dy \\
& + (1 - \epsilon) \int_{-\infty}^{\infty} \cdots \int_{-\infty}^{\infty} \mathbf{x}(-\mathbf{x}')\frac{\partial \mathbf{T}_\epsilon}{\partial \epsilon} h(\mathbf{x}, y)\, d\mathbf{x} dy \\
& + \int_{-\infty}^{\infty} \cdots \int_{-\infty}^{\infty} (y - \mathbf{x}'\mathbf{T}_\epsilon)\, d\Delta_{(\mathbf{x}_0, y_0)}(\mathbf{x}, y) + \epsilon B, \qquad (12.4.10)
\end{aligned}$$

where the expression for B is irrelevant because we are evaluating this partial at $\epsilon = 0$. Setting $\epsilon = 0$ and solving for the partial of $\mathbf{T}$, we see that the influence function is

$$IF(\mathbf{x}_0, y_0; \widehat{\boldsymbol{\beta}}_{LS}) = [E(\mathbf{x}\mathbf{x}')]^{-1}(y_0 - \mathbf{x}_0'\boldsymbol{\beta})\mathbf{x}_0. \qquad (12.4.11)$$

The LS influence function is unbounded in both the Y and the $\mathbf{x}$ spaces; hence, LS estimates are not robust in either space. It is also clear that it has breakdown point 0; see, for example, Exercise 12.2.6.

Influence Function for the Wilcoxon Estimator

We obtain the influence function for the Wilcoxon estimate by proceeding as in the LS case. For the probability model, (12.4.6), let $\mathbf{T}_W(H)$ denote the corresponding functional of the Wilcoxon estimator. Recall from Section 12.2, that the scores can be written as

$$a[R(Y_i - \mathbf{x}_{ci}'\boldsymbol{\beta})] = \varphi\left[\frac{n}{n+1}F_n(Y_i - \mathbf{x}_{ci}'\boldsymbol{\beta})\right],$$

where F_n is the empirical distribution of the residuals. Thus the functional solves

$$\int_{-\infty}^{\infty} \cdots \int_{-\infty}^{\infty} \mathbf{x}\varphi[y - \mathbf{x}'\mathbf{T}_W(H)]h(\mathbf{x}, y)\, d\mathbf{x}dy = \mathbf{0}. \qquad (12.4.12)$$

Now consider the equation that the Wilcoxon functional solves at the contaminated cdf $H_\epsilon(\mathbf{x}, y)$, (12.4.8). It is given by

$$\begin{aligned}
\mathbf{0} &= (1 - \epsilon) \int_{-\infty}^{\infty} \cdots \int_{-\infty}^{\infty} \mathbf{x}\varphi[y - \mathbf{x}'\mathbf{T}_\epsilon]h(\mathbf{x}, y)\, d\mathbf{x}dy \\
&\quad + \epsilon \int_{-\infty}^{\infty} \cdots \int_{-\infty}^{\infty} \mathbf{x}\varphi[y - \mathbf{x}'\mathbf{T}_\epsilon]\, d\Delta_{(\mathbf{x}_0, y_0)}(\mathbf{x}, y).
\end{aligned} \qquad (12.4.13)$$

We then implicitly differentiate this equation with respect to ϵ, taking the partial derivative of T_j with respect to ϵ, for $j = 1, \ldots, p$, and entering the results into a row vector. The derivation is quite close to the derivation in Section 12.2; hence, we proceed to give the result and leave the details for the reader in Exercise 12.4.2. The influence function for the Wilcoxon estimator is

$$IF(\mathbf{x}_0, y_0; \widehat{\boldsymbol{\beta}}_W) = \tau \left[E(\mathbf{x}\mathbf{x}') \right]^{-1} \varphi(y_0 - \mathbf{x}_0'\boldsymbol{\beta})\mathbf{x}_0, \qquad (12.4.14)$$

where $\tau = [\sqrt{12} \int f^2(t)\, dt]^{-1}$.

Remark 12.4.1. As in the simple linear case, the Wilcoxon influence function is bounded in the Y space but unbounded in the $\mathbf{x}$ space. Hence, the Wilcoxon estimator is robust in the Y space but, as with LS, it is not robust in the $\mathbf{x}$ space. However, simple, weighted versions of the Wilcoxon estimator have influence functions which are bounded in both the $\mathbf{x}$ and the Y spaces. There is a trade-off, though. While these estimators can achieve 50% breakdown, they can also be much less efficient than the Wilcoxon estimator. Discussions of these estimators are beyond the scope of this book; see Chang et al. (1999) and the fifth chapter of Hettmansperger and McKean (1998) for details. ∎

12.4.3 Asymptotic Distribution Theory

As in the simple linear case, we can read the asymptotic distribution of the estimators of $\boldsymbol{\beta}$ from their influence function. As in the simple linear case, we want this result for the regression model where the xs are thought of as fixed. An estimate of $E(\mathbf{x}\mathbf{x}')$ is the arthmetic average

$$\frac{1}{n} \sum_{i=1}^{n} \mathbf{x}_{ci}\mathbf{x}_{ci}' = \frac{1}{n}\mathbf{X}_c'\mathbf{X}_c.$$

For the theory discussed below, certain assumptions are required. Let $\mathbf{H}_c = \mathbf{X}_c(\mathbf{X}_c'\mathbf{X}_c)^{-1}\mathbf{X}_c'$ denote the projection matrix onto V_c. The **leverage** values of $\mathbf{H}_c$ are its diagonal entries; that is, the ith leverage value is

$$h_{cni} = \mathbf{x}_{ci}'(\mathbf{X}_c'\mathbf{X}_c)^{-1}\mathbf{x}_{ci}, \qquad (12.4.15)$$

where the subscript n is a reminder that the design depends on n. Recall that $\mathbf{x}_{ci}$ is the vector of deviations of the ith design point from the "center" of design space. If $\mathbf{X}'_c\mathbf{X}_c$ was the identity matrix, then h_{cni} would be the square of the distance of the ith design point from the center of design space. However, the matrix $\mathbf{X}'_c\mathbf{X}_c$ is positive definite; so it is the square of a distance relative to $\mathbf{X}'_c\mathbf{X}_c$, (this is often called a **Mahalanobis** squared distance).

Assumptions 12.4.1 (Assumptions for Asymptotic Distribution Theory). *Assumptions (a), (b), (d), and (e) are neeeded for LS, while Assumptions (a), (c), (d), and (e) are neeeded for the Wilcoxon estimates.*

(a): *The random errors $e_1, e_2, \ldots, e_n$ are iid with cdf $F(x)$ and pdf $f(x)$.*

(b): *$Var(e_i) = \sigma^2 < \infty$.*

(c): *The pdf $f(x)$ has finite Fisher information.*

(d): $\limsup h_{cni} \to 0$, *as $n \to \infty$.*

(e): *$n^{-1}\mathbf{X}'_c\mathbf{X}_c \overset{P}{\to} \mathbf{\Sigma}$, where $\mathbf{\Sigma}$ is positive definite.*

As Exercise 12.4.9 shows, the diagonal entries h_{cni} dominate the other entries of the projection matrix. So Assumption (d) essentially says that all design points are uniformly about the same. Part (e) is the usual standardization assumption.

Note further that $e_i = Y_i - \mathbf{x}'_{ci}\boldsymbol{\beta}$. Then the representation of the LS estimator corresponding to its influence function, (12.4.11), is

$$\sqrt{n}(\widehat{\boldsymbol{\beta}}_{LS} - \boldsymbol{\beta}) = \mathbf{\Sigma}^{-1}\sum_{i=1}^{n}\frac{1}{\sqrt{n}}\mathbf{x}_{ci}e_i + o_p(1). \qquad (12.4.16)$$

Note that the random vectors $\mathbf{W}_i = \frac{1}{\sqrt{n}}\mathbf{x}_{ci}e_i$ are independent and that, based on the assumptions,

$$E\left(\sum_{i=1}^{n}\mathbf{W}_{ni}\right) = \mathbf{0} \text{ and } \lim_{n\to\infty}\mathrm{Cov}\left(\sum_{i=1}^{n}\mathbf{W}_{ni}\right) = \lim_{n\to\infty}\sigma^2 n^{-1}\mathbf{X}'_c\mathbf{X}_c = \sigma^2\mathbf{\Sigma}; \qquad (12.4.17)$$

see Exercise 12.4.10. Using a multivariate Central Limit Theorem, the following theorem can be proved.

Theorem 12.4.1. *Under Parts (a), (b), (d), and (e) of Assumptions (12.4.1), the asymptotic distribution of the LS estimator can be stated as:*

$$\widehat{\boldsymbol{\beta}}_{LS} \text{ has a limiting } N_p\left(\boldsymbol{\beta}, \sigma^2(\mathbf{X}'_c\mathbf{X}_c)^{-1}\right) \text{ distribution.} \qquad (12.4.18)$$

We can read the asymptotic distribution of $\widehat{\boldsymbol{\beta}}_W$ from its influence function, (12.4.14). The asymptotic representation is given by

$$\sqrt{n}(\widehat{\boldsymbol{\beta}}_W - \boldsymbol{\beta}) = \mathbf{\Sigma}^{-1}\sum_{i=1}^{n}\frac{\tau}{\sqrt{n}}\mathbf{x}_{ci}\varphi[F(e_i)] + o_p(1). \qquad (12.4.19)$$

Note that the random vectors $\mathbf{W}_i = \frac{\tau}{\sqrt{n}} \mathbf{x}_{ci} \varphi[F(e_i)]$ are independent and that, based on the assumptions,

$$E\left(\sum_{i=1}^{n} \mathbf{W}_{ni}\right) = \mathbf{0} \text{ and } \lim_{n \to \infty} \text{Var}\left(\sum_{i=1}^{n} \mathbf{W}_{ni}\right) = \lim_{n \to \infty} \tau^2 n^{-1} \mathbf{X}_c' \mathbf{X}_c = \tau^2 \mathbf{\Sigma};$$
(12.4.20)

see Exercise 12.4.11. As with Theorem 12.4.1, a multivariate Central Limit Theorem can be used to prove the following theorem.

Theorem 12.4.2. *Under Parts (a), (c), (d) and (e) of Assumptions (12.4.1), the asymptotic distribution of the Wilcoxon estimator can be stated as:*

$$\widehat{\boldsymbol{\beta}}_W \text{ has a limiting } N_p\left(\boldsymbol{\beta}, \tau^2(\mathbf{X}_c'\mathbf{X}_c)^{-1}\right) \text{ distribution.}$$
(12.4.21)

Note that the asymptotic variance of $\widehat{\boldsymbol{\beta}}_W$ differs from the asymptotic variance of the LS only by the constant of proportionality, τ^2 for the Wilcoxon and σ^2 for LS. Thus the efficiency properties of the Wilcoxon regression estimates to the LS regression estimates are the same as in the simple location problems; see Remark 10.7.1.

Based on (12.4.21), asymptotic confidence intervals for the individual regression coefficients are formulated as,

$$\widehat{\beta}_{W,j} \pm t_{(\alpha/2, n-p-1)} \tau \sqrt{(\mathbf{X}_c'\mathbf{X}_c)_{jj}^{-1}},$$
(12.4.22)

where $(\mathbf{X}_c'\mathbf{X}_c)_{jj}^{-1}$ denotes the *jth* diagonal entry of the matrix $(\mathbf{X}_c'\mathbf{X}_c)^{-1}$. The parameter τ requires estimation, and such estimators are discussed in Hettmansperger and McKean (1998) along with the use of t-critical values. Furthermore, the packages and web sites discussed in Remark 12.2.1 compute these estimates. This approximate confidence interval is similar to the corresponding LS confidence interval. The only difference is that the LS confidence interval requires an estimate of σ.

12.4.4 Estimates of the Intercept Parameter

Usually the regression model has an intercept parameter α. Rewrite Model (12.4.1) as

$$\mathbf{Y} = \alpha \mathbf{1} + \mathbf{X}_c \boldsymbol{\beta} + \mathbf{e},$$
(12.4.23)

where, as usual, $\mathbf{1}$ denotes a column vector of length n all of whose entries are one. To complete the Wilcoxon fit, we need an estimator of the intercept parameter α. As discussed in Section 12.2, we use the median of the Wilcoxon residuals; that is,

$$\widehat{\alpha}_W = \text{med}_i\{Y_i - \mathbf{x}_{ci}'\widehat{\boldsymbol{\beta}}_W\}.$$
(12.4.24)

Let $\widehat{\mathbf{b}}_W = (\widehat{\alpha}_W, \widehat{\boldsymbol{\beta}}_W')'$. The asymptotic distribution of $\widehat{\mathbf{b}}_W$ can be shown to be:

$$N_{p+1}\left(\begin{pmatrix} \alpha \\ \boldsymbol{\beta} \end{pmatrix}, \begin{bmatrix} n^{-1}\tau_S^2 & \mathbf{0}' \\ \mathbf{0} & \tau^2(\mathbf{X}_c'\mathbf{X}_c)^{-1} \end{bmatrix}\right);$$
(12.4.25)

Table 12.4.1: Hald Data used in Exercise 12.4.5

x_1	x_2	x_3	x_4	Response
7	26	6	60	78.5
1	29	15	52	74.3
11	56	8	20	104.3
11	31	8	47	87.6
7	52	6	33	95.9
11	55	9	22	109.2
3	71	17	6	102.7
1	31	22	44	72.5
2	54	18	22	93.1
21	47	4	26	115.9
1	40	23	34	83.8
11	66	9	12	113.3
10	68	8	12	109.4

where the scale parameters τ_S and τ are given respectively by (10.2.15) and (12.2.25).

EXERCISES

12.4.1. Show that the LS estimator of β in expression (12.4.2) is the same as the LS estimator for β in expression (12.3.11).

12.4.2. Based on equation (12.4.13), derive the influence function for the Wilcoxon estimator given by (12.4.14).

12.4.3. Do Exercise 12.3.8 using the Wilcoxon estimates instead of the LS estimates. See Remark 12.2.1 for ways to perform the computations.

12.4.4. In Exercise 12.3.8 change the value of Y_{10} to -148.0. Now carry out Steps (a) and (b) of this exercise, using both LS and the Wilcoxon fits. Compare the fits.

12.4.5. These data consist of 13 observations and 4 predictors. They can be found in Hald (1952) but they are also discussed in Draper and Smith (1966) where they serve to illustrate a method of predictor subset selection based on R^2. The data are given in Table 12.4.1. The response is the heat evolved in calories per gram of cement. The predictors are the percents of weights of the ingredients used in the cement and are given by:

$$
\begin{aligned}
x_1 &= \text{percent of amount of tricalcium aluminate} \\
x_2 &= \text{percent of amount of tricalcium silicate} \\
x_3 &= \text{percent of amount of tetracalcium alumino ferrite} \\
x_4 &= \text{percent of amount of dicalcium silicate.}
\end{aligned}
$$

(a) Fit these data using the Wilcoxon estimates.

(b) Obtain the confidence intervals given in expression (12.4.22) for the regression coefficients.

12.4.6. Using expressions (12.4.16) and (12.4.19), determine the asymptotic covariance between $\widehat{\boldsymbol{\beta}}_{LS}$ and $\widehat{\boldsymbol{\beta}}_W$. Do this by first establishing the following result: Use independence to show that

$$E[(\widehat{\boldsymbol{\beta}}_{LS} - \boldsymbol{\beta})(\widehat{\boldsymbol{\beta}}_W - \boldsymbol{\beta})'] = \boldsymbol{\Sigma}^{-1} \sum_{i=1}^{n} \frac{1}{\sqrt{n}} \mathbf{x}_{ci} \tau E\{e_i \varphi[F(e_i)]\} \frac{1}{\sqrt{n}} \mathbf{x}'_{ci} \boldsymbol{\Sigma}^{-1}.$$

12.4.7. Using expression (12.4.19), show that this implies the asymptotic variance of $\widehat{\boldsymbol{\beta}}_W$ as given in expression (12.4.21).

12.4.8. Continuing with Exercise 12.4.7, obtain the asymptotic variance of the Wilcoxon fitted value $\mathbf{X}_c \widehat{\boldsymbol{\beta}}_W$.

12.4.9. Let $\mathbf{H}_c = \mathbf{X}_c (\mathbf{X}'_c \mathbf{X}_c)^{-1} \mathbf{X}'_c$ denote the projection matrix onto V_c. Let h_{cnil} denote the (i, l) entry of $\mathbf{H}_c$ and let h_{cni} denote its ith diagonal entry.

(a) Because $\mathbf{H}_c$ is idempotent, show that the following inequality is true:

$$h_{cni} = \sum_{j=1}^{n} h_{cnij}^2 \geq h_{cnil}^2, \quad \text{for all } i, l = 1, \ldots, n. \tag{12.4.26}$$

(b) Based on this last result, show that if (d) of Assumptions 12.4.1 is true then all design points get uniformly small.

   Hint: The range of the design matrix $\mathbf{X}_c$ is the same as the range of $\mathbf{H}_c$.

12.4.10. Prove the assertions found in expression (12.4.17).

12.4.11. Prove the assertions found in expression (12.4.20).

12.5 Tests of General Linear Hypotheses

In this section, we consider tests of general linear hypotheses. As in the previous sections, procedures are suggested from a simple geometric point-of-view. We will use the matrix formulation and take our model to be (12.3.2), namely

$$\mathbf{Y} = \mathbf{Xb} + \boldsymbol{\varepsilon}, \tag{12.5.1}$$

where $\mathbf{X}$ is an $n \times (p + 1)$ design matrix. Our hypotheses of interest are collections of independent, linear constraints on the regression parameters. More precisely, a general linear hypothesis and its alternative are given by

$$H_0 : \mathbf{Ab} = \mathbf{0} \text{ versus } H_1 : \mathbf{Ab} \neq \mathbf{0}, \tag{12.5.2}$$

where $\mathbf{A}$ is a $q \times (p+1)$ specified matrix of full row rank $q < p+1$. The rows of $\mathbf{A}$ provide the linear constraints.

For example, suppose we are predicting Y based on a second degree polynomial model in x_1 and x_2; that is, the expected value of Y is

$$E(Y) = \alpha + \beta_1 x_1 + \beta_2 x_2 + \beta_3 x_1^2 + \beta_4 x_2^2 + \beta_5 x_1 x_2. \tag{12.5.3}$$

Suppose our null hypothesis is that the first-order terms suffice to predict Y. The corresponding matrix $\mathbf{A}$ is

$$\mathbf{A} = \begin{bmatrix} 0 & 0 & 0 & 1 & 0 & 0 \\ 0 & 0 & 0 & 0 & 1 & 0 \\ 0 & 0 & 0 & 0 & 0 & 1 \end{bmatrix},$$

because, under H_0, $E(Y) = \alpha + \beta_1 x_1 + \beta_2 x_2$. As a second example, suppose for the model (12.5.3) we think that the slope parameters of x_1 and x_2 are the same. This null hypothesis can be described by the matrix

$$\mathbf{A} = \begin{bmatrix} 0 & 1 & -1 & 0 & 0 & 0 \end{bmatrix}.$$

In hypotheses testing, we consider Model (12.5.1) to be the **full model**. Let V_F, (where F stands for full), denote the column space of $\mathbf{X}$. For the hypotheses (12.5.2), the **reduced model** is the full model subject to H_0; that is, the subspace given by

$$V_R = \{\mathbf{v} \in V_F : \mathbf{v} = \mathbf{Xb} \text{ and } \mathbf{Ab} = \mathbf{0}\}, \tag{12.5.4}$$

where the R stands for reduced, (as Exercise 12.5.1 shows, V_R is a subspace of V_F). In Lemma 12.5.2 we show that V_R has dimension $(p+1) - q$.

Suppose we have a norm $\| \cdot \|$ for fitting models. Then based on geometry a simple test procedure can be described. Let $\widehat{\boldsymbol{\eta}}_F$ be the full model fit based on the norm; i.e., $\widehat{\boldsymbol{\eta}}_F = \text{Argmin} \| \mathbf{Y} - \boldsymbol{\eta} \|$, $\boldsymbol{\eta} \in V_F$. Then the distance between $\mathbf{Y}$ and the subspace V_F is

$$d(\mathbf{Y}, V_F) = \| \mathbf{Y} - \widehat{\boldsymbol{\eta}}_F \|. \tag{12.5.5}$$

Likewise, we next fit the reduced model. Let $\widehat{\boldsymbol{\eta}}_R$ denote the reduced model fit and let $d(\mathbf{Y}, V_R)$ denote the distance between $\mathbf{Y}$ and the reduced model space V_R. Because we are minimizing over a smaller subspace $d(\mathbf{Y}, V_R) \geq d(\mathbf{Y}, V_F)$. Figure 12.5.1 is a sketch of the geometry, here. It depicts the fits $\widehat{\boldsymbol{\eta}}_F$ and $\widehat{\boldsymbol{\eta}}_R$ and the distances of the response vector $\mathbf{Y}$ to the respective subspaces.

An intuitive test statistic is the difference in these distances given by

$$RD_{\|\cdot\|} = d(\mathbf{Y}, V_R) - d(\mathbf{Y}, V_F), \tag{12.5.6}$$

where $RD_{\|\cdot\|}$ denotes reduction in distance. Small values of $RD_{\|\cdot\|}$ indicate that H_0 is true while large values indicate that H_1 is true. Hence, the corresponding test is

$$\text{Reject } H_0 \text{ in favor of } H_1, \text{ if } RD_{\|\cdot\|} \geq c, \tag{12.5.7}$$

where c must be determined. Usually $RD_{\|\cdot\|}$ is standardized by an estimate of scale or variance.

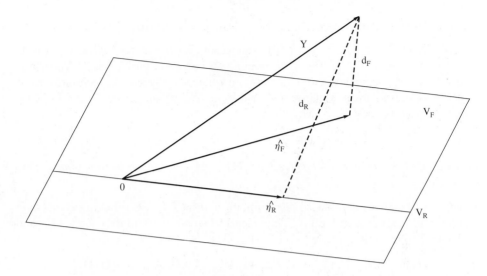

Figure 12.5.1: The sketch shows the geometry underlying the test statistic $RD_{\|\cdot\|}$, which is the difference in distances between $\mathbf{Y}$ and each of the subspaces V_F and V_R. On the plot, these respective distances are denoted by d_F and d_R and are the lengths of the dashed segments.

Suppose the norm is the Euclidean norm. In this case, it is more convenient to work with the squared norm. Thus the fitting is by LS. If we let $\mathbf{H}_F$ and $\mathbf{H}_R$ denote the projection matrices onto the subspaces V_F and V_R, respectively, then it follows from Section 12.4 that

$$
\begin{aligned}
d_{LS}^2(\mathbf{Y}, V_F) &= \|\mathbf{Y} - \mathbf{H}_F\mathbf{Y}\|_{LS}^2 = \mathbf{Y}'(\mathbf{I} - \mathbf{H}_F)\mathbf{Y}, \\
d_{LS}^2(\mathbf{Y}, V_R) &= \|\mathbf{Y} - \mathbf{H}_R\mathbf{Y}\|_{LS}^2 = \mathbf{Y}'(\mathbf{I} - \mathbf{H}_R)\mathbf{Y},
\end{aligned}
\tag{12.5.8}
$$

and hence, the LS reduction is

$$
RD_{LS} = d_{LS}^2(\mathbf{Y}, V_R) - d_{LS}^2(\mathbf{Y}, V_F) = \mathbf{Y}'(\mathbf{H}_F - \mathbf{H}_R)\mathbf{Y}.
\tag{12.5.9}
$$

Below we determine the distribution theory associated with this test.

Suppose the Wilcoxon norm $\|\cdot\|_W$ is selected. Because the Wilcoxon fit is invariant to the intercept, we assume that the matrix $\mathbf{A}$ does not involve an intercept parameter. Recall that the full model Wilcoxon fit is

$$
\widehat{\boldsymbol{\eta}}_{W,F} = \mathrm{Argmin}_{\boldsymbol{\eta} \in V_F} \|\mathbf{Y} - \boldsymbol{\eta}\|,
$$

and, hence,

$$
d_W(\mathbf{Y}, V_F) = \|\mathbf{Y} - \widehat{\boldsymbol{\eta}}_{W,F}\|_W.
$$

The expression $d_W(\mathbf{Y}, V_R)$ is defined similarly. The Wilcoxon test statistic is

$$
RD_W = d_W(\mathbf{Y}, V_R) - d_W(\mathbf{Y}, V_F).
$$

12.5.1   Distribution Theory for the LS Test for Normal Errors

In this subsection, we obtain the distribution theory for the LS test of the hypotheses
(12.5.2). We assume that the error random vector ε, (12.5.1), has a $N_n(\mathbf{0}, \sigma^2\mathbf{I})$
distribution. We begin with two lemmas from which the theory follows immediately.

First recall from linear algebra, if V_1 and V_2 are two subspaces of R^n and $V_1 \subset V_2$,
then the space V_2 mod V_1 is defined by

$$V_2|V_1 = \{\mathbf{v} \in V_2 : \mathbf{v} \perp \mathbf{w}, \text{ for all } \mathbf{w} \in V_1\}. \tag{12.5.10}$$

Lemma 12.5.1. *In expression (12.5.9), $\mathbf{H}_F - \mathbf{H}_R$ is the projection matrix onto
the space $V_F|V_R$.*

Proof: Let $\mathbf{U}_R$ be an o.n. basis matrix for V_R and let $[\mathbf{U}_R : \mathbf{U}_2]$ be an extension of
it to an o.n. basis matrix for V_F. Then clearly $\mathbf{U}_2$ is a basis matrix for $V_F|V_R$ and
$\mathbf{U}_2\mathbf{U}_2'$ is the projection matrix onto $V_F|V_R$. Also $\mathbf{H}_R = \mathbf{U}_R\mathbf{U}_R'$ and

$$\mathbf{H}_F = [\mathbf{U}_R : \mathbf{U}_2][\mathbf{U}_R : \mathbf{U}_2]' = \mathbf{U}_R\mathbf{U}_R' + \mathbf{U}_2\mathbf{U}_2' = \mathbf{H}_R + \mathbf{U}_2\mathbf{U}_2',$$

and hence, the result. ∎

Lemma 12.5.2. *Let $\mathbf{C} = \mathbf{X}(\mathbf{X}'\mathbf{X})^{-1}\mathbf{A}'$. Then $\mathbf{C}$ is a basis matrix for $V_F|V_R$.
Further, the dimension of $V_F|V_R$ is q and V_R is $p + 1 - q$.*

A proof of this lemma is sketched in Exercise 12.5.3; see, also, Arnold (1981).
Based on this lemma, because $\mathbf{C}$ is a basis matrix and has q columns, the dimension
of $V_F|V_R$ is q and hence, the dimension of V_R is $(p + 1) - q$. Furthermore, since
$\mathbf{H}_F - \mathbf{H}_R$ is a projection matrix for the subspace $V_F|V_R$, (Lemma 12.5.1), it follows
that $\mathbf{H}_F - \mathbf{H}_R = \mathbf{C}(\mathbf{C}'\mathbf{C})^{-1}\mathbf{C}'$. Using some simple matrix algebra, we get

$$
\begin{aligned}
RD_{LS} &= \mathbf{Y}'(\mathbf{H}_F - \mathbf{H}_R)\mathbf{Y} \\
&= \mathbf{Y}'\mathbf{X}(\mathbf{X}'\mathbf{X})^{-1}\mathbf{A}'[\mathbf{A}(\mathbf{X}'\mathbf{X})^{-1}\mathbf{A}']^{-1}\mathbf{A}(\mathbf{X}'\mathbf{X})^{-1}\mathbf{X}'\mathbf{Y} \\
&= (\mathbf{A}\widehat{\mathbf{b}}_{LS})'[\mathbf{A}(\mathbf{X}'\mathbf{X})^{-1}\mathbf{A}']^{-1}\mathbf{A}\widehat{\mathbf{b}}_{LS}. \tag{12.5.11}
\end{aligned}
$$

As stated above, usually the reduction in distance is standardized by an estimate
of scale or variance. The test statistic for the hypotheses (12.5.2) is given by

$$F_{LS} = \frac{(\mathbf{A}\widehat{\mathbf{b}}_{LS})'[\mathbf{A}(\mathbf{X}'\mathbf{X})^{-1}\mathbf{A}']^{-1}\mathbf{A}\widehat{\mathbf{b}}_{LS}/q}{\widehat{\sigma}^2}, \tag{12.5.12}$$

where $\widehat{\sigma}^2$ is the estimate of σ^2, (12.3.5). As Exercise 12.5.4 shows, this is the
likelihood ratio test statistic. The null distribution of this statistic is given in the
following theorem.

Theorem 12.5.1. *Under the Model (12.3.2) and under the assumption that the
random error vector ε has a $N_n(\mathbf{0}, \sigma^2\mathbf{I})$ distribution, the statistic F_{LS} has an F-
distribution with q and $n - p - 1$ degrees of freedom and noncentrality parameter*

$$\theta = \frac{(\mathbf{A}\mathbf{b})'[\mathbf{A}(\mathbf{X}'\mathbf{X})^{-1}\mathbf{A}']^{-1}\mathbf{A}\mathbf{b}}{\sigma^2}. \tag{12.5.13}$$

Proof: By Theorem 12.3.5, we know that $(n - p - 1)\widehat{\sigma}^2/\sigma^2$ has a $\chi^2(n - p - 1)$ distribution and that $\widehat{\sigma}^2$ is independent of $\widehat{\mathbf{b}}_{LS}$. Hence, the numerator and denominator of F_{LS} are independent. Also, from Theorem 12.3.5, we know that $\widehat{\mathbf{b}}_{LS}$ has a $N_{p+1}(\mathbf{b}, \sigma^2(\mathbf{X}'\mathbf{X})^{-1})$ distribution. Hence, by Theorem 3.5.1, $\mathbf{A}\widehat{\mathbf{b}}_{LS}$ has a $N_q(\mathbf{Ab}, \sigma^2\mathbf{A}(\mathbf{X}'\mathbf{X})^{-1}\mathbf{A}')$ distribution. It follows immediately from Theorem 3.5.4, that q/σ^2 times the numerator of F_{LS} has a χ^2 distribution with q degrees of freedom and noncentrality parameter θ. Putting these results together proves the theorem. ∎

Note that the noncentrality parameter of distribution of F_{LS} is 0 if and only if H_0 is true, in which case F_{LS} has a central F-distribution. Therefore, a level α test is

$$\text{Reject } H_0 \text{ in favor of } H_1 \text{ if } F_{LS} \geq F_{\alpha,q,n-p-1}. \tag{12.5.14}$$

12.5.2 Asymptotic Results

If the random vector of errors does not have a normal distribution then we can appeal to asymptotic results. This development, though, is beyond the scope of this book; see, for instance, Arnold (1981). It can be shown, though, that under Parts (a), (b), (d), and (e) of Assumptions 12.4.1 and assuming H_0 is true that

$$\frac{(\mathbf{A}\widehat{\mathbf{b}}_{LS})'[\mathbf{A}(\mathbf{X}'\mathbf{X})^{-1}\mathbf{A}']^{-1}\mathbf{A}\widehat{\mathbf{b}}_{LS}}{\sigma^2} \xrightarrow{D} \chi^2(q). \tag{12.5.15}$$

Further, $\widehat{\sigma}^2$ is a consistent estimator of σ^2. Hence, substituting $\widehat{\sigma}^2$ for σ^2 in expression (12.5.15) leads to an approximate χ^2 test statistic. In practice, though, the test is usually based on F_{LS} and F-critical values are used even though in the nonnormal case F_{LS} does not have an F-distribution under H_0 this provides a more conservative test.

Turning to the Wilcoxon test, an estimate of τ is used to standardize RD_W. The usual test statistic is of the form

$$F_W = \frac{RD_W/q}{\widehat{\tau}/2}. \tag{12.5.16}$$

LS theory is facilitated by the equivalence between the drop in the square of the distance and the quadratic form in $\widehat{\boldsymbol{\beta}}_{LS}$. While not algebraically true for the Wilcoxon test, it is asymptotically true. Using this asymptotic equivalence it can be shown that under Parts (a), (c), (d), and (e) of Assumptions 12.4.1 and assuming H_0 is true that

$$qF_W \xrightarrow{D} \chi^2(q). \tag{12.5.17}$$

Small sample studies, though, have indicated that F-critical values lead to levels closer to true levels. Hence, for a specified level α, we will use the decision rule given by

$$\text{Reject } H_0 \text{ in favor of } H_1, \text{ if } F_W \geq F_{\alpha,q,n-p-1}, \tag{12.5.18}$$

considering this to be an approximate test.

For local alternatives, we can obtain the noncentrality parameter for F_W. It is the same as θ, (12.5.13), except τ^2 replaces σ^2. Based on this, we have the same ARE between F_W and F_{LS} as we have for the regression estimates. The robust properties of the tests are also the same as for the regression estimators. For LS, the test is not robust in either the Y or $\mathbf{x}$ spaces, while for the Wilcoxon F_W is robust in the Y space but not robust in the $\mathbf{x}$ space.

Remark 12.5.1 (On Computations). There are many computer packages which obtain the LS test based on F_{LS} for many different models and hypotheses. It is not our purpose to discuss these. For the examples and exercises in this text, a few simple lines of R or S-PLUS code can be used to obtain the calculations. For example, suppose the hypothesis matrix is in the matrix `am`, the response is in the vector `y`, and the full model design matrix is in the matrix `x`, then the function `lslinhypoth(x,y,am)` of Appendix C returns the F_{LS} test statistic and p-value.

There are several ways to compute the test statistic F_W and the corresponding analysis. The web site `www.stat.wmich.edu/slab/RGLM` offers side-by-side results for both LS and Wilcoxon analyses. Also, Terpstra and McKean (2004) have developed R or S-PLUS code which obtains the test statistic F_W. This code can be downloaded from the web site `www.stat.wmich.edu/mckean/HMC/Rcode`. ∎

12.5.3   Examples

As we stated in the introduction to this chapter, linear models are the most widely used models in statistics. They encompass not only regression models but all the analysis of variance (ANOVA) models. Most of our examples so far have utilized regression models, so in this subsection we will consider ANOVA type models. This is a large area which we briefly touch on. In particular, we look at two-way ANOVA models and a simple analysis of covariance model. We illustrate our discussion with two examples drawn from real studies. The interested reader can consult many excellent texts in this area, including Neter et al. (1996), Scheffé (1959), Arnold (1981), Stapleton (1995), and Graybill (1976).

Two-Way ANOVA

Consider the two-way ANOVA model which was discussed in Section 9.5. Recall that this model was for an experimental design where the response can be affected by two factors, for example, A and B. Factor A has a levels while Factor B has b levels. The easiest way to table the data is in a table with a rows and b columns. Hence, the resulting table has ab cells. Suppose we have n_{ij} responses in cell (i,j). Let Y_{ijk} denote the response to the kth experimental unit (subject) at level i of Factor A and level j of Factor B. Let μ_{ij} denote the mean of Y_{ijk}. Then the two-way model can be written as

$$Y_{ijk} = \mu_{ij} + e_{ijk}, \quad k = 1, \ldots, n_{ij}; \, i = 1, \ldots, a; \, j = 1, \ldots, b; \quad (12.5.19)$$

where the random errors e_{ijk} are iid with pdf $f(t)$. Let $n = \sum \sum n_{ij}$ denote the total sample size and let $\mathbf{Y}$ be the $n \times 1$ vector of responses. For convenience, we assume that the responses Y_{ijk} are entered into this vector with the subscript k running the fastest, j running the second fastest, and i running the slowest. Let $\mathbf{x}_{ij}$ be a $n \times 1$ vector whose kth component is defined as

$$x_{ijk} = \begin{cases} 1 & \text{if } y_k \text{ is an observation from cell } (i,j) \\ 0 & \text{elsewhere,} \end{cases}$$

i.e., $\mathbf{x}_{ij}$ is the incidence vector for cell (i,j). Let $\mathbf{W}$ be the $n \times ab$ matrix whose columns are the $\mathbf{x}_{ij}$s; that is, $\mathbf{W} = [\mathbf{x}_{11} \ \mathbf{x}_{12} \ \cdots \mathbf{x}_{ab}]$. Then Model (12.5.19) can be written in matrix form as

$$\mathbf{Y} = \mathbf{W}\boldsymbol{\mu} + \mathbf{e}, \tag{12.5.20}$$

where $\mathbf{e}$ is the vector of e_{ijk}s and $\boldsymbol{\mu}$ is the vector of μ_{ij}s. This is a linear model and for our discussion on two-way ANOVA we take it to be the full model. We call the matrix $\mathbf{W}$ the **incidence matrix**.

It is easy to obtain the LS estimators of the full model in closed form. Because of the way the data are entered into $\mathbf{Y}$, $\mathbf{W}'\mathbf{W}$ is a diagonal matrix with the main diagonal $\text{diag}\{n_{11}, n_{12}, \ldots, n_{ab}\}$ and $\mathbf{W}'\mathbf{Y} = (\sum_{k=1}^{n_{11}} Y_{11k}, \sum_{k=1}^{n_{12}} Y_{12k}, \ldots, \sum_{k=1}^{n_{ab}} Y_{abk})'$. Hence, the LS estimator of $\boldsymbol{\mu}$ is $\widehat{\boldsymbol{\mu}} = (\overline{Y}_{11.}, \overline{Y}_{12.}, \ldots, \overline{Y}_{11.})'$, i.e., the cell sample means. Further, the estimator of σ^2 is given by

$$\widehat{\sigma}^2 = (n - ab)^{-1} \sum_{i=1}^{a} \sum_{j=1}^{b} \sum_{k=1}^{n_{ij}} (Y_{ijk} - \overline{Y}_{ij.})^2; \tag{12.5.21}$$

see Exercise 12.5.2. The Wilcoxon estimators cannot be obtained in closed form but the model is easily fitted by the algorithms cited in Remark 12.5.1.

Next consider the additive submodel where the the cell means are given by

$$\mu_{ij}^* = \overline{\mu}_{..} + (\overline{\mu}_{i.} - \overline{\mu}_{..}) + (\overline{\mu}_{.j} - \overline{\mu}_{..}). \tag{12.5.22}$$

The additive model is an attractive reduced model. Notice that under the additive model the null hypothesis $H_{0A} : \overline{\mu}_{i.} - \overline{\mu}_{..} = 0$, for $i = 1, \ldots a$ implies that the individual levels of Factor A do not affect the response. Likewise for Factor B, the null hypothesis $H_{0B} : \overline{\mu}_{.j} - \overline{\mu}_{..} = 0$, for $j = 1, \ldots b$, implies that the individual levels of Factor B do not affect the response. These two null hypotheses are called **main effect** hypotheses.

While the additive model is attractive, it is itself a null hypothesis. As in Section 9.5, define the interaction parameters as the difference between the full model and the additive model; that is, define the parameters

$$\gamma_{ij} = \mu_{ij} - \mu_{ij}^* = \mu_{ij} - \overline{\mu}_{i.} - \overline{\mu}_{.j} + \overline{\mu}_{..}, \quad i = 1, \ldots, a; j = 1, \ldots, b. \tag{12.5.23}$$

Hence, the additive model is equivalent to the null hypothesis $H_{0AB} : \gamma_{ij} = 0$, for all i and j. Notice that the hypotheses H_{0A}, H_{0B}, and H_{0AB} are given in terms of linear constraints on the full model parameters μ_{ij}. Hence, these are linear hypotheses. Recall from Chapter 9 that the number of linearly independent constraints for them are $a - 1$, $b - 1$ and $(a-1)(b-1)$, respectively. We offer the following data example to help clarify this discussion.

Table 12.5.1: Data for Example 12.5.1, Lifetimes of Motors, (hours). Note in the example the *log*s of these lifetimes were used.

	Insulation		
Temp.	1	2	3
	1176	2856	3528
	1512	3192	3528
200° F	1512	2520	3528
	1512	3192	
	3528	3528	
	624	816	720
	624	912	1296
225° F	624	1296	1488
	816	1392	
	1296	1488	
	204	300	252
	228	324	300
250° F	252	372	324
	300	372	
	324	444	

Example 12.5.1 (Lifetime of Motors). This problem is an unbalanced two-way design which is discussed on page 471 of Nelson (1982). The responses are lifetimes (in hours) of electric motors. The two factors are: Factor A, type of insulation (1, 2, and 3), and Factor B, temperature at which motor was kept until it died, (200° F, 225° F, and 250° F). Hence, both factors had three levels.

The data are displayed in Table 12.5.1. Following Nelson, as the response variable we considered the logs of the lifetimes. Let Y_{ijk} denote the log of the lifetime of the kth replicate at temperature level i and motor insulation j. As our full model we use the cell mean model (12.5.19). We enter the data into the response vector as discussed above; hence, $\boldsymbol{\mu}' = (\mu_{11}, \mu_{12}, \mu_{13}, \mu_{21}, \mu_{22}, \mu_{23}, \mu_{31}, \mu_{32}, \mu_{33})$.

To write the hypotheses matrices, consider first the main effect for Factor A. For our cell mean model H_{0A} means that the row averages are the same. This is true if and only if the following two constraints are true: $3(\bar{\mu}_{1.} - \bar{\mu}_{3.}) = 0$ and $3(\bar{\mu}_{2.} - \bar{\mu}_{3.}) = 0$. Hence, a hypothesis matrix for H_{0A} is given by

$$\mathbf{A}_1 = \begin{bmatrix} 1 & 1 & 1 & 0 & 0 & 0 & -1 & -1 & -1 \\ 0 & 0 & 0 & 1 & 1 & 1 & -1 & -1 & -1 \end{bmatrix}. \tag{12.5.24}$$

Similarly, a hypothesis matrix for H_{0B} is given by

$$\mathbf{A}_2 = \begin{bmatrix} 1 & 0 & -1 & 1 & 0 & -1 & 1 & 0 & -1 \\ 0 & 1 & -1 & 0 & 1 & -1 & 0 & 1 & -1 \end{bmatrix}. \tag{12.5.25}$$

To test that the interaction parameters are 0 requires four linearly independent constraints. We have selected $\gamma_{11}, \gamma_{12}, \gamma_{21}$, and γ_{22}. Upon simplification, this leads to the hypothesis matrix

$$
\mathbf{A}_3 = \begin{bmatrix}
4 & -2 & -2 & -2 & 1 & 1 & -2 & 1 & 1 \\
-2 & 4 & -2 & 1 & -2 & 1 & 1 & -2 & 1 \\
-2 & 1 & 1 & 4 & -2 & -2 & -2 & 1 & 1 \\
1 & -2 & 1 & -2 & 4 & -2 & 1 & -2 & 1
\end{bmatrix}. \tag{12.5.26}
$$

In Exercise 12.5.5, the reader is asked to verify these last two matrices are hypothesis matrices for H_{0B} and H_{0AB}, respectively.

Using the web site discussed in Remark 12.5.1, the following results of the tests based on F_{LS} and F_W were obtained.

Hypothesis	F_{LS}	p-value	F_W	p-value
H_{0A}	214.1	0.000	121.7	0.000
H_{0B}	12.51	0.000	17.15	0.000
H_{0AB}	1.297	0.293	2.854	0.041

Notice that there is a discrepancy between the LS and Wilcoxon results for the interaction hypothesis. This is investigated in Exercise 12.5.6. ∎

Analysis of Covariance

Often in an experimental design there are extraneous variables, other than the response variables, which cannot be controlled. For instance, consider a one-way experiment on weight reduction over several diets. The response is the amount of weight loss, but initial weight of the subject is an important often uncontrollable variable which would seem to be influential in the amount of weight loss. Hopefully, these variables explain some of the noise in the data. We call such a variable a **covariate** or a **concomitant variable** and the traditional analysis of such data is called **analysis of covariance**.

Covariates can occur in any experimental design. To keep matters simple, we consider a one-way ANOVA model with a single covariate, with a levels for a single factor A, and with n_i observations at level i. Let Y_{ij} and x_{ij} denote the jth response and the jth value of the covariate, respectively, at level i, for $j = 1, 2, \ldots, n_i$ and $i = 1, 2, \ldots, a$. A first-order model is given by

$$
Y_{ij} = \mu_i + \beta x_{ij} + e_{ij}, \quad j = 1, 2, \ldots, n_i; i = 1, 2, \ldots, a. \tag{12.5.27}
$$

The parameters μ_i in this model have a natural interpretation. Suppose for the two responses Y_{ij} and $Y_{i'j'}$, we have $x_{ij} = x_{i'j'}$. Then $E(Y_{ij}) - E(Y_{i'j'}) = \mu_i - \mu_{i'}$. In this sense, the parameters μ_i are measuring average effect. Hence, a null hypothesis of interest is given by $H_{0A} : \mu_1 = \cdots = \mu_a$, for under H_{0A}, the individual levels of Factor A do not affect the response.

Model (12.5.27), however, may be oversimplistic because it assumes that the covariate behaves the same within each level. A more general model is

$$
Y_{ij} = \mu_i + \beta_i x_{ij} + e_{ij}, \quad j = 1, \ldots, n_i, \; i = 1, \ldots, a. \tag{12.5.28}
$$

Thus the slope and the intercept at the *ith* level are β_i and μ_i, respectively. Thus, each level has its own linear model. A natural null hypothesis for this model is, of course, $H_{0C} : \beta_1 = \cdots = \beta_k$, because under H_{0C} the simple model (12.5.27) is true. If H_{0C} is not true then the covariate and the levels interact. For example, whether one level is better than another may depend on where in factor or x space the responses are measured. Thus as in the two-way design, the interpretation of main effect hypotheses may not be clear. These hypotheses are linear in the parameters of the model and, hence, can be tested as discussed above. We discuss the corresponding hypothesis matrices in the following data example.

Example 12.5.2 (Snake Data). As an example of an analysis of covariance problem consider the data set discussed by Afifi and Azen (1972). Forty subjects were given a behavior approach test to determine how close they could walk to a snake without feeling uncomfortable. This distance was taken as the covariate. Next they were randomly assigned to one of the four levels of a treatment. The first level was a control (placebo) while the other three levels were different methods intended to reduce a human's fear of snakes. The response was a subject's distance on the behavior approach test after treatment. Ten subjects were assigned to each level; hence, the total sample size is 40 and the number of parameters in Model (12.5.28) is 8. The data are given by, (where Initial distance is the covariate and final distance is the response),

Placebo		Level 2		Level 3		Level 4	
Initial Dist.	Final Dist.	Initial Dist.	Final Dist.	Initial Dist.	Final Dist.	Initial Dist.	Final Dist.
25	25	17	11	32	24	10	8
13	25	9	9	30	18	29	17
10	12	19	16	12	2	7	8
25	30	25	17	30	24	17	12
10	37	6	1	10	2	8	7
17	25	23	12	8	0	30	26
9	31	7	4	5	0	5	8
18	26	5	3	11	1	29	29
27	28	30	26	5	1	5	29
17	29	19	20	25	10	13	9

Let $\mathbf{Y}$ denote the vector of responses and let $\mathbf{z}$ denote the corresponding entries for the covariate. Let $\mathbf{w}_i$, $i = 1, \ldots, 4$, be the incidence vector for level i. Let $\mathbf{v}_i$ be the vector whose jth component is $w_{ij}\mathbf{z}_j$. Then the design matrix corresponding to Model (12.5.28) is

$$\mathbf{X} = [\mathbf{w}_1 \ \mathbf{v}_1 \ \mathbf{w}_2 \ \mathbf{v}_2 \ \mathbf{w}_3 \ \mathbf{v}_3 \ \mathbf{w}_4 \ \mathbf{v}_4]. \qquad (12.5.29)$$

Using the computer, we obtained the LS and Wilcoxon fits discussed below.

The plots of the response variable versus the covariate for each level are found in Panels A through D of Figure 12.5.2. It is clear from the plots that the relationship between the response and the covariate varies with the level from virtually no relationship for the first level (placebo) to a fairly strong linear relationship for the

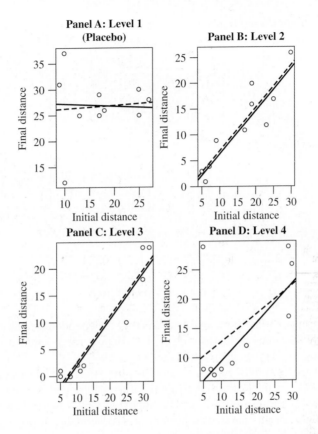

Figure 12.5.2: Panels A - D: For the Snake Data, scatterplots of Final Distance versus Initial Distance for the Placebo and Levels 2-4, overlaid with the Wilcoxon fit (solid line) and the LS fit (dashed line).

third level. Outliers are apparent in these plots also. These plots are overlaid with Wilcoxon and LS fits of the full model, Model (12.5.28).

The Wilcoxon and LS estimates of intercept and slope for each level are:

Level	Wilcoxon Estimates				LS Estimates			
	Int.	(SE)	Slope	(SE)	Int.	(SE)	Slope	(SE)
1	27.3	(3.6)	$-.02$	(.20)	25.6	(5.3)	.07	(.29)
2	-1.78	(2.8)	.83	(.15)	-1.39	(4.0)	.83	(.22)
3	-6.7	(2.4)	.87	(.12)	-6.4	(3.5)	.87	(.17)
4	2.9	(2.4)	.66	(.13)	7.8	(3.4)	.49	(.19)

As this table and Figure 12.5.2 shows, the larger discrepancies between the Wilcoxon and LS estimates occur for those levels which have large outliers. The estimates of τ and σ are 3.92 and 5.82, respectively; hence, as the table shows the estimated standard errors of the Wilcoxon estimates are lower than their LS counterparts.

To test the hypotheses H_{0A} and H_{0C}, we will formulate the hypothesis matrices corresponding to our formulation of the design matrix $\mathbf{X}$, (12.5.29). To test H_{0C}, the hypothesis matrix is

$$
\mathbf{A}_{0C} = \begin{bmatrix} 0 & 1 & 0 & 0 & 0 & 0 & 0 & -1 \\ 0 & 0 & 0 & 1 & 0 & 0 & 0 & -1 \\ 0 & 0 & 0 & 0 & 0 & 1 & 0 & -1 \end{bmatrix}, \tag{12.5.30}
$$

while the hypothesis matrix for H_{0A} is

$$
\mathbf{A}_{0A} = \begin{bmatrix} 1 & 0 & 0 & 0 & 0 & 0 & -1 & 0 \\ 0 & 0 & 1 & 0 & 0 & 0 & -1 & 0 \\ 0 & 0 & 0 & 0 & 1 & 0 & -1 & 0 \end{bmatrix}, \tag{12.5.31}
$$

The following table summarizes the tests based on F_{LS} and F_W:

Hypothesis	F_{LS}	p-value	F_W	p-value
H_{0C}	2.34	0.09	4.07	0.01
H_{0A}	9.63	0.00	12.67	0.00

Note for each hypothesis there are three linearly independent constraints. Further, there are eight parameters in the full model. Hence, the p-values in the table are based on 3 and 32 degrees of freedom.

Note that F_W strongly rejects H_{0C}, (p-value is .01), while the p-value of the test based on F_{LS} is 0.09. This confirms the discussion above on outliers as seen in Figure 12.5.2. The interpretation of the hypothesis H_{0A} is not clear when H_{0C} is rejected. Hence, for this particular problem the practical interpretations of the LS and Wilcoxon analyses would differ. ∎

EXERCISES

12.5.1. Using the defining equation of the reduced model space, (12.5.1), show that V_R is a subspace of V_F.

12.5.2. Verify the LS estimates of μ and σ^2 discussed around expression (12.5.21).

12.5.3. A proof of Lemma 12.5.2 is obtained by showing each of the following parts are true. Let $k = (p+1) - q$. Recall that if a matrix has full column rank, then its kernel is $\{\mathbf{0}\}$. Use this for both $\mathbf{X}$ and $\mathbf{A}'$ in the following.

   (a) Define the subspace $W = \{\mathbf{b} | \mathbf{Ab} = \mathbf{0}\}$. Show that W is the space orthogonal to the row space of $\mathbf{A}$. Hence, W has dimension k.

   (b) Let $\mathbf{b}_1, \ldots, \mathbf{b}_k$ be a basis for W. Show that $\mathbf{v}_1, \ldots, \mathbf{v}_k$, where $\mathbf{v}_i = \mathbf{Xb}_i$, is a basis for V_R. Hence, V_R has dimension k. Hence, $V_F | V_R$ has dimension $p + 1 - k = q$.

   (c) Let $\mathbf{C} = \mathbf{X}(\mathbf{X}'\mathbf{X})^{-1}\mathbf{A}'$. Show that range $\mathbf{C} \subset V_F | V_R$.

(d) Show that $\ker(\mathbf{C}) = \mathbf{0}$. Thus, the range $\mathbf{C}$ is a subspace of $V_F|V_R$ which has the same rank as $V_F|V_R$. Therefore, range $\mathbf{C} = V_F|V_R$.

12.5.4. Assume that Model (12.3.2) holds and that the random error vector $\boldsymbol{\varepsilon}$ has a $N_n(\mathbf{0}, \sigma^2 \mathbf{I})$ distribution. Show that the test based on the test statistic F_{LS}, (12.5.12), is the likelihood ratio test for the hypotheses (12.5.2). *Hint:* Use the identity (12.5.11).

12.5.5. For Example 12.5.1, verify that the matrices in expressions (12.5.25) and (12.5.26) are hypothesis matrices for the hypotheses H_{0A} and H_{0AB}, respectively.

12.5.6. Consider Example 12.5.1. Recall that the responses are the logs of the lifetimes in Table 12.5.1.

(a) Obtain the cell median profile plots that are discussed in Section 9.5. Plot cell median versus Factor A (motor insulation) and connect the points for each level of Factor B (temperature). Notice that the profiles do intersect, indicating possible interaction.

(b) Plot the LS (or Wilcoxon) residuals versus fitted values. Examine this plot for outliers.

(c) Based on Part (b), what can you say about the discrepancy between the LS and Wilcoxon tests for interaction.

12.5.7. Consider Example 12.5.2.

(a) Obtain the residual plots for the LS and Wilcoxon for the fits of Model (12.5.28). These plots show that the variance of the residuals is not constant across the fit.

(b) Apply the square root transform to the responses and refit the model. Comment on the resulting residual plots.

12.5.8. In this Section, the two-way ANOVA model of Section 9.5 was set up in a linear model (matrix) formulation.

(a) Set up the linear model formulation for the one-way ANOVA model in Section 9.2. Obtain the hypothesis matrix to test the hypothesis $H_{0A} : \mu_1 = \mu_2 = \cdots = \mu_a$.

(b) Use Part (a), to obtain the LS and Wilcoxon analyses for the data given in Exercise 9.2.7.

(c) For the data in Exercise 9.2.7, change the value of the last observation in Brand C from 40.3 to 4.3. Now obtain the LS and Wilcoxon analyses. Comment on how the analyses changed.

12.5.9. Set up the data in Exercise 9.5.7 as a two-way ANOVA model as described in this section. Obtain the hypotheses matrices for main effects and interaction. Use a computer package to conduct the LS and Wilcoxon analyses.

12.5.10. For the data in Exercise 9.5.7, change the value of the last observation in cell $(3, 4)$ from 4.2 to 0.2. Now obtain the LS and Wilcoxon analyses. Comment on how the analyses changed.

12.5.11. The following data set is taken from Shirley (1981) discussed from a robust point-of-view in McKean and Vidmar (1994). The response is the time it takes a rat to enter a chamber after receiving a treatment designed to delay the time of entry. There were 30 rats in the experiment and they were divided evenly into three groups. The rats in Groups 2 and 3 received an antidote to the treatment. The covariate is the time taken by the rat to enter the chamber prior to its treatment. The data are:

Group 1		Group 2		Group 3	
Initial time	Final time	Initial time	Final time	Initial time	Final time
1.8	79.1	1.6	10.2	1.3	14.8
1.3	47.6	0.9	3.4	2.3	30.7
1.8	64.4	1.5	9.9	0.9	7.7
1.1	68.7	1.6	3.7	1.9	63.9
2.5	180.0	2.6	39.3	1.2	3.5
1.0	27.3	1.4	34.0	1.3	10.0
1.1	56.4	2.0	40.7	1.2	6.9
2.3	163.3	0.9	10.5	2.4	22.5
2.4	180.0	1.6	0.8	1.4	11.4
2.8	132.4	1.2	4.9	0.8	3.3

(a) Obtain scatterplots of the data by group.

(b) Obtain the LS and Wilcoxon fits of the full model $y_{ij} = \mu_i + \beta_i x_{ij} + e_{ij}$, $i = 1, 2, 3$, $j = 1, \ldots, 10$, where y_{ij} denotes the response for the jth rat in Group i and x_{ij} denotes the corresponding covariate.

(c) Overlay the LS and Wilcoxon fits on your scatterplots in (a). Comment on the effect of outliers in groups 2 and 3.

(d) Form the hypothesis matrix to test the homogeneity of the slopes, i.e., $H_0 : \beta_1 = \beta_2 = \beta_3$.

(e) Carry out the LS and Wilcoxon tests of the hypothesis in Part (d).

Appendix A

Mathematics

A.1 Regularity Conditions

These are the regularity conditions referred to in Sections 6.4 and 6.5 of the text. A discussion of these conditions can be found in Chapter 6 of Lehmann and Casella (1998).

Let X have pdf $f(x; \boldsymbol{\theta})$, where $\boldsymbol{\theta} \in \Omega \subset R^p$. For these assumptions X can be either a scalar random variable or a random vector in R^k. As in Section 6.4, let $\mathbf{I}(\boldsymbol{\theta}) = [I_{jk}]$ denote the $p \times p$ information matrix given by expression (6.4.4). Also, we will denote the true parameter $\boldsymbol{\theta}$ by $\boldsymbol{\theta}_0$.

Assumptions A.1.1. *Additional Regularity Conditions for Sections 6.4 and 6.5.*

(R6): *There exists an open subset $\Omega_0 \subset \Omega$ such that $\boldsymbol{\theta}_0 \in \Omega_0$ and all third partial derivatives of $f(x; \boldsymbol{\theta})$ exist for all $\boldsymbol{\theta} \in \Omega_0$.*

(R7) *The following equations are true (essentially, we can interchange expectation and differentiation):*

$$
E_{\boldsymbol{\theta}} \left[\frac{\partial}{\partial \theta_j} \log f(x\,; \boldsymbol{\theta}) \right] = 0, \quad for\ j = 1, \ldots, p,
$$

$$
I_{jk}(\boldsymbol{\theta}) = E_{\boldsymbol{\theta}} \left[-\frac{\partial^2}{\partial \theta_j \partial \theta_k} \log f(x\,; \boldsymbol{\theta}) \right], \quad for\ j, k = 1, \ldots, p.
$$

(R8) *For all $\boldsymbol{\theta} \in \Omega_0$, $\mathbf{I}(\boldsymbol{\theta})$ is positive definite.*

(R9) *There exist functions $M_{jkl}(x)$ such that*

$$
\left| \frac{\partial^3}{\partial \theta_j \partial \theta_k \theta_l} \log f(x\,; \boldsymbol{\theta}) \right| \le M_{jkl}(x), \quad for\ all\ \boldsymbol{\theta} \in \Omega_0,
$$

and

$$
E_{\boldsymbol{\theta}_0}[M_{jkl}] < \infty, \quad for\ all\ j, k, l \in 1, \ldots, p. \quad \blacksquare
$$

A.2 Sequences

Let $\{a_n\}$ be a sequence of real numbers. Recall from calculus that $a_n \to a$ ($\lim_{n\to\infty} a_n = a$) if and only if

for every $\epsilon > 0$, there exists an N_0, such that $n \geq N_0 \implies |a_n - a| < \epsilon$. (A.2.1)

Let A be a set of real numbers which is bounded from above; that is, there exists and $M \in R$ such that $x \leq M$ for all $x \in A$. Recall that a is the **supremum** of A, if a is the least of all upper bounds of A. From calculus, we know that the supremum of a set bounded from above exists. Furthermore, we know that a is the supremum of A if and only if for all $\epsilon > 0$, there exists an $x \in A$ such that $a - \epsilon < x \leq a$. Similarly, we can define the **infimum** of A.

We need three additional facts from calculus. First is the sandwich theorem.

Theorem A.2.1 (Sandwich Theorem). *Suppose for sequences $\{a_n\}$, $\{b_n\}$, and $\{c_n\}$ that $c_n \leq a_n \leq b_n$, for all n, and that $\lim_{n\to\infty} b_n = \lim_{n\to\infty} c_n = a$. Then $\lim_{n\to\infty} a_n = a$.*

Proof: Let $\epsilon > 0$ be given. Because both $\{b_n\}$ and $\{c_n\}$ converge, we can choose N_0 so large that $|c_n - a| < \epsilon$ and $|b_n - a| < \epsilon$, for $n \geq N_0$. Because $c_n \leq a_n \leq b_n$, it is easy to see that

$$|a_n - a| \leq \max\{|c_n - a|, |b_n - a|\},$$

for all n. Hence, if $n \geq N_0$, then $|a_n - a| < \epsilon$. ∎

The second fact concerns subsequences. Recall that $\{a_{n_k}\}$ is a subsequence of $\{a_n\}$, if the sequence $n_1 \leq n_2 \leq \cdots$ is an infinite subset of the positive integers. Note that $n_k \geq k$.

Theorem A.2.2. *The sequence $\{a_n\}$ converges to a if and only if every subsequence $\{a_{n_k}\}$ converges to a.*

Proof: Suppose the sequence $\{a_n\}$ converges to a. Let $\{a_{n_k}\}$ be any subsequence. Let $\epsilon > 0$ be given. Then there exists an N_0 such that $|a_n - a| < \epsilon$, for $n \geq N_0$. For the subsequence, take k' to be the first index of the subsequence beyond N_0. Because for all k, $n_k \geq k$, we have that $n_k \geq n_{k'} \geq k' \geq N_0$ which implies that $|a_{n_k} - a| < \epsilon$. Thus, $\{a_{n_k}\}$ converges to a. The converse is immediate because a sequence is also a subsequence of itself. ∎

Finally, the third theorem concerns monotonic sequences.

Theorem A.2.3. *Let $\{a_n\}$ be a nondecreasing sequence of reals numbers, i.e., $a_n \leq a_{n+1}$, for all n. Suppose $\{a_n\}$ is bounded from above, i.e., for some $M \in R$, $a_n \leq M$, for all n. Then the limit of a_n exists.*

Proof: Let a be the supremum of $\{a_n\}$. Let $\epsilon > 0$ be given. Then there exists an N_0 such that $a - \epsilon < a_{N_0} \leq a$. Because the sequence is nondecreasing, this implies that $a - \epsilon < a_n \leq a$, for all $n \geq N_0$. Hence by definition, $a_n \to a$. ∎

Let $\{a_n\}$ be a sequence of real numbers and define the two subsequences,

$$b_n = \sup\{a_n, a_{n+1}, \ldots\}, \quad n = 1, 2, 3 \ldots, \tag{A.2.2}$$

$$c_n = \inf\{a_n, a_{n+1}, \ldots\}, \quad n = 1, 2, 3 \ldots. \tag{A.2.3}$$

It is obvious that $\{b_n\}$ is a nonincreasing sequence. Hence, if $\{a_n\}$ is bounded from below, then the limit of b_n exists. In this case, we call the limit of $\{b_n\}$ the **limit supremum**, (limsup), of the sequence $\{a_n\}$ and write it as

$$\overline{\lim_{n \to \infty}} \, a_n = \lim_{n \to \infty} b_n. \tag{A.2.4}$$

Note that if $\{a_n\}$ is not bounded from below then $\overline{\lim}_{n \to \infty} a_n = -\infty$. Also, if $\{a_n\}$ is not bounded from above, we define $\overline{\lim}_{n \to \infty} a_n = \infty$. Hence, the $\overline{\lim}$ of any sequence always exists. Also, from the definition of the subsequence $\{b_n\}$, we have

$$a_n \leq b_n, \quad n = 1, 2, 3, \ldots. \tag{A.2.5}$$

On the other hand, $\{c_n\}$ is a nondecreasing sequence. Hence, if $\{a_n\}$ is bounded from above then the limit of c_n exists. We call the limit of $\{c_n\}$ the **limit infimum**, (liminf), of the sequence $\{a_n\}$ and write it as

$$\underline{\lim_{n \to \infty}} \, a_n = \lim_{n \to \infty} c_n. \tag{A.2.6}$$

Note that if $\{a_n\}$ is not bounded from above, then $\underline{\lim}_{n \to \infty} a_n = \infty$. Also, if $\{a_n\}$ is not bounded from below, $\underline{\lim}_{n \to \infty} a_n = -\infty$. Hence, the $\underline{\lim}$ of any sequence always exists. Also, from the definition of the subsequences $\{c_n\}$ and $\{b_n\}$, we have

$$c_n \leq a_n \leq b_n, \quad n = 1, 2, 3, \ldots. \tag{A.2.7}$$

Also, because $c_n \leq b_n$, for all n, we have

$$\underline{\lim_{n \to \infty}} \, a_n \leq \overline{\lim_{n \to \infty}} \, a_n. \quad \blacksquare \tag{A.2.8}$$

Example A.2.1. Here are two examples. More are given in the exercises.

1. Suppose $a_n = -n$, for all $n = 1, 2, \ldots$. Then $b_n = \sup\{-n, -n-1, \ldots\} = -n \to -\infty$ and $c_n = \inf\{-n, -n-1, \ldots\} = -\infty \to -\infty$. So, $\underline{\lim}_{n \to \infty} a_n = \overline{\lim}_{n \to \infty} a_n = -\infty$.

2. Suppose $\{a_n\}$ is defined by

$$a_n = \begin{cases} 1 + \frac{1}{n} & \text{if } n \text{ is even} \\ 2 + \frac{1}{n} & \text{if } n \text{ is odd.} \end{cases}$$

Then $\{b_n\}$ is the sequence $\{3, 2 + (1/3), 2 + (1/3), 2 + (1/5), 2 + (1/5), \ldots\}$ which converges to 2, while $\{c_n\} \equiv 1$ which converges to 1. Thus $\underline{\lim}_{n \to \infty} a_n = 1$ and $\overline{\lim}_{n \to \infty} a_n = 2$. $\blacksquare$

It is useful that the $\underline{\lim}_{n\to\infty}$ and $\overline{\lim}_{n\to\infty}$ of every sequence exists. Also, the sandwich effects of expressions (A.2.7) and (A.2.8) lead to the following theorem.

Theorem A.2.4. *Let $\{a_n\}$ be a sequence of real numbers. Then the limit of $\{a_n\}$ exists if and only if $\underline{\lim}_{n\to\infty} a_n = \overline{\lim}_{n\to\infty} a_n$. In which case, $\lim_{n\to\infty} a_n = \underline{\lim}_{n\to\infty} a_n = \overline{\lim}_{n\to\infty} a_n$.*

Proof: Suppose first that $\lim_{n\to\infty} a_n = a$. Because the sequences $\{c_n\}$ and $\{b_n\}$ are subsequences of $\{a_n\}$, Theorem A.2.2 implies that they converge to a, also. Conversely, if $\underline{\lim}_{n\to\infty} a_n = \overline{\lim}_{n\to\infty} a_n$, then expression (A.2.7) and the Sandwich Theorem, A.2.1, imply the result. ∎

Based on this last theorem, we have two interesting applications which are frequently used in statistics and probability. Let $\{p_n\}$ be a sequence of probabilities and let $b_n = \sup\{p_n, p_{n+1}, \ldots\}$ and $c_n = \inf\{p_n, p_{n+1}, \ldots\}$. For the first application, suppose we can show that $\overline{\lim}_{n\to\infty} p_n = 0$. Then because $0 \le p_n \le b_n$ and the sandwich theorem, we have that $\lim_{n\to\infty} p_n = 0$. For the second application, suppose we can show that $\underline{\lim}_{n\to\infty} p_n = 1$. Then because $c_n \le p_n \le 1$ and the sandwich theorem, we have that $\lim_{n\to\infty} p_n = 1$.

We list some other properties in a theorem and ask the reader to provide the proofs in Exercise A.2.2.

Theorem A.2.5. *Let $\{a_n\}$ and $\{d_n\}$ be sequences of real numbers. Then*

$$\overline{\lim_{n\to\infty}} (a_n + d_n) \;\le\; \overline{\lim_{n\to\infty}} a_n + \overline{\lim_{n\to\infty}} d_n, \tag{A.2.9}$$

$$\overline{\lim_{n\to\infty}} a_n \;=\; -\underline{\lim_{n\to\infty}} (-a_n). \tag{A.2.10}$$

EXERCISES

A.2.1. Calculate the $\underline{\lim}$ and $\overline{\lim}$ of each of the following sequences:

(a) For $n = 1, 2, \ldots$, $a_n = (-1)^n \left(2 - \frac{4}{2^n}\right)$.

(b) For $n = 1, 2, \ldots$, $a_n = n^{\cos(\pi n/2)}$.

(c) For $n = 1, 2, \ldots$, $a_n = \frac{1}{n} + \cos\frac{\pi n}{2} + (-1)^n$.

A.2.2. Prove properties (A.2.9) and (A.2.10).

A.2.3. Let $\{a_n\}$ and $\{d_n\}$ be sequences of real numbers. Show that

$$\underline{\lim_{n\to\infty}} (a_n + d_n) \ge \underline{\lim_{n\to\infty}} a_n + \underline{\lim_{n\to\infty}} d_n.$$

A.2.4. Let $\{a_n\}$ be a sequence of real numbers. Suppose $\{a_{n_k}\}$ is a subsequence of $\{a_n\}$. If $\{a_{n_k}\} \to a_0$, as $k \to \infty$, show that $\underline{\lim}_{n\to\infty} a_n \le a_0 \le \overline{\lim}_{n\to\infty} a_n$.

Appendix B

R and S-PLUS Functions

Below in alphabetical order are the R and S-PLUS routines referenced in the text.

1. **boottestonemean**. Bootstrap test for

$$H_0 : \ \theta = \theta_0 \text{ versus } H_A : \ \theta > \theta_0 \ .$$

The test is based on the sample mean but can easily be changed to another test.

```
boottestonemean<-function(x,theta0,b){
#
#   x = sample
#   theta0 is the null value of the mean
#   b is the number of bootstrap resamples
#
#   origtest contains the value of the test statistics
#            for the original sample
#   pvalue is the bootstrap p-value
#   teststatall contains the b bootstrap tests
#
n<-length(x)
v<-mean(x)
z<-x-median(x)+theta0
counter<-0
teststatall<-rep(0,b)
for(i in 1:b){xstar<-sample(z,n,replace=T)
             vstar<-mean(xstar)
             if(vstar >= v){counter<-counter+1}
             teststatall[i]<-vstar}
pvalue<-counter/b
list(origtest=v,pvalue=pvalue,teststatall=teststatall)
}
```

2. **boottesttwo.** Program which obtains the bootstrap test for two samples as discussed in Exercise 5.9.6. The test statistic is the difference in sample means. To change to another test statistic simply substitute the appropriate call in place of the call to means.

```
boottesttwo<-function(x,y,b){
#
#   x vector containing first sample.
#   y vector containing first sample.
#   b number of bootstrap replications.
#
#   origtest: value of test statistic on original samples
#   pvalue: bootstrap p-value
#   teststatall: vector of bootstrap test statistics
#
n1<-length(x)
n2<-length(y)
v<-mean(y) - mean(x)
z<-c(x,y)
counter<-0
teststatall<-rep(0,b)
for(i in 1:b){xstar<-sample(z,n1,replace=T)
              ystar<-sample(z,n2,replace=T)
              vstar<-mean(ystar) - mean(xstar)
              if(vstar >= v){counter<-counter+1}
              teststatall[i]<-vstar}
pvalue<-counter/b
list(origtest=v,pvalue=pvalue,teststatall=teststatall)
#list(origtest=v,pvaule=pvalue)
}
```

3. **condsim1.** This algorithm generates observations from the pdf (11.4.5).

```
condsim1<-function(nsims){
collect<-rep(0,nsims)
for(i in 1:nsims)
   {y<--.5*log(1-runif(1))
    collect[i]<--log(1-runif(1))+y
   }
collect
}
```

4. **empalphacn.** Obtains the empirical level of the test discussed in Example 5.8.5.

```
empalphacn<-function(nsims){
#
#   Obtains the empirical level of the test discussed
#   in Example 5.8.5.
#
#   nsims is the number of simulations
#
sigmac<-25
eps<-.25
alpha<-.05
n<-20
tc<-qt(1-alpha,n-1)
ic<-0
for(i in 1:nsims){
    samp<-rcn(n,eps,sigmac)
    ttest<-(sqrt(n)*mean(samp))/var(samp)^.5
    if(ttest > tc){ic<-ic+1}
    }
empalp<-ic/nsims
err<-1.96*sqrt((empalp*(1-empalp))/nsims)
list(empiricalalpha=empalp,error=err)
}
```

5. **hieracrh1**. This R program performs the Gibbs sampler given in the Example 11.5.2.

```
hieracrh1<-function(nsims,x,tau,kstart){
bold<-1
clambda<-rep(0,(nsims+kstart))
cb<-rep(0,(nsims+kstart))
for(i in 1:(nsims+kstart))
    {clambda[i]<-rgamma(1,shape=(x+1),scale=(bold/(bold+1)))
    newy<-rgamma(1,shape=2,scale=(tau/(clambda[i]*tau+1)))
    cb[i]<-1/newy
    bold<-1/newy}
gibbslambda<-clambda[(kstart+1):(nsims+kstart)]
gibbsb<-cb[(kstart+1):(nsims+kstart)]
list(clambda=clambda,cb=cb,gibbslambda=gibbslambda,gibbsb=gibbsb)
}
```

6. **lslinhypoth**. Returns the F_{LS} test statistic and p-value based on the hypothesis matrix am, where the response is in y, and the full model design matrix is in x.

```
lslinhypoth<-function(x,y,am){
n<-length(x[,1])
p<-length(x[1,])
q<-length(am[,1])
beta<-lsfit(x,y)$coef
sig<-sum((lsfit(x,y)$resid)^2)/(n-p)
mid<-am%*%solve(t(x)%*%x)%*%t(am)
top<-t(am%*%beta)%*%solve(mid)%*%am%*%beta/q
fls<-top/sig
pvalue<-1-pf(fls,q,n-p)
list(fls=fls,pvalue=pvalue)
}
```

7. **mixnormal**. This R/S-Plus function returns one iteration of the EM step for Exercise 6.6.8 of Chapter 6. The initial estimate for the step is the input vector theta0.

```
mixnormal = function(x,theta0){
part1=(1-theta0[5])*dnorm(x,theta0[1],theta0[3])
part2=theta0[5]*dnorm(x,theta0[2],theta0[4])
gam = part2/(part1+part2)
denom1 = sum(1 - gam)
denom2 = sum(gam)
mu1 = sum((1-gam)*x)/denom1
sig1 = sqrt(sum((1-gam)*((x-mu1)^2))/denom1)
mu2 = sum(gam*x)/denom2
sig2 = sqrt(sum(gam*((x-mu2)^2))/denom2)
p = mean(gam)
mixnormal = c(mu1,mu2,sig1,sig2,p)
mixnormal
}
```

8. **piest**. Obtains the estimate of pi and its standard error for the simulation discussed in Example 5.8.1.

```
piest<-function(n){
#
#  Obtains the estimate of pi and its standard
#  error for the simulation discussed in Example 5.8.1
#
#  n is the number of simulations
#
```

```
u1<-runif(n)
u2<-runif(n)
cnt<-rep(0,n)
chk<-u1^2 + u2^2 - 1
cnt[chk < 0]<-1
est<-mean(cnt)
se<-4*sqrt(est*(1-est)/n)
est<-4*est
list(estimate=est,standard=se)
}
```

9. **piest2**. Obtains the estimate of pi and its standard error for the simulation discussed in Example 5.8.3.

```
piest2<-function(n){
#
#  Obtains the estimate of pi and its standard
#  error for the simulation discussed in Example 5.8.3
#
#  n is the number of simulations
#
samp<-4*sqrt(1-runif(n)^2)
est<-mean(samp)
se<-sqrt(var(samp)/n)
list(est=est,se=se)
}
```

10. **percentciboot**. Program which obtains a percentile confidence interval for the mean. To change this to a parameter other than the mean, simply substitute the appropriate function at both calls to the mean.

```
percentciboot<-function(x,b,alpha){
# x is a vector containing the original sample.
# b is the desired number of bootstraps.
# alpha: (1 - alpha) is the confidence coefficient.
#
# theta is the point estimate.
# lower is the Lower end of the percentile confidence interval.
# upper is the Upper end of the percentile confidence interval.
# thetastar is the vector of bootstrapped theta^*s.
#
theta<-mean(x)
thetastar<-rep(0,b)
n<-length(x)
```

```
for(i in 1:b){xstar<-sample(x,n,replace=T)
                thetastar[i]<-mean(xstar)
                }
thetastar<-sort(thetastar)
pick<-round((alpha/2)*(b+1))
lower<-thetastar[pick]
upper<-thetastar[b-pick+1]
list(theta=theta,lower=lower,upper=upper,thetastar=thetastar)
#list(theta=theta,lower=lower,upper=upper)
}
```

11. **rcn**. Returns a random sample of size n drawn from a contaminated normal distribution with percent contamination ϵ and standard deviation ratio σ_c.

```
rcn<-function(n,eps,sigmac){
#
#   returns a random sample of size n drawn from
#   a contaminated normal distribution with percent
#   contamination eps and variance ratio sigmac
#
ind<-rbinom(n,1,eps)
x<-rnorm(n)
rcn<-x*(1-ind)+sigmac*x*ind
rcn
}
```

Appendix C

Tables of Distributions

In this appendix, tables for the following distributions are presented:

Table I Cumulative distribution functions for selected Poisson distributions.

Table II Selected quantiles for chi-square distributions.

Table III Cumulative distribution function for the standard normal random variable.

Table IV Selected quantiles for t-distributions.

Table V Selected quantiles for F-distributions.

These tables were generated using the R software. Most statistical computing packages have functions which obtain probabilities and quantiles for these distributions as well as many other distributions. Furthermore, many hand calculators have such functions.

Table I
Poisson Distribution

The following table presents selected Poisson distributions. The probabilities tabled are

$$P(X \le x) = \sum_{w=0}^{x} e^{-m} \frac{m^w}{w!},$$

for the values of m selected.

						$m = E(X)$						
x	0.5	1.0	1.5	2.0	3.0	4.0	5.0	6.0	7.0	8.0	9.0	10.0
0	0.607	0.368	0.223	0.135	0.050	0.018	0.007	0.002	0.001	0.000	0.000	0.000
1	0.910	0.736	0.558	0.406	0.199	0.092	0.040	0.017	0.007	0.003	0.001	0.000
2	0.986	0.920	0.809	0.677	0.423	0.238	0.125	0.062	0.030	0.014	0.006	0.003
3	0.998	0.981	0.934	0.857	0.647	0.433	0.265	0.151	0.082	0.042	0.021	0.010
4	1.000	0.996	0.981	0.947	0.815	0.629	0.440	0.285	0.173	0.100	0.055	0.029
5	1.000	0.999	0.996	0.983	0.916	0.785	0.616	0.446	0.301	0.191	0.116	0.067
6	1.000	1.000	0.999	0.995	0.966	0.889	0.762	0.606	0.450	0.313	0.207	0.130
7	1.000	1.000	1.000	0.999	0.988	0.949	0.867	0.744	0.599	0.453	0.324	0.220
8	1.000	1.000	1.000	1.000	0.996	0.979	0.932	0.847	0.729	0.593	0.456	0.333
9	1.000	1.000	1.000	1.000	0.999	0.992	0.968	0.916	0.830	0.717	0.587	0.458
10	1.000	1.000	1.000	1.000	1.000	0.997	0.986	0.957	0.901	0.816	0.706	0.583
11	1.000	1.000	1.000	1.000	1.000	0.999	0.995	0.980	0.947	0.888	0.803	0.697
12	1.000	1.000	1.000	1.000	1.000	1.000	0.998	0.991	0.973	0.936	0.876	0.792
13	1.000	1.000	1.000	1.000	1.000	1.000	0.999	0.996	0.987	0.966	0.926	0.864
14	1.000	1.000	1.000	1.000	1.000	1.000	1.000	0.999	0.994	0.983	0.959	0.917
15	1.000	1.000	1.000	1.000	1.000	1.000	1.000	0.999	0.998	0.992	0.978	0.951
16	1.000	1.000	1.000	1.000	1.000	1.000	1.000	1.000	0.999	0.996	0.989	0.973
17	1.000	1.000	1.000	1.000	1.000	1.000	1.000	1.000	1.000	0.998	0.995	0.986
18	1.000	1.000	1.000	1.000	1.000	1.000	1.000	1.000	1.000	0.999	0.998	0.993
19	1.000	1.000	1.000	1.000	1.000	1.000	1.000	1.000	1.000	1.000	0.999	0.997
20	1.000	1.000	1.000	1.000	1.000	1.000	1.000	1.000	1.000	1.000	1.000	0.998
21	1.000	1.000	1.000	1.000	1.000	1.000	1.000	1.000	1.000	1.000	1.000	0.999
22	1.000	1.000	1.000	1.000	1.000	1.000	1.000	1.000	1.000	1.000	1.000	1.000

Table II
Chi-square Distribution

The following table presents selected quantiles of chi-square distribution; i.e, the values x such that

$$P(X \leq x) = \int_0^x \frac{1}{\Gamma(r/2)2^{r/2}} w^{r/2-1} e^{-w/2} \, dw,$$

for selected degrees of freedom r.

r	\multicolumn{8}{c}{$P(X \leq x)$}							
	0.010	0.025	0.050	0.100	0.900	0.950	0.975	0.990
1	0.000	0.001	0.004	0.016	2.706	3.841	5.024	6.635
2	0.020	0.051	0.103	0.211	4.605	5.991	7.378	9.210
3	0.115	0.216	0.352	0.584	6.251	7.815	9.348	11.345
4	0.297	0.484	0.711	1.064	7.779	9.488	11.143	13.277
5	0.554	0.831	1.145	1.610	9.236	11.070	12.833	15.086
6	0.872	1.237	1.635	2.204	10.645	12.592	14.449	16.812
7	1.239	1.690	2.167	2.833	12.017	14.067	16.013	18.475
8	1.646	2.180	2.733	3.490	13.362	15.507	17.535	20.090
9	2.088	2.700	3.325	4.168	14.684	16.919	19.023	21.666
10	2.558	3.247	3.940	4.865	15.987	18.307	20.483	23.209
11	3.053	3.816	4.575	5.578	17.275	19.675	21.920	24.725
12	3.571	4.404	5.226	6.304	18.549	21.026	23.337	26.217
13	4.107	5.009	5.892	7.042	19.812	22.362	24.736	27.688
14	4.660	5.629	6.571	7.790	21.064	23.685	26.119	29.141
15	5.229	6.262	7.261	8.547	22.307	24.996	27.488	30.578
16	5.812	6.908	7.962	9.312	23.542	26.296	28.845	32.000
17	6.408	7.564	8.672	10.085	24.769	27.587	30.191	33.409
18	7.015	8.231	9.390	10.865	25.989	28.869	31.526	34.805
19	7.633	8.907	10.117	11.651	27.204	30.144	32.852	36.191
20	8.260	9.591	10.851	12.443	28.412	31.410	34.170	37.566
21	8.897	10.283	11.591	13.240	29.615	32.671	35.479	38.932
22	9.542	10.982	12.338	14.041	30.813	33.924	36.781	40.289
23	10.196	11.689	13.091	14.848	32.007	35.172	38.076	41.638
24	10.856	12.401	13.848	15.659	33.196	36.415	39.364	42.980
25	11.524	13.120	14.611	16.473	34.382	37.652	40.646	44.314
26	12.198	13.844	15.379	17.292	35.563	38.885	41.923	45.642
27	12.879	14.573	16.151	18.114	36.741	40.113	43.195	46.963
28	13.565	15.308	16.928	18.939	37.916	41.337	44.461	48.278
29	14.256	16.047	17.708	19.768	39.087	42.557	45.722	49.588
30	14.953	16.791	18.493	20.599	40.256	43.773	46.979	50.892

Table III
Normal Distribution

The following table presents the standard normal distribution. The probabilities tabled are

$$P(X \leq x) = \Phi(x) = \int_{-\infty}^{x} \frac{1}{\sqrt{2\pi}} e^{-w^2/2} \, dw.$$

Note that only the probabilities for $x \geq 0$ are tabled. To obtain the probabilities for $x < 0$, use the identity $\Phi(-x) = 1 - \Phi(x)$.

x	0.00	0.01	0.02	0.03	0.04	0.05	0.06	0.07	0.08	0.09
0.0	.5000	.5040	.5080	.5120	.5160	.5199	.5239	.5279	.5319	.5359
0.1	.5398	.5438	.5478	.5517	.5557	.5596	.5636	.5675	.5714	.5753
0.2	.5793	.5832	.5871	.5910	.5948	.5987	.6026	.6064	.6103	.6141
0.3	.6179	.6217	.6255	.6293	.6331	.6368	.6406	.6443	.6480	.6517
0.4	.6554	.6591	.6628	.6664	.6700	.6736	.6772	.6808	.6844	.6879
0.5	.6915	.6950	.6985	.7019	.7054	.7088	.7123	.7157	.7190	.7224
0.6	.7257	.7291	.7324	.7357	.7389	.7422	.7454	.7486	.7517	.7549
0.7	.7580	.7611	.7642	.7673	.7704	.7734	.7764	.7794	.7823	.7852
0.8	.7881	.7910	.7939	.7967	.7995	.8023	.8051	.8078	.8106	.8133
0.9	.8159	.8186	.8212	.8238	.8264	.8289	.8315	.8340	.8365	.8389
1.0	.8413	.8438	.8461	.8485	.8508	.8531	.8554	.8577	.8599	.8621
1.1	.8643	.8665	.8686	.8708	.8729	.8749	.8770	.8790	.8810	.8830
1.2	.8849	.8869	.8888	.8907	.8925	.8944	.8962	.8980	.8997	.9015
1.3	.9032	.9049	.9066	.9082	.9099	.9115	.9131	.9147	.9162	.9177
1.4	.9192	.9207	.9222	.9236	.9251	.9265	.9279	.9292	.9306	.9319
1.5	.9332	.9345	.9357	.9370	.9382	.9394	.9406	.9418	.9429	.9441
1.6	.9452	.9463	.9474	.9484	.9495	.9505	.9515	.9525	.9535	.9545
1.7	.9554	.9564	.9573	.9582	.9591	.9599	.9608	.9616	.9625	.9633
1.8	.9641	.9649	.9656	.9664	.9671	.9678	.9686	.9693	.9699	.9706
1.9	.9713	.9719	.9726	.9732	.9738	.9744	.9750	.9756	.9761	.9767
2.0	.9772	.9778	.9783	.9788	.9793	.9798	.9803	.9808	.9812	.9817
2.1	.9821	.9826	.9830	.9834	.9838	.9842	.9846	.9850	.9854	.9857
2.2	.9861	.9864	.9868	.9871	.9875	.9878	.9881	.9884	.9887	.9890
2.3	.9893	.9896	.9898	.9901	.9904	.9906	.9909	.9911	.9913	.9916
2.4	.9918	.9920	.9922	.9925	.9927	.9929	.9931	.9932	.9934	.9936
2.5	.9938	.9940	.9941	.9943	.9945	.9946	.9948	.9949	.9951	.9952
2.6	.9953	.9955	.9956	.9957	.9959	.9960	.9961	.9962	.9963	.9964
2.7	.9965	.9966	.9967	.9968	.9969	.9970	.9971	.9972	.9973	.9974
2.8	.9974	.9975	.9976	.9977	.9977	.9978	.9979	.9979	.9980	.9981
2.9	.9981	.9982	.9982	.9983	.9984	.9984	.9985	.9985	.9986	.9986
3.0	.9987	.9987	.9987	.9988	.9988	.9989	.9989	.9989	.9990	.9990
3.1	.9990	.9991	.9991	.9991	.9992	.9992	.9992	.9992	.9993	.9993
3.2	.9993	.9993	.9994	.9994	.9994	.9994	.9994	.9995	.9995	.9995
3.3	.9995	.9995	.9995	.9996	.9996	.9996	.9996	.9996	.9996	.9997
3.4	.9997	.9997	.9997	.9997	.9997	.9997	.9997	.9997	.9997	.9998
3.5	.9998	.9998	.9998	.9998	.9998	.9998	.9998	.9998	.9998	.9998

Table IV
t-Distribution

The following table presents selected quantiles of the t-distribution; i.e, the values x such that

$$P(X \leq x) = \int_{-\infty}^{x} \frac{\Gamma[(r+1)/2]}{\sqrt{\pi r}\Gamma(r/2)(1+w^2/r)^{(r+1)/2}} \, dw$$

for selected degrees of freedom r. The last row gives the standard normal quantiles.

	$P(X \leq x)$					
r	0.900	0.950	0.975	0.990	0.995	0.999
1	3.078	6.314	12.706	31.821	63.657	318.309
2	1.886	2.920	4.303	6.965	9.925	22.327
3	1.638	2.353	3.182	4.541	5.841	10.215
4	1.533	2.132	2.776	3.747	4.604	7.173
5	1.476	2.015	2.571	3.365	4.032	5.893
6	1.440	1.943	2.447	3.143	3.707	5.208
7	1.415	1.895	2.365	2.998	3.499	4.785
8	1.397	1.860	2.306	2.896	3.355	4.501
9	1.383	1.833	2.262	2.821	3.250	4.297
10	1.372	1.812	2.228	2.764	3.169	4.144
11	1.363	1.796	2.201	2.718	3.106	4.025
12	1.356	1.782	2.179	2.681	3.055	3.930
13	1.350	1.771	2.160	2.650	3.012	3.852
14	1.345	1.761	2.145	2.624	2.977	3.787
15	1.341	1.753	2.131	2.602	2.947	3.733
16	1.337	1.746	2.120	2.583	2.921	3.686
17	1.333	1.740	2.110	2.567	2.898	3.646
18	1.330	1.734	2.101	2.552	2.878	3.610
19	1.328	1.729	2.093	2.539	2.861	3.579
20	1.325	1.725	2.086	2.528	2.845	3.552
21	1.323	1.721	2.080	2.518	2.831	3.527
22	1.321	1.717	2.074	2.508	2.819	3.505
23	1.319	1.714	2.069	2.500	2.807	3.485
24	1.318	1.711	2.064	2.492	2.797	3.467
25	1.316	1.708	2.060	2.485	2.787	3.450
26	1.315	1.706	2.056	2.479	2.779	3.435
27	1.314	1.703	2.052	2.473	2.771	3.421
28	1.313	1.701	2.048	2.467	2.763	3.408
29	1.311	1.699	2.045	2.462	2.756	3.396
30	1.310	1.697	2.042	2.457	2.750	3.385
∞	1.282	1.645	1.960	2.326	2.576	3.090

Table V
F-Distribution

The following table presents selected quantiles of the F-distribution; i.e, the values x such that

$$P(X \leq x) = \int_0^x \frac{\Gamma[(r_1 + r_2)/2](r_1/r_2)^{r_1/2} w^{r_1/2-1}}{\Gamma(r_1/2)\Gamma(r_2/2)(1 + r_1 w/r_2)^{(r_1+r_2)/2}}\, dw,$$

for selected numerator and denominator degrees of freedom r_1 and r_2, respectively.

$P(X \leq x)$	r_2	1	2	3	4	r_1 5	6	7	8
0.950	1	161.450	199.500	215.710	224.580	230.160	233.990	236.770	238.880
0.975	1	647.790	799.500	864.160	899.580	921.850	937.110	948.220	956.660
0.990	1	4052.180	4999.500	5403.350	5624.580	5763.650	5858.990	5928.360	5981.070
0.950	2	18.510	19.000	19.160	19.250	19.300	19.330	19.350	19.370
0.975	2	38.510	39.000	39.170	39.250	39.300	39.330	39.360	39.370
0.990	2	98.500	99.000	99.170	99.250	99.300	99.330	99.360	99.370
0.950	3	10.130	9.550	9.280	9.120	9.010	8.940	8.890	8.850
0.975	3	17.440	16.040	15.440	15.100	14.880	14.730	14.620	14.540
0.990	3	34.120	30.820	29.460	28.710	28.240	27.910	27.670	27.490
0.950	4	7.710	6.940	6.590	6.390	6.260	6.160	6.090	6.040
0.975	4	12.220	10.650	9.980	9.600	9.360	9.200	9.070	8.980
0.990	4	21.200	18.000	16.690	15.980	15.520	15.210	14.980	14.800
0.950	5	6.610	5.790	5.410	5.190	5.050	4.950	4.880	4.820
0.975	5	10.010	8.430	7.760	7.390	7.150	6.980	6.850	6.760
0.990	5	16.260	13.270	12.060	11.390	10.970	10.670	10.460	10.290
0.950	6	5.990	5.140	4.760	4.530	4.390	4.280	4.210	4.150
0.975	6	8.810	7.260	6.600	6.230	5.990	5.820	5.700	5.600
0.990	6	13.750	10.920	9.780	9.150	8.750	8.470	8.260	8.100
0.950	7	5.590	4.740	4.350	4.120	3.970	3.870	3.790	3.730
0.975	7	8.070	6.540	5.890	5.520	5.290	5.120	4.990	4.900
0.990	7	12.250	9.550	8.450	7.850	7.460	7.190	6.990	6.840
0.950	8	5.320	4.460	4.070	3.840	3.690	3.580	3.500	3.440
0.975	8	7.570	6.060	5.420	5.050	4.820	4.650	4.530	4.430
0.990	8	11.260	8.650	7.590	7.010	6.630	6.370	6.180	6.030
0.950	9	5.120	4.260	3.860	3.630	3.480	3.370	3.290	3.230
0.975	9	7.210	5.710	5.080	4.720	4.480	4.320	4.200	4.100
0.990	9	10.560	8.020	6.990	6.420	6.060	5.800	5.610	5.470
0.950	10	4.960	4.100	3.710	3.480	3.330	3.220	3.140	3.070
0.975	10	6.940	5.460	4.830	4.470	4.240	4.070	3.950	3.850
0.990	10	10.040	7.560	6.550	5.990	5.640	5.390	5.200	5.060
0.950	11	4.840	3.980	3.590	3.360	3.200	3.090	3.010	2.950
0.975	11	6.720	5.260	4.630	4.280	4.040	3.880	3.760	3.660
0.990	11	9.650	7.210	6.220	5.670	5.320	5.070	4.890	4.740
0.950	12	4.750	3.890	3.490	3.260	3.110	3.000	2.910	2.850
0.975	12	6.550	5.100	4.470	4.120	3.890	3.730	3.610	3.510
0.990	12	9.330	6.930	5.950	5.410	5.060	4.820	4.640	4.500
0.950	13	4.670	3.810	3.410	3.180	3.030	2.920	2.830	2.770
0.975	13	6.410	4.970	4.350	4.000	3.770	3.600	3.480	3.390
0.990	13	9.070	6.700	5.740	5.210	4.860	4.620	4.440	4.300
0.950	14	4.600	3.740	3.340	3.110	2.960	2.850	2.760	2.700
0.975	14	6.300	4.860	4.240	3.890	3.660	3.500	3.380	3.290
0.990	14	8.860	6.510	5.560	5.040	4.690	4.460	4.280	4.140
0.950	15	4.540	3.680	3.290	3.060	2.900	2.790	2.710	2.640
0.975	15	6.200	4.770	4.150	3.800	3.580	3.410	3.290	3.200
0.990	15	8.680	6.360	5.420	4.890	4.560	4.320	4.140	4.000
0.950	16	4.490	3.630	3.240	3.010	2.850	2.740	2.660	2.590
0.975	16	6.120	4.690	4.080	3.730	3.500	3.340	3.220	3.120
0.990	16	8.530	6.230	5.290	4.770	4.440	4.200	4.030	3.890

Table 5 (continued)
F-Distribution

$P(X \leq x)$	r_2	9	10	11	12	13	14	15	16
0.950	1	240.540	241.880	242.980	243.910	244.690	245.360	245.950	246.460
0.975	1	963.280	968.630	973.030	976.710	979.840	982.530	984.870	986.920
0.990	1	6022.470	6055.850	6083.320	6106.320	6125.860	6142.670	6157.280	6170.100
0.950	2	19.380	19.400	19.400	19.410	19.420	19.420	19.430	19.430
0.975	2	39.390	39.400	39.410	39.410	39.420	39.430	39.430	39.440
0.990	2	99.390	99.400	99.410	99.420	99.420	99.430	99.430	99.440
0.950	3	8.810	8.790	8.760	8.740	8.730	8.710	8.700	8.690
0.975	3	14.470	14.420	14.370	14.340	14.300	14.280	14.250	14.230
0.990	3	27.350	27.230	27.130	27.050	26.980	26.920	26.870	26.830
0.950	4	6.000	5.960	5.940	5.910	5.890	5.870	5.860	5.840
0.975	4	8.900	8.840	8.790	8.750	8.710	8.680	8.660	8.630
0.990	4	14.660	14.550	14.450	14.370	14.310	14.250	14.200	14.150
0.950	5	4.770	4.740	4.700	4.680	4.660	4.640	4.620	4.600
0.975	5	6.680	6.620	6.570	6.520	6.490	6.460	6.430	6.400
0.990	5	10.160	10.050	9.960	9.890	9.820	9.770	9.720	9.680
0.950	6	4.100	4.060	4.030	4.000	3.980	3.960	3.940	3.920
0.975	6	5.520	5.460	5.410	5.370	5.330	5.300	5.270	5.240
0.990	6	7.980	7.870	7.790	7.720	7.660	7.600	7.560	7.520
0.950	7	3.680	3.640	3.600	3.570	3.550	3.530	3.510	3.490
0.975	7	4.820	4.760	4.710	4.670	4.630	4.600	4.570	4.540
0.990	7	6.720	6.620	6.540	6.470	6.410	6.360	6.310	6.280
0.950	8	3.390	3.350	3.310	3.280	3.260	3.240	3.220	3.200
0.975	8	4.360	4.300	4.240	4.200	4.160	4.130	4.100	4.080
0.990	8	5.910	5.810	5.730	5.670	5.610	5.560	5.520	5.480
0.950	9	3.180	3.140	3.100	3.070	3.050	3.030	3.010	2.990
0.975	9	4.030	3.960	3.910	3.870	3.830	3.800	3.770	3.740
0.990	9	5.350	5.260	5.180	5.110	5.050	5.010	4.960	4.920
0.950	10	3.020	2.980	2.940	2.910	2.890	2.860	2.850	2.830
0.975	10	3.780	3.720	3.660	3.620	3.580	3.550	3.520	3.500
0.990	10	4.940	4.850	4.770	4.710	4.650	4.600	4.560	4.520
0.950	11	2.900	2.850	2.820	2.790	2.760	2.740	2.720	2.700
0.975	11	3.590	3.530	3.470	3.430	3.390	3.360	3.330	3.300
0.990	11	4.630	4.540	4.460	4.400	4.340	4.290	4.250	4.210
0.950	12	2.800	2.750	2.720	2.690	2.660	2.640	2.620	2.600
0.975	12	3.440	3.370	3.320	3.280	3.240	3.210	3.180	3.150
0.990	12	4.390	4.300	4.220	4.160	4.100	4.050	4.010	3.970
0.950	13	2.710	2.670	2.630	2.600	2.580	2.550	2.530	2.510
0.975	13	3.310	3.250	3.200	3.150	3.120	3.080	3.050	3.030
0.990	13	4.190	4.100	4.020	3.960	3.910	3.860	3.820	3.780
0.950	14	2.650	2.600	2.570	2.530	2.510	2.480	2.460	2.440
0.975	14	3.210	3.150	3.090	3.050	3.010	2.980	2.950	2.920
0.990	14	4.030	3.940	3.860	3.800	3.750	3.700	3.660	3.620
0.950	15	2.590	2.540	2.510	2.480	2.450	2.420	2.400	2.380
0.975	15	3.120	3.060	3.010	2.960	2.920	2.890	2.860	2.840
0.990	15	3.890	3.800	3.730	3.670	3.610	3.560	3.520	3.490
0.950	16	2.540	2.490	2.460	2.420	2.400	2.370	2.350	2.330
0.975	16	3.050	2.990	2.930	2.890	2.850	2.820	2.790	2.760
0.990	16	3.780	3.690	3.620	3.550	3.500	3.450	3.410	3.370

Appendix D

References

Abebe, A., Crimin, K., McKean, J. W., Haas, J. V., and Vidmar, T. J. (2001), Rank-based procedures for linear models: applications to pharmaceutical science data, *Drug Information Journal*, **35**, 947-971.

Afifi, A. A. and Azen, S. P. (1972), *Statistical Analysis: A Computer Oriented Approach*, New York: Academic Press.

Arnold, S. F. (1981), *The Theory of Linear Models and Multivariate Analysis*, New York: John Wiley and Sons.

Box, G. E. P. and Muller, M. (1958), A note on the generation of random normal variates, *Annals of Mathematical Statistics*, 29, 610-611.

Breiman, L. (1968), *Probability*, Reading, MA: Addison-Wesley.

Buck, R. C. (1965), *Advanced Calculus*, New York: McGraw-Hill.

Casella, G. and George, E. I. (1992), Explaining the Gibbs sampler, *The American Statistician*, **46**, 167-174.

Chang, W. H., McKean, J. W., Naranjo, J. D., and Sheather, S. J. (1999), High breakdown rank-based regression, *Journal of the American Statistical Association*, 94, 205-219.

Chung, K. L. (1974), *A Course in Probability Theory*, New York: Academic Press.

Conover, W. J. and Iman, R. L. (1981), Rank transform as a bridge between parametric and nonparametric statistics, *The American Statistician*, **35**, 124-133.

D'Agostino, R. B. and Stephens, M. A. (1986), *Goodness-of-Fit Techniques*, NY: Dekker.

Davison, A. C. and Hinkley, D. V. (1997), *Bootstrap Methods and Their Applications*, Cambridge, MA: Cambridge University Press.

Draper, N. R. and Smith, H. (1966), *Applied Regression Analysis*, New York: John Wiley & Sons.

DuBois, C., ed. (1960) *Lowie's Selected Papers in Anthropology*. Berkeley: University of California Press.

Efron B. and Tibshirani, R. J. (1993), *An Introduction to the Bootstrap*, New York: Chapman and Hall.

Graybill, F. A. (1969), *Introduction to Matrices with Applications in Statistics*, Belmont, CA: Wadsworth Publishing Co.

Graybill, F. A. (1976), *Theory and Application of the Linear Model*, North Scituate, MA: Duxbury.

Hald, A. (1952), *Statistical Theory with Engineering Applications*, New York: John Wiley & Sons.

Haldane, J. B. .S. (1948), The precision of observed values of small frequencies, *Biometrika*, **35**, 297-303.

Hampel, F. R. (1974), The influence curve and its role in robust estimation, *Journal of the American Statistical Association*, 69, 383-393.

Hardy, G. H. (1992), *A Course in Pure Mathematics*, Cambridge, England: Cambridge University Press.

Hettmansperger, T. P. (1984), *Statistical Inference Based on Ranks*, New York: John Wiley & Sons.

Hettmansperger, T. P. and McKean, J. W. (1998), *Robust Nonparametric Statistical Methods*, London: Arnold.

Hewitt. E. and Stromberg, K. (1965), *Real and Abstract Analysis*, New York: Springer-Verlag.

Hodges, J. L., Jr., and Lehmann, E. L. (1961), Comparison of the normal scores and Wilcoxon tests, In: *Proceedings of the Fourth Berkeley Symposium on Mathematical Statistics and Probability*, 1, 307-317, Berkeley: University of California Press.

Hodges, J. L., Jr., and Lehmann, E. L. (1963), Estimates of location based on rank tests. *Annals of Mathematical Statistics*, 34, 598-611.

Hogg, R. V. and Craig, A. T. (1958), On the decomposition of certain chi-square variables, *Annals of Mathematical Statistics*, **29**, 608.

Hogg, R. V., Fisher, D. M., and Randles, R. H. (1975), A two-sample adaptive distribution-free test, *Journal of the American Statistical Association*, **70**, 656-661.

Hollander, M. and Wolfe, D. A. (1999), *Nonparametric Statistical Methods, 2nd Edition*, New York: John Wiley & Sons.

Hsu, J. C. (1996), *Multiple Comparisons*, London: Chapman Hall.

Huber, P. J. (1981), *Robust Statistics*, New York: John Wiley & Sons.

Ihaka, R. and Gentleman, R. (1996), R: A language for data analysis and graphics, *Journal of Computational and Graphical Statistics*, 5, 229-314.

Jeffreys, H. (1961), *The Theory of Probability*, Oxford: Oxford University Press.

Kendall, M. G. (1962), *Rank Correlation Methods, 3rd Edition*, London: Griffin.

Kennedy, W. J. and Gentle, J. E. (1980), *Statistical Computing*, New York: Marcel Dekker, Inc.

Lehmann, E. L. (1983), *Theory of Point Estimation*, New York: John Wiley & Sons.

Lehmann, E. L. (1986), *Testing Statistical Hypotheses, 2nd Edition*, London, England: Chapman & Hall.

Lehmann, E. L. (1999), *Elements of Large Sample Theory*, New York: Springer-Verlag.

Lehmann, E. L. and Casella, G. (1998), *Theory of Point Estimation, 2nd Edition*, New York: Springer.

Lehmann, E. L. and Scheffé, H. (1950), Completeness, similar regions, and unbiased estimation, *Sankhya*, **10**, 305-340.

Marsaglia, G. and Bray, T. A. (1964), A convenient method for generating normal variables, *SIAM Review*, **6**, 260-264.

McKean, J. W. and Vidmar, T. J. (1994), A comparison of two rank-based methods for the analysis of linear models, *The American Statistician*, 48, 220-229.

McLachlan, G. J. and Krishnan, T. (1997), *The EM Algorithm and Extensions*, New York: John Wiley & Sons.

Minitab (1991), MINITAB Reference Manual, Valley Forge, PA: Minitab, Inc.

Mosteller, F. and Tukey, J. W. (1977), *Data Reduction and Regression*, Reading, MA: Addison-Wesley.

Naranjo, J. D. and McKean, J. W. (1997), Rank regression with estimated scores, *Statistics and Probability Letters*, **33**, 209-216.

Nelson, W. (1982), *Applied Lifetime Data Analysis*, New York: John Wiley & Sons.

Neter, J., Kutner, M. H., Nachtsheim, C. J., and Wasserman, W. (1996), *Applied Linear Statistical Models, 4th Ed.*, Chicago: Irwin.

Randles, R. H. and Wolfe, D. A. (1979), *Introduction to the Theory of Nonparametric Statistics*, New York: John Wiley and Sons.

Rao, C. R. (1973), *Linear Statistical Inference and Its Applications , 2nd Edition*, New York: John Wiley & Sons.

Robert, C. P. and Casella, G. (1999), *Monte Carlo Statistical Methods*, New York: Springer.

Rousseeuw, P. J. and Leroy, A. M. (1987), *Robust Regression and Outlier Detection*, New York: John Wiley & Sons.

Scheffé. H. (1959), *The Analysis of Variance*, New York: John Wiley & Sons.

Seber, G. A. F. (1984), *Multivariate Observations*, New York: John Wiley & Sons.

Serfling, R. J. (1980), *Approximation Theorems of Mathematical Statistics*, New York: John Wiley & Sons.

Shirley, E. A. C. (1981), A distribution-free method for analysis of covariance based on rank data, *Applied Statistics*, 30, 158-162.

S-PLUS (2000), *S-PLUS 6.0 Guide to Statistics, Volume 2*, Seattle WA: Data Analysis Division, MathSoft.

Stapleton, J. H. (1995), *Linear Statistical Models*, New York: John Wiley & Sons.

Terpstra, J. T. and McKean, J. W. (2004), Rank-Based Analyses of Linear Models using R, *Technical Report #151*, Statistical Computation Lab, Western Michigan University.

Tucker, H. G. (1967), *A Graduate Course in Probability*, New York: Academic Press.

Tukey, J. W. (1977), *Exploratory Data Analysis*, Reading, MA: Addison-Wesley.

Venables, W. N. and Ripley, B. D. (2002), *Modern Applied Statistics with S, 4th Edition*, New York: Springer.

Appendix E

Answers to Selected Exercises

Chapter 1

1.2.1 (a) $\{0, 1, 2, 3, 4\}$, $\{2\}$; (b) $(0, 3)$, $\{x : 1 \leq x < 2\}$;
(c) $\{(x, y) : 1 < x < 2, 1 < y < 2\}$.

1.2.2 (a) $\{x : 0 < x \leq 5/8\}$.

1.2.3 $C_1 \cap C_2 = \{mary, mray\}$.

1.2.8 (a) $\{x : 0 < x < 3\}$,
(b) $\{(x, y) : 0 < x^2 + y^2 < 4\}$.

1.2.9 (a) $\{x : x = 2\}$, (b) ϕ,
(c) $\{(x, y) : x = 0, y = 0\}$.

1.2.10 (a) $\frac{80}{81}$, (b) 1.

1.2.11 $\frac{11}{16}, 0, 1$.

1.2.12 $\frac{8}{3}, 0, \frac{\pi}{2}$.

1.2.13 (a) $\frac{1}{2}$, (b) 0, (c) $\frac{2}{9}$.

1.2.14 (a) $\frac{1}{6}$, (b) 0.

1.2.16 10.

1.3.2 $\frac{1}{4}, \frac{1}{13}, \frac{1}{52}, \frac{4}{13}$.

1.3.3 $\frac{31}{32}, \frac{3}{64}, \frac{1}{32}, \frac{63}{64}$.

1.3.4 0.3.

1.3.5 $e^{-4}, 1 - e^{-4}, 1$.

1.3.6 $\frac{1}{2}$.

1.3.11 (a) $\binom{6}{4}/\binom{16}{4}$, (b) $\binom{10}{4}/\binom{16}{4}$.

1.3.12 $1 - \binom{990}{5}/\binom{1000}{5}$.

1.3.14 (b) $1 - \binom{10}{3}/\binom{20}{3}$.

1.3.16 (a) $1 - \binom{48}{5}/\binom{50}{5}$.

1.3.19 $13 \cdot 12\binom{4}{3}\binom{4}{2}/\binom{52}{5}$.

1.3.24 (a) ≤ 1, (b) No.

1.4.3 $\frac{9}{47}$.

1.4.4 $2\frac{13}{52}\frac{12}{51}\frac{26}{50}\frac{25}{49}$.

1.4.6 $\frac{111}{143}$.

1.4.8 (a) 0.022, (b) $\frac{5}{11}$.

1.4.9 $\frac{5}{14}$.

1.4.10 $\frac{3}{7}, \frac{4}{7}$.

1.4.12 (c) 0.88.

1.4.14 (a) 0.1764.

1.4.15 $4(0.7)^3(0.3)$.

1.4.16 0.75.

1.4.18 (a) $\frac{6}{11}$.

1.4.20 $\frac{1}{7}$.

1.4.21 (a) $1 - \left(\frac{5}{6}\right)^6$, (b) $1 - e^{-1}$.

1.4.23 $\frac{3}{4}$.

1.4.25 $\frac{43}{64}$.

1.4.26 $\frac{3}{5}$.

1.4.28 $\frac{5 \cdot 4 \cdot 5 \cdot 4 \cdot 3}{10 \cdot 9 \cdot 8 \cdot 7 \cdot 6}$.

1.4.29 $\frac{13}{4}$.

1.4.30 $\frac{2}{3}$.

1.4.31 $0.518, 0.491$.

1.4.32 No.

1.5.1 $\frac{9}{13}, \frac{1}{13}, \frac{1}{13}, \frac{1}{13}, \frac{1}{13}$.

1.5.2 (a) $\frac{1}{2}$, (b) $\frac{1}{21}$.

1.5.3 $\frac{1}{5}, \frac{1}{5}, \frac{1}{5}$.

1.5.5 (a) $\frac{\binom{13}{x}\binom{39}{5-x}}{\binom{52}{5}}, x = 0, 1, 2, 3, 4, 5$,
(b) $\left[\binom{39}{5} + \binom{13}{1}\binom{39}{4}\right] / \binom{52}{5}$.

1.5.7 $\frac{3}{4}$.

1.5.8 (a) $\frac{1}{4}$, (b) 0, (c) $\frac{1}{4}$, (d) 0.

1.6.2 (a) $p_X(x) = \frac{1}{10}, x = 1, 2, \ldots, 10$,
(b) $\frac{4}{10}$.

1.6.3 (a) $\left(\frac{5}{6}\right)^{x-1} \frac{1}{6} \ x = 1, 2, 3, \ldots$,
(c) $\frac{6}{11}$.

1.6.4 $\frac{6}{36}, x = 0; \frac{12-2x}{36}, x = 1, 2, 3, 4, 5$.

1.6.7 $\frac{1}{3}, y = 3, 5, 7$.

1.6.8 $\left(\frac{1}{2}\right)^{\sqrt[3]{y}}, y = 1, 8, 27, \ldots$.

1.7.1 $F(x) = \frac{\sqrt{x}}{10}, 0 \le x < 100$;
$f(x) = \frac{1}{20\sqrt{x}}, 0 < x < 100$.

1.7.3 $\frac{5}{8}; \frac{7}{8}; \frac{3}{8}$.

1.7.5 $e^{-2} - e^{-3}$.

1.7.6 (a) $\frac{1}{27}, 1; (b) \frac{2}{9}, \frac{25}{36}$.

1.7.8 (a)1; (b) $\frac{2}{3}$; (c) 2.

1.7.9 (b) $\sqrt[3]{1/2}$; (c) 0.

1.7.10 $\sqrt[4]{0.2}$.

1.7.12 (a) $1 - (1-x)^3, 0 \le x < 1$;
(b) $1 - \frac{1}{x}, 1 \le x < \infty$.

1.7.13 $xe^{-x}, 0 < x < \infty$; mode is 1.

1.7.14 $\frac{7}{12}$.

1.7.17 $\frac{1}{2}$.

1.7.19 $-\sqrt{2}$.

1.7.20 $\frac{1}{27}, 0 < y < 27$.

1.7.22 $\frac{1}{\pi(1+y^2)}, -\infty < y < \infty$.

1.7.23 cdf $1 - e^{-y}, 0 \le y < \infty$.

1.7.24 pdf $\frac{1}{3\sqrt{y}}, 0 < y < 1$,
$\frac{1}{6\sqrt{y}}, 1 < y < 4$.

1.8.3 $2, 86.4, -160.8$.

1.8.4 $3, 11, 27$.

1.8.5 $\frac{\log 100.5 - \log 50.5}{50}$.

1.8.6 (a) $\frac{3}{4}$;(b) $\frac{1}{4}, \frac{1}{2}$.

1.8.7 $\frac{3}{20}$.

1.8.8 \$7.80.

1.8.10 (a) 2; (b) pdf is $\frac{2}{y^3}, 1 < y < \infty$;
(c) 2.

1.8.11 $\frac{7}{3}$.

1.8.14 (a) $\frac{1}{2}$; (c) $\frac{1}{2}$.

1.9.1 (a) $1.5, 0.75$; (b) $0.5, 0.05$;
(c) 2, does not exist.

1.9.2 $\frac{e^t}{2-e^t}, t < \log 2; 2; 2$.

1.9.11 $10; 0; 2; -30$.

1.9.13 (a) $-\frac{2\sqrt{2}}{5}$; (b) 0 ;(c) $\frac{2\sqrt{2}}{5}$.

1.9.15 $\frac{1}{2p}$; $\frac{3}{2}$; $\frac{5}{2}$; 5 ; 50 .

1.9.17 $\frac{31}{12}$; $\frac{167}{144}$.

1.9.18 $E(X^r) = \frac{(r+2)!}{2}$.

1.9.22 $\frac{5}{8}$; $\frac{37}{192}$.

1.9.25 $(1-\beta t)^{-1}$, β , β^2 .

1.10.3 0.84 .

Chapter 2

2.1.1 $\frac{15}{64}$; 0 ; $\frac{1}{2}$; $\frac{1}{2}$.

2.1.2 $\frac{1}{4}$.

2.1.6 ze^{-z} , $0 < z < \infty$.

2.1.7 $-\log z$, $0 < z < 1$.

2.1.8 $\binom{13}{x}\binom{13}{y}\binom{26}{13-x-y}/\binom{52}{13}$,
x and y nonnegative integers
such that, $x + y \le 13$.

2.1.10 $\frac{15}{2}x_1^2(1-x_1^2)$, $0 < x_1 < 1$;
$5x_2^4$, $0 < x_2 < 1$.

2.1.13 $\frac{2}{3}$; $\frac{1}{2}$; $\frac{2}{3}$; $\frac{1}{2}$; $\frac{4}{9}$; yes ; $\frac{11}{3}$.

2.1.14 $\frac{e^{t_1+t_2}}{(2-e^{t_1})(2-e^{t_2})}$, $t_i < \log 2$.

2.1.15 $(1-t_2)^{-1}(1-t_1-t_2)^{-2}$, $t_2 < 1$,
$t_1 + t_2 < 1$; No .

2.2.2

	1	2	3	4	6	9
	$\frac{1}{36}$	$\frac{4}{36}$	$\frac{6}{36}$	$\frac{4}{36}$	$\frac{12}{36}$	$\frac{9}{36}$

2.2.3 $e^{-y_1-y_2}$, $0 < y_i < \infty$.

2.2.4 $8y_1y_2^3$, $0 < y_i < 1$.

2.2.6 (a) $y_1 e^{-y_1}$, $0 < y_1 < \infty$;
(b) $(1-t_1)^{-2}$, $t_1 < 1$.

2.3.1 $\frac{3x_1+2}{6x_1+3}$; $\frac{6x_1^2+6x_1+1}{2(6x_1^2+3)^2}$.

2.3.2 (a) $2, 5$;
(b) $10x_1x_2^2$, $0 < x_1 < x_2 < 1$;
(c) $\frac{12}{25}$; (d) $\frac{449}{1536}$.

2.3.3 (a) $\frac{3x_2}{4}$; $\frac{3x_2^2}{80}$;
(b) pdf is $7(4/3)^7 y^6$, $0 < y < \frac{3}{4}$;
(c) $E(X) = E(Y) = \frac{21}{32}$;
$\mathrm{Var}(X_1) = \frac{553}{15360} > \mathrm{Var}(Y) = \frac{7}{1024}$.

2.3.8 $x + 1$, $0 < x < \infty$.

2.3.9 (a) $\binom{13}{x_1}\binom{13}{x_2}\binom{26}{5-x_1-x_2}/\binom{52}{5}$, x_1, x_2
nonnegative integers, $x_1 + x_2 \le 5$;
(c) $\binom{13}{x_2}\binom{26}{5-x_1-x_2}/\binom{39}{5-x_1}$, $x_2 \le 5 - x_1$.

2.3.11 (a) $\frac{1}{x_1}$, $0 < x_2 < x_1 < 1$;
(b) $1 - \log 2$.

2.3.12 (b) e^{-1} .

2.4.1 (a) 1 ; (b) -1 ; (c) 0 .

2.4.2 (a) $\frac{7}{\sqrt{804}}$.

2.4.8 $1, 2, 1, 2, 1$.

2.4.9 $\frac{1}{2}$.

2.5.4 $\frac{5}{81}$.

2.5.5 $\frac{7}{8}$.

2.5.6 2 ; 2 .

2.5.8 $\frac{2(1-y^3)}{3(1-y^2)}$, $0 < y < 1$.

2.5.9 $\frac{1}{2}$.

2.5.12 $\frac{4}{9}$.

2.5.13 4 ; 4 .

2.6.1 (g) $\frac{2+3y+3z}{3+6y+6z}$.

2.6.2 (a) $\frac{1}{6}$; 0 ;
(b) $(1-t_1)^{-1}(1-t_2)^{-1}(1-t_3)^{-1}$; yes .

2.6.3 pdf is $12(1-y)^{11}$, $0 < y < 1$.

2.6.4 pmf is $\frac{y^3-(y-1)^3}{6^3}$.

2.6.6 $\sigma_1(\rho_{12} - \rho_{13}\rho_{23})/\sigma_2(1 - \rho_{23}^2)$;
$\sigma_1(\rho_{13} - \rho_{12}\rho_{23})/\sigma_3(1 - \rho_{23}^2)$.

2.6.8 (a) $\frac{3}{4}$.

2.7.1 joint pdf $y_2 y_3^2 e^{-y_3}$, $0 < y_1 < 1$, $0 < y_2 < 1$, $0 < y_3 < \infty$.

2.7.2 $\frac{1}{2\sqrt{y}}$, $0 < y < 1$.

2.7.3 $\frac{1}{4\sqrt{y}}$, $0 < y < 1$; $\frac{1}{8\sqrt{y}}$, $1 \le y < 9$.

2.7.7 $24 y_2 y_3^2 y_4^3$, $0 < y_i < 1$.

2.7.8 (a) $\frac{9}{16}$; $\frac{6}{16}$; $\frac{1}{16}$; (b) $\left(\frac{3}{4} + \frac{1}{4} e^t\right)^6$.

Chapter 3

3.1.1 $\frac{40}{81}$.

3.1.4 $\frac{147}{512}$.

3.1.6 5.

3.1.9 $\frac{3}{16}$.

3.1.11 $\frac{65}{81}$.

3.1.14 $\left(\frac{1}{3}\right)\left(\frac{2}{3}\right)^{x-3}$, $x = 3, 4, 5, \ldots$.

3.1.15 $\frac{5}{72}$.

3.1.18 $\frac{1}{6}$.

3.1.19 $\frac{24}{625}$.

3.1.21 (a) $\frac{11}{6}$; (b) $\frac{x_1}{2}$; (c) $\frac{11}{6}$.

3.1.22 $\frac{25}{4}$.

3.2.1 0.09.

3.2.4 $4^x e^{-4}/x!$, $x = 0, 1, 2 \ldots$.

3.2.5 0.84.

3.2.8 about 6.7.

3.2.10 8.

3.2.11 2.

3.2.13 (a) $e^{-2} \exp\{(1 + e^{t_1})e^{t_2}\}$.

3.3.1 0.05.

3.3.2 0.831; 12.8.

3.3.3 0.90.

3.3.4 $\chi^2(4)$.

3.3.6 pdf is $3e^{-3y}$, $0 < y < \infty$.

3.3.7 2; 0.95.

3.3.15 $\frac{11}{16}$.

3.3.16 $\chi^2(2)$.

3.3.18 $\frac{\alpha}{\alpha+\beta}$; $\frac{\alpha\beta}{(\alpha+\beta+1)(\alpha+\beta)^2}$.

3.3.19 (a) 20; (b) 1260; (c) 495.

3.3.20 $\frac{10}{243}$.

3.3.24 (a) $(1 - 6t)^{-8}$, $t < \frac{1}{6}$; (b) $\Gamma(\alpha = 8, \beta = 6)$.

3.4.2 0.067; 0.685.

3.4.3 1.645.

3.4.4 71.3; 189.7.

3.4.8 0.598.

3.4.10 0.774.

3.4.11 $\sqrt{\frac{2}{\pi}}$; $\frac{\pi-2}{\pi}$.

3.4.12 0.90.

3.4.13 0.477.

3.4.14 0.461.

3.4.15 $N(0, 1)$.

3.4.16 0.433.

3.4.18 0; 3.

3.4.23 $N(0, 2)$.

3.4.28 0.24.

3.4.29 0.159.

3.4.30 0.159.

3.4.32 $\chi^2(2)$.

3.5.1 (a) 0.574; (b) 0.735.

3.5.2 (a) 0.264; (b) 0.440; (c) 0.433; (d) 0.642.

3.5.5 $\frac{4}{5}$.

3.5.6 (38.2, 43.4).

3.5.17 0.05.

3.6.1 0.05.

3.6.2 1.761.

3.6.9 $\frac{1}{4.74}$; 3.33.

Chapter 4

4.1.4 $\frac{8}{3}$; $\frac{2}{9}$.

4.1.5 7.

4.1.7 2.5 ; 0.25.

4.1.9 -5 ; 30.6.

4.1.10 $\frac{\sigma_1}{\sqrt{\sigma_1^2+\sigma_2^2}}$.

4.1.12 12 ; 168.

4.1.13 0.265.

4.1.15 22.5 ; 65.25.

4.1.16 $\frac{\mu_2\sigma_1}{\sqrt{\sigma_1^2\sigma_2^2+\mu_1^2\sigma_2^2+\mu_2^2\sigma_1^2}}$.

4.1.18 0.801.

4.1.22 (a)$e^{\mu+(\sigma^2/2)}$; $e^{2\mu}\left(e^{2\sigma^2}-e^{\sigma^2}\right)$.

4.2.5 No ; $Y_n - \frac{1}{n}$.

4.3.1 Degenerate at μ.

4.3.2 Gamma($\alpha = 1, \beta = 1$).

4.3.3 Gamma($\alpha = 1, \beta = 1$).

4.3.4 Gamma($\alpha = 2, \beta = 1$).

4.3.7 Degenerate at β.

4.3.9 0.682.

4.3.10 (b) 0.815.

4.3.13 Degenerate at $\mu_2 + \frac{\sigma_2}{\sigma_1}(x - \mu_1)$.

4.3.14 (b) $N(0,1)$.

4.3.16 (b) $N(0,1)$.

4.3.19 $\frac{1}{5}$.

4.4.2 0.954.

4.4.3 0.604.

4.4.4 0.840.

4.4.5 0.728.

4.4.7 0.08.

4.4.9 0.267.

Chapter 5

5.1.1 (a) $\frac{1}{m(m-1)}$, $x_i = 1, 2, \ldots, m$, $x_j = 1, 2, \ldots, m$, $x_i \neq x_j$.

5.1.2 (a)$b(n,p)$; (c) $\frac{p(1-p)}{n}$.

5.1.3 (b)Gamma($\alpha = n, \beta = \theta/n$) ; (d) $c = 9.59, d = 34.2$.

5.1.5 9.5.

5.2.5 $1 - (1 - e^{-3})^4$.

5.2.6 (a) $\frac{1}{8}$.

5.2.10 Weibull.

5.2.11 $\frac{5}{16}$.

5.2.12 pdf: $(2z_1)(4z_2^3)(6z_3^5)$, $0 < z_i < 1$.

5.2.13 $\frac{7}{12}$.

5.2.17 (a) $48y_3^5 y_4$, $0 < y_3 < y_4 < 1$; (b) $\frac{6y_3^5}{y_4^6}$, $0 < y_3 < y_4$; (c) $\frac{6}{7}y_4$.

5.2.18 $\frac{1}{4}$.

5.2.19 $6uv(u + v)$, $0 < u < v < 1$.

5.2.24 14.

5.2.25 (a) $\frac{15}{16}$; (b) $\frac{675}{1024}$; (c) $(0.8)^4$.

5.2.26 0.824.

5.2.27 8.

5.2.28 (a) 1.13σ ; (b) 0.92σ.

5.3.2 8.

5.3.5 (a) Beta($n - j + 1, j$) ; (b) Beta($n - j + i - 1, j - i + 2$).

5.3.6 $\frac{10!}{1!3!4!}v_1 v_2^3 (1 - v_1 - v_2)^4$, $0 < v_2$, $v_1 + v_2 < 1$.

5.4.1 $(77.28, 85.12)$.

5.4.2 24 or 25.

5.4.3 $(3.7, 5.7)$.

5.4.4 160.

5.4.5 (a) 1.31σ; (b) 1.49σ.

5.4.6 $c = \sqrt{\frac{n-1}{n+1}}$; $k = 1.60$.

5.4.9 $\left(\frac{5\bar{x}}{24}, \frac{5\bar{x}}{16}\right)$.

5.4.11 7675.

5.4.12 $(3.19, 3.61)$.

5.4.14 (b) $(3.625, 29.101)$.

5.4.19 $(-3.0, 2.0)$.

5.4.24 135 or 136.

5.5.3 $1 - \left(\frac{3}{4}\right)^\theta + \theta \left(\frac{3}{4}\right)^\theta \log\left(\frac{3}{4}\right)$, $\theta = 1, 2$.

5.5.4 0.17; 0.78.

5.5.8 $n = 19$ or 20.

5.5.9 $\gamma\left(\frac{1}{2}\right) = 0.062$; $\gamma\left(\frac{1}{12}\right) = 0.920$.

5.5.10 $n \doteq 73$; $c \doteq 42$.

5.5.12 (a) 0.051; (c) 0.256; 0.547; 0.780.

5.5.13 (a) 0.154; (b) 0.154.

5.6.5 (a) Reject; (b) $p - \text{value} \doteq 0.005$.

5.6.6 (a) Do not reject;
(b) $p - \text{value} \doteq 0.056$.

5.7.1 $8.37 > 7.81$; reject.

5.7.3 $b \le 8$ or $b \ge 32$.

5.7.4 $2.44 < 11.3$; do not reject H_0.

5.7.5 $6.40 < 9.49$; do not reject H_0.

5.7.8 $k = 3$.

5.8.4 $F^{-1}(u) = \log[u/(1-u)]$.

5.8.7 $F^{-1}(u) = \log[-\log(1-u)]$.

5.8.17 (a) $F^{-1}(u) = u^{1/\beta}$;
(b) eg., dominated by a uniform pdf.

5.9.3 (a) $\beta \log 2$.

5.9.7 Use: $s_x = 20.41$; $s_y = 18.59$.

5.9.9 (a) $\bar{y} - \bar{x} = 9.67$;
20 possible permutations;
(c) P_n^n / n^n.

5.9.10 μ_0; $n^{-1} \sum_{i=1}^n (x_i - \bar{x})^2$.

Chapter 6

6.1.1 $\overline{X}$.

6.1.2 $\overline{X}/3$.

6.1.3 (b) $-n/\log(\prod_{i=1}^n X_i)$.
(d) $Y_1 = \min\{X_1, \ldots, X_n\}$.

6.1.5 (a) $Y_n = \max\{X_1, \ldots, X_n\}$.
(b) $(2n+1)/n$.
(c) $\sqrt[2n]{1/2} Y_n$.

6.1.6 $1 - \exp\{-2/\overline{X}\}$.

6.1.7 $\hat{p} = \frac{53}{125}$,
$\sum_{x=3}^5 \binom{5}{x} \hat{p}^x (1 - \hat{p})^{5-x}$.

6.1.9 $\bar{x} = 2.109$; $\bar{x}^2 e^{-\bar{x}}/2$.

6.1.10 $\max\left\{\frac{1}{2}, \overline{X}\right\}$.

6.2.7 (a) $\frac{4}{\theta^2}$.

6.2.8 (a) $\frac{1}{2\theta^2}$.

6.3.15 (a) $\left(\frac{1}{3\bar{x}}\right)^{n\bar{x}} \left(\frac{2}{3(1-\bar{x})}\right)^{n-n\bar{x}}$.

6.3.16 (a) $n\bar{x} \log(2/\bar{x}) - n(2 - \bar{x})$.

6.3.17 $\left(\frac{\bar{x}/\alpha}{\beta_0}\right)^{n\alpha}$
$\times \exp\left\{-\sum_{i=1}^n x_i \left(\frac{1}{\beta_0} - \frac{\alpha}{\bar{x}}\right)\right\}$.

6.4.1 $\frac{4}{25}, \frac{11}{25}, \frac{7}{25}$.

6.4.2 (a) $\bar{x}, \bar{y}$,
$\frac{1}{n+m}\left[\sum_{i=1}^n (x_i - \bar{x})^2 + \sum_{i=1}^m (y_i - \bar{y})^2\right]$.
(b) $\frac{n\bar{x} + m\bar{y}}{n+m}$,
$\frac{1}{n+m}\left[\sum_{i=1}^n (x_i - \hat{\theta}_1)^2 + \sum_{i=1}^m (y_i - \hat{\theta}_1)^2\right]$.

6.4.3 $\hat{\theta}_1 = \min\{X_i\}, \frac{1}{n} \sum_{i=1}^n (X_i - \hat{\theta}_1)$.

6.4.4 $\hat{\theta}_1 = \min\{X_i\}$,
$n / \log\left[\prod_{i=1}^n X_i / \hat{\theta}_1^n\right]$.

6.4.5 $(Y_1 + Y_n)/2$, $(Y_n - y_1)/2$; No.

6.4.6 (a) $\overline{X} + 1.282\sqrt{\frac{n-1}{n}}S$.

(b) $\Phi\left(\frac{c - \overline{X}}{\sqrt{(n-1)/nS}}\right)$.

6.4.7 If $\frac{y_1}{n_1} \le \frac{y_2}{n_2}$ then $\hat{p}_1 = \frac{y_1}{n_1}$ and
$\hat{p}_2 = \frac{y_2}{n_2}$; else $\hat{p}_1 = \hat{p}_2 = \frac{y_1 + y_2}{n_1 + n_2}$.

6.5.1 $t = 3 > 2.262$; reject H_0.

6.5.4 (b) $c\frac{\sum_{i=1}^n X_i^2}{\sum_{i=1}^m Y_i^2}$.

6.5.5 $c\frac{\overline{X}}{\overline{Y}}$.

6.5.6 $c\frac{[\max\{-X_1, X_{n_1}\}]^{n_1}[\max\{-Y_1, Y_{n_2}\}]^{n_2}}{[\max\{-X_1, -Y_1, X_{n_1}, Y_{n_2}\}]^{n_1 + n_2}}$,
$\chi^2(2)$.

6.6.8 The R function `mixnormal` found in Appendix B produced these results:
(First row are initial estimates, second row are the estimates after 500 iterations),

μ_1	μ_2	σ_1	σ_2	π
105.00	130.00	15.00	25.00	0.600
98.76	133.96	9.88	21.50	0.704

Chapter 7

7.1.4 $\frac{1}{3}$, $\frac{2}{3}$.

7.1.5 $\delta_1(y)$.

7.1.6 $b = 0$, does not exist.

7.1.7 does not exist.

7.2.8 $\prod_{i=1}^n [X_i(1 - X_i)]$.

7.2.9 (a) $\left(\frac{1}{\theta}\right)^r e^{-\frac{1}{\theta}\sum_{i=1}^r y_i} e^{-\frac{1}{\theta}(n-r)y_r}$
(b) $r^{-1}[\sum_{i=1}^r y_i + (n - r)y_r]$.

7.3.2 $60y_3^2(y_5 - y_3)/\theta^5$
$0 < y_3 < y_5 < \theta$;
$6y_5/5$; $\theta^2/7$; and $\theta^2/35$.

7.3.3 $\frac{1}{\theta^2}e^{-y_1/\theta}$, $0 < y_2 < y_1 < \infty$;
$y_1/2$; $\theta^2/2$.

7.3.5 $n^{-1}\sum_{i=1}^n X_i^2$; $n^{-1}\sum_{i=1}^n X_i$;
$(n + 1)Y_n/n$.

7.3.6 $6\overline{X}$.

7.4.2 (a) X; (b) X

7.4.3 Y/n.

7.4.5 $Y_1 - \frac{1}{n}$.

7.4.7 (a) Yes. (b) Yes.

7.4.8 (a) $E(\overline{X}) = 0$.

7.4.9 (a) $\max\{-Y_1, 0.5Y_n\}$; (b) Yes;
(c) Yes.

7.5.1 $Y_1 = \sum_{i=1}^n X_i$; $Y_1/4n$; yes.

7.5.4 $\overline{x}$.

7.5.9 $\overline{x}$.

7.5.11 (a) Y_1/n; (c) θ; (d) Y_1/n.

7.5.12 (a) $Y = \sum_{i=1}^n X_i$. (b) $\frac{n-1}{nY}$.

7.6.1 $\overline{X}^2 - \frac{1}{n}$.

7.6.2 $Y^2/(n^2 + 2n)$.

7.6.4 (a) $\left(\frac{n-1}{n}\right)^Y \left(1 + \frac{Y}{n-1}\right)$;

(b) $\left(\frac{n-1}{n}\right)^{n\overline{X}} \left(1 + \frac{n\overline{X}}{n-1}\right)$;

(c) $N\left(\theta, \frac{\theta}{n}\right)$.

7.6.7 $1 - e^{-2/\overline{X}}$; $1 - \left(1 - \frac{2/\overline{X}}{n}\right)^{n-1}$.

7.6.8 (b) $\overline{X}$; (c) $\overline{X}$; (d) $1/\overline{X}$.

7.7.3 Yes.

7.7.5 $\frac{\Gamma[(n-1)/2]}{\Gamma[n/2]}\sqrt{\frac{n-1}{2}}S$.

7.7.6 (b) $\frac{Y_1 + Y_n}{2}$; $\frac{(n+1)(Y_n - Y_1)}{2(n-1)}$.

7.7.9 (a) $\frac{1}{n-1}\sum_{h=1}^n (X_{ih} - \overline{X}_i)$
$\times (X_{jh} - \overline{X}_j)$;
(b) $\sum_{i=1}^n a_i\overline{X}_i$.

7.7.10 $\left(\sum_{i=1}^{n} x_i, \sum_{i=1}^{n} \frac{1}{x_i}\right)$.

7.8.3 Y_1 ; $\sum_{i=1}^{n}(Y_i - Y_1)/n$.

7.9.13 (a) $\Gamma(3n, 1/\theta)$; No ;
(c) $(3n-1)/Y$;
(e) Beta$(3, 3n-3)$.

Chapter 8

8.1.4 $\sum_{i=1}^{10} x_i^2 \geq 18.3$; yes ; yes.

8.1.5 $\prod_{i=1}^{n} x_i \geq c$.

8.1.6 $3\sum_{i=1}^{10} x_i^2 + 2\sum_{i=1}^{10} x_i \geq c$.

8.1.7 about 96 ; 76.7.

8.1.8 $\prod_{i=1}^{n}[x_i(1-x_i)] \geq c$.

8.1.9 about 39 ; 15.

8.1.10 0.08 ; 0.875.

8.2.1 $(1-\theta)^9(1+9\theta)$.

8.2.2 $1 - \frac{15}{16\theta^4}$, $1 < \theta$.

8.2.3 $1 - \Phi\left(\frac{3-5\theta}{5}\right)$.

8.2.4 about 54 ; 5.6.

8.2.7 Reject H_0 if $\bar{x} \geq 77.564$.

8.2.8 about 27 ; reject H_0 if $\bar{x} \leq 24$.

8.2.10 $\Gamma(n, \theta)$;
reject H_0 if $\sum_{i=1}^{n} x_i \geq c$.

8.2.12 (b)$\frac{6}{32}$; (c) $\frac{1}{32}$.
(d) Reject if $y = 0$;
if $y = 1$ reject with probability $\frac{1}{5}$.

8.3.1 $|t| = 2.27 > 2.145$; reject H_0.

8.3.9 Reject H_0 if $|y_3 - \theta_0| \geq c$.

8.3.11 (a) $\prod_{i=1}^{n}(1-x_i) \geq c$.

8.4.1 $5.84n - 32.42$; $5.84n + 41.62$.

8.4.2 $0.04n - 1.66$; $0.04n + 1.20$.

8.4.4 $0.025, 29.7, -29.7$.

8.5.5 $(9y - 20x)/30 \leq c \Rightarrow (x, y) \in$ 2nd.

8.5.7 $2w_1^2 + 8w_2^2 \geq c \Rightarrow (w_1, w_2) \in$ II.

Chapter 9

9.2.4 6.39.

9.2.6 $7.875 > 4.26$; reject H_0.

9.2.7 $10.224 > 4.26$; reject H_0.

9.3.2 $r + \theta$; $2r + 4\theta$.

9.3.3 Mean: $r_2(\theta + r_1)/[r_1(r_2 - 2)]$.

9.3.6 $\chi^2(\sum a_j - b, 0)$; $\chi^2(b-1, \theta_4)$;
$F(b-1, \sum a_j - b, \theta_4)$.

9.5.5 7.00 ; 9.98.

9.5.7 4.79 ; 22.82 ; 30.73.

9.5.9 (a) $7.624 > 4.46$,reject H_A ;
(b) $15.538 > 3.84$,reject H_B.

9.5.10 $8 ; 0 ; 0 ; 0 ; 0 ; -3 ; 1 ; 2 ; -2$
$; 2 ; -2 ; 2 ; 2 ; -2 ; 2 ; -2 ; 0 ; 0 ; 0 ; 0$.

9.6.1 (a) $4.483 + 6.483x$.

9.6.7 $\widehat{\beta} = \sum_{i=1}^{n}(X_i/nc_i^2)$,
$\sum_{i=1}^{n}[(X_i - \widehat{\beta}c_i)^2/nc_i^2]$.

9.6.12 $\widehat{a} = \frac{5}{3}$.

9.7.2 Reject H_0.

9.8.2 $2 ; \boldsymbol{\mu}'\mathbf{A}\boldsymbol{\mu}$; $\mu_1 = \mu_2 = 0$.

9.8.3 (b) $\mathbf{A}^2 = \mathbf{A}$; tr$(\mathbf{A}) = 2$;
$\boldsymbol{\mu}'\mathbf{A}\boldsymbol{\mu}/8 = 6$.

9.8.4 (a) $\sum \sigma_i^2/n^2$.

9.8.5 $[1 + (n-1)\rho](\sigma^2/n)$.

9.9.1 Dependent.

9.9.3 $0, 0, 0, 0$.

9.9.4 $\sum_{i=1}^{n} a_{ij} = 0$.

Chapter 10

10.2.3 (a) 0.1148; (b) 0.7836.

10.2.8 (a) $P(Z > z_\alpha - (\sigma/\sqrt{n})\theta)$,
where $E(Z) = 0$ and Var$(Z) = 1$;
(c) Use Central Limit Theorem
(d) $\left[\frac{(z_\alpha - z_{\gamma^*})\sigma}{\theta^*}\right]^2$.

10.4.2 $1 - \Phi[z_\alpha - \sqrt{\lambda_1\lambda_2}(\delta/\sigma)]$.

10.5.2 $\frac{n(n-1)}{n+1}$.

10.5.13 (a) $W_S^* = 9$; $W_{XS}^* = 6$; (b) 1.2; (c) 9.5.

10.8.3 $\widehat{y}_{LS} = 205.9 + 0.015x$; $\widehat{y}_W = 211.0 + 0.010x$.

10.8.4 (a) $\widehat{y}_{LS} = 265.7 - 0.765(x - 1900)$; $\widehat{y}_W = 246.9 - 0.436(x - 1900)$; (b) $\widehat{y}_{LS} = 3501.0 - 38.35(x - 1900)$; $\widehat{y}_W = 3297.0 - 35.52(x - 1900)$.

10.8.8 $r_{qc} = 16/17 = 0.941$, (zeroes were excluded).

10.8.9 $r_N = 0.835$; $z = 3.734$.

Chapter 11

11.1.1 $2.67, 8.00, 16.00, 21.33, 64$.

11.2.1 0.45; 0.55.

11.2.3 $[y\tau^2 + \mu\sigma^2/n]/(\tau^2 + \sigma^2/n)$.

11.2.4 $\beta(y + \alpha)/(n\beta + 1)$.

11.2.6 $\frac{y_1 + \alpha_1}{n + \alpha_1 + \alpha_2 + \alpha_3}$; $\frac{y_2 + \alpha_2}{n + \alpha_1 + \alpha_2 + \alpha_3}$.

11.2.8 (a) $\left(\theta - \frac{10 + 30\theta}{45}\right)^2 + \left(\frac{1}{45}\right)^2 30\theta(1 - \theta)$.

11.2.9 $\sqrt[6]{2}$, $y_4 < 1$; $\sqrt[6]{2}y_4$, $1 \le y_4$.

11.3.1 (a) $\frac{\theta_2^2}{[\theta_2^2 + (x_1 - \theta_1)^2][\theta_2^2 + (x_2 - \theta_1)^2]}$.

11.3.3 (a) 76.84; (b) $(76.19, 77.48)$.

11.3.5 (a) $I(\theta) = \theta^{-2}$; (d) $\chi^2(2n)$.

11.3.8 (a) beta$(n\overline{x} + 1, n + 1 - n\overline{x})$.

11.4.1 (a) Let U_1 and U_2 be iid uniform$(0,1)$:

    (1). Draw $Y = -\log(1 - U_1)$

    (2). Draw $X = Y - \log(1 - U_2)$.

11.4.3 (b) $F_X^{-1}(u) = -\log(1 - \sqrt{u})$, $0 < u < 1$.

11.4.7 (b) $f(x|y)$ is a $b(n, y)$ pmf; $f(y|x)$ is a beta$(x + \alpha, n - x + \beta)$ pdf.

11.5.1 (b) $\widehat{\beta} = \frac{1}{2\overline{x}}$; (d) $\widehat{\theta} = \frac{1}{\overline{x}}$.

11.5.2 (a) $\delta(y) = $

$$\frac{\int_0^1 \left[\frac{a}{1 - a\log p}\right]^2 p^y(1-p)^{n-y}\, dp}{\int_0^1 \left[\frac{a}{1 - a\log p}\right]^2 p^{y-1}(1-p)^{n-y}\, dp}.$$

Chapter 12

12.1.4 Cases: $t < y$ and $t > y$.

12.1.5 (c) $y^2 - \sigma^2$.

12.1.7 (a) $n^{-1}\sum_{i=1}^n (Y_i - \overline{Y})^2$; (c) $y^2 - \sigma^2$.

12.1.9 0; $[4f^2(\theta)]^{-1}$.

12.2.6 $\widehat{y}_{LS} = 3.14 + .028x$; $\widehat{y}_W = 0.214 + .020x$.

12.2.8 $\widehat{y}_W = -1.1 + 4.0x$.

12.3.7 (a) $\widehat{y}_{LS,1} = 49.54 + 1.25c_1$; $\widehat{y}_{LS,2} = 49.54 + 1.69c_2$; (b) $\widehat{y}_{LS} = 49.54 - 13.16c_1 + 16.34c_2$.

12.3.8 (a) $\widehat{y}_{LS,i} = -22.113 + 16.06i$; (b) $\widehat{y}_{LS,i} = -2.113 + 6.057i + 0.909i^2$.

12.3.13 $N\left(\alpha^*, \sigma^2\left[\frac{1}{n} + \overline{x}'(\mathbf{X}_c'\mathbf{X}_c)^{-1}\overline{x}\right]\right)$.

12.4.3 (a) $\widehat{y}_W = -22.4 + 16.1i$; (b) $\widehat{y}_W = -2.83 + 5.27i + 0.996i^2$.

12.4.4 (b) $\widehat{y}_{LS} = -90.91 + 63.91i - 5.818i^2$; (b) $\widehat{y}_W = -9.14 + 12.0i + .246i^2$.

12.4.5 (a) $\widehat{y}_W = 71.95 + 1.48x_1 + 0.403x_2 + 0.082x_3 - 0.245x_4$.

12.4.6 $\tau E\{e_1\varphi[F(e_1)]\}\mathbf{\Sigma}^{-1}$.

12.4.8 $\tau^2\mathbf{H}_c$.

12.5.8 (b) $\mathbf{A} = \begin{bmatrix} 1 & -1 & 0 \\ 1 & 0 & -1 \end{bmatrix}$; $F_{LS} = 10.22$; $F_W = 6.40$; (c) $F_{LS} = 0.749$; $F_W = 5.43$.

12.5.9

Effect	F_{LS}	F_W
A	30.726	31.312
B	22.825	23.268
$A \times B$	4.789	6.240

12.5.10

Effect	F_{LS}	F_W
A	3.597	13.019
B	2.803	9.909
$A \times B$	2.670	4.591

Index